Power Conversion of Renewable Energy Systems

Ewald F. Fuchs • Mohammad A.S. Masoum

Power Conversion of Renewable Energy Systems

Springer

Ewald F. Fuchs
Department of Electrical, Computer, and Energy Engineering
University of Colorado
Boulder, CO
USA
ewald.fuchs@colorado.edu
ewald.fuchs@gmail.com

Mohammad A.S. Masoum
Department of Electrical and Computer Engineering
Curtin University
Perth, WA
Australia
m.masoum@curtin.edu.au

(Corrected at 2nd printing 2012).

ISBN 978-1-4899-9831-6 ISBN 978-1-4419-7979-7 (eBook)
DOI 10.1007/978-1-4419-7979-7
Springer New York Heidelberg Dordrecht London

Printed on acid-free paper

Springer is part of Springer Science+Business Media (www.springer.com)

Preface

This book is intended for undergraduate and graduate students in electrical and other engineering disciplines as well as for professionals in related fields. It is assumed that the reader has already completed electrical circuit and electronics courses covering basic concepts such as Ohm's, Kirchhoff's, Ampere's and Faraday's laws, Norton and Thevenin equivalent circuits, Fourier analysis, and the characteristics of diodes and transistors. This text combines these technologies and can serve as an introduction for an undergraduate course – where the prerequisites are circuits and electronic analyses – and as a first graduate course on renewable energy. The top-down approach introduces the systems first in terms of block diagrams and then proceeds to analyze each component. Rudimentary mechanical, chemical, aeronautical, and electrical principles are assumed to be known to the reader. Application examples highlight conventional and renewable energy problems. Software programs such as Mathematica, Spice and Matlab are applied to solve component and system problems.

This book has evolved from the content of courses given by the authors at the University of Colorado at Boulder, the Iran University of Science and Technology at Tehran, and the Curtin University of Technology at Perth, Australia. It is suitable for both electrical and non-electrical engineering students and has been particularly written for students or practicing engineers who want to educate themselves through the inclusion of about 170 application examples with solutions. More than 350 references are cited, mostly journal and conference papers as well as national and international standards and guidelines. The International System (SI) of units has been used throughout with some reference to the American/English system of units.

Figure 1 below – indicating the reversal of Arctic cooling trend as published in the Science 325 article "Recent Warming Reverses Long-Term Arctic Cooling," by Darrell S. Kaufman et al. [1] – graphically captures our rationale for writing this book. Historical temperatures as found by Mann et al. and Moberg et al. are plotted together in graph G with recent findings as documented in [2, 3]. Trend line F indicates Arctic cooling for most of the last 2,000 years, explained by a shift in the Earth's elliptical orbit of the sun. However, the cooling has been overwhelmed in the last century by human-caused global warming.

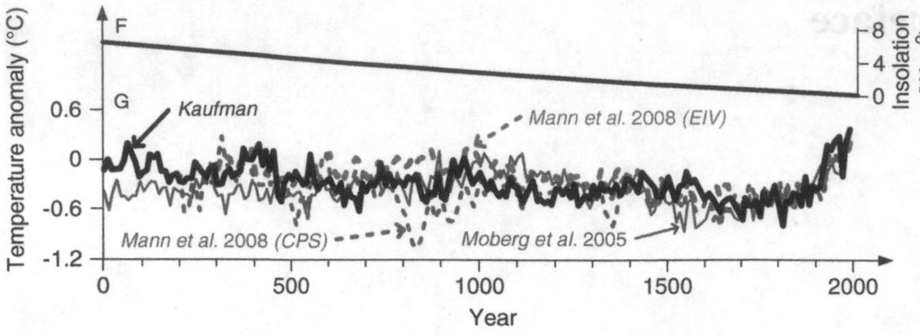

Fig. 1 Reversal of Arctic cooling trend during the past (AD) 2,000 years (From [1])

Key Features:

- Provides theoretical and practical insight into renewable energy problems.
- 170 practical application (example) problems with solutions, some implemented in PSpice, Mathematica, and Matlab.
- A total of 122 problems at the end of the chapters dealing with practical applications.

Boulder, CO — Ewald F. Fuchs
Perth, WA — Mohammad A.S. Masoum
August 2010

References

1. Kaufman, D. S. et al.: "Recent warming reverses long-term arctic cooling," *Science*, Vol. 325, No. 5945, Sept. 4, 2009, pp. 1236–1239, DOI: 10.1126/science.1173983.
2. M. E. Mann et al., *Proc. Natl. Acad. Sci. U.S.A. 105, 13252 (2008). doi:10.1073/pnas.0805721105 pmid: 18765811*
3. A. Moberg, D. M. Sonechkin, K. Holmgren, N. M. Datsenk, W. Karlén, *Nature 433, 613 (2005).*

Acknowledgments

The encouragement and support of Dipl.-Ing. Dietrich J. Roesler, formerly with the US Department of Energy, Washington, DC, one of the first professionals initiating the research of photovoltaic power plants [1] as part of the DOE mission, is greatly appreciated. The authors wish to express gratitude to their families: wives Wendy and Roshanak, sons Franz, Amir and Ali, and daughters Heidi and Maryam for their help in shaping and proofreading the manuscript. Lastly, the work on numerical field calculation [2] initiated by the late Professor Edward A. Erdelyi is reflected in part of this book. The authors thank the students of ECEN3170 (Energy Conversion 1), ECEN4167 (Energy Conversion 2) and ECEN5017 (Conventional and Renewable Energy Issues) taught at the University of Colorado during 2010 and 2011 for their constructive criticism and help.

References

1. Dugan, R. C.; Jewell, W. T.; Roesler, D. J.: "Harmonics and reactive power from line-commutated inverters in proposed photovoltaic subdivision," ID: 6637722; *DOE Report Contract Number W-7405-ENG-2*, Jan. 1, 1983.
2. Trutt, F. C.; Erdelyi, E. A.; Jackson, R. F.: "The non-linear potential equation and its numerical solution for highly saturated electrical machines," *IEEE Trans. on Aerospace*, Vol. 1, Issue 2, Aug, 1963, pp. 430–440.

Contents

Chapter 1
Introduction

1.1 Introduction

The rationale for this book is best characterized by a recent publication of the New IEEE-USA National Energy Policy Recommendation [1] stressing that the electric power system is an integral tool for the remaking of America. The clean and efficient production, transmission and use of electricity can play a key role in addressing the challenges of economic revitalization, climate change and "national security problem" of our addiction to foreign oil. Both established and new technologies must be applied at unprecedented scale and on an accelerated schedule.

On a regional and local basis, steps have been taken to achieve the abovementioned goal with the Midwestern utility Xcel's SmartGridCity Project [2].

The reversal of the global warming trend can be accomplished by energy conservation and efficiency increases of about 25%, the deployment of renewable-energy sources combined with short and long-term storage of about 50%, and continued reliance on natural gas, coal and nuclear power plants of about 25% which serve as frequency leaders [3, 4]. This menu of energy sources can be accomplished by combining old and new technologies, made possible by the invention of semiconductor switches [5, 6] resulting in photovoltaic [7, 8] and wind power plants [9], as well as short-term and long-term storage plants [10–12] with fast deployment time. Conventional technology is based on rotating machine analysis [13], wind power plants [14] in particular the GROWIAN-technology [15], hydro [16] and compressed-air [17] storage plants, and maintaining the quality of power [18].

1.2 Overview

This book first presents examples for entire systems using the block diagram approach, and then proceeds to discuss the individual block components. These systems encompass conventional constant speed drives, variable-speed drives, renewable energy systems such as photovoltaics, and wind power drive trains, as well as simple load flow within feeders, reactive power compensation,

E.F. Fuchs and M.A.S. Masoum, *Power Conversion of Renewable Energy Systems*,
DOI 10.1007/978-1-4419-7979-7_1, © Springer Science+Business Media, LLC 2011

transformers, linear and rotating machines. This text combines equivalent circuit theory, (power) electronics, magnetics, mechanics, and chemical energy storage devices such as fuel cells and batteries. The intent is to give the student a rather simplified overview of particular systems as they occur in industry applications. References leading to a more in-depth analysis are given; however, details such as the influence of saturation on magnetic devices, the impact of mechanical losses of bearings and gears on the efficiency of drives (rotating or linearly moving), will be largely neglected. Detailed examples complement the theoretical considerations.

The book is intended for undergraduate or graduate students in electrical and mechatronics engineering as well as for professionals in related fields. It is assumed that the reader has already completed electrical circuit analysis courses covering basic concepts such as Ohm's law, Kirchhoff's laws, source exchange/transformation, superposition, Thevenin's and Norton's theorems, phasor analysis, and the concept of Fourier analysis. In addition, knowledge of diodes and transistors and an introductory course on energy conversion (covering energy sources, transformers, simple control circuits, rudimentary power electronics, transformers, single- and three-phase systems as well as various rotating machine concepts such as brushless DC machines, induction and synchronous machines) is desirable.

Chapter 2 familiarizes the reader with the structure of an electromechanical system and detailed block diagrams for systems as they occur in industry applications.

Chapter 3 surveys energy sources such as power systems, fuel cells, batteries, and photovoltaics.

Chapter 4 reviews commonly used electronic controllers such as the P and PI types, used for closed-loop systems for current, voltage, speed and displacement control.

In Chap. 5, basic concepts relating to the design of (power) electronic components will be introduced: electronic switches, step-down and step-up DC-to-DC converters, phase-controlled rectifiers and inverters, pulse-width-modulated (PWM) rectifiers and inverters. These are preceded by the review of gating circuits commonly used for bi- and unipolar electronic switches such as BJTs, MOSFETs, thyristors, GTOs and IGBTs.

Chapter 6 discusses in a simplified manner fundamental laws as they relate to magnetic circuits, such as Ampere's law and Faraday's law. Concepts of self- and mutual inductances will be defined for linear systems.

Building on Chap. 6, Chap. 7 leads to the discussion of two-winding, single-phase transformers, including auto-transformers. Various computer-aided techniques, for the measurement of the losses and efficiencies of inductors and transformers will be introduced, respectively.

Chapter 8 extends the methods of Chap. 7 to three-phase systems and three-phase transformers.

Chapter 9 presents the computation of forces and torques based on the Lorentz force relation and the first law of thermodynamics: the conservation of energy for loss-less (conservative) systems.

In Chap. 10, rotating magnetic fields and the steady-state operating characteristics of the most important rotating machines will be discussed: machines of the longitudinal type such as three-phase induction, synchronous, reluctance and permanent-magnet machines. Some single-phase machines of the transversal type will be touched on, and the model laws for rotating machines will be relied on in application examples of Chap. 12.

In Chap. 11, mechanical components such as gear boxes and loads will be reviewed and the steady-state stability condition of a drive system will be examined. Energy conservation either requires constant-speed, variable-speed, constant-torque, or variable-torque operating constraints. These requirements then lead to the selection of an appropriate prime mover (e.g., motor) and the machine characteristics will then be matched with the given load characteristics for a certain (drive, generator) system.

Chapter 12 closes the loop of this text by returning to some of the electromechanical systems in Chap. 2, and mainly discussing the most important renewable energy sources and storage methods.

This text covers a wide range of topics; not all of them can be discussed in great detail. Therefore, for the sake of completeness, pertinent references are given with page numbers so that the reader can immerse him/herself in a more in-depth analysis.

Notes on units:

- This text uses exclusively the MKSAK (meter, kilogram, second, ampere, kelvin) system of units [19] and makes reference to English/American units only if absolutely necessary.
- The unit of mass is kg or lb while that of force or weight is kp in the US also called kg-force or lb-force. Here one should mention that 1 l (dm^3) of water has a mass m of 1 kg at 4°C, and exerts a (force or) weight of 1 kg-force (also 1 kilopond or 1 kp) on a scale [20].
- For some equations the units are attached in brackets, e. g., [T] = [teslas] or [N] = [newtons].
- In this text the same symbols for the time-domain quantities are used for phasor-domain quantities except that for the latter a tilde (~) sign is (e.g., $\tilde{i}(t)$) added. This removes any problems with identities of signals as defined in these two domains.
- Problems dealing with practical applications are provided at the end of each chapter.
- PSPICE and MATHEMATICA or MATLAB software will be used for some of the application examples and are also required to solve problems at the end of chapters.

1.3 The Role of Energy Conservation, Renewable Energy Systems, Fossil and Nuclear Power Plants

The role of energy efficiency in combating climate change cannot underestimated. Europeans, for example, consume on the average about half the energy per person as compared to Americans: sociological factors – smaller houses, smaller

Table 1.1 CO_2 emissions in (lbs-force) per mile and per passenger for various means of transportation [21]

Various means of transportation	CO_2 emissions (lbs-force per mile per passenger)
Sport utility vehicle (solo driver, 15 miles/gallon)	1.57
Average car (solo driver, 21.5 miles/gallon)	1.10
Airplane[a] (US average occupancy)	0.97
Bus (1/4 full)	0.75
Fuel-efficient car (solo driver, 40 miles/gallon)	0.59
Intercity train (US average occupancy)	0.45
Bus (3/4 full)	0.26

[a] Aircraft emissions are variable. Use an online calculator like Atmosfair.com to estimate impact of a trip

automobiles, and dense urban development play an important role. Nevertheless, it is estimated that about 25% of the energy presently consumed within the US can be saved. A significant part of this energy conservation can be achieved through more efficient transportation infrastructure resulting in reduced CO_2 emissions as indicated in Table 1.1 [21]. Half of the energy presently generated in fossil fuel plants can be replaced by renewable energy sources such as solar and wind. The remaining 25% of the energy required must be provided by fossil or nuclear plants due to the required frequency and load control. Remember that each and every generated (endowed) electron [22] must be utilized at the very same instant when it is generated. In existing power systems, the frequency and load control is achieved by drooping characteristics as will be explained in Application Examples 12.7–12.9. In future networks, this load sharing will have to be complemented by storage plants (e.g., hydro, compressed-air, flywheel, battery) discussed in Application Examples 12.22–12.28. Reference [23] presents a brief review of electric and hybrid cars manufactured at the present. These cars mostly rely on lithium-ion batteries [10, 24] with a stored energy content of 15–35 kWh.

1.4 Examples

To illustrate that not only electrical engineering concepts are employed in this book, a few examples demonstrate that principles of accounting – calculating the payback period of power plants – as well as mechanical, chemical engineering, and physics concepts represent a base for the understanding of the topics discussed in this book.

Application Example 1.1: Photovoltaic $P = 6.15\ kW_{DC}$ Plant on the Roof of a Residence

Colorado Amendment 37. Colorado voters passed Amendment 37 in November 2004 [25] which requires the state's largest utilities to obtain 3% of their electricity from renewable energy resources by 2007 and 10% by 2015 as well as establish a standard net metering system for homeowners and ranchers with small photovoltaic (PV) systems to connect to the power grid. The measure also calls for 4% of the mandated amount of renewable energy to come from solar resources, that is, 0.4% of its electricity must be generated through solar by 2015. Public Service Company (PSC) of Colorado (which is a part of Xcel Energy) had a peak power demand of 6,268 MW during the summer of 2003. This means at least 25 MW of electricity must originate from solar plants within the jurisdiction of PSC. To comply with this voter mandate, Xcel established a rebate program to provide financial incentives (e.g., \$4.5k/(kW_{DC}) installed power capacity) for installing solar electric systems (also called solar photovoltaic or solar PV) on residential buildings. In addition, tax benefits are available. The maximum installed power must not exceed 10 kW_{DC}.

Insolation or irradiance levels within contiguous US [26]. Figure 1.1 illustrates the insolation within the contiguous United States in terms of kWh/m^2 per day. In Boulder, Colorado, one can expect (5.5–6.0) kWh/m^2 per day.

Net metering [27]. Net metering programs adopted in many states offer the potential for individuals or businesses to realize financial benefits from installing renewable energy systems. Net metering allows consumers to offset the cost of electricity they buy from a utility by selling renewable electric energy generated at their homes or businesses back to the utility. In essence, a customer's electric meter can run both forward (on cloudy days) and backward when generating energy from photovoltaic system (on sunny days) during the same metering period, and the customer is charged only for the net amount of energy used. By definition, true net metering calls for the utility to purchase energy at the retail rate and use one meter. States have adopted a number of variations on this theme.

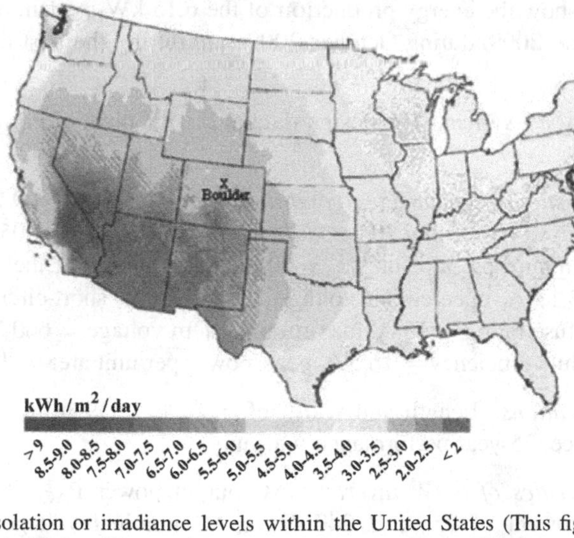

Fig. 1.1 Insolation or irradiance levels within the United States (This figure was created and prepared by the National Renewable Energy Laboratory for the U.S. Department of Energy [26])

Fig. 1.2 Residence with solar panels (Courtesy of Namasté Solar [28], 4571 North Broadway, Boulder, CO 80304) in Boulder, Colorado

As part of the Energy Policy Act of 2005, all public electric utilities are now required to offer net metering on request to their customers. Utilities have 3 years to implement this requirement.

6.15 kW_{DC} PV system in Boulder, Colorado. The photovoltaic system of Fig. 1.2 installed and put on line in September 2007 generated during the past 2 years the data of Table 1.2 indicating the total generated AC energy, $E_{generated}$ by the P_{DC} = 6.15 kW_{DC} plant. The AC energy consumed by the residence $E_{residence}$, the AC energy supplied to the utility Xcel $E_{supplied\ Xcel}$, and the CO_2 emissions avoided. The connection cost charged by the utility $C_{connection}$ is $7.66 per month. Figure 1.3 illustrates the circuit components which are required for net metering.

Figures 1.4–1.6 show the energy production of the 6.15 kW_{DC} plant of Fig. 1.2 during the entire year 2009, during October 2009, and during the 31st of October 2009, respectively.

Design of 6.15 kW_{DC} PV system. The design data of the PV plant of Fig. 1.2 are as follows:

Electrical characteristics at standard test conditions (irradiance of 1,000 W/m², cell temperature of 25°C) of residential PV module. Peak power P_{max} = 205 W, rated voltage at the maximum power point V_{mp} = 40 V, rated current at the maximum power point I_{mp} = 5.13 A, open-circuit voltage V_{oc} = 47.8 V, short-circuit current I_{sc} = 5.53 A, series fuse rating = 15 A, maximum system voltage = 600 V (UL) or 1,000 V (IEC), module efficiency = 16.5%, peak power per unit area = 165 W/m^2.

Mechanical specifications. Length and width of module 1.559 m and 0.798 m, weight = 15 kg-force, 25 year performance warranty.

Electrical characteristics of 6 kW inverter. AC output power P_{AC} = 6 kW, AC maximum output current of 25 A at 240 V_{AC}, AC nominal voltage range of (211–264) V_{AC} @ rated voltage of 240 V_{AC}, AC frequency range (59.3–60.5) Hz. Power factor of 1.0, peak inverter efficiency of 97%, weighted inverter efficiency

Table 1.2 Generated data of PV plant of Fig. 1.2 during a 2 year time frame

Time period	Cumulative net meter reading, $E_{net\ meter}$ (kWh)	Cumulative total kWh generated, $E_{generated}$ (kWh)	Excess energy to Xcel, $E_{supplied\ Xcel}$ (kWh)	Energy consumed by residence, $E_{residence}$ (kWh)	CO_2 emission avoided (lbs-force)
9/15/07	99,999				
9/15/07-12/31/07[a]	99,084	1,841	915	926	3,131
1/1/08-4/3/08	98,203	3,760	881	1,038	6,395
4/4/08-6/30/08	96,272	6,472	1,931	781	11,002
7/1/08-7/31/08	95,694	7,403	578	353	12,538
8/1/08-8/31/08	95,154	8,203	540	260	13,952
9/1/08-9/30/08	94,637	8,949	517	229	15,214
10/1/08-10/31/08	94,347	9,544	290	305	16,225
11/1/08-11/30/08	94,219	9,998	128	326	16,996
12/1/08-12/31/08	94,257	10,417	−38	457	17,710
1/1/09-1/31/09	94,201	10,844	56	371	18,434
2/1/09-2/28/09	93,955	11,440	246	350	19,449
3/1/09-3/31/09	93,532	12,257	423	394	20,837
4/1/09-4/30/09	93,201	12,972	331	384	22,052
5/1/09-5/31/09	92,678	13,775	523	280	23,417
6/1/09-6/30/09	92,118	14,600	560	265	24,819
7/1/09-7/31/09	91,479	15,513	639	274	26,372
8/1/09-8/31/09	90,857	16,361	622	226	27,814
9/1/09-9/30/09	90,420	17,053	437	255	28,990
10/1/09-10/31/09	90,287	17,525	133	339	29,792
11/1/09-11/30/09	90,109	18,057	178	354	30,697
12/01/09-12/31/09	90,206	18,370	−97	410	31,229

[a] Readings were recorded at the end of each time period

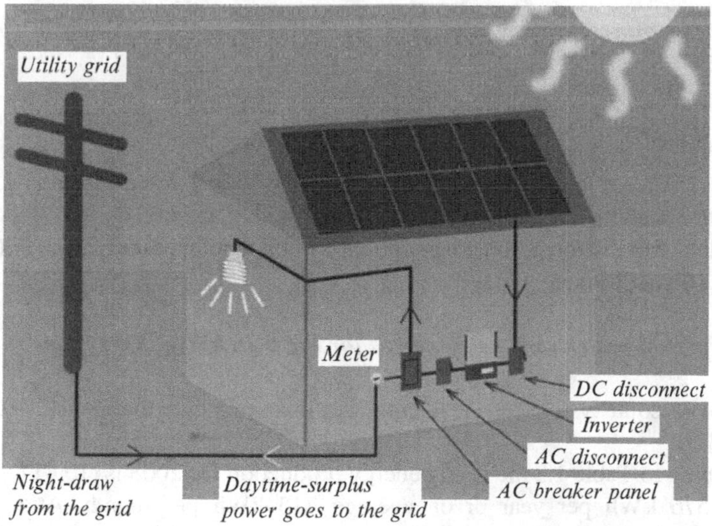

Fig. 1.3 Net metering (Courtesy of Namasté Solar [28], 4571 North Broadway, Boulder, CO 80304)

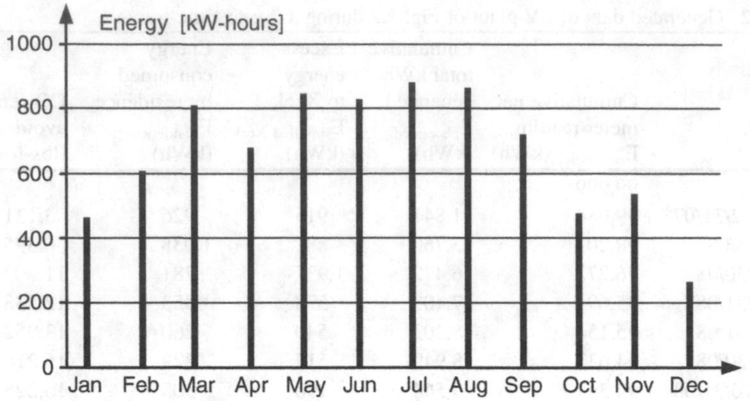

Fig. 1.4 Measured energy generation of 6.15 kW_{DC} of PV plant of Fig. 1.2 during 2009, see Table 1.2

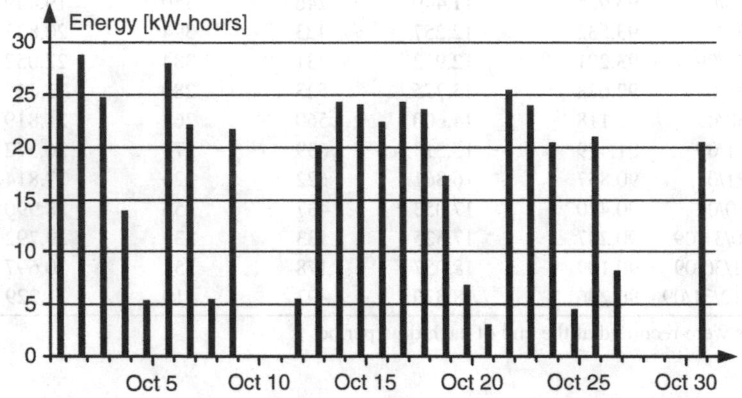

Fig. 1.5 Measured energy generation of 6.15 kW_{DC} of PV plant of Fig. 1.2 during October 2009, see Table 1.2

of 95.5%, recommended array input DC power of 6.4 kW, input DC voltage range of (250–600) V, peak power tracking voltage of (250–480) V, DC maximum input current = 25 A, DC ripple voltage ⟨5%, number of fused string inputs is 4, power consumption standby/nighttime ⟨7 W/0.25 W, grounding is of the positive ground type. Weight 143 lbs-force, forced-air cooling, ambient temperature range = (−13 to +113)°F, 10 year limited warranty.

Payback period under various scenarios for the 6.15 kW_{DC} PV system

A 6.15 kW_{DC} solar system for a residential application incurs the costs [28] listed in Table 1.3.

According to Table 1.2 the yearly energy production for 2008 is (10,417 – 1,841) kWh = 8,576 kWh per year or on average 715 kWh per month. At a cost of

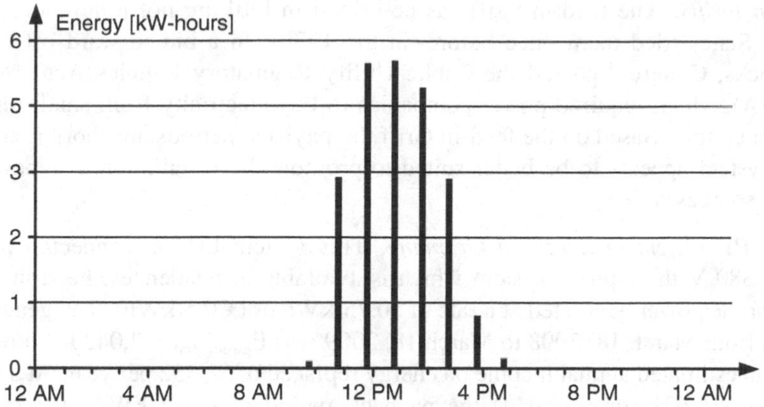

Fig. 1.6 Measured energy generation of 6.15 kW_{DC} of PV plant of Fig. 1.2 during October 31st, 2009, see Table 1.2

Table 1.3 Cost breakdown of a residential 6.15 kW_{DC} PV system in Boulder, Colorado

System component cost (installation in 2007)	\$38,354
City of Boulder sales tax (8.31%)	\$3,187
State of Colorado renewable energy sales tax waiver (−2.9%)	\$(1,112)
Installation labor	\$9,230
System value (before incentives)	\$49,658
Xcel energy rebate[a] (@ \$4.50 per watt DC for $P_{DC} = 6.15\ kW_{DC}$)	\$(27,675)
Total out-of-pocket costs	\$21,983
Federal tax credit (30% tax bracket, \$2k cap)	\$(2,000)
Net out-of-pocket cost	\$19,983

[a] The energy rebate of the utility requires that the PV plant be operated for 25 years
The underline following numbers indicates a subtotal

\$0.15/kWh the payback period is – if no interest is paid and 8,576 kWh per year are consumed by the residence at a connection fee of \$7.66/month – 16.7 years. At a cost of \$0.15/kWh the payback period is – if no interest is paid, 4,827 kWh per year are consumed by the residence and the remaining 3,749 kWh per year are fed into the grid at a reimbursement price of \$0.06/kWh at a connection fee of \$7.66/month – 23.3 years. At a cost of \$0.15/kWh – if 4.85% interest is paid and 8,576 kWh per year are consumed by the residence at a connection fee of \$7.66/month – no break-even point in time (payback period) exists.

Federal Incentives for Renewables and Efficiency [29] has changed in 2009. Based on this new regulation, the net-out-of pocket cost of the above 6.15 kW_{DC} PV plant would be reduced due to the 30% tax credit awarded to the system value (before incentives) of \$49,658 although the Xcel rebate was reduced to \$3.5 per watt DC. The net-out-of-pocket cost was now \$13,235.6. At a cost of \$0.15/kWh the payback period is – if no interest is paid and 8,576 kWh per year are consumed by the residence at a connection fee of \$7.66/month – 11.1 years.

Feed-in tariffs. The feed-in tariffs as described in [30] are not a new idea. The United States tried them once before, in the 1970s. In a bid to ward off future oil shocks, Congress passed the Public Utility Regulatory Policies Act of 1978 (PURPA), which required power companies to buy electricity from small renewable generators. Based on the feed-in tariff the payback periods are shorter, and no other system appears to be better suited to promote the installation of renewable energy sources.

2kW$_{AC}$ PV system in Southern Germany. This system [31] is connected to the $V_{L\text{-}L} = 380$ V three-phase system which is available in residences. Feed-in tariff pays for the power generated at a rate of \$0.75/kWh or (€0.5/kWh). The generated energy from March 18, 2008 to March 18, 2009 was $E_{generated} = 2{,}042$ kWh resulting in an estimated annual income on energy replaced of \$1,532 per year. At a retail price of \$18,621 (or €12,414) the payback period of the 2 kW$_{AC}$ PV system neglecting any interest payments is 12.15 years. This compares favorably with the payback period of the above-mentioned 6.15 kW$_{DC}$ system where the comparable payback period is 11.1 years based on the new regulations [29]. Note that the utility in Southern Germany charges \$0.30 (or €0.20) for 1 kWh this compares to \$0.15 in Colorado.

Application Example 1.2: Propulsion [32] of Magnetic or Conducting Body with Electromagnetic Force

A magnetic or conducting body (e.g., bullet, cannonball) can be propelled by electromagnetic force using a linear motor or accelerator as will be discussed in Chap. 9 (see Figs. 9.1a, b). Figure 1.7 shows the trajectory (parabola) of the propelled body in the x–y plane. You may assume that the acceleration in x-direction is $b_x = 0$ and that in y-direction $b_y = -g = -9.81$ m/s^2 – that is, independent of y- as indicated in Fig. 1.7.

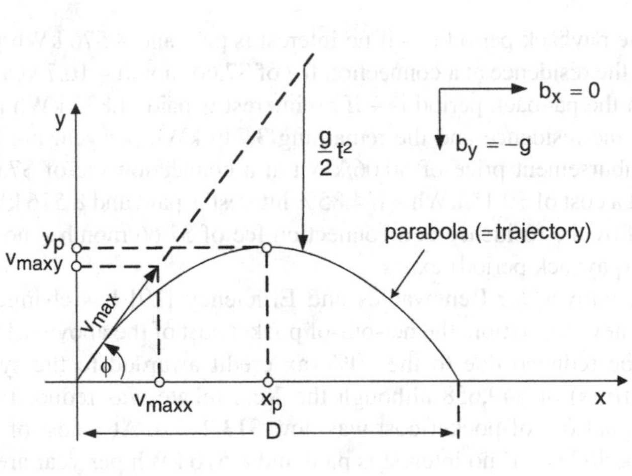

Fig. 1.7 Trajectory of propelled body

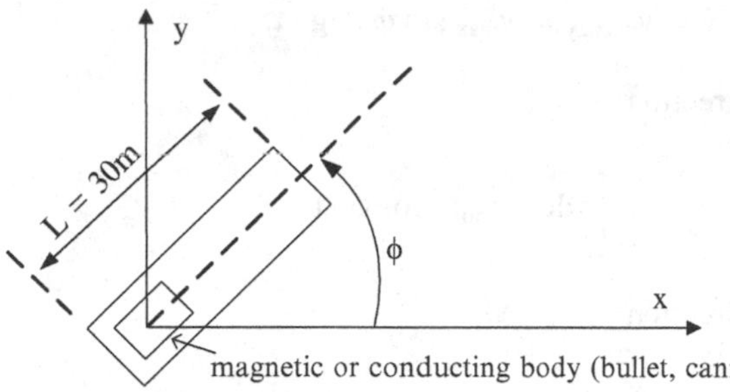

Fig. 1.8 Linear accelerator of length L = 30 m

(a) Find the velocity component in x-direction $v_x = \dot{x}$ as a function of b_x and Φ, the initial velocity v_{max} and time t.
(b) Find the velocity component in y-direction $v_y = \dot{y}$ as a function of b_y and Φ, the initial velocity v_{max} and time t.
(c) Find the displacements x and y as a function of v_{max}, Φ, g, and time t.
(d) What is the equation of the trajectory (parabola) in the x–y plane?
(e) How large is the angle Φ, if for a given v_{max} the distance D must be a maximum, that is, D_{max}?
(f) If the body must be propelled a distance $D_{max} = 370{,}000$ m, what is the maximum velocity, v_{max}, the body must have as it leaves the linear accelerator as shown in Fig. 1.8?
(g) Find the height $H = y_p$ the body reaches for $D_{max} = 2x_p = 370{,}000$ m and $\Phi = 45^{\circ}$.
(h) Calculate the required energy E_{requ} to propel the body to v_{max} as calculated in part (f) provided the propelled body has a mass of m = 3 kg.
(i) Calculate the required electrical force F_{requ} the linear accelerator must generate. You may assume that the force F_{requ} is uniformly applied to the accelerated body along the length L of the linear accelerator.
(j) How long does the acceleration of the body last (t_{accel}), if the body is accelerated from $v_{initial} = 0$ m/s to $v_{end} = v_{max}$ as calculated in part (f) and the accelerator has a length of L = 30 m?
(k) What is the required (average) power P_{avg} the linear accelerator absorbs during the acceleration of the body with a mass of m = 3 kg?
(l) Is this method suitable for transporting a satellite having a mass of m = 0.1 tons = 100 kg into space? What are the disadvantages of this method?

Solution

(a) Velocity component in x-direction $v_x = \dot{x} = v_{max\_x} + b_x t$; where $b_x = 0$, therefore,

$$\dot{x} = v_{max\_x} = v_{max} \cos \phi. \tag{1.1}$$

(b) Velocity component in y-direction $v_y = \dot{y} = v_{max\_y} + b_y t$; where $b_y = -g = -9.81\ \text{m/s}^2$, therefore,

$$\dot{y} = v_{max\_y} = v_{max} \sin \phi - g \cdot t. \tag{1.2}$$

(c) Displacement in x-direction

$$x = \int \dot{x} dt = v_{max} \cos \phi \cdot t. \tag{1.3}$$

Displacement in y-direction

$$y = \int \dot{y} dt = v_{max} \sin \phi \cdot t - g \frac{t^2}{2} \tag{1.4}$$

(d) Elimination of time t yields the parabolic function

$$\frac{2v_{max}^2 \cos^2 \phi}{g} y = \frac{v_{max}^2 \sin 2\phi}{g} x - x^2. \tag{1.5}$$

(e) The maximum velocity v_{max} is given, find angle Φ for $D = D_{max}$. For $y = 0$ and $x = D$ one obtains

$$D = \frac{v_{max}^2}{g} \sin 2\phi \tag{1.6}$$

with the differential

$$\frac{dD}{d\phi} = \frac{v_{max}^2}{g} \cos 2\phi = 0 \tag{1.7}$$

the angle $\Phi = \Phi_{max} = 45^{o}$ results in

$$D_{max} = \frac{v_{max}^2}{g}. \tag{1.8}$$

(f) From (1.8) $v_{max} = \sqrt{D_{max} \cdot g} = 1{,}905\,m/s = 6{,}858\,km/h$. This speed is about 5.8 times the speed of sound ($c = 331.6 + 0.6\,T$) m/s where T is the temperature in °C.

(g) For $x_p = D_{max}/2 = 185{,}000$ m one obtains

$$y_p = \frac{\frac{v_{max}^2}{g} \sin 2\phi \cdot x_p - x_p^2}{\frac{2v_{max}^2 \cos^2 \phi}{g}} = 92{,}370\,m. \tag{1.9}$$

(h) The kinetic energy of the propelled body is for m = 3 kg

$$E = \frac{1}{2} m v_{max}^2 = 5.44 \cdot 10^6\,\text{Nm} = 5.44\ 10^6\,\text{Ws} \tag{1.10}$$

(i) The average required force is

$$F_{required} = \frac{E}{L} = 181{,}510\,\text{N}. \tag{1.11}$$

(j) The acceleration time is

$$t_{accel} = \frac{2L}{v_{max}} = 31.49\,\text{ms}. \tag{1.12}$$

(k) The average required power is

$$P_{avg} = \frac{E}{t_{accel}} = 1.728 \cdot 10^8\,\text{W} = 172.8\,\text{MW}. \tag{1.13}$$

(l) Method is not feasible to put a satellite of mass m = 0.1 ton into space. Acceleration occurs during a very short time requiring a very high power. Once the satellite has left the rail gun no further control is possible.

1.5 Frequency/Load Control and Power Quality of Distribution System with High Penetration of Renewable Energy Sources and Storage Devices

The utility Xcel Energy has announced that the first SmartGrid City in its regional network will be Boulder, Colorado. Xcel Energy intends for this forward-thinking project to transform the way to do business. When fully deployed, the new system will provide customers with smart-grid technologies designed to provide environmental, financial, and operational benefits [2].

The rationale for the SmartGrid lies in the integrative analysis of wind and solar (photovoltaic and thermal) energy sources, micro-turbine power plants combined with cogeneration at the distribution voltage level, and the deployment of short-term and long-term storage plants [18]. This extensive reliance on renewable sources and storage plants will cause load sharing problems because of the peak-power operation of intermittently operating renewable sources, and the online response time of different storage plants (short-term storage devices such as battery, fuel cell, flywheel, variable-speed hydro plants, and supercapacitor versus long-term storage plants such as constant speed pump-hydro plants, and compressed air storage facility). The system's relatively high impedance at the distribution voltage level, resulting at acceptable current harmonics in unacceptably high voltage harmonics, single-time (e.g., spikes due to network switching and synchronization of renewable plants) and non-periodic but repetitive (e.g., flicker) events contribute to power quality problems. While the control of the renewable sources must occur at the distribution level that of large power plants occurs at the transmission level. Present-day load sharing is based on drooping characteristics, but this method cannot be employed if the intermittently operating renewable sources operate at peak power exploiting the renewable sources to the fullest. The reliability of such a smart grid will be greatly impacted by geographically dispersed intermittently operating energy sources, and the storage plants together with the spinning reserve and demand-side management will have to compensate for the above-mentioned intermittent operation. For the latter the plug-in vehicle – either receiving load or generating load – may play an important role. If the maximum load/source per vehicle is in the neighborhood of 6 kW installed power capacity many plug-in vehicles will be required. The network that is most efficient and reliable is unlikely to look like the present distribution system fed by a relatively small number of central power stations.

Present-day power and distribution systems employ base-load and peak-power plants with spinning reserve relying on drooping frequency-load characteristics which distribute additional load through load reference set points provided by the control and dispatch center. Future systems will rely on renewable energy sources such as wind and solar which operate at peak power in order to displace as much fuel (natural gas, oil and coal) as possible. This peak-power constraint [33] imposed by wind and insolation complicates the frequency/load control of the entire system. To compensate for this intermittent and changing power short-term and long-term energy storage plants are proposed. Storage plants can be charged during periods of low power demand and can supply power during high power demand. Short-term plants can be put online within a 60 Hz cycle for at least 10 min, while long-term plants can be put online within about 6–10 min and can supply power for a few hours. Thus the load sharing of present-day plants through drooping characteristics will be replaced by connecting additional storage plants if the peak power of renewable plants changes as a function of time. This deployment of storage plants necessitates frequent switching actions, cause flicker effects and not perfect synchronizations which may result in poor power quality [33–37]. While the design of renewable and storage plants is important, their concerted operation is very important and should be done first before refining the individual components in order to explore the approximate compatibility of the all components involved.

1.6 Summary

The approach of this text is the top-down approach where the systems are introduced first without relying on detailed information about their components. Photovoltaic plants demonstrate this approach and only high-school physics is required to understand these application examples. An overall description of the properties of a smart

Table 1.4 Standards boost efficiency [38], the efficiency reference index is 100% at 1972

Gas furnace
New state standards take effect: 1972: 100%; 1977: −8%; 1978: −11%; 1979: −12%; 1982: −10%
New federal standards take effect: 1992: −25%; 2006: −27%
Central air conditioner
New state standards take effect: 1972: 100%; 1978: −3%; 1979: −9%; 1982: −20%; 1983: −23%;
New federal standards take effect: 1992: −35%; 2006: −50%
Refrigerator
New state standards take effect: 1972: 100%; 1977: −13%; 1979: −22%; 1987: −44%
New federal standards take effect: 1990: −47%; 1993: −62%; 2001: −65%; 2006: −71%

grid with all its conventional (e.g., hydro-plant) and new (photovoltaic plant, fuel cell, supercapacitor) components is provided: A smart grid has to be very efficient through the minimization of losses, permit the use of intermittently operating renewable energy sources, short-term and long-term storage facilities, guarantee voltage/reactive and load/frequency control, minimize voltage and current harmonics, outages, and have superior protection, reliability and security. Short-term and long-term storage facilities must be brought online within a few cycles and within about 10 min, respectively. Through smart grid efforts – minimizing the losses and improving power quality – power consumption can be reduced through energy conservation and load management by "filling up the load valleys" during off-peak consumption employing possibly power line communication and the internet as a control mechanism.

Table 1.4 [38] illustrates how state and federal regulations boost efficiencies: It shows that state regulations initiate energy conservation and follow-up federal initiatives result in very large conservation measures. Such boosts in efficiency can be accomplished for most electric devices including plug-in hybrid electric vehicles (PHEV). State-of-the-art papers with respect to solar photovoltaic and thermal power plants in the USA, Europe and Japan are presented in the May/June 2009 issue of IEEE Power & Energy [39–44]. Reference [39] gives an overview of solar technologies. The utility experience, challenges and opportunities in photovoltaic power are described in [40]. The planning for large-scale solar power [41] and in [42] the greening of the grid with concentrating solar power are addressed next. The German [43] and the Japanese [44] experience with solar power applications are discussed at the end of this series of publications. These papers provide a good starting point for understanding the impact of renewable energy sources in particular intermittently operating [45] solar and wind power plants.

1.7 Problems

Problem 1.1: *Payback period of a 50 kW_{AC} solar power plant on private/commercial premises*

Derive, justify and explain the data of Application Example 1.1 to a potential buyer of a 50 kW_{AC} PV system if the electricity rate is \$0.15/kWh.

Problem 1.2: *CO_2 generation of a coal-fired power plant*

A coal-fired power plant has an efficiency of $\eta_{plant} = 30\%$. It is known that 1 (kg-force) = 1 kp coal contains 8.2 kWh of energy.

(a) Calculate the amount of CO_2 (measured in kg-force) released for each 1 kWh energy generated by the coal-fired plant. You may assume that coal consists 100% of carbon (C).

(b) On the average one person in the USA uses 200 kWh per month. How many (tons-force = 1,000 kg-force) CO_2 are produced by 300 million persons per year?

Problem 1.3: *Comparison of internal combustion (IC) engine car and electric car*

The average operating range of a compact car is 320 miles. The weight of this car is 1.5 (tons-force) = 1,500 (kg-force) = 1,500 kp = 3,300 (lbs-force), its IC engine including its mechanical gear has an overall energy efficiency of 16%. The gasoline consumption is 40 miles/gallon (1 gallon ≈ 3.8 l).

(a) How much gasoline does this car with IC engine use for 320 miles = (320 miles· 1.6 km/mile) = 520 km, and what is the weight of the gasoline ($W_{gasoline}$) used?

(b) What is the energy utilized ($E_{utilized}$) when driving the car at an IC engine efficiency of 16%?

(c) Figure 1.9 indicates [46] that for the GM EV1 (Impact) the vehicle-specific energy consumption (figure of merit, FM) is FM=(tons-force)-miles/(energy used at the wheel in kWh) = 8.2. What is the energy required at the wheel E_{wheel}

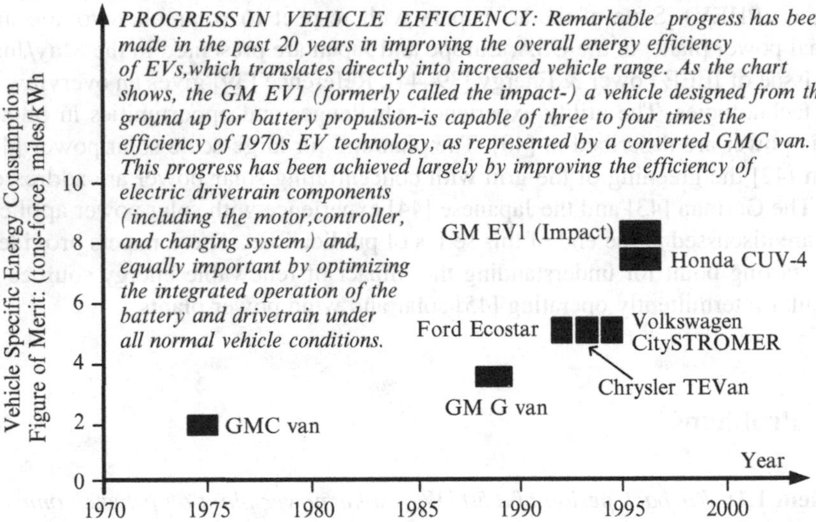

Fig. 1.9 Vehicle-specific energy consumption for past developed electric cars. For example the Honda CUV-4 has a figure of merit FM={[(tons-force) miles]/(energy used at the wheel in kWh)} = 7.5. (From [46])

and the energy stored in the battery $E_{battery}$ provided the depth of discharge (DoD) is DoD = 0.5? What is the energy available ($E_{available}$) while traveling 320 miles, provided the electric car has an energy efficiency of $\eta_{drive} = 80\%$ but can recover 20% of its energy used through regenerative braking (E_{regen})?

(d) What is the lead-acid battery weight ($W_{battery}$) required to store the energy $E_{battery}$ (in kWh) of part (c)? You may use an energy density of 50 Wh/(kg-force).

Problem 1.4: *Comparison of fuel costs and CO_2 generation for IC engine car and electric car*

(a) Provided an IC engine has an energy efficiency of 18%. How many liters of gasoline must be in the tank to supply 20 kWh of energy at the wheels?

(b) How heavy (kg-force) will be a lithium-ion battery to supply 20 kWh at the wheel at a depth of discharge of DoD = 0.5? You may assume an energy density of 100 Wh/(kg-force). You may assume a battery efficiency of $\eta_{bat} = 0.9$ and that of the drive train is $\eta_{drive} = 0.9$.

(c) What are the total fuel costs for both cases, provided 1 kWh costs $0.15? You may assume that 1 l of gasoline costs $1.1.

(d) Compute the CO_2 generation of an IC car and that of an electric car per 100 km (62.5 miles), if the IC car has a mileage of 36 miles/gallon and the electric car has a figure of merit of FM = 8.2. You may assume a car weight of 1.5 tons-force and the electricity consumed by the electric car is generated by a coal-fired power plant. Are the CO_2 emissions of the IC and electric cars below 100 (g-force)/km?

References

1. http://www.ieeeusa.org/policy/positions/energypolicy.pdf
2. http://birdcam.xcelenergy.com/sgc/media/pdf/smartgridcityhypothesisWhitePaper_July2008.pdf
3. Lamme, B. G.: "The technical story of the frequencies," *Proceedings of the Meeting of the American Institute of Electrical Engineers*, Washington, DC Jan. 18, 1918, pp. 65–89.
4. Neidhöfer, G.: "Early three-phase power," *IEEE Power & Energy Magazine*, Sept/Oct 2007, pp. 88–100.
5. Bardeen J. et al.: "Three-electrode circuit element utilizing semiconductive materials", US patent 2524035, 26.02.1948.
6. Pollack, J. J.: "Advanced pulse width modulated inverter techniques," *IEEE Trans. on Industry Applications*, vol. IA-8, no. 2, March 1972, pp. 145–154.
7. Dugan, R. C.; Jewell, W. T.; Roesler, D. J.: "Harmonics and reactive power from line-commutated inverters in proposed photovoltaic subdivision," ID: 6637722; *DOE Report Contract Number W-7405-ENG-2*, Jan. 1, 1983.
8. Masoum, M. A. S.; Mousavi Badejani, S. M.; Fuchs, E. F.: "Microprocessor-controlled new class of optimal battery chargers for photovoltaic applications," *IEEE Transactions on Energy Conversion*, vol. 19, no. 3, Sept 2004, pp. 599–606.
9. Yildirim, D.; Fuchs, E. F.; Batan, T.: "Test results of a 20 kW, direct-drive, variable-speed wind power plant," *Proceedings of the International Conference on Electrical Machines, Istanbul*, Turkey, Sept 2–4, 1998, pp. 2039–2044.
10. Hall, J. C.; Lin, T.; Brown, G.; Biesan, Li. Ph.; Bonhomme, F.: "Decay processes and life predictions for lithium ion satellite cells," 4th International Energy Conversion Engineering

Conference and Exhibit (IECEC), 26–29 June 2006, San Diego, CA, Paper No. AIAA 2006-4078, 11 pages.
11. http://www.mpoweruk.com/supercaps.htm
12. Ribeiro, P. F.; Johnson, B. K.; Crow, M. L.; Arsoy, A.; Liu, Y.: "Energy storage systems for advanced power applications," *Proceedings of the IEEE*, vol. 89, no. 12, Dec. 2001, pp. 1744–1756.
13. Trutt, F. C.; Erdelyi, E. A.; Jackson, R. F.: "The non-linear potential equation and its numerical solution for highly saturated electrical machines," *IEEE Transactions on Aerospace*, vol. 1, no. 2, Aug. 1963, pp. 430–440.
14. Stein, D. R.: "Utilization of wind power in the USSR," (in German), *Elektrizitätswirtschaft*, vol. 40, 1941, pp. 54–56.
15. Jarass, L.; Hoffmann, L.; Jarass, A.; Obermair, G.: *Windenergie* (in German), Springer, Berlin, 1980.
16. http://www.tva.gov/sites/raccoonmt.htm
17. Mattick, W.; Haddenhorst, H. G.; Weber, O.; Stys, Z. S.: "Huntorf- the world's first 290 MW gas turbine air storage peaking plant," *Proceedings of the American Power Conference*, vol. 37, 1975, pp. 322–330.
18. Fuchs, E. F.; Masoum, M. A. S.: *Power Quality in Power Systems and Electrical Machines*, Elsevier/Academic Press, Feb. 2008, 638 pages. ISBN: 978-0-12-369536-9.
19. http:/encarta.msn.com/encyclopedia_761561345_2/Metric_system.html#p3
20. http://www.unitconversion.org/unit_converter/force-ex.html
21. http://features.csmonitor.com/environment/2008/08/01/high-gas-prices-boost-bus-travel
22. Case number: 2006CV12445, District Court, City and County of Denver, Colorado, 1437 Bannock Street, Denver, Colorado 80202, October 9–10, 2007.
23. Voelcker, J.: "Top 10 tech cars," *IEEE Spectrum*, vol. 46, no. 4, April 2009, pp. 42–49.
24. Tichy, R.: "Battery advancements," *2007 Embedded Systems Conference*, April 05, 2007, 7 pages.
25. http://www.dora.state.co.us/occ/Cases/05R112E_Amendment37_Rulemaking/InitialComments Final.pdf
26. http://www.nrel.gov/gis/solar.html
27. http://apps3.eere.energy.gov/greenpower/markets/netmetering.shtml
28. Namasté Solar, 4571 North Broadway, Boulder, CO 80304.
29. http://www.dsireusa.org/library/includes/incentive2.cfm?Incentive_Code=US37F&State=federal&currentpageid=1&ee=1&re=1
30. Blake, M.: "The rooftop revolution," *Washington Monthly*, vol. 41, no. 2, March/April 2009, pp. 32–36.
31. http://www.elektrofuchs-munderkingen.de/
32. http://www.n-tv.de/913309.html
33. Masoum, M. A. S.; Dehbonei, H.; Fuchs, E. F.: "Theoretical and experimental analyses of photovoltaic systems with voltage- and current-based maximum power point tracking," *IEEE Transactions on Energy Conversion*, vol. 17, no. 4, Dec. 2002, pp. 514–522.
34. Yildirim, D.: *Commissioning of a 30 kVA Variable-Speed, Direct-Drive Wind Power Plant*, PhD Dissertation, University of Colorado at Boulder, 1999.
35. Masoum, M. A. S.; Ladjevardi, M.; Jafarian, A.; Fuchs, E. F.: "Optimal placement, replacement and sizing of capacitor banks in distorted distribution networks by genetic algorithms," *IEEE Transactions on Power Delivery*, vol. 19, no. 4, Oct. 2004, pp. 1794–1801.
36. Fuchs, E. F.; Fuchs, F. S.: "Intentional islanding – maintaining power systems operation during major emergencies," *Proceedings of the 10th IASTED International Conference on Power and Energy Systems (PES 2008)*, Baltimore, MD, April 16–18, 2008.
37. IEEE Standard 519.
38. Clayton, M.: "Bright ideas on energy efficiency," (Appliance Standards Awareness Project), *Christian Science Monitor*, April 26, 2009, pp. 36–37.
39. Kroposki, B.; Margolis, R.; Tan, D.: "Harnessing the sun," *IEEE Power & Energy*, vol. 7, no. 3, May/June 2009, pp. 22–33.

40. Key, T.: "Finding a bright spot," *IEEE Power & Energy*, vol. 7, no. 3, May/June 2009, pp. 34–44.
41. Bebic, J.; Walling, R.; O'Brien, K.; Kroposki, B.: "The sun also rises," *IEEE Power & Energy*, vol. 7, no. 3, May/June 2009, pp. 45–54.
42. Mehos, M.; Kabel, D.; Smithers, P.: " Planting the seed," *IEEE Power & Energy*, vol. 7, no. 3, May/June 2009, pp. 55–62.
43. Braun, M.; Arnold, G.; Laukamp, H.: "Plugging into the zeitgeist," *IEEE Power & Energy*, vol. 7, no. 3, May/June 2009, pp. 63–76.
44. Hara, R.; Kita, H.; Tanabe, T.; Sugihara, H.; Kuwayama, A.; Miwa, S.: "Testing the technologies, *IEEE Power & Energy*, vol. 7, no. 3, May/June 2009, pp. 77–85.
45. http://www.nrel.gov/midc/ss1/
46. Moore, T.: The road ahead for EV batteries, *EPRI Journal*, March/April 1996, pp. 7–15.

Chapter 2
Block Diagrams of Electromechanical Systems

The structure of an electromechanical drive system is given in Fig. 2.1. It consists of energy/power source, reference values for the quantities to be controlled, electronic controller, gating circuit for converter, electronic converter (e.g., rectifier, inverter, power electronic controller), current sensors (e.g., shunts, current transformer, Hall sensor), voltage sensors, (e.g., voltage dividers, potential transformer), speed sensors (e.g., tachometers), and displacement sensors (e.g., encoders), rotating three-phase machines, mechanical gear box, and the application-specific load (e.g., pump, fan, automobile). The latter component is given and one has to select all remaining components so that desirable operational steady-state and dynamic performances can be obtained: that is, the engineer has to match these electronic components with the given mechanical (or electronic) load. To be able to perform such a matching analysis, the performance of all components of a (drive) system must be understood. In Fig. 2.1 all but the mechanical gear are represented by transfer functions -that is, output variables X_{out} as a function of time t. The mechanical gear is represented by the transfer characteristic – that is X_{out} as a function of X_{in}.

Most constant-speed drives operate within the first quadrant I of the torque-angular velocity ($T-\omega_m$) plane of Fig. 2.2. Whenever energy conservation via regeneration is required, a transition from either quadrant I to II or quadrant III to IV is required. Most variable-speed traction drives accelerating and decelerating in forward and backward directions operate within all four quadrants I–IV.

Application Example 2.1: Motion of a Robot Arm

For accelerating a robot arm in forward direction the torque (T) and angular velocity (ω_m) are positive, that means the power is $P = T \cdot \omega_m > 0$: the power required for the motion must be delivered from the source to the motor. If deceleration in forward direction is desired $T < 0$ and $\omega_m > 0$, that is, the power required is $P = T\ \omega_m < 0$: the braking power to slow down the motion of the robot arm has to be generated by the motor and absorbed by the source of the drive.

For acceleration in reverse direction the torque $T < 0$ and angular velocity $\omega_m < 0$, that is $P = T \cdot \omega_m > 0$, the power has to be delivered by the source of the drive. If deceleration in reverse direction is required, the braking power must be absorbed by the source of the drive (Fig. 2.3).

E.F. Fuchs and M.A.S. Masoum, *Power Conversion of Renewable Energy Systems*,
DOI 10.1007/978-1-4419-7979-7_2, © Springer Science+Business Media, LLC 2011

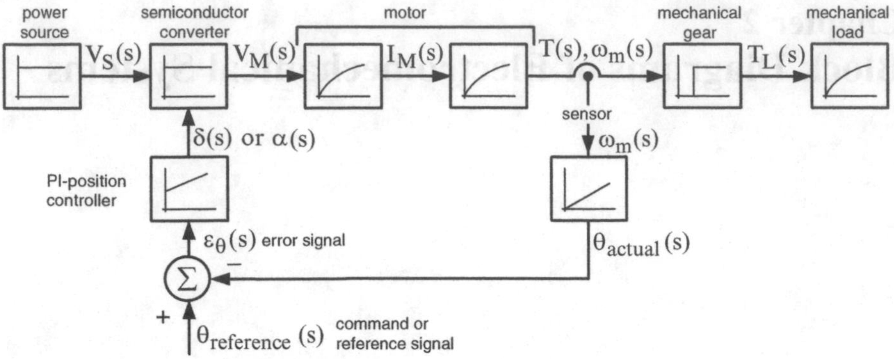

Fig. 2.1 Generic structure of an electromechanical drive system

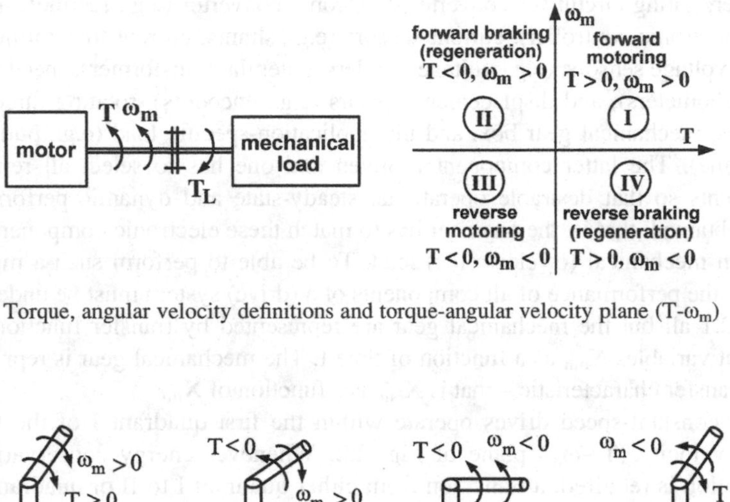

Fig. 2.2 Torque, angular velocity definitions and torque-angular velocity plane (T-ω_m)

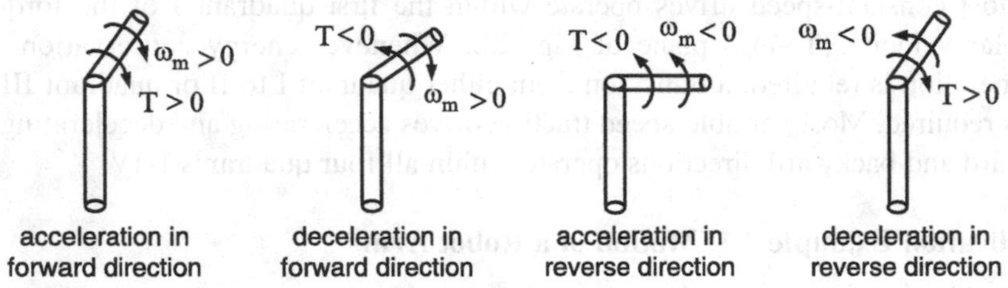

Fig. 2.3 Motion of a robot arm

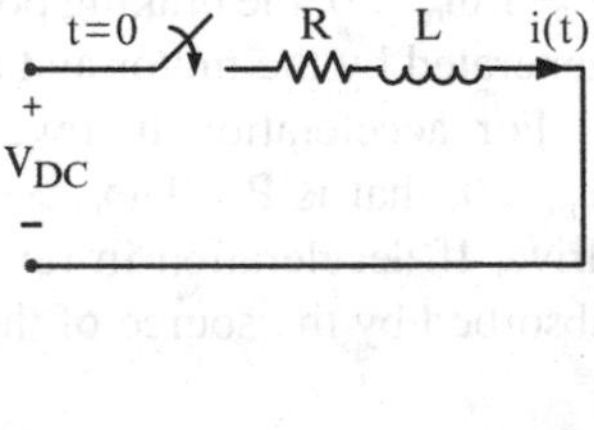

Fig. 2.4 R-L circuit with initial condition $i_{init}(t=0)=0$

2.1 Relation Between Differential Equations of Motion and Transfer Functions

In general, the equations governing a motion are complicated differential equations with non-constant (time-dependent), nonlinear coefficients. Sometimes approximations permit us to represent these equations of motion by ordinary differential equations with constant coefficients. These can be solved either via Laplace solution techniques [1] or other mathematical methods. One of the latter is where the general solution of a differential equation with constant coefficients is obtained by the superposition of the (unforced, natural) solution of the homogeneous differential equation and a particular (forced) solution of the inhomogeneous differential equation.

Application Example 2.2: Solution of Differential Equation with Constant Coefficients

The R-L circuit of Fig. 2.4 will be used to derive the solution of differential equation with constant coefficients and zero initial condition $i(t=0)=i_{init}(t=0)=0$. Inhomogeneous (or forced) first-order differential equation is

$$Ri + L\frac{di}{dt} = V_{DC} \tag{2.1}$$

or

$$i + \tau\frac{di}{dt} = \frac{V_{DC}}{R}, \tag{2.2}$$

where $\tau = L/R$ is the time constant.

Homogeneous (or unforced, natural) first-order differential equation is

$$i + \tau\frac{di}{dt} = 0. \tag{2.3}$$

1. Solution of the homogeneous (or unforced, natural response) differential equation: assume the solution to be of the form

$$i_{hom}(t) = Ae^{-(\alpha t)}, \tag{2.4}$$

$$\frac{di_{hom}(t)}{dt} = -A\alpha e^{-(\alpha t)}, \tag{2.5}$$

introducing these terms into the homogeneous differential equation, one obtains

$$Ae^{-(\alpha t)} - \tau A\alpha e^{-(\alpha t)} = 0, \tag{2.6}$$

with $e^{-(\alpha t)} \neq 0$ and $A \neq 0$ follows $\tau = 1/\alpha$ or the homogeneous (unforced, natural) solution is

$$i_{hom}(t) = Ae^{-\left(\frac{t}{\tau}\right)}. \tag{2.7}$$

2. Particular solution of inhomogeneous differential equation: assume the solution to be of the form

$$i_{particular} = C, \tag{2.8a}$$

$$\frac{di_{particular}}{dt} = 0, \tag{2.8b}$$

with

$$i + \tau\frac{di}{dt} = \frac{V_{DC}}{R}, \tag{2.9}$$

follows the forced solution

$$i_{particular} = V_{DC}/R. \tag{2.10}$$

3. General solution consists of the superposition of homogeneous and particular solutions:

$$i_{general} = i_{hom} + i_{particular}, \tag{2.11}$$

$$i_{general}(t) = Ae^{-\left(\frac{t}{\tau}\right)} + \frac{V_{DC}}{R}. \tag{2.12}$$

The coefficient A will be found by introducing the initial condition at $t = 0$, e.g., $i_{init}(t) = 0$, therefore

$$0 = Ae^{-0} + \frac{V_{DC}}{R} \tag{2.13a}$$

or

$$A = -\frac{V_{DC}}{R}. \tag{2.13b}$$

The solution for the given initial condition is now (Fig. 2.5)

$$i(t) = \frac{V_{DC}}{R}\left(1 - e^{-\left(\frac{t}{\tau}\right)}\right). \tag{2.14}$$

Fig. 2.5 Solution for i(t)

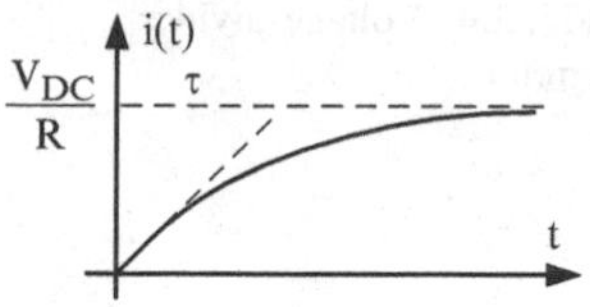

The solution for the general initial condition $i(t=0) = i_{init}(t=0)$ is

$$i(t) = \frac{V_{DC}}{R}\left(1 - e^{-\left(\frac{t}{\tau}\right)}\right) + i_{init}(t=0)e^{-\left(\frac{t}{\tau}\right)}. \tag{2.15}$$

2.1.1 *Circuits with a Constant or P (Proportional) Transfer Function*

The following circuits have constant transfer functions.

(a) Voltage-divider circuit (Fig. 2.6)

$$\frac{v_{out}(s)}{v_{in}(s)} = \frac{R_1}{(R_1 + R_2)} = G_1 \langle 1. \tag{2.16}$$

The ratio $G_1 = R_1/(R_1 + R_2)$ is called the gain of the circuit which is less than 1, where s is the Laplace operator.

(b) Operational amplifier with negative (resistive) feedback: inverting [2] configuration (Fig. 2.7a).

$$\frac{v_{out}(s)}{v_{in}(s)} = -\frac{R_2}{R_1} = G_2. \tag{2.17}$$

As can be seen, the gain G_2 is negative and can be larger or smaller than 1 (Fig. 2.7b).

The constant or proportional transfer function of (2.16) represents a gain G_1 which is independent of time. For this reason the transfer function of Fig. 2.7b with gain G can be also represented as a transfer characteristic as illustrated in Fig. 2.7c. The slope of the linear input-output characteristic $v_{in}(t)$ and $v_{out}(t)$ is the gain $G = G_1$ of the voltage divider (Fig. 2.6).

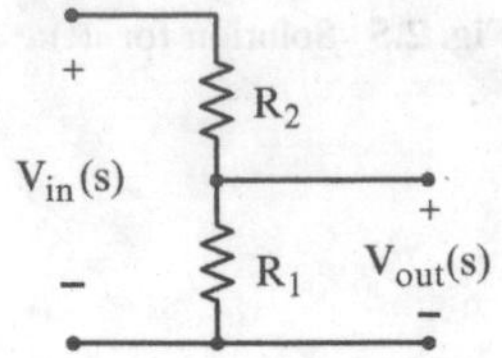

Fig. 2.6 Voltage-divider circuit

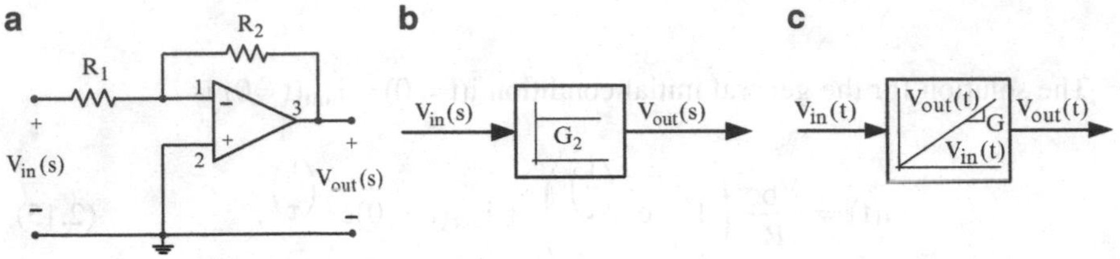

Fig. 2.7 (**a**) Inverting configuration of operational amplifier. (**b**) Representation of transfer function in a block diagram. (**c**) Representation of transfer characteristic in a block diagram

2.1.2 *Circuits Acting as Integrators*

The following circuits act as integrators, that is, the output signal is the integral of the input signal.

(a) R-C circuit (Fig. 2.8):

$i = C(dv_c/dt)$, with $v_c \ll v_R$ follows $v_{in} = v_c + v_R$ or $v_{in} \approx v_R$, $C(dv_c/dt) = v_{in}/R$, $v_c = v_{out} = (1/RC)\int v_{in}dt$, with $s = (d\ldots/dt)$

$$\frac{V_{out}(s)}{V_{in}(s)} = \frac{1}{s\tau}, \tag{2.18}$$

where $\tau = RC$ is a time constant.

(b) Operational amplifier with negative (capacitive) feedback: inverting configuration (Fig. 2.9a).

$$\frac{v_{out}(s)}{v_{in}(s)} = -\frac{Z_2}{R_1} = -\frac{\left(\frac{1}{sC_2}\right)}{R_1} \tag{2.19a}$$

or

$$\frac{v_{out}(s)}{v_{in}(s)} = -\frac{1}{sR_1C_2}, \tag{2.19b}$$

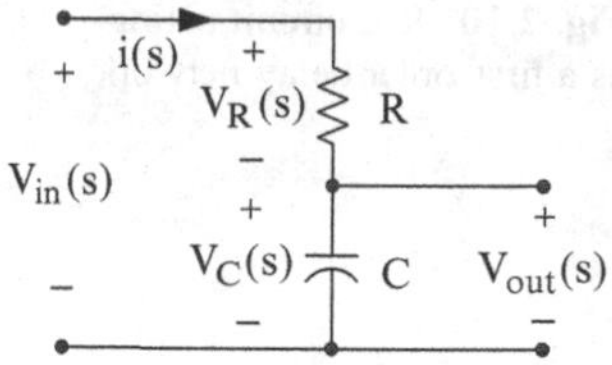

Fig. 2.8 R-C circuit as an integrator (I)

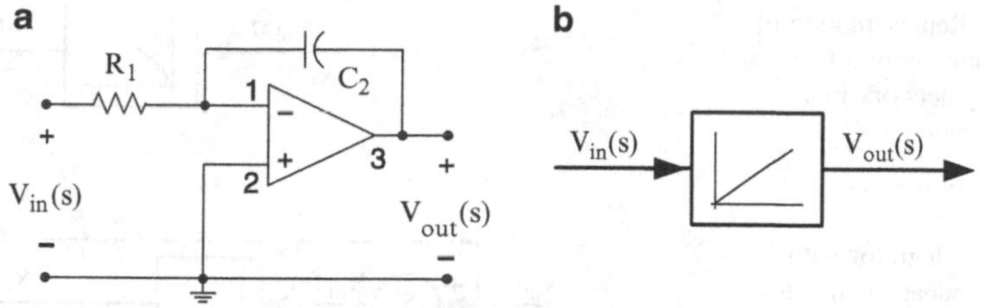

Fig. 2.9 (**a**) Operational amplifier with capacitive feedback acting as integrator (I). (**b**) Representation of transfer function of an integrator in a block diagram

where $\tau = R_1C_2$ is a time constant, thus

$$\frac{v_{out}(s)}{v_{in}(s)} = -\frac{1}{s\tau}. \tag{2.20}$$

Figure 2.9b shows the block diagram representation of an integrator circuit.

2.1.3 *First-Order Delay Transfer Function*

The following circuits have first-order delay transfer functions.

(a) R-L circuit (Fig. 2.10)

First-order differential equation with constant coefficients in time domain reads $i(t)R + L(di/dt) = v_{in}(t)$ or $i(t) + (L/R)(di/dt) = v_{in}(t)/R$, where $\tau = L/R$ is a time constant.

In Fig. 2.10 the voltage $v_{in}(t)$ is the input of the network, and the current i(t) is its response (or output). That is, the R-L circuit can be represented by a transfer function within a block diagram (Fig. 2.11) as before: $\frac{i(s)(1+s\tau)}{(v_{in}(s)/R)} = 1$ or

$$\frac{x_{out}(s)}{x_{in}(s)} = \frac{i(s)}{v_{in}(s)} = \frac{1}{R}\frac{1}{(1+s\tau)} = \frac{G_3}{(1+s\tau)}, \tag{2.21}$$

with $\tau = L/R$ and gain $G_3 = (1/R) < 1$.

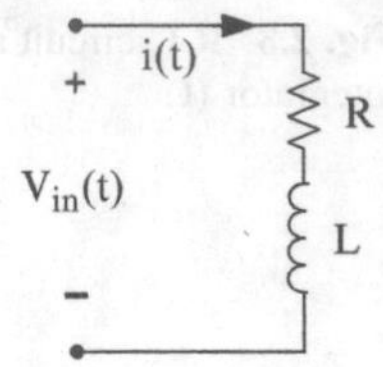

Fig. 2.10 R-L circuit acting as a first-order delay network

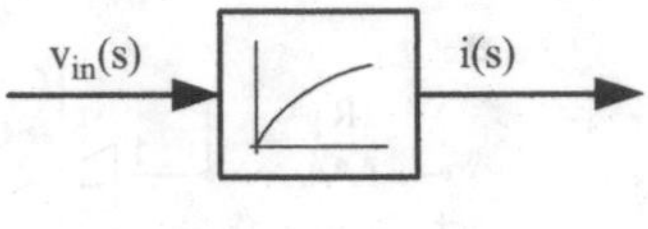

Fig. 2.11 Representation of transfer function of a first-order delay network in a block diagram

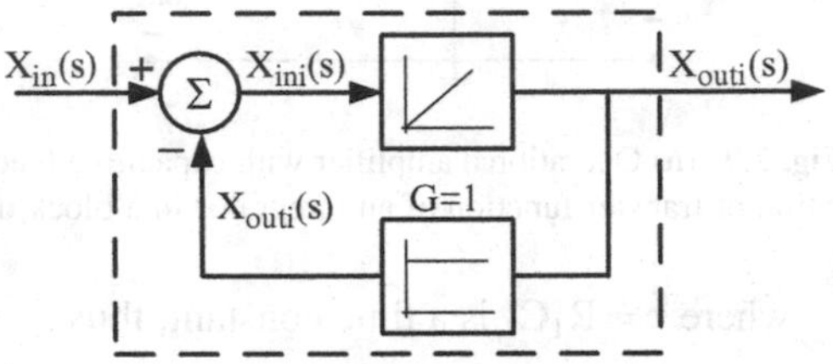

Fig. 2.12 Integrator with negative feedback generates a first-order delay network

(b) Integrator with negative feedback (Fig. 2.12).

$$x_{outi}(s) = x_{ini}(s)\frac{1}{s\tau},\ x_{in}(s) - x_{outi}(s) = x_{ini}(s),\ x_{outi}(s) = (x_{in}(s) - x_{outi}(s))\frac{1}{s\tau}$$

or

$$x_{outi}(s)\left(1 + \frac{1}{s\tau}\right) = \frac{x_{in}(s)}{s\tau}$$

$$\frac{x_{outi}(s)}{x_{in}(s)} = \frac{1}{s\tau(1+\frac{1}{s\tau})} = \left(\frac{1}{s\tau + 1}\right), \tag{2.22}$$

with gain G = 1.

2.1.4 Circuit with Combined Constant (P) and Integrating (I) Transfer Function

The circuit of Fig. 2.13 has a constant (P) transfer function combined with an integrating (I) transfer function.

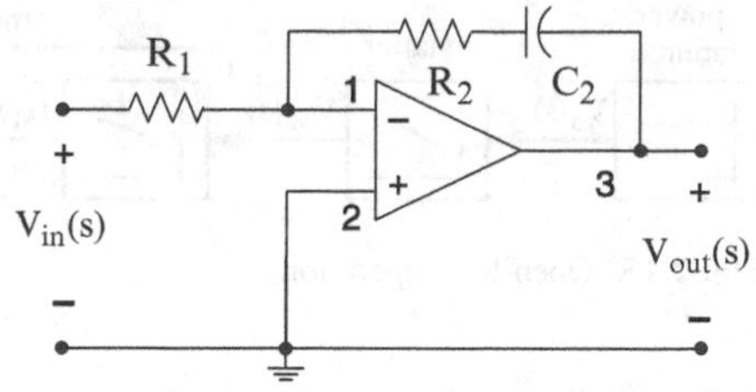

Fig. 2.13 Operational amplifier with resistive/capacitive feedback

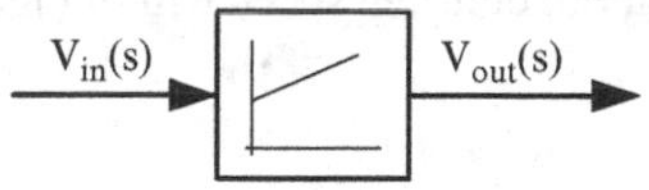

Fig. 2.14 Block diagram for proportional/integrator (PI) network

$$\frac{V_{out}(s)}{V_{in}(s)} = -\frac{Z_2}{R_1} = -\frac{\left(R_2 + \frac{1}{sC_2}\right)}{R_1} \tag{2.23a}$$

or

$$\frac{V_{out}(s)}{V_{in}(s)} = -\frac{R_2}{R_1} - \frac{1}{sR_1C_2}, \tag{2.23b}$$

where $\tau = R_1C_2$ is a time constant, thus

$$\frac{V_{out}(s)}{V_{in}(s)} = -\frac{R_2}{R_1} - \frac{1}{s\tau} = -\frac{\left(\frac{R_2}{R_1}\right)s\tau + 1}{s\tau} = G_4\frac{(1 + s\tau_1)}{s}, \tag{2.24}$$

where $G_4 = -(1/\tau) = (-1/R_1C_2)$, and $\tau_1 = (R_2/R_1)\tau = R_2C_2$.

Figure 2.14 shows the block diagram representation of a proportional/integrator (PI) circuit.

2.2 Commonly Occurring (Drive) Systems

Some of the more commonly occurring drive systems are presented using linear transfer functions within each and every block of these diagrams. In real designs, nonlinear elements frequently occur. However, such nonlinear components cannot be approximated by linear differential equations with constant coefficients (e.g., Laplace solution technique) and, therefore, this introductory text limits the analysis to mostly linear systems.

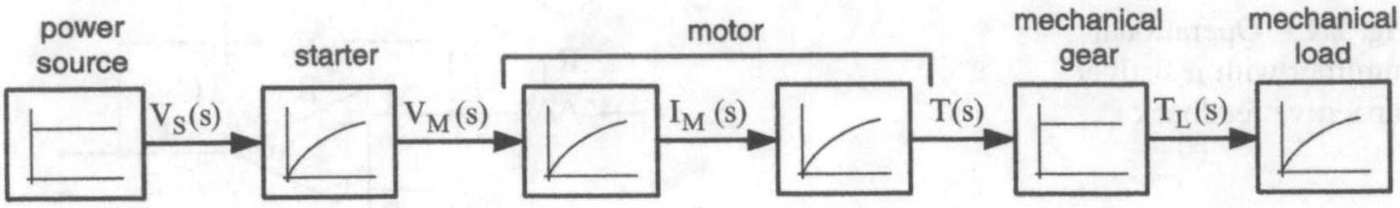

Fig. 2.15 Open-loop operation

2.2.1 Open-Loop Operation

The most commonly used electromechanical system is that of Fig. 2.15 operating at about constant speed without feedback control. The speed regulation is defined as [3]

$$\text{speed regulation} = \frac{(\text{no} - \text{load speed}) - (\text{full} - \text{load speed})}{(\text{full} - \text{load speed})}. \qquad (2.25)$$

2.2.2 Closed-Loop Operation

If the speed regulation is less than a few percent (< 3%) and constant speed is sufficient, then open-loop operation is acceptable for the majority of drive applications. However, if the speed regulation must be small and variable-speed is required, a closed-loop speed control, with negative feedback, must be chosen. Such a system is represented in Fig. 2.16a.

In this system, the (inner) current control loop is desirable because it improves the dynamic behavior of the drive and the current can be limited to a value below a given maximum value if a nonlinear element (see Application Example 2.8) is included.

Application Example 2.3: Description of Start-Up Behavior of a Current- and Speed-Controlled Drive

The dynamic response during start-up of the drive of Fig. 2.16a with inner current feedback and outer speed feedback loop – where the internal feedback loop of the electric machine is not shown in Fig. 2.16a, but shown in Fig. 2.16b – is as follows. Provided the reference value of the mechanical angular velocity is $\omega_m^* \cong \omega_{mrat}$ and the measured mechanical angular velocity at $t = t_0 = 0$ is $\omega_m \cong 0$ then the error angular velocity is $\varepsilon_{\omega m} \cong \omega_{mrat}$. If the gain of the speed controller is assumed to be $G_{\omega m} = 1.0$, then the reference signal of the current is $I_M^* \cong I_{M\,max}$. Because the filtered current is $I_{Mf} \cong 0$ the error signal of the current controller is $\varepsilon_{IM} \cong I_{M\max}$. Therefore, the gating circuit provides a gating signal $V_G \cong V_{Gmin}$ for the thyristor rectifier resulting in $\alpha \approx 0°$ and a maximum motor current $I_M \cong I_{M\max}$ corresponding to the maximum torque $T \cong T_{M\max}$ at $\omega_m \cong \omega_{mt1}$ (see Figs. 2.16c, d at time t_1).

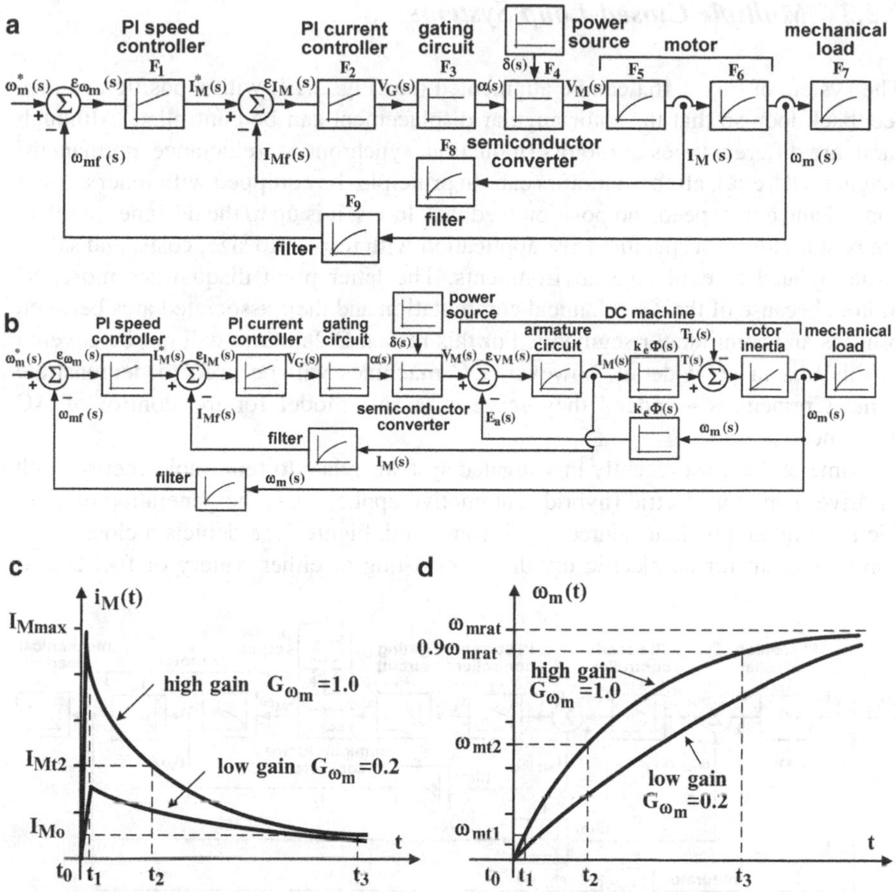

Fig. 2.16 (**a**) Closed-loop operation with current- and speed control where the load torque input of the electric machine is not shown. * indicates reference quantities. (**b**) Closed-loop operation with current- and speed control where the internal feedback loop of a DC machine is shown. * indicates reference quantities. Motor current $i_M(t)$ (**c**), motor (mechanical) angular velocity $\omega_m(t)$ (**d**)

Due to this large torque the motor gains speed quickly, say $\omega_m \cong \omega_{mt2}$ at $t = t_2$: the speed error becomes $\varepsilon_{\omega m} \cong \omega_{m\,max} - \omega_{m\,t2}$, at the same time the motor current reduces to $I_M \cong I_{Mt2}$ (see Figs. 2.16c, d at time t_2). After some time the current error signal becomes $\varepsilon_{IM} \cong 0.1 I_{Mmax}$; the gating voltage is increased to $V_G \cong V_{Gmax}$ corresponding to $\alpha = 175°$ and the thyristor rectifier delivers to the motor the no-load current corresponding to $I_M \cong I_{Mo}$ at $t = t_3$. In the meantime no-load speed has about been reached at $\omega_m \cong 0.9\ \omega_{m\,rat}$ (see Figs. 2.16c, d at time t_3).

2.2.3 *Multiple Closed-Loop Systems*

The system of Fig. 2.16a can be augmented (see Fig. 2.17) with a position control feedback loop so that the rotor angular displacement can be controlled. Although there are different types of motors (induction, synchronous, reluctance, permanent-magnet, DC, etc.), all these motors can -in principle- be equipped with inner current control and outer speed and position feedback loops. It is up to the designer to select the best motor for a specific drive application with respect to size, costs, and safety issues related to explosive environments. The latter point disqualifies most DC motors because of their mechanical commutation and their associated arcs between brushes and commutator segments. For this reason DC motors will not be covered in this text in great detail, however, DC machines will be used to demonstrate control principles – indeed they serve as a role model for the control of AC machines.

Some of the most recently investigated systems relate to renewable energy, such as drive trains for electric (hybrid) automotive applications, the generation of electricity from photovoltaic sources, and from wind. Figure 2.18 depicts a closed-loop control system for an electric car drive consisting of either battery or fuel cell as

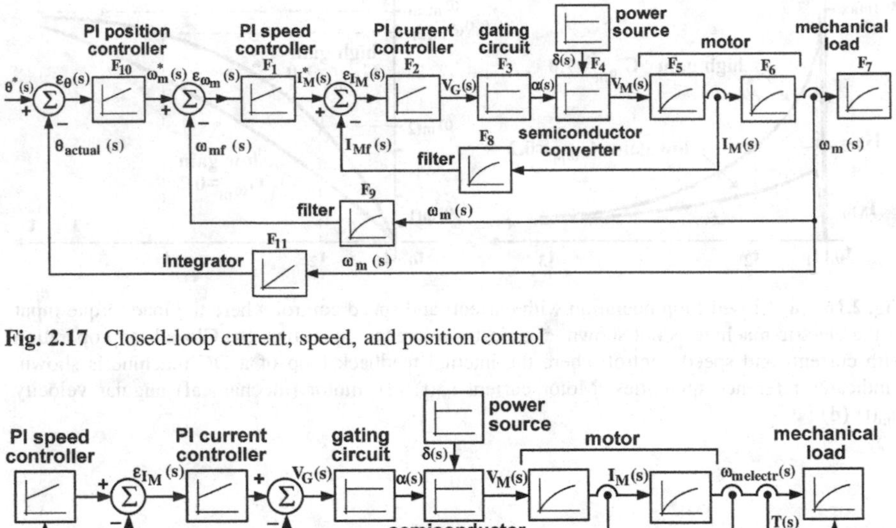

Fig. 2.17 Closed-loop current, speed, and position control

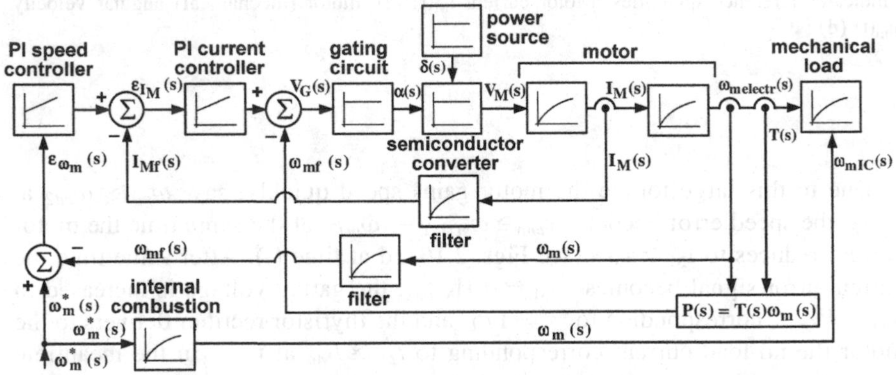

Fig. 2.18 Drive train for a parallel hybrid electric-internal combustion engine automobile drive

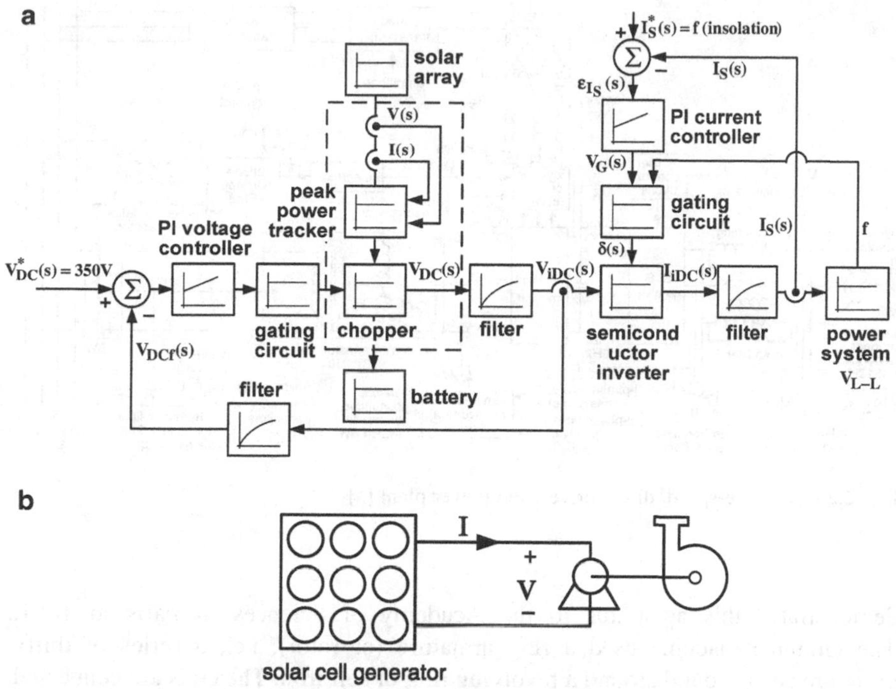

Fig. 2.19 (**a**) Photovoltaic power generation with maximum power tracker and inverter. (**b**) Photovoltaic system supplying power to DC motor and pump

energy source, a six-step or pulse-width-modulated (PWM) inverter, and either a permanent-magnet motor (where the combination of an inverter and a permanent-magnet machine/motor/generator is called a brushless DC machine/motor/generator) or an induction motor as well as an internal combustion (IC) engine.

A commonly employed solar power generation system is illustrated in Fig. 2.19. It consists of an array of solar cells, a peak-power tracking component, a DC-to-DC converter which is able to operate either in a step-down or step-up mode, and an inverter feeding AC power into the utility system. Less sophisticated photovoltaic drive applications also exist, where a solar array directly feeds a DC motor driving a pump [4], as shown in Fig. 2.19b; in this case the costs of the entire drive are reduced. Figure 2.20 illustrates the block diagram of a variable-speed, direct-drive permanent magnet 20 kW wind power plant [5].

2.3 Principle of Operation of DC Machines

The principle of operation of a DC machine can be best explained based on the Gramme ring named for its Belgian inventor, Zénobe Gramme. It was the first generator to produce power on a commercial scale for industry. Gramme

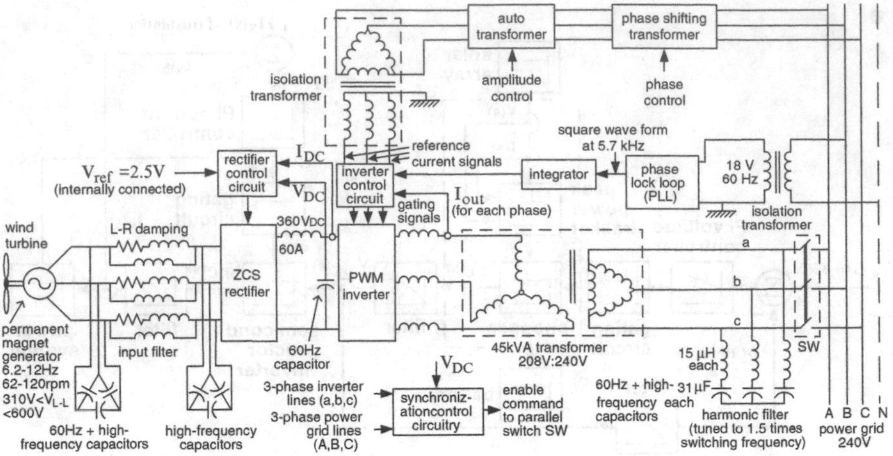

Fig. 2.20 Variable-speed, direct-drive wind power plant [5]

demonstrated this apparatus to the Academy of Sciences in Paris in 1871. The Gramme machine used a ring armature (or rotor), i.e., a series of thirty armature coils, wound around a revolving ring of soft iron. The coils are connected in series, and the junction between each pair is connected to a commutator on which two brushes run. The field excitation winding with resistance R_f and self inductance L_f or permanent magnets magnetize the stator core and the rotor (armature) iron ring, producing a magnetic field with the magnetomotive force (mmf) vector $\mathbf{F_f}$ which is stationary with respect to the stator or field iron core. The rotor coils (having resistance R_a and self inductance L_a) reside on the Gramme ring and rotate due the torque developed through the interaction of the armature current I_a with the field flux density $\vec{B}_f$ – the armature current I_a sets up a rotor or armature mmf vector $\mathbf{F_a}$ and interacts with the stator mmf $\mathbf{F_f}$ – I_a is supplied to the commutator segments via the two brushes. That is, the stator magnetic field $\vec{B_f}$ or mmf $\mathbf{F_f}$ and the rotor magnetic field or mmf $\mathbf{F_a}$ are stationary with respect to one another. The interaction of the stationary stator and rotor magnetic fields produce a torque $\vec{T} = R \times \vec{F}$ due to on the Lorentz force relation

$$\vec{F} = (\vec{I}_a \times \vec{B}_f)\ell, \tag{2.26}$$

where $\vec{I}_a$ is proportional to the mmf vector $\mathbf{F_a}$ and $\vec{B}_f$ is proportional to the mmf vector $\mathbf{F_f}$, and ℓ is the length of the DC machine (depth of the Gramme ring/cylinder), and R is the radius of the Gramme ring as indicated in Fig. 2.21a.

In Fig. 2.21a the brushes reside in the quadrature (interpolar) axis. The field mmf vector $\mathbf{F_f}$ and the armature mmf vector $\mathbf{F_a}$ are orthogonal, that is, the angle between both vectors is $\delta = 90°$ which is called the torque angle. For the configuration of Fig. 2.21a the torque can be calculated from

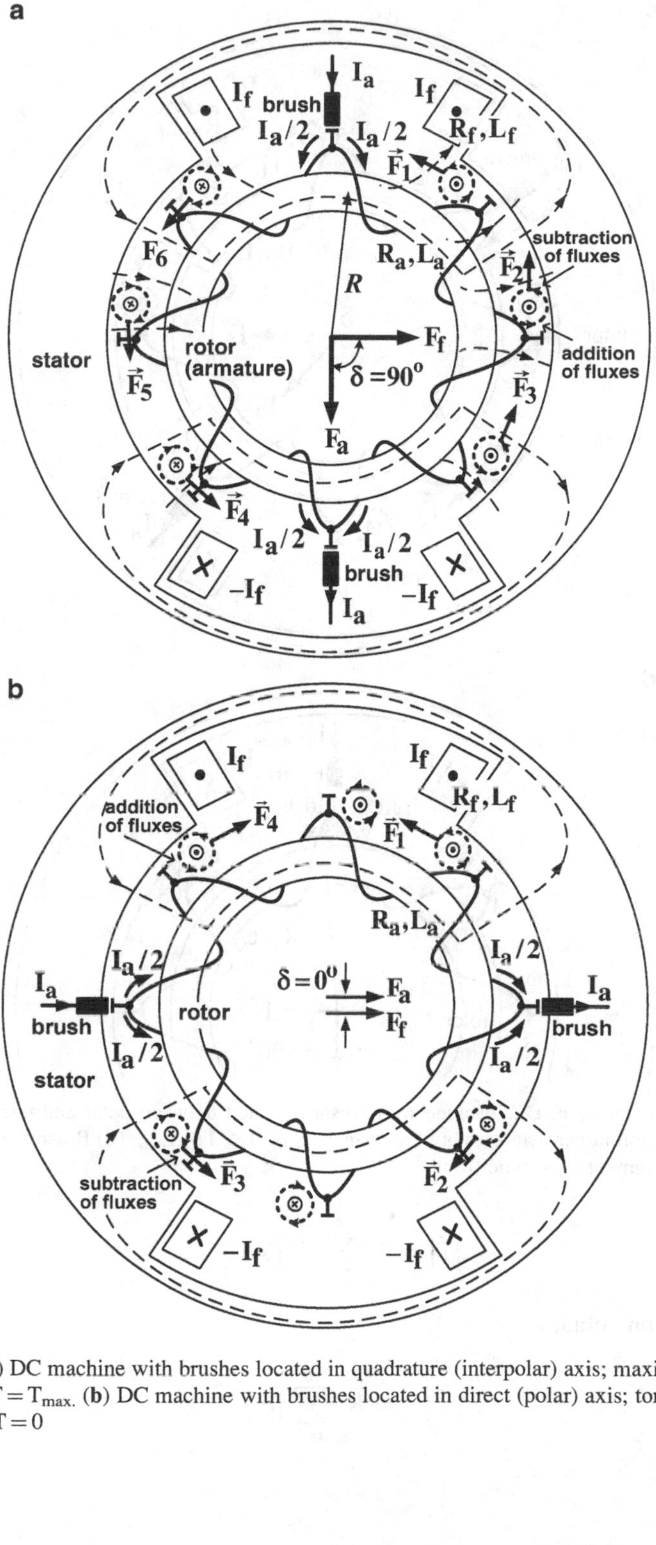

Fig. 2.21 (**a**) DC machine with brushes located in quadrature (interpolar) axis; maximum torque production, $T = T_{max.}$ (**b**) DC machine with brushes located in direct (polar) axis; torque production is zero, $T = 0$

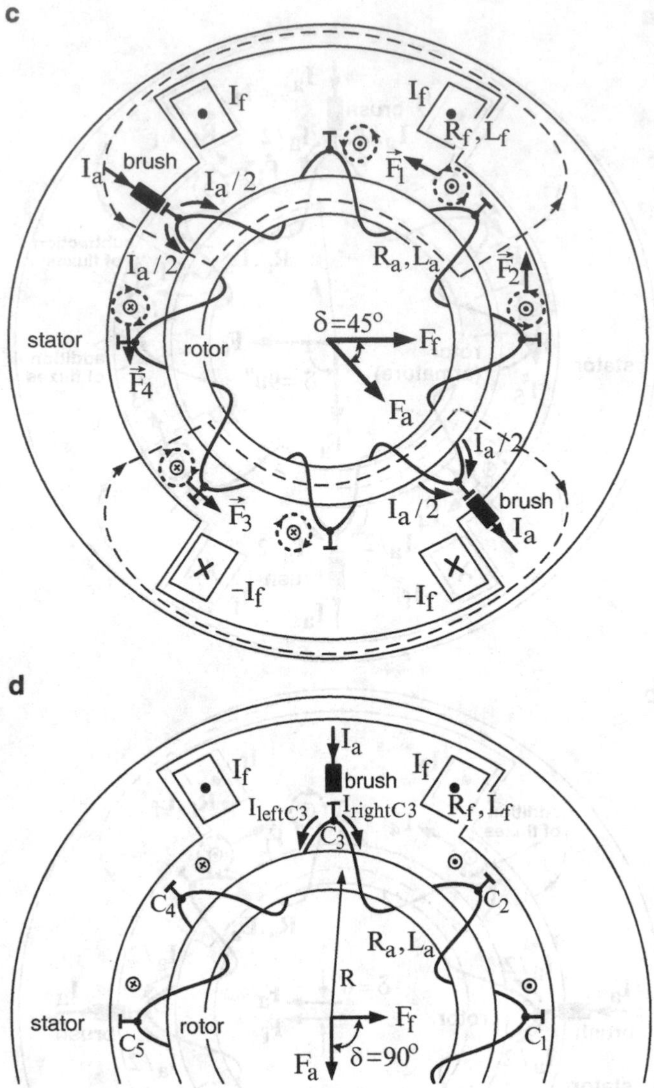

Fig 2.21 (continued) (**c**) DC machine with brushes located between polar and interpolar axes; torque production larger than zero but less than T_{max} or $0 \leq T \leq T_{max}$. (**d**) Brush is connected to commutator segment C_3 at time t_1

$$|\vec{T}| = R \cdot \sum_{i=1}^{6} |\vec{F}_i| \sin\delta, \tag{2.27}$$

for $\delta = 90°$ one obtains

$$|\vec{T}| = R \cdot \sum_{i=1}^{6} |\vec{F}_i|. \tag{2.28}$$

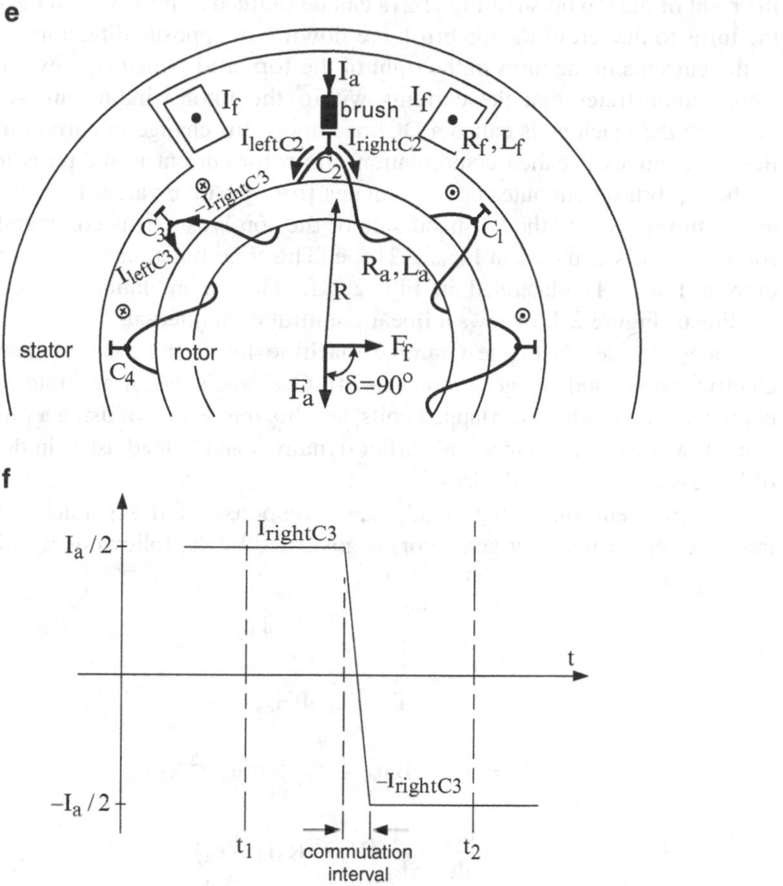

Fig 2.21 (continued) (**e**) Brush is connected to commutator segment C_2 at time t_2. (**f**) Commutation (change from positive current $I_{rightC3} = I_a/2$ at time t_1 when brush is connected to segment C_3 to negative current $(-I_{rightC3}) = -I_a/2$ at time t_2 when brush is connected to segment C_2) of rotor current. The linear commutation interval is indicated

In Fig. 2.21b the brushes reside in the direct (polar) axis. The field mmf vector $\mathbf{F_f}$ and the armature mmf vector $\mathbf{F_a}$ are collinear or parallel, that is, the angle between both vectors is δ=0°. For the configuration of Fig. 2.21b the torque can be calculated from (2.27) and is $|\vec{T}| = 0$. In Fig. 2.21c the brushes are in between the direct (d) and quadrature (q) axes. The angle between the field mmf vector $\mathbf{F_f}$ and the armature mmf vector $\mathbf{F_a}$ is δ, and the torque can be calculated from (2.27). This equation indicates that the torque $\vec{T}$ can be controlled as a function of δ by moving the brushes around the circumference of the machine. This concept will be exploited through electronic switching for the torque control of permanent-magnet machines and brushless DC machines.

When the rotor moves in counterclockwise direction, the currents in the turns to the right of the top brush of Fig. 2.21a can be defined as positive and the currents in the turns to the left of the top brush are flowing in opposite direction as compared to the currents in the turns to the right of the top brush – that is, they are negative. This demonstrates that the currents within the rotor winding are AC currents although the machine is called a DC machine. This change in current direction as the rotor moves is called commutation: the rotor current in the turns to the right of the top brush commutes or commutates from positive values to negative values as the turn moves to the left-hand side of the top brush. This commutation of the rotor current is depicted in Figs. 2.21d, e. The time functions of the commutating current $I_{rightC3}$ is illustrated in Fig. 2.21f. The commutation can be linear or nonlinear. Figure 2.21f shows a linear commutation interval.

Today the design of the Gramme machine forms the basis of nearly all DC electric motors and DC generators [6–24]. Gramme's use of multiple commutator contacts with multiple overlapped coils, and his innovation of using a ring armature (rotor), was an improvement on earlier dynamos and helped usher in development of large-scale electrical devices.

The transient (including steady-state) response of a separately excited DC machine (either motor or generator) is governed by the following equations:

$$\frac{d\omega_m}{dt} = \frac{1}{J}[T - T_L], \tag{2.29}$$

$$T = (k_e\ \Phi)i_a, \tag{2.30}$$

$$T_L = T_m + B\omega_m + T_c + c(\omega_m)^2 + T_s, \tag{2.31}$$

$$\frac{di_a}{dt} = \frac{1}{L_a}(V_a - R_a i_a - e_a), \tag{2.32}$$

$$e_a = (k_e\ \Phi)\omega_m. \tag{2.33}$$

In (2.29) ω_m is the mechanical angular velocity (measured in rad/s) of the rotor (armature) of the DC machine, J is the polar moment of inertia of the rotor, T is the magnitude of the motor torque [see (2.27)], and T_L is the magnitude of the load torque.

Equation 2.30 defines the magnitude of the motor torque as a function of the resultant flux Φ within the machine taking into account field flux and rotor flux (armature reaction), i_a is the instantaneous value of the rotor current whose steady-state value is I_a, and k_e is a constant. Equation 2.31 defines the load torque components: T_m is the mechanical torque doing useful work, B is a viscous damping coefficient, T_c is the Coulomb frictional torque, c is a constant, and T_s is the standstill torque.

In (2.32) V_a is the terminal voltage, and e_a the induced voltage within the rotor winding.

Application Example 2.4: Identify Transfer Functions of a Block Diagram

In this example, define all transfer functions of the block diagram of Fig. 2.16a and find the transfer function $\{\omega_m(s)/\omega_m^*(s)\}$.

Solution.

Individual transfer functions

$$F_1 = \frac{G_1(1+s\tau_1)}{s},\ F_2 = \frac{G_2(1+s\tau_2)}{s},\ F_3 = G_3,\ F_4 = G_4,\ F_5 = \frac{G_5}{(1+s\tau_5)},$$

$$F_6 = \frac{G_6}{(1+s\tau_6)},\ F_7 = \frac{G_7}{(1+s\tau_7)},\ F_8 = \frac{G_8}{(1+s\tau_8)},\ F_9 = \frac{G_9}{(1+s\tau_9)}.$$

$$I_M(s) = F_2F_3F_4F_5(I_M^*(s) - F_8I_M(s)),\ I_M(s)(1+F_2F_3F_4F_5F_8)$$

$$= F_2F_3F_4F_5I_M^*(s),\ \text{or}\ \frac{I_M(s)}{I_M^*(s)} = \frac{F_2F_3F_4F_5}{(1+F_2F_3F_4F_5F_8)},\ \omega_m(s) = F_6I_M(s),\ \omega_m(s)$$

$$= \frac{F_2F_3F_4F_5F_6}{(1+F_2F_3F_4F_5F_8)}I_M^*(s),\ I_M^*(s) = F_1(\omega_m^*(s) - F_9\omega_m(s)),$$

$$\frac{\omega_m(s)}{\omega_m^*(s)} = \frac{F_1F_2F_3F_4F_5F_6}{(1+F_2F_3F_4F_5F_8+F_1F_2F_3F_4F_5F_6F_9)} = F_{12}.$$

2.4 Central Air-Conditioning System for Residence

The central air-conditioning unit of a residence is shown in the block diagram of Fig. 2.22.

Air conditioning systems are used for space cooling (e.g., rooms, residences, etc.). They are similar to the cooling unit of a refrigerator.

The indoor heat (power) Q_L will be transported by an air handling ($P_{\text{air handlers}}$) and a compressor ($P_{\text{compressor}}$) system from the indoors to the outdoors.

- The constant-speed compressor(s) of the air conditioning unit is (are) located outdoors.
- The coil with the warm cooling medium is located outdoors, where the coil is cooled by a variable-speed outdoor fan (air handler), which dissipates the heat energy to the surrounding air, or there is a coil embedded in the ground and dissipates the heat to the soil or ground water. No fan will be required for the outside unit in this latter case.
- There is also a variable-speed indoor fan (air handler), which circulates the cool air originating from the coil with the cool medium within the residence.

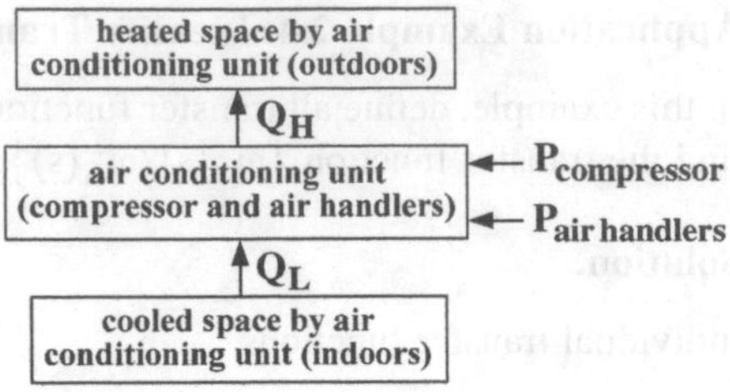

Fig. 2.22 Block diagram of a central air conditioning unit for a residence, where L and H means low and high temperature, respectively

It is the advantageous feature of an air conditioning system that for the given input electrical power P_{in}=($P_{compressor}$+ $P_{air\ handlers}$) one can provide cooling power Q_L for the indoors, where P_{in} is significantly smaller than Q_L. The coefficient of performance of an air conditioning unit is defined as

$$COP_L = Q_L/(P_{compressor} + P_{air\ handlers}).$$

Air conditioning systems are classified by the Seasonal Energy Efficiency Ratio (SEER). The relation between SEER and COP_L is:

$$SEER = 3.413 \cdot COP_L.$$

According to federal law [25], SEER=13 is the minimum seasonal energy efficiency ratio which is commercially available for residential air conditioning systems. Central air-conditioning systems are most efficient if they operate all the time at either high or low speed, and on-off operation (that is the start-up phase) should be avoided for high-efficient central air-conditioning systems. This is also good for extending the lifetime of compressors. One can classify air conditioning systems according to Table 2.1.

The size of an air conditioning unit is specified in terms of "tons", where 1 ton ≡ 12,000 BTU/h.

Additional useful units are: 1 BTU ≡ 1.055 kJ, 1 J ≡ 1 Ws, 1 kWh ≡ $3.6 \cdot 10^3$ kJ, and 1 quad ≡ 10^{15} BTU.

A new evaporative cooling technology [26, 27] can deliver cooler supply air temperatures than either direct or indirect evaporative cooling systems, without increasing humidity. The technology, known as the Coolerado Cooler™, has been described as an "ultra cooler" because of its performance capabilities relative to other evaporative cooling products. The Coolerado Cooler evaporates water in a secondary (or working) air stream, which is discharged in multiple stages. No water or humidity is added to the primary (or product) air stream in the process. This approach takes advantage of the thermodynamic properties of air, and it applies to both direct and indirect cooling technologies in an innovative cooling system that is drier than direct evaporative cooling and cooler than indirect cooling. The technology also uses much

Table 2.1 Classification of air conditioning units

Low efficiency	SEER = 13–14	Two (outdoor and indoor) constant-speed fans (air handlers) with either one or two speed ranges (high and low speed), one constant-speed compressor
Medium efficiency	SEER = 15–16	Two variable-speed (outdoor and indoor) fans (air handlers),
High efficiency	SEER = 17–19	two (one small rating and one large rating) constant-speed compressors. The difference between SEER = 15–16 and SEER =17–19 are the cooling surfaces of the outdoor and indoor coils of the air handlers: the larger the cooling surface the higher the efficiency

less energy than conventional vapor compression air-conditioning systems. Performance tests have shown that the efficiency of the Coolerado Cooler is 1.5–4 times higher than that of conventional vapor compression cooling systems, while it provides the same amount of cooling. It is suitable for climates having low to average humidity, as is the case in much of the Western half of the United States. This technology can also be used to pre-cool air in conventional heating, ventilating, and air-conditioning systems in more humid climates because it can lower incoming air temperatures without adding moisture.

Application Example 2.5: A Central Air-Conditioning Unit with SEER = 13

Central air-conditioning unit of rated capacity of $Q_L = 3$ tons (average residence of 2,500 $(ft)^2$), SEER = 13, one constant-speed compressor, and two constant-speed air handlers. Input power of two (outdoor and indoor fans) air handlers at 50% of rated cfm (cubic feet per minute): $P_{air\ handlers} = 2 \cdot I_{rms} \cdot \cos\Phi \cdot V_{rms} = 2 \cdot 7A \cdot 0.9 \cdot 120V = 0.864$ kW. Coefficient of performance is a function of SEER: $COP_L = SEER/3.413 = 3.81$. Cooling power provided to the cooled space (indoors) corresponding to 3 tons: $Q_L = 10.55$ kW; for given Q_L, COP_L and $P_{air\ handlers}$ values: $P_{compressor} = 1.91$ kW; $P_{in} = (P_{compressor} + P_{air\ handlers}) = 2.77$ kW

Application Example 2.6: A Central Air-Conditioning Unit with SEER = 19

Central air-conditioning unit of rated capacity of $Q_L = 3$ tons, SEER = 19, two constant-speed compressors (only one is on at any time, but not both) and two variable-speed air handlers: $P_{air\ handlers} = 2 \cdot 1.03A \cdot 120V = 0.247$ kW at 50% of rated cfm; $COP_L = SEER/3.413 = 5.57$; $Q_L = 10.55$ kW; $P_{compressor} = 1.65$ kW; $P_{in} = (P_{compressor} + P_{air\ handlers}) = 1.897$ kW. For a high-efficient air conditioner, the fan and coil of the outdoor air handler is rather large to increase the cooling surface.

The heat pump is a similar system with a reverse heat (power) flow Q_H from the outdoors to the indoors, where the heat is being released. A similar coefficient of performance can be defined for heat pumps: $COP_H = Q_H/(P_{compressor} + P_{air\ handlers})$.

Application Example 2.7: Emergency Standby Generating Set for Commercial Building

An emergency standby generating set for a commercial building is to be sized. The building has an air conditioning unit (10 tons, SEER=14) and two air handlers, two high-efficient refrigerators/freezers (18 (ft)^3, each), 100 light bulbs, three fans, two microwave ovens, three drip coffee makers, one clothes washer (not including energy to heat water), one clothes dryer, one electric range (with oven), TV and VCR, three computers, and one 3 hp motor for a water pump.

(a) Find the wattage of the required standby emergency power equipment.
(b) What is the estimated price for such standby equipment (without installation)?

Solution.

Typical wattages of appliances are listed in Table 2.2.

Table 2.2 Typical wattage of appliances

Clothes dryer: 4,800 W
Clothes washer (not including energy to heat hot water): 500 W
Drip coffee maker: 1,500 W
Fan (ceiling): 100 W
Fan(box/window): 200 W
Microwave oven: 1,450 W
Dishwasher: 1,200 W
Vacuum cleaner: 700 W
Range (with conventional oven): 3,200 W
Range (with self-cleaning oven): 4,000W
Refrigerator/freezer 18 (ft)^3: 150 W (new high-efficient) to 700 W (old)
VCR (including operation of TV): 120 W
Television: 250 W
Toaster: 1,100 W
Hair dryer: 1,250 W
Compact fluorescent light bulb (cfl) and incandescent light bulbs: 13 W (cfl) to 200 W (incandescent)
Computer: 360 W
Air conditioning unit, compressor (8 tons, SEER = 14): 6,000 W.
Two air handlers of air conditioning unit: 3,000 W.

(a) Required wattage

10 ton, SEER=14 air conditioning unit: 6,000·(10/8)= 7,500 W
air handlers for 10 ton, SEER=14 air conditioner: 3,000·(10/8)= 3,750 W
2 refrigerators/freezers (high-efficient): 2·150= 300 W
100 light bulbs (cfl): 100·65= 6,500 W
3 fans (box/window): 3·200= 600 W
2 microwave ovens: 2·1,450= 2,900 W
3 drip coffee makers: 3·1,500= 4,500 W
1 clothes washer (not including energy to heat hot water): 500 W
1 clothes dryer: 4,800 W
1 range (with oven): 3,200 W

TV and VCR: 370 W
2 computers: 3·360= 720 W
3 one-horse power motors (note 1 hp=746 W): 3·746= 2,238 W
Total wattage required: 37,878 W

(b) Price for a 40 kW standby generator according to [28] is about $10,210.

Application Example 2.8: Open-and Closed-Loop Operation of DC Machine Drive

Open-loop operation: Compute the transient response of a separately excited DC machine for changes of the armature voltage (V_a), of the load torque (T_L) doing the mechanical work (T_m), and of the flux ($k_e\Phi$) (see Fig. 2.23).

The transient response of a separately excited DC machine is governed by the following two first-order differential equations:

$$\frac{di_a}{dt} = \frac{1}{L_a}[V_a - R_a i_a - (k_e\Phi)\omega_m], \tag{2.34}$$

where the induced voltage is $e_a = (k_e\Phi)\omega_m$.

$$\frac{d\omega_m}{dt} = \frac{1}{J}\left[(k_e\Phi)i_a - T_m - B\omega_m - T_c - c(\omega_m)^2 - T_s\right], \tag{2.35}$$

where the mechanical load torque (including frictional torques $B\omega_m$, T_c, $c(\omega_m)^2$, T_s) is $T_L = T_m + B\omega_m + T_c + c(\omega_m)^2 + T_s$, and ω_m is the angular velocity related to the speed of the motor.

The electrical DC machine torque is

$$T = (k_e\Phi)i_a, \tag{2.36}$$

and the DC machine output power (neglecting windage and frictional losses) is

$$P = T\omega_m. \tag{2.37}$$

The constant machine parameters are defined as follows:

$L_a = 3.6$ mH, $R_a = 0.153\ \Omega$, $(k_e\Phi) = 2.314$ Vs/rad, $J = 0.56$ kgm^2, $T_c = 5$ Nm, $B = 0.1$ Nm/(rad/s), $c = 0.002$ Nm/(rad/s^2), and $T_s = 0$ Nm.

L_a is the armature inductance, R_a the armature resistance, $(k_e\Phi)$ corresponds to the exciting flux of the machine, J is the polar moment of inertia, T_c is the Coulomb frictional torque, B is the damping coefficient of the viscous torque, c is a

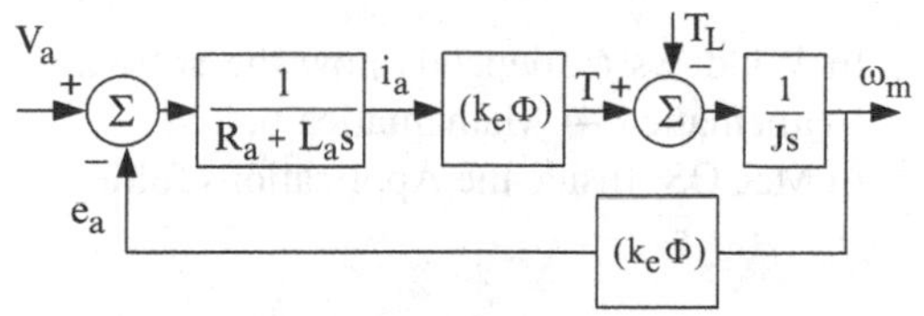

Fig. 2.23 Equivalent circuit of separately excited DC machine in time domain

Table 2.3 Variable input parameters for start-up based on armature voltage control, rated operation at rated speed, increased speed via flux weakening, and reversing the speed under reduced flux conditions

Time t [s]	0	2	4	6	8	10	12
Terminal voltage of DC motor V_a [V]	110	220	220	220	220	−220	
Field flux of DC motor ($k_e\phi$) [Vs/rad]	2.314	2.314	2.314	2.314	1.067	1.067	
Mechanical torque doing the work T_m [Nm]	0	0	150	190	75	−75	
Coulomb frictional torque T_c [Nm]	5	5	5	5	5	5	
Coefficient of windage torque c [Nm/(rad/s)^2]	0.002	0.002	0.002	0.002	0.002	−0.002	

constant for the windage torque, T_s is the torque at standstill, and in Fig. 2.23 "s" is the Laplace operator.

The input parameters V_a, T_m, ($k_e\Phi$) can be variable (see Table 2.3).

(a) Compute and plot ($0 \leq t \leq 12$ s) using either Matlab or Mathematica the armature current i_a (t), the angular velocity $\omega_m(t)$, the electrical machine torque T(t), and the machine output power P(t) as a function of time t for the initial conditions : $i_a(t=0)=0$, $\omega_m(t=0)=0$, and for the variations of the input parameters as detailed in Table 2.3. Print either the Matlab or Mathematica program. The start-up of the machine from 0 to 4 s is achieved by armature-voltage control, and the speed increase above rated speed from 8 to 10 s is accomplished based on flux reduction (below rated flux) or flux-weakening control. Reversing of speed is performed from 10 to 12 s under reduced-flux conditions.

Solution

Remove[] "removes symbols completely, so that their names are no longer recognized by Mathematica" rather than merely setting variables to zero, and "all of its properties and definitions are removed as well". More common than removing variables one at a time as in the remove statement is Remove["Global'*"], which removes everything except explicitly protected symbols in the Global (default) context. As an even more effective alternative, the Quit [] command can be used, which quits the kernel from the command line. All definitions are lost.

A way is to set at the very beginning of each and every program run all variables to zero push the button "Evaluation" and then choose "Quit Kernel" resulting in the response "local" and the question: "Do you really want to quit the kernel?". Then select "Quit".

How do we run the software Mathematica?:

1. In Windows (it depends how the software is installed): Start → Programs → Mathematica → Mathematica 6.
   In Mac OS: Inside the Applications folder, there should be a Mathematica folder

2. All Mathematica files have the extension .nb or notebook.
3. When entering a line or equation, a single equal sign (=) sets the value of a parameter, whereas a double equal sign (= =) means "equal" and relates to equations in Mathematica.
4. All functions of t are defined by x[t_]:= where the two important parts are t_ and := these are required whenever defining a function.
5. To run a program, one can execute the program one instruction at a time or have the computer go through all instructions for you.
   - To execute instructions one at a time, put the cursor on the line and press the return key while holding down the shift key. Or go to Kernel → Evaluation → Evaluate Cells.
   - To run the whole program at once, go to kernel → Evaluation → Evaluate Notebook.
6. Any blue messages from Mathematica signify an error in the notebook somewhere.

```
(*Remove statement*)

   Remove[T,TL,Va,Wm,sol,eqn1,eqn2,ic1,ic2,kphi,Tm,Tic,icky,P,Ia]

(*Definition of constant input parameters:*)

   La=0.0036;
   Ra=0.153;
   J=0.56;
   B=0.1;
   Ts=0;

(*Definition of time-dependent input parameters (see Table 2.3):*)

   Va[t_]:=If[t<10.0,If[t<2,110,220],-220];
   kphi[t_]:=If[t<8.0,2.314,1.067];
   Tm[t_]:=If[t<10.0,If[t<8.0,If[t<6.0,If[t<4.0,0,150],190],75],-75];
   Tic[t_]:=If[t<10.0,5,-5];
   icky[t_]:=If[t<10.0,0.002,-0.002];

(*Initial conditions at time t=0:*)

   ic1=Ia[0]==0;
   ic2=Wm[0]==0;

(*Parameters as a function of the independent variables (e.g.,Ia[t],T[t],Wm[t]):*)

   T[t_]:=kphi[t]*Ia[t];
   P[t_]:=T[t]*Wm[t];
```

```
(*Definition of the set of differential equations:*)
  eqn1=Ia'[t]==(1/La)*(Va[t]-Ra*Ia[t]-kphi[t]*Wm[t]);
  eqn2=Wm'[t]==(1/J)*(kphi[t]*Ia[t]-Tm[t]-B*Wm[t]-Tic[t]-icky[t]*
  (Wm[t]^2)-Ts);
(*Numerical solution of differential equation system:*)
sol=NDSolve[{eqn1,eqn2,ic1,ic2},{Ia[t],Wm[t]},{t,0,12}, MaxSteps->10000];
(*Plotting statements: Figs. 2.24-2.28*)
Plot[Evaluate[Tm[t]],{t,0,12},PlotRange->All,AxesLabel->{"t (s)","
Tm (Nm)"}]
Plot[Ia[t]/.sol,{t,0,12},PlotRange->All,AxesLabel-> {"t (s)","Ia (A)"}]
```

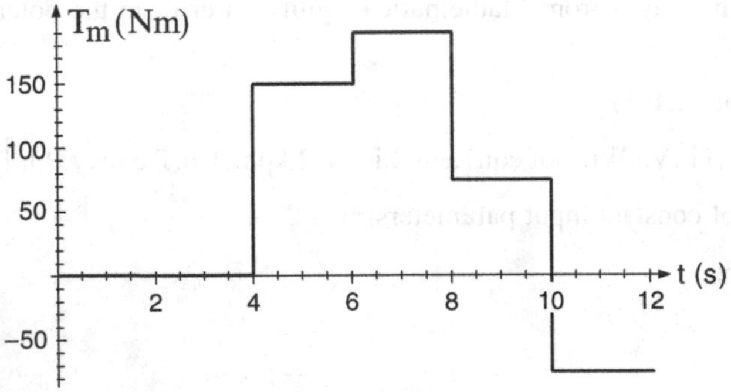

Fig. 2.24 Mechanical shaft torque in Nm

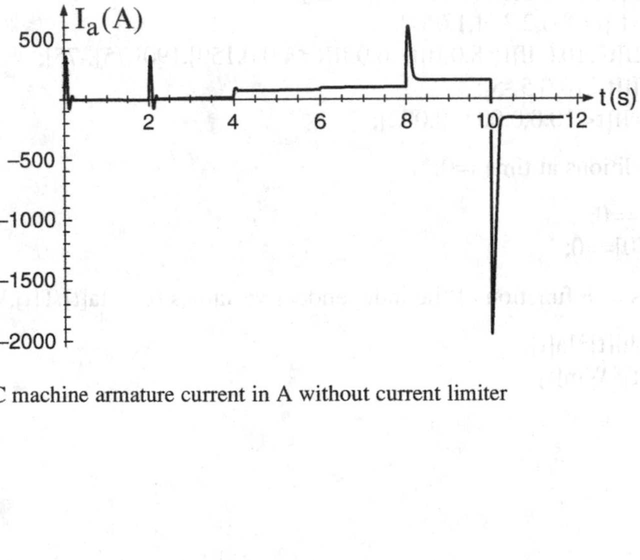

Fig. 2.25 DC machine armature current in A without current limiter

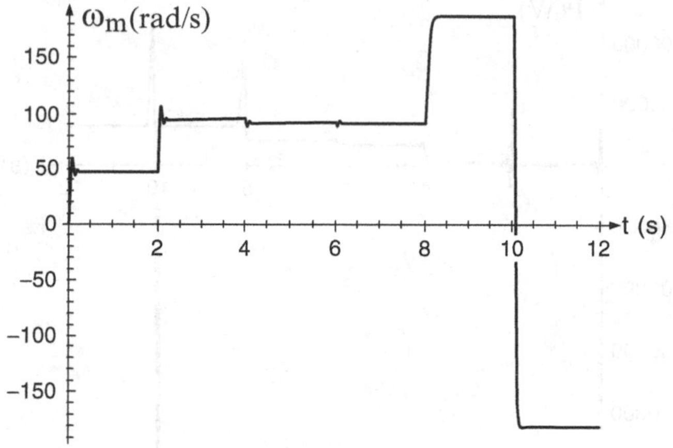

Fig. 2.26 Mechanical angular velocity in rad/s without current limiter

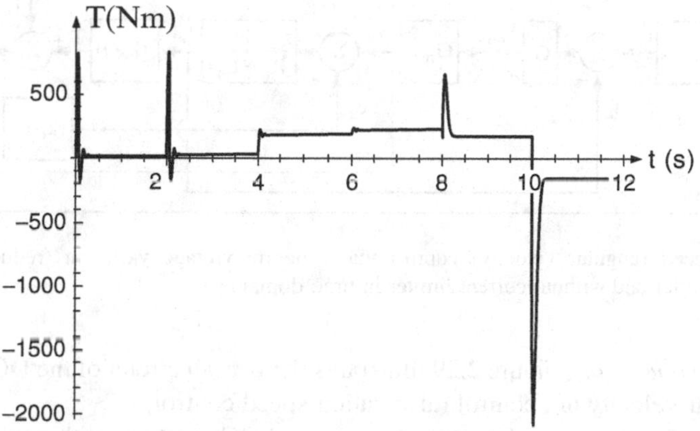

Fig. 2.27 DC machine torque in Nm without current limiter

```
Plot[Wm[t]/.sol,{t,0,12},PlotRange->All,AxesLabel->{"t (s)","Wm (rad/s)"}]
Plot[T[t]/.sol,{t,0,12},PlotRange->All,AxesLabel->{"t (s)","T (Nm)"}]
Plot[P[t]/.sol,{t,0,12},PlotRange->All,AxesLabel->{"t (s)","P (W)"}]
```

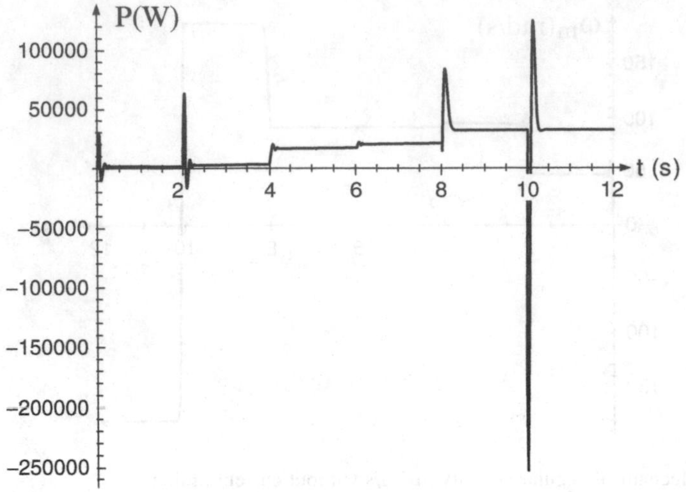

Fig. 2.28 DC machine output power in W without current limiter

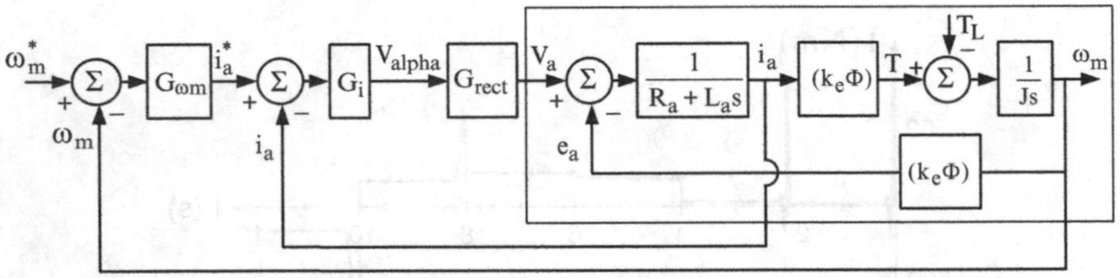

Fig. 2.29 Speed (angular velocity) control via armature voltage variation (reduction) with P-speed controller and without current limiter in time domain

Closed-loop operation: Figure 2.29 illustrates the overall circuit of the DC machine with angular velocity ω_m control (also called speed control).

The transient response of a separately excited DC machine with speed control, using armature–voltage variation, is governed by the following equations:

$$\frac{d\omega_m}{dt} = \frac{1}{J}[T - T_L], \tag{2.38}$$

$$T = (k_e\Phi)i_a, \tag{2.39}$$

$$T_L = T_m + B\omega_m + T_C + c\omega_m{}^2 + T_S, \tag{2.40}$$

$$\frac{di_a}{dt} = \frac{1}{L_a}(V_a - R_a i_a - e_a), \tag{2.41}$$

$$e_a = (k_e\Phi)\omega_m, \tag{2.42}$$

$$V_a = G_{rect}V_{alpha}, \tag{2.43}$$

$$V_{alpha} = G_i(i_a{}^* - i_a), \tag{2.44}$$

$$i_a{}^* = G_{\omega m}(\omega_m{}^* - \omega_m). \tag{2.45}$$

The constant machine parameters are defined as follows:

$L_a = 3.6$ mH, $R_a = 0.153\ \Omega$, $(k_e\Phi) = 2.314$ Vs/rad, $J = 0.56$ kgm^2, $T_c = 5$ Nm, B = 0.1 Nm/(rad/s),
c = 0.002 Nm/(rad/s^2), $T_s = 0$ Nm, $G_{rect} = 20$, $G_i = 1.0$, $G_{\omega m} = 10.0$.

Note all above values are maintained constant throughout closed-loop control and Table 2.3 does not apply. G_{rect} is the gain of the rectifier, G_i is the gain of the P-current controller, and $G_{\omega m}$ is the gain of the P-speed controller

(b) Using either Matlab or Mathematica, compute for $T_m = 190$ Nm and plot ($0 \leq t \leq 2$ s) the armature current $i_a(t)$, the angular velocity $\omega_m(t)$, the machine torque T(t), and the output power P(t) as a function of time for the initial conditions $i_a[0] = 0$, $\omega_m[0] = 0$, where at time $t = t_0 = 0$ $\omega_m{}^*$ changes from 0 to 100 rad/s.
(c) Replace at time $t_1 = 2$ s the P-speed controller by a PI-controller (where the final values of part b) serve as initial conditions). This means replace equation (2.45) by

$$\frac{di_a{}^*}{dt} = \frac{(\omega_m{}^* - \omega_m)}{0.2} - \frac{d\omega_m}{dt} \tag{2.46}$$

Mathematica program for the time interval t = 0 s to 2 s, and the speed P-controller gain of Gwm = 10.0:

(*A "remove" statement can be used to set variables to zero:*)

```
Remove[T,TL,Ea,Va,Iastar,Wmstar,Valpha,Wm,sol,eqn1,eqn2,ic1,ic2,kphi,
Tm,Tic,icky,P,Ia]
```

(*Definition of constant input parameters:*)

```
La=0.0036;
Ra=0.153;
J=0.56;
B=0.1;
Ts=0;
kphi=2.314;
Tc=5;
c=0.002;
Grect=20;
Gi=1;
Gwm=10;
Tm=190;
Wmstar=100;
```

(*Definition of time-dependent input parameters:*)
(*There are no time-dependent input parameters.*)

```
(*Initial conditions:*)

  ic1=Ia[0]==0;
  ic2=Wm[0]==0;

(*Parameters as a function of the independent variables:*)

  T[t_]:=kphi*Ia[t];
  P[t_]:=T[t]*Wm[t];
  TL[t_]:=Tm+B*Wm[t]+Tc+c*(Wm[t]*Wm[t])+Ts;
  Ea[t_]:=kphi*Wm[t];
  Va[t_]:=Grect*Valpha[t];
  Valpha[t_]:=Gi*(Iastar[t]-Ia[t]);
  Iastar[t_]:=Gwm*(Wmstar-Wm[t]);

(*Definition of the set of differential equations:*)

  eqn1=Ia'[t]==(1/La)*(Va[t]-Ra*Ia[t]-Ea[t]);
  eqn2=Wm'[t]==(1/J)*(T[t]-TL[t]);

(*Numerical solution of differential equation system:*)

  sol=NDSolve[{eqn1,eqn2,ic1,ic2},{Ia[t],Wm[t]},{t,0,2},MaxSteps->10000];

(*Plotting statements: Figs. 2.30-2.34*)

  Plot[Evaluate[Tm],{t,0,2},PlotRange->All,AxesLabel->{"t (s)","Tm
  (Nm)"}]
  Plot[Ia[t]/.sol[[1]],{t,0,2},PlotRange->All,AxesLabel->{"t (s)","Ia (A)"}]
  Plot[Wm[t]/.sol[[1]],{t,0,2},PlotRange->All,AxesLabel->{"t  (s)","Wm  (rad/
  s)"}]
  Plot[Evaluate[T[t]]/.sol[[1]],{t,0,2},PlotRange->All,  AxesLabel->{"t  (s)","T
  (Nm)"}]
Plot[Evaluate[P[t]]/.sol[[1]],{t,0,2},PlotRange->All,AxesLabel->{"t (s)","P (W)"}]
```

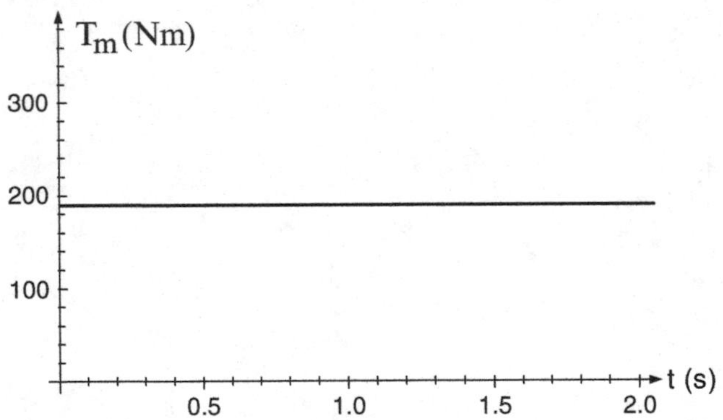

Fig. 2.30 Mechanical shaft torque in Nm

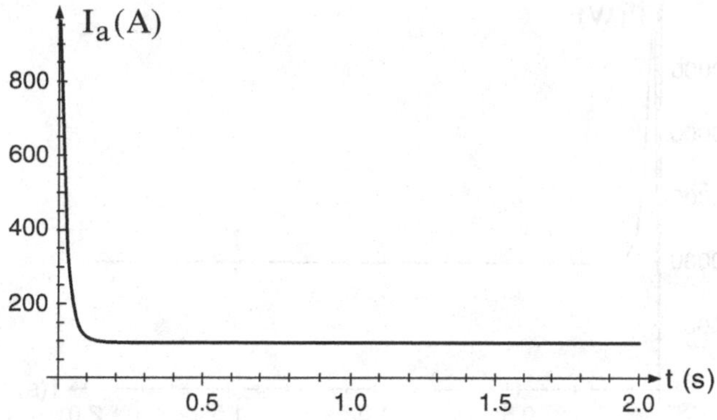

Fig. 2.31 DC machine armature current in A without current limiter

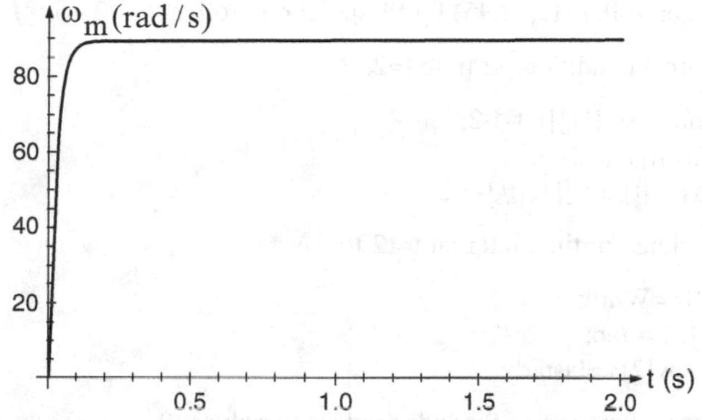

Fig. 2.32 Mechanical angular velocity in rad/s without current limiter

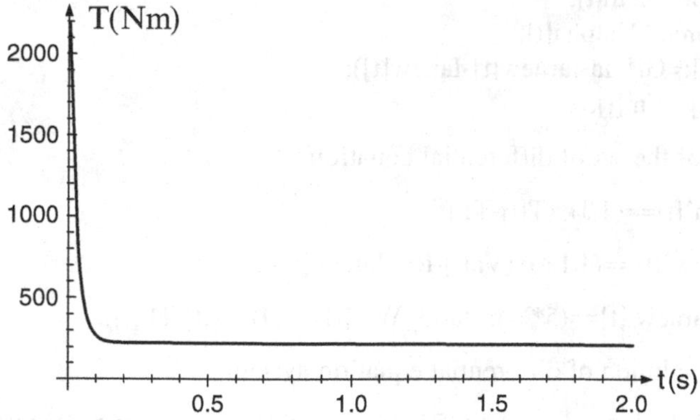

Fig. 2.33 DC machine torque in Nm without current limiter

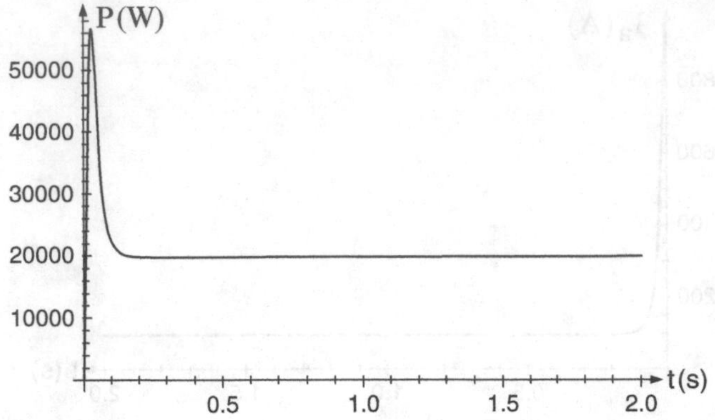

Fig. 2.34 DC machine output power in W without current limiter

(*Continuation of the Mathematica program for the time interval from t=2 s to 5 s replacing P-controller (Eq .2.45) by PI-speed controller (Eq .2.46):*)

(*Storing the final conditions at time t=2s:*)

```
Wmb=(Wm[t]/.sol[[1]])/.t->2;
Iab=(Ia[t]/.sol[[1]])/.t->2;
Iastarb=(Iastar[t]/.sol[[1]])/.t->2;
```

(*Initial conditions for time interval t=(2 to 5)s:*)

```
ic3=Wm[2]==Wmb;
ic4=Ianew[2]==Iab;
ic5=Iastarnew[2]==Iastarb;
```

(*Parameters as a function of the independent variables:*)

```
T[t_]:=kphi*Ianew[t];
TL[t_]:=Tm+B*Wm[t]+Tc+c*(Wm[t])^2+Ts;
Ea[t_]:=kphi*Wm[t];
Va[t_]:=Grect*Valpha[t];
Valpha[t_]:=Gi*(Iastarnew[t]-Ianew[t]);
P[t_]:=T[t]*Wm[t];
```

(*Definition of the set of differential equations:*)

```
eqn3=Wm'[t]==(1/J)*(T[t]-TL[t]);

eqn4=Ianew'[t]==(1/La)*(Va[t]-Ra*Ianew[t]-Ea[t]);

eqn5=Iastarnew'[t]==(5*Wmstar-5*Wm[t])-(1/J)*(T[t]-TL[t]);
```

(*Numerical solution of differential equation system:*)

```
sol2=NDSolve[{eqn3,eqn4,eqn5,ic3,ic4,ic5},{Wm[t], Ianew[t],Iastarnew[t]},
{t,2,5},MaxSteps->10000];
```

(*Plotting statements: Figs. 2.35–2.40*)

Plot[Evaluate[Tm],{t,2,5},PlotRange->All,AxesLabel->{"t (s)","Tm (Nm)"}]
Plot[Ianew[t]/.sol2[[1]],{t,2,5},PlotRange->{0,200}, AxesLabel->{"t (s)","Ianew (A)"}]
Plot[Wm[t]/.sol2[[1]],{t,2,5},PlotRange->{0,150}, AxesLabel->{"t (s)","Wm (rad/s)"}]

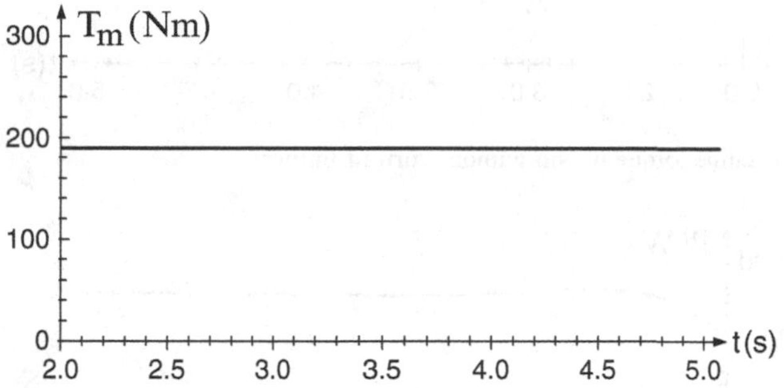

Fig. 2.35 Mechanical shaft torque in Nm

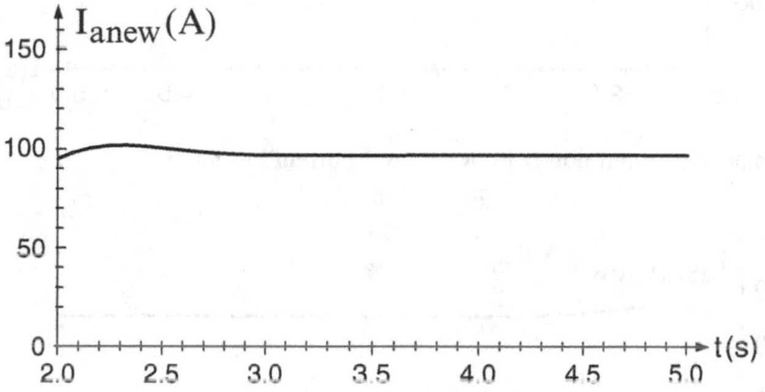

Fig. 2.36 DC machine armature current in A without current limiter

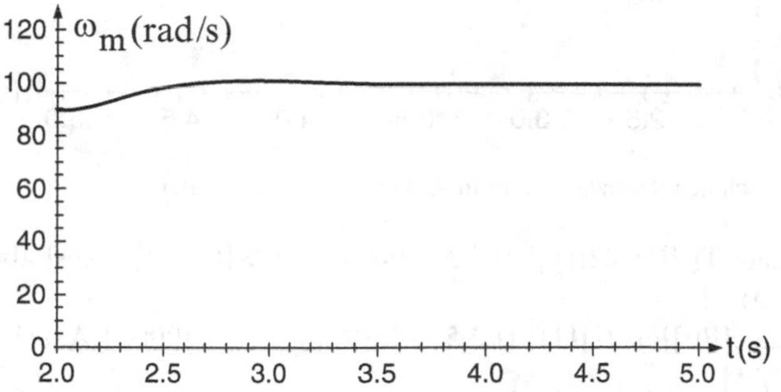

Fig. 2.37 Mechanical angular velocity in rad/s without current limiter

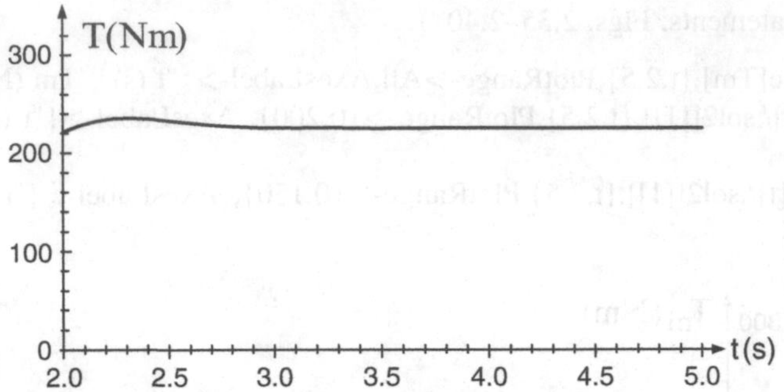

Fig. 2.38 DC machine torque in Nm without current limiter

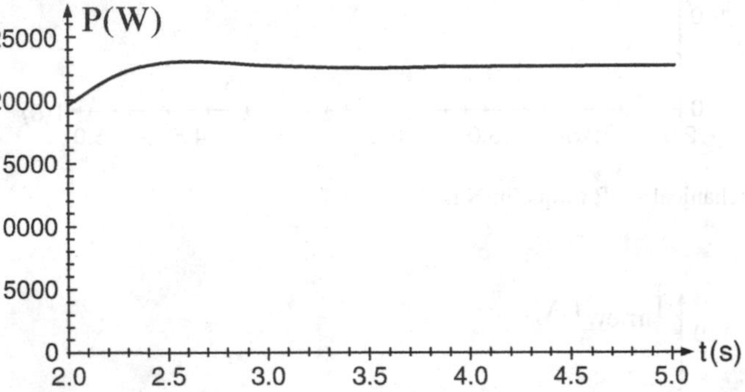

Fig. 2.39 DC machine output power in W without current limiter

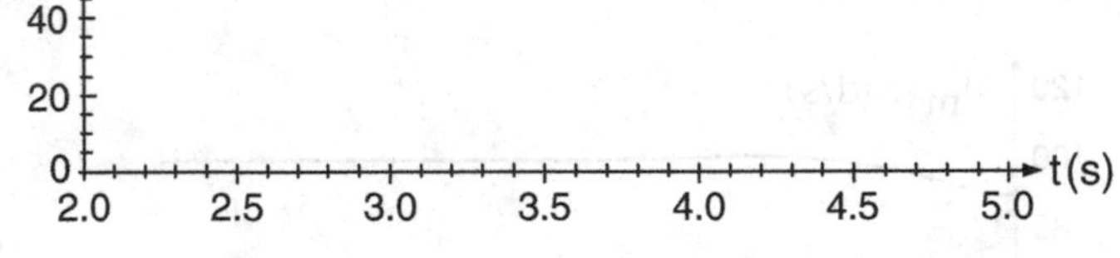

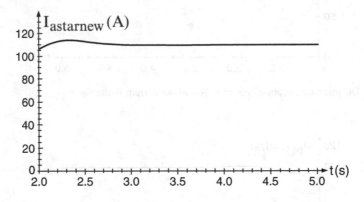

Fig. 2.40 DC machine reference current in A without current limiter

```
Plot[Evaluate[T[t]]/.sol2[[1]],{t,2,5},PlotRange->{0,400},AxesLabel->{"t (s)","T (Nm)"}]
Plot[Evaluate[P[t]]/.sol2[[1]],{t,2,5},PlotRange->{0,30000},AxesLabel->{"t (s)","P (W)"}]
Plot[Iastarnew[t]/.sol2[[1]],{t,2,5},PlotRange->{0,150},AxesLabel->{"t (s)","Iastarnew (A)"}]
```

Computation of the steady-state error $\varepsilon_{\omega m}$ of the angular velocity based on final-value theorem:

observe that there is a steady-state error due to the P-speed controller (see Problem 2.5), which is why the final value of the angular velocity ω_m is not 100 rad/s!

We can compute this steady-state error of the angular velocity from: $\varepsilon_{\omega m}(s) = (\omega_m^*(s) - \omega_m(s))/\omega_m^*(s)$: Considering Fig. 2.29, we know that

$$\omega_m(s) = \frac{1}{Js}\left[\frac{\left(\left[(\omega_m^*(s) - \omega_m(s))G_{\omega m} - i_a\right]G_i G_{rect} - e_a\right)}{R_a + L_a s}k_e\,\Phi - T_L\right], \tag{2.47}$$

$$\omega_m(s) = \frac{k_e \Phi i_a(s) - T_L}{Js}, \tag{2.48}$$

and

$$i_a(s) = \frac{Js\omega_m(s) + T_L}{k_e \Phi} \tag{2.49}$$

by substitution in the first equation, we obtain:

$$Js\,\omega_m(s) = \frac{\left(G_{\omega m}\omega_m^*(s) - G_{\omega m}\omega_m(s) - \frac{Js\omega_m(s)+T_L}{k_e\Phi}\right)G_i G_{rect} - k_e\Phi\omega_m(s)}{R_a + L_a s}k_e\,\Phi - T_L \tag{2.50}$$

or

$$\begin{aligned}(R_a + L_a s)Js\,\omega_m(s) = &\Big[G_i G_{rect} G_{\omega m}\,\omega_m^*(s) - G_i G_{rect} G_{\omega m}\,\omega_m(s)\\ &- \frac{G_i G_{rect}}{k_e\,\Phi}(Js\,\omega_m(s) + T_L) - k_e\,\Phi\,\omega_m(s)\Big]k_e\,\Phi\\ &- T_L(R_a + L_a s)\end{aligned} \tag{2.51}$$

or

$$\begin{aligned}&\omega_m(s)\left(R_a Js + JL_a s^2 + G_i G_{rect} G_{\omega m} k_e\,\Phi + G_i G_{rect} Js + (k_e\,\Phi)^2\right)\\ &\quad = k_e\,\Phi G_i G_{rect} G_{\omega m}\omega_m^*(s) - T_L(G_i G_{rect} + R_a + L_a s)\end{aligned} \tag{2.52}$$

resulting in the angular velocity as a function of the Laplace operator s

$$\omega_m(s) = \frac{G_i G_{rect} G_{\omega m}\omega_m^*(s)k_e\Phi - T_L(G_i G_{rect} + R_a + L_a s)}{R_a Js + JL_a s^2 + G_i G_{rect} G_{\omega m} k_e\,\Phi + G_i G_{rect}\,Js + (k_e\Phi)^2}. \tag{2.53}$$

The final (steady-state) value of $\omega_{mss} = \{\lim s \to 0$ applied to $\omega_m(s)\}$ is now for a $\omega_m(s)$ step function at the input:

$$\lim_{t\to\infty}\omega_m(t) = \omega_{mss} = \frac{\omega_m^*\,G_i\,G_{rect}\,G_{\omega m}\,k_e\,\Phi - T_L(R_a + G_i\,G_{rect})}{G_i\,G_{rect}G_{\omega m}\,k_e\,\Phi + (k_e\Phi)^2}. \tag{2.54}$$

For the specific case when the gain of the speed controller is $G_{\omega m} = 10$ one obtains (see given circuit parameters) with $\omega_m^* = 100$ rad/s and

$$T_L = T_m + B\omega_m + T_C + c{\omega_m}^2 + T_S = 190 + 0.1\omega + 0.002\omega^2 + 5 \qquad (2.55)$$

and if one approximately takes (we do not exactly know the speed $\omega_{mss} \approx \omega_{estimate} =$ 90 rad/s) $T_L \approx 220.2$Nm, one gets $\omega_{mss} \approx 89.38$rad/s or the steady-state error becomes

$$\varepsilon_{\omega m} = (\omega_m * - \omega_{mss})/\omega_m * = 0.1062 = 10.62\%. \qquad (2.56)$$

This steady-state error will become zero when we use a PI-speed controller. This is confirmed by the Mathematica simulation (see Fig. 2.37).

Closed-loop operation of a speed-controlled drive with current limiter or clipper using Mathematica and Matlab Software:

(d) Repeat the closed-loop analysis of parts b) and c) by inserting a current limiter or clipper circuit in Fig. 2.29. This modified block diagram is shown below in Fig. 2.41. The current limiter has a nonlinear characteristic, where the input is i_{in} and the output is $i_{out} = i_a^*$. The output current of this clipper must be limited to $i_{amax} = \pm 200$ A. This improves the safety of this drive. The block diagram of the current limiter is depicted in detail in Fig. 2.42.
(e) Use Mathematica to solve for the given set of linear differential equations together with the set of given algebraic linear equations as specified in parts b) and c) taking into account the nonlinear characteristic of the current limiter of Fig. 2.42.
(f) Use Matlab to solve for the given set of linear differential equations together with the given set of algebraic linear equations as specified in parts b) and c) taking into account the nonlinear characteristic of the current limiter of Fig. 2.42.
(g) What is the effect of the nonlinear current limiter on the dynamic performance of the drive?

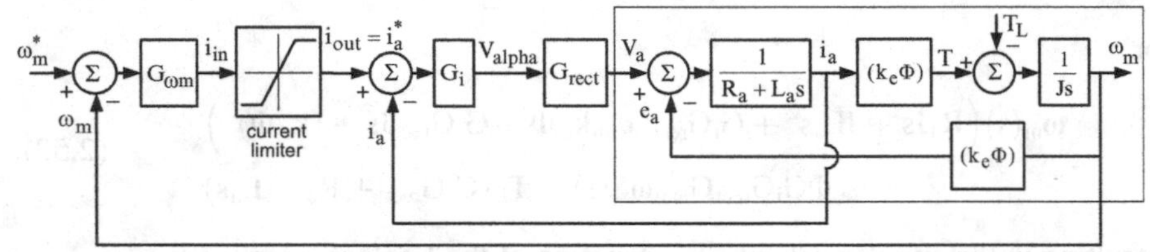

Fig. 2.41 Speed (angular velocity) control via armature-voltage variation with P and PI-controllers and a current limiter in time domain

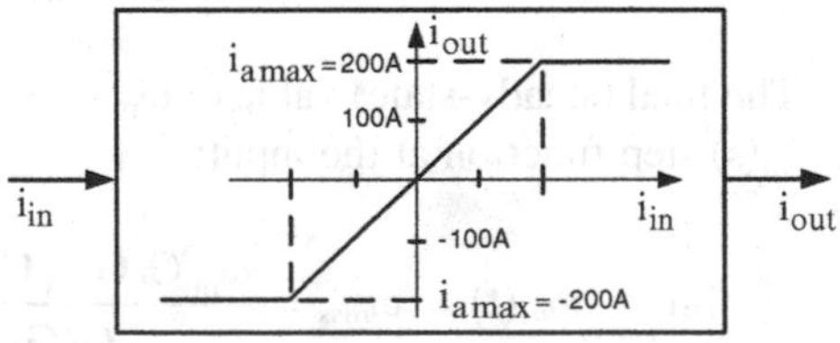

Fig. 2.42 Transfer characteristic of current limiter as used in Fig. 2.41

(*Remove statement: Figs. 2.43-2.50*)

```
Remove[T,Wmstar,eqn1,eqn2,eqn3,eqn4,eqn5,ic1,ic2,ic3,ic4,ic5]
La=0.0036;
Ra=0.153;
kphi=2.314;
J=0.56;
Tc=5;
B=0.1;
c=0.002;
Ts=0;
Grect=20;
Gi=1.0;
Gwm=10.0;
Tm=190;
Wmstar=100;
ic1=Ia[0]==0;
ic2=Wm[0]==0;
Valpha[t_]:=Gi*(Iastar[t]-Ia[t]);
Iastar[t_]:=Clip[Gwm*(Wmstar-Wm[t]),{-200,200}, {-200,200}];
Va[t_]:=Grect*Valpha[t];
T[t_]:=kphi*Ia[t];
P[t_]:=T[t]*Wm[t];
eqn1=Ia'[t]==(1/La)*(Va[t]-Ra*Ia[t]-kphi*Wm[t]);
eqn2=Wm'[t]==(1/J)*(kphi*Ia[t]-Tm-B*Wm[t]-Tc-c* (Wm[t]^2)-Ts);
sol1=NDSolve[{eqn1,eqn2,ic1,ic2},{Ia[t],Wm[t]},{t,0,2},MaxSteps->10000];
Plot[Ia[t]/.sol1[[1]],{t,0,2},PlotRange->All,AxesLabel->{"t (s)","Ia (A)"}]
```

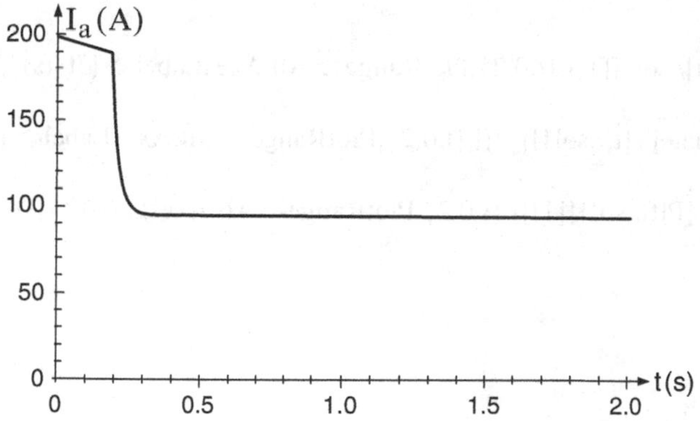

Fig. 2.43 DC machine armature current in A with current limiter

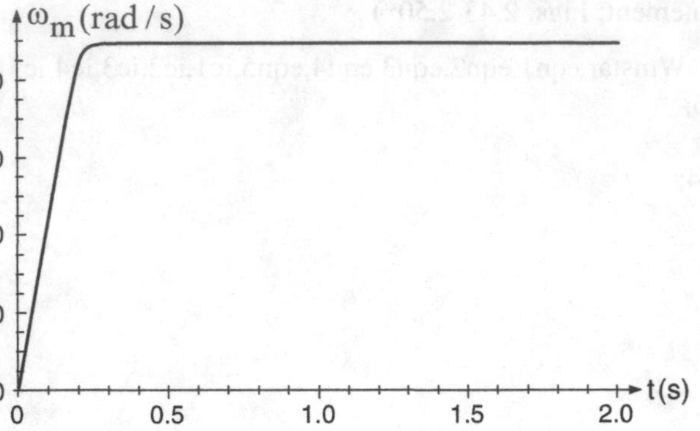

Fig. 2.44 Mechanical angular velocity in rad/s with current limiter

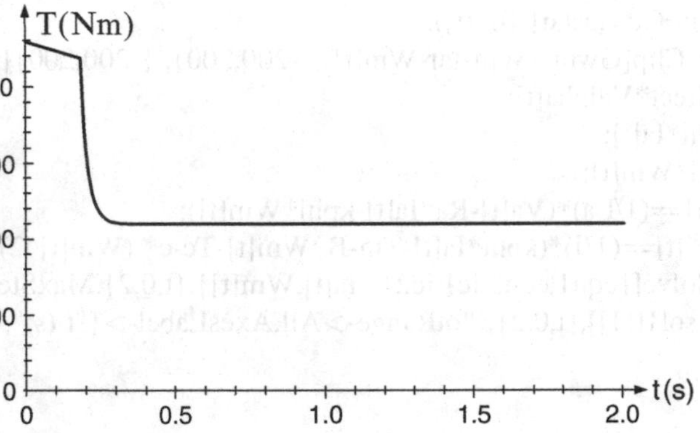

Fig. 2.45 DC machine torque in Nm with current limiter

```
Plot[Wm[t]/.sol1[[1]],{t,0,2},PlotRange->All,AxesLabel->{"t (s)","Wm (rad/s)"}]
Plot[Evaluate[T[t]/.sol1[[1]]],{t,0,2},PlotRange->All,AxesLabel->{"t (s)","T (Nm)"}]
Plot[Evaluate[P[t]/.sol1[[1]]],{t,0,2},PlotRange->All,AxesLabel->{"t (s)","P (W)"}]
```

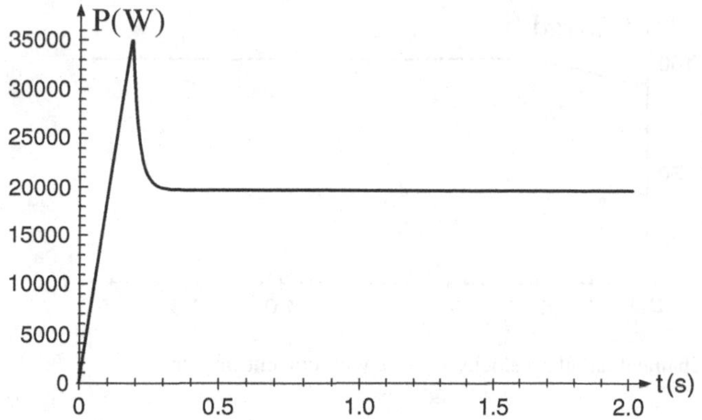

Fig. 2.46 DC machine output power in W with current limiter

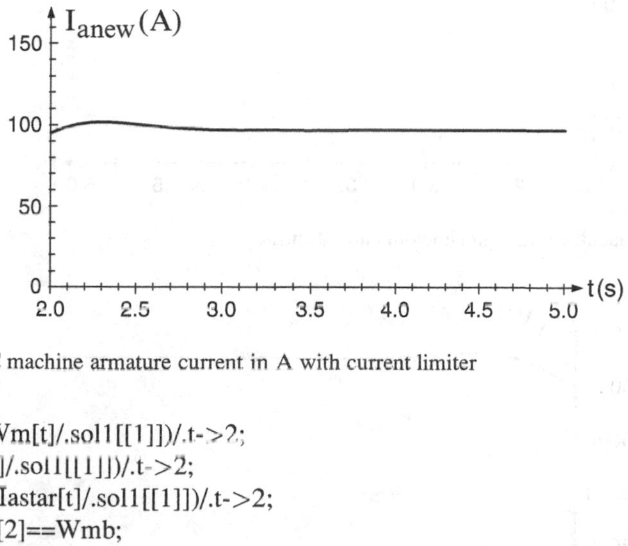

Fig. 2.47 DC machine armature current in A with current limiter

```
Wmb=(Wm[t]/.sol1[[1]])/.t->2;
Iab=(Ia[t]/.sol1[[1]])/.t->2;
Iastarb=(Iastar[t]/.sol1[[1]])/.t->2;
ic3=Wm[2]==Wmb;
ic4=Ianew[2]==Iab;
ic5=Iastarnew[2]==Iastarb;
Va[t_]:=Grect*Valpha[t];
Valpha[t_]:=Gi*(Iastarnew[t]-Ianew[t]);
T[t_]:=kphi*Ianew[t];
P[t_]:=T[t]*Wm[t];
eqn3=Wm'[t]==(1/J)*(kphi*Ianew[t]-Tm-B*Wm[t]-Tc-c*(Wm[t]^2)-Ts);
eqn4=Ianew'[t]==(1/La)*(Va[t]-Ra*Ianew[t]-kphi*Wm[t]);
eqn5=Iastarnew'[t]==(Wmstar-Wm[t])/0.2-Wm'[t];
sol2=NDSolve[{eqn3,eqn4,eqn5,ic3,ic4,ic5},{Wm[t],   Ianew[t],Iastarnew[t]},
{t,2,5},MaxSteps->10000];
Plot[Ianew[t]/.sol2[[1]],{t,2,5},PlotRange->{0,150}, AxesLabel->{"t (s)","Ianew
(A)"}]
```

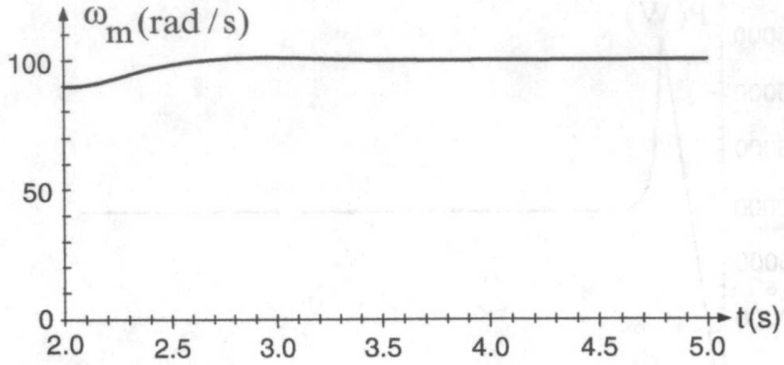

Fig. 2.48 Mechanical angular velocity in rad/s with current limiter

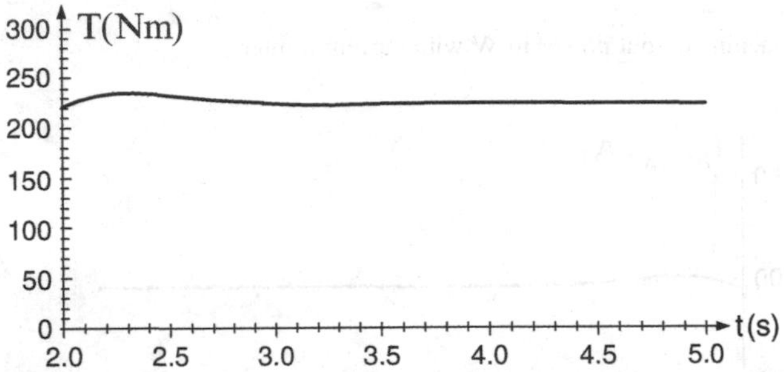

Fig. 2.49 DC machine torque in Nm with current limiter

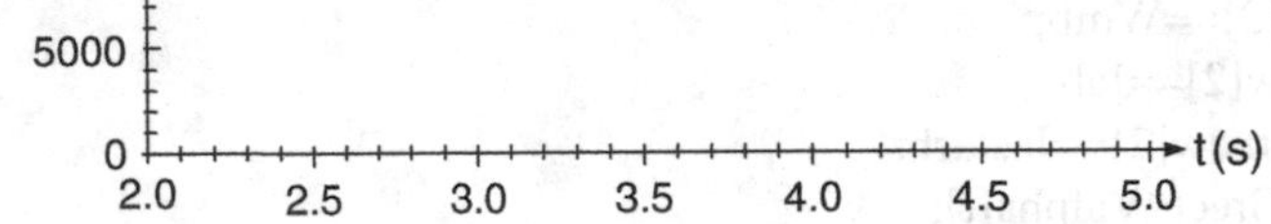

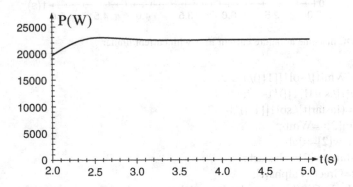

Fig. 2.50 DC machine output power in W with current limiter

```
Plot[Wm[t]/.sol2[[1]],{t,2,5},PlotRange-> {0,150},
AxesLabel->{"t (s)","Wm (rad/s)"}]
Plot[Evaluate[T[t]/.sol2[[1]]],{t,2,5},PlotRange->{0,300},AxesLabel->{"t
(s)","T (Nm)"}]
Plot[Evaluate[P[t]/.sol2[[1]]],{t,2,5},PlotRange-> {0,25000},AxesLabel->{"t
(s)","P (W)"}]
```

(h) The Matlab program is not listed, however, the results are identical with those obtained with Mathematica software.
(i) Without the current limiter, the DC machine starts with a maximum current of about 900 A (Fig. 2.31), which permits a large starting torque above 2,000 Nm (Fig. 2.33). The mechanical angular velocity reaches its steady-state value of 89.4 rad/s within about 0.1 s, Fig. 2.32. When the current limiter is added, the maximum current becomes limited to 200 A at start up (Fig. 2.43) and limits the maximum starting torque to about 460 Nm, Fig. 2.45. This in turn delays the start up, and the angular velocity reaches its steady-state value of approximately 89.4 rad/s after about 0.25 s, Fig. 2.44. Thus there is a tradeoff when implementing a current limiter: the acceleration of the DC machine is reduced through the current limiter and the circuit is now protected against dangerously high levels of current.

References [6–24, 29] provide a basic understanding of rotating machines and must be consulted if detailed operation and design parameters must be obtained.

2.5 Summary

Block diagrams illustrate wind power plants, air conditioning systems and drives for electric/hybrid automobiles. Block diagrams are used to analyze the response of electromechanical systems based on either Mathematica or Matlab. From an electrical viewpoint, the DC machine is relatively simple, however, it is complicated from a mechanical point of view due to the commutator. The DC machine principle is explained based on the Gramme ring. The field winding of a DC machine is excited by DC while the rotor winding is excited by AC. The mechanical commutator makes the conversion from DC to AC in motoring mode and from AC to DC in generating mode. Indeed in motoring mode the Gramme ring works like a DC to AC inverter and in the generating mode the Gramme ring works like an AC to DC rectifier. The start-up, rated operation, flux weakening operation, and the speed reversal of a DC machine has been demonstrated. Nonlinear current limiters are discussed.

2.6 Problems

Problem 2.1: *Solution of differential equation with constant coefficients*

Find the solution of the differential equation ($v_c(t)$, $i(t)$) as described by the circuit of Fig. 2.51 provided $v_c(t=0) = V_{max}$, and the terminal voltage applied at time $t=0$ is $v_T(t)=V_{max}\cos\omega t$. Hint: Use the Laplace transformation or solve the differential equation (see Application Example 2.2) for the state-variable $v_c(t)$, use the initial condition, and then find $i(t)$.

Fig. 2.51 R-C circuit

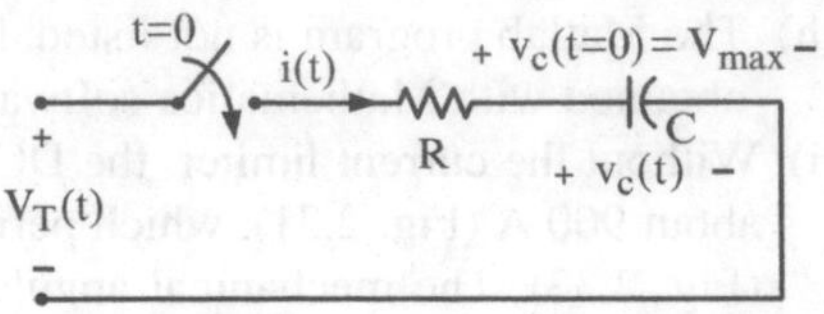

Problem 2.2: *Design of an electric drive for an automobile*

An electric drive for a 1.4 (tons-force) automobile consists of battery, step-up/step-down DC-DC converter, and a DC motor, as shown in Fig. 2.52. The available maximum power at the wheels of the car is (output power) $P_{max\_rated} = 50$ kW, the maximum velocity of the car is $v_{max} = 55$ miles/h, and the operating range is $d_{max} = 160$ miles.

(a) Express P_{max_rated} in hp, v_{max} in m/s, and d_{max} in m.
(b) How many hours t_h does one trip last (55 miles/h, 160 miles)?
(c) What is the total maximum energy utilized at the wheels (E_{wheels}) if the car has an output power of $P_{out} = 30$ kW during the entire time it travels 160 miles at the constant velocity of $v_{max} = 55$ miles/h?
(d) If the energy efficiencies of the battery (η_{bat}) DC-DC chopper (η_{con}) and the DC motor (η_{mot}) are 95% each, how much energy must be provided by the battery at the wheels (E_{wheels}) during one trip ($P_{out} = 30$kW, $v_{max} = 55$miles/h, $d_{max} = 160$ miles)?
(e) What is the figure of merit $\left(\mathrm{FM} = \frac{[(\text{tons} - \text{force}) \cdot \text{miles}]}{[\text{Energy used at the wheels } E_{wheels}]}\right)$ for this case [30]?
(f) What is $E_{battery}$? How much does the nickel-metal hydride battery [30, 31] weigh, if its specific energy is 100 Wh/(kg-force) and its depth of discharge is DoD=0.5 ?
(g) How much does the energy supplied to the wheels (E_{wheel}) cost if the price for 1 kWh is $0.15?
(h) What is the inherent disadvantage of an electric automobile where the energy is supplied by a battery?

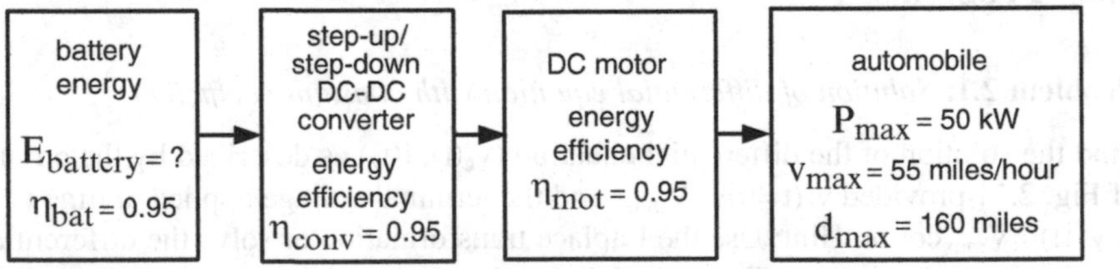

Fig. 2.52 Block diagram of a drive for an electric car at rated conditions

Problem 2.3: *Parallel hybrid drive vehicle*

A parallel hybrid vehicle has a $P_{IC\_rated} = 100$ hp internal combustion (IC) engine and a $P_{electric\_rated} = 15$ kW electric motor variable-speed drive consisting of brushless DC machine, solid-state converter (rectifier, inverter) and a 4 kWh nickel-metal hydride battery. You may assume that the motor/generator, converter and battery have energy efficiencies of $\eta_{bat} = \eta_{conv} = \eta_{mot} = 95\%$ each.

The mileage of the hybrid vehicle (without electric drive) is 40 miles/gallon and during braking 30% of the energy can be recovered by regeneration. Note that the energy efficiency of the IC engine is about 20% due to its constant-speed operation.

(a) How much gasoline (1 gallon = 3.8 ℓ) must be in the tank provided one drives 400 miles with the IC engine and without use of the electric drive?
(b) What is the energy content ($E_{provided}$) of the gasoline residing in the tank?
(c) What is the energy content (E_{used}) used by the IC engine (energy efficiency of 20%)?
(d) Provided the car travels at an average speed of 50 miles/h how long will the trip take to travel 400 miles?
(e) How many (equivalent) kWh are used during 1 h of travel ($E_{used\ per\ hour}$)?
(f) Let us assume that during 1 h 30% of the energy provided by the IC engine is recycled (see Fig. 2.53) by regeneration, how much energy is stored in the battery ($E_{battery\ stored}$)?
(g) The energy stored ($E_{battery\ stored}$) is then reused: how much of this energy ($E_{battery\ reused}$) can be reused at the wheels?
(h) How much gasoline (Gal_{gas}) must be in the tank provided one drives 400 miles with the IC engine and with use of the electric drive?
(i) How many percent does the electric drive increase the energy efficiency ($\Delta\eta$) of this hybrid vehicle?
(j) What is the weight of the gasoline (tank is full), and that of the nickel-metal hydride battery? For the battery you may assume a specific energy density of 100 Wh/(kg-force) and a depth of discharge DoD ≈ 0.4.
Note: During the uphill climb the electric drive contributes power to vehicle, and during downhill travel the 4 kWh battery is being recharged due to braking/regeneration.

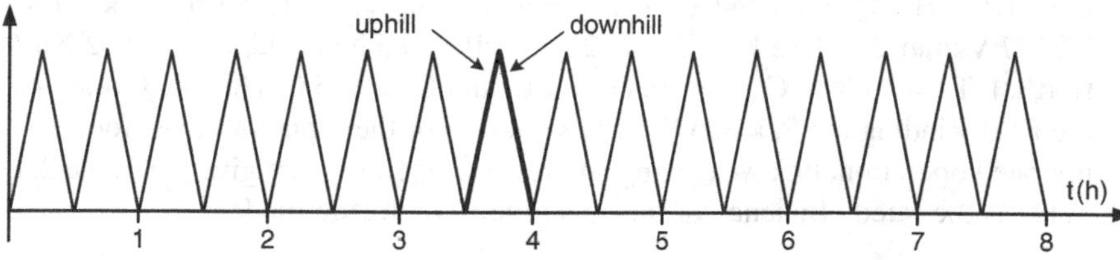

Fig. 2.53 Uphill-downhill diagram as a function of distance

Problem 2.4: *Find transfer function*

Identify all transfer functions of the block diagram of Fig. 2.17 and find $\{\theta_{actual}(s)/\ \theta^{*}(s)\}$. Hint: see Application Example 2.4.

Problem 2.5: *Steady-state errors using the final-value theorem*

(a) Compute for the block diagram of Fig. 2.54 the steady-state error $\varepsilon_{IM}(s)$ if $I_M^*(s)$ is a unit-step function. You may assume G_1=5, G_2=2, τ_2=0.2 s, G_3=10, and τ_3=0.1 s.

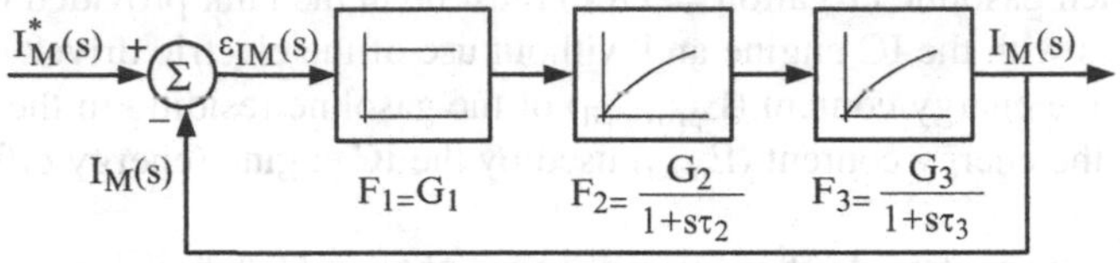

Fig. 2.54 Current-control loop of an electric drive with P-controller

(b) Show that for the block diagram of Fig. 2.55 the steady-state error $\varepsilon_{IM}(s)$ is zero if $I_M^*(s)$ is a unit-step function.

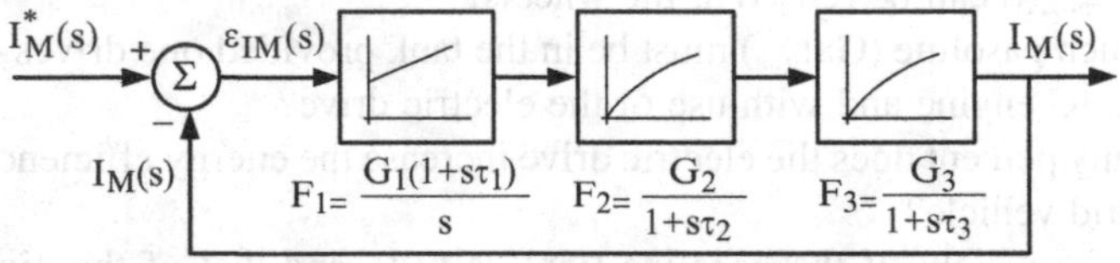

Fig. 2.55 Current-control loop of an electric drive with PI-controller

Problem 2.6: *Rated operation, rated efficiency, flux weakening, and reversal of DC machine*

(a) Repeat the Mathematica/Matlab analysis of Application Example 2.8 for a P_{out_rated}=100 hp, V_a=230 V DC machine. The machine parameters are [16] L_a= 1.1 mH, R_a = 0.0144 Ω, I_{a_rated}=349 A, n_{m_rated}=1,750 rpm, $(k_e\Phi)$ = 1.2277 Vs/rad, J = 1.82 kgm^2, T_c = 21 Nm, B = 0.5 Nm/(rad/s), c = 0.02 Nm/(rad/s^2), T_s = 10 Nm, G_{rect} = 10, G_i = 1.0, and $G_{\omega m}$ = 5.0. The rated losses of the field winding are P_f =325 W. The variation of the input variables for start-up, rated operation, flux weakening and reverse operation are given in Table 2.4

(b) What is the rated efficiency of this machine in motoring mode?

Table 2.4 Variable input parameters

Time t [s]	0	2	4	6	8	10	12
Terminal voltage of DC motor V_a [V]	115	230	230	230	230	−230	
Field flux of DC motor ($k_e\phi$) [Vs/rad]	1.2277	1.2277	1.2277	1.2277	0.614	0.614	
Mechanical torque doing the work T_m [Nm]	0	0	214	428	214	−214	
Coulomb frictional torque T_c [Nm]	21	21	21	21	21	−21	
Standstill frictional torque T_s [Nm]	10	10	10	10	10	−10	
Coefficient of windage torque c [Nm/(rad/s)^2]	0.02	0.02	0.02	0.02	0.02	−0.02	

Problem 2.7: *Speed control via flux weakening of a DC motor*

A $P_{rated} = 40$ kW, $V_{a\_rated} = 240$ V, $n_{rated} = 1{,}100$ rpm separately excited DC motor is to be used in a speed control system which may be represented by the block diagram of Fig. 2.56. For $k_e\Phi = 2.0$ Vs/rad, $R_a = 0.1\ \Omega$, $B = 0.3$ Nms/rad, $c_1 = 0.2$ V/rpm, and $c_2 = 0.10$ rpm/rad/s:

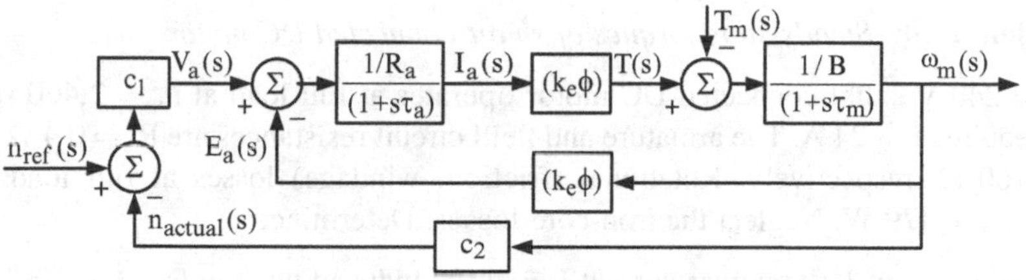

Fig. 2.56 Block diagram of separately excited DC motor with speed/angular velocity control via armature voltage

(a) Determine the steady-state angular velocity at no load ω_{mo} if $V_a = 240$ V. What is the no-load speed n_{mo}?
(b) What is the reference speed n_{ref} for the condition of (a) ?
(c) What is $(k_e\Phi)_{new}$ if the steady-state no-load angular velocity $(\omega_{mo})_{new}$ must be two times the rated angular velocity (see Fig. 2.57)?
(d) For $R_f = 1\ \Omega$, $V_f = 240$ V, and $k_e\Phi = 2.0$ Vs/rad determine c_3. What is the value of V_f for part c)? You may assume linear conditions (Fig. 2.57).

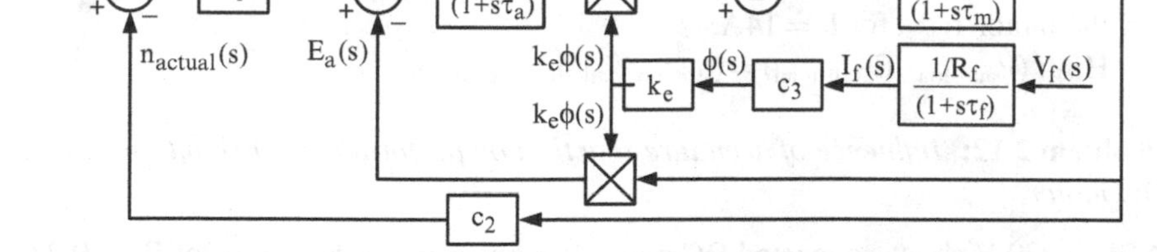

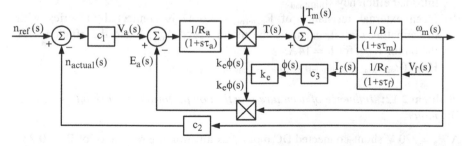

Fig. 2.57 Block diagram of separately excited DC motor with speed/angular velocity control via flux weakening

Problem 2.8: *Steady-state performance of shunt-connected DC motor*

A $P = 8.8$ kW, $V_a = 220$ V, $n_m = 1{,}200$ rpm shunt-connected DC motor has armature resistance of $R_a = 0.1\ \Omega$ and field resistance of $R_f = 100\ \Omega$. The rotational (friction, windage) loss is $P_{fr\_wi} = 2{,}200$ W. Neglect the iron-core loss. Compute the rated motor output torque T, the armature current I_a, and the efficiency η.

Problem 2.9: *Steady-state performance of series-connected DC motor*

When operated from a $V_a = 230$ V DC supply, a series-connected DC motor operates at n_m=975 rpm with an armature current of $I_a = 90$ A. The armature and the series field resistances are R_a= 0.12 Ω, and R_{sf} = 0.09 Ω, respectively.

(a) Compute the induced voltage E_a
(b) Compute the new value of the induced voltage E_{a_new} when the armature current is reduced to $I_{a\_new} = 30$ A.
(c) Find the new motor speed n_{m_new} when the armature voltage is $V_a = 230$ V and armature current is $I_{a\_new} = 30$ A. Assume that due to saturation, the flux produced by an armature current of 30 A is 48% of that at an armature current of 90 A.

Problem 2.10: *Steady-state torques of shunt-connected DC motor*

A V_a=240 V shunt-connected DC motor operates at full load at $n_m = 2{,}400$ rpm, and requires I_t= 24 A. The armature and field circuit resistances are $R_a = 0.4\ \Omega$ and R_f= 160 Ω, respectively. Rotational (friction, windage) losses at full load are $P_{rotational}$ = 479 W. Neglect the iron-core losses. Determine:

(a) Field current I_f, armature current I_a and the induced voltage E_a.
(b) Copper losses P_{cu}, developed internal motor power $P_d = E_a \cdot I_a$, output power P, and efficiency η.
(c) Developed internal motor torque $T_d = P_d/\omega_m$, and output load torque T.

Problem 2.11: *Speed control of shunt-connected DC motor*

A V_a=120 V, n_m=2,400 rpm shunt-connected DC motor requires $I_{t\_no\text{-}load} = 2$ A at no load, and I_{t_load}=14.75 A at full load. The armature circuit resistance is $R_a = 0.4\ \Omega$ and the field winding resistance is R_f =160 Ω.

(a) Determine the no-load speed $n_{no\text{-}load}$, the rotational and core loss P_{r+fe}, and the full-load efficiency $\eta_{full\text{-}load}$.
(b) If an external resistance of $R_{external} = 3.6\ \Omega$ is connected in series with the armature circuit, calculate the motor speed n_{m_new} and the efficiency of the motor η_{new} for $I_a = 14$A.
Hint: $(E_{ano\text{-}load}/E_{afull\text{-}load}) = (n_{m\_no\text{-}load}/n_{m\_full\text{-}load})$.

Problem 2.12: *Influence of armature reaction on performance of shunt-connected DC motor*

A $V_a = 220$ V shunt-connected DC motor has an armature resistance of $R_a = 0.2\ \Omega$ and field-circuit resistance of R_f = 110 Ω. At no-load, the motor runs at $n_{m\_no\text{-}load}$ =1,000 rpm and it draws a line current of $I_{t\_no\text{-}load} = 7$ A. At full load, the input to the motor is $P_{in\_full\text{-}load} = 11$ kW. Consider that the air-gap flux remains constant at its no-load value; that is, neglect armature reaction.

(a) Find the speed, speed regulation, and developed internal torque at full load $T_{d\_full\text{-}load}$.
(b) Find the starting torque if the starting armature current is limited to 150% of the full load current.
(c) Consider that the armature reaction reduces the air-gap flux by 5 % when full-load current flows in the armature. Repeat part (a).

References

1. Nilsson, J.W.: *Electric Circuits*, 2nd Edition, 1986, Addison-Wesley Series in Electrical Engineering, Addison-Wesley, Englewood Cliffs NJ, p. 596.
2. Sedra, A.S.; Smith, K.C.: *Microelectronic Circuits,* 4th Edition, 1998, Oxford University Press, Oxford, p. 60.
3. Dubey, G.K.: *Power Semiconductor Controlled Drives*, 1989, Prentice Hall, Englewood Cliffs NJ, p. 17.
4. Appelbaum, J.: "Starting and steady-state characteristics of DC motors powered by solar cell generators," *IEEE Transactions on Energy Conversion*, Vol. EC-1, No.1, March 1986.
5. Yildirim, D.; Fuchs, E.F.; Batan, T.: "Test results of a 20 kW variable-speed direct-drive wind power plant," *Proceedings of the ICEM 1998 International Conference on Electric Machines*, Istanbul, Turkey, 2–4 September 1998.
6. Fitzgerald, A.E.; Kingsley, Jr., C.; Umans, S.D.: *Electric Machinery*, 1990, 5th Edition, McGraw-Hill Publishing Company, New York.
7. Matsch, L.W.; Morgan, J.D.: *Electromagnetic and Electromechanical Machines*, 1986, 3rd Edition, Harper &Row Publishers, New York.
8. Chapman, S.J.: *Electric Machinery Fundamentals*, 1998, 3rd Edition, McGraw-Hill Publishing Company, New York.
9. Sarma, M.S.: *Electric Machines*, 1994, 2nd Edition, West Publishing Company, Minneapolis MN.
10. Ong, C.M.: *Dynamic Simulation of Electric Machinery Using Matlab/Simulink*, 1997, Prentice Hall, Englewood Cliffs NJ
11. Hiziroglu, B.S.G.H.R.: *Electric Machinery and Transformers*, 1988, Harcourt Brace Jovanovich Publishers, San Diego CA, Oxford University Press, 3rd Edition, 2001.
12. Kosow, I.L.: *Electric Machinery and Transformers*, 1991, 2nd Edition, Prentice Hall, Englewood Cliffs NJ.
13. Del Toro, V.: *Basic Electric Machines*, 1990, Prentice Hall, Englewood Cliffs, NJ.
14. McPherson, G.; Laramore, R. D.: *An Introduction to Electrical Machines and Transformers*, 1990, 2nd Edition, Wiley, New York.
15. R. Krishnana, *Electric Motor Drives*, 2001, Prentice Hall, Englewood Cliffs NJ.
16. Dewan, S.B.; Slemon, G.R.; Straughen, A.: *Power Semiconductor Drives*, 1984, Wiley, New York.
17. Kovacs, P.K.: *Transient Phenomena in Electrical Machines*, 1984, Akademiai Kiado, Budapest, Hungary
18. Wildi, Th.: *Electrical Machines, Drives, and Power Systems*, 2002, 5th Edition, Prentice Hall, Englewood Cliffs, NJ
19. Lindsay, J.F.; Rashid, M.H.: *Electromechanics and Electrical Machinery*, 1985, Prentice Hall, Englewood Cliffs, NJ

20. Ramshaw, R.; van Heeswijk, R.G.: *Energy Conversion, Electric Motors and Generators*, 1989, Saunders College Publishing, Orlando, FL
21. Leonhard, W.: *Control of Electrical Drives*, 1985, Springer, Berlin
22. Mohan, N.: *Electric Drives an Integrated Approach*, 2000, MNPERE, MN
23. El-Sharkawi, M.A.: *Fundamentals of Electric Drives*, 2000, Brooks/Cole Thomson Learning, Orlando, FL.
24. Krause, P.C.; Wasynczuk, O.: *Electromechanical Motion Devices*, 1989, McGraw-Hill Book Company, New York.
25. U. S. Department of Energy, *Press Release*, April 2, 2004.
26. http://www1.eere.energy.gov/femp/pdfs/tir_coolerado.pdf
27. http://www.coolerado.com/
28. Grainger, *Wholesale Net Price Catalog*, W. W. Grainger, Inc.
29. Nasar, S.A.; Unnewehr, L.E.: *Electromechanics and Electric Machines*, 1979, Wiley, New York
30. Moore, T.: "The road ahead for EV batteries," *EPRI Journal*, March/April 1996, pp. 6–15.
31. Tichy, R.: "Battery advancements," *Proceedings of the 2007 Embedded Systems Conference*, April 5, 2007, 7 pages.

Chapter 3
Electric Energy Sources

3.1 Public Utility System

The most common electric source of electrical energy is in the USA the public three-phase, 60 Hz [1–4] utility system. Within the continental US there are three distinct separate AC power pools: the Eastern, Western, and Texan system representing a total of 1,000 GW installed capacity. These separate systems are interconnected by relatively low power (100 MW) DC transmission lines in order to permit some exchange of electric power between the Eastern Interconnection (System), Western Interconnection (System), and Texan ERCOT System – where ERCOT stands for Electric Reliability Council of Texas – as depicted in Fig. 3.1 [5, 6]. The DC transmission line in California has a rating of about 1,000 MW.

The reason why these three separate power systems exist is of technical and regulatory nature: it is not possible to operate an AC power system in a synchronized manner (at a nominal frequency of 60 Hz) in New York and San Francisco due to inherent stability problems associated with long AC transmission lines (more than 1,000 miles). The separation of the Texan grid from the Eastern and Western Interconnection is due to a difference in regulatory procedures in Texas and the remaining 47 lower continental states. DC transmission lines do not exhibit stability problems and, therefore, their length is not limited. However, they require AC-to-DC voltage/current conversion at one end of the high-voltage line terminal and a DC-to-AC voltage/current conversion at the other end of the high-voltage line terminal.

The Western Interconnection is coordinated by the WECC – Western Electricity Coordinating Council – formed 14 August 1967. WECC is responsible for coordinating and promoting electric system reliability. On a national basis since 1968, the North American Electric Reliability Corporation (NERC) has been committed to ensuring the reliability of the bulk power system in the US. On the federal level the Federal Energy Regulatory Commission (FERC) is in charge of the rules valid for the Extra-High-Voltage System (EHV) $\geq$ 345 kV. In January 1994 the Eastern Interconnection experienced significant power outages due to extreme cold weather, inadequate energy resources, and insufficient fuel (deliveries hampered).

Any AC power system can be either represented by Thevenin's or Norton's equivalent circuit, as shown in Figs. 3.2a, b [7].

E.F. Fuchs and M.A.S. Masoum, *Power Conversion of Renewable Energy Systems*,
DOI 10.1007/978-1-4419-7979-7_3,

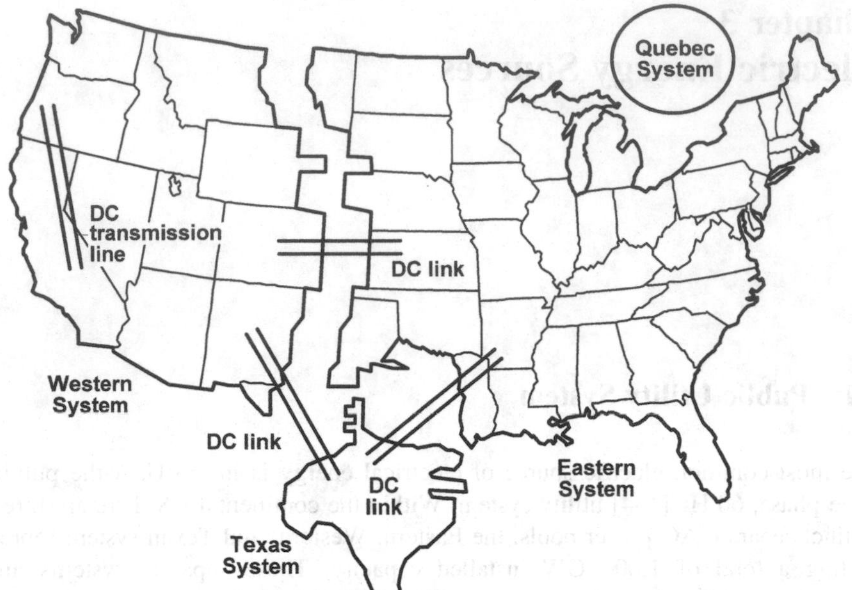

Fig. 3.1 Three separate power systems within the continental US

For ideal systems $Z_t \rightarrow 0$ and $Z_n \rightarrow \infty$. One should note that to operate the public system reliably there must be a so-called "spinning reserve" of about 10% of the installed power capacity (1,000 GW) in order to cover high-power demands during certain times of the day, especially when air conditioning units are operated. Remember that an average size of a steam-driven power plant (see Fig. 3.2c) can output about 600 MW (efficiency of about 35%), whereas combined-cycle plants (e.g., gas and steam turbines in series, Fig. 3.4, with efficiency of about 60%) output up to 200 MW. That is, nationwide more than 100 steam power plants must be operating in a standby mode to supply power for all conceivable eventualities (e.g., hot days, failures of some stations).

Here it might be useful to remind the reader that one human being can generate on a steady-state basis an average power of about 75 W. Serving 300 million people within the continental US the installed power capacity per capita is 1,000 GW/300 M = 3.33 kW, that is, 3,333 W/75 W $\approx$ 44 "energy/power servants" are required to cover the needs of one US citizen. This compares to about half as many "energy servants" required in European countries, which indicates that there is still a great potential for energy savings within the US and it can be achieved without sacrificing the level of standard of living by switching from less efficient components e.g., incandescent lighting, constant-speed drives, to more efficient components e.g., compact fluorescent light (CFL) bulbs, light-emitting diodes (LEDs), variable-speed drives, conservation, and regeneration techniques discussed later within this text. Figure 3.2c illustrates a typical electrical three-phase power system.

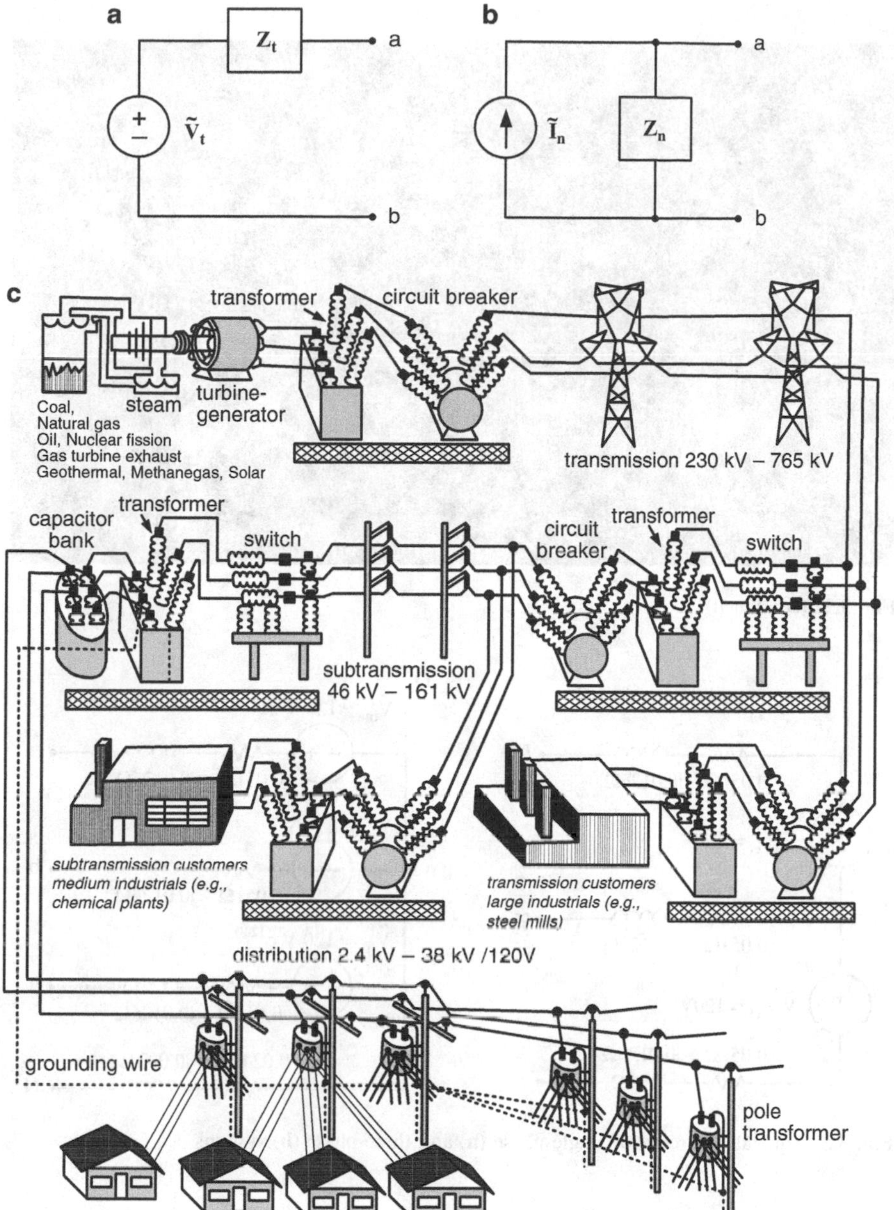

Fig. 3.2 (**a**) Thevenin's equivalent circuit for a single-phase system, (**b**) corresponding Norton's equivalent circuit, (**c**) typical electrical three-phase power system, (**d**) Bright lights. A satellite photo shows America at night (Courtesy of National Aeronautics and Space Administration, NASA)

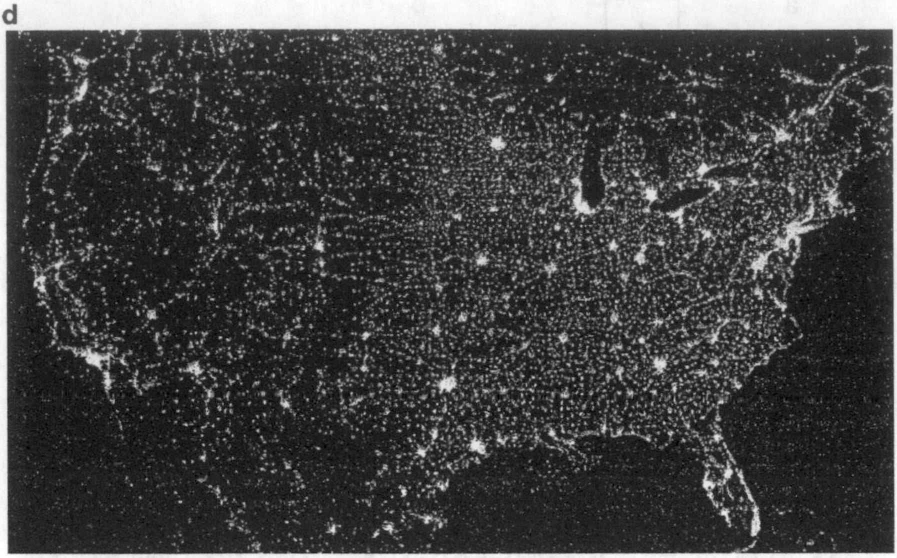

Fig. 3.2 (continued)

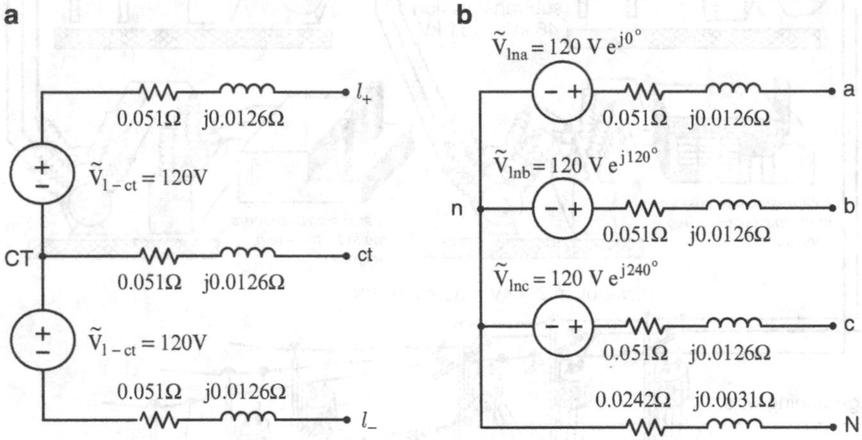

Fig. 3.3 Equivalent circuits of single-phase (**a**) and three-phase (**b**) systems

A satellite photo is reproduced in Fig. 3.2d, of North America at night. About 20% of 1,000 GW, that is, 200 GW of power consumption is used for lighting installations. It is estimated that without any sacrifice in the standard of living at least 20–30% of 1,000 GW can be saved through energy conservation and a reduction of electricity usage. The latter of course requires the cooperation of each and every individual living within the US.

The North American Electric Reliability Council (NERC) located in Princeton, NJ, estimates that US demand for electricity will grow just under 2% per year over the next decade, while supply will grow at 1.5% per year for the next 5 years. Among areas particularly vulnerable to power disruptions are the West Coast, New England, New York, Arizona, New Mexico, and southern Nevada [8].

Application Example 3.1: Single-Phase and Balanced Three-Phase Systems

Equivalent circuits of single-phase and balanced three-phase systems are shown in Figs. 3.3a, b [9]. In these figures ℓ_+, ℓ_-, ct, and CT stand for positive line, negative line, center tap at the load, and center tap at the transformer, respectively; a, b, c, n, and N stand for phase a, phase b, phase c, neutral at the transformer, and neutral at the load, respectively.

Application Example 3.2: Computation of Short-Circuit Current of Single-Phase Power System

The short-circuit current of the 240 V single-phase system (line-to-line) is according to Fig. 3.3a

$$I_{sc}^{\ell-\ell} = \frac{240\,\text{V}}{\sqrt{(0.102)^2 + (0.0252)^2}} = 2{,}284.26\,\text{A} \tag{3.1}$$

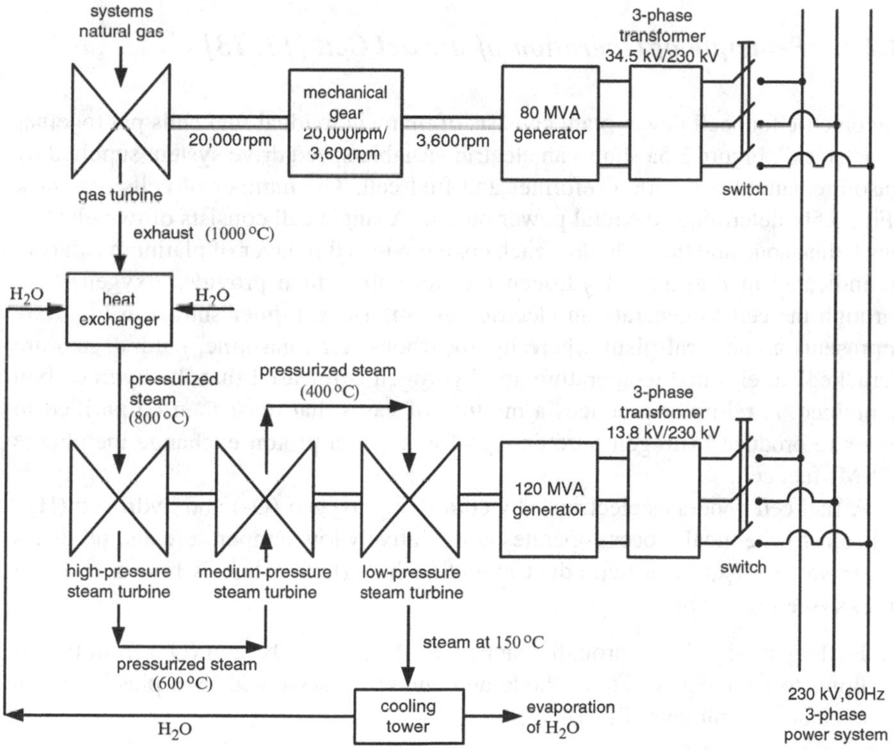

Fig. 3.4 Block diagram of combined-cycle (gas- and steam turbines) power plant

3.2 Fuel Cells [10]

About 100 years ago, before the advent of the internal combustion engine as used today in automobiles, electric cars were relied on having a few horsepower and lead-acid batteries as an energy source. At that time regenerative braking (that is, operation in quadrant II of Fig. 2.2 ($T < 0$, $\omega_m > 0$) was not feasible due to the lack of electronic switches. Since the invention of the transistor (BJT) in 1947 many attempts have been made to revive the electric car to reduce pollution in urban areas: however, it was not very successful due to the lack of an energy source with sufficiently large energy content per unit weight. From [11] one gathers that one liter of gasoline ideally contains 8.8 kWh of energy. Although the internal combustion engine has a low efficiency (e.g., 16%) this high energy density of gasoline per liter permits the automobile to travel at least 300 miles without refilling the tank. Recent advances in fuel cell designs, and the introduction of a reformer permits the replacement of the "hot combustion" of an internal combustion engine with the "cold combustion" of a fuel cell. Cold combustion has the advantage of producing less heat losses as compared to the internal combustion engine, and the efficiency of such a cold combustion drive train will be significantly increased (e.g., 50% as compared with 16%), over that of the internal combustion engine.

3.2.1 Principle of Operation of a Fuel Cell [12, 13]

An electric fuel cell power plant consists of many individual fuel cells put together in a "stack". Figure 3.5a shows an electric variable-speed drive system supplied by gasoline/natural gas with a reformer and fuel cell. The number of cells per stack (Fig. 3.5b) determines the total power output. A single cell consists of two electrodes – the anode and the cathode – each coated with a thin layer of platinum catalyst, as indicated in Fig. 3.5b. Hydrogen fuel and air (which provides oxygen) flow through the cell to generate an electric current. The reformer shown in Fig. 3.5a represents a chemical plant where hydrocarbons (e.g., gasoline, natural gas) are "cracked" at elevated temperature and hydrogen is released that the hydrocarbon contained. A reformer produces a mixture of gases that must then be purified in order to produce hydrogen pure enough for use in a proton exchange membrane (PEM) fuel cell.

A fuel cell generates electricity by combining oxygen (O_2) and hydrogen (H_2). The electrochemical process operates at a relatively low temperature and produces water vapor (H_2O) as a byproduct including heat (loss). Here is how a fuel cell works (see Fig. 3.5b):

1. Hydrogen (H_2) flows through channels to the anode. Oxygen (O_2) from the air flows to the cathode. The cathode and anode are separated by a plastic proton exchange membrane (PEM).

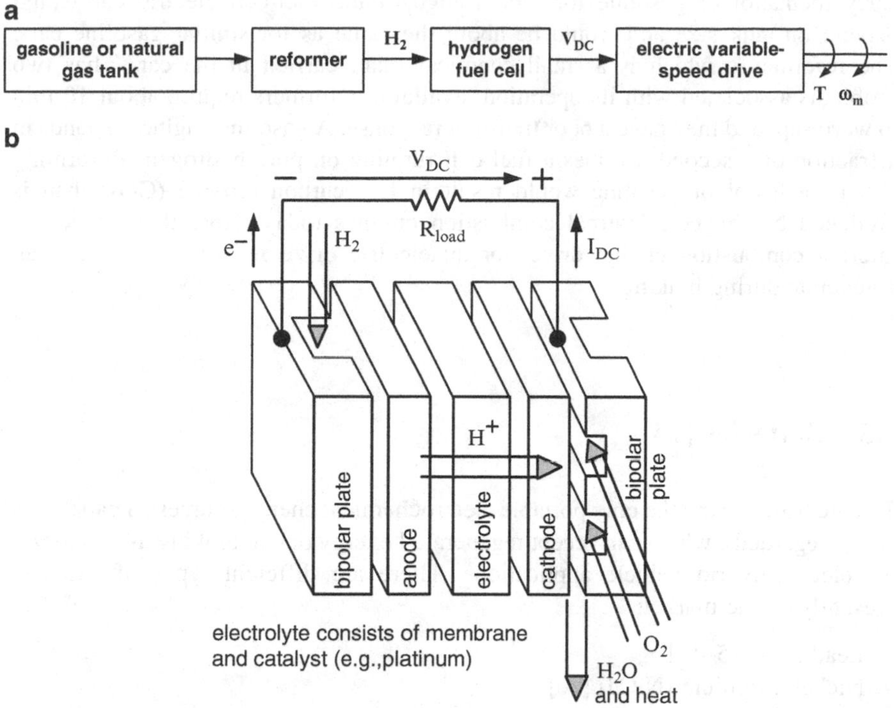

Fig. 3.5 (**a**) Fuel-cell electric variable-speed drive system. (**b**) PEM hydrogen fuel cell and generation of electricity through electrochemical process. PEM is also an abbreviation for polymer electrolyte membrane

2. At the anode, hydrogen is dissociated into electrons and protons by reacting with the platinum catalyst. Protons pass through the membrane to the cathode.
3. Free electrons flow as an electric current through a load (e.g., resistance R_{load}, motor) to the cathode.
4. At the cathode, electrons and protons combine again and form water.

3.2.2 *Fuel for a Fuel Cell*

The fuel required for a fuel cell is for example hydrogen. The problems associated with carrying hydrogen within a tank of an automobile (extreme explosiveness as compared with gasoline or methanol also called methane gas) and the lack of an infrastructure for the generation of hydrogen from electricity by reversing the process of the fuel cell's electrochemical process led to the idea to extract pure hydrogen from methanol or gasoline via a so-called "reformer". Vehicles can carry methanol or gasoline for fuel. Range of the fuel-cell electric car would depend on tank size and would be about the same as for similar gasoline cars. The reformer – which is a small chemical plant carried in the car – has two problems associated with its operation: available reformers require about 10 min to warm-up, and they have a poor transient response. A gasoline engine responds in a fraction of a second, as does a fuel cell running on pure hydrogen. Reforming either methanol or gasoline would result in less carbon dioxide (CO_2) than is produced by the best internal combustion engines today. Note that neither an internal combustion engine drive nor an electric drive fed by a fuel cell can regenerate during braking.

3.3 Batteries [14]

To date batteries are the only portable electrochemical energy sources (if capacitors are disregarded), which can accept regenerated energy due to braking as it is useful for electric/hybrid vehicle applications. There are different types of batteries presently on the market:

- Lead acid [15–19]
- Nickel–cadmium (NiCd) [20]
- Nickel-metal hydride (NiMH) [20]
- Lithium-ion (Li-ion) [20, 21]
- Lithium polymer (Li-polymer) [20]
- Sodium–sulfur
- Sodium-nickel chloride
- Flow batteries [22]

Surprisingly, most of them have very similar properties with respect to specific energy per unit of weight (Wh/kg-force), peak specific power per unit of weight (W/kg-force), and lifetime (cycles), as can be seen from Fig. 3.6 [23]. The United States Advanced Battery Consortium (USABC) [24] established its mid-and long-term criteria (Table 3.1) in 1992 for manufacturers to benchmark their development of advanced battery for vehicle. An intermediate criterion, commercial goal, was established in 1997. Figure 3.6 compares the performance of advanced batteries for electric vehicle applications [23]. Unfortunately, none of the presently available battery types meet these goals. Lithium-ion batteries appear to be the most promising: they have 3.3 V per cell. It is well-known that Li-ion battery cells charge and discharge in a non-uniform manner and therefore each and every cell must be controlled (e.g., voltage, current) separately to prevent overheating and explosion of the battery. Some of the automobiles described in [24] use 100 Li-ion 3.3 V cells, which are based on iron phosphate. That chemistry has less energy than the cobalt alternative, but it is far less prone to internal short-circuits of the sort that have caused laptops to burst into flames.

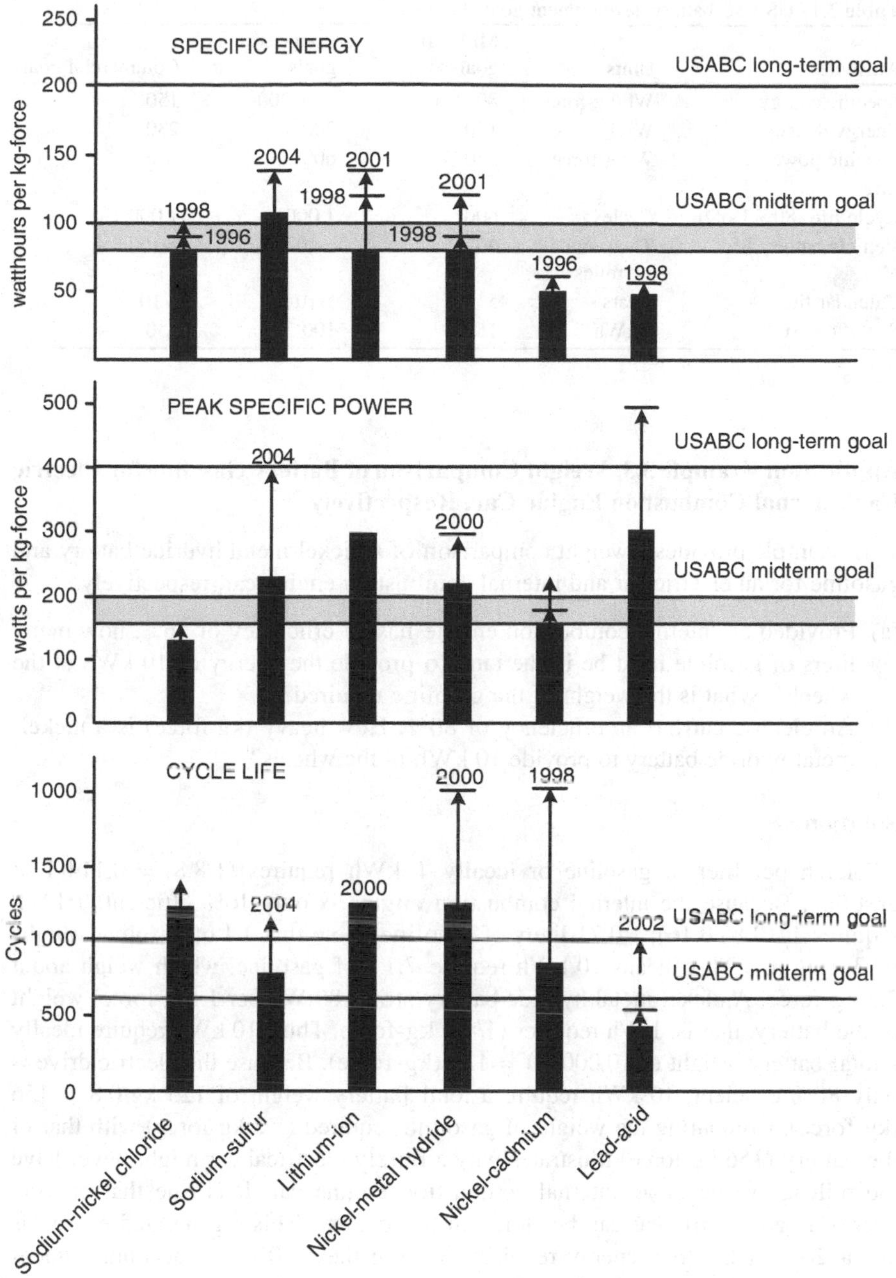

Fig. 3.6 Comparison of performance of advanced batteries for electric vehicle applications (From Moore, T.: "The road ahead for EV batteries," *EPRI Journal*, March/April 1996, pp. 7–15, [23])

Table 3.1 USABC battery development goals [24]

Attributes	Units	Mid-term goals	Long-term goals	Commercial goals
Specific energy	Wh/kg-force	80–100	150–200	150
Energy density	Wh/l	130	300	230
Specific power	W/kg-force	250	600	460
Life				
Cycle life (80% DoD)	Cycles	600	1,000	1,000
Vehicle battery life	Thousand miles	40	>100	>100
Calendar life	Years	5	>10	>10
Specific cost	$/kWh	150	100	150

Application Example 3.3: Weight Comparison of Battery/Gasoline for Electric Car/Internal Combustion Engine Car, Respectively

This example provides a weight comparison of a nickel-metal hydride battery and gasoline for an electric car and internal combustion engine car, respectively.

(a) Provided an internal combustion engine has an efficiency of 16%, how many liters of gasoline must be in the tank to provide the energy of 10 kWh at the wheels? What is the weight of the gasoline required?
(b) An electric car has an efficiency of 80%. How heavy (kg-force) is a nickel-metal hydride battery to provide 10 kWh at the wheels?

Solution

8.8 kWh per liter of gasoline or ideally 1 kWh requires $(1/8.8) = 0.114$ l of gasoline. Because the internal combustion engine is only 16% efficient, 1 kWh requires $1/(8.8 \cdot 0.16) = 0.71$ liters of gasoline. Note that 1 l of gasoline weighs 0.73 kg-force. This means 10 kWh require 7.1 l of gasoline, which weigh about 5.2 kg-force. A nickel-metal hydride battery stores 80 Wh per 1 (kg-force) weight of the battery, that is, 1 Wh requires (1/80) kg-force. Thus, 10 kWh require ideally a total battery weight of $10{,}000/80 = 125$ (kg-force). Because the electric drive is only 80% efficient, 10 kWh require a total battery weight of $125\ \text{kg}/0.8 = 156$ (kg-force). Comparing the weight of gasoline required (5.2 kg-force) with that of the battery (156 kg-force) illustrates why a purely electrical car might never have the mileage range of an internal combustion engine car. It is true that braking energy in an electric car can be stored in the battery. This regenerated energy is about 20% of the total energy required. Taking these 20% into account reduces the battery weight to $156 \cdot 0.8 = 124.8$ kg-force. In addition a battery cannot be completely discharged and this requires the battery to be even heavier than 124.8 kg-force. A depth of discharge (DoD) of about 50–70% [21] is acceptable depending on the type of battery. For DoD = 50% one obtains a battery weight of 249.6 kg-force.

3.3.1 *Principle of Operation of Batteries*

3.3.1.1 Galvanic Elements and Voltaic Series

A galvanic element consists, for example, of a Zn (zinc) electrode, an electrolyte (e.g., $CuSO_4$), and a Cu (copper) electrode. If one connects the Zn and Cu electrodes by a conducting wire (Fig. 3.7) the Zn^{2+} ions migrate from the Zn electrode into the electrolyte, and electrons (e) wander from the Zn electrode to the Cu electrode: that is, the Zn electrode becomes smaller losing Zn atoms ($Zn \rightarrow Zn^{2+}$ ions + 2e). Correspondingly, the electrons flowing from the Zn electrode combine with Cu^{2+} ions residing within the electrolyte and form copper atoms, which increase the volume of the copper electrode (Cu^{2+} ions + 2e $\rightarrow$ Cu).

The direction the electrons travel (here from Zn electrode to Cu electrode) depends upon the potential difference $\Delta\varphi_{\text{metal-electrolyte}}$ which exists between the metal and electrolyte: the electron flow is such that the metal (here Zn) separates into ions and electrons having a lower potential difference $\Delta\varphi_{\text{metal-electrolyte}}$ between metal (e.g., Zn) and electrolyte than Cu. The metal which has the higher potential serves as the positive pole of the galvanic element.

One can arrange the metals of the voltaic sequence or series such that all metals to the right (e.g., Fe, Cd, Ni, Pb, H_2, Cu, ...) of a certain metal (e.g., Zn) form the positive pole if a combination of electrodes is chosen from the voltaic series of Table 3.2. For example, Zn is the negative pole with respect to Fe, Cd, ... Au electrodes, and a Li electrode forms the negative pole with respect to K, Na, Mg, Zn, Fe, ...Au electrodes.

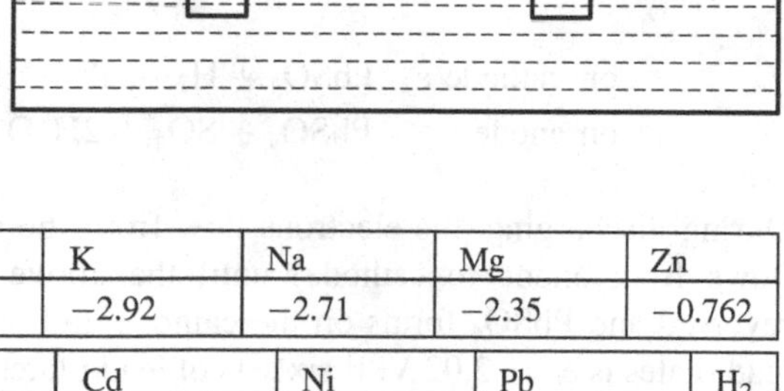

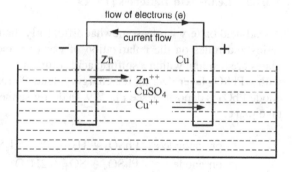

Fig. 3.7 Current flow of a galvanic element consisting of Zn, Cu electrodes and $CuSO_4$ electrolyte

Table 3.2 Voltaic series

Metal electrode	Li	K	Na	Mg	Zn
$\Delta\varphi_{\text{metal-electrolyte}}$ (V)	−3.02	−2.92	−2.71	−2.35	−0.762

Metal electrode	Fe	Cd	Ni	Pb	H_2
$\Delta\varphi_{\text{metal-electrolyte}}$ (V)	−0.44	−0.402	−0.25	−0.126	0

Metal electrode	Cu	Ag	Hg	Au
$\Delta\varphi_{\text{metal-electrolyte}}$ (V)	+0.345	+0.80	+0.86	+1.5

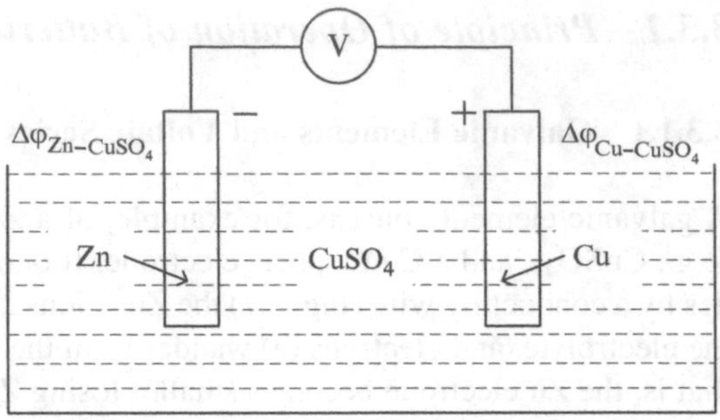

Fig. 3.8 Potential distribution across a galvanic element consisting of Zn, Cu electrodes and $CuSO_4$ electrolyte

The voltage measured between the two electrodes of Fig. 3.8 is the difference in voltage which exists between the Zn electrode and the electrolyte ($CuSO_4$) and the Cu electrode and the electrolyte. It is not possible to measure the potential difference between an electrode and the electrolyte ($CuSO_4$) directly: that is, only $\Delta\varphi_{metal\_1-metal\_2}$ can be measured.

Application Example 3.4: Li and Au Electrodes

Note that a Li electrode and an Au electrode generate the largest potential difference

$$\Delta\varphi_{Li-Au} = -3.02\,V - 1.5\,V = -4.52\,V.$$

3.3.1.2 Lead-Acid Batteries [15–19]

A lead-acid battery works somewhat differently than batteries based on the voltaic series and relies on the polarization of positive and negative charges. Two lead plates are residing within a sulfuric acid solution (H_2SO_4) and each plate is covered with a $PbSO_4$ layer. During the charging, lead Pb forms on the cathode and on the anode PbO_2 is deposited, see Fig. 3.9. In doing so the following chemical processes take place,

$$\text{on cathode}: \quad PbSO_4 + H_2 \rightarrow Pb + H_2SO_4,$$
$$\text{on anode}: \quad PbSO_4 + SO_4 + 2H_2O \rightarrow PbO_2 + 2H_2SO_4.$$

During discharging, the electrons flow from the cathode to anode (or the current flows from anode to cathode) until the above chemical processes have been reversed and $PbSO_4$ forms on the cathode and anode: the voltage across the two lead plates is $e_0 = 2.02$ V. If six sets of lead plates are connected in series, then the total terminal voltage is $E_0 = 6e_0 = 12.12$ V; the positive pole (anode) is covered with PbO_2 which gets reduced as the discharging progresses. The energy efficiency of a lead-acid battery (η_B) is such that one can recover about $\eta_B = (83–90\%)$ of the energy with which the battery has been charged.

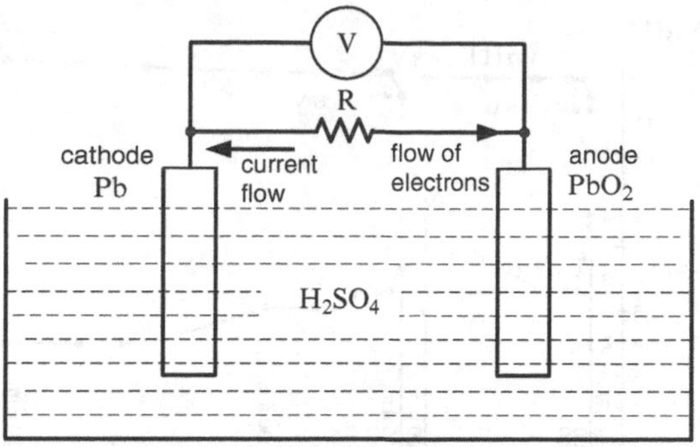

Fig. 3.9 Configuration of a (fully charged) lead-acid battery with discharge current direction

3.3.2 Charging and Discharging of Lead-Acid Batteries

Lead-acid batteries are fully charged if one can measure an open-circuit voltage of fully "discharged" battery cell(s) of $E_0 = 12.6$–12.8 V. The term "discharged" means that all free charges within the battery are zero (battery capacitors are discharged, see Fig. 3.11) and the only voltage source is the cell(s) voltage E_o. One can measure E_0 after the battery has been neither charged nor discharged for a few hours.

In the following all numerical values pertain to the charging of a deep-cycle 500 Wh, 12 V battery with the weight of 27 kg-force as depicted in Fig. 3.10. At the start of the charging process, at 2:45 p.m., the measured battery terminal voltage is $E_0 = 11.3$ V – which is less than 11.6 V, that is, the battery is completely empty resulting in an initial charging current of 4 A. At 8:00 p.m. battery terminal voltage is 12.3 V at a battery charging current of 2.5 A. The following day, at 7:15 a.m., charging is continued, whereby at the beginning of the charging process the voltage of $E_0 = 11.8$ V at a current of 3.2 A is measured. At 7:30 p.m. the voltage has risen to 12.8 V at a current of 2 A. The next day at 7:00 a.m. the charging is continued at a voltage of $E_0 = 12.2$ V and a charging current of 2.5 A etc. At the end of the fourth day at 6:00 p.m. the terminal voltage is 13.4 V at 0.5 A. This low charging current indicates that the charging process is about complete. Indeed, at 7:00 p.m., the terminal voltage reduced to $E_0 = 12.7$ V after the equivalent capacitors (C_b, C_B of Fig. 3.11) of the battery are completely discharged. At these latter values the battery can be considered to be fully charged.

A discharge test consists of connecting a 12.7 Ω resistance across the battery terminals, resulting in a discharge current of about 1 A. As a consequence of this discharge the battery voltage is $V_B = 11.8$ V, indicating that the series resistance (Fig. 3.11) of an equivalent circuit for the battery is $R_{series} = 1$ Ω. This resistance becomes smaller with increasing battery capacity, and 0.05 Ω values are not uncommon for larger batteries.

The internal discharge resistance $R_{discharge}$ of a battery (Fig. 3.11) is in the range of 5 kΩ. The differences in E_0, V_{Boc}, and V_B indicate that the equivalent circuit of a battery for low frequencies (less than 25 kHz) consists of a cell voltage, resistors

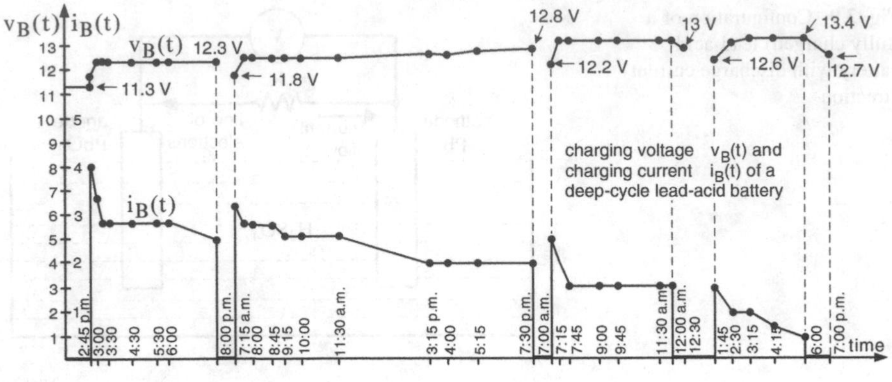

Fig. 3.10 Charging of a deep-cycle 500 Wh lead-acid battery

Fig. 3.11 Nonlinear lead-acid battery equivalent circuit

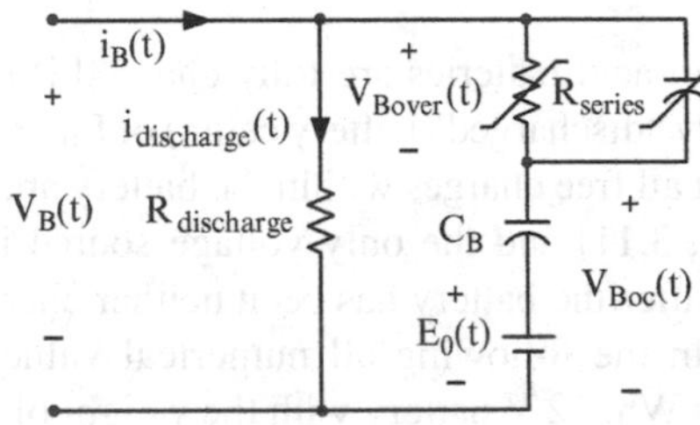

Table 3.3 Charging time chart

	Battery's starting percent of charge (%)			
	75	50	25	0
Switch setting	Hours to reach full charge (h)			
2A/12 V	6.5	12	18	23
6A/12 V	2.3	4.6	7	9

and capacitors, as will be presented in the next subsection, see Fig. 3.11. As a guideline one should remember that lead-acid batteries must not be completely discharged (depth of discharge, DoD $\approx$ 50%) and considerable time will be needed to charge a battery, as is demonstrated in Fig. 3.10 and outlined in Table 3.3. Note, that some battery chargers tend to overcharge, reducing the lifetime of a battery.

3.3.3 Lead-Acid Battery Care

Lead-acid batteries should be recharged every few weeks. Do not expose them to extreme low temperatures so that they freeze. The charging state of batteries can

Table 3.4 Assessment of charging state of lead-acid batteries

No-load voltage E_0 (V)	Hydrometer specific gravity	Approx. state of charge (%)	Approx. charge time @ 6A (h)
12.6	1.255	100	None needed
12.45	1.230	75	3.5–5.5
12.15	1.180	50	7.5–9.5
11.90	1.135	30	12.0–14
11.7 – down	1.100	Very low	15.5–17.5

either be checked with a hydrometer or a voltmeter. Table 3.4 presents detailed information on how the charging state of a battery can be assessed.

Adequately long charge time will result in much lower reduction of battery lifetime (e.g., 4 years). Right after charge time is over, one notices battery voltage will be considerably higher than 12.6 V and will be as much as 13.5–14.5 V (see Fig. 3.10). This increase in terminal battery voltage is due to the presence of electrochemical capacitance of battery cell(s) (C_B). After the battery is idled and is neither charged nor discharged for some time, the voltage across this capacitance C_B will approach zero and the battery terminal voltage will be identical to the open-circuit voltage E_0 of the fully "discharged" battery cell(s) (see Sect. 3.3.4). This open-circuit voltage for a fully charged battery is at about $E_0 = 12.6$ V.

Application Example 3.5: Charging State and Charging Time of a Lead-Acid Battery

The no-load voltage E_0 of a lead-acid battery is 12.15 V. What is the charging state of this battery and how long will it take to charge the battery to $E_0 = 12.6$ V?

Solution

Charging state is 50%. It will take 7.5–9.5 h to charge the battery, depending upon available charger.

3.3.4 Equivalent Circuit of Lead-Acid Battery

A lead-acid battery equivalent circuit for static and dynamic operations – corresponding to changes in charging and discharging currents – is illustrated in Fig. 3.11, where

$R_{discharge}$ – self-discharge resistance (kΩ range)
C_B – electrochemical capacitance of battery cell(s)
E_0 – open-circuit voltage of fully "discharged" battery cell(s) (that means C_b and C_B are discharged)
R_{series} – nonlinear series resistance (1–0.05 Ω)
C_b – nonlinear capacitance

$v_{Boc}(t)$ – open-circuit battery cell(s) voltage
$v_{Bover}(t)$ – battery cell(s) over voltage
$v_B(t)$ – battery terminal voltage
$i_B(t)$ – battery terminal current
$i_{discharge}(t)$ – battery discharge current.

This equivalent circuit shows that the discharging and charging behavior of a (lead-acid) battery is quite complicated, and the prediction of the charging state of a battery is not straightforward. Nevertheless, the information about the charging state of a battery is vital for the safe operation of a battery so that too much discharge or even overcharging can be prevented – both of which reduce the lifetime of a battery. As a rule of thumb a lead-acid car battery must be replaced after about 4 years of service. The energy-specific weight of commercially available lead-acid batteries is about 20 Wh/(kg-force). Note, there is a difference of the equivalent circuit for short-term and long-term operating conditions. Also, some batteries exhibit a memory effect [24–26] which can be mitigated by proper charging and discharging.

3.3.5 Performance Properties of Automobile Batteries Used for Starting the Internal Combustion (IC) Engine

Deep-cycle batteries used in recreational vehicles (RVs) and electric cars are rated either in ampere-hours (Ah) or in watt-hours (Wh), see Fig 3.10, and characterize the long-term use of batteries. IC-car batteries must deliver for starting of the IC engine during a short time (e.g., 30 s) a large current (e.g., 550 A) and their performance is defined in terms of cold cranking amperes (CCA) and reserve capacity (RC). In colder climates higher CCA ratings are more important than in hot climates, while in hot climates higher RC ratings are more important than CCA.

3.3.5.1 Cold Cranking Amperes

CCAs relate to a short-term use of a battery, and they are the discharge load measured in amperes that a new, fully charged battery, operating at 0°F (−17.8°C) can deliver for 30 s and while maintaining the voltage above 7.2 V. To start a 4-cylinder gasoline IC engine about 600–700 CCA are required. Four-cylinder IC Diesel engines require about 700–800 CCA. A higher cylinder number requires higher CCA values. As batteries age, they are also less capable of producing CCAs. Winter starting requires 140–170% more current than during summer. Assume that at 80°F (26.7°C) the power required to crank the engine is 100% and the available power from the battery is 100% as well. At −20°F (−29°C) the power required to crank the engine is 350% and the available power from the battery is only 25%. This explains why in a cold climate even a well maintained battery may not start the car due to increased power required to crank a sluggish engine and the inefficiency of a cold battery.

3.3.5.2 Reserve Capacity

The second most important consideration is the RC rating because of the effects of increased parasitic ("key off") loads, e.g., servos, lights, locks, and emergencies. RC is the number of minutes a fully charged battery at 80°F (26.7°C) can be discharged at 25 A until the voltage falls below 10.5 V. RC relates to the long-term use of a battery and thus the RC can be expressed in terms of ampere-hours or watt-hours. To convert RC to approximate ampere-hours, multiply RC by 0.4. For example, a battery with a 100 min RC will have approximately 40 Ah or $40 \cdot 12.6$ Wh = 504 Wh of capacity at the 25 A discharge rate. The capacity of a battery is temperature dependent. If at 68°F (20°C) the capacity is 97% then at 32°F (0°C) the capacity is 80% and at −40°F (−40°C) the capacity is only 30%.

3.3.5.3 Difference Between IC-Car Batteries and Deep-Cycle Batteries

IC-car batteries are specially designed for high initial current delivery during a very short time and shallow discharges. IC engines usually start in 5–15 s; to start an IC engine typically consumes 5–10% of the battery's capacity. Car batteries should not be discharged below 90% rated state-of-charge, that is DoD = 10%. A car battery is not designed for deep discharges and will have a very short life if it is abused by deep discharges. By contrast, deep-cycle batteries are designed for prolonged (long-term) discharges at lower amperage that typically consume between 20 and 80% of the battery's capacity, that is, DoD = 20–80%. A "dual" battery is a compromise between an IC starter battery and a deep-cycle battery. Such a battery will work as a (short-term) starting battery and a (long-term) drive battery as used in a hybrid/electric vehicle.

3.3.5.4 Starter and Charging System Tests

Figures 3.12a, b illustrate the discharging (a) and the charging (b) performances of a car battery. The rated values of this battery are CCA = 550 and RC = 100 at a purchase price of \$100–150. As can be noted the battery voltage starts in Fig. 3.12a at 12 V (which is below the rated voltage of 12.6 V) and can deliver at t = 0 s a peak $CCA_{t=0s} \approx 400$ which reduces after t = 30 s to $CCA_{t=30s} \approx 150$. Clearly this battery is unable to start an IC engine for which CCA = 550 are required. The charging of the battery occurs between upper and lower limits on Fig. 3.12b at about 14.4 V with a sufficiently small voltage ripple.

A flow battery [22] is a form of rechargeable battery in which the electrolyte containing one or more dissolved electro-active species flows through an electrochemical cell that converts chemical energy directly to electricity. Various classes

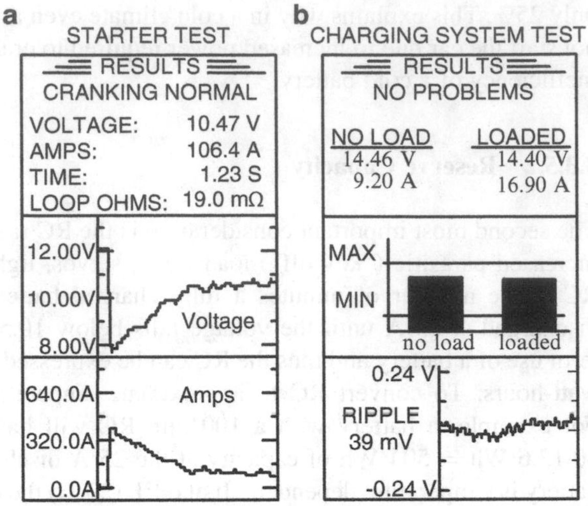

Fig. 3.12 (**a**) Starter test and (**b**) charging test of a car battery

of flow batteries exist including the *redox (reduction–oxidation) flow* battery, in which all electro-active components are dissolved in the electrolyte. If one or more electro-active component is deposited as a solid layer the system is known as a *hybrid flow* battery. The main difference between these two types of flow battery is that the energy of the *redox flow* battery can be determined fully independently of the battery power, because the energy is related to the electrolyte volume (tank size) and the power to the reactor size. The *hybrid flow* battery, similar to a conventional battery, is limited in energy to the amount of solid material that can be accommodated within the reactor. In practical terms this means that the discharge time of a *redox flow* battery at full power can be varied, as required, from several minutes to many days, whereas a *hybrid flow* battery may be typically varied from several minutes to a few hours. This enables the flow batteries to be used as storage elements which have a fast response time as a buffer for intermittently operating renewable sources such as wind and photovoltaic [27] power plants.

3.4 Photovoltaic Array Model [28–33]

3.4.1 Solar Radiation and Solar Cells

The amount of solar radiation incident at a location (insolation or irradiance) on the earth varies daily and annually. The daily variations are caused by the spinning of the earth, while the solar declination and the earth–sun distance change annually. Insolation depends also on the geographical position of the location, on the weather conditions (e.g., cloudiness), on the physical conditions (e.g., dust cover, snow, leaves), and the geometry of the collecting surface.

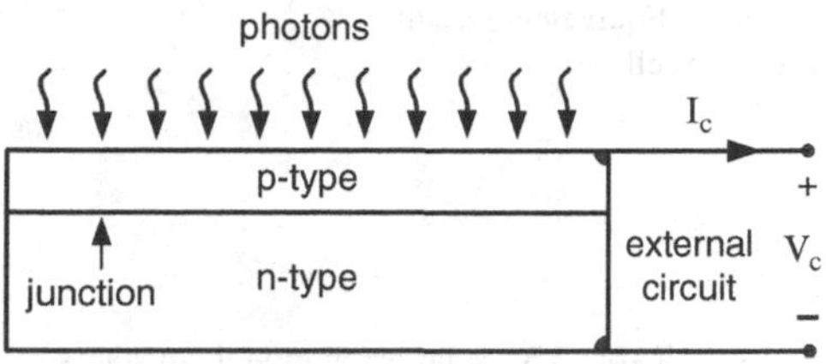

Fig. 3.13 Schematic representation of a solar cell with cell terminal voltage V_c, and cell terminal current I_c

The main component of the total radiation is the global shortwave radiation (0–3 μm). This component itself consists of three parts: the direct shortwave radiation, the diffused shortwave radiation, and the shortwave radiation reflected from nearby surfaces. The common practice of solar collector installations is to keep an optimal angle of incidence throughout the entire year.

The incident sunlight can be converted into electricity by photovoltaic conversion using a solar cell [29], which uses the photoelectric properties of a semiconductor (e.g., silicon or cadmium/telluride based). A solar cell is a large-area semiconductor diode, so constructed that light can penetrate into the region of a p-n junction diode. The junction formed between the n-type silicon wafer and the p-type surface layer provides the electric fields which cause the diode's characteristics as well as the photovoltaic effect. Light is absorbed in the silicon, generating both excess holes and electrons. These excess charges generated by the p-n junction can be made to flow through an external circuit. The operating principle of the solar cell is illustrated in Fig. 3.13.

3.4.2 Voltage–Current Characteristics and Equivalent Circuit

The photon-generated electron–hole pairs determine the cell photon-current I_{ph}. This current is proportional to the incident photon flux and thus proportional to the insolation. The solar cell equivalent circuit is shown in Fig. 3.14, from which the nonlinear voltage–current (V_c–I_c) [28, 29]:

$$V_c = -R_s I_c + \frac{1}{\Lambda}\ln\left[\frac{I_{ph} - (1+\frac{R_s}{R_{sh}})I_c - \frac{V_c}{R_{sh}}}{I_r} + 1\right], \tag{3.2}$$

where R_{sh} is the cell shunt resistance [Ω], R_s is the cell series resistance [Ω], I_r is the cell reverse saturation current [A], and

$$\Lambda = \frac{q}{A_0 k T_{kelvin}}. \tag{3.3}$$

In this relation q is the electron charge (1.6×10^{-19} coulombs), k is the Boltzmann constant ($1.3708 \cdot 10^{-23}$ J/K), T_{kelvin} is the cell temperature [K], A_0 [W/m^2] is a cell constant (see Table 3.5).

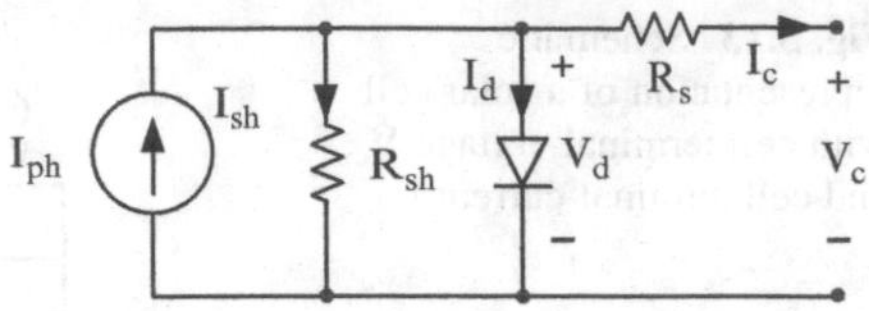

Fig. 3.14 Equivalent circuit of a solar cell

Table 3.5 Solar array model constants

Constant	Value	Constant	Value
Rs	0.0001 Ω	D2	2.805×10^{-3} V/°C
Rx	0.007 Ω	D3	0.1197 V
Ns	Variable	D4	3.979 m²/V
Np	Variable	D5	−1.042 A/°C
A0	$1.95 + 0.02(Q_s)$ W/m²	D6	1.978×10^{-4} V/°C
D1	3.973 V	D7	−1220.0°C

Usually $R_{sh} \to \infty$ and, therefore,

$$V_c = -R_s I_c + \frac{1}{\Lambda}\ln\left[\frac{I_{ph} - I_c}{I_r} + 1\right]. \quad (3.4)$$

If there are N_s cells connected in series and N_p parallel strings in the array (Fig. 3.15a), then

$$V_{DC} = N_s V_c, \quad (3.5)$$

and

$$I_{DC} = N_p I_c. \quad (3.6)$$

Equation (3.4) can now be modified to give:

$$V_{DC} = -R_s\frac{N_s}{N_p}I_{DC} - R_x I_{DC} + \frac{N_s}{\Lambda}\ln\frac{I_{ph}}{I_r} + \frac{N_s}{\Lambda}\ln\left[\left(1 + \frac{I_r}{I_{ph}}\right) - \frac{1}{N_p I_{ph}}(I_{DC})\right], \quad (3.7)$$

where R_x is the array to inverter resistance.

Following the development of Otterbein [30], I_{ph} and I_r are given by

$$I_{ph} = I_{sc}\frac{e^{Eoc\Lambda} - 1}{e^{Eoc\Lambda} - e^{IreRs\Lambda}}, \quad (3.8)$$

and

$$I_r = I_{sc}\frac{1}{e^{Eoc\Lambda} - e^{IreRs\Lambda}}, \quad (3.9)$$

where I_{sc} is the cell short-circuit current [A] and E_{oc} is the cell open-circuit voltage [V]. The remaining parameters of (3.8) and (3.9) are defined in [30]. Equations 3.7–3.9

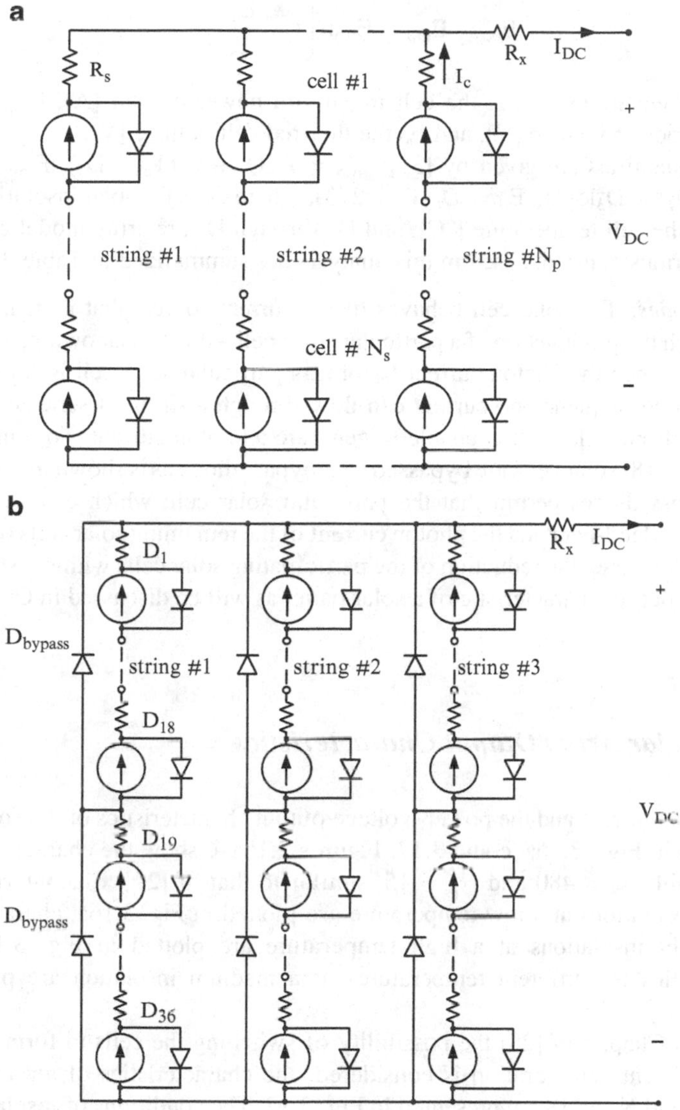

Fig. 3.15 (**a**) Equivalent circuit of a solar (photovoltaic) array or panel without bypass diodes. (**b**) Equivalent circuit of a solar (photovoltaic) array or panel with bypass diodes

require values for R_s, N_s, N_p, R_x, A_0, I_{sc}, and E_{oc}. The values for R_s and A_0 are determined numerically by adjusting them until the V_{DC}–I_{DC} curve, approximated by (3.7), matches an experimentally recorded V_{DC}–I_{DC} curve [30]. R_x, N_s, and N_p are determined by the array configuration. I_{sc} and E_{oc} are estimated using [31]:

$$I_{sc} = I_{mp}\left[1 + \frac{E_0}{E_{mp}}\right], \tag{3.10}$$

$$E_{oc} = E_{mp} + E_0 \ln\left[1 + \frac{E_{mp}}{E_0}\right]. \qquad (3.11)$$

In these equations, I_{mp} is the cell maximum-power current [A], E_{mp} the cell maximum-power voltage [V], and E_0 the thermal cell voltage [V].

These quantities are given by [31]: $I_{mp} = D_4Q_s + D_5(T_c - D_7)$, $E_{mp} = D_1 - D_2(T_c - D_7) + D_3\log Q_s$, $E_0 = D_6(T_c + 273)$, where Q_s is the solar insolation [kW/m^2], T_c is the cell temperature [°C], and D_1 through D_7 are array model constants [-]. The various constants used in this analysis are summarized in Table 3.5.

Bypass Diodes: The solar cell behaves like a current source, that is, if insolation cannot reach the p-n junction of a particular solar cell – due to shadowing, dirt, snow and leaves – then the photon current I_{ph} of this particular solar cell is zero or very small. As a consequence no current can flow through a string of series connected solar cells if one solar cell is unable to generate a photon current. To remedy this several (e.g., 18) solar cells are bypassed by a bypass diode as is shown in Fig. 3.15b. These bypass diodes permit that the particular solar cell, which cannot conduct current, is bridged-over and the photon current of the remaining solar cells of a string can flow. Of course, the reduction of the participating solar cells within a string will change the output characteristic of a solar panel as will be discussed in Chap 12.

3.4.3 Solar Array Output Characteristics

The voltage–current and the power–voltage output characteristics of the solar array are plotted in Figs. 3.16a–c and 3.17. Figures 3.16a–c show the characteristics of an array with $N_s = 480$ and $N_p = 15$, similar to that of [29]. Characteristics for different insolations at a low temperature are plotted in Fig. 3.16a, characteristics for different insolations at a high temperature are plotted in Fig. 3.16b, and characteristics for different temperatures at a medium insolation are plotted in Fig. 3.16c.

Since in Chap. 5 of [28] the possibility of switching the cells to form an array with a different configuration is considered, the characteristics of an array with $N_s = 400$ and $N_p = 18$ are presented in Fig. 3.17. The conditions of insolation and temperature are similar to those of Fig. 3.16a. Note that the voltage–current curves approach asymptotically the short-circuit current, determined by the solar insolation level and the configuration of the array.

Application Example 3.6: Voltage–Current Characteristics and Maximum Power of a Solar Cell

For $R_s = 0.05\ \Omega$, $R_{sh} = 10\ k\Omega$, $I_r = 0.0005$ A, $T_c = 313.2$ K corresponding to 40°C, and the open circuit cell voltage $V_c = V_{coc} = 1.7$ V for maximum insolation $Q_s = 1{,}000$ W/m^2 compute the cell voltage V_c as a function of I_c, and find the maximum output power of this solar cell.

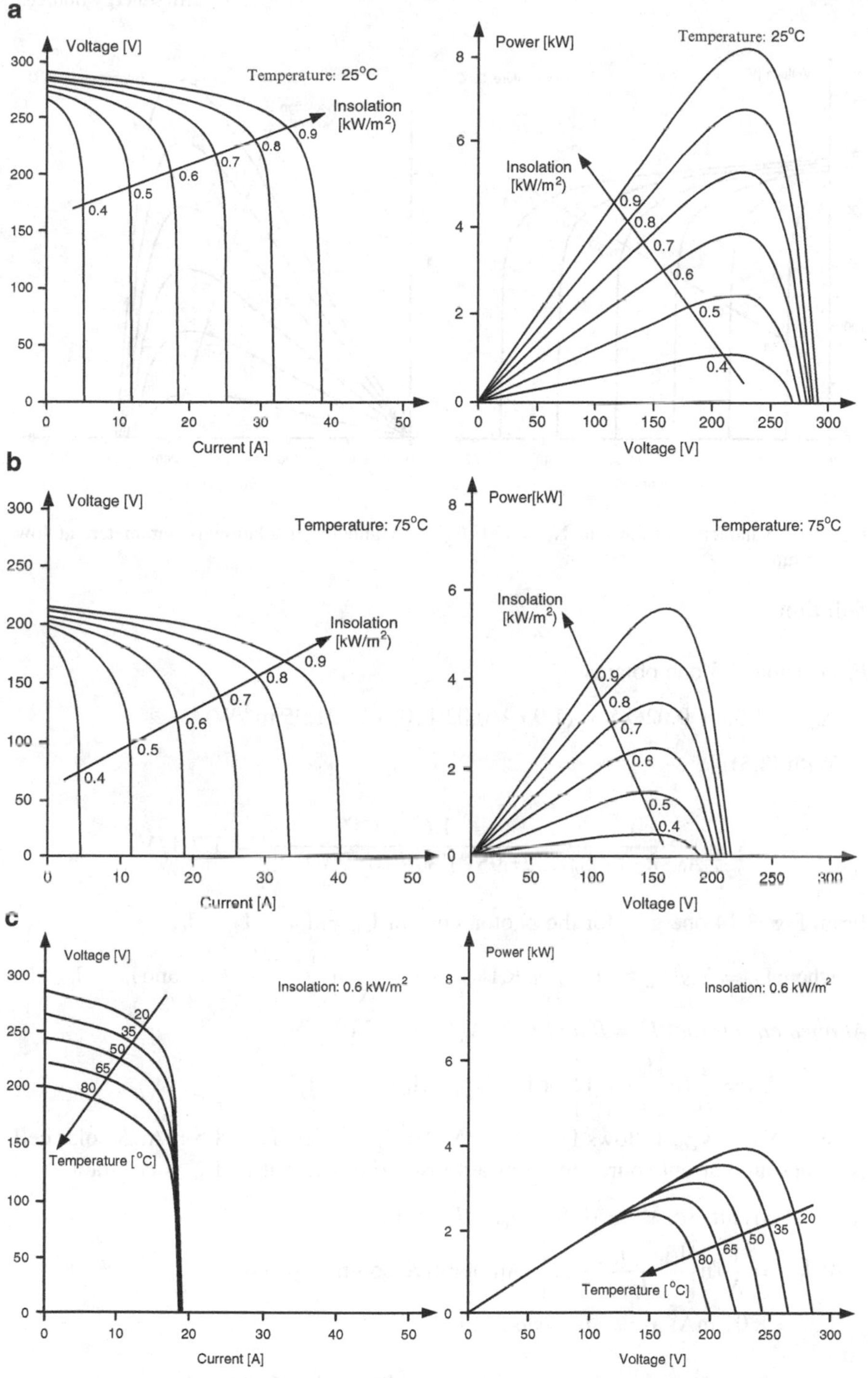

Fig. 3.16 (**a**) Solar array characteristics where $N_s = 480$, $N_p = 15$, and the insolation is parameter; at low temperature. (**b**) Solar array characteristics where $N_s = 480$, $N_p = 15$, and the insolation is parameter; at high temperature. (**c**) Solar array characteristics where $N_s = 480$, $N_p = 15$, and the temperature is parameter; at medium insolation

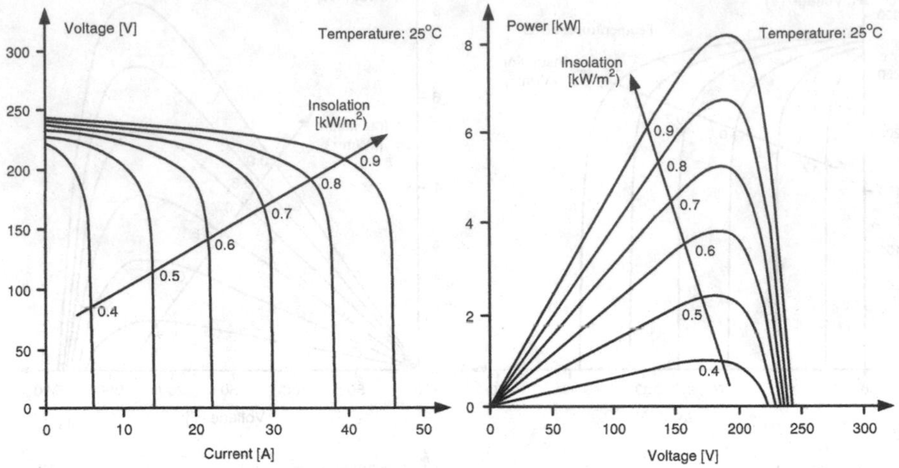

Fig. 3.17 Characteristics where $N_s = 400$, $N_p = 18$ and the insolation is parameter; at low temperature

Solution

From Table 3.5 one obtains

$A_0 = (1.95 + 0.02Q_s) = (1.95 + 0.02 \cdot 1000) = 21.95\ m^2/W$.

With (3.3)

$$\Lambda = \frac{q}{A_0 \cdot k \cdot T_{kelvin}} = \frac{1.6 \cdot 10^{-19}}{21.95 \cdot 1.37 \cdot 10^{-23} \cdot 313.2} = 1.7\ 1/V.$$

From Fig. 3.14 one gets for the photon current $I_{ph} = I_{sh} + I_d + I_c$,

where $I_{sh} = V_d/R_{sh} \approx 0$, $V_d = R_s I_c + V_c = \frac{1}{\Lambda}\ell n\left[\frac{I_{ph} - I_c}{I_r} + 1\right]$ and $I_d = I_{ph} - I_c$.

At open circuit (oc) $I_c = 0$ and $V_c = V_{coc} = 1.7\ V$:

$V_c = V_d = \frac{1}{\Lambda}\ell n\left[\frac{I_{ph}}{I_r} + 1\right]$ or $I_d = I_{ph} = I_r[e^{\Lambda V_d} - 1]$.

With $V_d = V_{coc}$ follows $I_d = I_{ph} = 5 \cdot 10^{-4}[e^{1.7 \cdot 1.7} - 1] = 8.5$ mA. A solar cell is a constant current source for constant insolation Q_s, that is, I_{ph} = constant.

At short circuit (sc) $V_c = 0$, $I_c = I_{csc}$, $I_{sh} \approx 0$:

$R_s I_{csc} = \frac{1}{\Lambda}\ell n\left[\frac{I_{ph} - I_{csc}}{I_r} + 1\right]$. An iterative solution yields

$I_{csc} = 8.01$ mA; as can be seen from

$$0.05\Omega \cdot 8.01\text{mA} = \frac{1}{1.7}\ell n\left[\frac{8.5\text{mA} - 8.01\text{mA}}{5 \cdot 10^{-4}\text{A}} + 1\right]$$

or 0.400 mV ≈ 0.401 mV, this nonlinear relation is approximately satisfied.

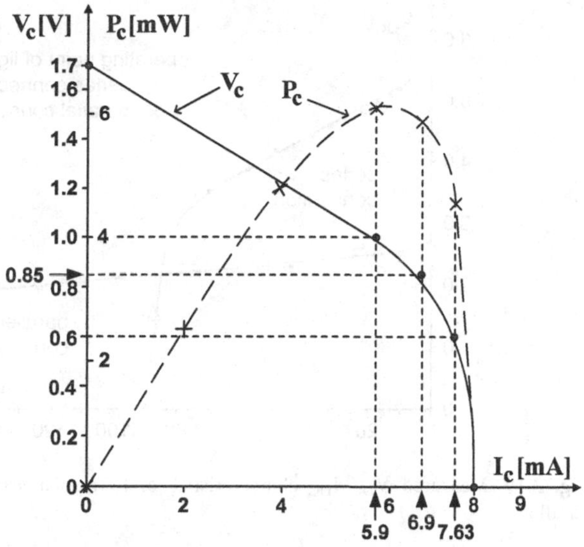

Fig. 3.18 V_c-I_c and P_c-I_c output characteristics of a solar cell

At $V_{c0.85V} = 0.85$ V, the current is $I_{c0.85V} = 6.9$ mA:

As can be noted from

$$0.05\Omega \cdot 6.9\text{mA} + 0.85\text{V} = \frac{1}{1.7}\ell n\left[\frac{8.5\text{mA} - 6.9\text{mA}}{5 \cdot 10^{-4}\text{A}} + 1\right]$$

or 0.85 V ≈ 0.844 V, this nonlinear relation is approximately satisfied.

At $V_{c1.03V} = 1.03$ V, the current is $I_{c1.03V} = 5.9$ mA:

As can be seen from

$$0.05\Omega \cdot 5.9\text{mA} + 1.03\text{V} = \frac{1}{1.7}\ell n\left[\frac{8.5\text{mA} - 5.9\text{mA}}{5 \cdot 10^{-4}\text{A}} + 1\right]$$

or 1.0303 V ≈ 1.07 V, this nonlinear relation is approximately satisfied.

At $V_{c0.6V} = 0.6$ V, the current is $I_{c0.6V} = 7.63$ mA:

As can be found from

$$0.05\Omega \cdot 7.63\text{mA} + 0.6\text{V} = \frac{1}{1.7}\ell n\left[\frac{8.5\text{mA} - 7.63\text{mA}}{5 \cdot 10^{-4}\text{A}} + 1\right]$$

or 0.6004 V ≈ 0.593 V, this nonlinear relation is approximately satisfied.

The above calculated data are represented by the graph $V_c = f(I_c)$ in Figure 3.18.

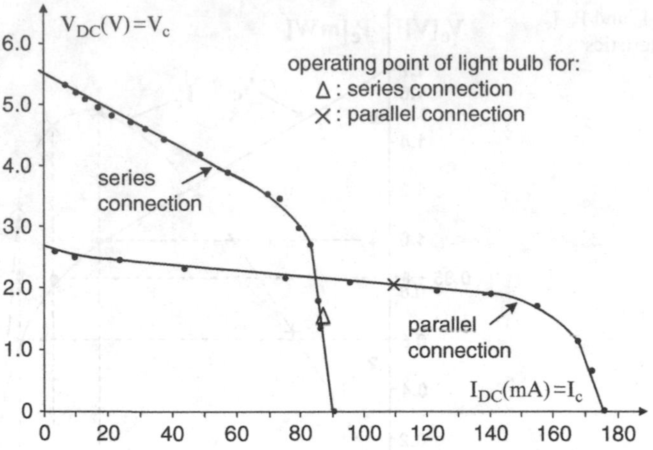

Fig. 3.19 Measured V_{DC}–I_{DC} characteristics of four solar arrays connected in series and in parallel

The $P_c = V_c \cdot I_c = f(I_c)$ function is also shown in Fig. 3.18, where the maximum output power occurs at the knee of the $V_c = f(I_c)$ characteristic, and is $P_{c\_max} \approx V_{c\_knee} \cdot I_{c\_knee} = 1.03\,\text{V} \cdot 5.9\,\text{mA} = 6.08\,\text{mW}$. The determination of the maximum output power of a solar cell or a solar array requires a maximum power tracker [33].

This example shows that a nonlinear characteristic model including the diode nonlinear $I_d = f(V_d)$ relation generates a convex $V_c = f(I_c)$ characteristic for a solar cell as illustrated in Fig. 3.18.

It is advisable to solve the above nonlinear equations with software programs such as either Matlab or Mathematica, whereby the influence of the shunt resistance R_{sh} can be taken into account resulting in more complicated nonlinear equations.

Application Example 3.7: Measured Voltage–Current Characteristic of a Solar Array

Figure 3.19 shows the measured V_{DC}–I_{DC} characteristic of a solar array. Four panels were purchased: In the "series" connection two panels were connected in series and connected in parallel with the other two in series-connected panels, while in the "parallel" connection all four panels were connected in parallel. A light bulb was operated in series (Δ) and in parallel (X) connection as indicated in Fig. 3.19: the light bulb is operating in both connections not at the maximum power point [33] of the solar array.

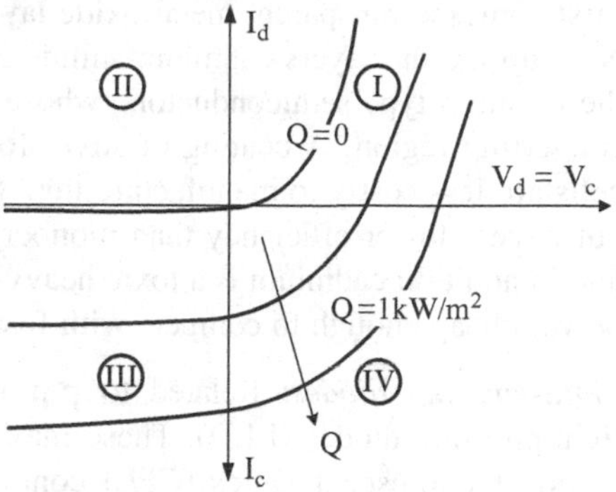

Fig. 3.20 Solar-cell or photo-diode characteristic, where insolation $Q_S = Q$ is parameter

Application Example 3.8: Identity of Diode Equation and Solar-Cell Equation

Show that the equation for a photovoltaic cell reduces to that of a diode if the insolation (Q_S) is zero. You may assume that $R_S = 0$ and $R_{sh} \to \infty$.

Background information: Figure 3.20 illustrates the characteristic of a solar cell under the influence of light. For an insolation of $Q_S = Q = 0$ one obtains the well known diode characteristic (first and third quadrants), and for $Q_S = Q$ from 0 to 1 kW/m^2 one recognizes the solar cell characteristic (fourth quadrant).

Solution

With $I_c = -I_d$; $V_c = V_d$; $R_s = 0$; $R_{sh} = \infty$; $I_{ph} = 0$ follows

$$V_d = \frac{1}{\Lambda} \ln\left[\frac{I_d}{I_r} + 1\right] = \frac{1}{\Lambda} \ln\left[\frac{I_d + I_r}{I_r}\right],$$

or $e^{V_d \Lambda} = [(I_d + I_r)/I_r]$; $I_r e^{V_d \Lambda} = I_d + I_r$.

One obtains for the solar cell $I_d = I_r(e^{V_d \Lambda} - 1)$.
The diode relation is $I_d = I_s(e^{(V_d/nV_T)} - 1)$.
Note $V_T = kT/q$, $\Lambda = 1/nV_T$, and $I_r = I_s$.

Types of solar cells: There are three basic types of solar cells. The *monocrystalline* cells are cut from a silicon ingot grown from a single large crystal of silicon whilst the *polycrystalline* cells are cut from an ingot made up of many smaller crystals. The third type is the *amorphous* or *thin-film* solar cell, resulting in the cadmium/telluride (CdTe) thin-film solar cell technology. In a conventional CdTe cell, the layers are deposited on a glass sheet, which is flipped over to face the sun [34, 35].

First comes a transparent metal-oxide layer that will form the cell's top electrode. Next up are the layers cadmium-sulfide followed by cadmium-telluride that form the p- and n-type semiconductors, whose junction forms the cell's active, energy-converting region. A coating of silver forms the bottom electrode. The thin-film cells are less costly to manufacture than the monocrystalline/polycrystalline cells, but have a lower efficiency than monocrystalline/ polycrystalline ones (18–21%) and in addition cadmium is a toxic heavy-metal material. The aim is to make solar power cheap enough to compete with fossil fuels, perhaps within a decade.

Light-emitting diodes: Related to p-n junction diodes and solar cells are the light-emitting diodes (LED). These may replace in the not too distant future the compact fluorescent lamps (CFL) containing mercury. In particular LEDs may replace traditional street lights such as high pressure sodium and metal halogen lamps. The advantages of LEDs compared with traditional street lights are:

1. High color rendering, where the white LEDs light spectrum covers the entire visible light.
2. Low power consumption and operating cost. An LED street light of 130 W corresponds (has the same lighting effect) to the metal halide lamp or the high pressure sodium lamp of 400–500 W.
3. LEDs rely on less copper cable and construction investment.
4. LEDs are resource-conserving and environment-friendly.
5. LEDs are easy to control by time, sound, light, and networks can be wireless controlled which is impossible for traditional lights.

Application Example 3.9: Design of an Electric Storage Plant for a Stand-Alone Residential Power System Based on Lead-Acid Batteries

The storage of electricity in batteries is feasible for short-term storage (that is, a few days). To investigate this feasibility, answer the following preliminary questions based on the battery-charging characteristic of Fig. 3.21.

(a) What is the charging state (e.g., empty, 50% charged, fully charged) of the battery at the beginning of the charging period?
(b) What is the charging state of the battery at the end of the charging period at $t_{charging} = 15$ h?

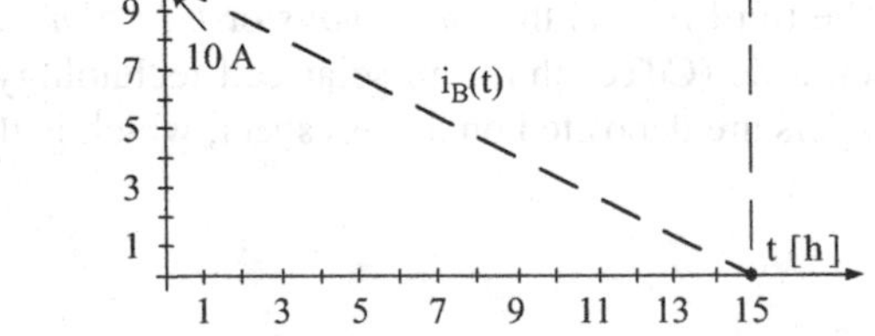

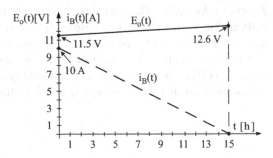

Fig. 3.21 Linear charging diagram of a lead-acid battery

(c) Compute the total charge of the battery expressed in ampere-hours (Ah) at the end of the charging period at $t_{charging} = 15$ h.
(d) Compute the total charge stored (Q_B) in the battery expressed in coulombs at the end of the charging period at $t_{charging} = 15$ h.
(e) Compute as a function of time the instantaneous power $p_B(t) = \{E_0(t) \cdot i_B(t)\}$ delivered to the battery during the charging period. You may approximate this function by a few points. Note that $E_0(t)$ is the battery voltage, and $i_B(t)$ is the battery charging current as defined in Fig. 3.21. Compute the average power of the battery.
(f) Compute the total energy stored ($Energy_B$) in the battery at the end of the charging period.
(g) What is the weight (in kg-force) of this battery pack?
(h) How much of the stored energy can be utilized?

Solution

(a) The charging state at the beginning of charging period is with $V_B = 11.5$ V: empty.
(b) The charging state at the end of the charging period is with $V_B = 12.6$ V: fully charged.
(c) The total charge of the battery at the end of the charging period (after 15 h of charging)

$$\int_{t=0}^{t=15h} i_B(t)dt = \frac{10A}{2} \cdot 15h = 75Ah. \tag{3.12}$$

(d) The total charge stored expressed in coulombs (C)

$$Q_B = \int_{t=0}^{t=15h} i_B(t)dt = \frac{10\,A}{2} \cdot 15h \cdot 3,600s/h = 270\,kAs = 270\,kC. \tag{3.13}$$

(e) The average battery power is defined as

$$P_B = \int_{i_B=0A}^{i_B=10A} E_o(t) \cdot di_B(t) = 60.25\,W, \tag{3.14}$$

(f) The total energy stored in the battery is defined as

$$Energy_B = \int_{t=0}^{t=15h} E_o(t) \cdot i_B(t)dt = 906.30\,Wh, \tag{3.15}$$

see Table 3.6

Table 3.6 Approximate hand calculations based on three 5h-segments yields for the total stored $Energy_B$ = 906.30 Wh, while a more accurate numerical calculation based on Mathlab/Simulink (with many segments) the total stored energy is 890 Wh

t (h)	$E_o[t]$ (V)	i(t) (A)	$Energy_B(t)$ (Wh)
0	11.5	10	0
5	≈12	≈6.8	≈493.5[a]
10	≈12.5	≈3.3	≈309.3
15	12.6	0	≈103.5

[a] At the end of the first 5 h interval

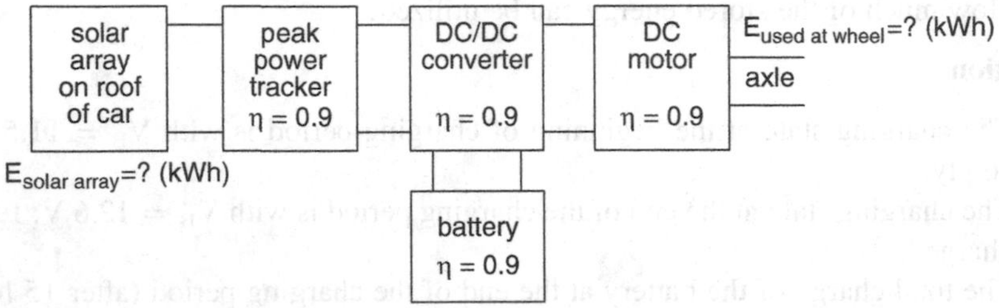

Fig. 3.22 Drive of an automobile with solar array and battery

Although the battery characteristics of Fig. 3.21 are linear, the characteristics are divided in 3 time intervals as shown in Table 3.6 in order to keep the numerical integration as flexible as possible, and nonlinear characteristics could be approximated by step functions.

(g) The weight of the lead-acid battery – based on a specific energy per kg-force of 50 Wh/(kg-force) – is 18.08 kg-force.
(h) A lead-acid battery cannot be completely discharged. Provided the permissible depth of discharge (DoD) is 50% then 453.2 Wh are available.

Application Example 3.10: Design of an Automobile Drive with Solar Array and Battery

A car with a weight of 500 (kg-force) = 0.5 (ton-force) is equipped with a photovoltaic array on the roof (1.5 m × 2.3 m), and a lithium-ion battery with the stored energy content of $E_{battery}$ = 10 kWh is used for storage as indicated in Fig. 3.22. It can be assumed that this car is driven for 75 miles for 1.5 h in the early morning (per day) from 7:00 to 8:30 a.m., and thereafter parked for 8 h from 8:30 a.m. to 4:30 a.m. so that the battery can be recharged via the solar array. At 5:00 p.m. the vehicle will be used again for the 1.5 h return trip. It can be assumed that the vehicle's battery is fully charged in the garage during the previous night at a cost of \$0.20/kWh.

The vehicle specific energy consumption is (see Fig. 1.9)

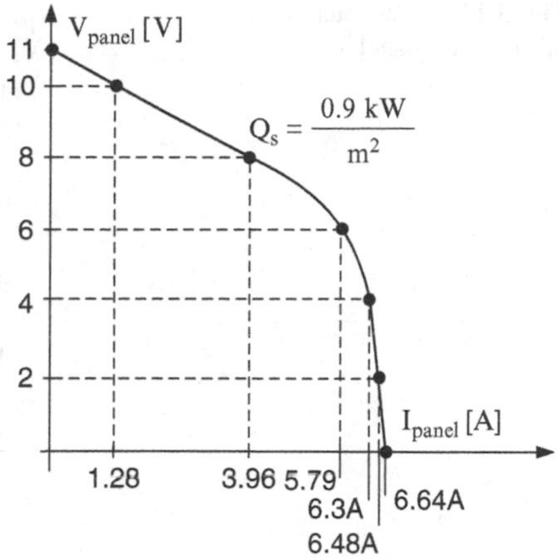

Fig. 3.23 V-I characteristic of one solar panel

$$\text{figure of merit} = \text{FM} = \frac{\{\text{weight [tons} - \text{force]} \times \text{distance [miles]}\}}{\{\text{energy used at wheel } E_{used\_at\_wheel}\ [\text{kWh}]\}} = 10.$$

(a) For a commercially available solar panel the V–I characteristic of Fig. 3.23 was measured at an insolation of $Q_s = 0.9\ kW/m^2$. Plot the power curve of this solar panel: $P_{panel} = f(I_{panel})$.
(b) At which point of the power curve $P_{panel} = f(I_{panel})$ would you operate the solar panel assuming $Q_s = 0.9\ kW/m^2$ is constant? What values for power, voltage and current correspond to this point?
(c) What is the energy required at the wheel $E_{used\_at\_wheel}$ for the (2·75) miles = 150-mile roundtrip?
(d) What is the energy generated by the solar array (1.5 m × 2.3 m = 3.45 m^2) E_{solar_array}, and the energy supplied by the solar array to the car drive at the wheel (at the output of DC motor) $E_{solar\_array\_wheel}$ during 8 h of operation at 0.9 kW/m^2 if the solar cell efficiency is 15%, and the energy efficiencies η of peak-power tracker, DC-to-DC converter, battery, and DC motor are 90% each?
(e) How many solar panels are required for the entire rooftop solar array?
(f) What is the depth of discharge (DoD) of the lithium-ion battery after the 150 mile roundtrip has been completed?
(g) What is the weight of the lithium-ion battery pack if it has a specific weight of 160 (Wh)/(kg force)?
(h) What is the average output power (at the wheels) developed?
(i) How much is the purchase price of this solar power plant, $cost_{electric\_drive}$, if 1 kW installed output power capacity of the drive P_{wheel} (entire solar array + peak power tracker + Li-ion battery + DC-to-DC converter + DC motor) costs $11,200?
(j) What is the purchase price of an IC drive (e.g., two or three cylinder engine), $cost_{IC\_drive}$, which could replace the electric drive if 1 kW installed output power capacity costs $2,000?

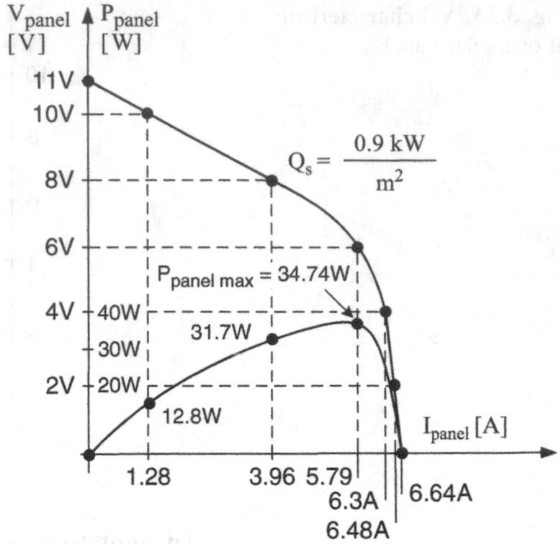

Fig. 3.24 P-I characteristic of one solar panel

(k) What is the payback period (in years) neglecting interest payments for this electric drive, if a comparable internal combustion engine drive has a mileage of 60 miles/gallon, and 1 gallon of gasoline costs \$4.00? You may assume that the car is used every day for two 75 mile-1.5 h trips.

Solution

(a) Figure 3.24 shows the power curve P_{panel} as a function of the panel current I_{panel}.
(b) The V–I characteristic is operated at the knee corresponding to $V_{panel} = 6$ V, $I_{panel} = 5.79$ A resulting in a peak (max) power of $P_{panel\_max} = 34.74$ W.
(c) The energy required at the wheels for the 150-mile roundtrip is

$$E_{used\_at\_wheel} = \left[\frac{0.5\,(\text{ton} - \text{force}) \cdot 150\,(\text{miles})}{10}\right] = 7.5\,\text{kWh}.$$

(d) The generated energy of the solar array is $E_{solar\_array} = 0.9 \cdot 3.45 \cdot 8 \cdot 0.15 = 3.726$ kWh, and the energy supplied to the wheels by the solar array is $E_{solar\_array\_wheel} = E_{solar\_array} \cdot (\eta)^6 = 3.726 \cdot (0.9)^6 = 1.98$ kWh, or $P_{solar\_array} = (E_{solar\_array}/8h) = 466$ W. One concludes that the energy $E_{solar\_array\_wheel}$ is smaller than the energy $E_{used\_at\_wheel}$ required for the $(2 \cdot 1.5)$ h = 3 h drive. This means that for each 150 mile roundtrip the battery energy $E_{enery\_supplied\_by\_battery} = E_{used\_at\_wheel} - E_{solar\_array\_wheel} = (7.5 - 1.98)$ kWh = 5.52kWh, must be supplied from the energy of the battery which is stored on the battery during the previous night.
(e) The number of panels required for the solar array are $N_{panels} = (P_{solar\_array}/P_{panel\_max}) = 466\ \text{W}/34.74\ \text{W} \approx 14$ panels.

(f) The depth of discharge is $DoD = (7.5\,kWh - 1.98\,kWh)/10\,kWh = 0.552$, that is the battery energy is reduced to 44.8% after the 150 mile roundtrip; this DoD of about 50% is acceptable.
(g) The weight of the lithium-ion battery is (10 kWh)/(0.160 kWh)/(kg-force) = 62.5 kg-force, which amounts to 12.5% of the entire weight of the car of 500 (kg-force).
(h) The average power developed at the wheels is $P_{wheel} = E_{used\_at\_wheel}/3h = 7.5\,kWh/3h = 2.5\,kW = 3.35\,hp$.
(i) The purchase price of the electric drive is:

$$cost_{electric\_drive} = (\$\ 11,200/kW) \cdot 2.5\,kW \ = \$28,000.$$

(j) The purchase price of an internal combustion (IC) engine drive is: $cost_{IC\_drive}$ = (\$2,000/kW)·2.5 kW = \$5,000.
(k) The operating cost of the electric drive is (including purchasing cost) for (y) years:

$$\begin{aligned} cost_{electric\_operating} &= cost_{electric\_drive} + \left(E_{enery\_supplied\_by\_battery}\right) \cdot 365 \cdot (y) \cdot 0.20 \\ &= \$28,000 + \$5.52 \cdot 365 \cdot (y) \cdot 0.20 = \$28,000 + \$402.96(y). \end{aligned}$$

The operating cost of the IC drive is (including purchasing cost) for (y) years:

$$\begin{aligned} cost_{IC\_operating} &= \$5,000 + (y) \\ &\quad \cdot [365\ days\ (150\,miles/day) \cdot (\$4/gallon)]/(60\,miles/gallon) \\ &= \$5,000 + \$3,650\ (y). \end{aligned}$$

The payback period in (y) years is obtained from the condition $cost_{electric\_operating} = cost_{IC\_operating}$ or \$28,000 + 402.96 (y) = \$5,000 + 3,650 (y) resulting in (y) = 7.08 years.
The disadvantages of the electric drive are

- The low average output power of the drive $P_{wheel} = 2.5$ kW or 3.35 hp
- The fact that the solar panel cannot provide the required energy for the 150 mile roundtrip
- That a Li-ion battery must be replaced about every 8 years

Application Example 3.11: Final Values After a Transient Response

Figure 3.25 represents the block diagram and the transfer functions of a separately excited DC machine, where R_a, B, and K are machine constants and τ_a and τ_m are time constants.

(a) Find the transfer function of the angular velocity $\omega_m(s)$ as a function of the mechanical torque $T_M(s)$and the terminal voltage $V_t(s)$, that is $\omega_m(s) = f\{T_M(s), V_t(s)\}$.
(b) What is the final value of $\omega_m(s)$ if the terminal voltage $V_t(s)$ is a step function changing from 0 to V_t and $T_M(s) = 0$?

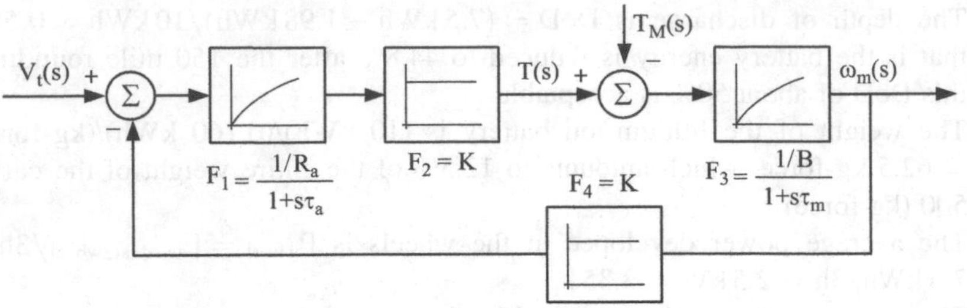

Fig. 3.25 Block diagram of separately excited DC machine with transfer functions

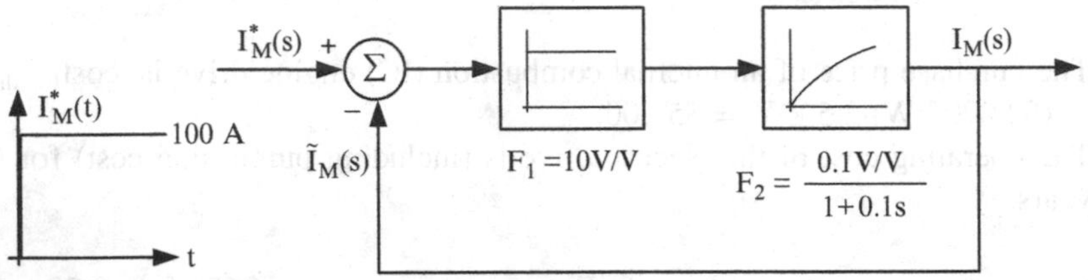

Fig. 3.26 Current–control loop for an electric drive

(c) What is the final value of $\omega_m(s)$ if the mechanical torque $T_M(s)$ is a step function changing from 0 and to T_M and $V_t(s) = 0$?

Solution

(a) By inspection one obtains from the block diagram $\omega_m(s) = F_3[T(s) - T_M(s)]$, where $T(s) = F_1F_2[V_t(s) - F_4\,\omega_m(s)]$, therefore, $\omega_m(s) = F_1F_2F_3V_t(s) - F_1F_2F_3F_4\,\omega_m(s) - F_3T_M(s)$, or

$$\omega_m(s) = \left\{\frac{F_1KF_3V_t(s) - F_3T_M(s)}{(1 + F_1F_3K^2)}\right\}$$

(b) The final-value theorem $\lim\limits_{s\to\infty} \omega_m(t) = \lim\limits_{s\to 0} s\,\omega_m(s)$ yields

$$\lim_{t\to\infty} \omega_m(t) = \lim_{s\to 0} s\,\omega_m(s) = \frac{V_tK}{(R_aB + K^2)}.$$

(c) Correspondingly,

$$\lim_{t\to\infty} \omega_m(t) = \lim_{s\to 0} s\omega_m(s) = \frac{-T_MR_a}{(R_aB + K^2)}.$$

Application Example 3.12: Transient Response of an Electric Drive

(a) For the block diagram of Fig. 3.26 compute the final value (at $t \to \infty$) of $I_M(s)$ provided $I^*_M(s)$ is a step function of 100 A.

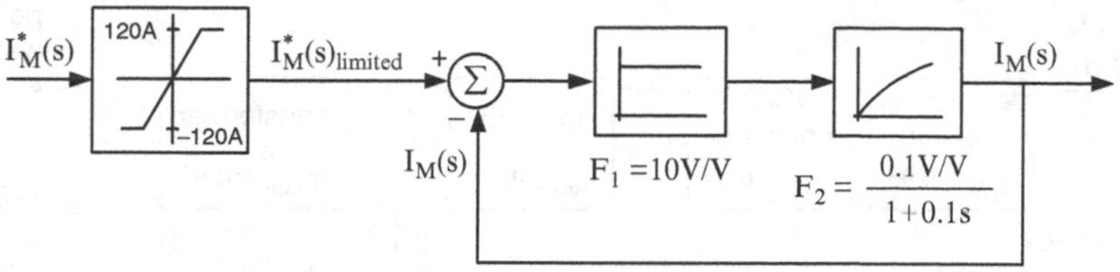

Fig. 3.27 Current-control loop with current limiter

Fig. 3.28 Implementation of the limiting circuit using two zener diodes

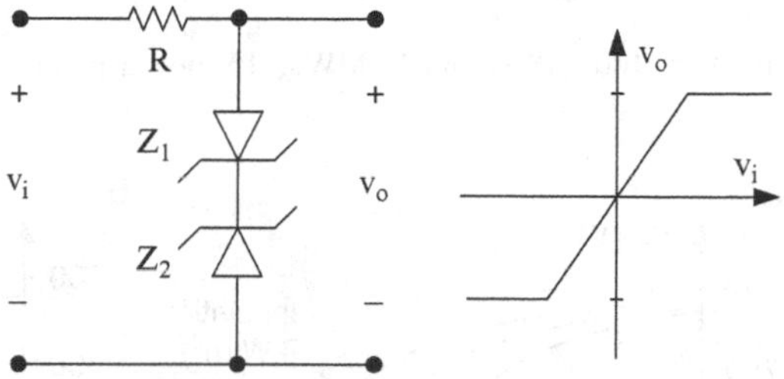

(b) For the block diagram of Fig. 3.26 compute the initial value (at $t = 0$) of $I_M(s)$ provided $I^*_M(s)$ is a step function of 100 A.

(c) Between the time $t = 0$ and time $t \rightarrow \infty$ the current $I_M(s)$ may have large instantaneous excursions (e.g., 200 A). How could you limit such large excursions to ±120 A using one additional circuit element? Indicate in the block diagram where you would place such an additional circuit element and identify the circuit of such an element.

Solution

(a) $I_M(s) = (F_1F_2I^*_M(s))/(1+F_1F_2)$. The final value theorem yields $\lim_{t\rightarrow\infty} I_m(t) = \lim_{s\rightarrow 0} sI_m(s) = 50\,A$.

(b) The initial-value theorem gives $\lim_{t\rightarrow 0} \omega_m(t) = \lim_{s\rightarrow\infty} s\,\omega_m(s) = 0$.

(c) Figure 3.27 illustrates the addition of a limiting circuit. This limiting circuit can be assembled using two zener diodes (see Fig. 3.28) and possibly two batteries if the zeners do not have sufficient breakdown voltages.

Application Example 3.13: Design of a $P_{out} = 4\ MW_{AC}$ Photovoltaic (PV) Power Plant

A PV power plant consists of solar panels, peak-power tracker, step-down/step-up DC-to-DC converter(s) (20 connected in parallel), three-phase inverter(s) (20 connected in parallel), Y–Δ three-phase transformer, and a three-phase power system absorbing the rated (nominal) output power $P_{out\ inverter}$ of the 20 inverters

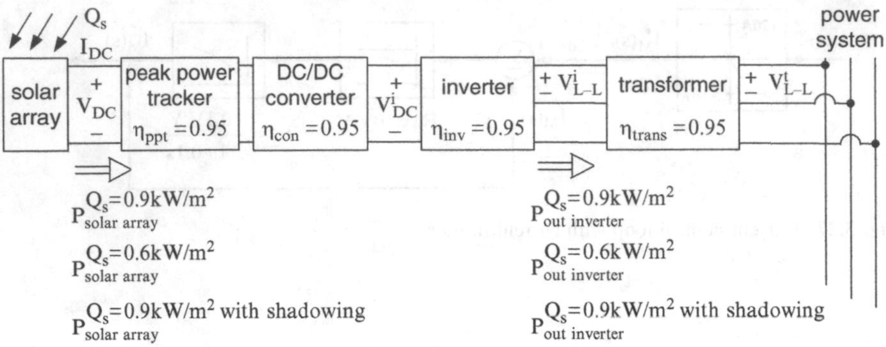

Fig. 3.29 Block diagram of 4 MW_{AC} PV power plant

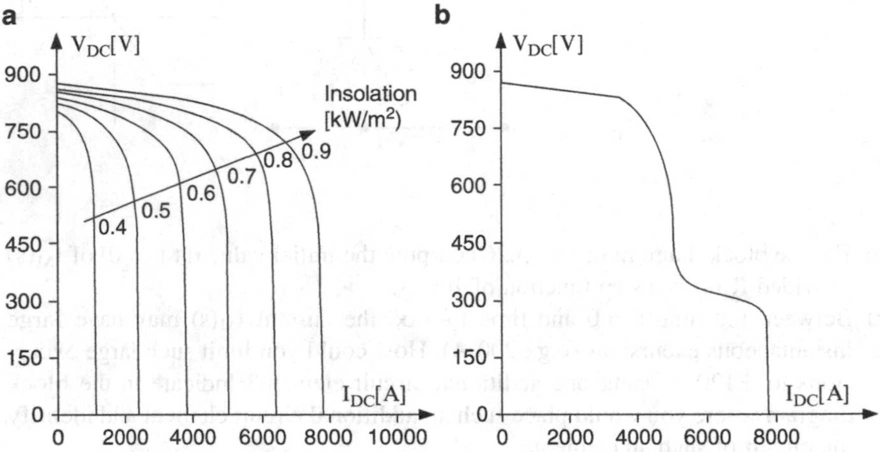

Fig. 3.30 Voltage–current characteristics of solar array; (**a**) without shadowing effect, (**b**) with shadowing effect at insolation $Q_s = 0.9\ kW/m^2$

at rated operation and rated (maximum) insolation $Q_s = 0.9\ kW/m^2$. The block diagram of this plant is shown in Fig. 3.29. The power efficiencies of the peak-power tracker, step-down/step-up DC-to-DC converters, inverters, and transformer are $\eta = 95\%$ each.

(a) Using the output characteristics of the solar array of Fig. 3.30a determine at an insolation of $Q_s = 0.9\ kW/m^2$ the output (peak) power of the solar array $P^{Qs=0.9\,kW/m^2}_{solar\,array}$, and the resulting output power of the 20 parallel-connected inverters $P^{Qs=0.9\,kW/m^2}_{out\,inverter}$.

(b) If the line-to-line voltage of the power system – which is identical to the output line-to-line voltage of the Δ-secondary of the Y–Δ transformer (t) – is $V^t_{L-L} = 13.2\,kV$ determine the input line-to-line voltage of the Y-primary of the transformer – which is identical with the output line-to-line voltage of the 20 (parallel-connected) inverters (i) V^{i}_{L-L} – provided the DC input voltage of one inverter is $V^{i}_{DC} = 1\,kV$ and the modulation index m = 1 (see modulation index m for inverters in Chap. 5). What is the turns ratio of the transformer N_p/N_s where the primary (p) is the Y-winding and the secondary (s) is the Δ-winding of the Y–Δ transformer (see 3-phase transformer connections in Chap. 8)?

(c) Repeat the analysis of part (a) at an insolation of $Q_s = 0.6\ kW/m^2$, that is, determine the output (peak) power of the solar array $P^{Qs=0.6\,kW/m^2}_{solar\ array}$ and the resulting output power of the 20 (parallel-connected) inverters $P^{Qs=0.6\,kW/m^2}_{out\ inverter}$.

(d) The output characteristic of the solar array changes due to shadowing [36] caused by clouds, snow, dirt and leaves even though there are bypass diodes mitigating this effect. Repeat the analysis of part (a) for the output characteristic of Fig. 3.30b, that is, determine the output (peak) power of the solar array $P^{Qs=0.9\,kW/m^2\ with\ shadowing}_{solar\ array}$ and the resulting output power of the 20 parallel-connected inverters $P^{Qs=0.9\,kW/m^2\ with\ shadowing}_{out\ inverter}$.

(e) The solar array consists of 335,000 solar panels each having an area of $(0.8 \times 1.60)\ m^2$. What is the total area of all solar panels ($Area_{total}$)?

(f) What is the payback period (in years) if 1 kW installed output power capacity of the 4 MW PV plant costs $4,500 provided the average wholesale price of 1 kWh is $0.06? Note: The fuel costs are zero and the operational costs are negligible; for the payback period calculation you may assume on the average a 6 h operation of the plant per day. No rebates and tax breaks are available, and no interest must be paid.

Solution

Figure 3.31a shows the operating points at 0.9 and 0.6 kW/m² insolation without shadowing.

(a) The solar array output power is for 0.9 kW/m² irradiation/insolation: $P^{Qs=0.9\,kW/m^2}_{solar\ array} = 7,000\,A \cdot 720\,V = 5.04\,MW$, and the corresponding inverter output power is $P^{Qs=0.9\,kW/m^2}_{out\ inverter} = (0.95)^3 \cdot 5.04\,MW = 4.32\,MW$.

(b) The Y–Δ transformer with its number of turns N_p and N_s as well as its voltages is shown in Fig. 3.32 (see 3-phase transformer connections in Chap. 8). The output voltage of one inverter $|\tilde{V}^{i}_{L-N}|$(see modulation index m for inverters in Chap. 5) is for a DC input voltage of the inverter $V_{DC} = V^{i}_{DC}$ at about unity power factor of the output inverter current

$$|\tilde{V}^{i}_{L-N}| = m \cdot \frac{V^{i}_{DC}}{2\sqrt{2}} = 1 \cdot \frac{1,000\,V}{2\sqrt{2}} = 354\,V$$

or the line-to-line output voltage of one inverter is $|\tilde{V}^{i}_{L-L}| = \sqrt{3}|V^{i}_{L-N}| = 612.4\,V$. Thus the transformer turns ratio is

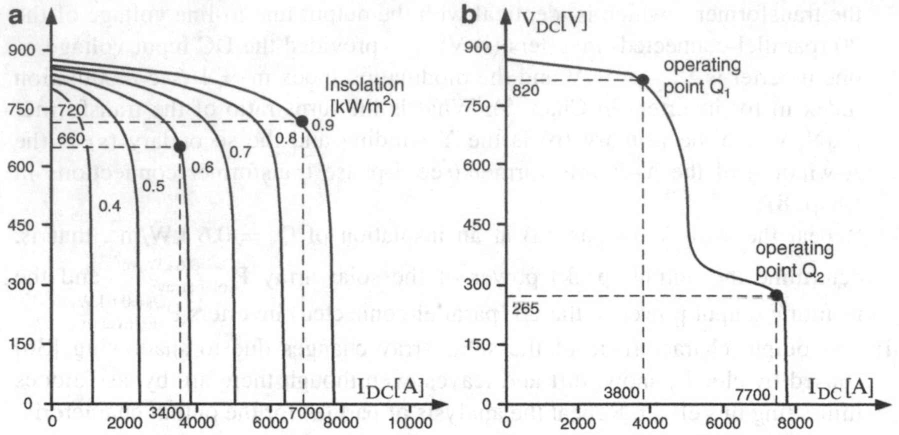

Fig. 3.31 Operating points; (a) without shadowing, (b) with shadowing at an insolation of 0.9 kW/m^2

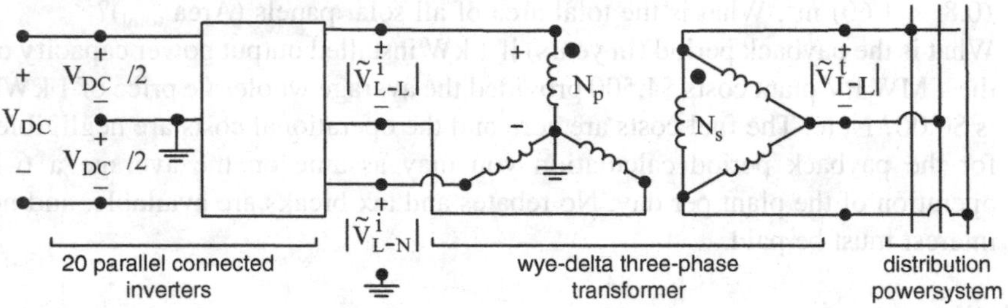

Fig. 3.32 Y–Δ transformer, where the grounding of the inverters may be either an actual ground or a virtual ground

$$\frac{N_p}{N_s} = \frac{|\tilde{V}^i_{L-N}|}{|\tilde{V}^t_{L-L}|} = \frac{354}{13,200} = 0.0268$$

or $N_s/N_p = 37.29$.

(c) The solar array output power is for 0.6 kW/m^2 irradiation:
$P^{Qs=0.6\,kW/m^2}_{solar\,array} = 3,400\,A \cdot 660\,V = 2.24\,MW$, and the corresponding output power of the 20 inverters is $P^{Qs=0.6\,kW/m^2}_{out\,inverter} = (0.95)^3 \cdot 2.24\,MW = 1.92\,MW$.

(d) Figure 3.31b shows the operating points Q_1 and Q_2 at 0.9 kW/m^2 insolation with shadowing. The solar and inverter powers are for operating point Q_1

$$P^{Qs=0.9\,kW/m^2\,with\,shadowing}_{solar\,array} = 3,800\,A \cdot 820\,V = 3.11\,MW,$$

and

$$P_{\text{out inverter}}^{\text{Qs=0.9 kW/m}^2\text{with shadowing}} = (0.95)^3 \cdot 3.11\,\text{MW} = 2.67\,\text{MW}.$$

The solar and inverter powers are for operating point Q_2

$$P_{\text{solar array}}^{\text{Qs=0.9 kW/m}^2\text{with shadowing}} = 7,700\,\text{A} \cdot 265\,\text{V} = 2.04\,\text{MW},$$

and

$$P_{\text{out inverter}}^{\text{Qs=0.9 kW/m}^2\text{with shadowing}} = (0.95)^3 \cdot 2.04\,\text{MW} = 1.75\,\text{MW}.$$

One concludes that the power generated at operating point Q_1 is larger than that of operating point Q_2 and a peak-power tracker is required to sense this optimal maximum [33].

(e) The total required area is $\text{Area}_{\text{total}} = 335{,}000 \cdot 0.8 \cdot 1.6 = 42.88 \cdot 10^4\ \text{m}^2$.

(f) *The payback period:* If interest payments are neglected the expenses are expenses = 4,000 kW·$4,500/kW = $\$18 \cdot 10^6$, and
the income is for y years
income = 4,000 kW·($0.06/kWh)·6 h·(y)·365 = $525.6 \cdot 10^3 \cdot (\text{y})$.
The payback condition is obtained from the condition "income = expenses" resulting in the payback period of $\text{y} = 18 \cdot 10^6 / 525.6 \cdot 10^3 = 34.24$ years.

3.5 Summary

The three-phase 60 Hz power system was established about 1885 [1–4]. Today's system is the most complicated and largest man-made machine, and is about to be further improved by the incorporation of renewable energy sources and extensive demand-side management based on two-way communications. This chapter gives an overview of the three-phase power system, fuel cells, batteries, and intermittently [27] operating photovoltaic energy sources.

Table 3.7 lists the costs of various available energy sources. According to this table the least expensive source of electricity in the next decade will be nuclear power, followed by hydroelectric and biomass generating plants, according to federal forecasts. Those technologies are among the least used today. Cost estimates could be greatly affected by such factors as advances in scientific research, government subsidies, and availability of raw materials, and dealing with spent nuclear fuel. Utilities price energy in terms of MWh, while consumers use the unit kWh. Note that $104.80 per MWh correspond to $0.1048 per kWh. Today's cost for generating 1 kWh in a coal fired plant is about $(0.04–0.06), that is, Table 3.7 indicates that the generation cost of electricity might increase by about a factor of 2 within the next 8 years [37].

Table 3.7 The price of our electrical future (from http://chronicle.com [37])

Energy source	Future cost (per megawatt-hour estimate for new plants in 2016, in 2007 dollars)	Current use in electrical production (Trillion Btu)
Nuclear	$104.80	8,415
Hydroelectric	$112.80	2,463
Biomass	$113.00	824
Natural gas	$114.80[a]	7,716
Wind	$115.50	319
Coal	$120.40[a]	20,990
Solar	$385.40 for photovoltaic; $257.50 for thermal	6
Petroleum	NA	715

[a] These prices include the cost of advanced technologies to reduce pollution

3.6 Problems

Problem 3.1: *Short-circuit currents of power system*

Compute for Fig. 3.3b of Application Example 3.1 the short-circuit currents, provided the three-phase voltage sources have an rms value of 120 V_{rms}.

(a) If a line-to-line short-circuit occurs.
(b) If a line-to-neutral short-circuit takes place.

Problem 3.2: *Determination of the peak power of a solar array*

Construct for the measured V_c–I_c characteristics of the solar array of Application Example 3.7 (Fig. 3.19) the power P_c as a function of the current I_c. Determine the point of maximum power (P_{cmax}) for both the series and the parallel connections.

Problem 3.3: *Parallel and series operation of photovoltaic cells*

A solar cell power plant configuration consisting of ten solar arrays (panels) which can be either connected in parallel (see Fig. 3.33a) or in series (Fig. 3.33b). What happens to the output power when one solar array is covered with snow and does not receive any light for both above-mentioned configurations? You may assume that there are no bypass diodes.

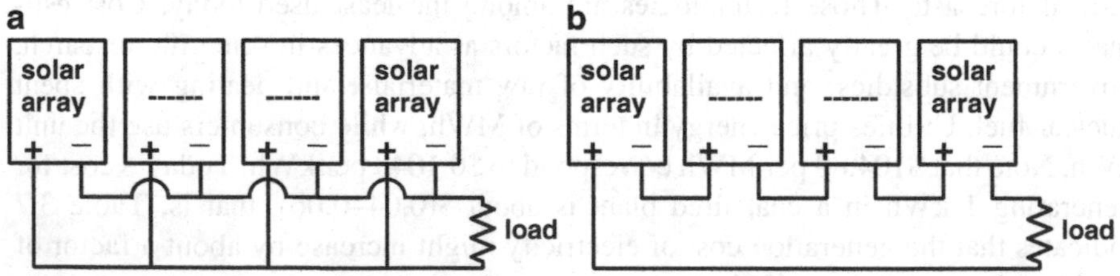

Fig. 3.33 Solar arrays connected in parallel and in series

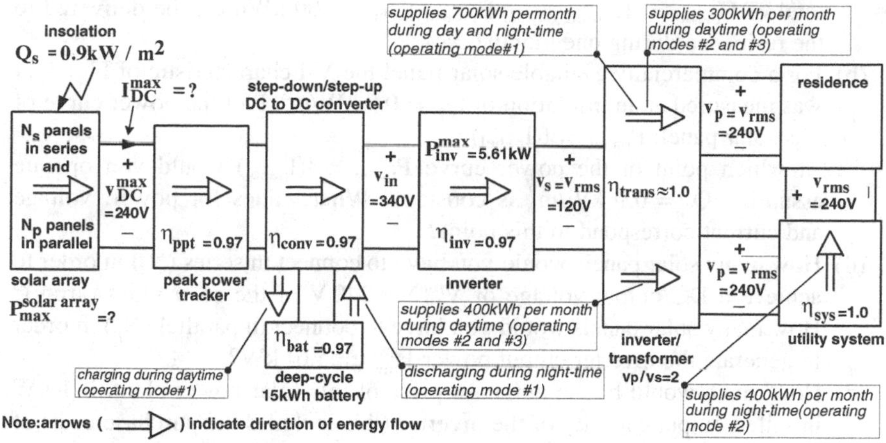

Fig. 3.34 Block diagram of a photovoltaic plant power plant for a residence. Note: Solar array consists of N_s panels in series and N_p panels in parallel

Problem 3.4: *Design of a photovoltaic power plant for a residence*

A photovoltaic power plant consists of solar array, peak (maximum)-power tracker [33], a step-up/step-down DC-to-DC converter, a deep-cycle battery for part f) only, a single-phase inverter, single-phase transformer, and residence requires a maximum inverter AC output power of $P_{inv}^{max} = 5.61$ kW as shown in Fig. 3.34. Note, the maximum inverter output AC power has been specified because the entire power must pass through the inverter for all operating modes as is explained below. In addition, inverters cannot be overloaded even for a short time due to the low heat capacity of the semiconductor switches.

Three operating modes will be investigated:

1. In part (f) the operating mode #1 is a stand-alone configuration (700 kWh are consumed per month)
2. In part (g) the operating mode #2 is a configuration, where the entire energy (700 kWh) is consumed by the residence and the utility system is used as storage device only
3. In part (h) the operating mode #3 is a configuration, where 300 kWh are consumed by the residence and 400 kWh are sold to the utility

   (a) The power efficiencies of the maximum power tracker, the step-up/step-down DC to DC converter, the battery, and the inverter are 97% each, while that of the transformer is about 1.00. What maximum power $P_{max}^{solar\ array}$ must be generated by the solar plant (array), provided during daytime E_{month_day} = 300 kWh will be delivered via the inverter to the residence (without storing this energy in battery), and sufficient energy will be stored in the battery so that the battery energy of E_{month_night} = 400 kWh can be delivered during nighttime by the battery via the inverter to the residence: that is a total of $E_{month} = E_{month\_day} + E_{month\_night} = 700$ kWh can be delivered to the residence during one month?

(b) For a commercially available solar panel the V-I characteristic of Fig. 3.24 was measured at an insolation of $Q_s = 0.9\ kW/m^2$. Plot the power curve of this solar panel: $P_{panel} = f(I_{panel})$.

(c) At which point of the power curve $P_{panel} = f(I_{panel})$ would you operate assuming $Q_s = 0.9\ kW/m^2$ is constant? What values for power, voltage and current correspond to this point?

(d) How many solar panels would you have to connect in series (N_s) in order to achieve a DC output voltage of $V_{DC}^{max} = 240$ V of the solar plant (array)? How many solar panels would you have to connect in parallel (N_p) in order to generate the inverter output power $P_{inv}^{max} = 5.61$ kW?

(e) How much would be the purchase price of this solar power plant, if 1 kW installed output capacity of the inverter (this includes the purchase costs of solar cells + peak-power tracker + DC-to-DC converter + inverter) costs $3,000 (after utility rebates and state/federal government tax-related subsidies) ? Without tax rebates and subsidies the buyer would have to pay about $4,500 per 1 kW installed inverter capacity.

(f) *Operating mode #1*
What is the payback period (in years, without taking into account interest payments) of this solar plant if you use in your residence 700 kWh per month at an avoided cost of $0.20/kWh (includes service fees and tax)? (You may assume that this solar plant can generate every month 700 kWh and there is no need to buy electricity from the utility: 300 kWh per month will be used in the residence during daytime and during nighttime 400 kWh per month will be supplied via inverter from the battery to the residence). However, there is a need for the use of a 15 kWh deep-cycle battery as a storage element so that electricity will be available during hours after sunset. This battery must be replaced every 4 years at a cost of $3,000.

(g) *Operating mode #2*
What is the payback period (in years, without taking into account interest payments) of this solar plant if you use in your residence 300 kWh per month at an avoided cost of $0.20 per kWh? You may assume that this solar plant can generate every month 700 kWh and every month the solar plant feeds 400 kWh into the power system of the utility company which reimburses you $0.20 per kWh (so-called "net metering", see [27] of Chap. 1). In this case there is no need for a battery as a storage element because during the hours after sunset the electricity can be supplied by the utility: 400 kWh at $0.20 per kWh. There is a connection charge of $8.55 per month.

(h) *Operating mode #3*
What is the pay-back period (in years, without taking into account interest payments) of this solar plant if you use in your residence 300 kWh per month at an avoided cost of $0.20 per kWh? You may assume that this solar plant can generate every month 700 kWh of which every month the solar plant feeds 400 kWh per month into the power system of the utility company which reimburses you $0.06 per kWh. There is a connection charge of $8.55 per month.

(i) Which power plant configuration (e.g., f, g or h) is more cost effective (e.g., has the shortest payback period)?
(j) What is the total surface of the solar panels provided the efficiency of solar cells is 15% at $Q_s = 0.9\ kW/m^2$?
(k) Instead of obtaining tax rebates and state/federal government subsidies the owner of a photovoltaic power plant obtains a higher price (feed-in tariff, see Chap. 1) for the electricity delivered to the utility: Provided 700 kWh are fed into the utility grid at a reimbursement cost of \$0.75/kWh and the utility supplies 300 kWh to the residence at a cost of \$0.20/kWh, what is the payback period if the entire plant generating 5.61 kW_{AC} (there are no batteries required for storage) costs \$30 k. You may neglect interest payments, and there is a connection charge of \$8.55 per month.
(l) Repeat part (k) taking into account interest payments of 4.85%.

Problem 3.5: *Energy efficiency of proton exchange membrane (PEM) fuel cell*

Pertinent data of a PEM fuel cell are as follows:

Output power: $P_{rat} = 1{,}200$ W
Output current: $I_{rat} = 46$ A
DC voltage range: $V_{rat} = 22$–50 V
Operating lifetime: $T_{life} = 1{,}500$ h
Hydrogen volume: V = 18.5 standard liters per minute (SLPM)
Weight: W = 13 kg-force
Emission: 0.87 l of water per hour

(a) The energy efficiency is defined by η_{energy} = (rated electrical energy output)/(hydrogen energy input). Calculate the energy efficiency of the fuel cell. Note that the nominal energy density of hydrogen is 28 kWh/(kg force), significantly larger than that of gasoline. This makes hydrogen a desirable fuel for automobiles.
(b) Find the specific energy density (energy per unit of weight) of this fuel cell expressed in Wh/(kg-force).
(c) How does this specific energy density compare with that of a nickel-metal hydride battery?

Problem 3.6: *Emergency power supply*

An emergency standby generating unit is to be sized for a residence. The building has an air conditioning unit (3 tons, SEER = 14) consisting of compressor and two air handlers, two refrigerators/freezers (18 cubic feet, 150 W each), 35 compact fluorescent light (CFL) bulbs @ 23 W each, four fans, two microwave ovens, two drip coffee makers, two dishwashers, one clothes washer (not including energy to heat hot water), one clothes dryer, one range (with oven), two 2 hp sump pumps, two TV sets, two computers, two VCRs, and one vacuum cleaner.

Find the size of the emergency power supply required provided all devices are operated at the same time (worst case).

Hint: See Application Example 2.7.

References

1. Lamme, B.G.: "The technical story of the frequencies," *Presented at the section meeting of the American Institute of Electrical Engineers*, Washington, DC, January 18, 1918, pp. 65–89.
2. Mixon, P.: "Technical origins of 60 Hz as the standard AC frequency in North America," *IEEE Power Engineering Review*, Vol. 19, 1999, pp. 35–37.
3. Neidhöfer, G.: "Early three-phase power," *IEEE Power & Energy Magazine*, September/October 2007, pp. 88–100.
4. Neidhöfer, G.: "Der Weg zur Normfrequenz 50 Hz" Bulletin SEV/AES 17/2008, pp. 29–34.
5. Hingorani, N.G.: "High-voltage DC transmission: A power electronic workhorse," *IEEE Spectrum*, Vol. 33, 1996, pp. 63.
6. Brumhagen, H.; Schwarz, J.: "The European power systems on the threshold of a new east-west cooperation," *IEEE Transactions on Energy Conversion*, Vol. 11, No. 2, 1996, pp. 462.
7. Nilsson, J.W.: *Electric Circuits, Second Edition*, 1986, Addison-Wesley Publishing Series in Electrical Engineering, pp. 343.
8. "Summertime ... and Electricity is in short supply," *The Christian Science Monitor*, July 5, 2000, p. 4.
9. Fuller, J.; Fuchs, E. F.; Roesler, D. J.: "Influence of harmonics on power distribution system protection," *IEEE Transactions on Power Delivery*, Vol. 3, No. 2, 1988, pp. 549.
10. Cook, W.J.:"Piston Engine R.I.P.?" *U.S. News & World Report*, May 11, 1998, pp. 46.
11. Kassakian, J.G.; Wolf, H.C.; Miller, J.M.; Hurton, C.J.: "Automotive electrical systems circa 2005," *IEEE Spectrum*, Vol. 33, 1996, pp. 23
12. Wang, C.; Nehrir, M.H.; Shaw, S.R.: "Dynamic models and model validation for PEM fuel cells using electrical circuits," *IEEE Transactions on Energy Conversion*, Vol. 20, No. 2, 2005, pp. 442–451.
13. O'Hayre, R.; Cha, S.-W.; Colella, W.; Prinz, F.B.: *Fuel Cell Fundamentals*, Wiley, Hoboken, NJ, 2006.
14. O'Hayre, R.; Cha, S.-W.; Colella, W.; Prinz, F.B.: "The road ahead for EV batteries," *EPRI Journal*, Vol. 21, 1996, pp. 7.
15. Bode, H.: *Lead-Acid Batteries*, Wiley, New York, 1977.
16. Weiss, R.; Appelbaum, J.: "Hardware simulation of a scaled-down photovoltaic power system," *1980 Photovoltaic Solar Energy Conference*, pp. 531–534, 27–31 October, 1980, Cannes, France.
17. Weiss, R.; Appelbaum, J.: "Battery state of charge determination in photovoltaic system," *Journal of the Electrochemical Society*, Vol. 129, No. 9, 1982, pp. 1928–1933.
18. Appelbaum, J.; Weiss, R.: "An electrical model of the lead-acid battery," *International Telecommunications Energy Conference, 1982. INTELEC 1982*. 3–6 October, 1982, pp. 304–307.
19. Zhan, C.-J.; Wu, X.G.; Kromlidis, S.; Ramachandaramurthy, V.K.; Barnes, M.; Jenkins, N.; Ruddell, A.J.: "Two electrical models of the lead-acid battery used in a dynamic voltage restorer," *IEE Proceedings of the Generation, Transmission and Distribution*, Vol. 150, No. 2, 2003, pp. 175–182.
20. Tichy, R.: "Battery advancements," *2007 Embedded Systems Conference*, April 05, 2007, 7 pages.
21. Hall, J.C.; Lin, T.; Brown, G.; Biesan, Li. Ph.; Bonhomme, F.: "Decay processes and life predictions for lithium ion satellite cells," *4th International Energy Conversion Engineering Conference and Exhibit (IECEC)*, 26–29 June 2006, San Diego, CA, Paper No. AIAA 2006-4078, 11 pages.
22. http://en.wikipedia.org/wiki/Flow_battery
23. Moore, T.: "The road ahead for EV batteries," *EPRI Journal*, vol. 21, 1996, pp. 7–15.
24. http://www.powerzinc.com/en/index-3-usabc.htm
25. Voelker, J.: "Top 10 tech cars", *IEEE Spectrum*, Vol. 46, No. 4, 2009, pp. 42–49.
26. http://www.zbattery.com/Battery-Memory-Effect

27. http://www.nrel.gov/midc/ss1/
28. Montu, P.: "Impact of dispersed photovoltaic system on the design of residential distribution networks," *PhD. Dissertation*, University of Colorado at Boulder, 1986.
29. Rauschenbach, H.S.: *Solar Cell Array Design Handbook*, Van Nostrand Reinhold Company, New York, 1980.
30. Otterbein, R.T.; Evans, D.L.; Facinelli, W.A.: "A modified single-phase model for high illumination solar cells for simulation work," *13th Photovoltaic Specialist Conference*, Washington, DC, June 1978.
31. Evans, D.L.; Facinelli, W.A.; Koehler, L.P.: "Simulation and simplified design studies of photovoltaic systems," *SAND 80-7013*, Arizona State University, Tempe, AZ, 1980.
32. McNeill, B.W.; Mirza, M.A.: "Estimated power quality for line-commutated photovoltaic residential system," *IEEE Transactions on Power Apparatus and Systems*, Vol. PAS-102, No. 10, 1983, pp. 3288–3295.
33. Masoum, M.A.S.; Dehbonei, H.; Fuchs, E.F.: "Theoretical and experimental analyses of photovoltaic systems with voltage- and current-based maximum power point tracking," *IEEE Transactions on Energy Conversion*, Vol. 17, No. 4, 2002, pp. 514–522.
34. Stevenson R. "First Solar: Quest for the $1 watt," *IEEE Spectrum*, August 2008.
35. Fairley, P.: "Heavy metal power," *IEEE Spectrum*, Issue 6, Vol. 47, pp. 12–13. June 2010.
36. Masoum, M.S.; Badejani, S.M.M.; Fuchs, E.F.: Closure of "Microprocessor-controlled new class of optimal battery chargers for photovoltaic applications," *IEEE Transactions on Energy Conversion*, Vol. 22, 2007, pp. 550–553.
37. http://chronicle.com Section: The Faculty, Volume 55, Issue 35, Page A1.

Chapter 4
Electronic Controllers for Feedback Systems

In any negative feedback system the measured actual control variable x_{actual} is compared with a given reference value x_{ref}. Ideally the actual instantaneous value of a control variable should follow the given reference value. However, this is not entirely possible and the quality of a control system depends to a large extent upon the type of controller used. In applications there are linear and nonlinear controllers. Here only the simplest linear controllers will be discussed: the proportional (P) controller and the proportional-integral (PI) controller.

4.1 P Controller

In Fig. 2.17 current, speed, and position controllers are indicated, and these essentially consist of a summation of the actual (inverted) signal with the reference signal $\varepsilon(s) = x_{ref}(s) - x_{actual}(s)$, and (proportional) amplification of the error signal $\varepsilon(s)$. Using ideal operational amplifiers one obtains the P controller of Fig. 4.1.

If the gain of the proportional amplifier is G, then one can show that the actual value of the control variable cannot be made identical to the reference value of the control variable, and a steady-state deviation/error exists. This steady-state deviation Δy is approximately inversely proportional to gain G (e.g., the larger the gain G is the smaller is $\varepsilon(s)$, and the smaller the gain G is the larger will be $\varepsilon(s)$ [1]):

$$\varepsilon(s) = 1/(1 + G). \tag{4.1}$$

If $G \to \infty$ then $\varepsilon(s) \to 0$; however, one obtains a large overshoot due to the large gain. For $G = 1$ follows $\varepsilon(s) = 1/2$ which is also undesirable, because the feedback control loop is not very effective and the actual control variable does not closely follow its reference value. Therefore, for most practical applications the gain G is $1.0 \leq G \leq 10.0$.

E.F. Fuchs and M.A.S. Masoum, *Power Conversion of Renewable Energy Systems*,
DOI 10.1007/978-1-4419-7979-7_4, © Springer Science+Business Media, LLC 2011

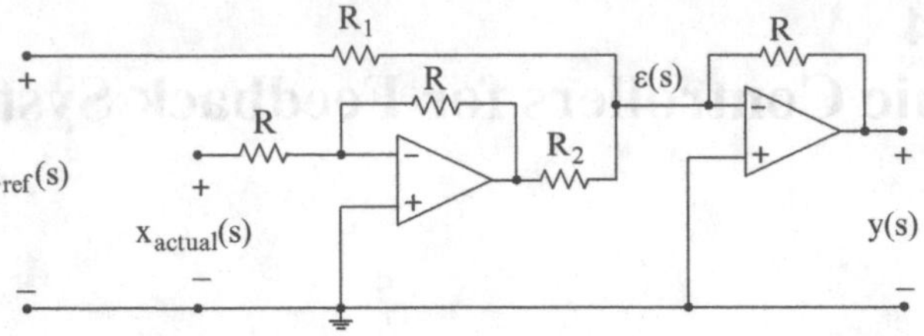

Fig. 4.1 Proportional (P) controller including summing network

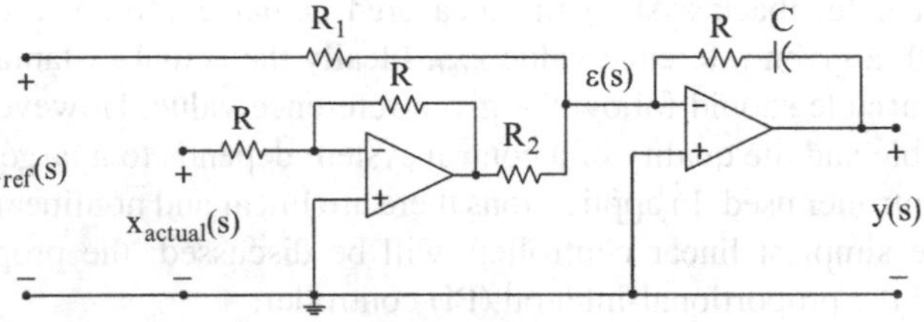

Fig. 4.2 Proportional-integral (PI) controller including summing network

4.2 PI Controller

In Fig. 2.17 the current, speed, and position controllers can be designed to be of the PI type, as illustrated in Fig. 4.2.

A Bode plot analysis [2, 3] reveals that the stability margin of the PI controller is somewhat smaller than that of the P controller: the control is less stable. However, the integral (I) component of this controller makes the above-mentioned steady-state error zero. For this reason – when designing a controller – it is advisable to start out with a P controller and then progress to a PI controller if the dynamic stability of the system is acceptable. There are pure I and D (differential) controllers which are seldom used. However, there are applications where proportional-integral-differential (PID) controllers are beneficial.

Instead of operational amplifiers one can use in the circuits of Figs. 4.1 and 4.2 differential amplifiers, which have a very high input resistance, a large voltage gain, and excellent noise suppressing properties due to their large common-mode rejection ratio (CMRR) [4].

In these examples analog op-amp circuitry has been used. However, digital equivalents maybe used as well. For such systems the differential equations are converted to difference equations and then implemented in embedded software chips.

4.3 Transfer Functions of Separately Excited DC Machine [5]

4.3.1 Speed Control via Armature-Voltage Variation (So-Called Armature-Voltage Control)

The differential equations governing a separately excited DC machine can be obtained from Fig. 4.3. From Chap. 2 follow for the armature circuit of the DC machine (4.2), machine load system (4.3), and the torque–current relationship (4.4) the equations

$$v_a = R_a i_a + L_a \frac{di_a}{dt} + k_e\ \Phi\omega_m \tag{4.2}$$

$$J\frac{d\omega_m}{dt} = T - T_m - B\ \omega_m = k_e\Phi i_a - T_m - B\ \omega_m, \tag{4.3}$$

where

$$T = k_e\Phi i_a. \tag{4.4}$$

The coefficient $k_e\Phi$ is a function of the excitation or field current – therefore, the subscript "e" is indicating excitation. Assuming zero initial conditions for current i_a and angular velocity ω_m one obtains in the Laplace domain the complex equations

$$sL_a I_a(s) + R_a I_a(s) + k_e\Phi\omega_m(s) = V_a(s), \tag{4.5}$$

$$sJ\omega_m(s) + B\omega_m(s) + T_m(s) = k_e\Phi I_a(s), \tag{4.6}$$

where $I_a(s)$, $V_a(s)$, $\omega_m(s)$, and $T_m(s)$ are Laplace transforms of the variables i_a, v_a, ω_m, and T_m, respectively. With the definitions of the time constant $\tau_{m1} = JR_a/\{BR_a + (k_e\phi)^2\}$ and the gain $K_{m1} = B/(BR_a + (k_e\phi)^2)$ one obtains the current–voltage relation

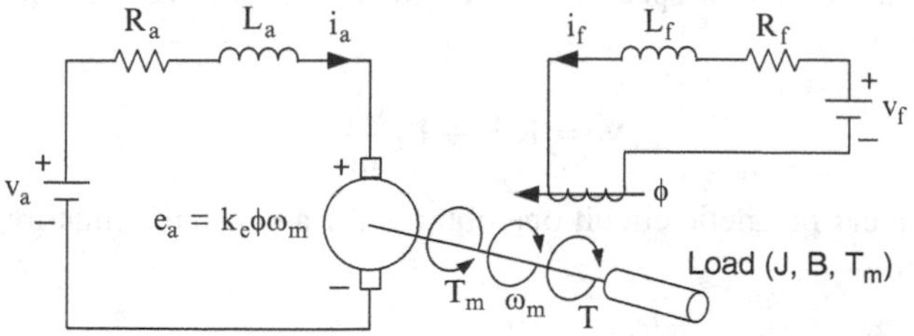

Fig. 4.3 Dynamic equivalent circuit of separately excited DC machine in time domain

$$\frac{I_a(s)}{V_a(s)}=\frac{K_{m1}(1+s\tau_m)}{\{1+\tau_{m1}(1+\frac{\tau_a}{\tau_m})s+\tau_{m1}\ \tau_a s^2\}}, \tag{4.7}$$

where $\tau_a = L_a/R_a$ and $\tau_m = J/B$. Alternatively one can write

$$\frac{I_a(s)}{V_a(s)}=\frac{K_{m1}(1+s\tau_m)}{(1+s\tau_1)(1+s\tau_2)}, \tag{4.8}$$

where $\tau_1 = \frac{\tau_{m1}}{\tau_2}\tau_a$, and

$$\tau_2=\frac{\tau_{m1}\left(1+\frac{\tau_a}{\tau_m}\right)\pm\sqrt{\tau_{m1}^2\left(1+\frac{\tau_a}{\tau_m}\right)^2-4\tau_{m1}\tau_a}}{2}.$$

The time constants τ_1 and τ_2 are complex conjugates, if

$$\left(4\frac{L_a}{R_a}\right)\rangle\left\{\frac{JR_a}{\{BR_a+(k_e\phi)^2\}}\left(1+\frac{L_aB}{R_aJ}\right)^2\right\},$$

or in other words, if J and L_a are large, and R_a is small as it applies to large DC machines. The transfer function between $\omega_m(s)$ and $V_a(s)$ is for $T_m(s) = 0$

$$\frac{\omega_m(s)}{V_a(s)}=\frac{(k_e\phi)/\{BR_a+(k_e\phi)^2\}}{(1+s\tau_1)(1+s\tau_2)} \tag{4.9}$$

4.3.2 Speed Control via Field–Voltage Variation (So-Called Flux-Weakening Control)

At above rated speed DC drives are operated with field–voltage $V_f(s)$ control at a constant armature current $I_a(s)$ and constant armature voltage $V_a(s)$. The armature current is maintained constant using closed-loop control. Because the armature time constant $\tau_a = L_a/R_a$ is very small compared to the field time constant $\tau_f = L_f/R_f$, the response time of the closed-loop system controlling the armature current can be considered zero, and thus the change in the armature current due to the variation of the field current and motor speed can be neglected. From Fig. 4.3 one obtains the voltage relation

$$v_f = R_f i_f + L_f\frac{di_f}{dt}. \tag{4.10}$$

Assuming a linear magnetic circuit one obtains for a constant armature current the torque relation

$$T = k_a i_f \tag{4.11}$$

The coefficient k_a is a function of the constant armature current, the subscript "a" indicating armature. From (4.3) one gets in terms of the field current

$$J\frac{d\omega_m}{dt} = T - T_m - B_m\omega_m = k_a i_f - T_m - B_m\,\omega_m. \tag{4.12}$$

Assuming zero initial conditions for current i_f and angular velocity ω_m one obtains in the Laplace domain the complex equations

$$sL_f I_f(s) + R_f I_f(s) = V_f(s), \tag{4.13}$$

$$sJ\omega_m(s) + B\omega_m(s) + T_m(s) = k_a I_f(s). \tag{4.14}$$

From these equations one obtains

$$I_f(s) = \frac{V_s(s)}{R_f(1+s\tau_f)}, \tag{4.15}$$

$$\omega_m(s) = \frac{(k_a/B)I_f(s)}{(1+s\tau_m)} - \frac{T_m(s)}{B(1+s\tau_m)} = \frac{T(s)-T_m(s)}{B(1+s\tau_m)}, \tag{4.16}$$

where $\tau_f = L_f/R_f$. Substituting (4.15) into (4.16) yields

$$\omega_m(s) = \frac{k_a/(R_f B)}{(1+s\tau_f)(1+s\tau_m)} V_f(s) - \frac{T_m(s)}{B(1+s\tau_m)}. \tag{4.17}$$

The output of the DC machine is $P(s) = T(s)\cdot\omega_m(s)$. Equation (4.11) indicates that if $I_f(s)$ is reduced then T(s) is reduced for a constant armature current resulting in k_a = constant. From (4.17) one concludes that if T(s) is reduced $T_m(s)$ must be reduced as well such that $\omega_m(s)$ increases in order to maintain the output power to be constant. This increase in ω_m and the associated decrease of T at steady-state is demonstrated in Chap. 2, Figs. 2.26–2.28.

Other more advanced control methods such as State Space Control are sometimes used for high precision motor control application.

4.4 Relative Stability: Gain Margin and Phase Margin

If the damping of the machine-load system of Fig. 4.3 is small oscillations may occur and the system may even become unstable. Nyquist plots [1, 2] can be used to demonstrate the stability of a control system based on the closed-loop transfer function

$$\frac{C(s)}{R(s)} = \frac{G(s)}{1+G(s)H(s)}, \tag{4.18}$$

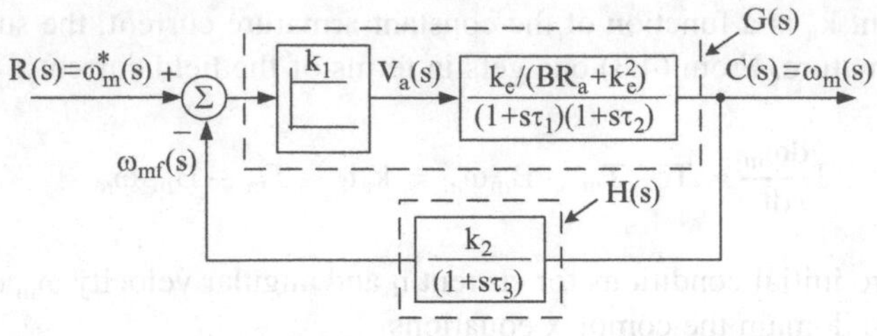

Fig. 4.4 DC motor controlled with P-controller and filter in the feedback path at mechanical torque $T_m = 0$

where C(s) is the output (controlled) variable of a control system, R(s) is the input (reference) variable, G(s) is the open-loop transfer function of a control system, and H(s) is the transfer function of the feedback loop of a control system. The denominator $F(s) = 1 + G(s)H(s)$ represents the characteristic equation of the feedback system and contains its stability information. If $F(s) = 0$ then $G(s)H(s) = -1$. First- and second-order transfer functions G(s)H(s) result always in a stable control because each first-order transfer function contributes a phase shift of $\leq 90°$. A third- or higher order transfer function G(s)H(s) may result in instability. The block diagram of Fig. 4.4 [with (4.9)] shows the speed control of a DC motor with a filter in the feedback path and a P-controller. This configuration results in a third-order G(s)H(s) transfer function for $T_m = 0$ and small R_a value as follows

$$G(s)H(s) = \frac{\text{gain}}{(1 + s\tau_1)(1 + s\tau_2)(1 + s\tau_3)}, \quad (4.19)$$

where gain $= k_1 k_2 k_e / (BR_a + k_e^2)$.

Figures 4.5a–d illustrate the relation between the Nyquist plots and their associated transient responses C(t) at an input step change R(t) for a third-order G(s)H(s). In Fig. 4.5a the gain G_1 is very large and instability results because at the negative real axis intercept $G(s)H(s) < -1$, that is $(1 + G(s)H(s))$ is negative, resulting in positive feedback. The gain G_2 of Fig. 4.5b is such that a sustained oscillation occurs. Gain G_3 in Fig. 4.5c results in a periodic response, and G_4 produces a brief overshoot with an acceptable control response (Fig. 4.5d).

4.4.1 Gain Margin

The gain margin is a measure of the closeness of the phase-crossover point to the critical point: the phase-crossover point is the point at which the phase of $G(j\omega)$ is 180°, which is called the critical point $(-1, j0)$. In other words it is the point where $G(j\omega)$ crosses the negative real axis as shown in Fig. 4.6. There can be more than one phase-crossover point.

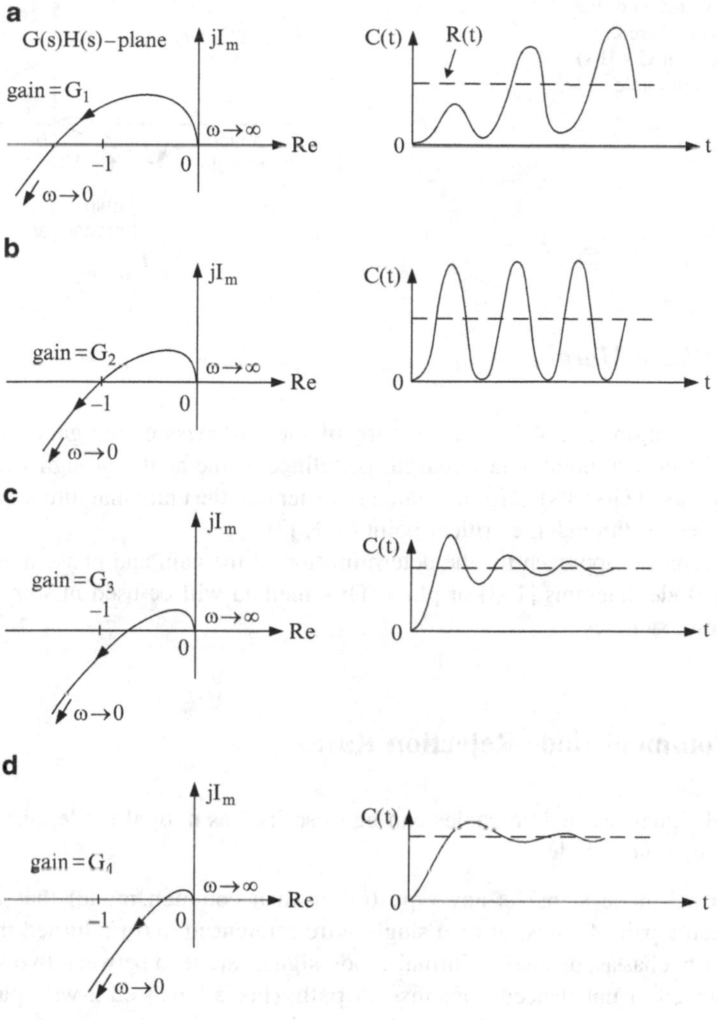

Fig. 4.5 **(a–d) Correspondence of Nyquist plots and transient responses**

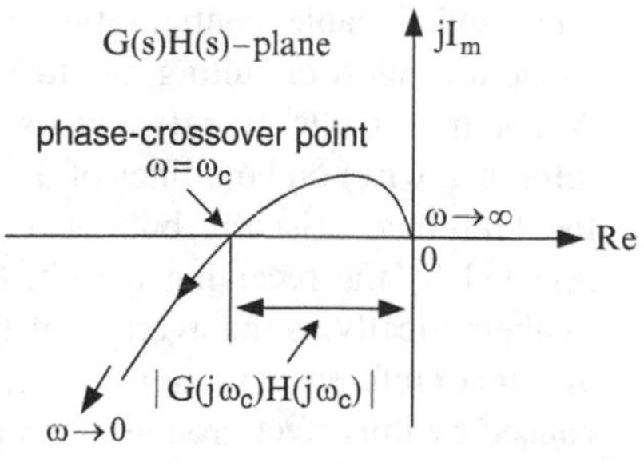

Fig. 4.6 Definition of gain margin $|G(j\omega)H(j\omega)|$ where the Nyquist locus of $G(s)H(s)$ crosses the negative real axis

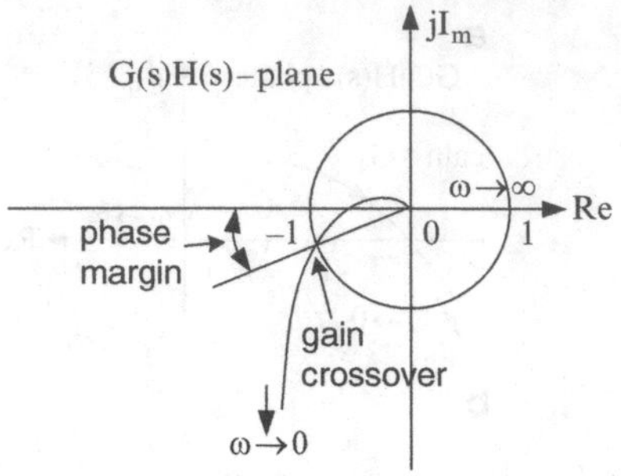

Fig. 4.7 Definition of the phase margin where the Nyquist locus of G(s)H(s) crosses the unit circle

4.4.2 Phase Margin

The phase margin (Fig. 4.7) is a measure of the closeness of the gain-crossover point to the critical point: phase margin is defined as the angle through which the Nyquist locus of G(s)H(s) must be rotated in order that the unity magnitude point on the locus passes through the critical point (−1, j0).

An alternative approach for the determination of the gain and phase margins is based on Bode diagrams [1–3] or plots. This method will be used in some of the application examples.

4.5 Common-Mode Rejection Ratio

Electrical signals carried on cables can be described as normal mode, differential mode, or common mode.

1. A normal-mode signal is any type (other than common mode) that appears between a pair of wires, or on a single wire referenced to (or returned through) the earth, chassis, or shield. Normal-mode signals are read between two wires in a balanced or unbalanced transmission path. (For a balanced 2-wire path, one wire is driven positive while the other is driven negative by an equal amount, both with respect to a static or no-signal condition in which both lines assume the same voltage level relative to a common reference wire).
2. A differential-mode signal appears differentially on a pair of wires in an ungrounded cable configuration and can be reduced or mitigated by filter capacitors short-circuiting the differential noise.
3. A common-mode signal appears equally (with respect to a local common reference wire) on both lines of a 2-wire cable not connected to earth, shield, or local reference. Usually, but not always, this is an unwanted signal that should be rejected by the receiving circuit. Common-mode voltage (V_{CM}) is expressed mathematically as the average of the two signal voltages with respect to local ground or reference: $v_{CM}(t) = \{v_A(t) + v_B(t)\}/2$. Common-mode noise is mostly caused by imperfect grounding. A common-mode signal can be reduced by the

use of optically coupled circuits, capacitively coupled differential circuits, inductively coupled circuits (e.g., transformer), resistively coupled differential circuits, and the use of high toroidal (doughnut-type) cores with high permeability on which the two wires – carrying a common-mode noise – will be wound (few to several turns of twisted or not twisted wires carrying the common-mode noise). The mechanism for reducing common-mode noise is based on the reduction of the magnetic field strength H within the toroid. If the permeability μ is approaching infinity and the flux density is finite then $H = B/\mu$ approaches zero resulting in the suppression of the common-mode current noise due to Ampere's law, as will be discussed in Chap. 6. Ideally the common-mode rejection ratio (CMRR) should be infinite.

Application Example 4.1: Analysis of a PI Controller

Show that the circuit of Fig. 4.8 is a PI (proportional-integral) controller.

Solution

The output of the R_2–R_1 operational amplifier is

$$- v_{in}(s)\frac{R_2}{R_1}, \tag{4.20}$$

and the output of the C R_3 operational amplifier is

$$-\frac{v_{in}(s)}{s\tau}, \tag{4.21}$$

where $\tau = CR_3$.

The output of the R–R amplifier is

$$v_{out}(s) = v_{in}(s)\frac{R_2}{R_1} + \frac{v_{in}(s)}{s\tau} = v_{in}(s)\left(\frac{R_2}{R_1} + \frac{1}{s\tau}\right), \tag{4.22}$$

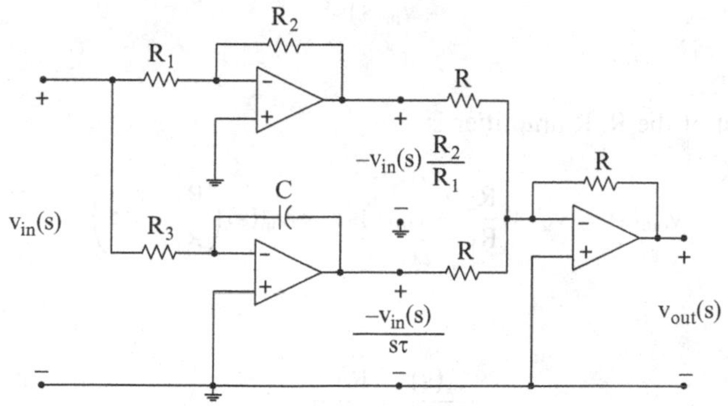

Fig. 4.8 PI controller

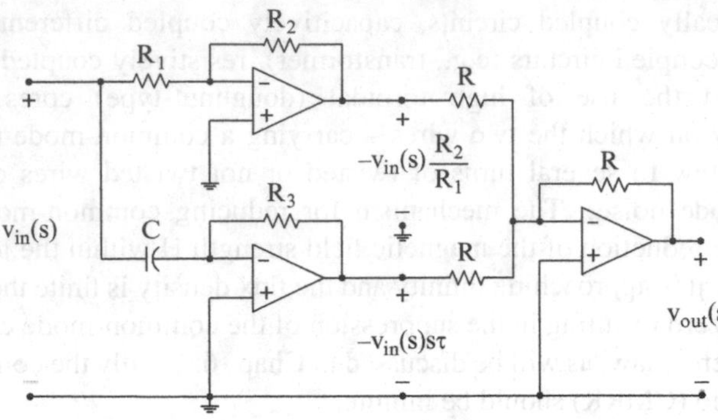

Fig. 4.9 PD controller

or

$$\frac{v_{out}(s)}{v_{in}(s)} = \frac{R_2}{R_1} + \frac{1}{s\tau}. \tag{4.23}$$

Application Example 4.2: Analysis of a PD Controller

Show that the circuit of Fig. 4.9 is a PD (proportional, differential) controller

Solution

The output of the R_2–R_1 operational amplifier is

$$-v_{in}(s)\frac{R_2}{R_1}, \tag{4.24}$$

and the output of the R_3–C operational amplifier is

$$-v_{in}(s)s\tau, \tag{4.25}$$

where $\tau = CR_3$.

The output of the R–R amplifier is

$$v_{out}(s) = v_{in}(s)\frac{R_2}{R_1} + v_{in}(s)s\tau = v_{in}(s)\left(\frac{R_2}{R_1} + s\tau\right), \tag{4.26}$$

or

$$\frac{v_{out}(s)}{v_{in}(s)} = \frac{R_2}{R_1} + s\tau. \tag{4.27}$$

Application Example 4.3: Bode Diagram for Transfer Function with Two Real Poles and One Zero

The numerical expression for an open-loop transfer function is

$$G(s) = \frac{10^5(s+5)}{(s+100)(s+5{,}000)}$$

(a) On the basis of a straight-line (asymptotes) approximation of $|G(j\omega)|_{dB}$ vs. ω, estimate the maximum magnitude of $G(j\omega)$ in dB.
(b) Approximate the phase angle $\angle G(j\omega)$ of the above transfer function by means of a straight-line (asymptotes) plot, that is, plot $\angle G(j\omega)$ vs. ω.

Solution

(a) The transfer function can be rewritten in the form

$$G(s) = \left.\frac{\left(\frac{s}{5}+1\right)}{\left(\frac{s}{100}+1\right)\left(\frac{s}{5{,}000}+1\right)}\right|_{s=j\omega}. \tag{4.28}$$

The standard, normalized form is now

$$G(j\omega) = \frac{\left(1+\frac{j\omega}{5}\right)}{\left(1+\frac{j\omega}{100}\right)\left(1+\frac{j\omega}{5{,}000}\right)}. \tag{4.29}$$

The corner angular frequencies are $\omega_{c1} = 5$ rad/s, $\omega_{c2} = 100$ rad/s, $\omega_{c3} = 5{,}000$ rad/s. The magnitude plot is given in Fig. 4.10 and one obtains the maximum magnitude $|G(j\omega)|_{dB\,max} = 26$ dB.
(b) The phase plot is shown in Fig. 4.11.

Application Example 4.4: Bode Diagram for Transfer Function with a Complex Conjugate Pair of Poles

The numerical expression for an open-loop transfer function G(s) is

$$G(s) = \frac{I_o(s)}{I_i(s)} = \frac{25 \cdot 10^8}{s^2 + 20{,}000\,s + 25 \cdot 10^8}.$$

(a) Compute the corner frequency ω_o.

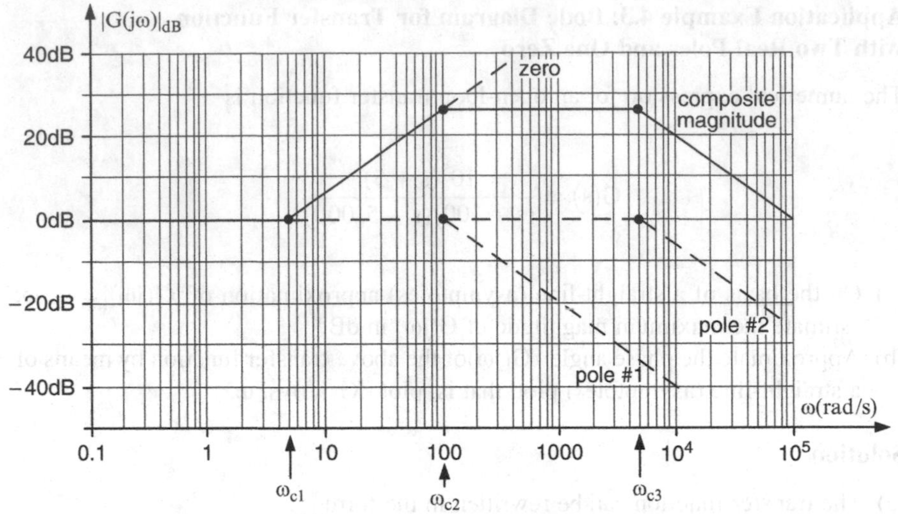

Fig. 4.10 Magnitude plot as a function of angular frequency ω

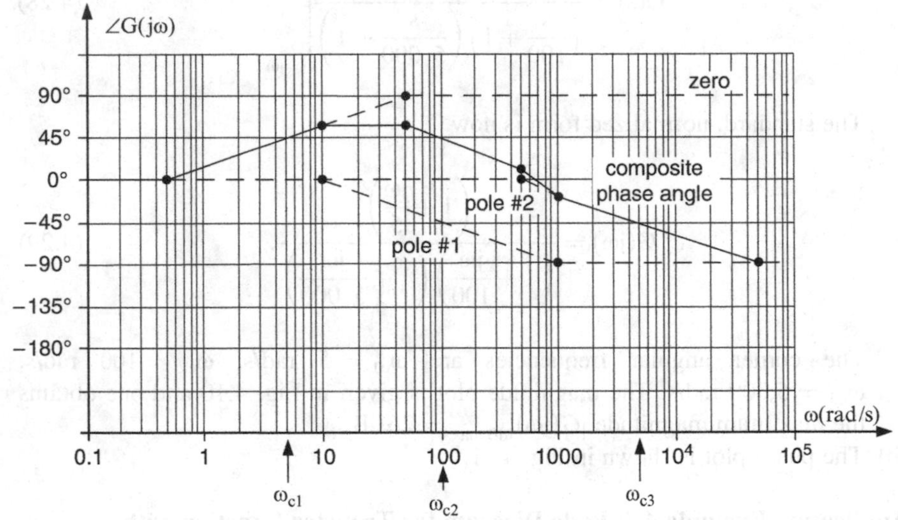

Fig. 4.11 Phase plot as a function of angular frequency ω

(b) Compute the damping coefficient ζ and the quality factor Q.
(c) Plot the magnitude $|G(j\omega)|_{dB}$ vs. the angular frequency ω.
(d) Calculate the frequency ω_{MAX} at which the peak magnitude of $G(j\omega) = G(\omega_{MAX})$ occurs.
(e) Calculate the peak magnitude of $G(j\omega)$ in dB, that is, $|G(j\omega_{MAX})|_{dB}$.
(f) Plot the phase angle $\angle G(j\omega)$ vs. the angular frequency ω.

Solution

(a) The transfer function can be rewritten in the form

$$G(s) = \frac{1}{1 + \frac{20 \cdot 10^3}{5 \cdot 10^4}\left(\frac{s}{5 \cdot 10^4}\right) + \left(\frac{s}{5 \cdot 10^4}\right)^2} \quad (4.30)$$

or in the standard, normalized form

$$G(s) = \frac{1}{1 + 0.4\left(\frac{s}{\omega_o}\right) + \left(\frac{s}{\omega_o}\right)^2} \quad (4.31)$$

where the corner frequency is $\omega_o = 5{\cdot}10^4$ rad/s.

(b) The damping coefficient is $2\varsigma = 0.4$, and the quality factor $Q = 2.5$.

(c) For $s = j\omega$ the transfer function becomes

$$G(s) = \frac{1}{1 + 0.4\left(\frac{j\omega}{\omega_o}\right) - \left(\frac{\omega}{\omega_o}\right)^2} \quad (4.32)$$

or the magnitude

$$|G(j\omega)| = \frac{1}{\sqrt{\left\{1 - \left(\frac{\omega}{\omega_o}\right)^2\right\}^2 + \left(\frac{0.4\omega}{\omega_o}\right)^2}} \quad (4.33)$$

which is represented in Fig. 4.12

(d) The peak magnitude occurs at $\omega_{MAX} = \omega_o\sqrt{1 - 2(\varsigma)^2} = 5 \cdot 10^4\sqrt{1 - 2 \cdot 0.04} = 4.79 \cdot 10^4$ rad/s.

(e) The peak magnitude of $G(j\omega)$ is

$$|G(j\omega_{MAX})| = \frac{G(\omega = 0)}{2\varsigma\sqrt{1 - \varsigma^2}} = \frac{1}{2 \cdot 0.2\sqrt{1 - 0.04}} = 2.55,$$

or in decibel (dB)$|G(\omega_{MAX})|_{dB} = 20\log_{10}(|G(\omega_{MAX}|) = 8.14\,\text{dB}$.

(f) The phase plot is shown in Fig. 4.13 with $\omega_a = \omega_o \cdot 10^{(-1/2Q)} = 5 \cdot 10^4 \cdot 10^{(-0.2)} = 31,547$ rad/s, and $\omega_b = \omega_o \cdot 10^{(+1/2Q)} = 5 \cdot 10^4 \cdot 10^{(0.2)} = 79,245$ rad/s.

Application Example 4.5: Gain and Phase Margins for a Transfer Function with Three Real Poles [1]

Consider the open-loop transfer function G(s)

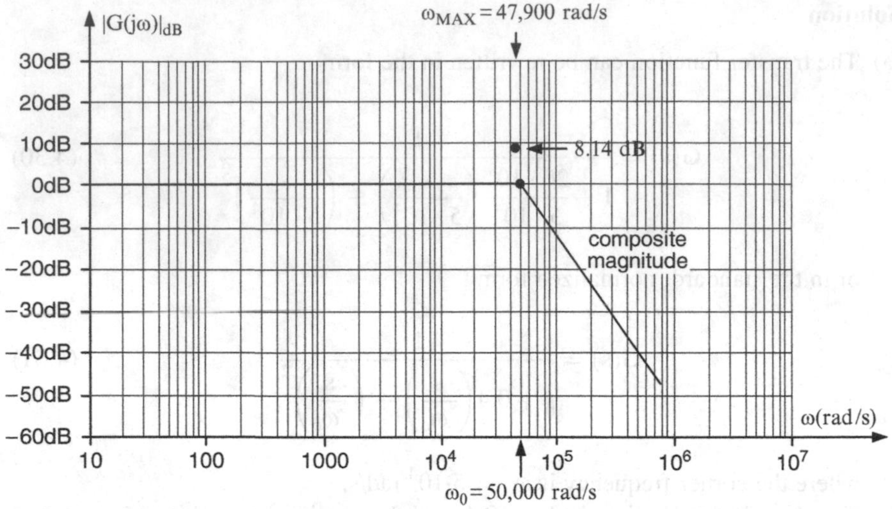

Fig. 4.12 Magnitude plot as a function of angular frequency ω

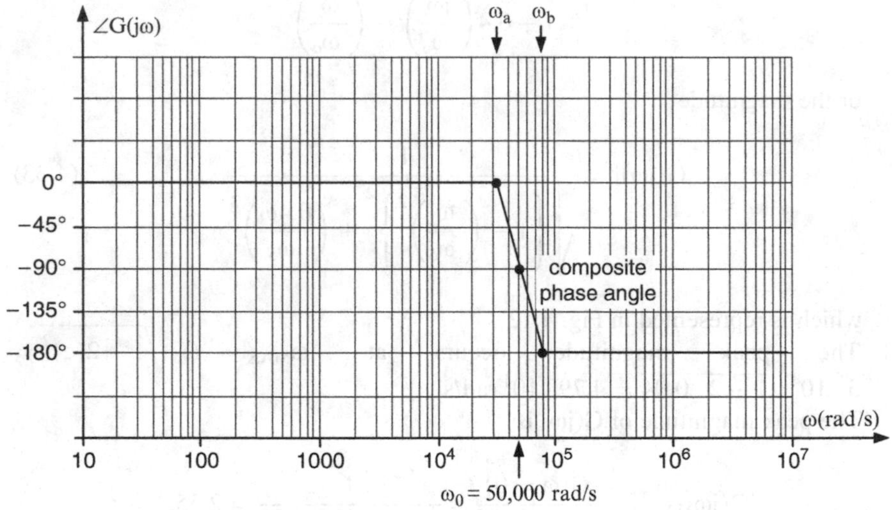

Fig. 4.13 Phase plot as a function of angular frequency ω

$$G(s) = \frac{K}{s(1 + 0.2s)(1 + 0.02s)},$$

where K = 10 dB Find the gain margin at the phase crossover ω_{c_phase} and the phase margin at the gain crossover ω_{c_gain}.

Solution

The corner frequencies of G(jω) are at $\omega_{c1} = 0$, $\omega_{c2} = (1/0.2)$ rad/s $= 5$ rad/s, and $\omega_{c3} = (1/0.02)$ rad/s $= 50$ rad/s. Straight line asymptotes are drawn for the magnitude curve $|G(j\omega)|/K$ (Fig. 4.14) and the phase curve $\angle G(j\omega)$ (Fig. 4.15) corresponding to the corner frequencies.

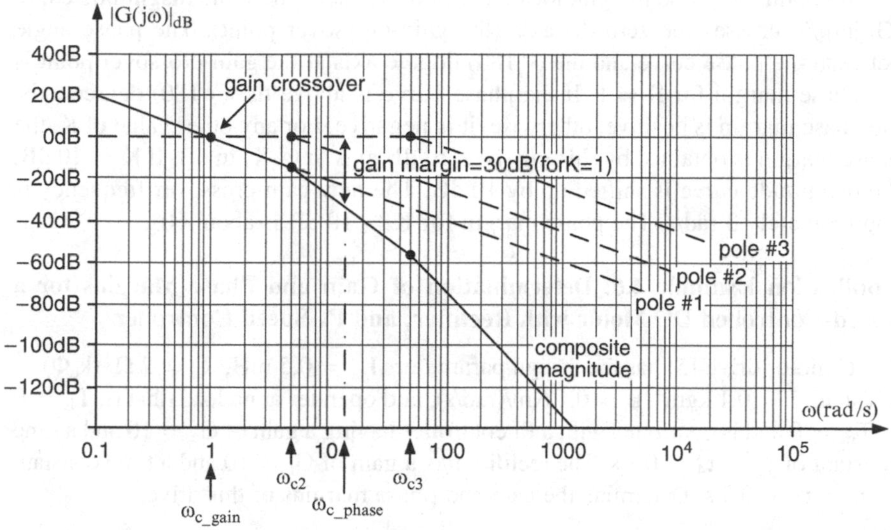

Fig. 4.14 Magnitude plot as a function of the angular frequency ω, where $\omega_{c\_gain} = 1$ rad/s, $\omega_{c\_phase} = 14$ rad/s at K = 1

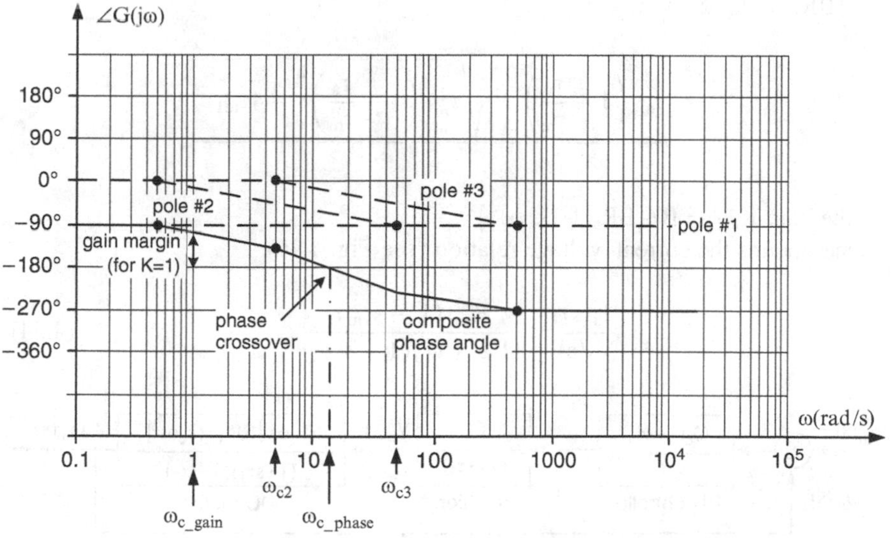

Fig. 4.15 Phase plot as a function of the angular frequency ω, where $\omega_{c\_gain} = 1$ rad/s, $\omega_{c\_phase} = 14$ rad/s

To obtain the gain margin locate the point ω_{c_phase} at which the phase curve crosses the (−180) degree axis (the phase crossover point). The magnitude of the $|G(j\omega)|/K$ curve in decibels (dB) at the phase – crossover point is the gain margin for K = 1. For any other value of K, the gain margin is simply the value of K in dB. If K = 10 dB, the gain margin is [30 dB(for K = 1) – 10 dB(for given K)] = 20 dB.

To obtain the phase margin, locate the point ω_{c_gain} where the magnitude curve $|G(j\omega)|/K$ crosses the zero dB axis (the gain-crossover point). The phase angle between the phase curve and the (−180) degree axis at the gain-crossover point is the phase margin for K = 1. If the phase curve is above the (−180) degree axis, the phase margin is positive, otherwise, it is negative. For any other value of K, the phase margin is obtained by shifting the zero dB axis to (−K) in dB. If K = 10 dB, the magnitude curve is shifted up by 10 dB. The new gain-crossover frequency is approximately 3 rad/s. The phase margin for K = 10 dB is about 60°.

Application Example 4.6: Determination of Gain and Phase Margins for a Speed-Controlled DC Motor with Rectifier, and PI-Speed Controller

A DC motor drive [5] has the design parameters $L_a = 0.5$ mH, $R_a = 2\ \Omega$, $(k_e\Phi) = 2$ Vs/rad, $J = 0.1\ \text{kgm}^2$, $B = 0.1$ Nm/(rad/s), and operates at no load, that is, $T_L = 0$ or $T_m = 0$ and is controlled with a PI controller having a gain of $G_c = 10$ and a time constant of $\tau_c = \tau_4 = 0.1$ s. The rectifier has a gain of $G_r = 10$ and a time constant of $\tau_r = \tau_3 = 0.3$ s. Determine the gain and phase margins of this drive.

Solution

With the definitions of the time constants $\tau_a = L_a/R_a$, $\tau_m = J/B$, $\tau_{m1} = JR_a/\{BR_a + (k_e\,\phi)^2\}$, $\tau_1 = (\tau_{m1}/\tau_2)\,\tau_a$,

$$\tau_2 = \frac{\tau_{m1}\left(1+\frac{\tau_a}{\tau_m}\right) \pm \sqrt{\tau_{m1}^2\left(1+\frac{\tau_a}{\tau_m}\right)^2 - 4\tau_{m1}\tau_a}}{2},$$

and the gain $K_{m1} = B/\{BR_a + (k_e\phi)^2\}$.

One obtains the current–voltage relation (see Fig. 4.16)

$$\frac{I_a(s)}{V_a(s)} = \frac{K_{m1}(1+s\tau_m)}{(1+s\,\tau_1)(1+s\,\tau_2)}. \tag{4.34}$$

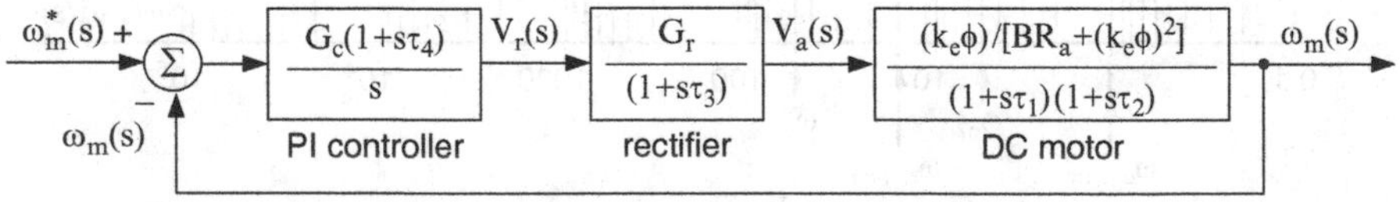

Fig. 4.16 Speed control of DC motor drive with rectifier and PI controller at no load $T_L = 0$

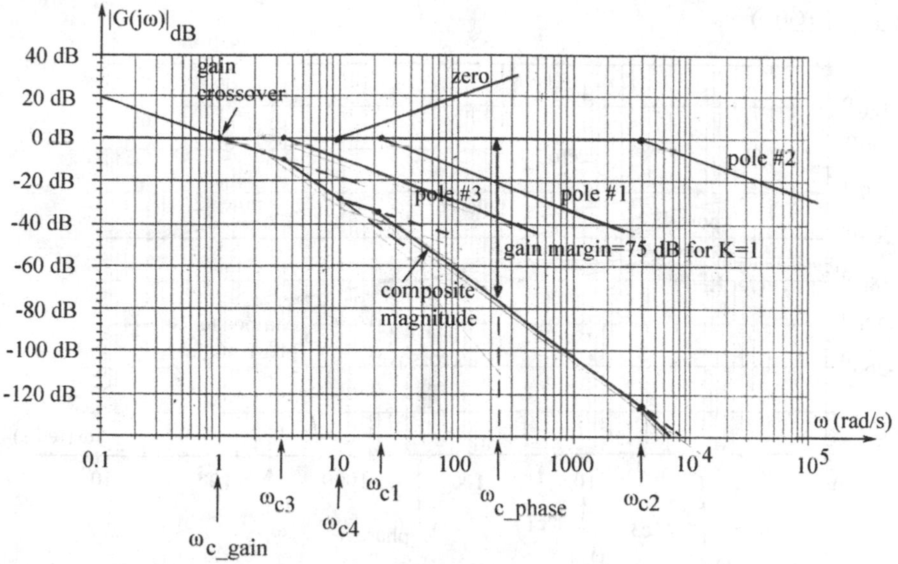

Fig. 4.17 Magnitude plot as a function of the angular frequency ω, where $\omega_{c\_gain} = 1$ rad/s, $\omega_{c\_phase} = 230$ rad/s at K = 1

The transfer function between $\omega_m(s)$ and $V_a(s)$ is for $T_m(s) = 0$ (at no load)

$$\frac{\omega_m(s)}{V_a(s)} = \frac{(k_e\phi)/\{BR_a + (k_e\phi)^2\}}{(1 + s\tau_1)(1 + s\tau_2)}. \tag{4.35}$$

The open-loop transfer function G(s) is

$$G(s) = \frac{K(1 + 0.1s)}{s(1 + 0.3s)(1 + 0.000251s)(1 + 0.0474s)}$$

$$G(s) = \frac{K\left(1 + \frac{s}{10}\right)}{s\left(1 + \frac{s}{3.33\ \text{rad/s}}\right)\left(1 + \frac{s}{3984\ \text{rad/s}}\right)\left(1 + \frac{s}{21.1\ \text{rad/s}}\right)} \tag{4.36}$$

where $\omega_{c1} = 21.1$ rad/s, $\omega_{c2} = 3984$ rad/s, $\omega_{c3} = 3.33$ rad/s, $\omega_{c4} = 10$ rad/s.

For K = 33.6 dB, the gain margin is [75 dB(for K = 1) −33.6 dB(for given K)] = 41.4 dB. If K = 33.6 dB, the composite magnitude curve is shifted up by 33.6 dB. The new gain-crossover frequency is approximately 10 rad/s. The phase margin for K = 33.6 dB is about 30° (Figs. 4.17 and 4.18).

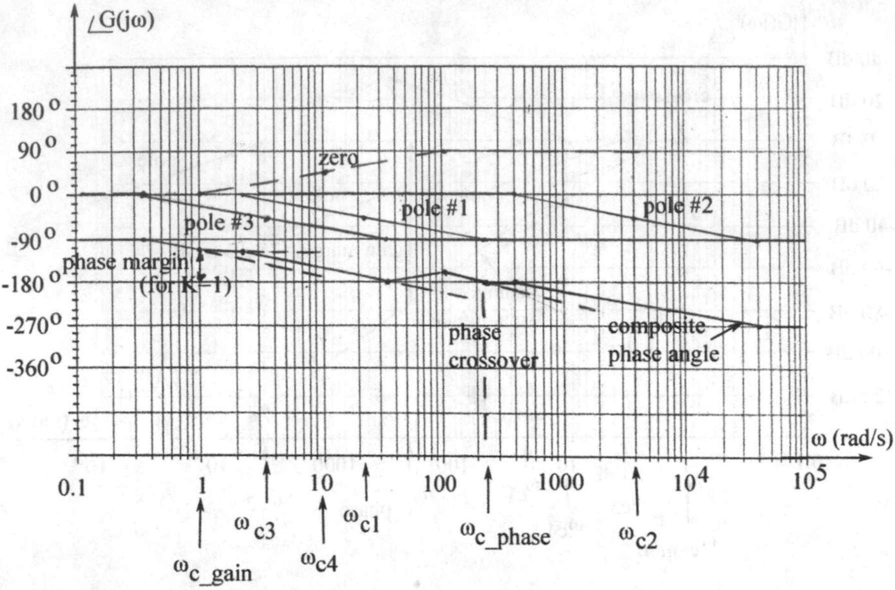

Fig. 4.18 Phase plot as a function of the angular frequency ω, where $\omega_{c\_gain} = 1$ rad/s, $\omega_{c\_phase} = 230$ rad/s at K = 1

4.6 Summary

This chapter addresses the proportional (P) and proportional-integral (PI) controllers: these are the most common controllers in drive applications. While the P-controller always has a steady-state error, the PI-controller makes the steady-state error to zero. After a review of the transfer functions of the separately excited DC machine, speed-control via armature-voltage variation and field–voltage variation (also called field-weakening control) are applied to DC motor drives. The relative stability principles – such as gain and phase margins – are examined. The common-mode rejection ratio is an important electric noise indicator for control applications. Lastly, Bode diagrams are drawn for the analysis of the stability of a DC motor drive.

4.7 Problems

Problem 4.1: *Computation of gain margin using G(s)H(s)*

The loop transfer function of a feedback control system is given by

$$G(s)H(s) = \frac{5}{(1+s)(1+2s)(1+3s)}. \tag{4.37}$$

Find the gain margin [1].

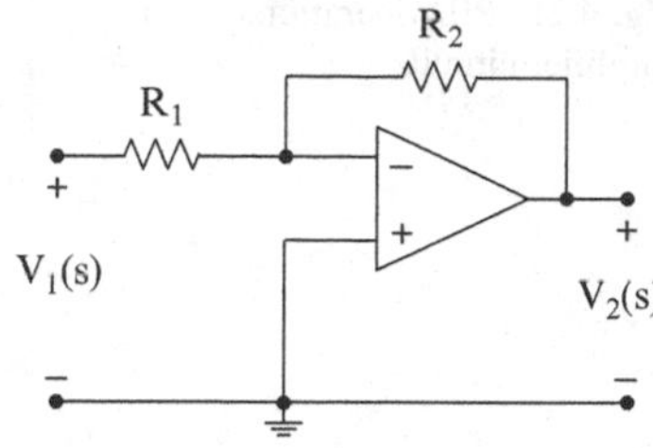

Fig. 4.19 P operational amplifier circuit

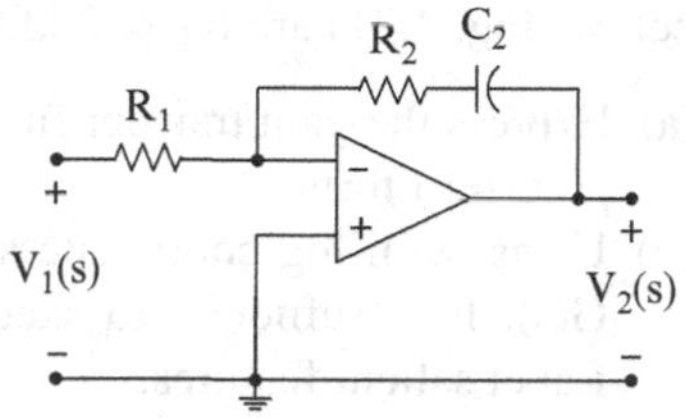

Fig. 4.20 PI operational amplifier circuit

Problem 4.2: *Frequency response of a P-operational amplifier circuit*

The component values for the P (proportional) circuit below (Fig. 4.19) are $R_1 - 2\ \text{k}\Omega$, and $R_2 = 5\ \text{k}\Omega$.

(a) Express the gain transfer function $G(s) = V_2(s)/V_1(s)$ in factored normalized pole-zero form.
(b) Using semi-log coordinate system, sketch the magnitude of the Bode plot of G(s). It is sufficient to sketch the asymptotes (straight-line approximation). Label salient features.
(c) On the same semi-log coordinate system, sketch the phase asymptotes of the Bode plot of G(s). Label salient features.

Problem 4.3: *Frequency response of a PI-operational amplifier circuit*

The component values for the PI (proportional, integrating) circuit below (Fig. 4.20) are $R_1 = 2\ \text{k}\Omega$, $R_2 = 5\ \text{k}\Omega$, and $C_2 = 100\ \text{nF}$.

(a) Express the gain transfer function $G(s) = V_2(s)/V_1(s)$ in factored normalized pole-zero form.
(b) Using semi-log coordinate system, sketch the magnitude of the Bode plot of G(s). It is sufficient to sketch the asymptotes (straight-line approximation). Label salient features.
(c) On the same semi-log coordinate system, sketch the phase asymptotes of the Bode plot of G(s). Label salient features.

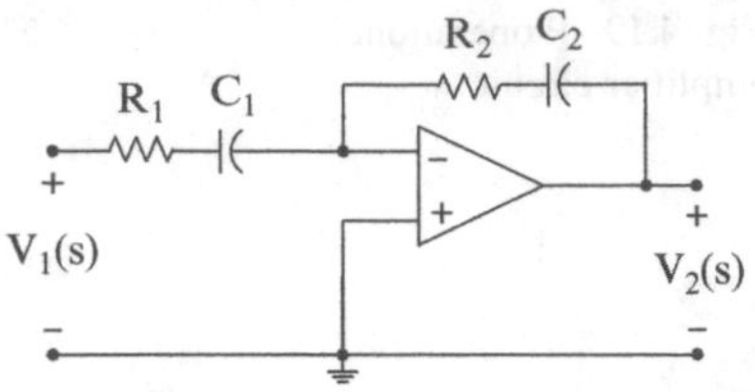

Fig. 4.21 PID operational amplifier circuit

Problem 4.4: *Frequency response of a PID operational amplifier circuit*

The component values for the PID (proportional, integrating, differentiating) circuit below (Fig. 4.21) are $R_1 = 2\ k\Omega$, $R_2 = 5\ k\Omega$, $C_1 = 1\ \mu F$, and $C_2 = 100\ nF$.

(a) Express the gain transfer function $G(s) = V_2(s)/V_1(s)$ in factored normalized pole-zero form.
(b) Using semi-log coordinate system, sketch the magnitude of the Bode plot of G(s). It is sufficient to sketch the asymptotes (straight-line approximation). Label salient features.
(c) On the same semi-log coordinate system, sketch the phase asymptotes of the Bode plot of G(s). Label salient features.

Problem 4.5: *Determination of gain and phase margins for a speed-controlled DC motor with rectifier, and P-speed controller*

A DC motor drive has the design parameters $L_a = 0.5$ mH, $R_a = 2\ \Omega$, $(k_e\Phi) = 2$ Vs/rad, $J = 0.1\ kgm^2$, $B = 0.1$ Nm/(rad/s), and operates at no load, that is, $T_L = 0$ or $T_m = 0$ and is controlled with a P controller having a gain of $G_c = 20$. The rectifier has a gain of $G_r = 15$ and a time constant of $\tau_r = \tau_3 = 0.3$ s. Determine the gain and phase margins of this drive. Hint: see Application Example 4.6.

References

1. Kuo, B. C.: *Automatic Control Systems*, Prentice Hall, Englewood Cliffs, NJ, 1967.
2. Horowitz, I. M.: *Synthesis of Feedback Systems*, Academic, New York, 1963.
3. Thomas, R. E.; Rosa, A. J.: *The Analysis and Design of Linear Circuits*, 4th Edition, Wiley, New York, 2004.
4. Sedra, A. S.; Smith, K. C.: *Microelectronic Circuits*, 4th Edition, Oxford University Press, New York, 1982.
5. Dubey, G. K.: *Power Semiconductor Controlled Drives*, Prentice Hall, Englewood Cliffs, NJ, 1989.

Chapter 5
Power Electronic Converters

5.1 Introduction

For many electric motor drives constant angular velocity (ω_m) and constant torque (T_m) operation is acceptable. Because of $P = T_m \cdot \omega_m$ the power delivered to the load will be constant as well. In many residential (e.g., air conditioner), commercial (e.g., elevator), and industrial (e.g., rolling mill, paper manufacturing) applications, the speed of drives must be variable and controllable from 0 to rated (or base) speed even up to maximum (or critical) speed, which can be about three times the base speed. The torque can vary as well from about zero at no-load to rated torque up to two times the rated torque. In some applications the speed is about constant at its rated value and the torque varies from no-load torque to rated torque. By varying speed and torque independently a high-performance drive evolves, resulting in high efficiency, conserving energy, and producing the highest yield in a manufacturing process.

To be able to vary speed and torque independently via closed-loop control, voltage and current sources (e.g., inverters) are required where voltage/current amplitudes and their frequency can be adjusted or controlled independently, and (E/f) = constant or (V/f) control can be achieved (see Chap. 10). Realizing that most available power/energy sources (Chap. 3) provide constant input voltage ($\tilde{V}_1$) and constant input frequency (f_1), semiconductor converters with a control input must be used for processing of power, resulting in variable voltage ($\tilde{V}_2$) or current amplitudes ($\tilde{I}_2$) and variable frequency (f_2) at the output port of such a converter, as shown in Fig. 5.1.

In the following subsections some of the important types and concepts of power semiconductor converters are introduced. The approach is a simplified and idealized one with the limited objective of presenting some of the major features that influence the performance of electromechanical drive systems.

E.F. Fuchs and M.A.S. Masoum, *Power Conversion of Renewable Energy Systems*,
DOI 10.1007/978-1-4419-7979-7_5, © Springer Science+Business Media, LLC 2011

Fig. 5.1 Power processing network with constant input AC voltage/current and frequency, and variable AC output voltage/current and frequency

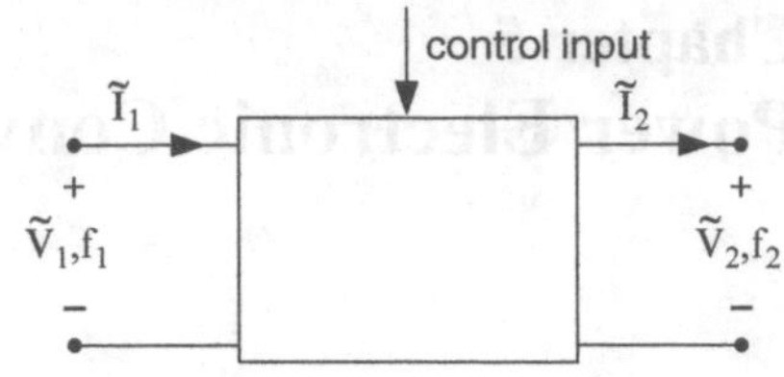

5.2 Power Electronic Switches and Their Gating Circuits

5.2.1 Diodes

The first group and the simplest semiconductor devices are diodes. These provide essentially a unidirectional flow of current and have the idealized voltage-current characteristic of Fig. 5.2, which also contains the circuit symbol for the diode. Diodes can be manufactured with large current and voltage ratings (e.g., $i_{max} = 5$ kA and $v_{maxprv} = 5$ kV, where "maxprv" stands for maximum peak reverse voltage), and if necessary they can be connected in series with a proper equalizing network (e.g., RC network) so that each and every series-connected diode has the same voltages. No gating circuits are necessary. The diode-surrounding network determines whether a diode is on or off.

5.2.2 Thyristors

The next group consists of SCR (silicon-controlled-rectifier) devices or thyristors (this name originates in the Greek language and means "current gate"). These are three-terminal devices that have properties similar to diodes when the applied voltage v_{TH} across the device is negative (reverse biasing). With positive applied voltage v_{TH} (forward biasing), the current i_{TH} is negligibly small until a positive low-power (current) pulse (i_G) is applied to the third terminal or gate, whereupon the device then acts much as a diode.

To re-establish the off state, the device current (i_{TH}) must be reduced to zero for a short time (e.g., 0.5 ms) by the external circuitry. An alternative term used for this class of device is "naturally commutated semiconductor switches", which reflects the property that they can be switched on but must be naturally (by line or load voltage) turned off by the associated external circuitry. The idealized voltage-current characteristics of a thyristor and its circuit symbol are illustrated in Fig. 5.3. If no gating current is provided ($i_G = 0$) then the forward breakover voltage is identical to the maximum peak forward blocking voltage. For a gating current of $i_G = 1$ A the forward breakover voltage is less than the maximum peak forward blocking voltage, and for $i_G = 2$ A the forward breakover voltage is practically zero. The latching, holding and forward leakage currents as indicated in Fig. 5.3 are very small.

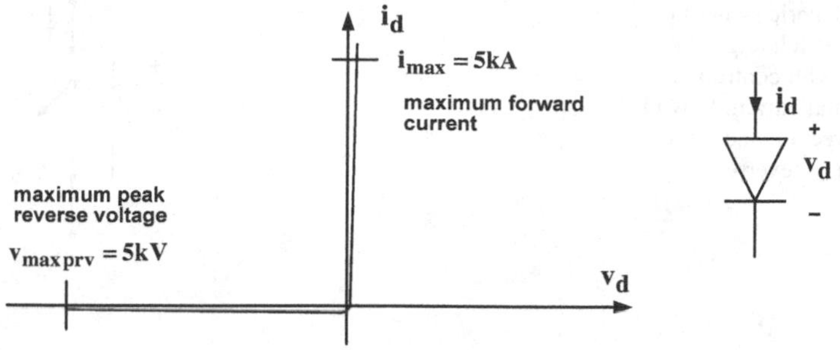

Fig. 5.2 i_d–v_d characteristic of idealized diode and its circuit symbol

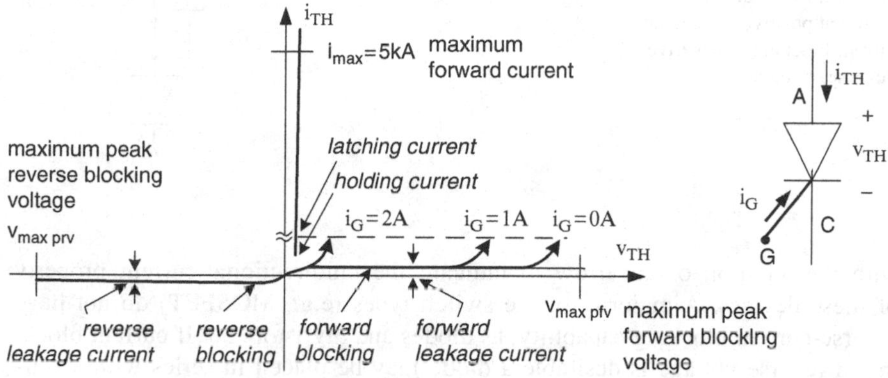

Fig. 5.3 i_{TH}–v_{TH} characteristics of an idealized thyristor including circuit symbol: characteristics for gating currents of $i_G = 2$ A, 1 A and 0 A

Thyristors can be manufactured with large current and voltage ratings ($i_{max} = 5$ kA and $v_{maxprv} = 5$ kV). If necessary they can be connected in series with proper equalizing network (e.g., RC network) so that each and every thyristor experiences the same voltage. Gating networks are required (see Application Example 5.4).

5.2.3 *Self-Commutated Switches*

The devices of the third category are denoted as self-commutated semiconductor switches. The general properties of devices in this class are the capability of being turned-on or turned-off at will by use of some form of low power (either voltage or current controlled) signal into a third (gate or base) terminal, and normally they have a capability of controlling current only in forward direction. Examples are BJTs, MOSFETs, IGBTs (insulated-gate-bipolar transistors), and many others.

In this text, we will represent all devices of this category by the generic symbol shown in Fig. 5.4, which is the classic symbol for an electronic switch

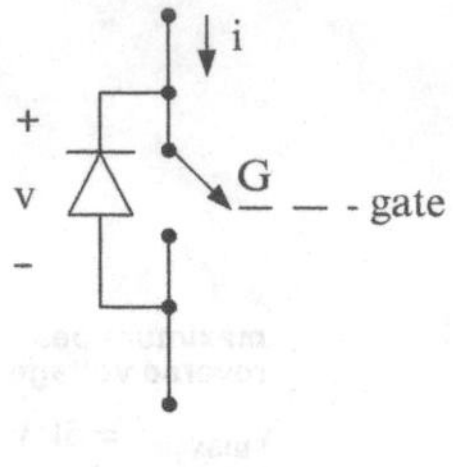

Fig. 5.4 Generic symbol of electronic switch (e.g., MOSFET) with controlled unidirectional current flow in forward direction and with conduction in reverse direction

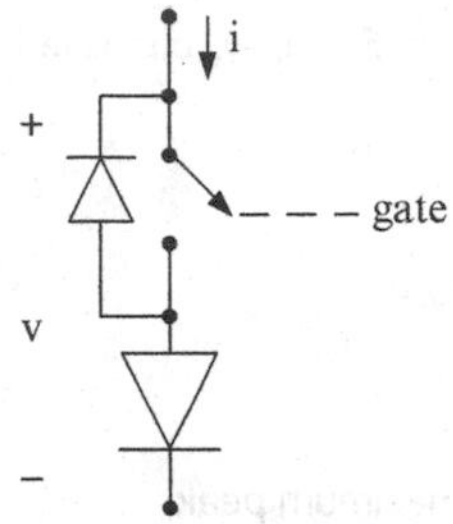

Fig. 5.5 Electronic switch (e.g., thyristor) with controlled unidirectional current flow (in direction of arrow) at positive voltage and current blocking at negative (reverse) voltage

with the addition of an arrow to indicate the unidirectional-current property of these devices. A majority of the switch types (e.g., MOSFET) do not have reverse-current blocking capability, as diodes and thyristors do. If current blocking at reverse voltage is desirable a diode may be placed in series with a self-commutated switch, as shown in Fig. 5.5.

If current is to be permitted to flow freely in the reverse direction, an additional diode can be placed in anti-parallel with the self-commutated switch (Fig. 5.4). Note that none of the electronic switches in the three categories has the simple properties of a mechanical switch, i.e., conduction in both directions when on and current-blocking ability in both directions when off.

Application Example 5.1: Uncontrolled, Single-Phase, Half-Wave Rectifier with Output Voltage Increase (Boost-Rectifier)

(a) Simulate the circuit of Fig. 5.6 using PSpice [1–3] for a diode saturation current of $I_s = 10^{-12}$ A, an inductance $L_1 = 1$ nH, a capacitance $C_1 = 100$ µF, and a load resistance $R_1 = 100\ \Omega$. List the PSpice input program. Plot the input voltage v_{in}, the diode current of D_1, and the output voltage $v_{out} = v(3,0)$.

(b) Repeat part (a) with $L_1 = 10$ mH and all other quantities remaining the same.

Solution

PSpice input program is listed below and the simulation plots are shown in Fig. 5.7. Note the extension of this input file must be .cir

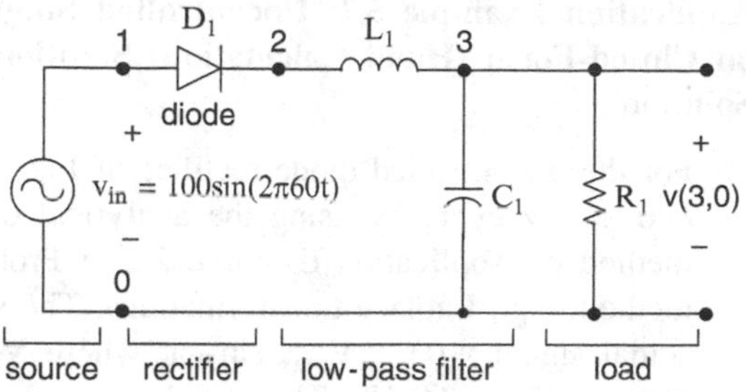

Fig. 5.6 Uncontrolled, single-phase, half-wave rectifier

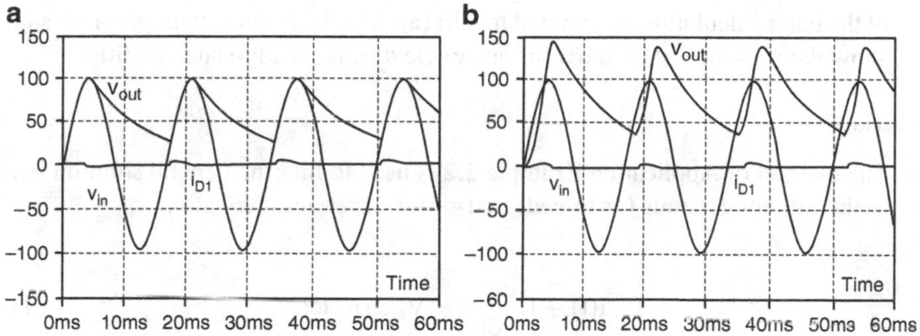

Fig. 5.7 Input voltage v_{in}, output voltage $v_{out} = v(3,0)$, and diode current i_{D1} for uncontrolled, single-phase half-wave rectifier; (**a**) at small boost inductance L = 1 nH, (**b**) at large boost inductance L = 10 mH

```
*Half-wave rectifier
Vin  1  0  sin (0 100 60)
R1   3  0  100
L1   2  3  1m
Cl   3  0  100μ
D1   1  2  dio
.model dio d (is=1p)
.tran 1n 60m
.probe
.four 60 v(3)
.end
```

Figures 5.7a, b illustrate the input voltage v_{in}, the diode current i_{D1} and the output voltage v_{out} for L = 1 nH and 10 mH, respectively. The boost action of the 10 mH inductance can be clearly seen.

Note that for any PSpice program the title line * *Half-wave rectifier* and the terminating line *.end* are important.

Application Example 5.2: Uncontrolled Single-Phase Diode Rectifier Based on Closed-Form (Hand Calculation) Solution and Comparison with PSpice Solution

(a) For the uncontrolled diode rectifier of Fig. 5.8 compute the current i(t) over one 60 Hz cycle by using the analytical or closed-form (hand calculation) method of Application Example 2.2 or Problem 2.1 (or any other analytical method, e.g., Laplace transformation). The voltage source consists of a sinusoidal signal $v_s(t) = V_{max} \sin \omega t$, where $V_{max} = \sqrt{2} \cdot 120$ V and has a frequency of f = 60 Hz. The initial condition for the current at time t = 0 is i(t = 0) = 0.

(b) Write a PSpice input program with the .cir extension for the circuit of Fig. 5.8 and obtain a PSpice solution. Compare the result of the PSpice solution with that of the hand calculation as obtained in part (a). List the PSpice input program and show the plots of $v_s(t)$ and i(t) for one cycle during steady-state condition.

Solution

(a) The method of Application Example 2.2 is used to find the general solution i(t) to the inhomogeneous (or forced) first-order differential equation

$$R \cdot i(t) + L\frac{di(t)}{dt} = V_{max} \sin \omega t \tag{5.1}$$

or

$$i(t) + \frac{L}{R}\frac{di(t)}{dt} = \frac{V_{max}}{R} \sin \omega t, \tag{5.2}$$

where τ = L/R is the time constant.

The homogeneous first-order differential equation is

$$i(t) + \frac{L}{R}\frac{di(t)}{dt} = 0 \tag{5.3}$$

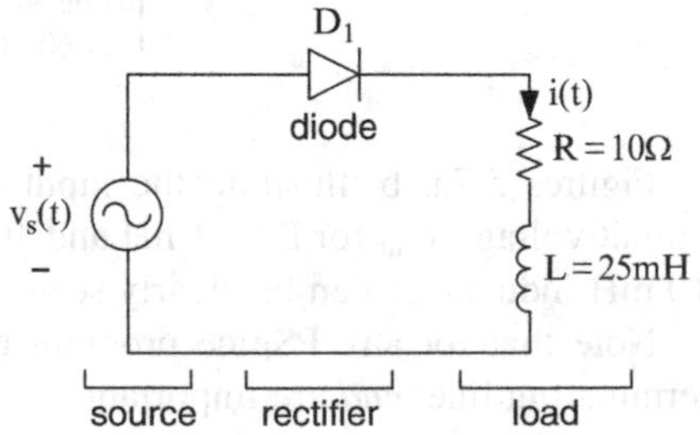

Fig. 5.8 Uncontrolled diode rectifier

1. Solution of the homogeneous (or unforced, natural response) differential equation: assume the solution to be of the form

$$i_{hom}(t) = C_1 e^{-(\alpha t)}, \tag{5.4}$$

$$\frac{di_{hom}(t)}{dt} = -C_1\ \alpha e^{-(\alpha t)}, \tag{5.5}$$

introducing these terms into the homogeneous differential equation, one obtains

$$C_1 e^{-(\alpha t)} - \ \tau C_1\ \alpha e^{-(\alpha t)} = 0, \tag{5.6}$$

with $e^{-(\alpha t)} \neq 0$, $C_1 \neq 0$, and $\tau = 1/\alpha$ follows the homogeneous solution

$$i_{hom}(t) = C_1 e^{-(\frac{t}{\tau})}. \tag{5.7}$$

2. Particular solution of inhomogeneous differential equation: assume the solution to be of the form

$$i_{particular}(t) = B_1 \cos\omega t + D_1 \sin\omega t \tag{5.8a}$$

$$\frac{di_{particular}(t)}{dt} = -B_1\ \omega \sin\omega t + D_1\ \omega \cos\omega t \tag{5.8b}$$

with

$$i(t) + \frac{L}{R}\frac{di(t)}{dt} = \frac{V_{max}}{R}\sin\omega t \tag{5.9}$$

follows

$$B_1 \cos\omega t + D_1 \sin\omega t - \ \tau B_1\ \omega \sin\omega t + \ \tau D_1\ \omega \cos\omega t = \frac{V_{max}}{R}\sin\omega t. \tag{5.10}$$

The coefficients of $\cos\omega t$ must add up to zero:

$$B_1 + \tau D_1 \omega = 0. \tag{5.11}$$

Similarly, the coefficients of $\sin\omega t$ must add up to zero:

$$D_1 - \ \tau B_1\ \omega - \frac{V_{max}}{R} = 0. \tag{5.12}$$

From (5.11) and (5.12) follows

$$D_1 = \frac{\frac{V_{max}}{R}}{(1 + \ \tau^2\ \omega^2)}, \tag{5.13}$$

$$B_1 = \frac{(-\tau\omega)\left(\frac{V_{max}}{R}\right)}{(1+\ \tau^2\ \omega^2)}. \tag{5.14}$$

The particular solution is now

$$i_{particular}(t) = \frac{(-\tau\omega)\left(\frac{V_{max}}{R}\right)}{(1+\ \tau^2\ \omega^2)}\cos\ \omega t + \frac{\left(\frac{V_{max}}{R}\right)}{(1+\ \tau^2\ \omega^2)}\sin\ \omega t. \tag{5.15}$$

3. General solution consists of the superposition of homogeneous and particular solutions:

$$i_{general}(t) = i_{hom}(t) + i_{particular}(t), \tag{5.16}$$

$$i_{general}(t) = C_1 e^{-\left(\frac{t}{\tau}\right)} - \left(\frac{\tau\omega\left(\frac{V_{max}}{R}\right)}{(1+\ \tau^2\ \omega^2)}\cos\ \omega t + \frac{\left(\frac{V_{max}}{R}\right)}{(1+\ \tau^2\ \omega^2)}\sin\ \omega t\right). \tag{5.17}$$

4. The coefficient C_1 will be found by introducing the initial condition at $t = 0$, e.g., $i_{init}(t) = 0$, therefore

$$0 = C_1 e^{-0} - \frac{\tau\omega\left(\frac{V_{max}}{R}\right)}{(1+\ \tau^2\ \omega^2)} \tag{5.18}$$

or

$$C_1 = \left(\frac{\tau\omega\left(\frac{V_{max}}{R}\right)}{(1+\tau^2\ \omega^2)}\right). \tag{5.19}$$

5. The general solution for the given initial condition is now

$$i_{general}(t) = \frac{\tau\omega\left(\frac{V_{max}}{R}\right)}{(1+\ \tau^2\ \omega^2)}\left(e^{-\left(\frac{t}{\tau}\right)} - \cos\ \omega t\right) + \frac{\left(\frac{V_{max}}{R}\right)}{(1+\tau^2\ \omega^2)}\sin\ \omega t$$

$$= \text{RHS}. \tag{5.20}$$

6. Find zero crossing of i(t) at time t_O for given parameters: $V_{max} = \sqrt{2}\cdot 120$ V, $R = 10\ \Omega$, $\tau = L/R = 2.5$ ms, $f = 60$ Hz, $\omega = 2\pi f = 377$ rad/s, $T = 1/f = 16.666$ ms, $V_{max}/R = 12\sqrt{2}$ A, $\tau\omega = 0.943$. Pointwise determination of i(t) as depicted in Figs. 5.9a, b: i(t = T/2 = 8.34 ms) = 8.77 A, i(t = 6 ms) = 13.08 A, i(t = 11 ms) = −2.945 A, i(t = 12 ms) = −10.33 A, i(t = 10 ms) = 1.71 A, i(t = 10.4 ms) = −0.159 A, i(t = 10.37 ms)= −0.0578 A.
7. Figure 5.10 shows the positive cycles of $v_s(t)$ and i(t) for the given parameters.

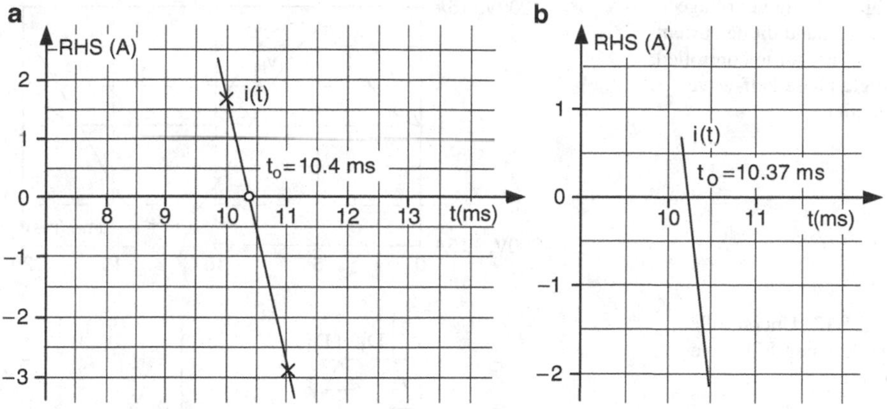

Fig. 5.9 Approximate solution (**a**) and detailed solution (**b**) of transcendental equation (5.20) for the zero crossing of the horizontal (time) axis

Fig. 5.10 Solution of transcendental equation (5.20), amplitude of $v_s(t)$ is not to scale

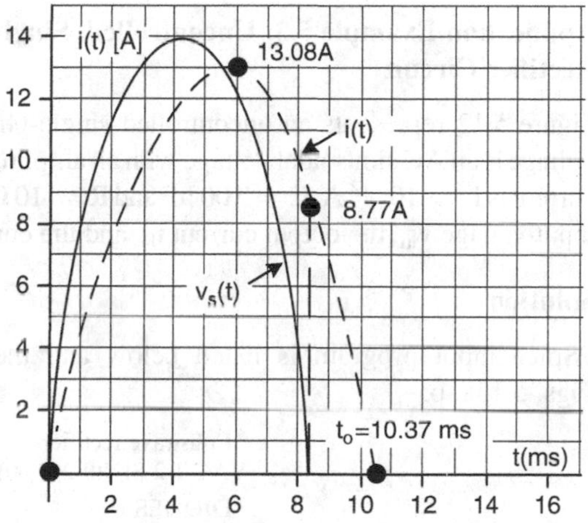

(b) PSpice input program is listed below and the simulation plots are shown in Fig. 5.11. Note the extension of this input file must be .cir. Figure 5.11 illustrates the input voltage $v_{in} = v_s$ and the diode current i_{D1}.

```
*Single-phase diode rectifier
Vs 1 0 sin(0 169.7 60)
D1 1 2 dio
R 2 3 10
L 3 0 25m
.model dio d(is=1p)
.tran 1u 20m
.probe
.end
```

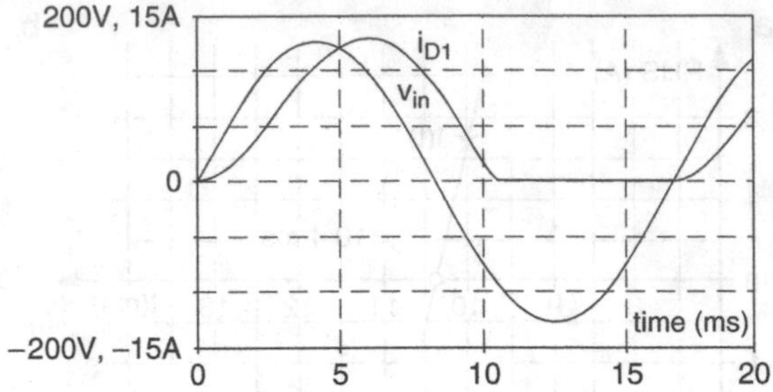

Fig. 5.11 Input voltage $v_{in} = v_s$ and diode current $i_{D1} = i(t)$ for uncontrolled, single-phase half-wave rectifier

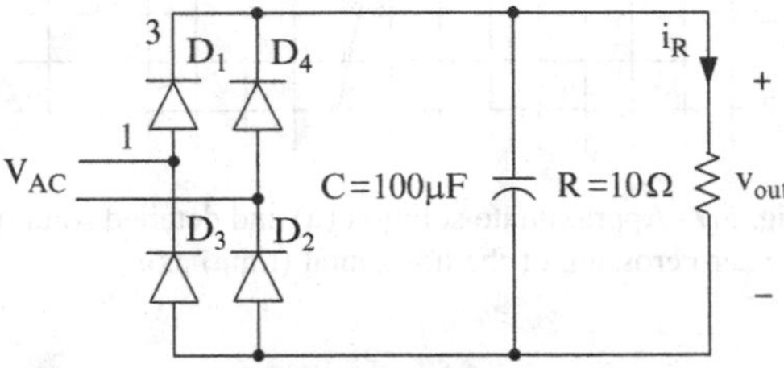

Fig. 5.12 Uncontrolled, single-phase full-wave rectifier

Application Example 5.3: Uncontrolled, Single-Phase, Full-Wave Rectifier Circuit

Figure 5.12 represents an uncontrolled single-phase, full-wave rectifier. The input voltage is an AC sinusoidal voltage with an amplitude value of 240 V, diode saturation current is $I_s = 10^{-12}$ A, C = 100 μF and R = 10 Ω. Use PSpice and compute/plot the input voltage v_{in}, the output current i_R, and the current through diode D_1.

Solution

PSpice input program is listed below and the simulation plots are shown in Figs. 5.13a, b:

```
*Full-wave rectifier
VAC 1 2 SIN(0 240 60)
*DIODES
D1 1 3 IDEAL
D2 0 2 IDEAL
D3 0 1 IDEAL
D4 2 3 IDEAL
*RESISTOR & CAPACITOR
R 3 0 10
C 0 3 1m
*MODELS
.MODEL IDEAL D(IS=1E-12)
.TRAN 0.5U 50m 0.5m
.PROBE
.end
```

Application Example 5.4: Gating Circuit for Thyristor

Thyristors are phase-controlled devices, where within one cycle the current is delayed from about 0 degrees ($\alpha_{min} = 3°$) to about 180 ($\alpha_{max} = 177°$) degrees.

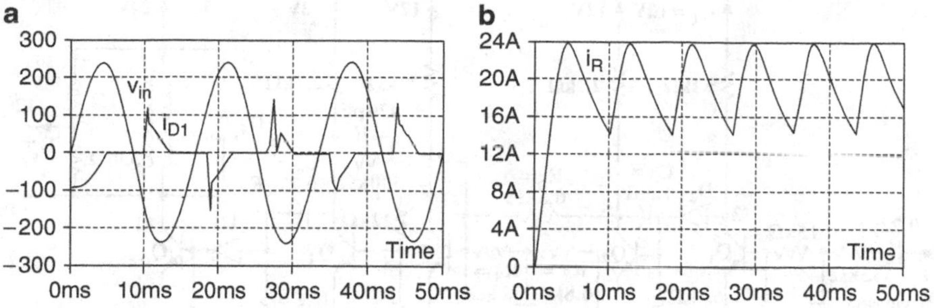

Fig. 5.13 Simulation plots for the uncontrolled, single-phase, full-wave rectifier of Fig. 5.12; (**a**) input voltage v_{in} [V] = V_{AC} [V] and current i_{D1} [A] through diode D_1, (**b**) current i_R [A] through load resistor R of rectifier

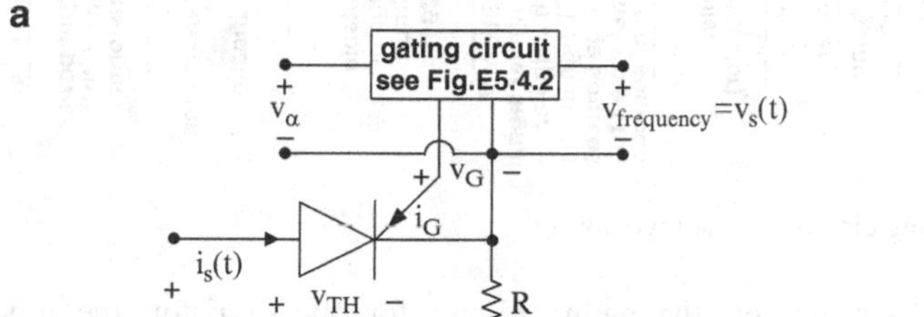

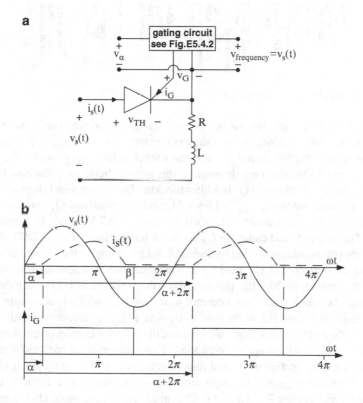

Fig. 5.14 Thyristor controls AC current through inductive-resistive load, (**a**) circuit, (**b**) signals

Figures 5.14a, b illustrate a single-phase circuit where the positive half-wave current can be delayed by the firing angle α and the half wave of negative current is always zero. The turn-off of the thyristor at β occurs through the line voltage being negative and the thyristor current reaching zero value, that is, turn-off occurs through line (voltage) commutation. The duration of the gating current i_G must be always somewhat larger than β as indicated in Fig. 5.14b. Figure 5.15 details the firing or gating circuit of one thyristor which is a current-controlled switch.

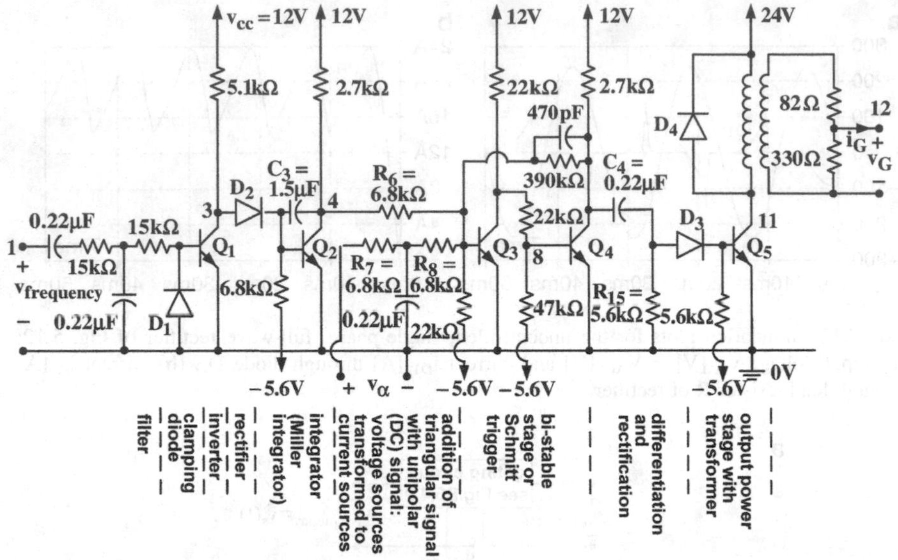

Fig. 5.15 Gating circuit for one thyristor

Figure 5.15 represents the gating circuit for one thyristor: the input voltage $v_{frequency}$ senses the frequency of the source voltage, the input voltage v_α is instrumental for controlling the firing angle α, and the output voltage v_G provides the gating power (current) for the thyristor. Inspecting this gating circuit, one can note that the network to the left of transistor Q_1 is a filter blocking DC currents and short-circuiting harmonic voltages within $v_{frequency}$. Diode D_1 protects transistor Q_1 and clamps the negative base voltage to a diode voltage drop of about −0.7 V. Diode D_2 represents a rectifier. Transistor Q_2 and capacitor C_3 act as an integrating network. BJTs (bipolar junction transistors) are controlled by currents, not by voltages; therefore, the voltage existing at the collector of Q_2 must be transformed into a current source, which is achieved by the large resistor R_6. The unipolar voltage source v_α controlling the firing angle α must be transformed into a current source as well, which is accomplished by the large resistors R_7 and R_8. At the base of transistor Q_3 the current originating at the collector voltage of Q_2 (triangular voltage), and the unipolar current originating at the variable DC (unipolar) voltage v_α are added. This superposition enables us to make Q_3 conducting as a function of v_α, and thus effectively introducing the delay of firing angle α. Transistors Q_3 and Q_4 represent a bi-stable stage or a Schmitt trigger. Capacitor C_4 and resistor R_{15} act as a differentiating network, diode D_3 is a rectifier, and transistor Q_5 is a part of the output power stage. Diode D_4 is a freewheeling diode, and the output transformer produces a floating ground. The output voltage v_G is the gating voltage for the thyristor of Fig. 5.14a.

Specific data

In the gating circuit of Fig. 5.15 the input voltage $v_{frequency}$ determines the frequency of the gating signal v_G, and the variable but unipolar input DC voltage $v_\alpha = v_{alpha}$ determines the phase-shift angle α or the delay of the gating signal with respect to the reference voltage $v_{frequency}$. In particular $v_{frequency} = \sqrt{2} \cdot 100 \sin 2\pi\, 60t$. The models for the transistors, diodes, and the output transformer are:

BJT transistors

```
.model Q2N2222  NPN( Is=3.108f  Xti=3 Eg=1.11 Vaf=131.5 Bf=217.5 Ne=1.541 Ise=190.7f
+ Ikf=1.296 Xtb=1.5 Br=6.18 Nc=2 Isc=0 Ikr=0 Rc=1 Cjc=14.57p Vjc=.75 Mjc=0.3333
+ Fc= 0.5 Cje=26.08p Vje=0.75 Mje=0.3333 Tr=51.35n Tf=451p Itf=0.1 Vtf=10 Xtf=2)
```

Diodes

```
.model D1N4001 D( Is=10^-12)
```

transformer:

```
L1  18  19  4H
L2  0   20  1H
K12  L1  L2  0.9999
```

Plot the time function $v_G(t)$ for the time range $0 < t < 8.333$ ms (at $f = 60$ Hz), and determine point-wise (e.g., 21 points) the thyristor firing angle α as a function of v_α. The PSpice input program is listed below and the simulation plots are shown in Figs. 5.16a, b:

```
*Gating circuit for thyristor
Vfrequency   1    0     Sin(0 141.4V 60Hz)
Vcc1         6    0     DC 12V
Vcc2         22   0     DC 12V
Vcc3         23   0     DC 12V
Vcc4         24   0     DC 12V
Vdd          19   0     DC 24V
Vee1         8    0     DC −5.6V
Vee2         25   0     DC −5.6V
Vee3         26   0     DC −5.6V
Vee4         27   0     DC −5.6V
* User defined source (for example Valpha=2.5V)
Valpha       12   0     2.5V
* Input filter
C1           1    2     0.22uF
R1           2    3     15k
C2           3    0     0.22uF
R2           3    4     15k
D1           0    4     D1N4001
Q1           5    4 0   Q2N2222
R3           5    6     5.1k
D2           5    7     D1N4001
* Integrator
R4           7    8     6.8k
C3           7    9     1.5uF
Q2           9    7 0   Q2N2222
R5           9    22    2.7k
* Transforming unipolar voltage Valpha to a current source
R6           9    10    6.8k
R7           11   12    6.8k
R8           11   10    6.8k
R9           10   25    22k
C4           11   0     0.22uF
```

```
* Schmitt trigger
Q3      13    10 0    Q2N2222
R10     13    23      22k
R11     10    15      390k
C5      10    15      470pF
R12     13    14      22k
R13     14    26      47k
R14     15    24      2.7k
Q4      15    14 0    Q2N2222
* Input of output transformer
C6      15    16      0.22uF
R15     16    0       5.6k
D3      16    17      D1N4001
R16     17    27      5.6k
Q5      18    17 0    Q2N2222
D4      18    19      D1N4001
L1      18    19      4H
* Floating ground output signal
L2      20    0       1H
R17     20    21      82
R18     21    0       330
* Definition of coupling of output transformer
K12     L1    L2      0.9999
* Definition of BJT model
.model Q2N2222 NPN(Is=3.108f Xti=3 Eg=1.11 Vaf=131.5 Bf=217.5 Ne=1.541 Ise=190.7f
+ Ikf=1.296 Xtb=1.5 Cjc=14.57p vjc=0.75 Mjc=0.3333  Fc=0.5 Cje=26.08p vje=0.75
+ Mje=0.333 Tr=51.35n Tf=451p  Itf=0.1 vtf=10 Xtf=2)
*Definition of diode model
.model D1N4001 D(IS=1E-12)
.tran 1.0u  0.033  0  0
.probe
.end
```

Plots of (a) the reference voltage $v(1) = v_{frequency}(t)$, the gating voltage $v_G(t)$ as a function of time t for a specific v_α (=2.5 V), and (b) the thyristor firing angle α as a function of v_α are shown in Figs. 5.16a, b, respectively. Note values for less than $\alpha_{min} \approx 3°$ and larger than $\alpha_{max} \approx 177°$ cannot be obtained because of the reverse-recovery currents of the switches (diodes, transistors). These parasitic currents and the associated reverse-recovery time due to the capacitances of the switches are discussed in Application Example 5.13 for MOSFETS.

5.2.4 *Triac*

A triac consists of two thyristors connected in anti-parallel, as depicted in Figs. 5.17a, b. Note that for the two anti-parallel connected thyristors two gating circuits are required (each carrying a positive current pulse), while for the triac one gating circuit is sufficient, carrying a positive and a 180°-delayed negative current

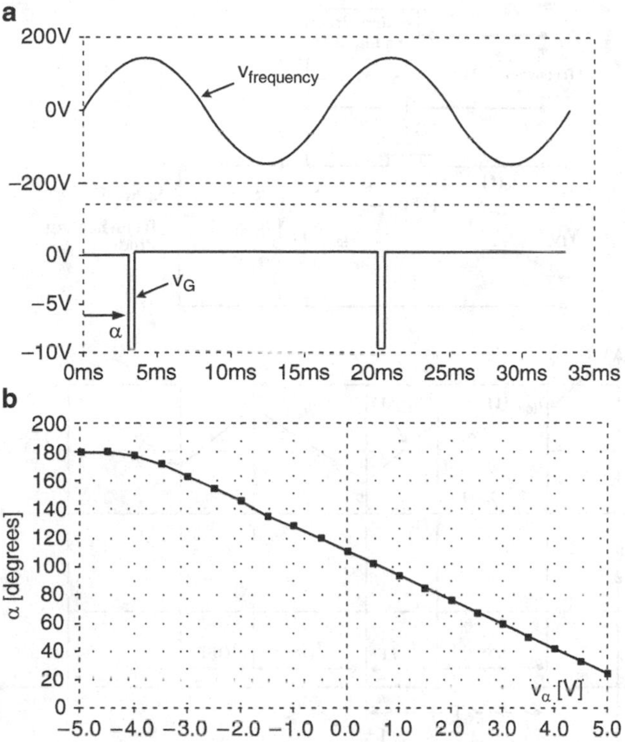

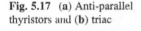

Fig. 5.16 PSpice plots; (**a**) the reference voltage $v(1) = v_{frequency}(t)$, the gating voltage $v_G(t)$ as a function of time t for a specific v_α (=2.5 V), (**b**) the thyristor firing angle α as a function of v_α

Fig. 5.17 (**a**) Anti-parallel thyristors and (**b**) triac

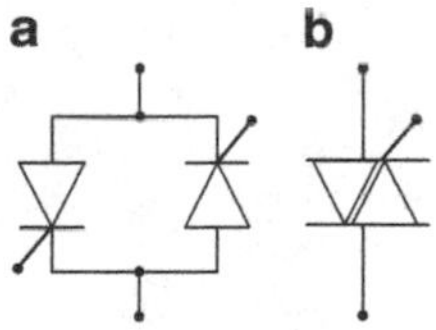

pulse. This means a thyristor or triac is a current-controlled switch which can be turned-on via the gate; turn-off must be initiated through a naturally occurring current zero crossing caused by the external (the thyristor/triac surrounding) circuit, whereby the triac voltage must be negative for positive triac current and the triac voltage must be positive for negative triac current.

5.2.5 *Gating Circuit for MOSFET*

In contrast to a thyristor which is a current-controlled switch, a MOSFET is a voltage-controlled device which can be turned-off and turned-on through the gating voltage v_{GS}, where G is the gate and S is the source of the MOSFET (see Fig. 5.18a). This gating voltage can be generated by an oscillator where the

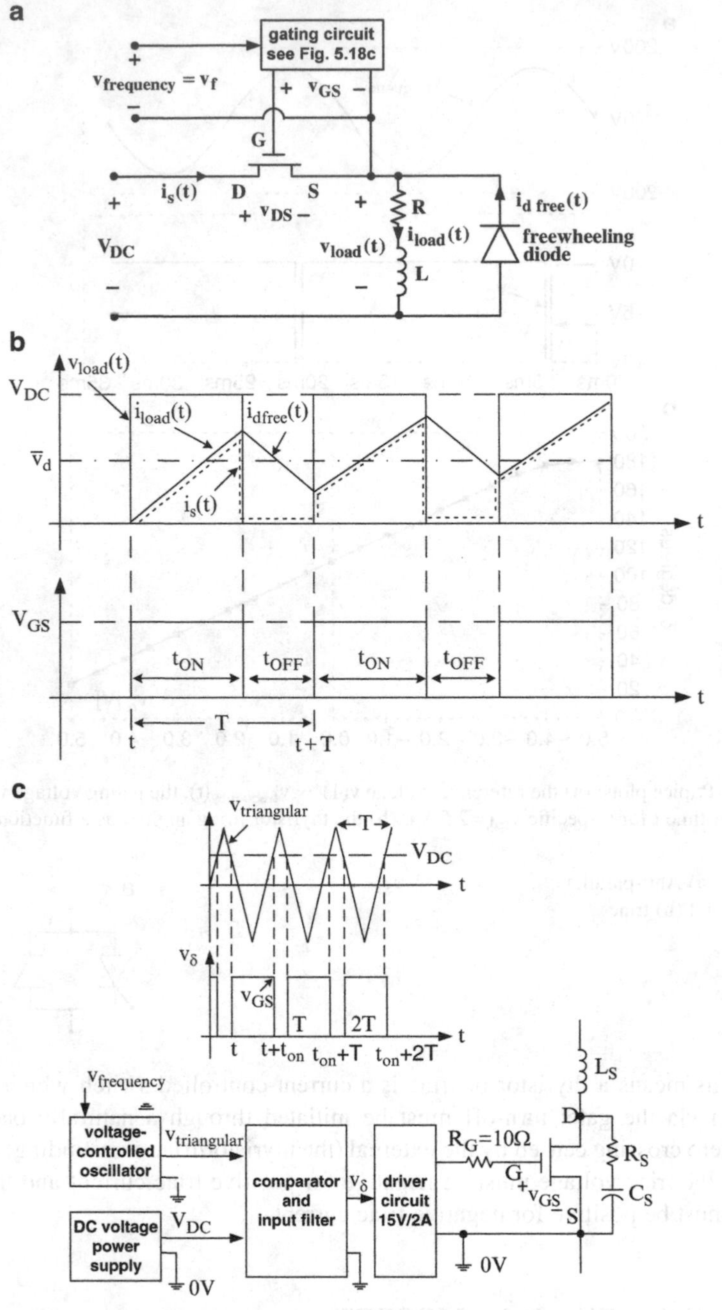

Fig. 5.18 MOSFET controls unipolar current through inductive-resistive R-L load: (**a**) circuit, (**b**) output voltage and current, (**c**) signals (the generation of the gating voltage v_{GS} comes about via the superposition of a triangular voltage with a unipolar voltage) and gating circuit. The inductance L_s limits di/dt levels acting as a current snubber, and the R_s–C_s network limits the dv/dt levels acting as a voltage snubber. Representative values for L_s, R_s and C_s are 10 nH, 10 Ω and 1 μF, respectively

on time t_{ON} and off time t_{OFF} within one period T can be varied (see Fig. 5.18c). Note that $\delta = t_{ON}/T$ is the duty cycle or the duty ratio and varies ideally between 0 and 1, because $t_{ON} + t_{OFF} = T$. The gate resistance $R_G = 10\ \Omega$ in Fig. 5.18c limits the gating current of the MOSFET.

Application Example 5.5: Controlled Single-Phase Rectifier with Self-Commutated MOSFET Switch

The circuit of Fig. 5.19 represents a controlled single-phase rectifier with a self-commutated switch (e.g., MOSFET). This switch provides the circuit with the necessary control of the output DC voltage by changing either the switching frequency (frequency control) or by changing the duty cycle (duty-cycle control). Use duty-cycle control ($\delta = D = 50\%$) to compute with PSpice the current through the load resister R which is proportional to the output voltage v_{out} without and with capacitor C for the input voltage $v_{ab}(t) = 240\ \sin(2\pi\cdot 60\ \text{Hz})$.

The PSpice input program is listed below and the simulation plots are shown in Figs. 5.20a, b:

```
* Supply Voltage
Vab va vb SIN(0 240 60)
*Gating (pulse) voltage of  MOSFET for duty cycle of D=50%
VG 5  4 PULSE (0 30 10u 0n 0n 0.83m 1.66m)
* Diodes
D1  1  3 IDEAL
D2  0  2 IDEAL
D3  0  1 IDEAL
D4  2  3 IDEAL
* MOSFET
MOS 3 5 4 4 SMM
* Resistor and capacitor (load)
R  4  0 100
* without capacitor
*C 0  4 500u
*Source parameters R and L
Ra va a 0.5
Rb vb b 0.5
La a 1 500n
Lb b 2 500n
*MOSFET model
.model SMM NMOS(level=3 gamma=0 delta=0 eta=0 theta=0 kappa=0 vmax=0
+xj=0 tox=100n uo=600 phi=0.6  rs=42.69m  KP=20.87u L=2u W=2.9 VTO=3.487
+RD=0.19  CBD=200n PB=0.8 MJ=0.5  CGSO=3.5n  CGDO=100p RG=1.2  IS=10f)
*Diode model
.model IDEAL D(is=1E-12)
.tran 0.5u 50m 0.5m
.options abstol=10u chgtol=10p reltol=0.1 vntol=100m
.probe
.end
```

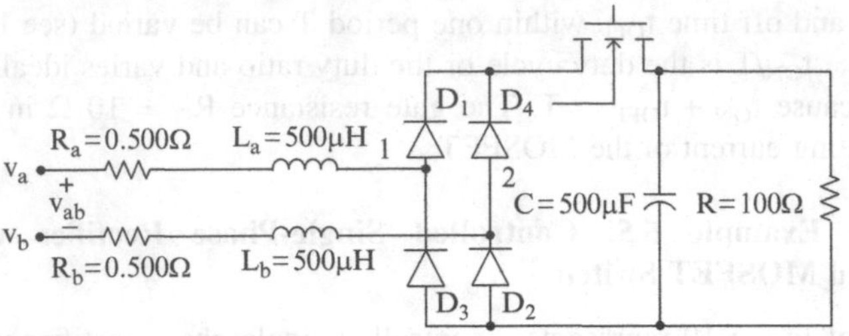

Fig. 5.19 Single-phase, full-wave rectifier with self-commutated switch

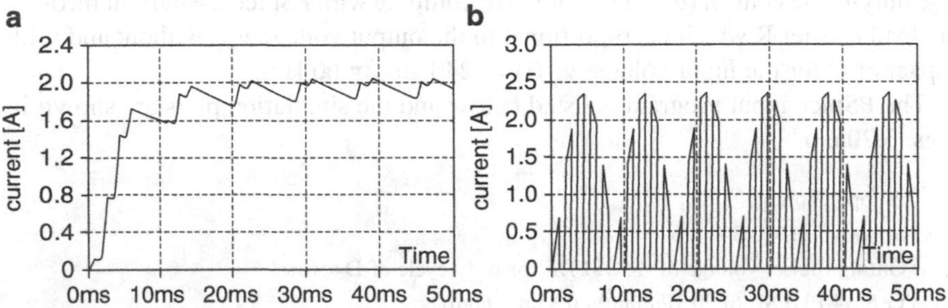

Fig. 5.20 PSpice plots for the current through the load resistor R of Fig. 5.19; (**a**) with capacitor C, (**b**) without capacitor C

5.2.6 *Representation of a Thyristor (Line-Commutated Switch) by a Self-Commutated Switch (e.g., MOSFET) and a Diode*

A line-commutated switch can approximately be represented by a self-commutated switch (e.g., MOSFET) and a diode connected in series, as is indicated in Fig. 5.21. Note that the MOSFET turns on the line-commutated switch and the diode is used for turn-off, that is, the MOSFET must be on until the diode gets reverse biased (diode voltage becomes negative) and the diode current becomes zero. There are more complicated models for thyristors available [3].

Principle of line (voltage) commutation: The principle of line commutation will be explained for a thyristor 3-phase rectifier [4] as illustrated in Figs. 5.22a–d.

1. For a thyristor to turn on (conduct) it must be forward biased with $V_{TH} > 0$ as depicted in Fig. 5.22b, and it must receive a positive gating current $i_G > 0$.
2. For a thyristor to turn off it must be reverse biased ($V_{TH} \leq 0$), see Fig. 5.22c, and the thyristor current i_{TH} must have a zero crossing as shown in Fig. 5.22d. The gating current i_G does not play any role for turn-off.

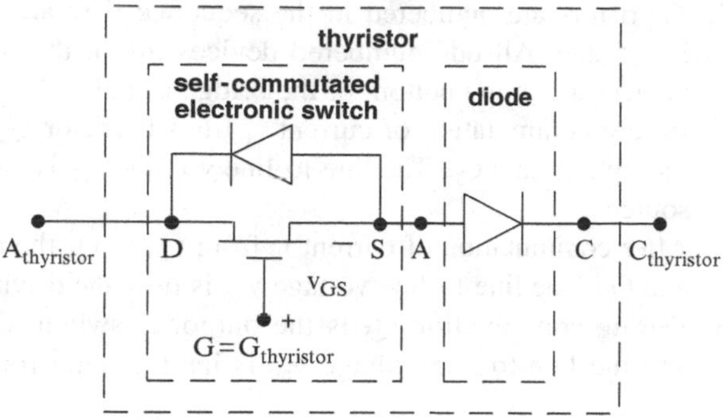

Fig. 5.21 Line-commutated switch

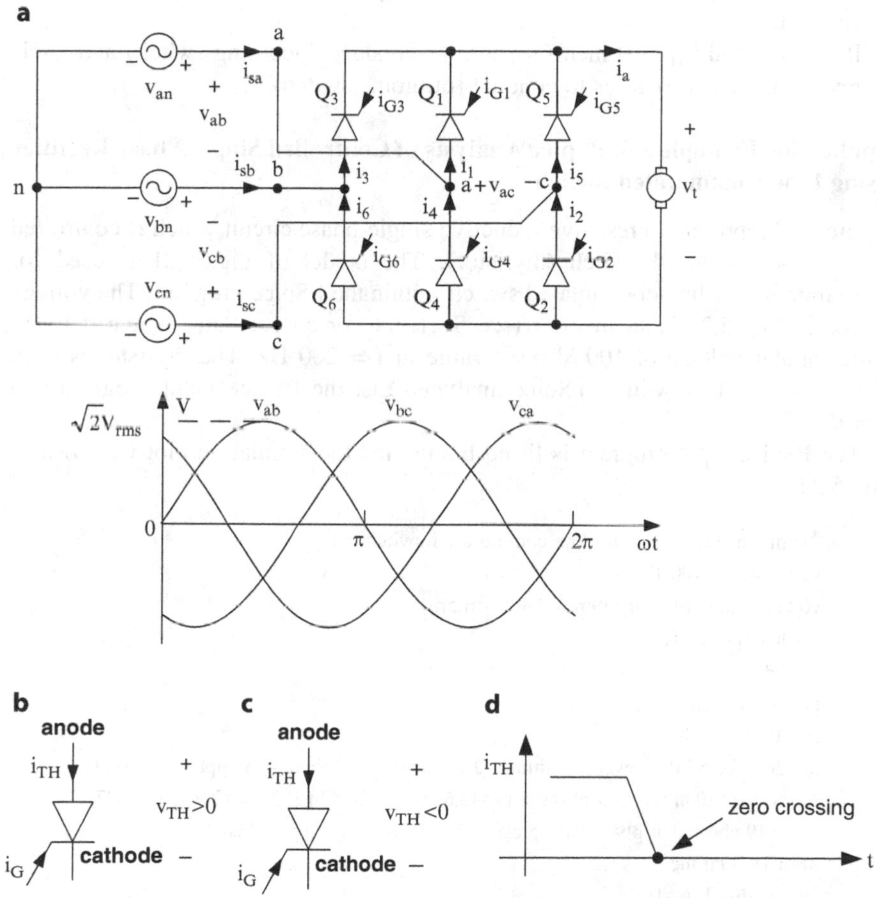

Fig. 5.22 (**a**) Three-phase, controlled bridge thyristor rectifier circuit and line-to-line terminal voltages. (**b**) Forward-biased thyristor. (**c**) Reverse-biased thyristor. (**d**) Zero crossing of thyristor current

3. Thyristors are numbered in the sequence they are conducting as indicated in Fig. 5.22a. All odd-numbered devices are at the top, and all even-numbered devices are at the bottom of the bridge circuit.
4. Before commutation of current i_a from thyristor Q_5 to Q_1 the current i_a flows through Q_5 and Q_6. The line-to-line voltage v_{cb} is the driving (positive) voltage source.
5. After commutation of current i_a from Q_5 to Q_1 the current i_a flows through Q_1 and Q_6. The line-to-line voltage v_{ab} is now the driving (positive) voltage.
6. During commutation Q_5 is the outgoing switch, Q_1 is the incoming switch, and the line-to-line voltage v_{ac} is instrumental for i_a to commutate from Q_5 to Q_1.
7. If $v_{ac} \langle 0$ and Q_5 conducts (and is forward biased) then Q_1 is reverse biased and, therefore, is not ready to conduct.
8. If $v_{ac} > 0$, $i_{G1} = 0$ and Q_5 still conducts, Q_1 is forward biased and is ready to conduct.
9. If $v_{ac} > 0$ and $i_{G1} > 0$ then Q_1 starts to conduct (incoming switch) and Q_5 is reverse biased and ceases to conduct (outgoing switch).

Application Example 5.6: PSpice Analysis of Controlled Single-Phase Rectifier Using Line-Commutated Switch

Figure 5.23 represents a resistive/inductive single-phase circuit which is controlled by a line-commutated switch (thyristor). The model of Fig. 5.21 is used for programming the line-commutated switch within the PSpice program. The voltage source in Fig. 5.23 is an inverter (see Sect. 5.6) or an oscillator which delivers a rectangular voltage of 100 V peak value at f = 200 Hz. The thyristor is fired at $\alpha = 10°$, find i(t) with a PSpice analysis. List the PSpice input program and plot i(t).

The PSpice input program is listed below and the simulation plot is shown in Fig. 5.24:

```
*Half-wave rectifier with line-commutated switch
vs 1 0 pulse(-100 100 0 0 0 2.5m 5m)
vfet sp 2 pulse(0 15 0.139m 0 0 4.861m 5m)
msch 1 sp 2 2 qfet
d1 2 3 dio
r1 3 4 10
l1 4 0 25m ic=0
.model qfet NMOS(level=3 gamma=0 delta=0 eta=0 theta=0 kappa=0 vmax=0
+xj=0 tox=100n uo=600 phi=0.6 rs=42.69m kp=20.87u l=2u w=2.9 vto=3.487
+rd=.19 cbd=200n pb=.8 mj=.5 cgso=3.5n cgdo=100p rg=1.2 is=10f)
.tran 1u 30m uic
.model dio d (is=1p)
.four 200 i(r1)
.probe
.end
```

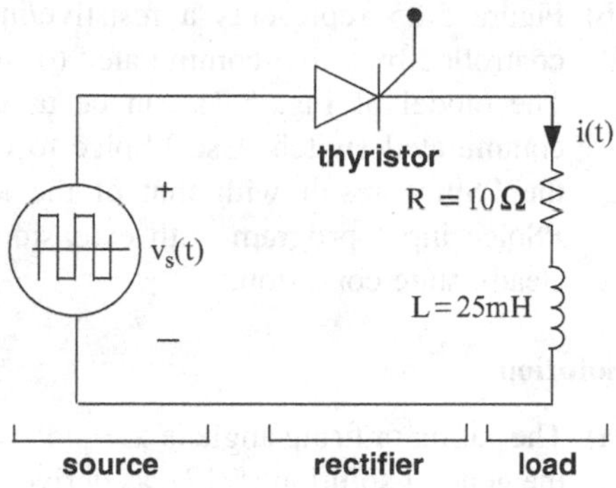

Fig. 5.23 Controlled half-wave rectifier employing a line-commutated switch

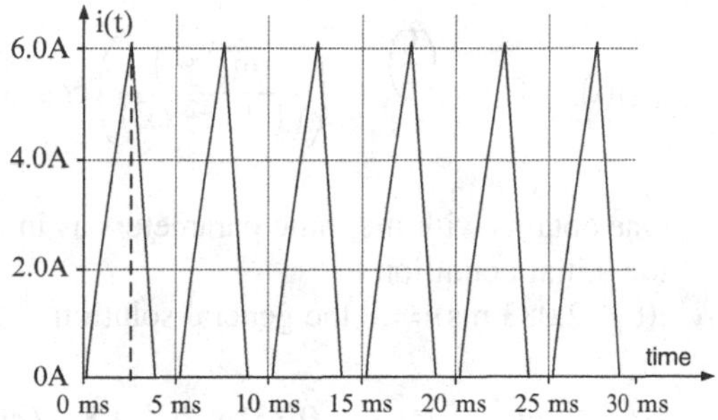

Fig. 5.24 PSpice plot for load current i(t) of Fig. 5.23

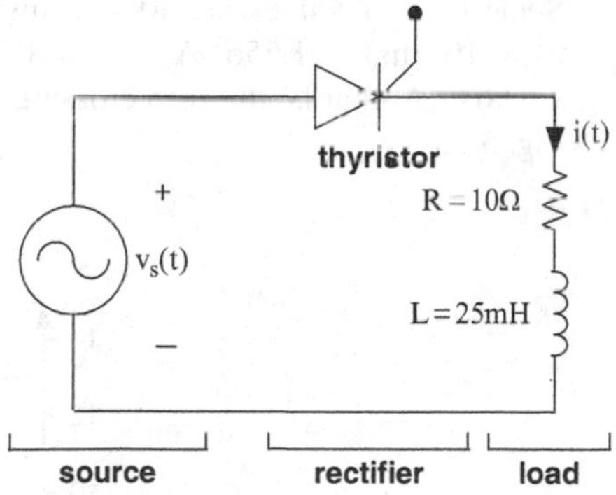

Fig. 5.25 Controlled single-phase thyristor rectifier

Application Example 5.7: Controlled Single-Phase Rectifier Using Line-Commutated Switch Based on Closed-Form (Hand Calculation) Solution and PSpice Analysis

(a) For the controlled thyristor rectifier circuit of Fig. 5.25, where $\alpha = 45°$, compute the current i(t) by using the analytical method of Application Example 2.2 or Problem 2.1, or any other analytical method, see also Application Example 5.2. The voltage source consists of a sinusoidal signal $v_s(t) = V_{max} \sin \omega t$, where $V_{max} = \sqrt{2} \cdot 120\,V$ and has a frequency of $f = 60$ Hz.

(b) Figure 5.25 represents a resistive/inductive single-phase rectifier which is controlled by a line-commutated (or naturally commutated) switch (thyristor). The model of Fig. 5.21 can be used for PSpice programming of the line-commutated switch. Use PSpice to obtain the solution for i(t) and compare the PSpice result with that of the analytical solution of part (a). List the PSpice input program with extension .cir and plot i(t) for one cycle during steady-state condition.

Solution

(a) The gating or firing angle is $\alpha = 45° = (45°/360°)\cdot 16.66$ ms $= 2.083$ ms. With the general solution (5.17), as derived in application Example 5.2

$$i_{general}(t) = C_1 e^{-\left(\frac{t}{\tau}\right)} - \left(\frac{\tau\omega\left(\frac{V_{max}}{R}\right)}{(1+\tau^2\omega^2)}\cos\omega t + \frac{\left(\frac{V_{max}}{R}\right)}{(1+\tau^2\omega^2)}\sin\omega t.\right) \quad (5.21)$$

One obtains with the same parameters as in Application Example 5.2 and with the initial condition
$i(t = 2.083\text{ ms}) = 0$ the general solution

$$i_{general}(t) = -0.831e^{(-t/\tau)} - 8.47\cos\omega t + 8.98\sin\omega t. \quad (5.22)$$

Some current values are $i(t = 6\text{ ms}) = 12.24$ A, $i(t = 8.33\text{ ms}) = 8.44$ A, $i(t = 10\text{ ms}) = 1.556$ A, $i(t = 10.3\text{ ms}) = 0.167$ A, $i(t = 10.35\text{ ms}) = -0.0647$ A, that is, the zero crossing is at about $t_0 = 10.34$ ms as indicated in Fig. 5.26

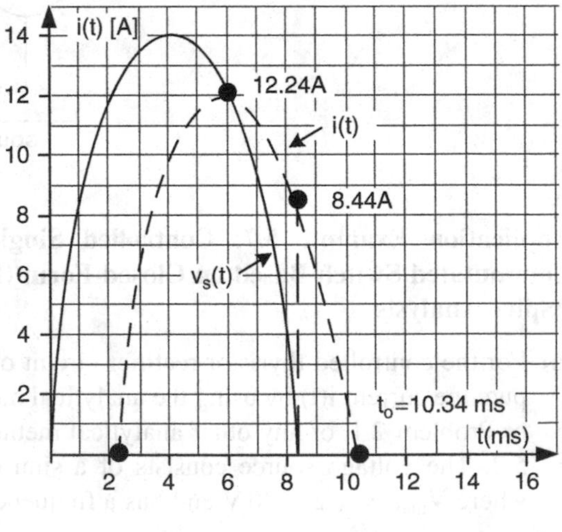

Fig. 5.26 Input voltage $v_s(t)$ for which the scale is not shown, and the solution of transcendental equation (5.22) for i(t)

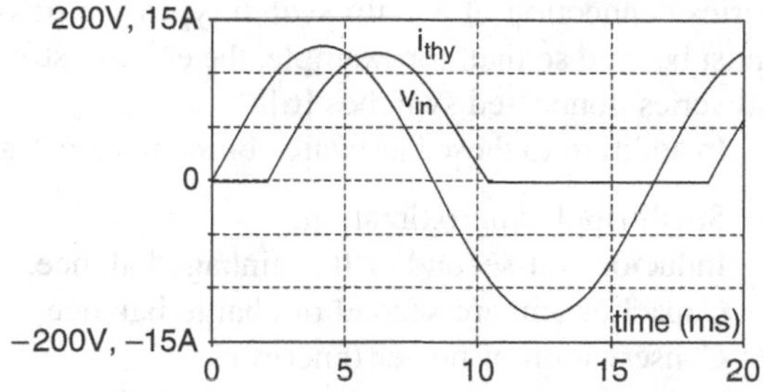

Fig. 5.27 Plot for input voltage $v_{in} = v_s(t)$, and PSpice plot for thyristor current $i(t) = i_{thy}$

(b) The PSpice input program is listed below and the thyristor current simulation plot is shown in Fig. 5.27:

```
*Half-wave rectifier with line-commutated switch
  vs 1  0  sin(0 169.7 60)
  vfet sp 2 pulse(0 15 2.083m 0 0 8.3m 16.66m)
  msch 1 sp 2 2 qfet
  d1 2 3 dio
  R1 3 4 10
  L1 4 0 25m
  .model qfet NMOS(level=3 gamma=0 delta=0 eta=0 theta=0 kappa=0 vmax=0
  +xj=0 tox=100n uo=600 phi=0.6 rs=42.69m kp=20.87u l=2u w=2.9 vto=3.487
  +rd=.19 cbd=200n pb=.8 mj=.5 cgso=3.5n cgdo=100p rg=1.2 is=10f)
  .tran 1u 30m uic
  .model dio d (is=1p)
  .four 200 i(r1)
  .probe
  .end
```

5.3 Basic Concepts Related to Electronic Converters

During the past 5 decades power supplies and converters with BJT transistors operating in the active (amplifying) region have been used and are still used for certain applications. The signals of such devices have a low harmonic voltage and current content (ripple). However, the fact that the transistor is operated in the active region entails relatively large transistor losses. During the past 30 years beginning with the invention of the MOSFET, switched-mode power supplies and converters have been employed. In these the transistors are operated in the ON and OFF regions of the $i_D - v_{DS}$ characteristic and relatively low losses occur in the switching transistor. However, the harmonic content (ripple) of voltages and currents can be significant and proper filtering techniques must be relied on.

In the following we are concerned only with converters based on ON–OFF switching actions. To make such a switching technique possible one has to rely on the use of inductors, transformers and capacitors as storage elements and so-called steady-state averaging techniques [5] are used for the analysis of such circuits. To improve performance and reduce voltage (dv/dt) and current stresses (di/dt), freewheeling diodes and snubber circuits are employed. The parallel and series connection of certain switch types are possible and symmetrizing networks must be used so that, for example, the electric stresses are equally distributed across all series-connected switches [6].

In addition to these hardware components certain basic principles are employed:

- Small ripple approximation.
- Inductor volt-second or flux-linkage balance.
- Capacitor ampere-second or charge balance, and
- Conservation of power (energy).

5.3.1 Small Ripple Approximation [5]

The DC-to-DC step-down (buck) converter of Fig. 5.28a, where $v_{load} < V_{bat}$, consists of a DC voltage source, a single-pole double-throw (SPDT) switch, an inductor L and capacitor C, as well as the load resistor R with the output voltage

$$v_{out}(t) = v_{load}(t) = V_{load} + v_{ripple}(t). \tag{5.23}$$

The starting transient current $i_L(t)$ is shown in Fig. 5.28b, where the duty ratio or duty cycle is $\delta = D = \frac{t_{ON}}{T_s}$, and $T_s = t_{ON} + t_{OFF}$ is the period length (at steady-state).

One requirement is that

$$|v_{ripple}(t)| \langle\langle V_{load}, \text{ or } v_{load}(t) \approx V_{load}, \tag{5.24}$$

therefore, the voltage across the inductor is

$$v_L(t) = V_{bat} - v_{load}(t) \approx V_{bat} - V_{load} = L\frac{di_L(t)}{dt}. \tag{5.25}$$

For the time duration when the SPDT is in position 1:

$$\frac{di_L(t)}{dt} = \frac{v_L(t)}{L} = \frac{V_{bat} - V_{load}}{L}. \tag{5.26}$$

For the time duration when the SPDT is in position 2:

$$\frac{di_L(t)}{dt} = \frac{v_L(t)}{L} = \frac{-V_{load}}{L}. \tag{5.27}$$

With

$$2\Delta i_L = \frac{V_{bat} - V_{load}}{L} t_{ON} \tag{5.28}$$

follows for the inductance required as a function of the ripple current Δi_L

$$L = \frac{V_{bat} - V_{load}}{2\Delta i_L} t_{ON}. \tag{5.29}$$

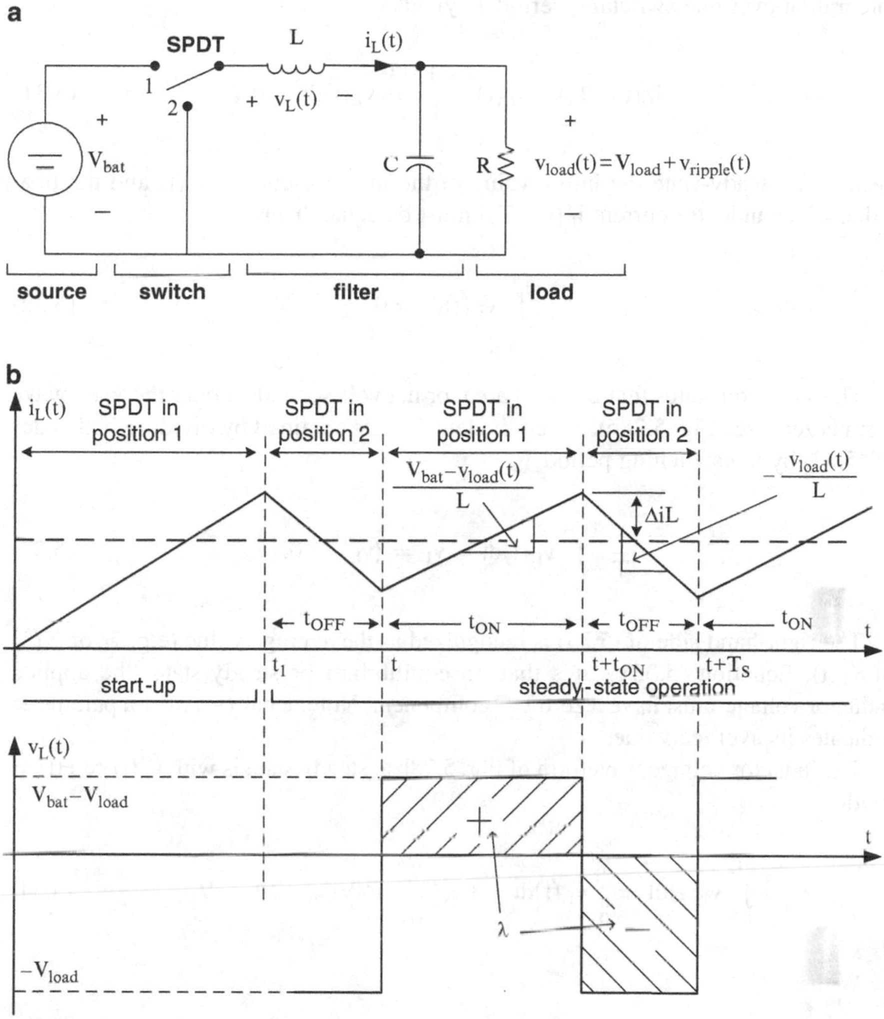

Fig. 5.28 (**a**) Step-down (buck) DC-to-DC converter with single-pole, double-throw (SPDT) switch and low-pass filter. (**b**) Start-up and steady-state performance of a step-down (buck) DC- to-DC converter

5.3.2 *Inductor Volt-Second Balance or Inductor Flux Linkage Balance*

At steady-state (equilibrium) operation, the requirement that the net change in inductor current over one switching period $T_s = 1/f_s$, where f_s is the switching frequency (e.g., $f_s = 20$ kHz) be zero leads us to a way to find steady-state conditions in any switching converter: the principle of inductor volt-second balance. Given the defining relation of an inductor

$$v_L(t) = L\frac{di_L(t)}{dt}, \tag{5.30}$$

integration over one switching period T_s yields

$$i_L(t+T_s) - i_L(t) = \frac{1}{L}\int_{t}^{t+T_s} v_L(t)dt = 0 \tag{5.31}$$

because at steady-state the initial value of the inductor current $i_L(t)$ and the final value of the inductor current $i_L(t+T_s)$ must be equal. Thus,

$$\int_{t}^{t+T_s} v_L(t)dt = 0. \tag{5.32}$$

This equation states that the total area, or net volt-seconds, under the $v_L(t)$ wave form is zero (see Fig. 5.28b). An equivalent form is obtained by dividing both sides of (5.32) by the switching period T_s

$$\frac{1}{T_S}\int_{t}^{t+T_s} v_L(t)dt = \bar{v}_L = \langle v_L \rangle = 0. \tag{5.33}$$

The right-hand side of (5.33) is recognized as the average value (e.g.,$\bar{v}_L$ or $\langle v_L \rangle$) of $v_L(t)$. Equation (5.33) states that, in equilibrium or steady-state, the applied inductor voltage must have a zero DC component. Note, a bar (–) over a parameter indicates its average value.

The inductor voltage waveform of Fig. 5.28b at steady state is with $v_L(t) \approx e(t) = d\lambda/dt$

$$\lambda = \int_{t}^{t+T_s} v_L(t)dt = \int_{0}^{T_s} v_L(t)dt = t_{ON}(V_{bat} - V_{load}) - t_{OFF}V_{load}. \tag{5.34}$$

With

$$t_{OFF} + t_{ON} = T_s \text{ or } t_{OFF} = T_s - t_{ON} \tag{5.35}$$

follows

$$\frac{\lambda}{T_s} = \langle v_L \rangle = \frac{t_{ON}}{T_s}(V_{bat} - V_{load}) - V_{load}\left(\frac{T_s - t_{ON}}{T_s}\right) = 0. \tag{5.36}$$

Therefore,

$$\frac{t_{ON}}{T_s}(V_{bat} - V_{load}) = V_{load}\left(1 - \frac{t_{ON}}{T_s}\right), \tag{5.37}$$

or with

$$\delta = D = \frac{t_{ON}}{T_s}, \tag{5.38}$$

$$V_{load} = \delta V_{bat}. \tag{5.39}$$

This principle of inductor volt-second balance allows us to derive an expression for the DC component of the converter output voltage V_{load}. An advantage of this approach is its generality; it can be applied to any converter [5]. One simply sketches the applied inductor voltage waveform (see Fig. 5.28b), and equates its average value to zero.

5.3.3 Capacitor Ampere-Second Balance or Capacitor Charge Balance

Similar arguments can be applied to capacitors. The defining equation of a capacitor is

$$i_C(t) = C\frac{dv_C(t)}{dt}. \tag{5.40}$$

Integration of this equation over one switching period yields in analogy to an inductor

$$v_C(T_s) - v_C(0) = \frac{1}{C}\int_{t}^{t+T_s} i_C(t)\,dt = \frac{1}{C}\int_{0}^{T_s} i_C(t)dt. \tag{5.41}$$

In steady-state, the net change over one switching period of the capacitor voltage (e.g., Eq. (2.26) of [5]) must be zero, so that the left-hand side of (5.41) is equal to zero. Therefore, in equilibrium the integral of the capacitor current over one switching period (having the units of charge, that is, ampere-seconds or coulombs) should be zero. There is no net change in capacitor charge in steady- state. An equivalent statement is

$$0 = \frac{1}{T_s}\int_{0}^{T_s} i_C(t)dt = \bar{i}_C = \langle i_C \rangle. \tag{5.42}$$

The average value, or DC component, of the capacitor current must be zero in equilibrium. If a DC current is applied to a capacitor, then the capacitor will charge continually and its voltage will increase without bound. Likewise, if a DC voltage is applied to an inductor, then the flux will increase continually and the inductor current will increase without bound, provided the ohmic resistance of the inductor is neglected. Equation (5.42) is called the principle of capacitor ampere-second balance or capacitor charge balance and can be used to find the steady-state currents of certain switching converters.

5.4 DC-to-DC Converters (Choppers)

In some situations, a supply of variable DC voltage may be available, and the requirement may be for establishing a source of constant DC voltage. An example would be the maximum or peak-power tracker of a solar array, where the DC input voltage of the peak-power tracker is changing due to the varying insolation and the DC output voltage of the peak-power tracker must be constant; this necessitates closed-loop control via a step-up/step-down DC-to-DC converter. Figure 5.29a illustrates the block diagram of a solar array, peak-power tracker, step-up/step-down DC-to-DC converter and an inverter feeding power at a constant voltage of $V_{rms}^{inverter}$ into the single-phase power system.

The relation between the DC input voltage of the inverter $V_{DC}^{inverter}$ and its AC output voltage $\tilde{V}_{rms}^{inverter}$ is illustrated in Fig. 5.29b without the use of a transformer.

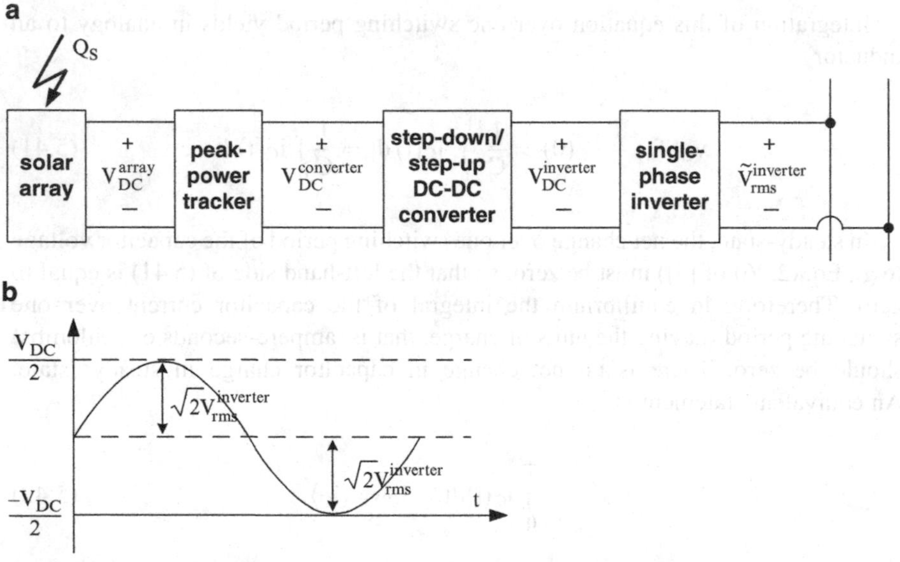

Fig. 5.29 **(a)** Block diagram of a solar power plant feeding power into a single-phase system. **(b)** Relationship between input DC voltage $V_{DC} = V_{DC}^{inverter}$ and output AC voltage magnitude of a single-phase inverter $V_{rms}^{inverter}$

The AC voltage wave shape must fit into the area defined by the DC voltage magnitudes $V_{DC}/2$ and $(-V_{DC}/2)$ and the time axis t. The modulation index m makes sure that the AC voltage $\tilde{V}_{rms}^{inverter}$ and the resulting AC current $\tilde{I}_{rms}^{inverter}$ can have leading or lagging phase shifts. For $m = 1$ the phase shift is about 0° and for $m = 0.5$ a leading (capacitive) power factor (consumer notation) can be permitted (e.g., 30°). In general $0 \leq m \leq 1.0$, and $V_{rms}^{inverter} = m \cdot V_{DC}^{inverter}/(2 \cdot \sqrt{2})$.

Now, we will introduce some simple switching circuits that

- Can convert power at a constant direct input voltage to power at a variable unipolar (DC) output voltage that is either less than or greater than the constant input DC voltage, and can also accommodate a reversal of power flow (from consumer to source);
- Can convert power at a variable direct input voltage to a power at a constant direct output voltage that is either less than or greater than the variable input voltage.

The general terms buck/boost choppers or step-down/step-up DC-to-DC converters are used for these circuits because the approach is to use self-commutated switches (e.g., MOSFET, IGBT) to "chop" sections out of the supply voltage (e.g., for a buck chopper) and thus provide for control of the average value of the output voltage of a DC-to-DC converter (Figs. 5.30a, b). The various types of choppers differ in the number of quadrants of the voltage-current plane in which they are capable of operating. Note, for a step-down converter $\bar{e}_d \leqslant V_{DC}$, where the bar (–) above e_d indicates the average voltage of e_d.

5.4.1 *Step-Down DC-to-DC Converter or Chopper (or Buck Converter)*

Figure 5.30a shows the circuit diagram of a step-down chopper capable of feeding power at a controllable voltage $\bar{v}_d$ to a generalized load circuit from a DC voltage V_{DC}, subject to the limitation that $\bar{v}_d \leqslant V_{DC}$, S_1 is a self-commutated switch capable of being turned ON or OFF by a timed gate signal; D_1 is a freewheeling diode.

5.4.1.1 Operation of Step-Down DC-to-DC Converter

Assume an initial condition with zero load current $i_d(t = 0) = 0$, S_1 is OFF, and $\bar{e}_d$ may be positive. When S_1 is turned ON, the load current $i_d(t)$ will increase according to the first-order differential equation

$$L\frac{di_d}{dt} + Ri_d + \bar{e}_d = V_{DC},$$

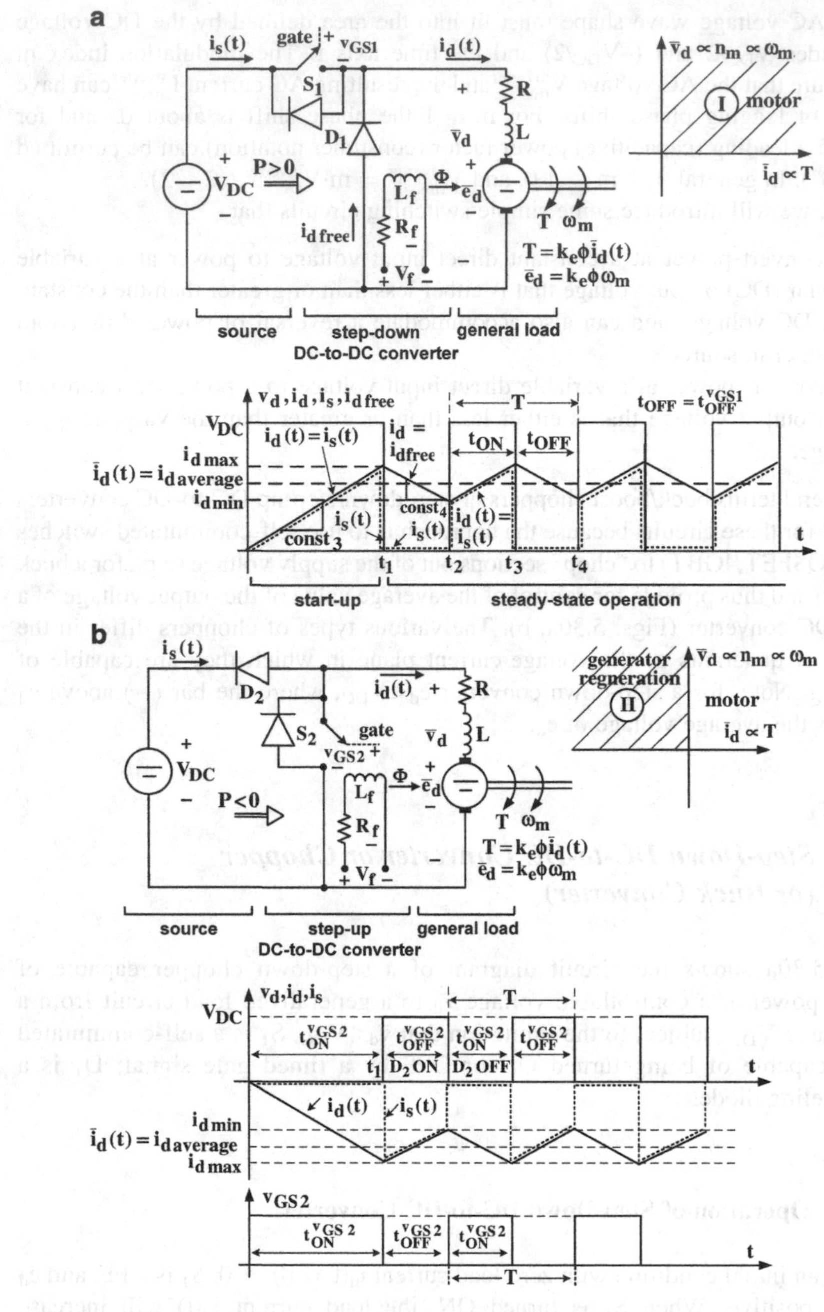

Fig. 5.30 (**a**) Basic circuit of a step-down DC-to-DC converter with operation in quadrant I of $v_d - i_d$ plane; typical load current $i_d(t)$ of a step-down DC-to-DC converter during starting and steady-state operation. Note: $\bar{i}_d(t) = i_{d\,average}$, $\bar{v}_d$ and v_{GS1} have about the same waveshape. (**b**) Step-up DC-to-DC converter with operation in quadrant II of $v_d - i_d$ plane; typical load current $i_d(t)$ of step-up DC-to-DC (boost) converter during starting and steady-state operation

or

$$\frac{di_d}{dt} + \frac{R}{L} i_d = \frac{di_d}{dt} + \frac{1}{\tau} i_d = \frac{(V_{DC} - \bar{e}_d)}{L},$$

where

$$\tau = \frac{L}{R}$$

or

$$\frac{di_d}{dt} = \frac{(V_{DC} - \bar{e}_d - Ri_d)}{L}. \tag{5.43}$$

If the resistance drop is small, this rate of increase is approximately constant because $V_{DC} = \text{const}_1 > 0$, $\bar{e}_d = \text{const}_2 > 0$, and $V_{DC} \rangle \bar{e}_d$. Therefore,

$$\frac{di_d}{dt} \approx \frac{(V_{DC} - \bar{e}_d)}{L} = \text{const}_3 > 0 \tag{5.44}$$

as is illustrated in Fig. 5.30a.

If S_1 is turned OFF now at $t = t_1$, the load current will continue to flow through diode D_1 because the current through the inductance L cannot change quickly. The above differential equation becomes now for $V_{DC} = 0$

$$\frac{di_d}{dt} \approx \frac{(-\bar{e}_d)}{L} = \text{const}_4 < 0 \tag{5.45}$$

which means the load current now will decrease at a constant negative rate of const_4.

Figure 5.30a presents a sequence of switching actions that keep the average value of the load current nearly constant by appropriate choice of the relation between the time t_{ON} of switch S_1 (at steady-state, switch S_1 is ON for a time t_{ON}) and t_{OFF} for the remainder of the repetition period

$$T = t_{ON} + t_{OFF}. \tag{5.46}$$

During the OFF period, the load voltage v_d is equal to zero as the load current flows through the freewheeling diode D_1 that is assumed to be ideal. From Fig. 5.30a it is seen that the average load voltage is determined by the proportion of the ON time within the period T, i.e.,

$$\bar{v}_d = \frac{1}{T}\int_t^{t+T} v_d(t)dt = \frac{1}{T}\int_t^{t+t_{ON}} V_{DC}dt = \frac{1}{T}V_{DC}t_{ON} = \delta V_{DC}, \tag{5.47}$$

where $\delta = \frac{t_{ON}}{T}$ is called the duty ratio or duty cycle. Ideally

$$0 \leq \delta \leq 1.0, \tag{5.48a}$$

and practically due to the reverse-recovery currents and the associated reverse-recovery time (see Application Example 5.13)

$$0.05 \leq \delta \leq 0.95. \tag{5.48b}$$

The steady-state load current wave form is illustrated in Fig. 5.30a. This wave form is made up of sections having the exponential form characteristic of the following first-order differential equations. For the charging current (charging the inductor, that is the inductor field is increasing)

$$\frac{di_d}{dt} = \frac{(V_{DC} - \bar{e}_d - Ri_d)}{L} > 0, \tag{5.49}$$

and the discharging current (discharging of inductor, that is, inductor field is decreasing)

$$\frac{di_d}{dt} = \frac{(-\bar{e}_d - Ri_d)}{L} < 0. \tag{5.50}$$

If the time constant $\tau = L/R$ is considerably greater than the period T, these sections of exponentials may be assumed to be nearly straight lines. In any case, the average load current will be given by

$$\bar{i}_d = \frac{(\bar{v}_d - \bar{e}_d)}{R}. \tag{5.51}$$

The source current $i_s(t)$ is equal to the load current $i_d(t)$ during the ON interval and is zero during the OFF interval, as shown in the wave form of Fig. 5.30a.

Its average value is (see Fig. 5.30a):

$$\bar{i}_s = \frac{1}{T}\int_t^{t+T} i_s(t)dt = \frac{1}{T}\int_t^{t+t_{ON}} i_d(t)dt = \frac{t_{ON}}{T}\bar{i}_d = \delta\bar{i}_d. \tag{5.52}$$

With $\bar{v}_d = \delta V_{DC}$ and $\bar{i}_s = \delta\bar{i}_d$ the output power of the step-down DC–DC converter is

$$P_{out} = \bar{v}_d\bar{i}_d = \delta V_{DC}\bar{i}_d, \tag{5.53}$$

and the input power is

$$P_{in} = V_{DC}\bar{i}_s = \delta V_{DC}\bar{i}_d. \tag{5.54}$$

One concludes $P_{in} = P_{out}$ for this idealized circuit where switching, ohmic ($R = 0$), and iron-core losses are neglected.

5.4.2 *Step-Up DC-to-DC Converter or Chopper (or Boost Converter)*

In some situations, the delivery of power from a low-voltage source to a higher-voltage system is desired. For example, a motor is operated in regeneration mode braking a variable-speed drive, or a solar array output voltage might be low (e.g., $V_{DC} = 50$ V) due to clouds but the output voltage of the DC-to-DC converter must be maintained at $v_{drated} = 120$ V. Figure 5.30b shows a circuit that is capable of this mode of operation.

Note that the reference directions of the variables $\bar{v}_d$, $\bar{i}_d$, $\bar{i}_s$ for the step-up converter have been chosen to be the same as for the step-down chopper (Fig. 5.30a) for reasons that will become evident in the next section (where it is shown that the same average voltage and current formulae apply for step-up as well as for step-down operation).

Because the load voltage $\bar{e}_d$ is assumed to be less than the source voltage $V_{DC} > \bar{e}_d$, nothing happens until the self-commutated switch S_2 is turned on. Turning S_2 on causes the current $i_d(t)$ to increase in a negative direction (see Fig. 5.30b).

Now, when S_2 is turned off, the presence of the inductance L ensures that the load current $i_d(t)$ will continue to flow through diode D_2 into the source. During the time interval when S_2 is ON ($V_{DC} = 0$), one obtains the charging current (see Fig. 5.30b)

$$\frac{di_d}{dt} = \frac{(-\bar{e}_d - Ri_d)}{L} < 0, \tag{5.55}$$

and the discharging current during the time interval when S_2 is OFF

$$\frac{di_d}{dt} = \frac{(V_{DC} - \bar{e}_d - Ri_d)}{L} > 0. \tag{5.56}$$

In the next section, we will want to combine this step-up chopper (Fig. 5.30b) with the step-down chopper (Fig. 5.30a) of Sect. 5.4.1 to achieve a two-quadrant chopper. For this reason, we will designate $t_{ON} = t_{ON}^{VGS1} = t_{OFF}^{VGS2}$, see Fig. 5.31.

Consider a starting point at time t_1 where the load current $i_d(t)$ is already flowing through D_2 to the source, as shown in Fig. 5.30b. During this interval (while D_2 is ON), the rate of change of current is positive, i.e., from the above (5.56) one notes that the magnitude of the negative current $i_d(t)$ decreases, that is $i_d(t)$ becomes more positive.

During t_{on}^{VGS2} the switch S_2 is closed and the rate of change of the current becomes negative, increasing the load current magnitude. The load voltage v_d is V_{DC} when diode D_2 is conducting and is zero with switch S_2 closed, as shown in Fig. 5.30b. Thus, the average load voltage is related to the source voltage by

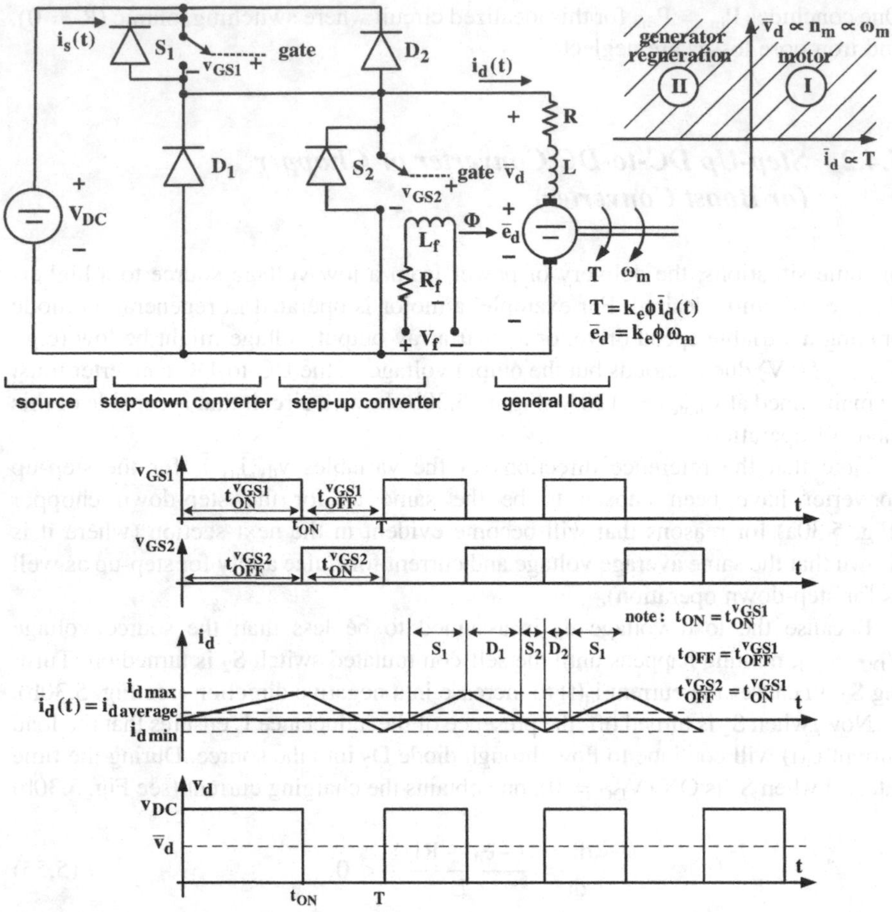

Fig. 5.31 Two-quadrant chopper with operation in quadrants I and II of $\bar{v}_d - \bar{i}_d$ plane

$$\bar{v}_d = \frac{t_{ON}}{T} V_{DC} = \delta V_{DC}, \tag{5.57}$$

and the average load current is

$$\bar{i}_d = \frac{(\bar{v}_d - \bar{e}_d)}{R}. \tag{5.58}$$

Because of our choices of the current direction in Fig. 5.30a, b and the designation of $t_{ON} = t_{ON}^{VGS1} = t_{OFF}^{VGS2}$, these expressions (5.57) and (5.58) are the same as for the step-down configuration (5.47) and (5.51). The average source current $\bar{i}_s$ is as given by (5.52).

5.4.3 Two-Quadrant (Step-Down, Step-Up) DC-to-DC Converters or Choppers (or Buck-Boost Converters)

Frequently, a combination of the properties of the step-up and the step-down DC-to-DC converters is required. This can be accomplished by the circuit of Fig. 5.31: this circuit is capable of providing a controllable positive average load voltage $\overline{v}_d$ while accommodating load current in either direction with a smooth transition through zero current, i.e., it operates in the first and second quadrants of the $\overline{v}_d - \overline{i}_d$ plane.

In this buck-boost converter, the switches S_1 and S_2 are turned on alternately, as shown in Fig. 5.31, with a very short time interval in between (few μs) to allow the switch being turned off to recover its forward blocking capability. This blocking capability is determined by the reverse-recovery time of the switch (see Application Example 5.13)

For operation when $i_d(t)$ is always positive, switch S_1 and diode D_1 act as a step-down chopper. Neither S_2 nor D_2 is operative with this current polarity. Similarly, when $i_d(t)$ is always negative, switch S_2 and diode D_2 act as a step-up chopper, allowing S_1 and D_1 to be ignored. Figure 5.31 illustrates a typical current waveform during the transition from positive to negative current. When the current is positive, it flows through S_1 or D_1 while negative current flows through S_2 or D_2.

The average voltage and current relations for this two-quadrant chopper are the same as for both the step-down and step-up choppers (see Sects. 5.4.1 and 5.4.2).

Application Example 5.8: Regenerative Braking of a Hybrid Automobile Consisting of Step-up DC-to-DC Converter, DC Machine and Battery

When a hybrid automobile travels downhill, its DC-to-DC converter is connected in the configuration shown in Fig. 5.32. The DC machine may be modeled as an induced direct voltage ($\overline{e}_d$ or E_a) in series with a resistance of $R_a = 0.2\ \Omega$ and an inductance of L_a which is sufficiently large so that the current ripple can be neglected. The battery voltage is $V_{bat} = 500$ V, and the switching frequency is $f_s = 5$ kHz.

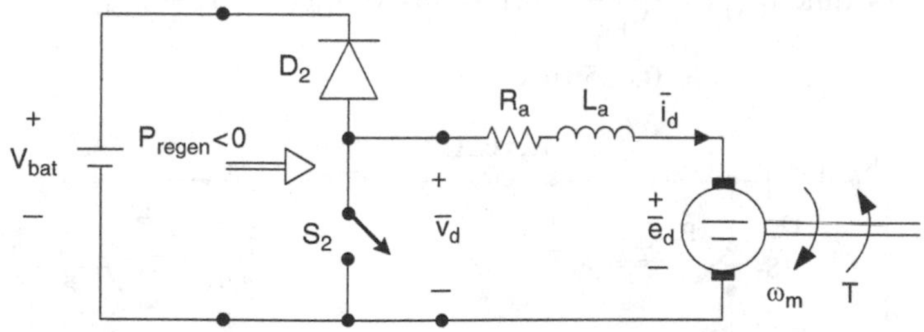

Fig. 5.32 Variable-speed drive consisting of battery, step-up DC-to-DC converter and DC machine

For what length of time t_{on}^{VGS2} should the chopper switch S_2 be in the conducting (ON) condition if the generated (induced) voltage of the DC machine is $\bar{e}_d = E_a = 300V_{DC}$ and a regenerated power of $P_{regen} = -20$ kW is to be fed back to the battery?

Solution

The switching frequency $f_s = 5$ kHz results in the switching period of $T_s = 1/f_s = 0.2$ ms. The terminal voltage is

$$\bar{v}_d = \frac{t_{ON}^{VGS1}}{T_s} V_{bat}, \tag{5.59}$$

and the terminal current is

$$\bar{i}_d = \frac{(\bar{v}_d - \bar{e}_d)}{R_a} \tag{5.60}$$

as indicated in Fig. 5.33.

From (5.59) and (5.60) one gets the regenerated power

$$P_{regen} = \bar{v}_d \cdot \bar{i}_d = \bar{v}_d \left(\frac{\bar{v}_d - \bar{e}_d}{R_a} \right), \tag{5.61}$$

or the second-order relation

$$(\bar{v}_d)^2 - 300\bar{v}_d + 4,000 = 0 \tag{5.62}$$

with the two solutions

$$\bar{v}_{d1} = 286.01\ \text{V}, \tag{5.63}$$

$$\bar{v}_{d2} = 13.98\ \text{V}. \tag{5.64}$$

For the first solution (subscript 1) one obtains

$$\text{the ON time } t_{ON1}^{VGS1} = \frac{\bar{v}_{d1} T_s}{V_{bat}} = 0.1144\ \text{ms or } t_{ON1}^{VGS2} = T_s - t_{ON1}^{VGS1} = 0.0856\ \text{ms} \tag{5.65}$$

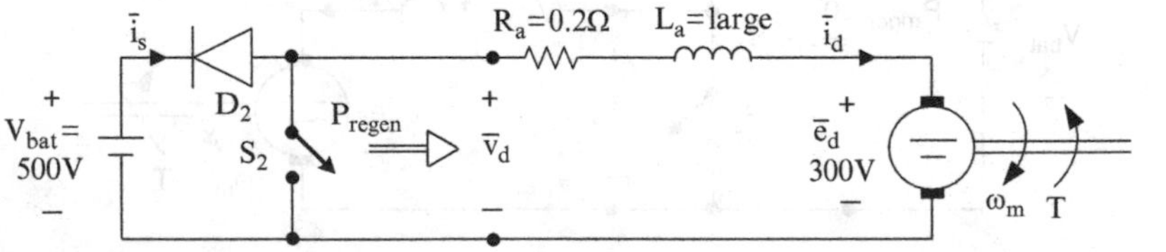

Fig. 5.33 Definition of currents, voltages, and power $P_{regen} < 0$ for regeneration mode

with the current

$$\bar{i}_{d1} = \frac{(\bar{v}_{d1} - \bar{e}_d)}{R_a} = -69.95\,\text{A}, \tag{5.66}$$

and the regenerated power

$$P_{regen1} = \bar{v}_{d1} \cdot \bar{i}_{d1} = -20{,}000\,\text{W}. \tag{5.67}$$

For the second solution (subscript 2) one obtains

$$\text{the ON time } t_{ON2}^{VGS1} = \frac{\bar{v}_{d2} T_s}{V_{bat}} = 0.0059\,\text{ms} \text{ or } t_{ON2}^{VGS2} = T_s - t_{ON2}^{VGS1} = 0.1941\,\text{ms} \tag{5.68}$$

with the current

$$\bar{i}_{d2} = \frac{(\bar{v}_{d2} - \bar{e}_d)}{R_a} = -1{,}430\,\text{A}, \tag{5.69}$$

and the regenerated power

$$P_{regen2} = \bar{v}_{d2} \cdot \bar{i}_{d2} = -20{,}000\,\text{W}. \tag{5.70}$$

The second solution is not feasible because the current is too high, resulting in excessive losses in R_a.

Application Example 5.9: Operation of a Variable-Speed Drive of an Electric Bus in Motoring and Regeneration Modes

A bus has an electric drive train consisting of a battery $V_B = 600$ V with battery resistance $R_B \approx 0$, a two-quadrant chopper, and a general load e.g., DC machine with $E_a = 450$ V, $R_a = 0.1\ \Omega$, L_a is large enough to produce approximately a DC armature current where the current ripple can be neglected, and $\bar{i}_d = 130$ A, E_a is proportional to the angular mechanical velocity ω_m measured at the wheel(s) of the bus, as illustrated in Figs. 5.34a, b.

1. Calculate for step-down (or motoring) operation:
   (a) If the bus moves at a (linear) speed of v = 50 miles/hour compute ω_m at the wheel(s).
   (b) The t_{ON}^{VGS1} of S_1 of the step-down chopper for $\bar{i}_d = 130$ A at $E_a = 450$ V at a switching frequency of $f_s = 5$ kHz.
   (c) Compute the power transferred from the battery to the motor P_{motor}.
2. Calculate for step-up (or regeneration) operation:
   (a) For $E_a = 450$ V the t_{ON}^{VGS2} of S_2 required for the step-up chopper to supply $\bar{i}_d = -130$ A to the battery.

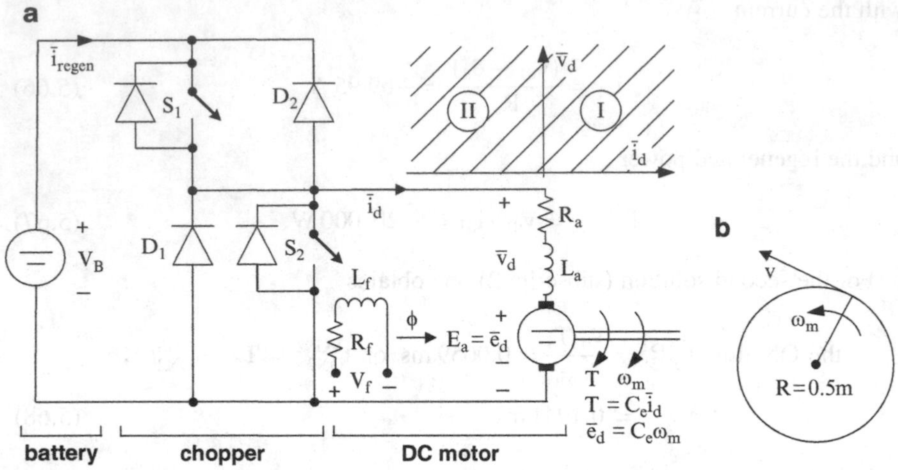

Fig. 5.34 (a) Two-quadrant chopper, (**b**) definition of angular velocity ω_m, velocity v and radius R at the wheel

(b) What is the power transferred P_{regen} from the DC machine acting as a generator to the battery for the condition of part (2a)?
(c) What t_{ON}^{VGS2} of S_2 and $(-\bar{i}_d)$ will be required to regenerate a power of 25 kW at $E_a = 100$ V?

Solution

(1a) Relation between linear speed and angular velocity: $v = R\,\omega_m = 50$ miles/h or 80 km/h (note: 1 mile = 1.6 km, and 1 h = 3,600 s), therefore, v = 80,000/3,600 = 22.22 m/s or

$$\omega_m = v/R = 22.22/0.5 = 44.44\,\text{rad/s}. \tag{5.71}$$

(1b)

$$\bar{v}_d = E_a + \bar{i}_d R_a = 450 + 130 \cdot 0.1 = 463\,\text{V}. \tag{5.72}$$

The switching period is $T_s = 1/f_s = 1/5$ kHz = 0.2 ms, and the ON time of switch S_1 is

$$t_{ON}^{VGS1} = t_{ON} = \frac{\bar{v}_d T_s}{V_B} = (463 \times 0.2\,\text{m})/600 = 0.154\,\text{ms}. \tag{5.73}$$

(1c) The motor input power is

$$P_{motor} = \bar{v}_d \bar{i}_d = 463 \cdot 130 = 60.20\,\text{kW}. \tag{5.74}$$

(2a)

$$\overline{v}_d = E_a + (-\overline{i}_d)R_a = E_a - I_aR_a = 450 - 130 \cdot 0.1 = 437\,V, \tag{5.75}$$

$$\overline{v}_d = \frac{t_{ON}^{VGS1}}{T_s} V_B, \tag{5.76}$$

or the ON time of switch S_1

$$t_{ON} = t_{ON}^{VGS1} = \frac{T_s \overline{v}_d}{V_B} = \frac{0.2\,m \cdot 437}{600} = 0.146\,ms. \tag{5.77}$$

This results in the ON time of switch S_2

$$t_{ON}^{VGS2} = T_s - t_{ON}^{VGS1} = 0.054\,ms. \tag{5.78}$$

(2b)

$$\begin{aligned} P_{regen} &= \overline{v}_d \overline{i}_d - R_B \overline{i}_d^2 = 437 \cdot (-130) + 0 \cdot (130)^2 \\ &= -56,810\,W + 0 = -56.81\,kW. \end{aligned} \tag{5.79}$$

(2c) $\overline{i}_d = \frac{(\overline{v}_d - E_a)}{R_a}$, and $\overline{i}_d = \frac{P_{regen}}{\overline{v}_d}$, where $P_{regen} = -25{,}000$ W

or

$$\frac{-25,000}{\overline{v}_d} = \frac{(\overline{v}_d - E_a)}{R_a},$$

$$\overline{v}_d^2 - 100\,\overline{v}_d + 2,500 = 0$$

$$\overline{v}_d = \frac{100 \pm \sqrt{(10,000 - 10,000)}}{2} = 50\,V. \tag{5.80}$$

The ON time for switch S_1 is

$$t_{ON} = t_{ON}^{VGS1} = \frac{\overline{v}_d T_s}{V_B} = \frac{50 \cdot 0.2\,m}{600} = 0.0166\,ms, \tag{5.81}$$

and the ON time for switch S_2 is

$$t_{ON}^{VGS2} = T_s - t_{ON}^{VGS1} = 0.1834\,ms. \tag{5.82}$$

Application Example 5.10: Efficiencies in Motoring and Regeneration Modes of a Variable-Speed Drive of an Electric Bus

A variable-speed drive of an electric bus consists of a battery, a two-quadrant chopper, and a DC machine, as shown in Fig. 5.35. When motoring the power

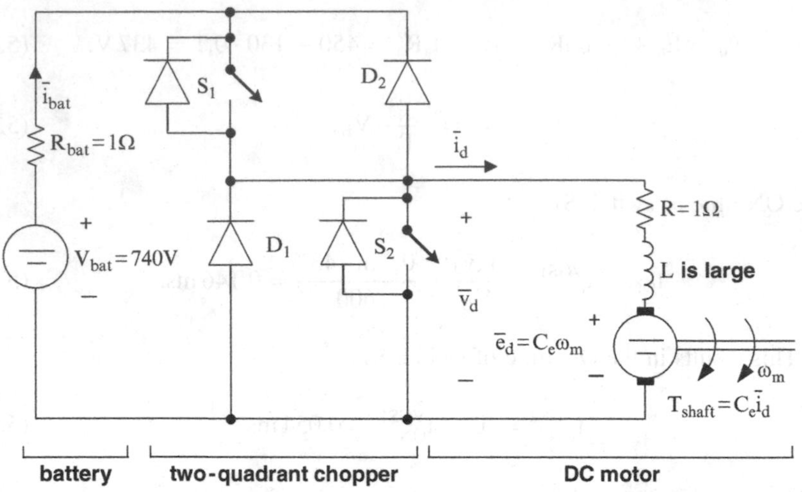

Fig. 5.35 Variable-speed drive train for electric bus

flows from the battery to the motor and when braking (regeneration) the power flows from the motor to the battery. The switching frequency of the chopper is $f_s = 10$ kHz.

(a) Compute for a rated shaft power of $P_{shaft} = 40$ kW the torque at the shaft of the motor (T_{shaft}) at a speed of 600 rpm. Compute the armature current $\bar{i}_d$ and the induced voltage $\bar{e}_d$ if $C_e = 9.69$ V/(rad/s); you may assume a duty cycle of $\delta = t_{ON}^{VGS1}/T_s \approx 1.0$. Note that $T_{shaft} = C_e\bar{i}_d$ and $\bar{e}_d = C_e\,\omega_m$. The battery internal resistance is $R_{bat} = 1\ \Omega$. Hint: use the relation $P = T{\cdot}\omega_m$ to relate shaft power and shaft torque.
(b) What is the output power and the efficiency η_{motor} of the drive train at $\bar{i}_d = 65.7$A and 600 rpm? You may neglect the frictional losses, iron-core losses of the DC motor, and the losses of the chopper.
(c) What is the maximum power that can be delivered from the battery to the motor shaft at $t_{ON}^{VGS1} = T_s$, and what is the efficiency for such a maximum power condition η_{motor_max}?
(d) The energy available at the wheel can be stored in the battery via regeneration. If the induced voltage is $\bar{e}_d = 225$ V and the duty ratio $\delta = t_{ON}^{VGS1}/T_s = 0.2$, how large is the regenerated power P_{regen} and how large is the efficiency η_{regen} for such an operating mode? What is the ON time of switch S_2?

Solution

(a) $P_{shaft} = 40{,}000$ W, $n^{rpm} = 600$ rpm, $n^{rps} = 10$ rps, $\omega_m = 2\pi{\cdot}n^{rps} = 62.83$ rad/s;

$$T_{shaft} = \frac{P_{shaft}}{\omega_m} = \frac{40{,}000}{62.83} = 636.6\,\text{Nm}, \tag{5.83}$$

$$I_a = \bar{i}_d = \frac{T_{shaft}}{C_e} = 65.7\,A, \tag{5.84}$$

$V_{bat} = 740$ V, $R_{bat} = 1$ Ω, $R = 1$ Ω, $\bar{e}_d = C_e \cdot \omega_m = 609\,V$ or $\bar{e}_{d=}V_{bat} - (R_{bat} + R)\,\bar{i}_d = 609$ V.

Checking of motor output power:

$$P_{motorout} = \bar{i}_d\bar{e}_d = \bar{i}_d(V_{bat} - (R_{bat} + R)\bar{i}_d = 65.7 \cdot (740 - 2 \cdot 65.7) = 39{,}985\,W \approx 40\,kW$$

(b)

$$\bar{v}_d = \bar{i}_d R + \bar{e}_d = 65.7 + 609 = 674.7\,V$$

$$\bar{v}_d = \frac{t_{ON}^{VGS1}}{T_s}(V_{bat} - \bar{i}_d R_{bat}),$$

or

$$t_{ON}^{VGS1} = \frac{\bar{v}_d T_s}{(V_{bat} - \bar{i}_d R_{bat})};$$

$$\delta = \frac{t_{ON}^{VGS1}}{T_s} = 674.7/(740 - 65.7) = 1.0.$$

The motor losses are

$$P_{loss\_motor} = (\bar{i}_d)^2 \cdot R = (65.7)^2 \cdot 1 = 4{,}316\,W. \tag{5.85}$$

The battery current is

$$\bar{i}_{bat} = \frac{t_{ON}^{VGS1}}{T_s}\bar{i}_d = 1.0 \cdot 65.7 = 65.7\,A,$$

and the battery losses become

$$P_{loss\_bat} = (\bar{i}_{bat})^2 \cdot R_{bat} = (65.7)^2 \cdot 1 = 4{,}316\,W. \tag{5.86}$$

The motor output power is now

$$P_{motor\_out} = \bar{e}_d\bar{i}_d = 609 \cdot 65.7 = 40\,kW. \tag{5.87}$$

The efficiency of the motor including (chopper and) battery is

$$\eta_{motor} = \frac{P_{shaft}}{P_{shaft} + P_{lossbat} + P_{lossmotor}} = \frac{40{,}000}{40{,}000 + 4{,}316 + 4{,}316} = 0.822$$
$$= 82.2\%. \tag{5.88}$$

(c)

$$P_{motor\_out} = \bar{e}_d \cdot \bar{i}_d = \frac{\bar{e}_d(\bar{v}_d - \bar{e}_d)}{R} = \frac{\bar{v}_d\bar{e}_d - \bar{e}_d^2}{R}. \tag{5.89}$$

To find maximum find the first derivative

$$\frac{\partial P_{motor\ out}}{\partial \bar{e}_d} = \frac{\bar{v}_d - 2\bar{e}_d}{R} = 0 \text{ or } \bar{e}_d = \frac{\bar{v}_d}{2}. \tag{5.90}$$

Choose $t_{ON}^{VGS1} = t_{ON} = T_s = 1/f_s = 0.1\,\text{ms}$, then $\bar{v}_d = V_{bat} - R_{bat}\bar{i}_{d\,max}$, or

$$\bar{i}_{d\_max} = \frac{\bar{v}_d - \frac{\bar{v}_d}{2}}{R} = \frac{\bar{v}_d}{2R} = \frac{V_{bat} - R_{bat}\bar{i}_{d\,max}}{2R},$$

or

$$\bar{i}_{d\_max} = \frac{V_{bat}}{(2R + R_{bat})} = \frac{740}{(2+1)} = 246.7\,\text{A},$$

$$\bar{v}_d = V_{bat} - R_{bat}\bar{i}_{d\,max} = 740\,\text{V} - 246.7\,\text{V} = 493.3\,\text{V}.$$

$$\bar{e}_d = \frac{\bar{v}_d}{2} = 246.65\,\text{V}.$$

The maximum motor output power is

$$P_{out\text{-}max} = \bar{i}_{d\,max}\bar{e}_d = 246.7 \cdot 246.65 = 60.85\,\text{kW}. \tag{5.91}$$

The efficiency at maximum power output becomes

$$P_{loss\_motor\_max} = (246.7)^2 \cdot 1 = 60.86\,\text{kW},$$

$$\bar{i}_{bat\_max} = \left(\frac{t_{ON}^{VGS1}}{T_s}\right)\bar{i}_{d\,max},$$

$$P_{loss\_bat\_max} = (\bar{i}_{d\,max})^2 R_{bat} = (246.7)^2 \cdot 1 = 60.86\,\text{kW},$$

$$P_{motor\_out} = P_{shaft} = \bar{e}_d\bar{i}_{dmax} = 246.65 \cdot 246.7 = 60.85\,\text{kW},$$

$$\begin{aligned}\eta_{motor\text{-}max} &= \frac{P_{shaft}}{P_{shaft} + P_{lossbat} + P_{lossmotor}} \\ &= \frac{60.85\,\text{kW}}{60.85\,\text{kW} + 60.85\,\text{kW} + 60.85\,\text{kW}} = 0.333.\end{aligned} \tag{5.92}$$

(d) $\overline{e}_d = 225\,V$ and $\delta = 0.2$ lead to $\overline{v}_d = \delta V_{bat} = 148\,V$, with $\overline{v}_d = \overline{e}_d + \overline{i}_d R$ follows $\overline{i}_d = \frac{\overline{v}_d - \overline{e}_d}{R} = (148 - 225)/1 = -77\,A$.

The regenerated (delivered to battery) power is:

$$P_{regen} = \overline{v}_d \overline{i}_d + (\overline{i}_d)^2 R_{bat} = 148\,V(-77\,A) + 5,929 \\ = -11,396\,W + 5,929 = -5,467\,W. \tag{5.93}$$

The motor loss becomes

$$P_{loss\_motor} = (\overline{i}_d)^2 R = (77)^2 \cdot 1 = 5,929W, \text{ and } P_{lossbat} = (\overline{i}_d)^2 R_{bat} = (77)^2 \cdot 1 \\ = 5,929\text{ W},$$

and the efficiency at regeneration is

$$\eta_{regen} = \frac{P_{regen}}{P_{regen} + P_{lossbat}} = \frac{5,467W}{5,467W + 5,929W} = 48\%. \tag{5.94}$$

The ON time for switch S_1 is

$$t_{ON}^{VGS1} = \delta \cdot T_s = 0.2 \cdot 0.1\text{ ms} = 0.02\text{ ms},$$

and the ON time of switch S_2 is for this regeneration

$$t_{ON}^{VGS2} = T_s - t_{ON}^{VGS1} = 0.1\text{ ms} - 0.02\text{ ms} = 0.08\text{ ms}. \tag{5.95}$$

5.4.4 Four-Quadrant (Step-Down, Step-Up) DC-to-DC Converters or Choppers (or Buck-Boost Converters)

Some applications require a four-quadrant chopper, i.e., the ability to provide voltage of reversible polarity as well as current of reversible polarity. An example would occur in a DC motor drive for a robot arm (see Fig. 2.3), where positive current is needed for acceleration and negative current for deceleration. Positive voltage is needed for forward velocity and negative voltage for backward velocity. This four-quadrant operation can be achieved by use of two of the systems of Fig. 5.31 connected, as shown in Fig. 5.36.

For operation with positive voltage $\overline{v}_{dforward}$, switch S_4 is held in the ON condition which, together with diode D_3, produces effectively the circuit configuration of Fig. 5.31. For negative load voltage $\overline{v}_{dreverse}$, switch S_2 is held in ON condition, allowing devices S_3, D_3, S_4, and D_4 to act as chopper in quadrants III and IV.

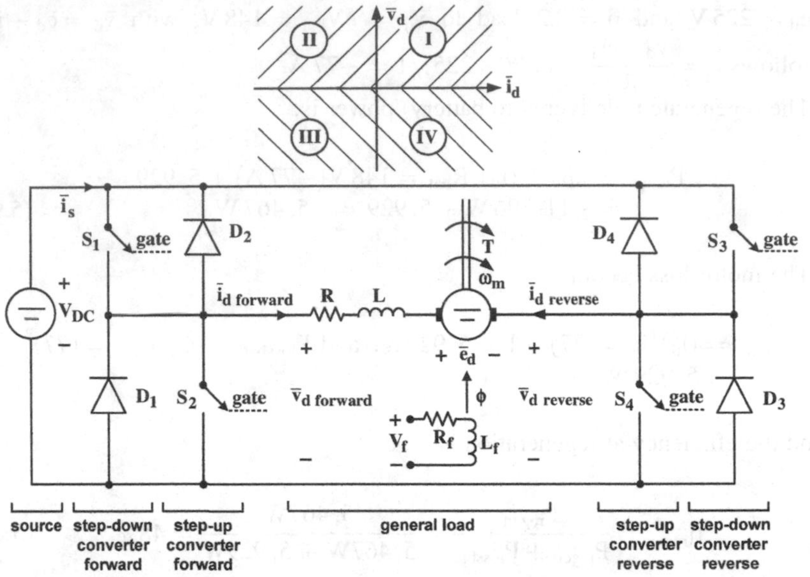

Fig. 5.36 Four-quadrant chopper with operation in quadrants I–IV in $\bar{v}_d - \bar{i}_d$ plane

5.4.5 The Need for Voltage and Current Control

Similarly to the speed regulation discussed in (2.25), the voltage of a converter must sometimes be maintained constant independent of the load. Voltage regulation can be defined, and the control circuit can be designed (P or PI controller) so that the steady-state deviation and transient overshoot is within given limits (see Chap. 2). Figure 5.37 illustrates the step-down DC-to-DC converter with its control circuit. The voltage error signal ε_{vd} is obtained from the difference of the reference voltage and the actual output voltage of the converter: $\varepsilon_{vd} = v_d{}^* - v_d$. An oscillator generates a triangular peak-to-peak voltage $v_{triangular}$ of about 20 V. This triangular voltage is compared with the output voltage V_{DC} of the PI controller which makes the steady-state deviation zero. The comparator must have an input filter in order to eliminate switching harmonics and electrical noise. The driver circuit for the MOSFET must be able to deliver about 2 A at $v_{GS} = 15$ V. It is advisable to introduce a low-pass filter at the voltage pick-up (R_{D1} and R_{D2}).

The efficiency of PWM converters (e.g., DC-to-DC buck/boost converters) can be increased at light load by decreasing the switching frequency say from 20 kHz at rated load to 5 kHz at light load. While at full-load the efficiency is about maximum say 95% depending upon the rating of the converter; the efficiency at light load is relatively low and in the range of 70%. This means the percentage of the switching ripple of the current can be increased at light load, where the average DC current magnitude is relatively small, while at about full-load the percentage of the switching ripple of the current must be small because the average magnitude of the DC current

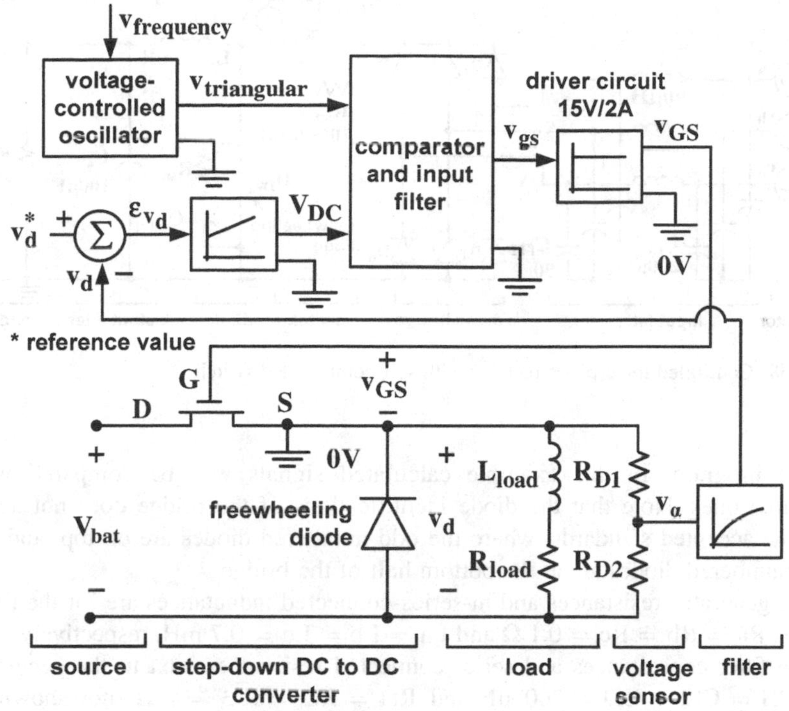

Fig. 5.37 Complete control circuit and power circuit for a step-down DC-to-DC converter

is large. This load-dependent variable PWM switching frequency calls for the measurement of the load current which, of course, complicates somewhat the control of the converter. Preliminary calculations indicate that the light-load efficiency can be increased by about 10% depending upon given power quality conditions.

5.5 AC-to-DC Converters (Rectifiers)

The literature of rectifiers feeding a general load consisting of a resistance, inductance and DC voltage is extensive [7]. The analytical expressions for the rectified DC voltage are complicated and in many cases the analytical treatment of harmonics on the DC bus is not satisfactory. In order not to spend too much time on complicated analytical expressions and their derivations, it is best to treat this subject by examples based on numerical PSpice simulations.

Application Example 5.11: $P_{out} = 6$ kW Controlled Three-Phase Rectifier with Self-Commutated Switch

Figure 5.38 shows a three-phase rectifier with six diodes and one self-commutated switch. The load resistance is $R = 15\ \Omega$. All other data can be taken from the

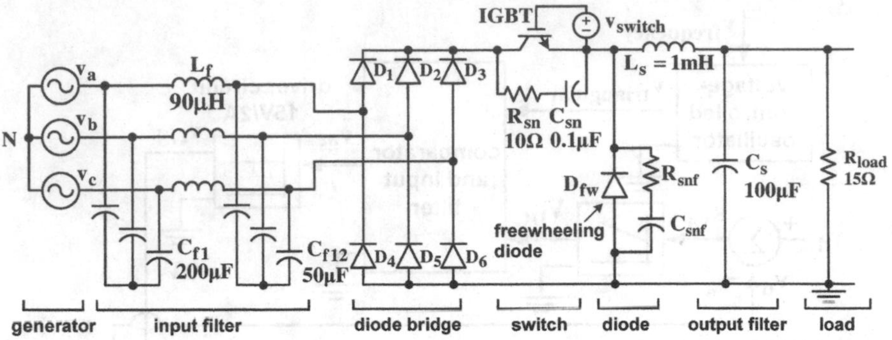

Fig. 5.38 Controlled three-phase rectifier with self-commutated switch

PSpice program. In this example calculated signals will be compared with measured ones. Note that the diode identifications of the bridge does not agree with the accepted standards, where the odd-numbered diodes are on top, and the even-numbered diodes are at the bottom half of the bridge.

The generator resistances and in-series-connected inductances are for the three phases: Ra = Rb = Rc = 0.1 Ω and La = Lb = Lc = 0.7 mH, respectively.

The filter capacitances and series-connected resistances next to the generator are: Cf1 = Cf2 = Cf3 = 200 μF and Rt1 = Rt2 = Rt3 = 4 Ω (not shown in Fig. 5.38), respectively.

The resistance between the generator neutral and the neutral of the filter next to the generator is Rtt = 1 GΩ (not shown in Fig. 5.38).

The filter inductances connected between the two filters (next to the generator and next to the diode bridge) are L_f = Lf1 = Lf2 = Lf3 = 90 μH.

The filter capacitances and series-connected resistances next to the diode bridge are: Cf12 = Cf22 = Cf32 = 50 μF and Ru1 = Ru2 = Ru3 = 2 Ω (not shown in Fig. 5.38), respectively.

The resistance between the generator neutral and the neutral of the filter next to the diode bridge is Rut = 1 GΩ (not shown in Fig. 5.38).

The diodes of the bridge including snubbers (not shown in Fig. 5.38) are:

D1 with model dio and snubbers Rsn1 = 20 Ω, Csn1 = 200 pF,
D2 with model dio and snubbers Rsn2 = 20 Ω, Csn2 = 200 pF,
D3 with model dio and snubbers Rsn3 = 20 Ω, Csn3 = 200 pF,
D4 with model dio and snubbers Rsn4 = 20 Ω, Csn4 = 200 pF,
D5 with model dio and snubbers Rsn5 = 20 Ω, Csn5 = 200 pF,
D6 with model dio and snubbers Rsn6 = 20 Ω, Csn6 = 200 pF, the series resistances of the diodes 1–6 are R1 = R2 = R3 = R4 = R5 = R6 = 10 mΩ.

The freewheeling diode has the model dfree and the snubber Rsnf = 10 Ω, Csnf = 100 pF.

The output filter has the parameters Ls = 1 mH, Cs = 100 μF with the series connected resistance RbL = 100 mΩ (not shown in Fig. 5.38).

The load resistance is Rload = 15 Ω.

The diode model of the bridge is:

.model dio d (is = 10p cjo = 100 p)

The diode model of the freewheeling diode is:

.model dfree d (is = 1n)

The NMOS model is:

```
.model mfet nmos(level=3 gamma=0 kappa=0 tox=100n rs=0 kp=20.87u l=2u w=2.9
+ delta=0 eta=0 theta=0 vmax=0 xj=0 uo=600 phi=0.6 vto=0 rd=0  cbd=200n pb=0.8
+ mj=0.5 cgso=3.5n cgdo=100p rg=0 is=10f)
```

Solution

The PSpice input program with .cir extension is listed below and the simulation plots are shown in Fig. 5.39:

```
*6kW controlled three-phase rectifier
*with self-commutated switch , duty cycle=
*50%, voltage amplitude per phase
*Vgen_ph=361V, frequency of generator voltage
*fgen=73Hz

va a st sin(0 361 73 0 0 0)
vb b st sin(0 361 73 0 0 240)
vc c st sin(0 361 73 0 0 120)
* Gating signal 3 kHz
Vmos gate 7 pulse(0 100 0 1u 1u 166u 334u)
* Generator impedances
La a aa .7m
Ra aa 1 .1
Lb b bb .7m
Rb bb 2 .1
Lc c cc .7m
Rc cc 3 .1
*Filter capacitors next to generator
Cf1 1 t1 200u
Cf2 2 t2 200u
Cf3 3 t3 200u
Rt1 t1 tt 4
Rt2 t2 tt 4
Rt3 t3 tt 4
Rtt tt st 1G
*Filter inductors between
*filter next to generator and that next to
*diode bridge
* continue program list in column #2
```

```
Lf1 1 11 90u
Lf2 2 22 90u
Lf3 3 33 90u
*Filter capacitors
*next to diode
* bridge
Cf12 11 u1 50u
Cf22 22 u2 50u
Cf32 33 u3 50u
Ru1 u1 ut 2
Ru2 u2 ut 2
Ru3 u3 ut 2
Rut ut st 1G
* Diode bridge
R1 b1 oben 10m
R2 b2 oben 10m
R3 b3 oben 10m
D1 11 b1 dio
Rsn1 11 sn1 20
Csn1 sn1 b1 200p
D2 22 b2 dio
Rsn2 22 sn2 20
Csn2 sn2 b2 200p
D3 33 b3 dio
Rsn3 33 sn3 20
Csn3 sn3 b3 200p
D4 b4 11 dio
Rsn4 b4 sn4 20
*continue program
   list in column #3
```

```
Csn4 sn4 11 200p
D5 b5 22 dio
Rsn5 b5 sn5 20
Csn5 sn5 22 200p
D6 b6 33 dio
Rsn6 b6 sn6 20
Csn6 sn6 33 200p
R4 0 b4 10m
R5 0 b5 10m
R6 0 b6 10m
* Output filter
Ls 7 last 1m
Cs last bL 100u
RbL bL 0 100m
* Load
Rload last 0 15
*MOSFET
mos oben gate 7 7 mfet
Rent oben ent 10
Cent ent 7 0.1u
* Freewheeling diode
Dfw 0 7 dfree
Rfw 7 fw 10
Cfw fw 0 100p
* Models for diodes
.model dio d(is=10p
+ cjo=100p)
.model dfree d(is=1n)
*continue program list
   below table
```

```
*Model for MOSFET
.model mfet nmos(level=3 gamma=0 kappa=0 tox=100n rs=0 kp=20.87u l=2u w=2.9
+ delta=0 eta=0 theta=0 vmax=0 xj=0 uo=600 phi=0.6 vto=0 rd=0  cbd=200n pb=0.8
+ mj=0.5 cgso=3.5n cgdo=100p rg=0 is=10f)
.tran 100u 50m 0 65u

.probe
.options abstol=10m chgtol=10m reltol=0.1 vntol=100m
.end
```

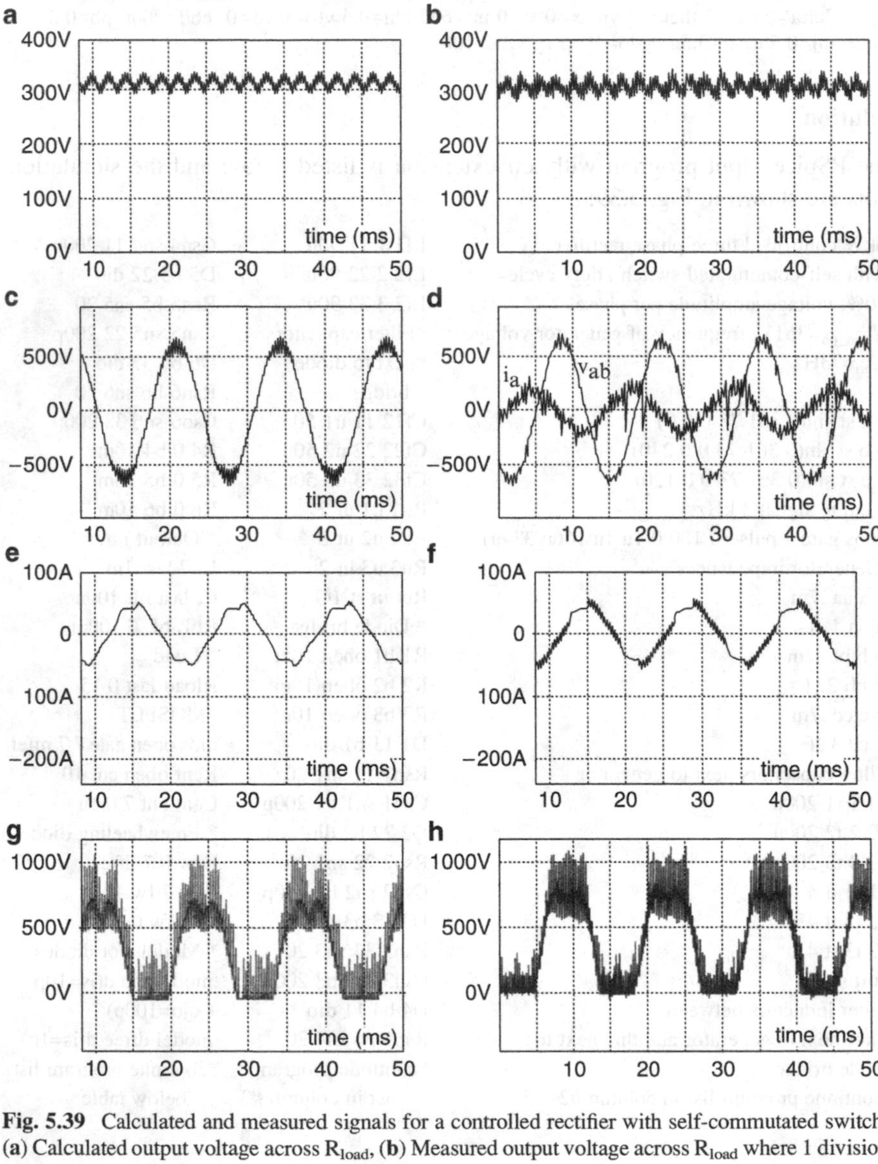

Fig. 5.39 Calculated and measured signals for a controlled rectifier with self-commutated switch: **(a)** Calculated output voltage across R_{load}, **(b)** Measured output voltage across R_{load} where 1 division

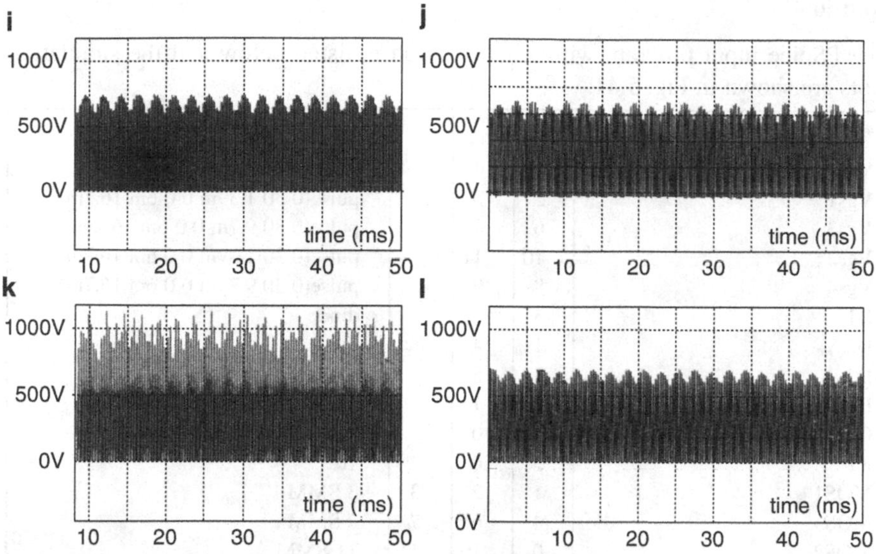

Fig. 5.39 (continued) corresponds to 5 ms, (**c**) Calculated input line-to-line voltage, (**d**) Measured input line-to-line voltage v_{ab} and phase current i_a where 1 division corresponds to 5 ms, the 500 V scale corresponds to 100 A for i_a , (**e**) Calculated phase current i_a at (generator inductance) $L_a = 700\ \mu H$, (**f**) Calculated phase current i_a at (generator inductance) $L_a = 200\ \mu H$, (**g**) Calculated reverse voltage of one diode, (**h**) Measured reverse voltage of one diode where 1 division corresponds to 5 ms, (**i**) Calculated reverse voltage of transistor, (**j**) Measured reverse voltage of transistor where 1 division corresponds to 5 ms, (**k**) Calculated reverse voltage of free-wheeling diode, (**l**) Measured reverse voltage of free-wheeling diode where 1 division corresponds to 5 ms

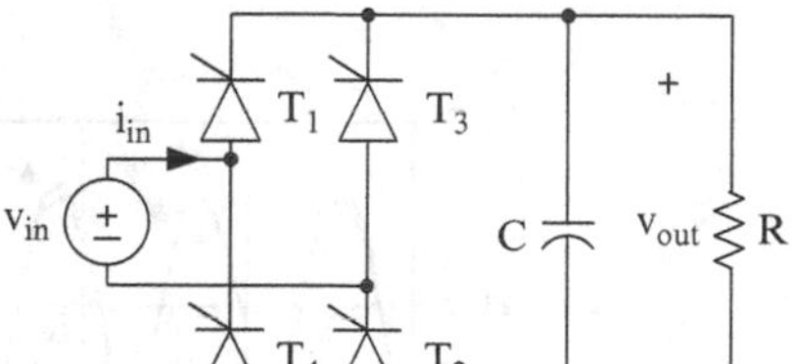

Fig. 5.40 Controlled single-phase rectifier with line-commutated switches

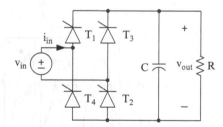

Application Example 5.12: Single-Phase, Bridge Thyristor Rectifier

For the circuit of Fig. 5.40 compute the input current i_{in}, and input and output voltages, where $R = 10\ \Omega$, $C = 100\ \mu F$, and firing angle $\alpha = 30^{\circ}$; all other data can-be found in the PSpice program. The large reverse-recovery currents (see Application Example 5.13) of the thyristors can be reduced by introducing small inductances in the range of nH between T_1 and T_4, as well as between T_3 and T_2 of Fig. 5.40.

Solution

The PSpice input program with .cir extension is listed below and the simulation plots are shown in Fig. 5.41:

*Single-phase thyristor rectifier				
Vs	1	4		sin(0 339.4 60 0)
Vgs1	2	3		pulse(0 30 1.39m 0 0 6m 16.7m)
Vgs3	6	7		pulse(0 30 9.7m 0 0 6m 16.7m)
Vgs2	10	11		pulse(0 30 1.39m 0 0 6m 16.7m)
Vgs4	8	9		pulse(0 30 9.73m 0 0 6m 16.7m)
D1	3	5		diode
D2	11	4		diode
D3	7	5		diode
D4	9	1		diode
C1	5	0		100u
R1	5	0		10
MOS1	1	2	3	3 SMM
MOS3	4	6	7	7 SMM
MOS2	0	10	11	11 SMM
MOS4	0	8	9	9 SMM
				* program list continues below table

```
.tran 60u 60m
.model SMM NMOS(level=3 gamma=0 delta=0 eta=0 theta=0 kappa=0 vmax=0
+xj=0 tox=100n uo=600 phi=0.6 rs=42.69m kp=20.87u L=2u w=2.9 vto=3.487
+rd=0.19 cbd=200n pb=0.8 mj=0.5 cgso=3.5n cgdo=100p rg=1.2 is=10f)
.model diode d(is=1e-12)
.options ITL4=75
.options ITL5=0
.options reltol=0.1
.probe
.end
```

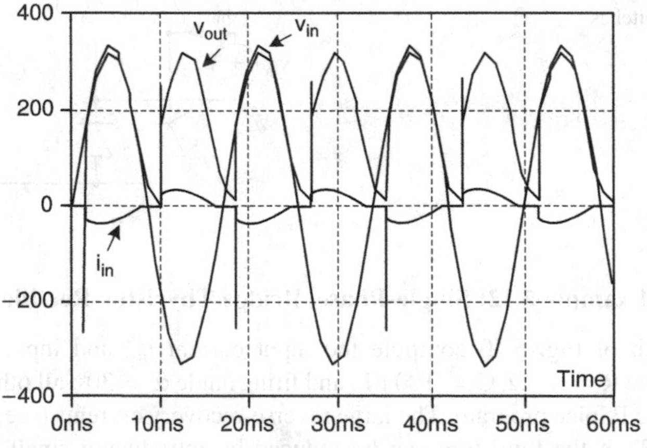

Fig. 5.41 PSpice plots of input voltage v_{in}, input current i_{in}, and output voltage v_{out} of Fig. 5.40

Application Example 5.13: Computation of Reverse-Recovery Current and Reverse-Recovery Time of Enhancement MOSFET

In the circuit of Fig. 5.42 the input voltage at the gate of MOSFET Q_2 consists of two pulses, each having an amplitude of 15 V and a duration (pulse width) of 15 μs; the gate resistance is $R_G = 100\,\Omega$. The time in between the two pulses is 15 μs as well. The rise and fall times (t_r and t_f) of the two pulses one may assume to be $t_r = t_f = 0.01$ μs. Simulate this circuit using PSpice and determine the magnitude $A_{reverse\text{-}recovery\ magnitude}$ and the duration $T_{reverse\text{-}recovery\ time}$ (see definition in Fig. 5.42) of the reverse-recovery current i_{DQ1} through enhancement MOSFET Q_1. Plot the gating currents i_{GQ1} and i_{GQ2}. The model of the MOSFETs Q_1, Q_2 and that of the inductor $L = 50\,\mu H$ are given below. Choose a maximum step size in the range of $\Delta t_{max} = 50$ ps.

MOSFET model:

```
.MODEL SMM NMOS(LEVEL=3 GAMMA=0 DELTA=0 ETA=0 THETA=0 KAPPA=0
+ VMAX=0 XJ=0 TOX=100N UO=600 PHI=0.6 RS=42.69M KP=20.87U L=2U W=2.9
+ VTO=3.487 RD=0.19 CBD=200N PB=0.8 MJ=0.5 CGSO=3.5N CGDO=100P RG=1.2
+ IS=10F)
inductor model:
L1 5 6 50U
```

Solution

The input PSpice .cir file is as follows:

```
*Computation of reverse-recovery current and reverse-recovery time
VIN 1 0 PULSE(0 15 100E-9 100E-9 15E-6 15E-6 30E-6)
RG 1 2 100
L1 5 6 50UH
MQ2 5 2 0 0 SMM
MQ1 6 5 55 SMM
VDD 6 0 DC 24V
*MOSFET model:
.MODEL SMM NMOS(LEVEL=3 GAMMA=0 DELTA=0 ETA=0 THETA=0 KAPPA=0
+ VMAX=0 XJ=0 TOX=100N UO=600 PHI=0.6 RS=42.69M KP=20.87U L=2U W=2.9
+ VTO=3.487 RD=0.19 CBD=200N PB=0.8 MJ=0.5 CGSO=3.5N CGDO=100P RG=1.2
+ IS=10F)
*TOTAL ITERATION LIMIT
.OPTIONS ITL%=0
*MAXIMUM POINTS ALLOWED
.OPTIONS LIMPTS=32000
*ITERATION LIMIT PER POINT
.OPTIONS ITL4=100
.TRAN 0.001u 200U
*RELATIVE TOLERANCE
.OPTIONS RELTOL=0.001
.probe
.end
```

Figure 5.43 shows the current i_{DQ1} through MOSFET Q_1 during 200 μs and Fig. 5.44 depicts the gating voltage v_{in} and the current i_{DQ1} through MOSFET Q_1

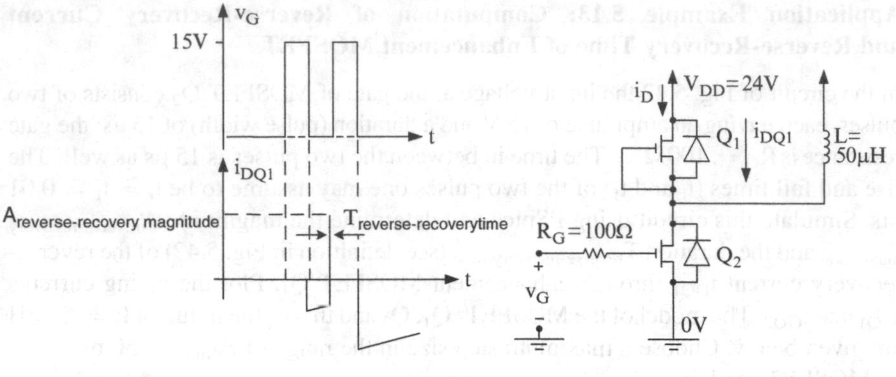

Fig. 5.42 Set up for computation/measurement of reverse-recovery current magnitude $A_{\text{reverse-recovery magnitude}}$ and reverse-recovery time $T_{\text{reverse-recovery time}}$

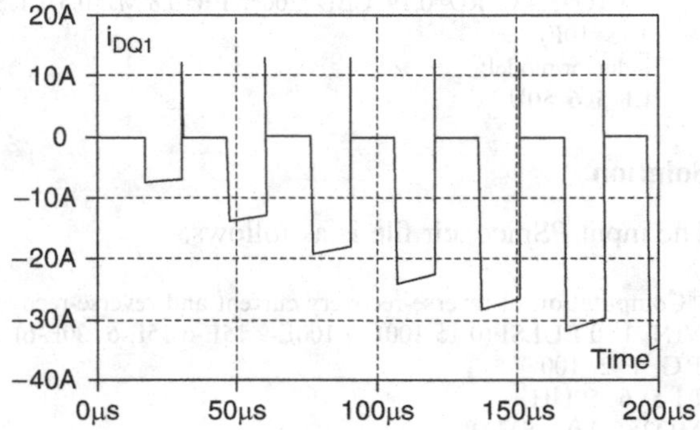

Fig. 5.43 Current i_{DQ1} through MOSFET Q_1 during 200 μs

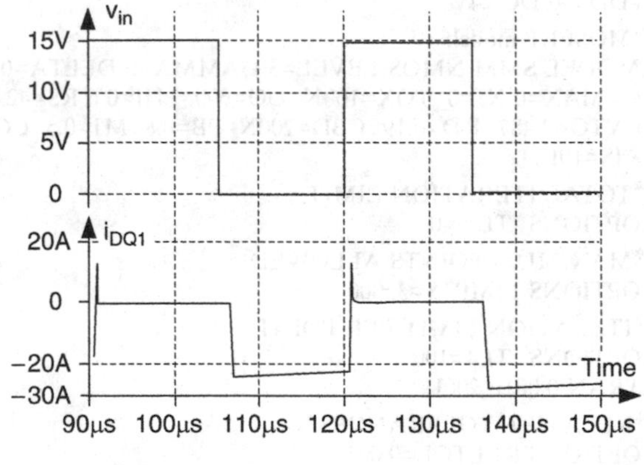

Fig. 5.44 Gating voltage v_{in} and the current i_{DQ1} through MOSFET Q_1 from 90 μs to150 μs

from 90 μs to150 μs. Figure 5.45 shows the reverse-recovery magnitude $A_{\text{reverse-recovery magnitude}}$ and the reverse-recovery time $T_{\text{reverse-recovery time}}$, and Fig. 5.46 represents the current in the gate of Q_1 and that of Q_2. From these plots one obtains $A_{\text{reverse-recovery magnitude}} = 12.5$ A, and $A_{\text{reverse-recovery time}} = 0.204$ μs.

Fig. 5.45 Reverse-recovery magnitude $A_{reverse\text{-}recovery\ magnitude}$, and reverse-recovery time $T_{reverse\text{-}recovery\ time}$ of MOSFET Q_1

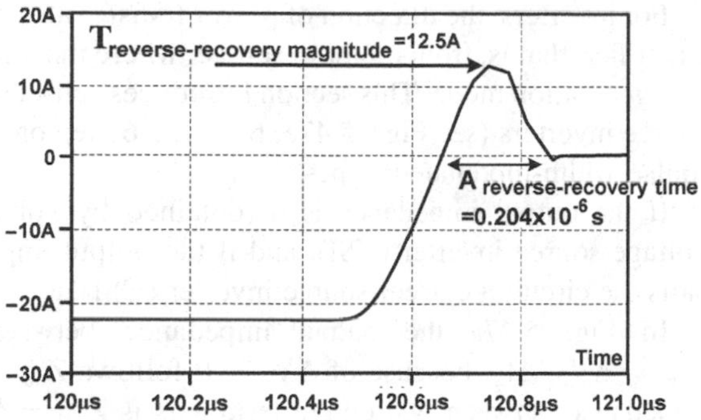

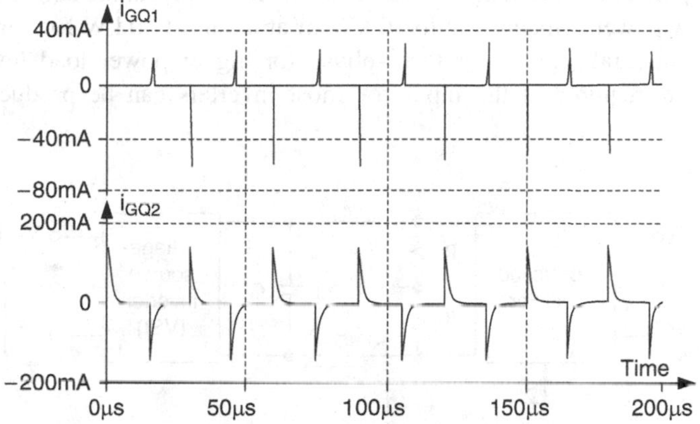

Fig. 5.46 Currents i_{GQ1} and i_{GQ2} of gates of Q_1 and Q_2, respectively

5.6 DC-to-AC Converters (Inverters)

The purpose of an inverter is to convert power from a direct voltage or direct current source to a controllable-frequency output with controllable voltage or current amplitude such that the ratio (see Chap. 10)

$$\frac{E}{f} = 4.44 \cdot N \cdot \phi_{max} = 4.44 \cdot N \cdot A \cdot B_{max} = \text{const.} \tag{5.96}$$

is satisfied, where E is the AC induced voltage within an electric machine, and f is the frequency of the applied voltages or currents. This relation must be satisfied in order to maintain the flux density within the motor/generator at its rated maximum flux density value B_{max_rated}. Failure to maintain rated maximum flux density results in burn-out of the machine if $B_{max} > B_{max\_rated}$ or low torque production if $B_{max} < B_{max\_rated}$. As a rule of thumb

$$0.3B_{max\_rated} \leq B_{max} \leq 1.1B_{max\_rated} \tag{5.97}$$

For inverters, the direction of power transfer also may be reversible operating as a rectifier; that is, from motor to source, where the motor is operating as a generator in regeneration mode. This section introduces self-commutated voltage and current-source inverters (see Figs. 5.47a, b) of the 6-step or 6-pulse type, and of the PWM (pulse-width-modulated) types.

If the output impedance is 0 (obtained by voltage control) one speaks of a voltage-source inverter (VSI), and if the output impedance is infinitely large one calls the circuit a current-source inverter (CSI).

In Fig. 5.47a the output impedance between two output terminals is $Z_{out} = \Delta V_2/\Delta I_2$; because of $\Delta V_2 = 0$ follows $Z_{out} = 0$. In Fig. 5.47b the output impedance between two output terminals is $Z_{out} = \Delta V_2/\Delta I_2$; because of $\Delta I_2 = 0$ follows $Z_{out} = \infty$.

In most applications the source of power is an AC supply, either single-phase for relatively low power demands (up to 10 kW or at the most 50 kW for very special applications in rural regions) or three-phase for higher power load levels. The voltage that is required at the input for most inverters can be produced using

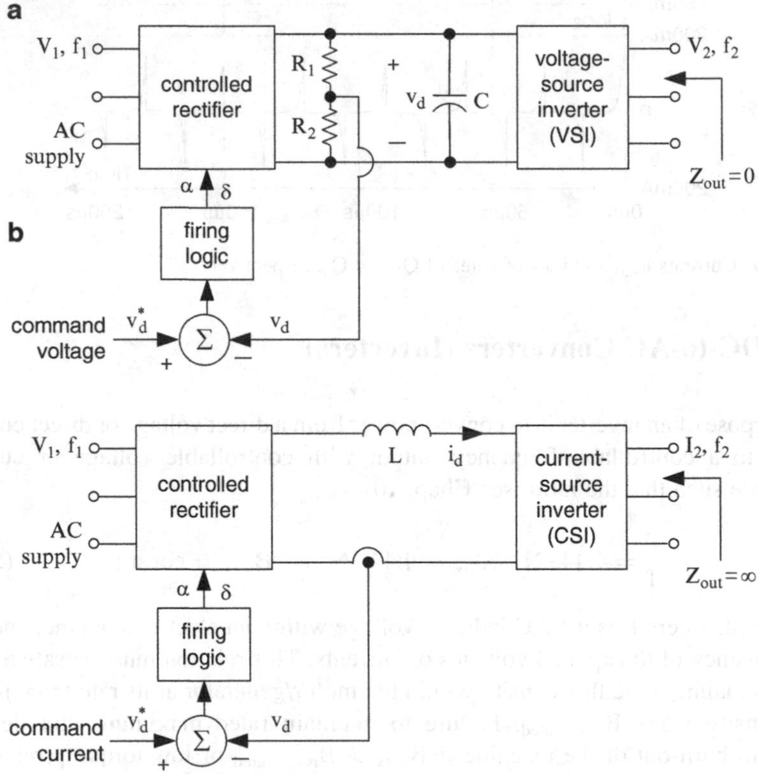

Fig. 5.47 Control circuits of (**a**) voltage-source inverter (VSI) and (**b**) current-source inverter (CSI)

rectifiers, as discussed in Sect. 5.5. In some instances a direct-voltage supply may be available, such as from batteries or supercapacitors in an electric vehicle, where regeneration is possible if battery/supercapacitor is not fully charged and from fuel cells or solar cells in a photovoltaic array, where regeneration is not possible.

5.6.1 A Simple (6-Step or 6-Pulse) Voltage-Source Inverter [7, 8]

A major concern is the provision of three-phase power at variable induced voltage E and frequency f to supply induction and synchronous/permanent-magnet motors maintaining the ratio

$$\frac{E}{f} = 4.44 \cdot N \cdot A \cdot B_{max} = \text{const.} \tag{5.98}$$

Figures 5.48a–c show the circuit of a simple three-phase voltage-source inverter, the so-called 6-step inverter. This system uses six self-commutated switches S_1 to S_6 to produce a variable-frequency output voltage, where the amplitude of the voltage is fixed. Therefore, it requires a controllable direct-voltage source (controlled rectifier) feeding the inverter to provide the variable alternating-voltage output of the inverter which satisfies the constraint of (5.98).

In Figs. 5.48a–c the controllable direct-voltage source is a controlled rectifier operating from AC supply. Typically a large capacitor C is connected across the direct-voltage link between the rectifier and the inverter. This can be considered as the approximate equivalent of the source voltage represented in the load circuits of the rectifiers in Sect. 5.5. In this inverter, the switches are turned on in the sequence in which they are numbered. The gate signals holding the switches in their ON condition are shown in Figs. 5.48a–c. Each is held (turned) on for just under one-half cycle ($\omega t \lesssim 178°$) and then is turned off. It will be seen that this normally causes three switches to be conducting at any time: two connected to one source terminal and one to the other source terminal (this, however, does not apply during the commutation interval where a total of four switches are simultaneously conducting). When any switch is on, it and the diode connected in anti-parallel with it constitute an effective short circuit.

Thus, with two switches in the ON condition connected to the source terminal, there is a short-circuit between the two corresponding output or load phases (see Figs. 5.48a–c).

The waveforms of the line-to-line output voltages of the inverter are illustrated in Figs. 5.48a–c. These consist during each half-cycle of a rectangular wave of voltage equal in magnitude to the direct-link voltage v_d and has a duration of about 120° (electrical). The load impedances in Figs. 5.48a–c are $Z_a = Z_b = Z_c = R + j\omega L$ and the neutral n of the three-phase load is virtually connected to point 0 of the output voltage of the controlled rectifier. The line-to line-voltage $v_{ab} = v_{an} - v_{bn} = v_{ao} - v_{bo}$, where 0 indicates the virtual ground or the imaginary middle point 0 of the DC source of Figs. 5.48a–c.

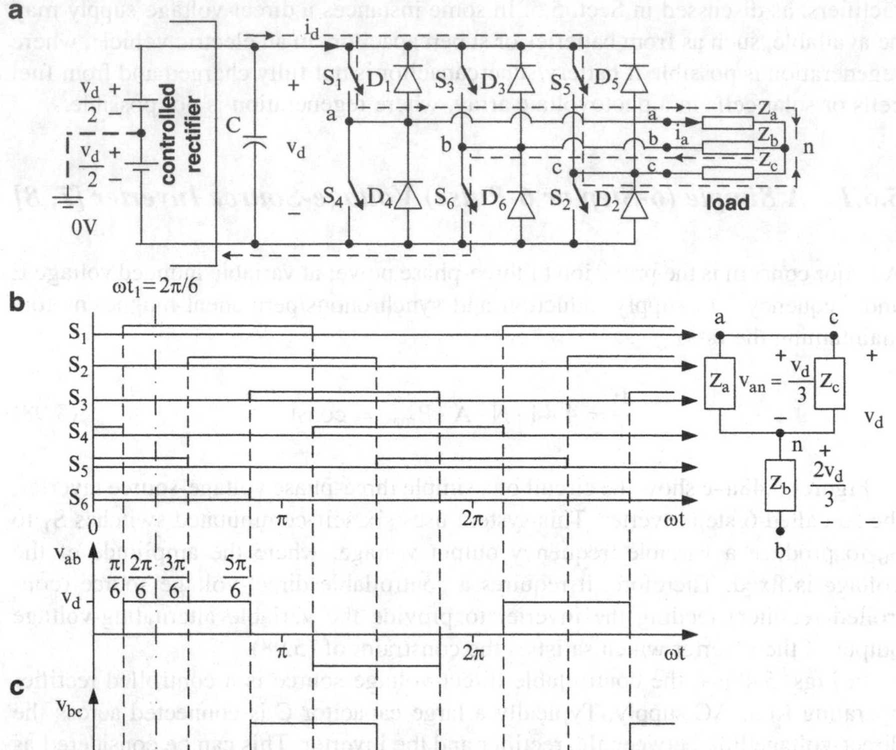

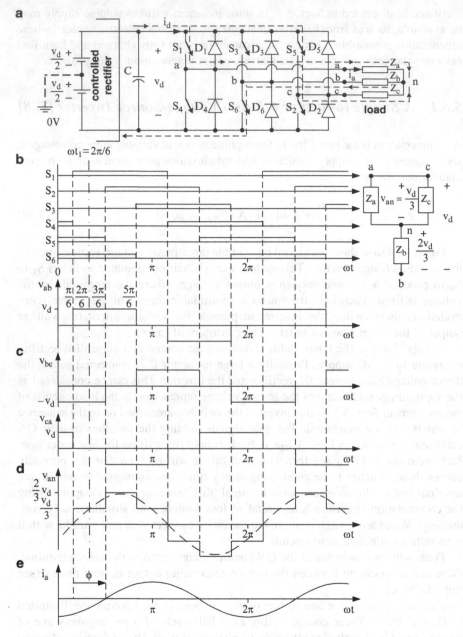

Fig. 5.48 (a) 6-pulse, voltage-source inverter with wave forms at time $t = t_1 = \frac{2\pi}{6\omega}$, where phases a and c are short-circuited. $Z_a = Z_b = Z_c = R + j\omega L$ with $R \ll \omega L$

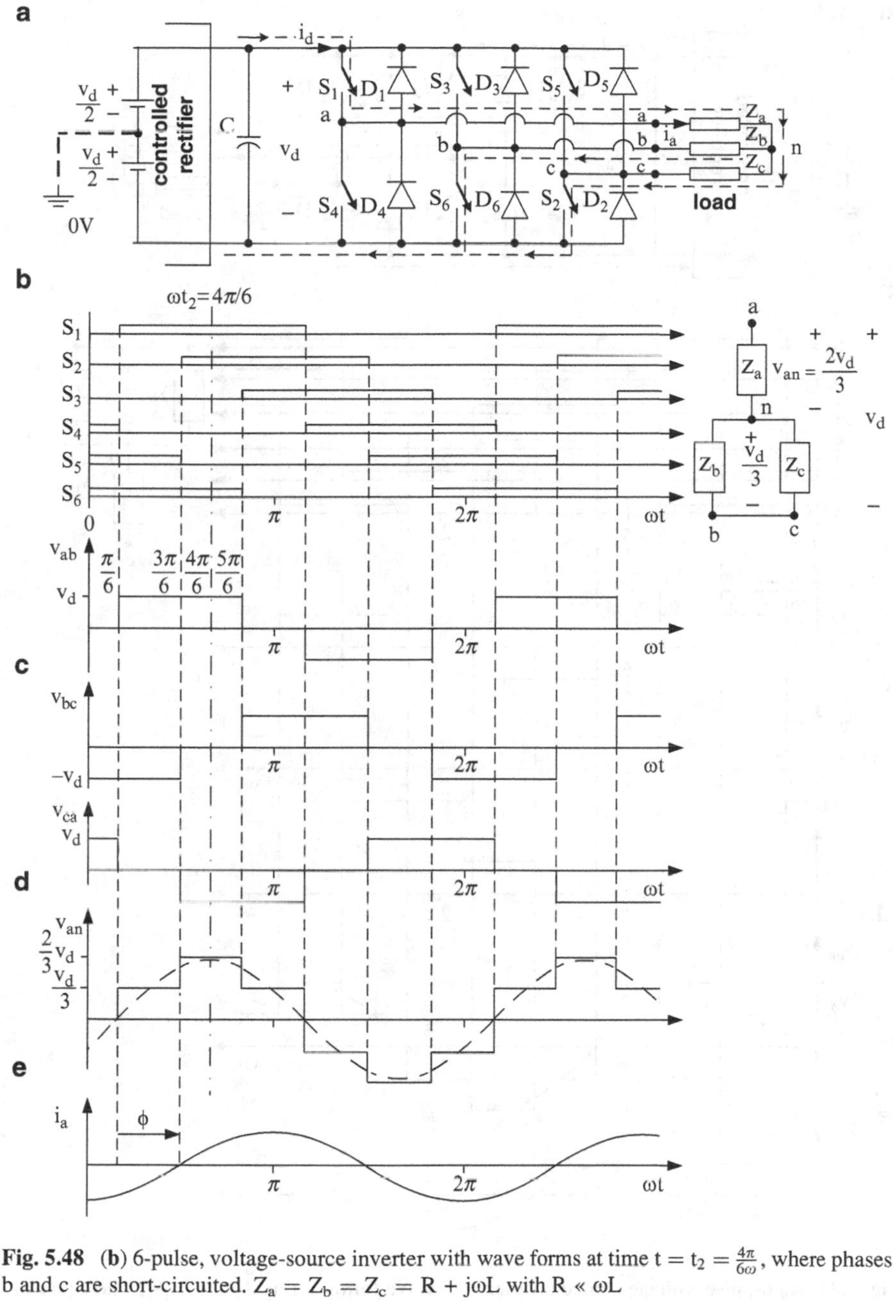

Fig. 5.48 (**b**) 6-pulse, voltage-source inverter with wave forms at time $t = t_2 = \frac{4\pi}{6\omega}$, where phases b and c are short-circuited. $Z_a = Z_b = Z_c = R + j\omega L$ with $R \ll \omega L$

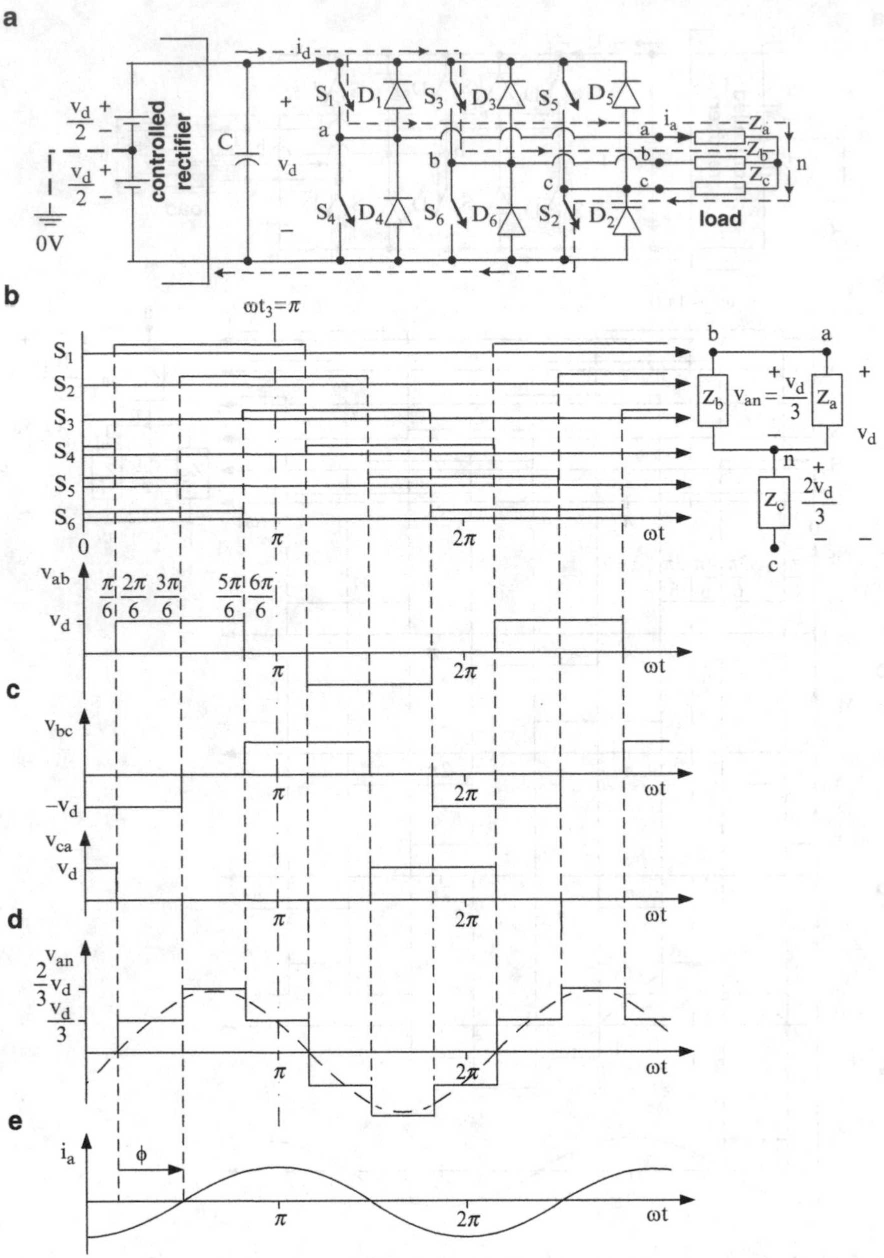

Fig. 5.48 **(c)** 6-pulse, voltage-source inverter with wave forms at time $t = t_3 = \frac{6\pi}{6\omega}$, where phases b and a are short-circuited. $Z_a = Z_b = Z_c = R + j\omega L$ with $R \ll \omega L$

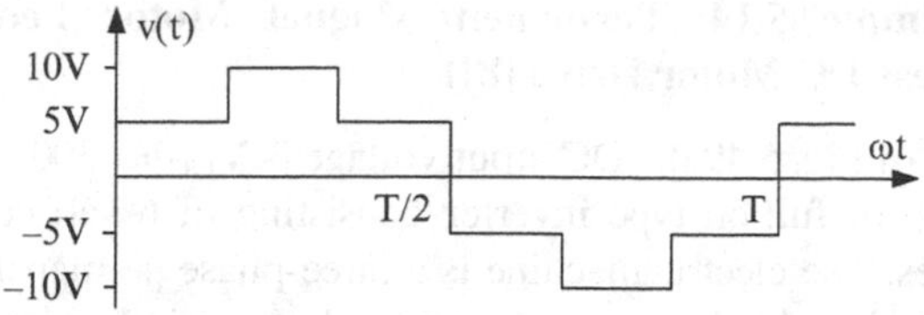

Fig. 5.48 (d) 6-step voltage waveform

In Figs. 5.48a–c the three-phase voltages have a 6-step waveform while the phase currents are nearly sinusoidal. This can be explained based on Fig. 5.48d and the fact that the 6-step inverter supplies an inductive/resistive load.

Consider the 6-step voltage v(t) of Fig. 5.48d. A Fourier analysis results in the complex representation

$\tilde{V}_h = -j\frac{10}{h\,\pi}\left(1 + \cos\left\{\frac{h\pi}{3}\right\}\right)$, with h = odd but no triplen (multiples of three) harmonics.

For the fundamental (h = 1) one obtains

$$\tilde{V}_1 = -j\frac{10}{\pi}(1.5) \tag{5.99}$$

for h = 5

$$\tilde{V}_5 = -j\frac{10}{5\,\pi}(1.5) \tag{5.100}$$

for h = 7

$$\tilde{V}_7 = -j\frac{10}{7\pi}(1.5) \tag{5.101}$$

If the load impedance is assumed to be $Z_L = R + jh\omega L$ and $R \ll h\omega L$ – which is true for most rotating machines-then the harmonic (h = 5, 7, 11, 13, ...) phase currents are in relation to the fundamental current (h = 1):

$$\tilde{I}_h = \frac{\tilde{V}_h}{jh\omega L} = \frac{-j\frac{15}{h\pi}}{jh\omega L} = \frac{-15}{(h)^2\pi\omega L} \propto \frac{1}{h^2}. \tag{5.102}$$

As the harmonic currents are inversely proportional to the square of the harmonic order h the resulting harmonic currents and associated ohmic losses within the rotating machine will be very small. However, due to the six-step phase voltages the harmonic iron-core losses will be increased [9].

Application Example 5.14: Permanent-Magnet Motor Fed by Six-Pulse Inverter (Brushless DC Motor Drive [8])

In the drive circuit of Fig. 5.49 the DC input voltage is $V_{DC} = 300$ V. The inverter is a 6-pulse or 6-step or full-on type inverter consisting of 6 self-commutated (e.g., MOSFET) switches. The electric machine is a three-phase permanent-magnet motor represented by an induced voltage, resistance and leakage inductance per phase. The induced voltage in the stator winding (phase A) of the permanent-magnet motor is

$$e_A = 160\sin(\omega t + \theta)[\text{V}], \tag{5.103}$$

where $\omega = 2\pi f_1$ and $f_1 = 1{,}500$ Hz; correspondingly,

$$e_B = 160\sin(\omega t + 240° + \theta)[\text{V}], \tag{5.104}$$

$$e_C = 160\sin(\omega t + 120° + \theta)[\text{V}]. \tag{5.105}$$

The resistance R_1 and the leakage inductance $L_{1\ell}$ of one of the motor phases are 0.5 Ω and 50 μH, respectively. The leakage inductance is that between two stator phases. The relation $V_{rms}^{inverter} = m \cdot V_{DC}^{inverter}/(2 \cdot \sqrt{2})$·(see Sect. 5.4) must be satisfied where $\sqrt{2}V_{rms}^{inverter} \approx 160\text{V}$, $V_{DC}^{inverter} = 300\text{V}$, and m is somewhat larger than 1, which is acceptable for full-on mode or six-step operation. The magnitude of the gating voltages of the 6 MOSFETs is $V_{Gmax} = 15$ V; the gating signals with their sequence are shown in Fig. 5.50. Note that the sequence of the induced voltages and that of the gating signals must be the same.

Using PSpice, compute and plot the current within MOSFET Q_{AU} (e.g., i_{QAU}) and the motor current of phase A (e.g., i_{MA}) for the angles $\theta = 0°$, $\theta = \pm 30°$, and $\theta = \pm 60°$. These phase angles permit to control the torque of the brushless DC motor drive while a change of the frequency f_1 permits to control the speed of the drive. Both controls are independent from one another. Note that the gating signal frequency of the MOSFETs corresponds to the frequency f_1, that is, full-on mode

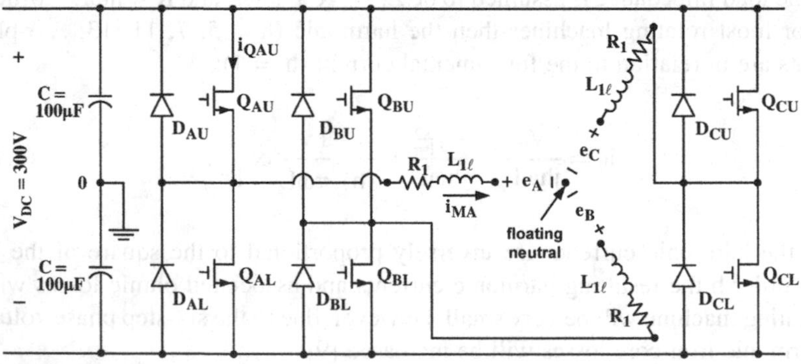

Fig. 5.49 Circuit of brushless DC motor

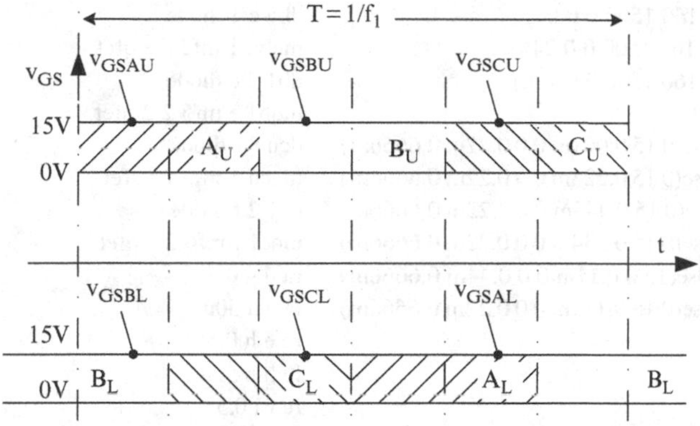

Fig. 5.50 Sequence of gating signals for brushless DC motor in 6-step (6-pulse or full-on mode) operation

operation exists. For switching sequence see Fig. 5.50, in this figure the upper (e.g., v_{GSAU}) and lower (e.g., v_{GSBL}) have been plotted separately in order to ease the understanding of the gating signals which overlap. The models of the enhancement metal-oxide semiconductor field effect transistors and those of the freewheeling diodes are as follows:

MOSFETs:

```
.MODEL SMM NMOS(LEVEL=3 GAMMA=0 DELTA=0 ETA=0 THETA=0
+KAPPA=0 VMAX=0 XJ=0 TOX=100N UO=600 PHI=0.6 RS=42.69M KP=20.87U
+L=2U W=2.9 VTO=3.487 RD=.19 CBD=200N PB=.8 MJ=.5 CGSO=3.5N
+CGDO=100P RG=1.2 IS=10F)
diodes:
.MODEL D1N4001 D(IS=1E-12)
```

Note

1. The step size for the numerical solution should be in the neighborhood of $\Delta t = 0.05\ \mu s$.
2. To eliminate computational transients due to inconsistent initial conditions, compute at least three periods of all quantities and plot the last (third) period of i_{QAU} and i_{MA} for all five cases where θ assumes the values given above.

Solution

The PSpice input program with .cir extension is listed below and the simulation plots are shown in Fig. 5.51:

```
* Brushless DC motor drive
* DC power supply
vplus 1 0 150
vminus 2 0 -150
*Induced voltages
```

```
* 6-step (6 pulse) inverter bridge
dau d 1 diode
mqau 1 mf1 d d qfet
dal 2 d diode
mqal d mf4 2 2 qfet
```

```
vea a st sin(0 160 1500 0 0 0)
veb b st sin(0 160 1500 0 0 240)
vec c st sin(0 160 1500 0 0 120)
*Gating signals
vau mf1 d pulse(0 15 0.001m 0 0 0.22m 0.6666m)
vbu mf2 e pulse(0 15 0.223m 0 0 0.22m 0.6666m)
vcu mf3 f pulse(0 15 0.445m 0 0 0.22m 0.6666m)
val mf4 2 pulse(0 15 0.334m 0 0 0.22m 0.6666m)
vbl mf5 2 pulse(15 0 0.11m 0 0 0.44m 0.6666m)
vcl mf6 2 pulse(0 15 0.112m 0 0 0.22m 0.6666m)
*Capacitors
c1 0 1 100u
c2 0 2 100u
* program list is continued in second column
dbu e 1 diode
mqbu 1 mf2 e e qfet
dbl 2 e diode
mqbl e mf5 2 2 qfet
dcu f 1 diode
mqcu 1 mf3 f f qfet
dcl 2 f diode
mqcl f mf6 2 2 qfet
ra d g 0.5
la g a 50u
rb e h 0.5
lb h b 50u
rc f i 0.5
lc i c 50u
*program list is continued below table

* Mosfet model
.model qfet nmos(level=3 gamma=0 kappa=0 tox=100n rs=42.69m kp=20.87u l=2u w=2.9
+ delta=0 eta=0 theta=0 vmax=0 xj=0 uo=600 phi=0.6
+ vto=3.487 rd=0.19 cbd=200n pb=0.8 mj=0.5 cgso=3.5n cgdo=100p rg=1.2 is=10f)

*diode model
.model diode d(is=1p)
.options abstol=1m reltol=10m vntol=10m chgtol=1m
.tran 0.05u 2m uic
.probe
.end
```

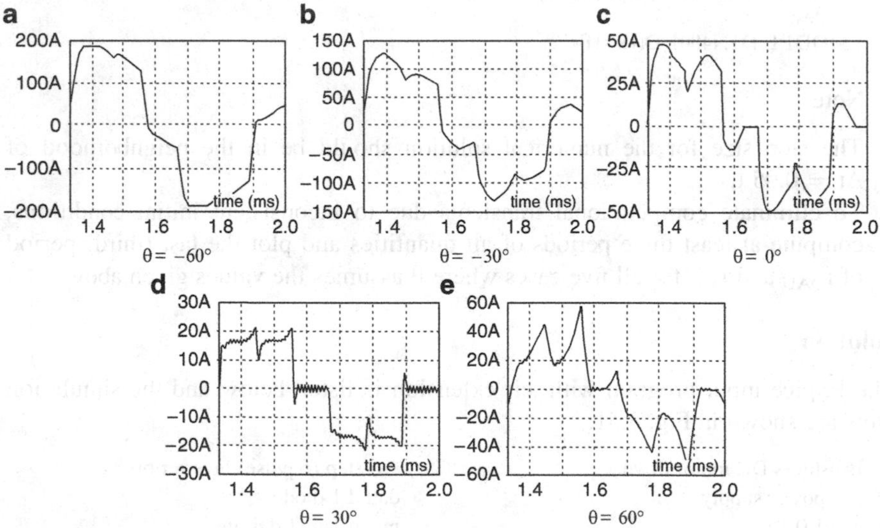

Fig. 5.51 Current waveforms of brushless DC motor for different phase angles θ controlling the torque of the drive

5.6.2 *Pulse-Width-Modulated (PWM) Voltage-Source Inverter*

The most common PWM technique for voltage source inverters is sinusoidal pulse-width modulation. In the following the principle of PWM will be explained. Three-phase reference voltages (or currents for a current-controlled voltage source inverter) v_a, v_b, and v_c of variable amplitude A are compared in three separate comparators with a common (isosceles) triangular carrier wave v_T of a fixed amplitude A_m, as shown in Fig. 5.52b. The outputs of comparators 1, 2, and 3 form the control signals for the three legs (phases) of the inverter formed by switch pairs (S_1, S_4), (S_3, S_6) and (S_5, S_2), respectively (see Fig. 5.52a). Let us consider the operation of the pair (S_1, S_4) which controls the voltage of the machine phase a with respect to the imaginary middle point 0 of the DC source of Figs. 5.48a–c. This is explained in Fig. 5.52b where the reference wave v_a and the carrier wave v_T are drawn on a common time axis for a positive half-cycle of v_a. Switch S_1 receives the control signal when $v_a > v_T$, and switch S_4 receives it when $v_a < v_T$. The resultant waveform of v_{ao} is shown in Fig. 5.52b. The waveforms of Fig. 5.52b are drawn for the case when a cycle of the reference wave consists of 12 cycles of the triangular wave. One can similarly draw voltages v_{bo} and v_{co} by considering the operation of switch pairs (S_3, S_6) and (S_5, S_2), respectively. The modulation is called sinusoidal PWM because the pulse width is a sinusoidal function of its angular position in the cycle. The modulation is also known as triangulation or PWM with natural sampling. The line voltage v_{ab} is obtained by subtracting v_{bo} from v_{ao}. Similarly the line voltages v_{bc} and v_{ca} are obtained. The line voltage waveform v_{ab} when each cycle of the reference wave has six cycles of triangular wave is shown in Fig. 5.53.

The frequency of the fundamental component of the motor terminal voltage is the same as that of the reference sinusoidal voltages. Hence, the frequency of the motor voltage can be changed by changing the frequency of the reference voltages.

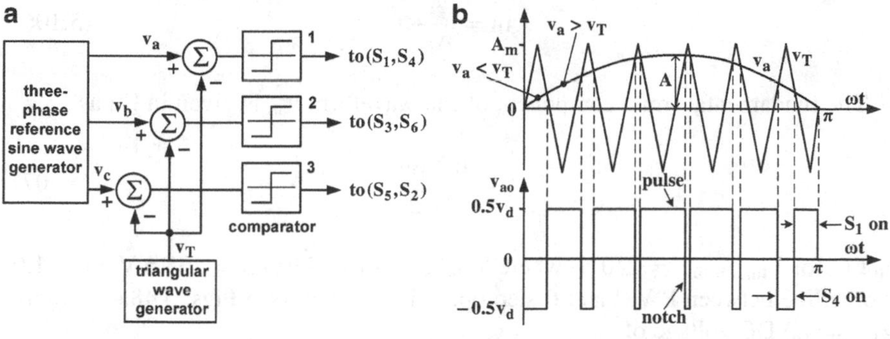

Fig. 5.52 Principle of sinusoidal pulse-width modulation: (**a**) generation of control signals for the circuit of Figs. 5.48a–c when operated in PWM mode; (**b**) modulated waveform for 12 switching cycles per fundamental period

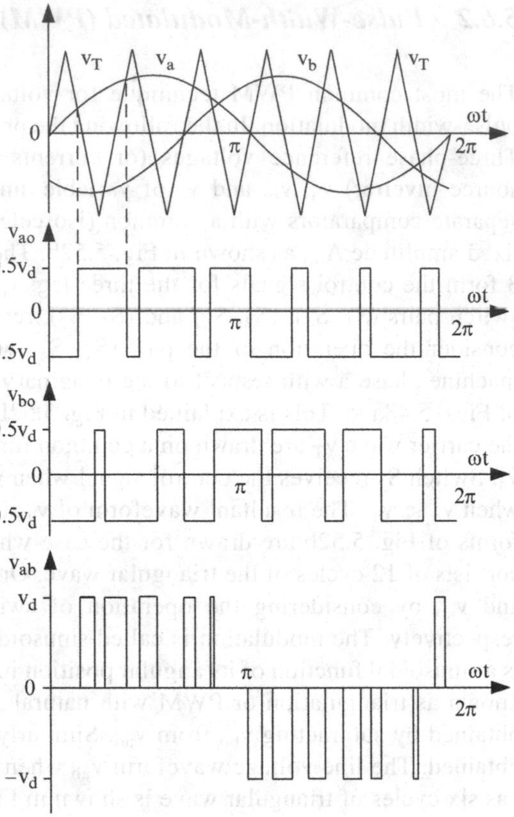

Fig. 5.53 Sinusoidal pulse-width modulation where $v_{ab} = (v_{ao} - v_{bo})$

The ratio of the amplitude of the reference wave (A) to that of the carrier wave (A_m) is called the modulation index m. Thus

$$m = \frac{A}{A_m}. \tag{5.106}$$

The fundamental (rms) component of the waveform v_{ao} is given in [7] as

$$V_1 = \frac{mV_{DC}}{2\sqrt{2}}, \tag{5.107}$$

that is, for $V_{line\text{-}to\text{-}line} = 240$ V, where $V_{phase} = V_1 = 240/\sqrt{3} = 138.6$ V, m = 1.0 (borderline between PWM and 6-step operation) a total (see Figs. 5.48a–c, where $V_{DC} = v_d$) DC voltage of

$$V_{DC} = \frac{V_1 2\sqrt{2}}{m} = 391.9 \text{ V} \tag{5.108}$$

is required. Unfortunately, this is not quite correct because references [6, 10, 11] teach us that at about 400 V_{DC} only leading (inductive-load operation for generator notation, that is, the inverter absorbs reactive power) and unity power factor operation can be achieved, requiring less DC voltage than lagging (capacitive-load operation for generator notation, that is, inverter supplies reactive power) power factor operation. For lagging power factor operation a substantially larger DC voltage is required in part due to the (di/dt) switching voltages, and the fundamental current-voltage relations are governed by the phasor diagram [9].

Overall one can see that the fundamental voltage increases linearly with m until m = 1 (that is, when the amplitude of the reference wave becomes equal to that of the carrier wave). For m > 1.0, the number of pulses in v_{ao} becomes less and the modulation ceases to be sinusoidal PWM. The wave form v_{ao} and the resulting current contain harmonics which are odd and even [6, 10] multiples of the carrier frequency f_c (that is, f_c, $2f_c$, $3f_c$, $4f_c$, $5f_c$, and so on). The existence of even switching harmonics in the current does not appear to be common knowledge [6, 10].

Application Example 5.15: Current-Controlled, Voltage-Source PWM Inverter [6, 10, 11]

In many renewable energy applications, e.g., solar photovoltaic power plants or in most wind power plants supplying electric power to the utility system, an inverter must be connected to the public utility system, and operate at unity-power factor or any other leading/lagging power factor supplying electric energy at phase-voltage levels of 120 V, 240 V, 480 V or higher voltages. Figure 5.54 [6, 10] shows such an inverter which can be connected to the three-phase utility system: the circuit

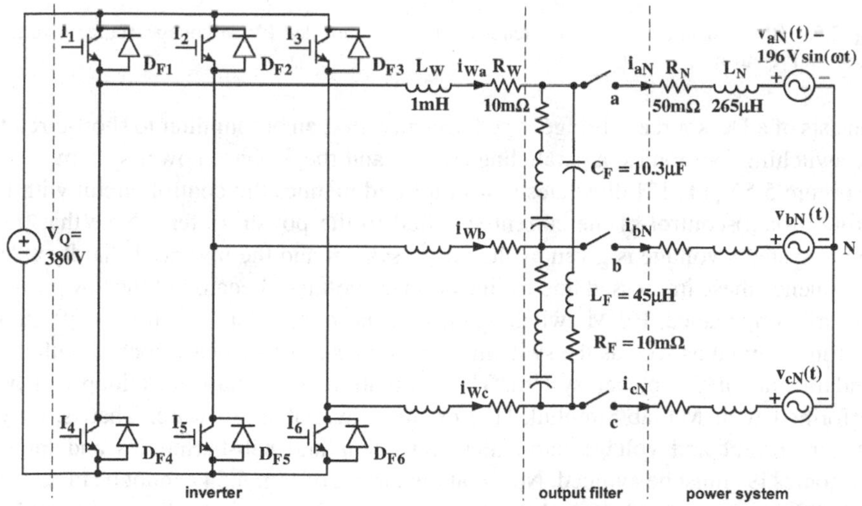

Fig. 5.54 Current-controlled PWM voltage-source inverter feeding power into utility system with $V_Q = 380\ V = V_{suppl}$

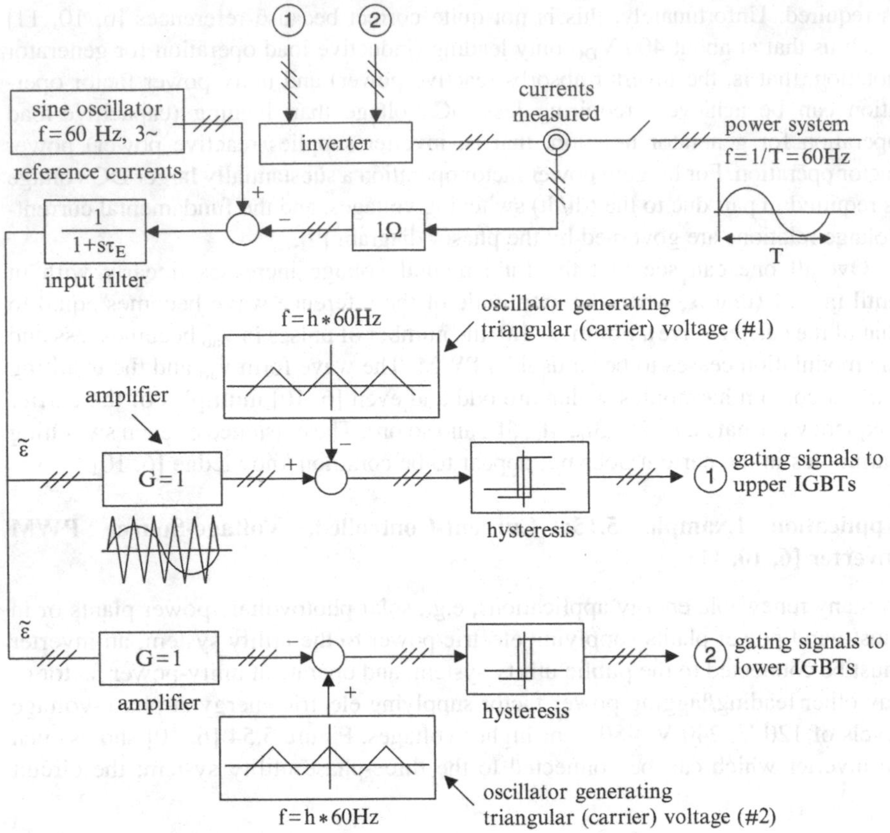

Fig. 5.55 Block diagram of control circuit for current-controlled PWM voltage–source inverter based on P-control

consists of a DC source, a bridge-type IGBT inverter, an output filter to short-circuit the switching harmonics, a paralleling switch, and the 3-phase power system.

Figure 5.55 [11, 12] illustrates in a simplified manner the control circuit with P (proportional)-control of the current supplied to the power system. Note that the power systems voltage is given by the utility system and the inverter is hardly able to influence these impressed line-to-line or phase voltages because of the low power system's impedance. PWM switching-current harmonics must be limited [9] and the fundamental as well as the switching harmonics must be phase-locked with the fundamental of the power system. The simulation of a phase-lock loop can be performed with Matlab/Simulink [13] or other available software. That is, only integer current and voltage harmonics can occur and sub-harmonics and inter-harmonics [9] must be avoided. Note that the circuit of Fig. 5.54 cannot be modeled with PSpice because the number of electrical nodes exceeds the capability of PSpice where the maximum number of nodes is 64. To reduce the number of nodes so that PSpice can be used it is recommended to delete the output filter between inverter and paralleling switch. The PSpice solutions of this application example take a few minutes of computation time; this time depends on the speed of the computer used. If the professional version of Spice is available then the entire circuit including the output filter can be analyzed.

Solution

The PSpice input program with .cir extension is listed below with a P-control and the simulation plots are shown in Figs. 5.56a, b:

```
*Current-controlled PWM voltage-source
* inverter with P-controller
*DC voltage supply
Vsuppl 2 0 380
* IGBT switches
msw1 2 11 10 10 qfet
dsw1 10 2 diode
msw2 2 21 20 20 qfet
dsw2 20 2 diode
msw3 2 31 30 30 qfet
dsw3 30 2 diode
msw4 10 41 0 0 qfet
dsw4 0 10 diode
msw5 20 51 0 0 qfet
dsw5 0 20 diode
msw6 30 61 0 0 qfet
dsw6 0 30 diode
L_W1 10 15 1m
L_W2 20 25 1m
L_W3 30 35 1m
* Resistors or voltage sources required
*to measure current resistors used as shunt
R_W1 15 16 10m
R_W2 25 26 10m
R_W3 35 36 10m
* Voltages represent reference currents
vref1 12 0 sin(0 56.6 60 0 0 0)
vref2 22 0 sin(0 56.6 60 0 0 -120)
vref3 32 0 sin(0 56.6 60 0 0 -240)
*Voltages represent load currents, measured
* with shunts
eout1 13 0 15 16 100
eout2 23 0 25 26 100
eout3 33 0 35 36 100
*Error signals are (vref – eout)
rdiff1 12 13a 1k
rdiff2 22 23a 1k
rdiff3 32 33a 1k
cdiff1 12 13a 1u
cdiff2 22 23a 1u
cdiff3 32 33a 1u
vtria1 5 0 pulse(-10 10 0 86.5u 86.5u
+ 0.6u 173.6u)
* Gating signals for upper MOSFETs
* as result of comparison between
* triangular waveform and error signal
xgs1 14 5 11 10 comp
xgs2 24 5 21 20 comp
xgs3 34 5 31 30 comp
* Gating for lower MOSFETs
egs4 41 0 poly(1) (11,10) 50 -1
egs5 51 0 poly(1) (21,20) 50 -1
egs6 61 0 poly(1) (31,30) 50 -1
*Filter removed, as node limit is 64
* rfi1 16 15b 0.1
*rfi2 26 25b 0.1
* rfi3 36 35b 0.1
* lfi1 15b 15c 45u
*lfi2 25b 25c 45u
*lfi3 35b 35c 45u
*cfi1 15c 26 10.3u
*cfi2 25c 36 10.3u
*cfi3 35c 16 10.3u
* Power system parameters
RM1 16 18 50m
LM1 18 19 265u
Vout1 19 123 sin(0 196 60 0 0 -30)
RM2 26 28 50m
LM2 28 29 265u
Vout2 29 123 sin(0 196 60 0 0 -150)
RM3 36 38 50m
LM3 38 39 265u
Vout3 39 123 sin(0 196 60 0 0 -270)
* Model of comparator: v1-v2, vgs
.subckt comp 1 2 9 10
rin 1 3 2.8k
r1 3 2 20meg
e2 4 2 3 2 50
r2 4 5 1k
d1 5 6 zenerdiode1
d2 2 6 zenerdiode2
e3 7 2 5 2 1
```

```
rdiff4 13a 13 1k
rdiff5 23a 23 1k
rdiff6 33a 33 1k
ecin1 14 0 12 13a 2
ecin2 24 0 22 23a 2
ecin3 34 0 32 33a 2
* Program list is continued in column 2
r3 7 8 10
c3 8 2 10n
r4 3 8 100k
e4 9 10 8 2 1
.model zenerdiode1 D (Is=1p BV=0.1)
.model zenerdiode2 D (Is=1p BV=50)
.ends comp
* Program list is continued below table

*Models
.model qfet nmos(level=3 gamma=0 kappa=0 tox=100n rs=42.69m kp=20.87u l=2u w=2.9
+ delta=0 eta=0 theta=0 vmax=0 xj=0 uo=600 phi=0.6 vto=3.487 rd=0.19 cbd=200n pb=0.8
+ mj=0.5 cgso=3.5n cgdo=100p rg=1.2 is=10f)
.model diode d(is=1p)

*Options
.options abstol=0.01m chgtol=0.01m reltol=50m vntol=1m itl5=0 itl4=200
.tran 5u 350m 16.67m 5u
.probe
.end
```

The use of the above program has to satisfy the (5.107) which is $V_1 = \frac{mV_{DC}}{2\sqrt{2}}$. In Fig. 5.56a $V_Q = V_{DC} = 380$ V $= V_{suppl}$ and $V_1 = 196/\sqrt{2} = 138.61$ V which results in a modulation index of $m = 138.13 \cdot 2 \cdot \sqrt{2}/380 = 1.028$. A modulation index of larger than 1.00 operates the inverter outside the PWM near the border to the 6-step mode. This is the reason why the actual current is nonsinusoidal and cannot follow the sinusoidal reference current. The DC input voltage $V_Q = V_{DC} = 380$ V $= V_{suppl}$ can be changed in the statement (see above table)

```
*DC voltage supply
Vsuppl 2 0 380
```

The reference current $I_{reference_amplitude} = 56.6$A can be changed in the statement (see above table)

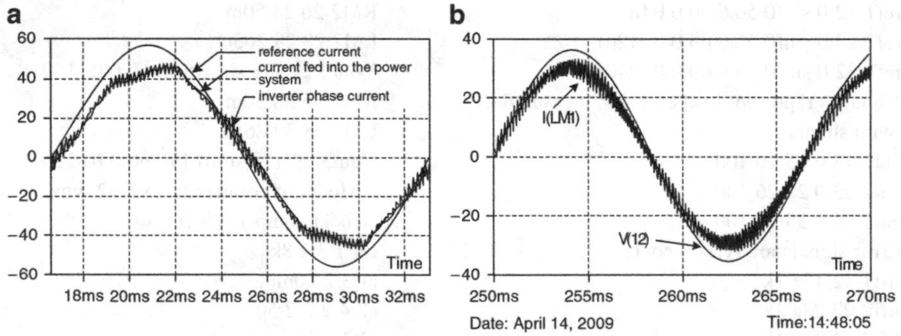

Fig. 5.56 PSpice plot of reference current, inverter phase current, and current fed into the power system based on P (proportional)-control at (**a**) $V_Q = V_{DC} = 380$ V $= V_{suppl}$ and $I_{reference\_amplitude} = 56.6$ A, (**b**) $V_Q = V_{DC} = 600$ V $= V_{suppl}$ and $I_{reference\_amplitude} = 35.82$ A

```
*Voltages represent reference currents
Vref1 12 0 sin(0 56.6 60 0  0  0)
Vref2 22 0 sin(0 56.6 60 0  0  -120)
Vref3 32 0 sin(0 56.6 60 0  0  -240)
```

and the power system voltage(s) $V_{system\_amplitude} = 196$ V can be changed in the statement (see above table)

```
*Power system parameters
RM1 16 18 50m
LM1 18 19 265u
Vout1 19 123 sin(0 196 60 0  0  -30)
RM2 26 28 50m
LM2 28 29 265u
Vout2 29 123 sin(0 196 60 0  0  -150)
RM3 36 38 50m
LM3 38 39 265u
Vout3 39 123 sin(0 196 60 0  0  -270)
```

For $V_Q = V_{DC} = 600$ V $= V_{suppl}$, $I_{reference\_amplitude} = 35.82$ A and $V_{system\_amplitude} = 196$ V a modulation index of $m = 0.65$ results and one obtains the sinusoidal current (neglecting the ripple) of Fig. 5.56b. Note that the waveform of the inverter current also depends upon the power factor, that is, the phase angle between inverter output phase voltage and current discussed in [9].

References [14–16] discuss some of the applications where PWM inverters are used such as vector analysis and control of advanced static VAR compensator, SSSC-static synchronous series compensator: theory, modeling and applications, UPFC-unified power flow controller: theory, modeling and applications – just to cite a few of the many papers available in the public domain. More detailed information is presented in [9].

The efficiency of PWM converters (e.g., inverters, rectifiers) can be increased at light load by decreasing the switching frequency say from 20 kHz at rated load to 5 kHz at light load. While at full-load the efficiency is about maximum say 95% depending upon the rating of the converter, the efficiency at light load is relatively low and in the range of 70%. This means the percentage of the switching ripple of the current can be increased at light load, where the fundamental current amplitude is relatively small, while at about full-load the percentage of the switching ripple of the current must be small because the fundamental amplitude of the current is large. This load-dependent variable PWM switching frequency calls for the measurement of the load current which, of course, complicates somewhat the control of the converter. Preliminary calculations indicate that the light-load efficiency can be increased by about 10% depending upon given power quality conditions.

5.7 Thyristor Controllers

Frequently inexpensive thyristor controllers are used for single- and three-phase induction motors, provided the speed of the motors is about constant and the load changes frequently from no-load to full-load torques. Such controllers do not

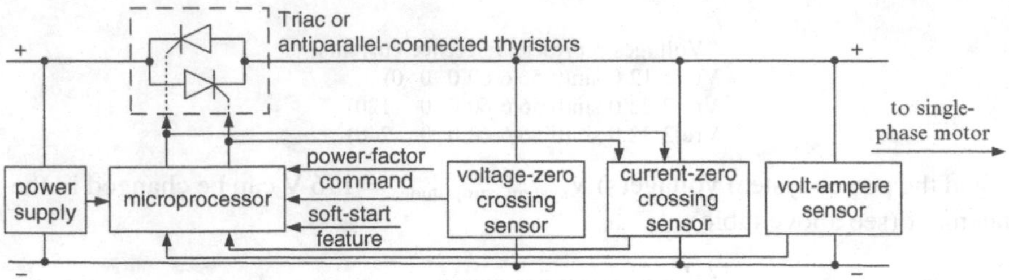

Fig. 5.57 Thyristor controller for single-phase induction motor

change the frequency but rather the terminal voltage amplitude. Thus at no-load operation the terminal voltage is only about 70% of its rated value and the phase angle between terminal voltage and terminal current is reduced: therefore, the efficiency and the power factor of the drive can be improved at light-load condition [17–23]. In most cases the efficiency at full-load condition is actually reduced due to the losses within the controller. Figure 5.57 represents the controller circuit for a single-phase configuration [17]. Representative plots of terminal voltage, and current are given in Figs. 5.58a–e for a single-phase induction motor and in Figs. 5.59a–e for a three-phase induction motor.

Application Example 5.16: Design of an Electric Drive for a Bicycle Fed by Solar Array

The drive train of an electric bicycle consists of photovoltaic array, deep-cycle battery, and variable-speed brushless DC motor-consisting of three-phase inverter and permanent-magnet motor- with regeneration capability (see Fig. 5.60).

A human being can output on the average a steady-state power of about 75 W. For this reason it is assumed that the brushless DC motor must output 77.76 W at an efficiency of $\eta_{motor} = 0.80$, and 3 h of biking are assumed with one battery charge – this includes the recovery of braking energy. In the worst case there will be no recovered braking energy. The three-phase inverter providing a variable AC voltage has an output power of 97.2 W at an efficiency of $\eta_{inverter} = 0.80$. The deep-cycle battery of 900 Wh has an efficiency of $\eta_{battery} = 0.90$ (note that the battery has an efficiency of 0.90 during charging and discharging which occur at different times, that is, a total efficiency of 0.81) and puts out a power of 121.5 W for about 3 h. The solar array must provide at an insolation of 0.75 kW/m^2 an output power of 75 W for 6 h of charging time while the bicycle is parked and replenishes the charge of the battery to a maximum of 900 Wh with a $\frac{(3 \cdot 121.5\text{Wh}/0.9)}{900\text{Wh}} = 45\%$ depth of discharge (DoD). At an insolation of $Q_s = 0.75$ kW/m^2 and a solar array efficiency of $\eta_{cell} = 0.10$ the solar array must have an area of 1 m^2. If the battery and the solar array characteristics are matched no peak-power tracker is required. The variable-frequency/voltage amplitude PWM (pulse-width-modulated) voltage-source inverter makes a DC-to-DC converter unnecessary.

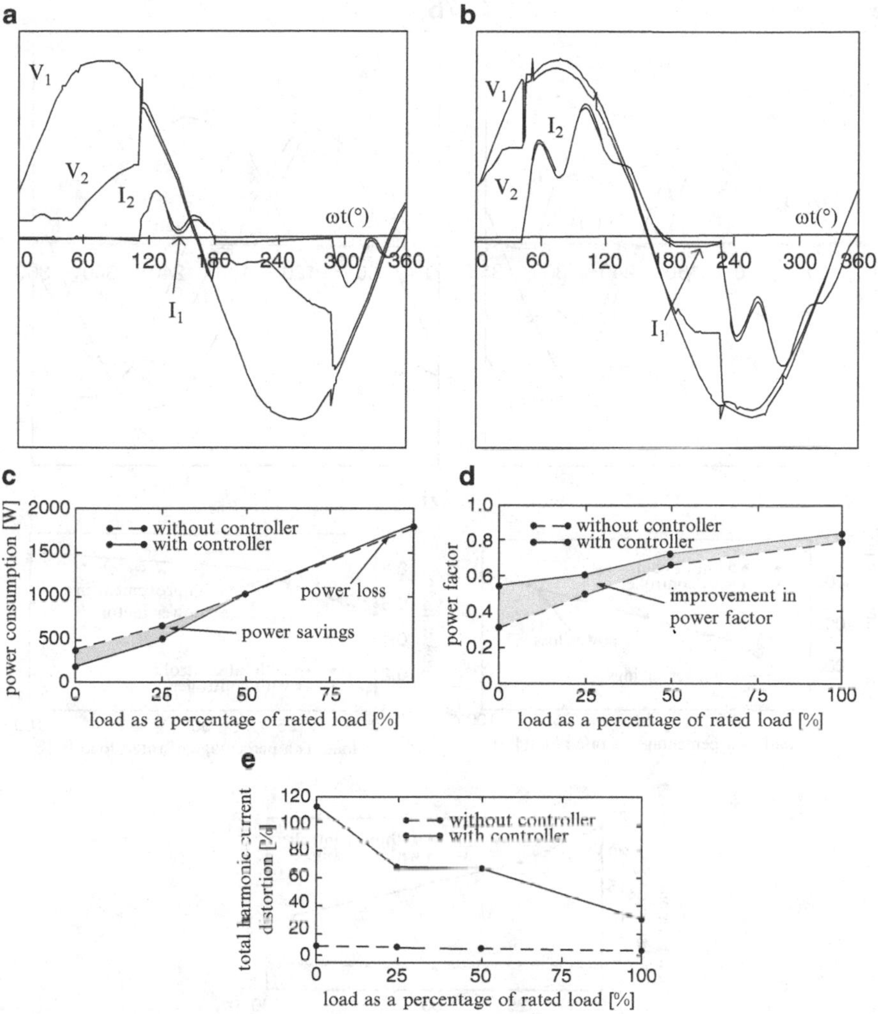

Fig. 5.58 Terminal voltage and current wave shapes for single-phase induction motor due to thyristor controller (**a**) at no-load, (**b**) at full-load where V_1 and V_2 are input and output voltages, respectively. I_1 and I_2 are input and output currents, respectively, (**c**) Efficiency improvement (power savings) of single-phase induction motor due to thyristor controller, (**d**) Power-factor improvement of single-phase induction motor due to thyristor controller, (**e**) Total harmonic current distortion (THD_i) due to single-phase thyristor controller

The weights of the solar array, lithium-ion battery, inverter, and motor are 4 (lbs-force), 9 (lbs-force), 3 (lbs-force), and 3 (lbs-force), respectively, that is, a total of 19 (lbs-force). This compares favorably with the weight of a mountain bike of 33 (lbs-force): the electric (mountain) bike weighs 52 (lbs-force).

The costs of the solar array, battery, inverter, and motor are $4,000, $900, $200, and $200, respectively. This would essentially triple the cost of a mountain bike.

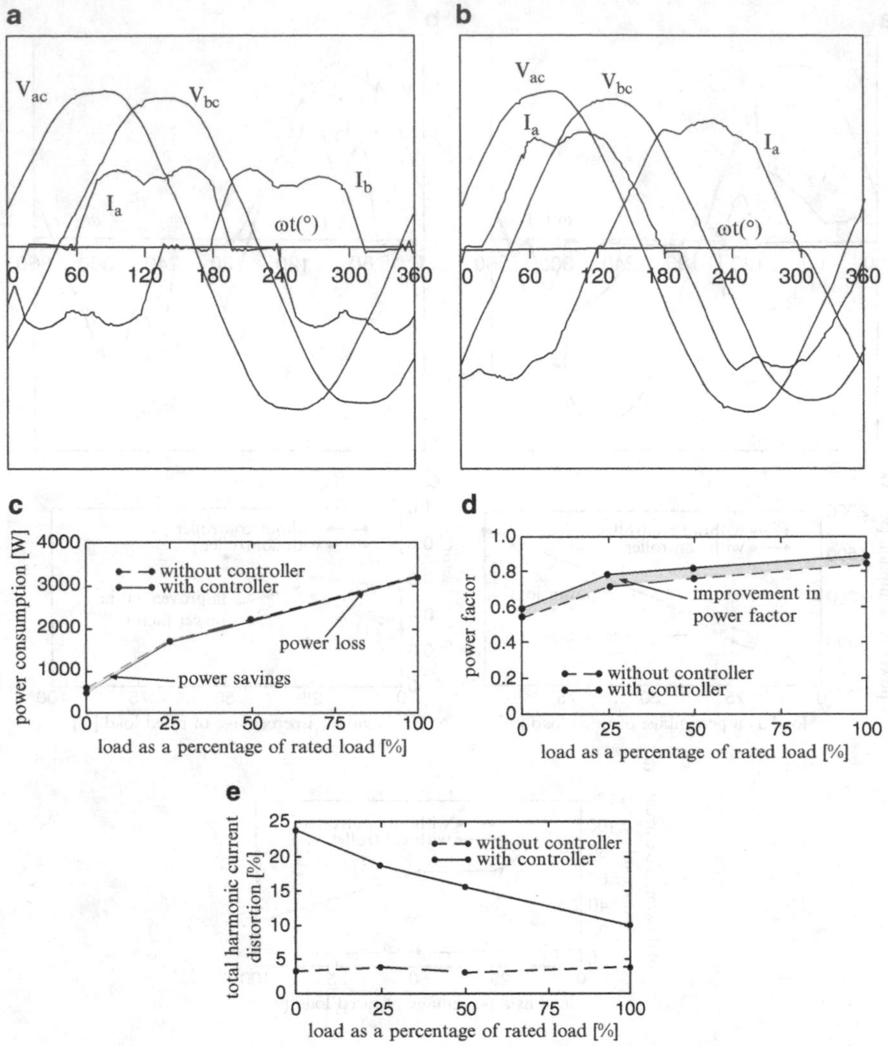

Fig. 5.59 Terminal voltage and current waveshapes for three-phase induction motor due to thyristor controller; (**a**) at no-load, (**b**) at full-load where V_{ac}, V_{bc} are line-to-line input voltages and I_a, I_b are phase input currents, (**c**) Efficiency (power consumption) improvement of three-phase induction motor due to thyristor controller, (**d**) Power-factor improvement of three-phase induction motor due to thyristor controller, (**e**) Total harmonic current distortion (THD_i) due to three-phase thyristor controller

Note that these costs do not include labor. The drawback of using a solar array is its size. This disadvantage can be mitigated by folding the solar panel while riding the bicycle. It is expected that the folded array will be a quarter of the unfolded panel. Provisions are available so that the battery can be charged from power-system outlets – a plug-in bike.

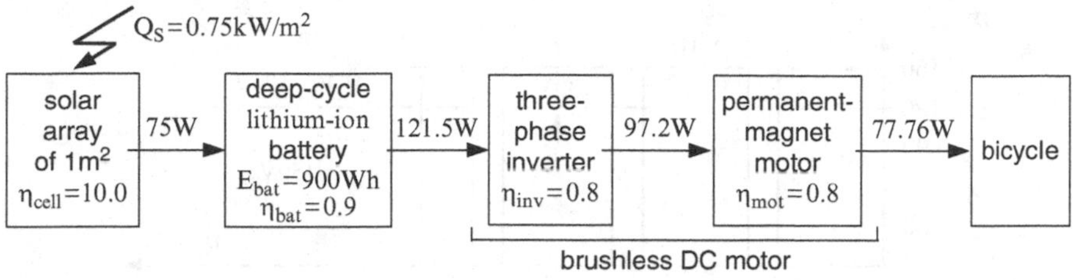

Fig. 5.60 Electric drive train for bicycle

Application Example 5.17: Speed Control of a Separately Excited DC Motor with a Switch-Mode DC-to-DC Converter

A switch-mode DC–DC converter with the switching frequency $f_{switching} = 20$ kHz and duty cycle $\delta = 0.8$ is supplying a DC machine with the following specifications: $E_a = 118$ V, $R_a = 0.3\ \Omega$, and $L_a = 5$ mH.

(a) If the peak-to-peak ripple of the output current I_o is $\Delta I_{p-p} = 0.12$ A, compute the supply (input) voltage of the converter V_{DC}, the average output voltage of the converter, V_o, and the average output current of the converter I_o.
(b) What is the operating mode of the machine? Plot the output voltage $v_o(t)$, and the output voltage ripple $\Delta v_o(t)$.
(c) Assume the motor operates in regenerative braking mode with an average current of -10 A, compute the average power flow into the converter.

Solution

(a)

$$\begin{cases} \dfrac{V_o}{V_{DC}} = \delta = 0.8 \\ v_{ind} = L\dfrac{di}{dt} \Rightarrow V_{DC} - V_o = L\dfrac{\Delta I_{p-p}}{\Delta(T_{s/2})} = 5\times 10^{-3}\dfrac{0.12}{0.8(5\times 10^{-5})/2} = 30\,\text{V} \end{cases}$$

$$\Rightarrow \begin{cases} V_o = 120\,\text{V} \\ V_{DC} = 150\,\text{V} \end{cases}$$

$$V_o = E_a + R_a I_o \ \Rightarrow\ I_o = \frac{V_o - E_a}{R_a} = \frac{120 - 118}{0.3} = 6.67\,\text{A}.$$

(b) Since the current is positive, the machine is operating in the motoring mode. Plots of the output voltage $v_o(t)$ and the output voltage ripple Δv_o are shown in Figs. 5.61a, b.

(c)

$$P_{Ea} = E_a(I_o)_{average} = 118(-10) = -1180\,\text{W}$$

$$P_{Ea} = P_{converter} + P_{dissipated\ in\ R} \Rightarrow P_{converter} = -P_{Ea} + I_o^2R_a$$

$$= -1180 + 10^2(0.3) = -1150\,\text{W}.$$

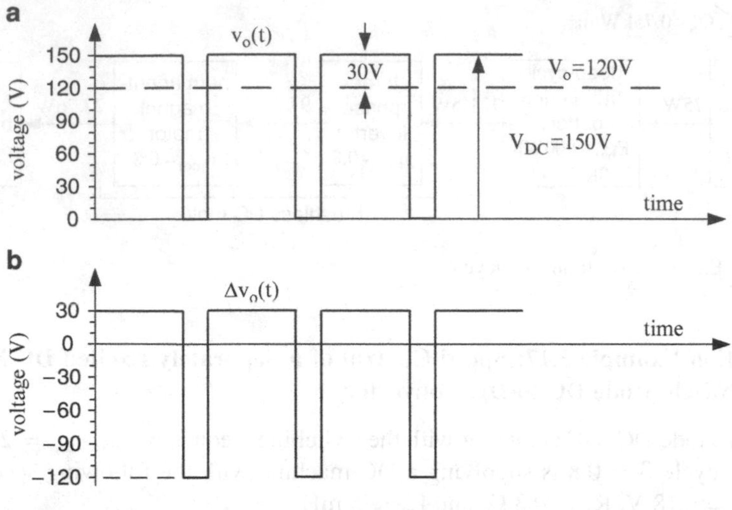

Fig. 5.61 Waveforms of switch-mode DC-to-DC converter, (**a**) output voltage $v_o(t)$, (**b**) output voltage ripple $\Delta v_o(t)$

The minus sign in the above relations indicates that regeneration occurs and the power flow is from the motor to the converter which is connected to an energy storage (e.g., battery).

5.8 Summary

In this chapter the basic electronic device characteristics are described and power electronic principles discussed. After the discussion of DC-to-DC converters, AC-to-DC converters or rectifiers DC-to-AC converters or inverters are introduced. Example applications ranging from very simple ones to more complicated ones such as six-step and PWM-inverter types are analyzed based on PSpice. The conditions under which P controlled inverters can operate are outlined. It should be noted that there is a certain relation between the AC output voltage and the DC input voltage of an inverter and this relation depends also on the power factor at the output of the inverter. The energy/power efficiency improvements of induction motor drives based on thyristor-voltage controllers show the potential for combining power electronics and electric machines provided the voltage and current harmonics do not result in excessive losses, that is, a decrease in efficiency and lifetime reduction [9]. The speed control of brushless DC motors is discussed.

5.9 Problems

Problem 5.1: *Design of a hydro-power plant for a residence*

A hydro-power plant consists of a (propeller) bulb-type turbine, a synchronous generator with a synchronous speed $n_s = 900$ rpm, a controlled rectifier, a single-

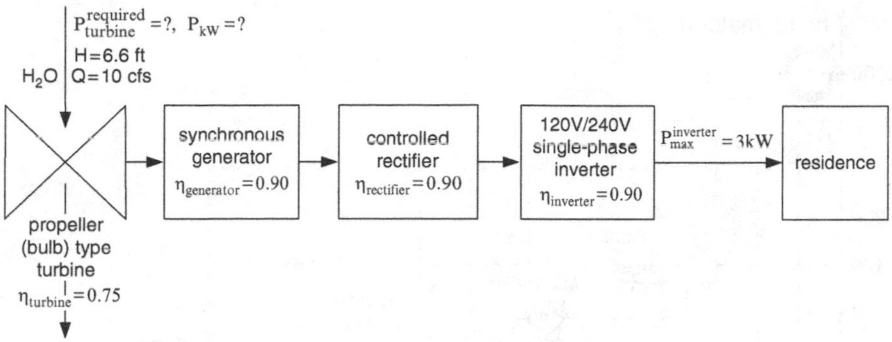

Fig. 5.62 Hydro-power plant for a residence

phase inverter (120/240 V_{AC} at f = 60 Hz) and consumers requiring a maximum power of $P_{max}^{residence} = P_{max}^{inverter} = 3$ kW as indicated in Fig. 5.62.

(a) If the power efficiencies of the water turbine, the synchronous generator, the controlled rectifier, and the single-phase inverter are $\eta_{turbine} = 0.75$, $\eta_{generator} = 0.90$, $\eta_{rectifier} = 0.90$, and $\eta_{inverter} = 0.90$, respectively, compute the required input power of the water turbine $P_{turbine}^{required}$.

(b) Provided the head of the water is H = 6.6 ft and the water flow measured in cubic feet per second (cfs) is Q = 10 cfs compute the mechanical power available at the turbine input [24–27]

$$P_{kW} = \frac{H \cdot Q \cdot W \cdot 0.746}{550}, \tag{5.109}$$

where

H is the head of water in feet (ft),
Q is the flow of water in cubic feet per second (cfs),
W is the weight of a cubic foot of water = 62.4 (lbs-force)/ft^3,
0.746 is a constant used to convert horsepower (hp) to kilowatts (kW),
550 is a constant used to convert {ft-(lb-force)}/s into horsepower.

(c) How does $P_{turbine}^{required}$ compare with P_{kW}?

(d) Compute the specific speed [25–27]

$$N_q = n_s \frac{Q^{0.5}}{H^{0.75}}, \tag{5.110}$$

where

n_s is the (synchronous mechanical) speed measured in rpm,
Q is the flow of water measured in (m^3/s),
H is the head of water measured in m.

(e) Is the selection of the bulb-type turbine according to Fig. 5.63 justified?

(f) What other types of water turbines exist?

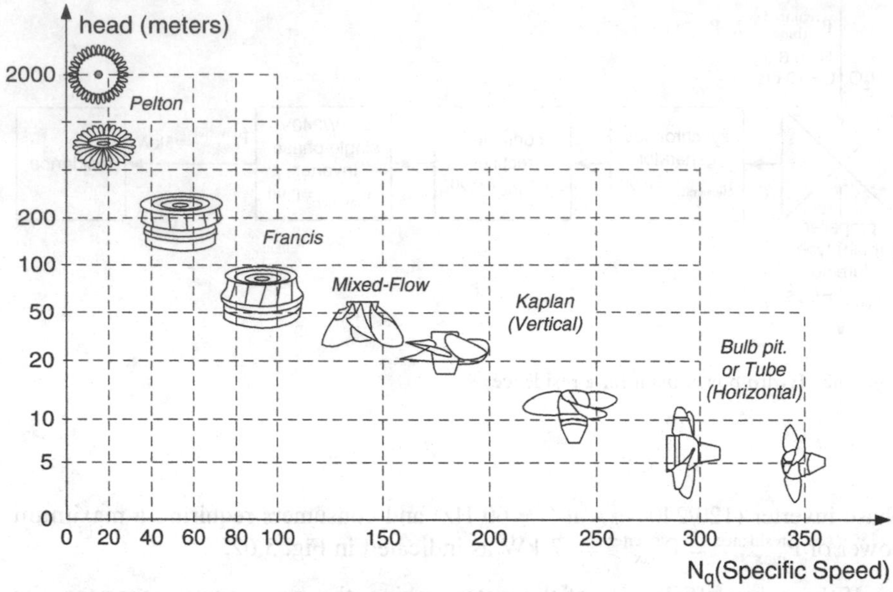

Fig. 5.63 Summary of the types of water turbines

(g) The power supply has to deliver the energy E = 500 kWh per month at an avoided cost of \$0.20/kWh. What is the payback period of this hydro-power plant if the purchase prices of the turbine, the generator, the controlled rectifier, and the single-phase inverter are \$2,000, \$1,000, \$2,000, \$2,000, respectively?

Note that the water inlet (input power) of the turbine can be controlled via wicket gates so that the input power of the turbine can be adjusted as a function of the consumed electrical power and no storage device is required.

Problem 5.2: *Uncontrolled half-wave diode rectifier*

(a) For the uncontrolled diode rectifier of Fig. 5.64 compute the current i(t) over one 60 Hz cycle by using the analytical (hand calculation) method of Application Example 2.2 or Problem 2.1 (or any other analytical method, e.g., Laplace transformation). The voltage source consists of a sinusoidal signal $v_s(t) = V_{max} \cos \omega t$, where $V_{max} = \sqrt{2} \cdot 120$ V and has a frequency of f = 60 Hz. The initial condition for the current at time t = 0 is i(t = 0) = 0.

(b) Write a PSpice program for the circuit of Fig. 5.64 and obtain a Pspice solution. Compare the result of the PSpice solution with that of the hand calculation as requested in part (a). List the PSpice input program and show the plots of $v_s(t)$ and i(t) for one cycle during steady-state condition.

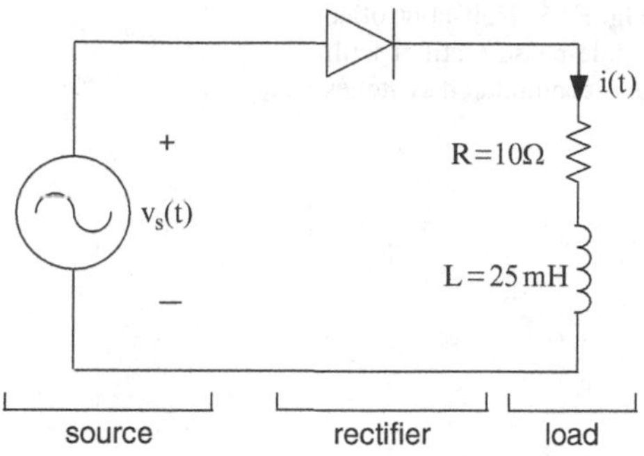

Fig. 5.64 Uncontrolled diode rectifier

Problem 5.3: *Gating circuit for thyristor*

In Application Example 5.4 a PSpice program is listed. The corresponding circuit is given in Fig. 5.15.

(a) Use "reverse" engineering and identify the nodes of Fig. 5.15 with the node numbers as used in the PSpice program listed in Application Example 5.4. For the thyristor gating circuit of Fig. 5.15 compute (simulate) based on PSpice v_G for $v_\alpha = 1.5$ V, and $v_{frequency}(t) = \sqrt{2}\cdot 120$ V $\sin 2\pi 60t$.
(b) Plot $v_{frequency}(t)$ and v_G (t).
(c) Plot the voltages at the bases and collectors of transistors Q_1 to Q_5.

Problem 5.4: *Analysis of a current-controlled switch*

(a) For the controlled thyristor rectifier of Fig. 5.23, where the firing angle is $\alpha = 60°$, compute the current i(t) by using the analytical method of Application Examples 2.2, 5.6, 5.7, and Problem 5.2 or any other analytical method. The voltage source consists of a rectangular signal with a magnitude of 100 V and a frequency of f = 200 Hz.
(b) Figure 5.23 represents a resistive/inductive single-phase rectifier which is controlled by a line-commutated (or naturally commutated) switch (thyristor). The model of Fig. 5.21 can be used for programming the line-commutated switch within the PSpice program. Use PSpice to obtain a numerical solution for i(t) and compare the numerical result with that of part (a).

Problem 5.5: *Half-controlled single-phase rectifier with 2 line-commutated switches (thyristors)*

For the circuit of Fig. 5.65 compute – using PSpice – the thyristor current, and output voltage, where R = 10 Ω, C = 100 μF, and the firing angle $\alpha = 45°$ at a line-to-line voltage of $v_{in}(t) = \sqrt{2}\cdot 240$ V $\sin \omega t$, where $\omega = 2\pi f$ with f = 60 Hz. List the input PSpice program. Plot the output voltage v_d, the input i_{in}, the input voltage v_{in}, and the current through thyristor T_1.

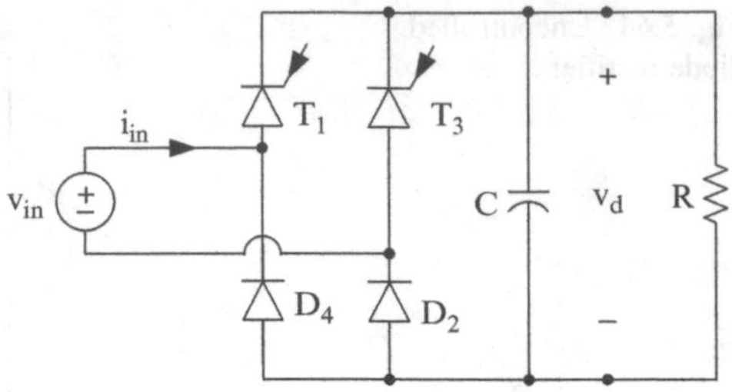

Fig. 5.65 Half-controlled single-phase rectifier with line-commutated switches

Problem 5.6: *Step-down (buck) DC-to-DC converter or chopper*

A step-down chopper, Fig. 5.30a, is to be used to supply power from a 1,000 V_{DC} overhead wire (catenary) to the DC motor of a commuter train in an airport. The motor may be modeled as an induced direct voltage ($\bar{e}_d$ or E_a) in series with a resistance of $R_a = 0.2\ \Omega$ and an inductance of $L_a = 25$ mH. The switching frequency of the chopper is set at $f_S = 1{,}000$ Hz.

(a) When the train is stopped, what should be the ON time t_{ON} of the chopper to provide a motor current of $\bar{i}_d = I_a = 150$ A?
(b) What is the maximum possible (at $\delta = 1.0$) induced voltage $\bar{e}_{d\,max}$ or E_{amax} of the motor for which a current of $I_a = \bar{i}_d = 200$ A results?
(c) What is the average value of the supply current $\bar{i}_s$ for the conditions of (a) and (b)?

Problem 5.7: *Step-up (boost) DC-to-DC converter or chopper*

When the commuter train of Problem 5.6 travels downhill, its chopper is reconnected in the configuration shown in Fig. 5.30b. For what length of time should the chopper switch be in the conducting (ON) condition if the generated (induced) voltage of the motor (now working as a generator) is $\bar{e}_d = E_a = 750$ V and a regenerated power of $P_{regen} = -50$ kW is to be fed back (note the minus sign!) to the DC supply?

Problem 5.8: *Step-down, step-up DC-to-DC converter or chopper*

A two-quadrant chopper of the configuration shown in Fig. 5.31 is used to control the speed of a small DC motor when accelerating in forward direction (motoring) and decelerating in forward direction (braking). The motor has a resistance of $R_a = 2.5\ \Omega$ and an inductance L_a that is sufficiently large to make the motor current essentially free of ripple. The supply direct voltage is 60 V. The chopper frequency is $f_S = 2$ kHz.

(a) Determine the generated (induced) voltage $\bar{e}_d$ or E_a of the motor when it is operating with a current of $\bar{i}_d = I_a = 4.0$ A and the chopper switch S_1 is set at $t_{ON} = 200\ \mu$s.
(b) Determine the time t_{ON} of switch S_2 when the machine is in the braking mode with a current of $\bar{i}_d = I_a = -4.0$ A and an induced voltage of $\bar{e}_d = E_a = 30$ V?

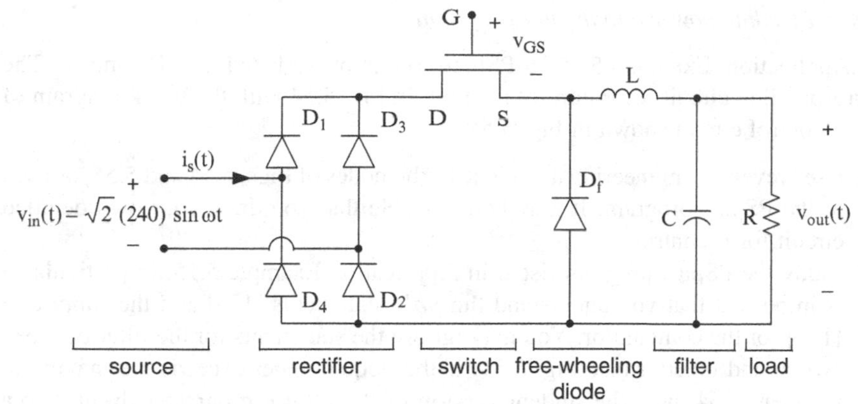

Fig. 5.66 Controlled, single-phase rectifier with self-commutated switch

Problem 5.9: *Controlled single-phase rectifier with self-commutated switch (MOSFET) (see application example 5.5)*

Figure 5.66 shows a single-phase rectifier with 4 diodes and one self-commutated switch (e.g., MOSFET). The load resistance is $R = 15\ \Omega$, $C = 100\ \mu F$, and $L = 1$ mH. There is no input filter and the voltage source can be considered to be ideal with $V_{in\_rms} = 240$ V, $f = 60$ Hz. The switching frequency of the MOSFET is $f_S = 500$ Hz. For the MOSFET switches you may assume a gating voltage magnitude of $v_{GS} = 30$ V, and the duty cycle is $\delta = 50\%$. Using PSpice compute and plot the output voltage $v_{out}(t)$, input voltage $v_{in}(t)$, and source current $i_s(t)$. List the input PSpice program.

Problem 5.10: *Brushless DC motor drive consisting of 6-step inverter and synchronous (permanent-magnet) motor*

In Application Example E5.14 a PSpice program is listed. The corresponding circuit is given in Fig. 5.49 with the gating signals of Fig. 5.50.

(a) Use "reverse" engineering and identify the nodes in Fig. 5.49, as used in the PSpice program listed in Application Example E5.14
(b) Modify this program and obtain a PSpice solution of the motor currents i_{MA}, i_{MB} and i_{MC} for a phase angle of $\theta = 15°$.
(c) Plot the current in MOSFET Q_{AU}, i_{QAU}, and the motor currents i_{MA}, i_{MB} and i_{MC}. List the modified input PSpice program.
Hint: compute a few periods so that the transient due to inconsistent initial conditions has decayed to zero.
(d) Compute the quantities as in part (b) with reversed-phase rotation and plot the motor currents i_{MA}, i_{MB} and i_{MC}.

Problem 5.11: *Current-controlled PWM (pulse-width-modulated) voltage–source inverter feeding power into the utility system*

In Application Example 5.15 a PSpice program is listed for P-control. The corresponding circuit for P-control is given in Fig. 5.54 with the block diagram of the P-control circuit shown in Fig. 5.55.

(a) Use "reverse" engineering and identify the nodes of Figs. 5.54 and 5.55, as used in the PSpice program. It may be advisable that you draw your own detailed circuit for P-control.
(b) Study the PSpice program listed in Application Example 5.15; in particular it is important that you understand the *poly* statements [1–3] and the *subcircuit* [1–3] for the comparator. You may ignore the statements for the filter between switch and inverter (see Fig. 5.54) if the node number exceeds the maximum number of 64 (note the student version of the PSpice program is limited to a maximum of 64 nodes).
(c) Run the program with inverter inductance values of $L_w = 1$ mH and 10 mH for a DC voltage of $V_Q = 360$ V. List the modified input PSpice program.
(d) Plot for both cases (1 mH and 10 mH) the current supplied by the inverter to the power system.
(e) Repeat parts (c) and (d) for a DC voltage $V_Q = 600$ V.

Problem 5.12: *Design of an emergency power supply for a residence*

An emergency power supply for a residence is to be designed. It consists of an exercise bicycle, a three-phase, permanent-magnet AC generator, three-phase rectifier, a deep-cycle battery, and a single-phase 120 V/240 V inverter, see Fig. 5.67. For each component the energy efficiencies are given. Compute the energy flow per day -when two human beings are pedaling for 1 h (each) per day-to the inputs of the AC generator, the three-phase rectifier, the deep-cycle battery, the single-phase inverter, and the residence $E_{day}^{residence}$. Before working on the design of this emergency power supply, review the electric bicycle drive drain of Application Example 5.16.

Problem 5.13: *Speed control of a separately excited DC motor with a switch-mode DC-to-DC converter*

A switch-mode DC-to-DC converter with the switching frequency $f_{switching} = 20$ kHz and supply voltage $V_{DC} = 140$ V is connected to a DC machine with the following specifications: $E_a = 110$ V, $R_a = 0.2\ \Omega$, and $L_a = 5$ mH. Calculate duty

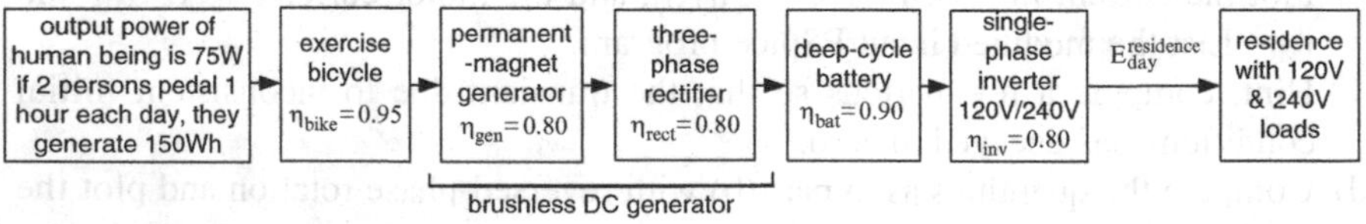

Fig. 5.67 Design of an emergency power supply for a residence

cycle δ, peak-to-peak ripple of the output current ΔI_{p-p}, and plot the output voltage $v_o(t)$ and output voltage ripple $v_{o\_ripple}(t)$ for the following cases:

(a) Motoring mode – DC motor draws an average current of 10 A.
(b) Regenerative braking mode – the average current being supplied by the motor to the converter during braking is 10 A.

References

1. Tuinenga, P. W.: *SPICE, A Guide to Circuit Simulation & Analysis Using Pspice*, 2nd Edition, Prentice Hall, Englewood Cliffs, NJ, 1992.
2. Roberts G. W.; Sedra, A. S.: *SPICE*, 2nd Edition, Oxford University Press, New York, 1997.
3. Rashid, M. H.: *SPICE for Power Electronics and Electric Power*, Prentice Hall, Englewood Cliffs, 1993.
4. Dewan, S. B.; Slemon, G. R. T.; Straughen, A.: *Power Semiconductor Drives*, John Wiley & Sons, New York, 1984.
5. Erickson, R. W.: *Fundamentals of Power Electronics*, Chapman & Hall, New York, 1997.
6. Yildirim, D.: *Commissioning of a 30 kVA Variable-Speed, Direct-Drive Wind Power P*lant, Ph.D. Thesis, University of Colorado, Boulder, 1999.
7. Dubey, G. K.: *Power Semiconductor Controlled Drives*, Prentice Hall, Englewood Cliffs, NJ, 1998.
8. Fardoun, A. A.; Fuchs, E. F.; Huang, H.: "Modeling and simulation of an electronically commutated permanent-magnet machine drive system using SPICE," *IEEE Transactions on Industry Applications*, Vol 30, No. 4, July/August 1994.
9. Fuchs, E. F.; Masoum, M. A. S.: *Power Quality in Power Systems and Electrical Machines*, Elsevier/Academic Press, 2008, 638 pages. ISBN: 978-0-12-369536-9.
10. Yildirim, D.; Fuchs, E. F.; Batan, T.: "Test results of a 20 kW variable-speed direct-drive wind power plant," *Proceedings of the ICEM'98 International Conference on Electrical Machines*, Istanbul, Turkey, 2–4 Sept. 1998.
11. Teltsch, M.: *Adjustable-Speed Drive for an Electric Car with Large Rated Torque at Low Rated Speed*," M.S. Thesis, University of Colorado at Boulder, August 27, 1997.
12. Schraud, J.: *Control of an Induction Motor by Electronically Switching its Windings with Different Pole Numbers*, Independent Study, University of Colorado, Boulder, 1999.
13. Matlab/Simulink software.
14. Schauder C. D.; Mehta, H.: "Vector analysis and control of advanced static VAR compensator," *IEE Proceedings-C*, Vol. 140, No. 4, July 1993.
15. Sen, K. K.: "SSSC-static synchronous series compensator: theory, modeling and applications," *IEEE Transactions on Power Delivery*, Vol.13, No.1, 1998.
16. Sen, K. K.; Stacey, E. J.: "UPFC-unified power flow controller: theory, modeling and applications," PE-282-PWRD-0-12-1997, *IEEE PES Winter Meeting*, Tampa, Florida.
17. Fuchs, E. F.; Lin, D.; Yildirim, D.: "Measured efficiency and power-factor improvements of single-phase and three-phase controllers," *Proceedings of the ICEM'98 International Conference on Electrical Machines*, Istanbul, Turkey, 2–4 Sept. 1998.
18. Fuchs, E. F.; Roesler, D. J.; Masoum, M. A. S.: "Are harmonic recommendations according to IEEE and IEC too restrictive?" *IEEE Transactions on Power Delivery*, Vol. 19, No. 4, 2004, pp. 1775–1786.
19. Fuchs, E. F.; Hanna, W. J.: "Measured efficiency improvements of induction motors with thyristor/triac controllers", *IEEE Transactions on Energy Conversion*, Vol. 17, No. 4, 2002, pp. 437–444.

20. Fuchs, E. F.; Roesler, D. J.; Kovacs, K. P.: "Aging of electrical appliances due to harmonics of the power system's voltage", *IEEE Transactions on Power Delivery*, Vol. TPWRD-1, No. 3, 1986, pp. 301–307.
21. Fuchs, E. F.; Roesler, D. J.; Alashhab, F. S.: "Sensitivity of electrical appliances to harmonics and fractional harmonics of the power system's voltage, Part I", *IEEE Transactions on Power Delivery*, Vol. TPWRD-2, No.2, 1987, pp. 437–444.
22. Fuchs, E. F.; Roesler, D. J.; Kovacs, K. P.: "Sensitivity of electrical appliances to harmonics and fractional harmonics of the power system's voltage, Part II", *IEEE Transactions on Power Delivery*, Vol. TPWRD-2, No.2, 1987, pp. 445–453.
23. Fuchs, E. F.; Poloujadoff, M.; Neal, G. W.: "Starting performance of saturable three-phase induction motors", *IEEE Transactions on Energy Conversion*, Vol. EC-3, No. 3, 1988, pp. 624–635.
24. Cowdrey, J. M.: "Making small hydro work well on a water supply system", *Hydro Review*, Vol. 15, No. 1, 1996, HCI Publications, 410 Archibald Street, Kansas City, MO 64111.
25. Cowdrey, J. M.: "Hydroelectric power in a municipal water system", City of Boulder website http://www.ci.boulder.co.us.
26. Zipparro, V. J.; Hasen, H.: *Davis' Handbook of Applied Hydraulics*, 4th Edition, McGraw-Hill, Inc. 1993, ISBN 0-70-073002-4.
27. American Society of Mechanical Engineers, *The Guide to Hydropower Mechanical Design*, HIC Publications, 1996, ISBN 0-9651765-0-9.

Chapter 6
Magnetic Circuits: Inductors and Permanent Magnets

Static magnetic devices used in electromechanical energy conversion processes (e.g., rotating machines, rectifiers, inverters) are inductors and (permanent) magnets: their operation is based on Maxwell's equations [1] and in particular for low-frequency ($f < 50$ kHz) applications on Ampere's and Faraday's laws neglecting displacement currents. Although the nonlinear characteristics of iron cores will be discussed in one of the following subsections, mostly ideal characteristics, that is, no iron-core losses ($P_{Fe} = 0$) and an infinitely large relative permeability ($\mu_r = \infty$) will be assumed in this chapter. Computer-aided measurement techniques are introduced for measuring, as a function of frequency, small loss levels (few watts), impedances, resistances and inductances of inductors as used in solid-state energy conversion devices.

6.1 Ampere's Law and Ohm's Law Applied to Magnetic Circuits

The complete behavior of the magnetic field within magnetic and nonmagnetic devices (e.g., inductors) is described by Maxwell's equations either in differential or integral form [1] supplemented by the constituent relation

$$\vec{B} = \mu_r \mu_0 \vec{H} = \mu \vec{H}. \tag{6.1}$$

For low-frequency applications the displacement-current terms of Maxwell's equations can be neglected and the "quasi-static" form is employed. One of Maxwell's equations is also called Ampere's law which reads in integral form

$$\text{LHS} = \int_S \vec{J} \cdot d\vec{S} = \oint_C \vec{H} \cdot d\vec{\ell} = \text{RHS}, \tag{6.2}$$

where $\vec{J}$ is the current density, $\vec{H}$ the magnetic-field intensity, S the surface and C the contour path as identified in Fig. 6.1. Note that Ampere's law relates a surface integral to a line integral. LHS means left-hand side, and RHS means right-hand side.

E.F. Fuchs and M.A.S. Masoum, *Power Conversion of Renewable Energy Systems*, DOI 10.1007/978-1-4419-7979-7_6, © Springer Science+Business Media, LLC 2011

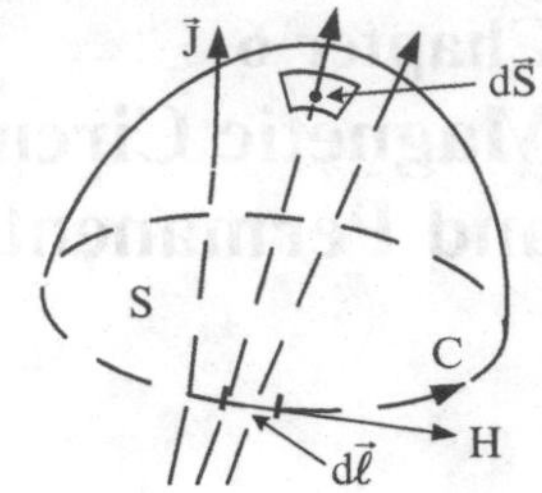

Fig. 6.1 Graphical interpretation of Ampere's law

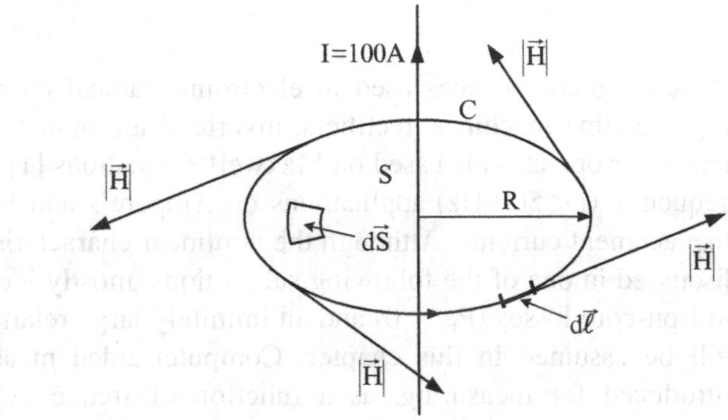

Fig. 6.2 Single wire in free space carrying a current of I = 100 A

Application Example 6.1: Computation of Magnetic-Field Intensity $\left|\vec{H}\right|$

Compute the magnetic-field intensity $\left|\vec{H}\right|$ of a wire at a distance R = 1 m in free space as is illustrated in Fig. 6.2.

Solution

Ampere's law is

$$\int_S \vec{J}\cdot d\vec{S} = \oint_C \vec{H}\cdot d\vec{\ell} \quad \text{or} \quad I = 2\pi R\left|\vec{H}\right| \tag{6.3}$$

The magnetic-field intensity is

$$\left|\vec{H}\right| = \frac{I}{2\pi R}, \tag{6.4}$$

and the magnetic-flux density is then

$$\left|\vec{B}\right| = \mu_r\mu_0\left|\vec{H}\right| = \frac{4\pi 10^{-7}}{2\pi R} I. \tag{6.5}$$

At R = 1 m, I = 100 A, $\mu_r = 1.0$, and $\mu_0 = 4\pi\cdot 10^{-7}$ H/m

$$\left|\vec{B}\right| = 200 \cdot 10^{-7}\mathrm{T} = 0.00002\mathrm{T} = 0.2\mathrm{G}. \tag{6.6}$$

Note that 1 G (gauss) = 10^{-4} T (tesla); for the sake of comparison the magnetic flux density of the earth is

$$\left|\vec{B}\right|_{\mathrm{earth}} = 0.7\mathrm{G}\ (\mathrm{gauss}). \tag{6.7}$$

6.1.1 Magnetic Circuit with One Winding and an Ideal (B–H) Characteristic ($\mu_r = \infty$)

Inductors are operated in the linear region of the B–H characteristic of Fig. 6.3. As a matter of convenience one assumes $\mu_r = \infty$, neglecting leakage fields and fringing flux. To limit the current for a given flux density B (e.g., $B_{max} = 1$ T) one has to introduce into the iron core an air gap of length g, as is depicted in Fig. 6.4.

In Fig. 6.4 an N-turn coil (or winding) is wound on a "C" shaped iron core containing an air gap g. The surface S and the contour path C are identified. Therefore, the left-hand side of Ampere's law of (6.2) becomes

$$\mathrm{LHS} = \int_S \vec{J} \bullet d\vec{S} = N \cdot i, \tag{6.8}$$

where N is the number of turns and i can be either DC or AC current. The right-hand side of (6.2) is

$$\mathrm{RHS} = \oint_C \vec{H} \bullet d\vec{l} = H_g \cdot g. \tag{6.9}$$

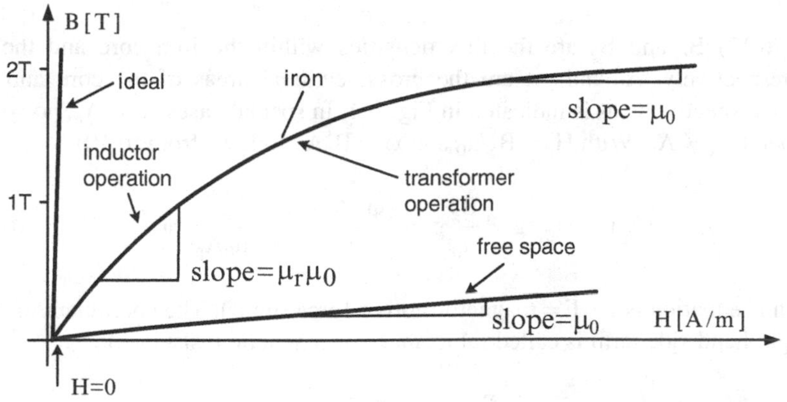

Fig. 6.3 Ideal ($\mu_r = \infty$), actual and free-space B-H characteristics

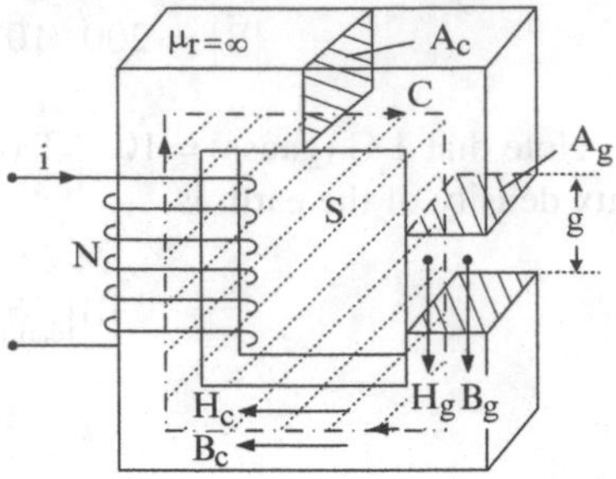

Fig. 6.4 Inductor circuit with ideal "C" iron core ($\mu_r = \infty$) and finite air-gap length g

According to (6.2) one notes that (6.8) = (6.9), therefore,

$$N \cdot i = H_g \cdot g, \tag{6.10}$$

where H_g is the magnetic-field intensity within the air gap. Note that the magnetic-field intensity within the iron core is zero, because $B_c = \mu_c H_c$ = finite and as a consequence for $\mu_c = \infty$ one obtains $H_c = 0$.

For any magnetic circuit (e.g., "C" core), the magnetic flux ϕ must be continuous, that is, there is neither a sink nor a source of flux. The flux ϕ is defined by the integral

$$\phi = \int_S \vec{B} \bullet d\vec{S}, \tag{6.11}$$

and the flux must be continuous, that is divergence of B must be 0: div(B) = 0. If fringing is neglected one obtains

$$\phi = B_c A_c = B_g A_g. \tag{6.12}$$

In (6.12) B_c and B_g are the flux densities within the iron core and the air gap, respectively. A_c and A_g are the cross-sectional areas of the core and the air gap, respectively, as is indicated in Fig. 6.4. In special cases $A_c = A_g$, however, in general $A_c \neq A_g$. With $H_g = B_g/\mu_0$ and $\phi = B_g A_g$ follows from (6.10)

$$N \cdot i = H_g \cdot g = \frac{B_g}{\mu_0} g = \frac{(\phi/A_g)}{\mu_0} g = \frac{g}{\mu_0 A_g}\ \phi. \tag{6.13}$$

In this equation $N \cdot i = F$ = magneto-motive force (mmf). The coefficient of ϕ of the right-hand side term is called reluctance (or magnetic resistance)

$$\Re_g = g/(\ \mu_0 A_g). \tag{6.14}$$

Thus (6.13) becomes

$$F = N \cdot i = \Re_g\ \phi. \qquad (6.15)$$

In this equation F represents the mmf driving the flux ϕ through the magnetic circuit represented by $\Re_g$. In analogy to Ohm's law of an electric circuit (Fig. 6.5a) one can draw an equivalent for magnetic circuits (Fig. 6.5b).

The following Application Example 6.2 demonstrates the linear approximation of the iron-core permeability and the nonlinear approximation of the iron-core permeability is demonstrated in reference [2].

Application Example 6.2: Magnetic Circuit with One Winding and a Linear (B–H) Characteristic: Power and Energy of R-L Circuit

An electromagnet (see Fig. 6.6) is to be used to lift cast-iron bars in a foundry. The surface roughness of the bars is such that when a bar and the electromagnet are in contact, there is a minimum average air gap of g = 0.5 cm in each leg.

(a) Draw the magnetic equivalent circuit. Calculate the self inductance of the magnet provided the relative permeability of the magnetic circuit is $\mu_r = 1{,}000$, and the equivalent core length is (electromagnet length and iron bar length) $\ell_c = 2\,\text{m}$.

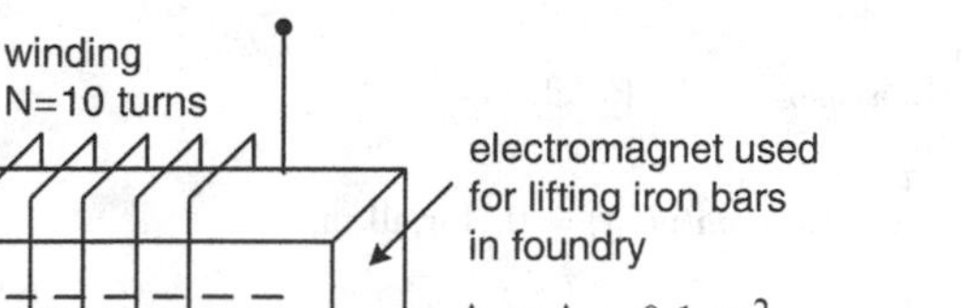

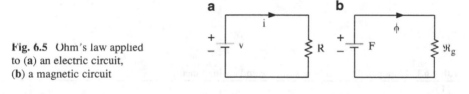

Fig. 6.5 Ohm's law applied to (**a**) an electric circuit, (**b**) a magnetic circuit

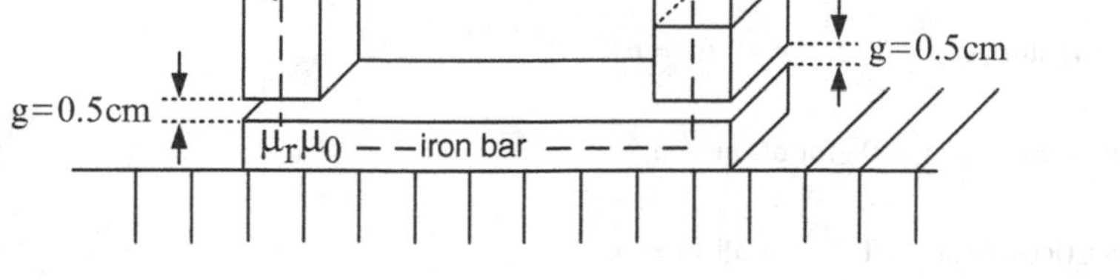

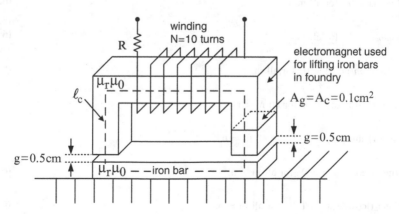

Fig. 6.6 Magnet as used in a foundry

(b) Draw the electric equivalent circuit for the magnet-bar configuration if the magnet is powered by an AC voltage of $v(t) = V_0 \cos\omega t = \sqrt{2}V_{rms} \cos\omega t$, where $V_{rms} = 600$ V at $\omega = 2\pi f$ and $f = 60$ Hz.

(c) Derive the current $\tilde{i}(t)$ in the phasor domain and the current i(t) in the time domain (no numerical calculations are required).

(d) Derive the time-averaged magnetic stored energy in the inductor $<W_{fld}>$. For the solution of the integral encountered you may remember the so-called orthogonality conditions of sine and cosine as listed in Table 6.1.

(e) Find the instantaneous power $p(t) = v(t)\, i(t)$. Use the trigonometric function $\cos\alpha \cos\beta = (1/2)\cos(\alpha - \beta) + (1/2)\cos(\alpha + \beta)$.

(f) Derive the time-averaged power $<p(t)>$. It is useful to employ for the evaluation of the integral the orthogonality conditions listed below. Also, it will be useful to use the relation

$$\cos\theta = \frac{1}{\sqrt{1 + (\tan\theta)^2}}. \tag{6.16}$$

(g) Show that the above time-averaged power $<p(t)>$ is identical with the power dissipated $<p(t)_{dissipated}>$ in the ohmic resistance R.

Solution

In Sect. 6.1.1 the relations for a magnetic circuit with one winding and an ideal (B–H) characteristic ($\mu_r = \infty$) are derived. In the following the finite value of the relative permeability of the iron core, $\mu_c = \mu_r \neq \infty$, is taken into account.

Table 6.1 Orthogonality conditions of sine and cosine functions

$$\int_{t_o}^{t_o+T} \sin(m\omega_o t) = 0, \text{ for all m}, \tag{6.17}$$

$$\int_{t_o}^{t_o+T} \cos(m\omega_o t) = 0, \text{ for all m}, \tag{6.18}$$

$$\int_{t_o}^{t_o+T} \cos(m\omega_o t)\sin(n\omega_o t) = 0, \text{ for all m, n}, \tag{6.19}$$

$$\int_{t_o}^{t_o+T} \sin(m\omega_o t)\sin(n\omega_o t) = 0, \text{ for all } m \neq n, \tag{6.20}$$

$$\int_{t_o}^{t_o+T} \sin(m\omega_o t)\sin(n\omega_o t) = T/2, \text{ for all } m = n, \tag{6.21}$$

$$\int_{t_o}^{t_o+T} \cos(m\omega_o t)\cos(n\omega_o t) = 0, \text{ for all } m \neq n, \tag{6.22}$$

$$\int_{t_o}^{t_o+T} \cos(m\omega_o t)\cos(n\omega_o t) = T/2, \text{ for all } m = n, \tag{6.23}$$

Applying Ampere's law (6.10) becomes

$$N \cdot i = H_g \cdot g + H_c \cdot \ell_c \tag{6.24}$$

or (6.13) becomes now with $H_c = B_c/(\mu_c \cdot \mu_o)$

$$N \cdot i = \frac{B_g}{\mu_0} g + \frac{B_c}{\mu_c \, \mu_o} \ell_c. \tag{6.25}$$

With the definition of the air-gap and core reluctances $\Re_g = g/(\mu_0 A_g)$ and $\Re_c = \ell_c/(\mu_c \mu_0 A_c)$, respectively, one obtains for the magneto-motive force (mmf)

$$F = N \cdot i = (\Re_g + \Re_c)\, \phi. \tag{6.26}$$

Equation (6.15) is obtained from (6.26) for $\Re_c = 0$.

(a) The magnetic equivalent circuit of Fig. 6.6 is shown in Fig. 6.7 with $\Re_g = 7.96 \cdot 10^4 \frac{\text{At}}{\text{Wb}}$ and $\Re_c = 1.59 \cdot 10^4 \frac{\text{At}}{\text{Wb}}$.
The inductance is defined now as

$$L = \frac{N^2}{(\Re_g + \Re_c)} = 1.05 \text{ mH} \tag{6.27}$$

(b) The electrical equivalent circuit in the phasor domain is depicted in Fig. 6.8.
(c) For $v(t) = V_0 \cos \omega t = \sqrt{2} V_{rms} \cos \omega t$ or $\tilde{v}(t) = V_o \angle 0°$ one obtains the current in the phasor domain

$$\tilde{i}(t) = \frac{V_o \angle 0° - \theta}{\sqrt{R^2 + (\omega L)^2}} \quad \text{and} \quad \theta = \tan^{-1}\left(\frac{\omega L}{R}\right). \tag{6.28}$$

The current in the time domain is

$$i(t) = \frac{V_o}{\sqrt{R^2 + (\omega L)^2}} \cos(\omega t - \theta). \tag{6.29}$$

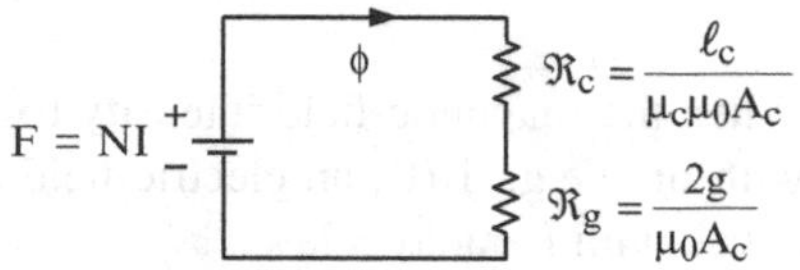

Fig. 6.7 Magnetic equivalent circuit with $F = N \cdot i = (\Re_g + \Re_c)\phi$

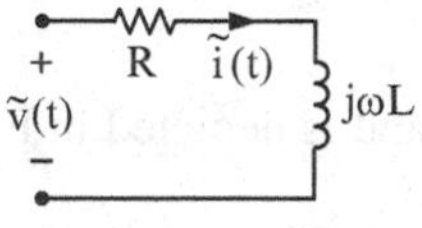

Fig. 6.8 Electrical equivalent circuit in phasor domain

(d) The time-averaged magnetic stored energy in the inductor $<W_{fld}>$ is

$$<W_{fld}> = \frac{1}{2}\cdot L\cdot\frac{1}{T}\int_0^T [i(t)]^2 dt = \frac{1}{2}\cdot L\cdot\frac{1}{T}\cdot\frac{V_o^2}{R^2+(\omega L)^2}\int_0^T \cos^2(\omega t-\theta)dt$$
$$= \frac{L}{4}\frac{V_o^2}{R^2+(\omega L)^2} = \frac{1}{2}L\frac{V_{rms}^2}{R^2+(\omega L)^2}. \tag{6.30}$$

(e) The instantaneous power is

$$p(t) = v(t)\, i(t) = V_o\cos(\omega t)\cdot\frac{V_o\cos(\omega t-\theta)}{\sqrt{R^2+(\omega L)^2}}$$
$$= \frac{V_o^2}{2\sqrt{R^2+(\omega L)^2}}\cos\,\theta + \frac{1}{2}\cdot\frac{V_o^2\cos(2\omega t-\theta)}{2\sqrt{R^2+(\omega L)^2}}. \tag{6.31}$$

(f) The time-averaged power $<p(t)>$ is

$$<p(t)> = \frac{V_o^2}{2\sqrt{R^2+(\omega L)^2}}\cos\,\theta = \frac{V_o^2}{2\sqrt{R^2+(\omega L)^2}}\frac{R}{\sqrt{R^2+(\omega L)^2}}$$
$$= \frac{V_{rms}^2 R}{R^2+(\omega L)^2}. \tag{6.32}$$

(g) The above time-averaged power $<p(t)>$ is identical with the power dissipated $<p(t)_{dissipated}>$ provided

$$I_{rms} = \frac{V_{rms}}{\sqrt{R^2+(\omega L)^2}},\quad \text{and } <p(t)_{dissipated}> = I_{rms}^2 R. \tag{6.33}$$

This application example will be relied on in Chap. 9 to demonstrate the calculation of magnetic forces for AC and DC excitation.

6.2 Flux Linkage, Inductance and Magnetic Stored Energy

When the magnetic-field intensity (or magnetic flux density) $\vec{B}$ is allowed to vary with time e.g., $\vec{B}(t)$, an electric-field strength $\vec{E}(t)$ is generated in space in accordance with Faraday's law

$$e(t) = \oint_C \vec{E}(t)\bullet d\vec{\ell} = \frac{d}{dt}\left(N\int_S \vec{B}(t)\bullet d\vec{S}\right) \tag{6.34}$$

and is depicted in Fig. 6.9.

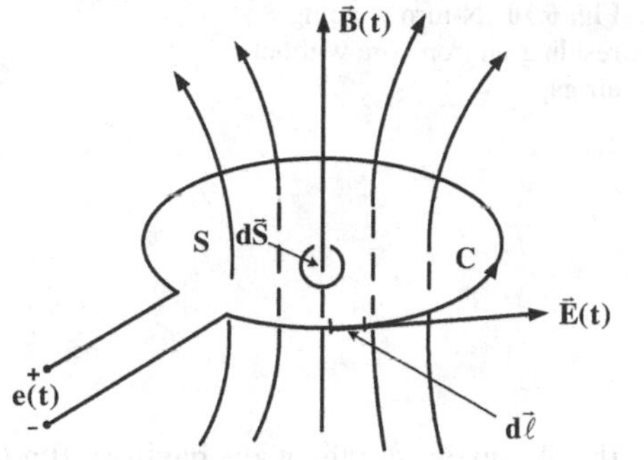

Fig. 6.9 Graphical illustration of Faraday's law (for N = 1 turn)

For N turns and

$$\phi = \int_S \vec{B} \bullet d\vec{S}, \tag{6.35}$$

follows

$$e(t) = N\frac{d\phi}{dt} = \frac{d\lambda}{dt}, \tag{6.36}$$

where the flux linkages are defined as

$$\lambda = N\,\phi. \tag{6.37}$$

For a magnetic circuit with $\mu_r \to \infty$, one can define the self inductance

$$L = \frac{\lambda}{i} = \frac{N\phi}{i} = N\frac{(F/\Re_g)}{i} = N\frac{(Ni/\Re_g)}{i} = \frac{N^2\mu_0 A_g}{g} \tag{6.38}$$

This equation shows the dimensional form of the expression for inductance.

Application Example 6.3: Application of Faraday's Law

(a) A sinusoidal voltage $e(t) = E_{max}\cos\omega t$ has a frequency $f = (\omega/2\pi) = 60$ Hz which is applied to a winding with zero ohmic resistance consisting of N = 1,000 turns surrounding a closed (without air gap) iron core of $A_c = 1.25 \cdot 10^{-3}$ m^2 cross-section (see Fig. 6.10).

1. Sketch the induced voltage and the core flux as a function of time.
2. Find the maximum permissible value of E_{max} if the maximum flux density is not to exceed $B_{max} = 1$ T.

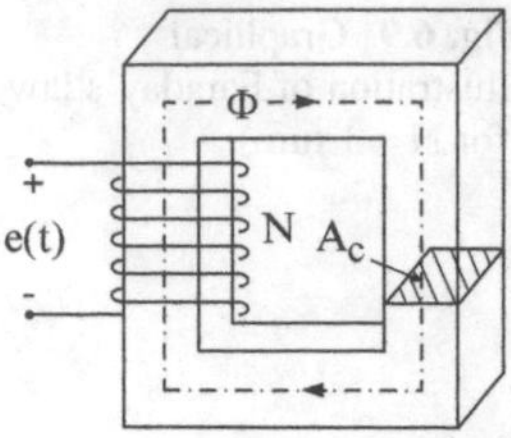

Fig. 6.10 N-turn winding residing on iron core without air gap

(b) A square-voltage wave having a fundamental frequency of f = 60 Hz and equal positive and negative half cycles – that is, there is no DC component – of magnitude E_{max} is applied to the same winding and iron core as in part (a).

1. Sketch the voltage, the core flux as a function of time, and the flux density; assume that the core flux does not have any DC component.
2. Find the maximum permissible value of E_{max} if the maximum flux density is not to exceed $B_{max} = 1$ T.

Solution

(a) The induced voltage is

$$e(t) = N\frac{d\Phi}{dt} \quad \text{or} \quad \Phi = \frac{1}{N}\int e(t)dt = \frac{1}{N}\frac{E_{max}}{\omega}\sin\omega t = \Phi_{max}\sin\omega t \tag{6.39}$$

where

$$\Phi_{max} = \frac{E_{max}}{N\,\omega} = A_c B_{max} \tag{6.40}$$

from this follows

$$E_{max} = N\cdot\omega\cdot A_c\cdot B_{max} = 1,000\cdot 377\cdot 1.25\cdot 10^{-3}\cdot 1 = 471\,\text{V}. \tag{6.41}$$

Figure 6.11 illustrates the induced voltage e(t), and core flux Φ(t) as a function of time.

(b) With $e(t) = N\frac{d\Phi}{dt}$ or

$$\Delta e(t) = N\frac{\Delta\Phi}{\Delta t} = N\frac{B_{max}A_c}{T/4} = \frac{1,000\cdot 1\cdot 1.25\cdot 10^{-3}}{16.666\,\text{m}/4} = 300\,\text{V}. \tag{6.42}$$

Figure 6.12 illustrates the induced voltage e(t), core flux Φ(t), and flux density B(t) as a function of time.

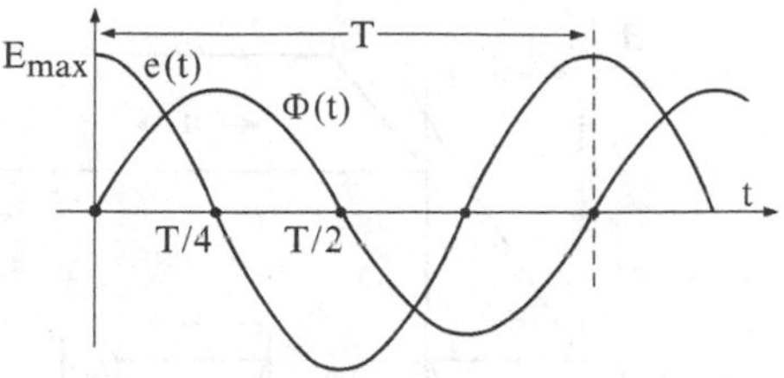

Fig. 6.11 Induced voltage e(t), and core flux Φ(t) as a function of time

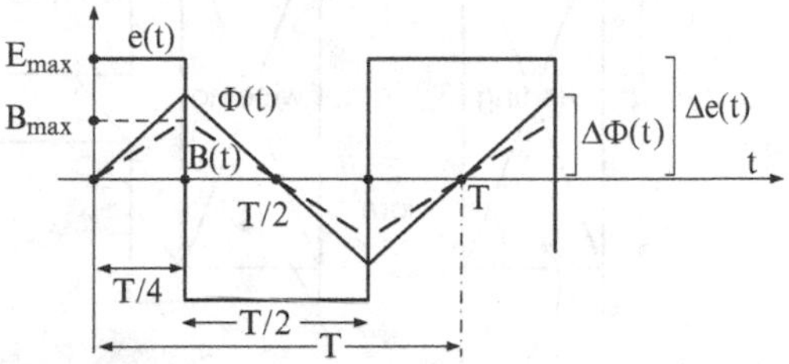

Fig. 6.12 Induced voltage e(t) and flux density B(t) as a function of time

This example demonstrates that sinusoidal signals utilize the material (winding, iron core) to the fullest as compared with square-wave signals. This is one of the reasons why most electromagnetic machines operate with the best efficiency if sinusoidal signals are present.

Application Example 6.4: Design of an Inductor

Figure 6.13 shows an inductor wound on a high-permeability laminated core of rectangular cross-section. Assume that the permeability of the iron is infinite. Neglect magnetic leakage and fringing around the air gap g. The winding consists of insulated copper wire whose resistivity is ρ [Ωm]. Assume that the fraction f_w (called winding fill factor) of the winding space is available for copper; the rest of the space being used for insulation.

(a) Estimate the mean length ℓ of one turn of the winding.
(b) Derive an expression for the electric power dissipation in the coil for a specified steady-state flux density B. This expression should be in terms of B, ρ, μ_o, ℓ, f_w, and the given dimensions. Note that the expression is independent of the number of turns N if the winding factor f_w is assumed to be independent of the turns.
(c) Derive an expression for the magnetic stored energy in terms of B and the given dimensions.
(d) From parts (b) and (c) derive an expression for the time constant τ = L/R of the inductor (coil).
(e) For the specific numerical values $a = h = w = 1.5$ cm $= 1.5 \cdot 10^{-2}$ m, $b = 2$ cm $= 2 \cdot 10^{-2}$ m, $g = 0.3$ cm $= 3 \cdot 10^{-3}$ m, winding factor $f_w = 0.70$, resistivity of copper $\rho = 1.73 \cdot 10^{-6}$ Ωcm $= 1.73 \cdot 10^{-8}$ Ωm, constant applied terminal voltage $V_t = 40$ V, and air-gap flux density $B = B_{max} = 1.2$ T, find numerical values for the power dissipated in the coil, coil current, number of turns, coil resistance, inductance, time constant, and wire size [3] to the nearest standard size.

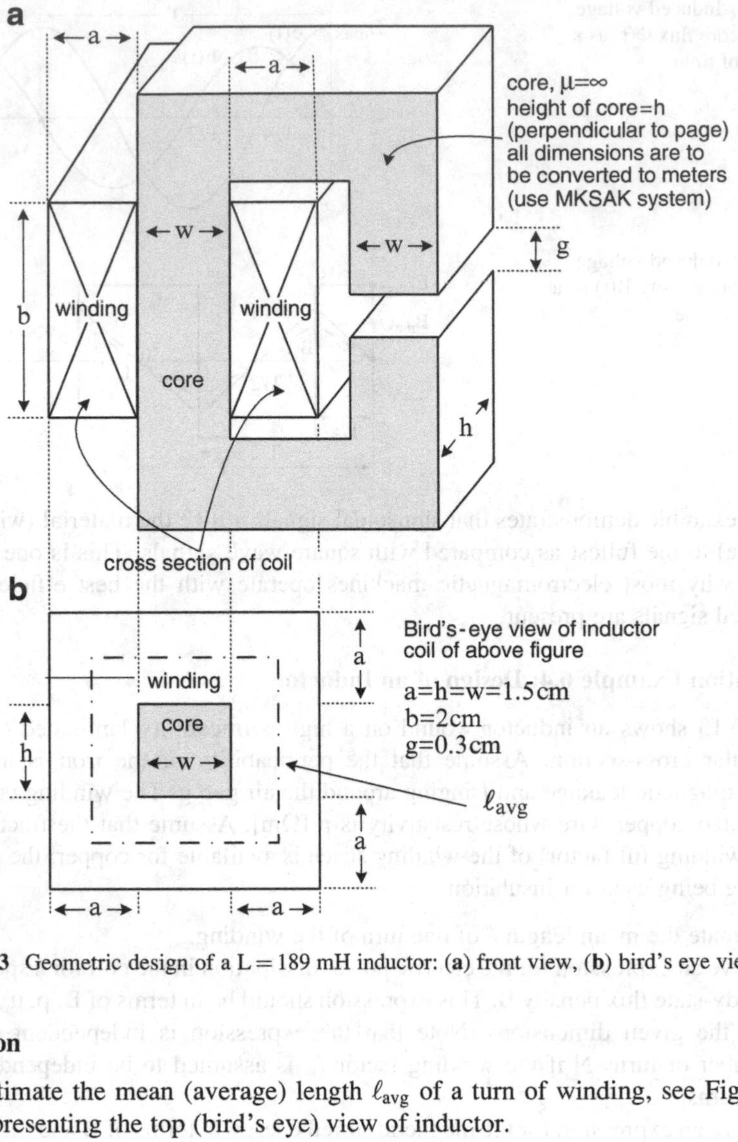

Fig. 6.13 Geometric design of a L = 189 mH inductor: (**a**) front view, (**b**) bird's eye view

Solution

(a) Estimate the mean (average) length ℓ_{avg} of a turn of winding, see Fig. 6.13b representing the top (bird's eye) view of inductor.

From geometric considerations

$$\ell_{avg} = 2(h + w + 2a) \text{ or } \ell_{avg} = 2(h + w + \pi(a/2)). \qquad (6.43)$$

(b) Electric power input to the coil (assume DC excitation):
The total ampere turns are $F = N{\cdot}i = g\ H_g$, or with $B_g = \mu_o H_g = B$ follows $F = g(B/\mu_o)$ or $i = g\ B/(\mu_o N)$.

The power is

$$p = i^2R = \frac{B^2g^2}{\mu_o^2N^2}\frac{\rho\ell_{avg}N}{area}, \tag{6.44}$$

where the wire cross-section is

$$area = \frac{abf_w}{N}, \tag{6.45}$$

and f_w is the (copper) winding fill factor.
Therefore, the power is now

$$p = \frac{B^2g^2\rho\ell_{avg}}{\mu_o^2abf_w}. \tag{6.46}$$

(c) Magnetic energy stored

$$W_{magnetic} = \frac{1}{2}\int_{volume\ of\ air\ gap} B_gH_g dV, \tag{6.47}$$

with $B_g = \mu_o H_g = B$ follows

$$W_{magnetic} = \frac{B^2}{2\mu_o}(\text{volume of air gap}) = \frac{B^2}{2\mu o}gwh. \tag{6.48}$$

(d) Time constant of coil $\tau = \frac{L}{R}$, where $L = \frac{\lambda}{i}$, $\lambda = N\phi = N{\cdot}B{\cdot}h{\cdot}w$, and $i = \frac{B}{\mu_o N}g$, therefore,

$$L = \frac{N^2h \cdot w \cdot \mu_o}{g}, \text{ and } R = \frac{\rho \cdot \ell_{avg} \cdot N^2}{a \cdot b \cdot f_w}, \tag{6.49}$$

and the time constant is

$$\tau = \frac{L}{R} = \frac{\mu_o}{\rho}\frac{a \cdot b \cdot h \cdot w \cdot f_w}{g \cdot \ell_{avg}}. \tag{6.50}$$

(e) For the given values one can calculate:
The average turn length

$$\ell_{avg} = 2(h + w + 2a) = 12 \cdot 10^{-2}\text{m}, \tag{6.51}$$

the power dissipated is

$$p_{diss} = \frac{B^2 g^2 \rho \ell_{avg}}{\mu_o^2 abf_w} = \frac{(1.2)^2 9 \cdot 10^{-6} 1.73 \cdot 10^{-8} 12 \cdot 10^{-2}}{(4\pi)^2 \cdot 10^{-14} 3 \cdot 10^{-4} 0.7} = 81\,\text{W}. \tag{6.52}$$

The coil current is now

$$i = \frac{p_{diss}}{V_t} = \frac{81}{40} = 2.02\text{A}, \tag{6.53}$$

and the coil resistance is

$$R_{coil} = \frac{V_t}{i} = \frac{40}{2.02} = 19.8\ \Omega. \tag{6.54}$$

The number of turns is

$$N = \frac{J(a \cdot b \cdot f_w)}{i} = \frac{B \cdot g \cdot a \cdot b \cdot f_w}{\mu_o a \cdot b \cdot f_w \cdot i} = \frac{B \cdot g}{\mu_o \cdot i} = 1{,}418 \ \ \text{turns}. \tag{6.55}$$

Note in the above equation J is the current density

$$J = \frac{B \cdot g}{\mu_o a \cdot b \cdot f_w}. \tag{6.56}$$

The inductance is

$$L = \frac{N^2 \cdot h \cdot w \cdot \mu_o}{g} = 0.189\ \text{H, and the time constant } \tau = \frac{L}{R} = \frac{0.189\,\text{H}}{19.8\ \Omega}$$
$$= 9.54\,\text{ms}. \tag{6.57}$$

Wire cross-section

area $= \frac{a \cdot b \cdot f_w}{N} = \frac{1.5 \cdot 2 \cdot 10^{-4} 0.7}{1{,}418} = 14.81 \cdot 10^{-6} \text{m}^2 = 0.1481\,\text{mm}^2$, the wire cross-section is related to the wire diameter or wire radius as area $= \pi R_{wire}^2 = \frac{\pi}{4} D_{wire}^2$, or

$$D_{wire} = \left(\text{area} \cdot \frac{4}{\pi}\right)^{1/2} = 0.434\ \text{mm} = \frac{0.434}{25.4}\text{in.} = 0.0171\ \text{in.} \tag{6.58}$$

From the wire table, Table 6.2 one gets now the wire size #25.

Table 6.3 presents additional information about wire sizes. A large proportion of electrical signals are transmitted through solid electrical conductors. Most such signal-carrying conductors are in the form of wires or cables. A *wire* is a single

Table 6.2 American wire gauge (AWG) sizes (Adapted from *Electrical Engineering Pocket Handbook*, Electrical Apparatus Service Association, Inc., 1331 Baur Blvd, St. Louis, Missouri 63132, [3])

Dimensions, weight and resistance of round copper and aluminum wire

Wire size	Diameter		DC Resistance at 29°C (68 °F) at 100% conductivity		DC Resistance at 29°C (68 °F) at 61.8% conductivity		Weight				Wire size
			Copper		Aluminum		Copper		Aluminum		
AWG	Nom. inch	Circular mils	Ohms per 1,000 feet	Ohms per lb-force	Ohms per 1,000 feet	Ohms per lb-force	lb-force per 1,000 feet	Feet per lb-force	lb-force per 1,000 feet	Feet per lb-force	AWG
4/0	0.4600	211,600	0.04901	0.00007652	0.07930	0.0004072	640.5	1.561	194.7	5.135	4/0
3/0	0.4096	167,800	0.06182	0.0001218	0.1000	0.0006482	507.8	1.969	154.4	6.478	3/0
2/0	0.3648	133,100	0.07793	0.0001935	0.1261	0.001030	402.8	2.482	122.5	8.163	2/0
1/0	0.3249	105,600	0.09825	0.0003075	0.1590	0.001637	319.5	3.130	97.14	10.29	1/0
1	0.2893	83,690	0.1239	0.0004891	0.2005	0.002603	253.3	3.947	77.02	12.98	1
2	0.2576	66,360	0.1563	0.0007781	0.2529	0.004141	200.9	4.978	61.08	16.37	2
3	0.2294	52,620	0.1971	0.001237	0.3189	0.006584	159.3	6.278	48.44	20.65	3
4	0.2043	41,740	0.2485	0.001967	0.4021	0.01047	126.3	7.915	38.40	26.04	4
5	0.1819	33,090	0.3134	0.003130	0.5072	0.01666	100.2	9.984	30.47	32.82	5
6	0.1620	26,240	0.3952	0.004975	0.6395	0.02647	79.44	12.59	24.15	41.41	6
7	0.1443	20,820	0.4981	0.007902	0.8060	0.04206	63.03	15.87	19.16	52.19	7
8	0.1285	16,510	0.6281	0.01257	1.016	0.06688	49.98	20.01	15.20	55.79	8
9	0.1144	13,090	0.7925	0.02000	1.282	0.1065	39.62	25.24	12.05	82.99	9
10	0.1019	10,380	0.9988	0.03180	1.616	0.1691	31.43	31.82	9.56	105	10
11	0.0907	8,230	1.26	0.0506	2.04	0.269	24.9	40.2	7.57	132	11
12	0.0808	6,530	1.59	0.0804	2.57	0.428	19.8	50.6	6.02	166	12
13	0.0720	5,180	2.00	0.127	3.24	0.679	15.7	63.7	4.77	210	13
14	0.0641	4,110	2.52	0.203	4.08	1.08	12.4	80.4	3.77	265	14
15	0.0571	3,260	3.18	0.322	5.15	1.72	9.87	101	3.00	333	15
16	0.0508	2,580	4.02	0.514	6.50	2.74	7.81	128	2.37	422	16
17	0.0453	2,050	5.05	0.814	8.18	4.33	6.21	161	1.89	529	17
18	0.0403	1,620	6.39	1.30	10.3	6.91	4.92	203	1.50	666	18
19	0.0359	1,290	8.05	2.06	13.0	11.0	3.90	256	1.19	840	19
20	0.0320	1,020	10.1	3.27	16.4	17.4	3.10	323	0.943	1,060	20

(continued)

Table 6.2 (continued)

Dimensions, weight and resistance of round copper and aluminum wire

Wire size	Diameter		DC Resistance at 29°C (68 °F) at 100% conductivity		DC Resistance at 29°C (68 °F) at 61.8% conductivity		Weight				Wire size
			Copper		Aluminum		Copper		Aluminum		
AWG	Nom. inch	Circular mils	Ohms per 1,000 feet	Ohms per lb-force	Ohms per 1,000 feet	Ohms per lb-force	lb-force per 1,000 feet	Feet per lb-force	lb-force per 1,000 feet	Feet per lb-force	AWG
21	0.0285	812	12.8	5.19	20.7	27.6	2.46	407	0.749	1,340	21
22	0.0253	640	16.2	8.36	26.2	44.5	1.94	516	0.590	1,690	22
23	0.0226	511	20.3	13.1	32.9	69.9	1.55	647	0.471	2,120	23
24	0.0201	404	25.7	21.0	41.5	112	1.22	818	0.371	2,700	24
25	0.0179	320	32.4	33.4	52.4	178	0.970	1,030	0.295	3,390	25
26	0.0159	253	41.0	53.6	66.4	285	0.765	1,310	0.233	4,290	26
27	0.0142	202	51.4	84.3	83.2	449	0.610	1,640	0.185	5,410	27
28	0.0126	159	65.3	136	106	724	0.481	2,080	0.146	6,850	28
29	0.0113	128	81.2	210	131	1120	0.387	2,590	0.118	8,470	29
30	0.0100	100	104	343	168	1,830	0.303	3,300	0.0921	10,900	30
31	0.0089	79.2	131	546	212	2,910	0.240	4,170	0.0730	13,700	31
32	0.0080	64.0	162	836	262	4,450	0.194	5,160	0.0590	16,900	32
33	0.0071	50.4	206	1,350	333	7,180	0.153	6,550	0.0465	21,500	33
34	0.0063	39.7	261	2,180	422	11,600	0.120	8,330	0.0365	27,400	34
35	0.0056	31.4	331	3,480	536	18,500	0.0949	10,500	0.0289	34,600	35
36	0.0050	25.0	415	5,480	671	29,200	0.0757	13,200	0.0230	43,500	36
37	0.0045	20.2	512	8,360	–	–	0.0613	16,300	–	–	37
38	0.0040	16.0	648	13,400	–	–	0.0484	20,600	–	–	38
39	0.0035	12.2	874	22,800	–	–	0.0371	27,000	–	–	39
40	0.0031	9.61	1,080	37,100	–	–	0.0291	34,400	–	–	40
41	0.0028	7.84	1,320	55,700	–	–	0.0237	42,100	–	–	41
42	0.0025	6.25	1,660	87,700	–	–	0.0189	52,900	–	–	42
43	0.0022	4.84	2,140	146,000	–	–	0.0147	68,300	–	–	43
44	0.0020	4.00	2,590	214,000	–	–	0.0121	82,600	–	–	44

Table 6.3 American wire gauge (AWG) sizes of copper wire

Application	AWG number	Area (circular[a] mils)	Ohms per 1,000 ft at (20°C)	Maximum allowable current (A) at (90°C)
Power distribution	500 MCH	25×10^{10}	0.0216	427
	0000	211,600	0.049	253
	00	133,100	0.078	186
	1	83,690	0.124	137
	4	41,470	0.240	89
House main power carriers	6	26,240	0.395	65
	8	16,510	0.620	48
Lighting, outlets, general home use	12	6,530	1.588	20
	14	4,110	2.52	15
Television, radio	20	1021.5	10.1	–[b]
	22	642.4	16.1	–[b]
Telephone instruments	28	159.8	64.9	–[b]
	35	31.5	329.0	–[b]
	40	9.9	1049.0	–[b]

[a]1 circular mil = 1CM = (diameter of wire in mils)^2 = d^2
[b]Current rating must be calculated per NEC® section 310–315

conductor. A *cable* is a configuration of insulated wires bound together with a plastic sleeve or ties, and is color coded for identification purposes. A single large power wire is also referred to as a cable.

Usually, the best electrical conductor is most suitable as a carrier of electrical signals. In other words, the better the electrical conductor, the lower the resistance losses that occur during the transmission of electrical signals. Therefore, the conductors in wires and cables are usually made of copper or aluminum. Gold, tin, and nickel are used extensively as conductors in semiconductors, relays, and sensors. The earth and the ocean are used as conductors for large transmission systems. Wire manufactured in the United States is sized according to the American Wire Gauge (AWG) convention. Wire numbers larger than 0000 are expressed in circular mils [4], because the wires are stranded rather than solid.

The properties of the wire shown in Table 6.3 are based upon the National Electric Code® [5] resistance and current carrying capacity. The National Electrical Code®, NEC® is a trademark of the National Fire Protection Association, Inc. Note, 1 mil = 1 inch/1,000.

6.2.1 Magnetic Circuit with Two Windings

Figure 6.14a illustrates a magnetic circuit ($\mu_r \rightarrow \infty$) with an air gap g and two windings (coils) with N_1 and N_2 number of turns. The winding sense of the coils is indicated by dot (•) polarity markings. Note that the reference directions of the

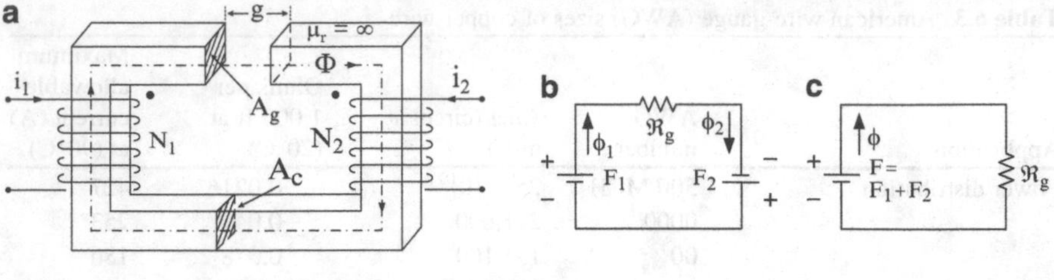

Fig. 6.14 (**a**) Inductor with "C" core, air gap g and two windings with turns N_1 and N_2. (**b** and **c**) Magnetic equivalent circuits, (**a**) separate mmfs F_1 and F_2, (**c**) combined mmf F with $\phi = \phi_1 = \phi_2$

currents i_1 and i_2 have been chosen to produce (for the given winding sense) flux Φ in the same direction. Figure 6.14b, c represent the magnetic equivalent circuits of Fig. 6.14a.

The total mmf is

$$F = N_1 \cdot i_1 + N_2 \cdot i_2. \tag{6.59}$$

With $\Re_g = g/(\mu_0 A_g)$ follows for the flux

$$\phi = F/\Re_g = (N_1 \cdot i_1 + N_2 \cdot i_2)\ \mu_0 A_g/g, \tag{6.60}$$

where ϕ is the resultant core flux due to i_1 and i_2. If the above relation is broken up into terms attributable to the individual currents, the resultant flux linkages with coil 1 can be expressed as

$$\lambda_1 = N_1 \phi = \left(N_1^2 \frac{\mu_0 A_g}{g}\right) i_1 + \left(N_1 N_2 \frac{\mu_0 A_g}{g}\right) i_2, \tag{6.61}$$

or

$$\lambda_1 = L_{11} i_1 + L_{12} i_2, \tag{6.62}$$

where the self inductance is

$$L_{11} = \left(N_1^2 \frac{\mu_0 A_g}{g}\right), \tag{6.63}$$

and the mutual inductance

$$L_{12} = \left(N_1 N_2 \frac{\mu_0 A_g}{g}\right). \tag{6.64}$$

Note that L_{12} corresponds to the inductance caused by the flux linkage with coil 1 (index 1 of L_{12}) due to the current i_2 in coil 2 (index 2 of L_{12}). Similarly, the flux linkage with coil 2 is

$$\lambda_2 = N_2\ \phi = \left(N_1 N_2 \frac{\mu_0 A_g}{g}\right) i_1 + \left(N_2^2 \frac{\mu_0 A_g}{g}\right) i_2, \tag{6.65}$$

or

$$\lambda_2 = L_{21} i_1 + L_{22} i_2, \tag{6.66}$$

where $L_{21} = L_{12}$ (law of reciprocity) is the mutual inductance. The self inductance of winding (coil) 2 is

$$L_{22} = \left(N_2^2 \frac{\mu_0 A_g}{g}\right). \tag{6.67}$$

Induced Voltage: Substitution of $\lambda = L \cdot i$ into $e = d\lambda/dt$ yields for a magnetic static circuit with a single winding and constant inductance L

$$e = d(Li)/dt = L(di/dt). \tag{6.68}$$

However, in electromechanical energy-conversion devices inductances are often time varying and one obtains in this case for linear motion where the inductance is a function of the linear displacement x

$$e(t) = \frac{d}{dt}\{L(x)i(t)\} = L(x)\frac{di}{dt} + i\frac{dL(x)}{dx}\frac{dx}{dt}, \tag{6.69}$$

or for rotary motion where the inductance is a function of the angular displacement θ

$$e(t) = \frac{d}{dt}\{L(\theta)i(t)\} = L(\theta)\frac{di}{dt} + i\frac{dL(\theta)}{d\,\theta}\frac{d\theta}{dt}, \tag{6.70}$$

where θ is, for example, the circumferential coordinate of an electric machine.

The first term of the induced voltage (6.69 and 6.70) is called the “transformer voltage” occurring predominantly in transformers, as will be discussed in Chap. 7. The second term is denoted the “speed voltage”, as it occurs mainly in linear and rotating machines, discussed in Chap. 10. Both voltage components occur in homopolar machines as discussed in Chap. 10.

Power: The power at the terminals of a winding of a magnetic circuit (e.g., inductor) is a measure of the rate of energy flow into the circuit through that particular winding. The power is determined from the product of the voltage and the current

$$p(t) = e(t)i(t) = i(t)\frac{d\lambda(t)}{dt}. \tag{6.71}$$

Magnetic Stored Energy: The change in magnetic stored energy ΔW in the magnetic circuit in the time interval from t_1 to t_2 is (see Application Example 6.5)

$$\Delta W = \int_{t1}^{t2} p(t)dt = \int_{\lambda 1}^{\lambda 2} i(\lambda)d\lambda. \tag{6.72}$$

For a single-winding system of constant inductance (linear circuit and there is no motion), the change in magnetic stored energy can be written as

$$\Delta W = \int_{\lambda 1}^{\lambda 2} i(\lambda)d\lambda = \int_{\lambda 1}^{\lambda 2} \frac{\lambda}{L}d\lambda = \frac{1}{2L}(\lambda_2^2 - \lambda_1^2). \tag{6.73}$$

The total magnetic stored energy at any given value of λ can be found from setting λ_1 equal to zero and $\lambda_2 = \lambda$.

$$W = \frac{1}{2L}\lambda^2 = \frac{1}{2}Li^2. \tag{6.74}$$

With (6.13)

$$i = \frac{g}{\mu_0 A_g N}\phi, \text{ and } \phi = B_g A_g \tag{6.75}$$

and (6.38)

$$L = \frac{N^2 \mu_0 A_g}{g} \tag{6.76}$$

follows

$$W = \frac{1}{2}\frac{B_g^2}{\mu_0}A_g\, g = \frac{1}{2}\int_{\text{volume of air gap}} B_g H_g dV. \tag{6.77}$$

This means the entire energy W is stored in the air gap and none will be stored in the core because $\mu_c = \infty$ results in $H_c = 0$.

Application Example 6.5: Differential Energy ΔW as a Function of i and λ

Show that

$\Delta W = \int_{t1}^{t2} p(t)dt = \int_{t1}^{t2} eidt = \int_{\lambda 1}^{\lambda 2} i(\lambda)d\lambda$ for given e(t) and i(t) functions, e.g., $e(t) = E_{max}\cos\omega t$, $R = 0$, and linearity may be assumed, that is, saturation can be neglected (Fig. 6.15).

Solution

From R = 0 follows e(t) = v(t),

$$e(t) = d\lambda/dt, \tag{6.78}$$

or

$$\lambda = \int_0^t e dt = \frac{E_{max}}{\omega} \sin \omega t = \lambda_{max} \sin \omega t. \tag{6.79}$$

Instantaneous power is defined as

$$p = e(t)\ i(t), \tag{6.80}$$

and differential energy absorbed (stored because of R = 0) is

$$\Delta W = \int_{t1}^{t2} p(t)dt = \int_{t1}^{t2} e(t)i(t)dt.$$

With

$$i(t) = \frac{\lambda}{L} = \frac{\lambda_{max}}{L} \sin \omega t = I_{max} \sin \omega t \tag{6.81}$$

follows

$$\Delta W = \int_{t1}^{t2} p(t)dt = \int_{t1}^{t2} eidt = \int_{t1}^{t2} E_{max} \cos \omega t\, I_{max} \sin \omega t\, dt,$$

or

$$\Delta W = E_{max} I_{max} \left| \frac{1}{2\omega} \sin^2 \omega t \right|_{t1}^{t2} = \frac{E_{max} I_{max}}{2\omega} \left| \sin^2 \omega t_2 - \sin^2 \omega t_1 \right|. \tag{6.82}$$

We have to show that

$$\Delta W = \int_{t1}^{t2} eidt = \int_{t1}^{t2} \frac{d\lambda}{dt} idt = \int_{\lambda 1}^{\lambda 2} i(\lambda) d\lambda, \tag{6.83}$$

this operation corresponds to a transformation of the variable of integration (Fig. 6.15).

In the above equation

$$\lambda_1 = \lambda_{max} \sin \omega t_1, \tag{6.84}$$

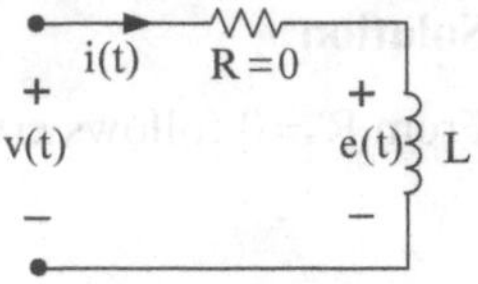

Fig. 6.15 Inductive circuit

$$\lambda_2 = \lambda_{max} \sin \omega t_2. \tag{6.85}$$

$$\Delta W = \int_{\lambda 1}^{\lambda 2} i(\lambda) d\lambda = \int_{\lambda 1}^{\lambda 2} \frac{\lambda}{L} d\lambda = \left| \frac{1}{2L} \lambda^2 \right|_{\lambda 1}^{\lambda 2} = \frac{1}{2L} \left[\lambda_2^2 - \lambda_1^2 \right]. \tag{6.86}$$

For specific values $t_1 = 0$, $t_2 = \frac{\pi/2}{\omega} = \frac{1}{4f}$ one obtains from (6.84) and (6.85) $\lambda_1 = 0$ and $\lambda_2 = \lambda_{max}$. With (6.86)

$$\Delta W_{Eq.\,6.86} = \frac{1}{2L} \lambda_{max}^2. \tag{6.87}$$

With (6.82)

$$\Delta W_{Eq.\,6.82} = \frac{E_{max} I_{max}}{2\omega} [1 - 0] = \frac{E_{max} I_{max}}{2\omega}. \tag{6.88}$$

Because of $\lambda_{max} = \frac{E_{max}}{\omega}$, and $L = \frac{\lambda_{max}}{I_{max}}$ one obtains from (6.86)

$$\Delta W_{Eq.\,6.86} = \frac{1}{2L} \lambda_{max}^2 = \frac{I_{max}}{2\,\lambda_{max}} \lambda_{max}^2 = \frac{\lambda_{max} I_{max}}{2} = \frac{E_{max} I_{max}}{2\omega},$$

thus for linear circuits

$$\Delta W_{Eq.\ 6.86} = \Delta W_{Eq.\ 6.82}. \tag{6.89}$$

Figure 6.16 shows measured quantities: $i(t)$ and $\lambda(t)$ are in phase if hysteresis is neglected. Also, they have the same waveshape provided saturation is neglected.

Application Example 6.6: Calculation of Mutual Inductances and Induced Voltages

The symmetric magnetic circuit of Fig. 6.17 has three windings A, B, and C, with N-turns each, air gap lengths $g_1 = g_2 = g_3 = g$ and air-gap cross sections $A_1 = A_2 = A_3 = A$.

(a) Draw an equivalent magnetic circuit identifying reluctances $\Re_1$, $\Re_2$, $\Re_3$, mmfs F_A, F_B, F_C, and fluxes ϕ_A, ϕ_B, ϕ_C.
(b) Find the self inductance of each of the 3 windings.

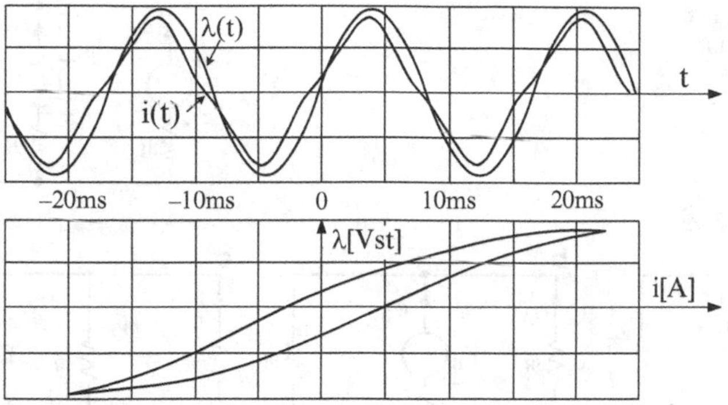

Fig. 6.16 Measured current i(t), λ(t) and (λ–i) characteristic, where the unit of λ is [V·s·turns] or [V·s·t] corresponds to Wb [weber · turns]

Fig. 6.17 Symmetric magnetic circuit

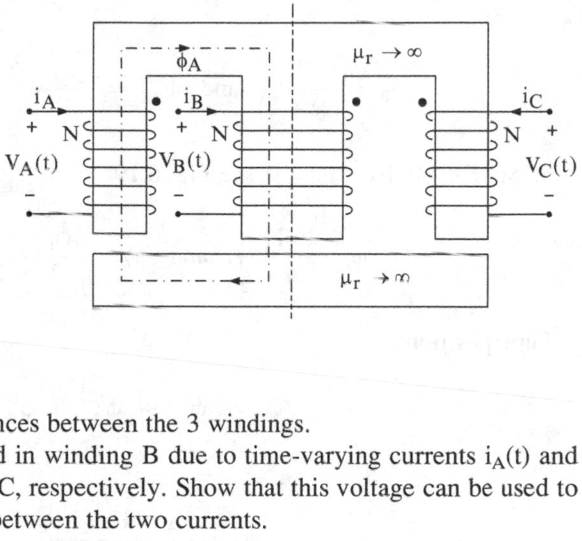

(c) Find the mutual inductances between the 3 windings.
(d) Find the voltage induced in winding B due to time-varying currents $i_A(t)$ and $i_C(t)$ in windings A and C, respectively. Show that this voltage can be used to measure the difference between the two currents.

Solution

(a) Equivalent circuit is shown in Fig. 6.18.

(b) and (c) Find self- and mutual inductances:

Calculation of ϕ_A via superposition using Figs. 6.19a–c

1. Set $F_A \neq 0$, $F_B = F_C = 0$, see Fig. 6.19a

$$\phi_A^{(1)} = \frac{F_A}{\frac{3}{2}\Re}. \tag{6.90}$$

2. Set $F_B \neq 0$, $F_A = F_C = 0$, see Fig. 6.19b

Fig. 6.18 Magnetic equivalent circuit

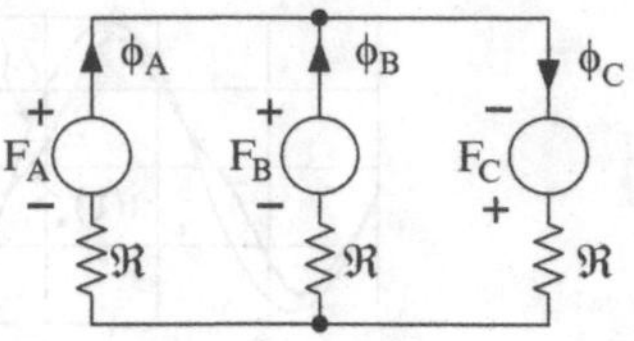

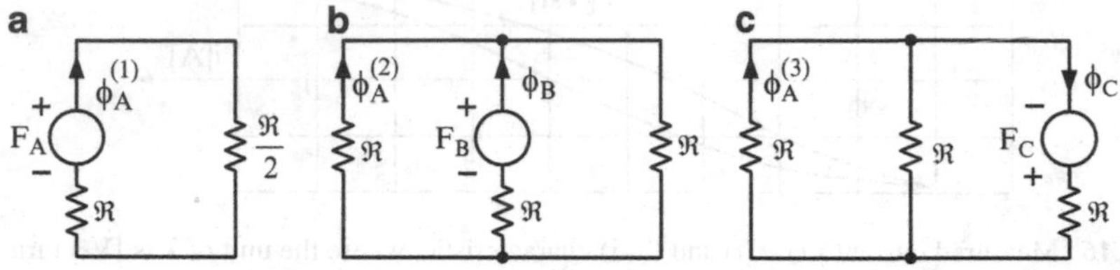

Fig. 6.19 Equivalent circuits of Fig. 6.18, (**a**) for calculation of $\phi_A^{(1)}$: $F_A \neq 0$, $F_B = F_C = 0$, (**b**) for calculation of ϕ_B, and $\phi_A^{(2)}$: $F_B \neq 0$, $F_A = F_C = 0$, and (**c**) for calculation of ϕ_C, and $\phi_A^{(3)}$: $F_C \neq 0$, $F_A = F_B = 0$

$$\phi_B = \frac{F_B}{\left(\Re + \frac{\Re}{2}\right)}, \text{and } \phi_A^{(2)} = -\frac{\phi_B}{2} = -\frac{F_B}{(2\Re + \Re)}. \tag{6.91}$$

3. Set $F_C \neq 0$, $F_A = F_B = 0$, see Fig. 6.19c

$$\phi_C = \frac{F_C}{\Re + \frac{\Re}{2}}, \text{ and } \phi_A^{(3)} = \frac{\phi_C}{2} = \frac{F_C}{2\Re + \Re}. \tag{6.92}$$

Superposition:

$$\phi_A = \phi_A^{(1)} + \phi_A^{(2)} + \phi_A^{(3)} \tag{6.93}$$

or

$$\phi_A = \frac{F_A}{\frac{3\Re}{2}} - \frac{F_B}{3\Re} + \frac{F_C}{3\Re}, \tag{6.94}$$

with

$$\lambda_A = N\,\phi_A = \frac{N^2 i_A}{\frac{3\Re}{2}} - \frac{N^2 i_B}{3\Re} + \frac{N^2 i_C}{3\Re} = L_{AA} i_A + L_{AB} i_B + L_{AC} i_C, \tag{6.95}$$

therefore,

$$L_{AA} = \frac{N^2}{\frac{3\Re}{2}};\ L_{AB} = -\frac{N^2}{3\Re};\ L_{AC} = \frac{N^2}{3\Re}. \tag{6.96}$$

Calculation of ϕ_B via superposition using Figs. 6.20a–c

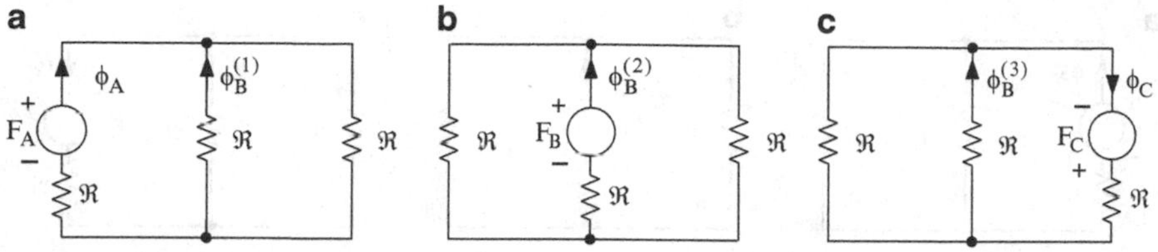

Fig. 6.20 Equivalent circuits of Fig. 6.18, (**a**) for calculation of ϕ_A, and $\phi_B^{(1)}$: $F_A \neq 0$, $F_B = F_C = 0$, (**b**) for calculation of $\phi_B^{(2)}$: $F_B \neq 0$, $F_A = F_C = 0$, and (**c**) for calculation of ϕ_C, and $\phi_B^{(3)}$: $F_C \neq 0$, $F_A = F_B = 0$

4. Set $F_A \neq 0$, $F_B = F_C = 0$, see Fig. 6.20a

$$\phi_A = \frac{F_A}{\Re + \frac{\Re}{2}} = \frac{F_A}{\frac{3}{2}\Re}, \tag{6.97}$$

$$\phi_B^{(1)} = -\frac{\phi_A}{2} = -\frac{F_A}{3\Re}. \tag{6.98}$$

5. Set $F_B \neq 0$, $F_A = F_C = 0$, see Fig. 6.20b

$$\phi_B^{(2)} = \frac{F_B}{\left(\Re + \frac{\Re}{2}\right)} = \frac{F_B}{\frac{3\Re}{2}}. \tag{6.99}$$

6. Set $F_C \neq 0$, $F_A = F_B = 0$, see Fig. 6.20c

$$\phi_C = \frac{F_C}{\Re + \frac{\Re}{2}}, \text{ and } \phi_B^{(3)} = \frac{\phi_C}{2} = \frac{F_C}{2\Re + \Re} = \frac{F_C}{3\Re}. \tag{6.100}$$

Superposition:

$$\phi_B = \phi_B^{(1)} + \phi_B^{(2)} + \phi_B^{(3)} \tag{6.101}$$

or

$$\phi_B = -\frac{F_A}{3\Re} + \frac{F_B}{\frac{3\Re}{2}} + \frac{F_C}{3\Re}, \tag{6.102}$$

with

$$\lambda_B = N\,\phi_B = -\frac{N^2 i_A}{3\Re} + \frac{N^2 i_B}{\frac{3\Re}{2}} + \frac{N^2 i_C}{3\Re} = L_{BA} i_A + L_{BB} i_B + L_{BC} i_C, \tag{6.103}$$

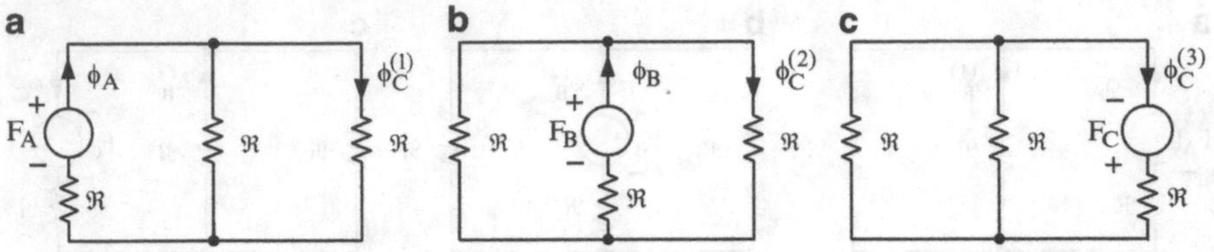

Fig. 6.21 Equivalent circuits of Fig. 6.18, (**a**) for calculation of ϕ_A, and $\phi_C^{(1')}$: $F_A \neq 0$, $F_B = F_C = 0$, (**b**) for calculation of ϕ_B, and $\phi_C^{(2)}$: $F_B \neq 0$, $F_A = F_C = 0$, and (**c**) for calculation of $\phi_C^{(3)}$: $F_C \neq 0$, $F_A = F_B = 0$

therefore,

$$L_{BA} = -\frac{N^2}{3\Re};\ \ L_{BB} = \frac{N^2}{\frac{3\Re}{2}};\ \ L_{BC} = \frac{N^2}{3\Re}. \tag{6.104}$$

Calculation of ϕ_C via superposition using Figs. 6.21a–c

7. Set $F_C = F_B = 0$, $F_A \neq 0$, see Fig. 6.21a

$$\phi_A = \frac{F_A}{\Re + \frac{\Re}{2}} = \frac{F_A}{\frac{3}{2}\Re}, \tag{6.105}$$

$$\phi_C^{(1)} = \frac{\phi_A}{2} = \frac{F_A}{3\Re}. \tag{6.106}$$

8. Set $F_A = F_C = 0$, $F_B \neq 0$, see Fig. 6.21b

$$\phi_B = \frac{F_B}{\left(\Re + \frac{\Re}{2}\right)} = \frac{F_B}{\frac{3\Re}{2}}, \tag{6.107}$$

$$\phi_C^{(2)} = \frac{F_B}{3\Re}. \tag{6.108}$$

9. Set $F_A = F_B = 0$, $F_C \neq 0$, see Fig. 6.21c

$$\phi_C^{(3)} = \frac{F_C}{\Re + \frac{\Re}{2}} = \frac{F_C}{\frac{3\Re}{2}}, \tag{6.109}$$

Superposition:

$$\phi_C = \phi_C^{(1)} + \phi_C^{(2)} + \phi_C^{(3)} \tag{6.110}$$

or

$$\phi_C = \frac{F_A}{3\Re} + \frac{F_B}{3\Re} + \frac{F_C}{\frac{3\Re}{2}}, \tag{6.111}$$

with

$$\lambda_C = N\phi_C = \frac{N^2 i_A}{3\Re} + \frac{N^2 i_B}{3\Re} + \frac{N^2 i_C}{\frac{3\Re}{2}} = L_{CA} i_A + L_{CB} i_B + L_{CC} i_C, \tag{6.112}$$

therefore,

$$L_{CA} = \frac{N^2}{3\Re};\ L_{CB} = \frac{N^2}{3\Re};\ L_{CC} = \frac{N^2}{\frac{3\Re}{2}}. \tag{6.113}$$

(d) Calculation of induced voltage $v_B(t)$

$$\begin{aligned} v_B(t) &= \frac{d}{dt}(L_{BA} i_A(t) + L_{BC} i_C(t)) = \frac{d}{dt}\left(-\frac{N^2}{3\Re} i_A(t) + \frac{N^2}{3\Re} i_C(t)\right) \\ &= \frac{N^2}{3\Re}\frac{d}{dt}(i_C(t) - i_A(t)). \end{aligned} \tag{6.114}$$

6.3 Magnetic Properties of Materials

There are three types of magnetic materials:

- Ferromagnetic ($\mu_r > 1$) such as iron (Fe),
- Paramagnetic ($\mu_r = 1$) such as free space,
- Diamagnetic ($\mu_r < 1$) such as water (H_2O), and bismuth (Bi)

In the context of electromechanical energy-conversion devices, the importance of ferromagnetic materials is twofold:

- Through their use it is possible to obtain relatively large magnetic maximum flux densities (e.g., B_{max} ranges from 0 to 2 T) with relatively low levels of magnetizing currents. Because magnetic forces (see Chaps. 9 and 10) are increased with increasing flux density, this effect plays a large role in the performance of energy-conversion devices.
- Ferromagnetic materials can be used to constrain and direct magnetic fields in well-defined paths.

Ferromagnetic Materials: They are composed of iron and alloys of iron with cobalt, tungsten, nickel, aluminum and other metals. If one takes a microscope one can discern a large number of domains within the structure of ferromagnetic materials. These are regions in which the dipole magnetic moments (see Fig. 6.22a with magnetic moment $= q_o\omega r^2/2$) of all the atoms within one domain are parallel, giving rise to a net magnetic moment for that domain. In a sample of

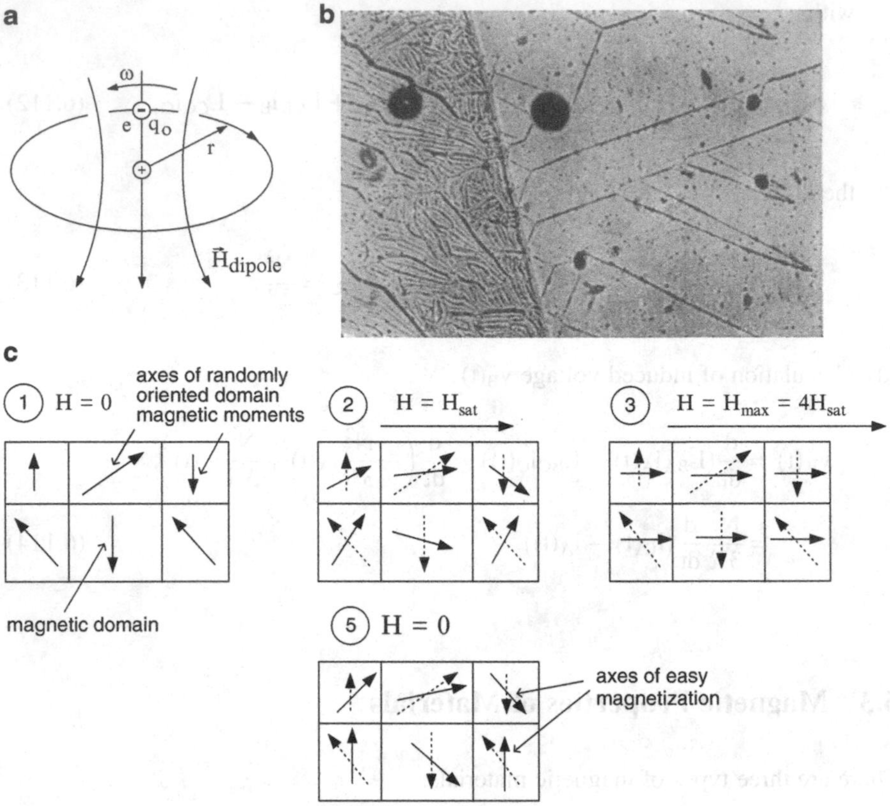

Fig. 6.22 (**a**) Magnetic dipole moment of hydrogen atom, where q_o is the electron charge and ω is the angular velocity of the electron. (**b**) Magnetic domains of silicon electrical steel (From Gerthsen, Ch.; Kneser, H. O.: *Physik*, 6th Edition, Springer-Verlag, Berlin, 1960, [6]). (**c**) Magnetic domains of ferromagnetic materials under the influence of the external magnetic-field strength H

ferromagnetic material which has not been exposed yet to any magnetic field (iron fresh from foundry), the domain magnetic moments are randomly oriented (see Figs. 6.22b [6], and 6.22c for H = 0, point 1 of Fig. 6.23a) and the net magnetic flux in the material is zero. As a consequence the operating point on the flux/flux density-current/magnetic field strength function of Fig. 6.23a is in the origin.

In the following we perform a set of "experiments", where we increase the external magnetizing current i (or magnetic field strength H) of a coil wrapped around a ferromagnetic material in a stepwise (see points 1–5 in Fig. 6.23a) manner. Let us assume that at current i_{sat} the ferromagnetic material starts to saturate, that is, the linear B–H characteristic becomes nonlinear with currents above values larger than i_{sat}. In general, when an external magnetizing force (current) is applied to ferromagnetic material the domain magnetic moments tend to align with the externally applied field H. As a result, the dipole magnetic moments of the domains add to the applied field H, resulting in a much larger value of the flux density B within the ferromagnetic material

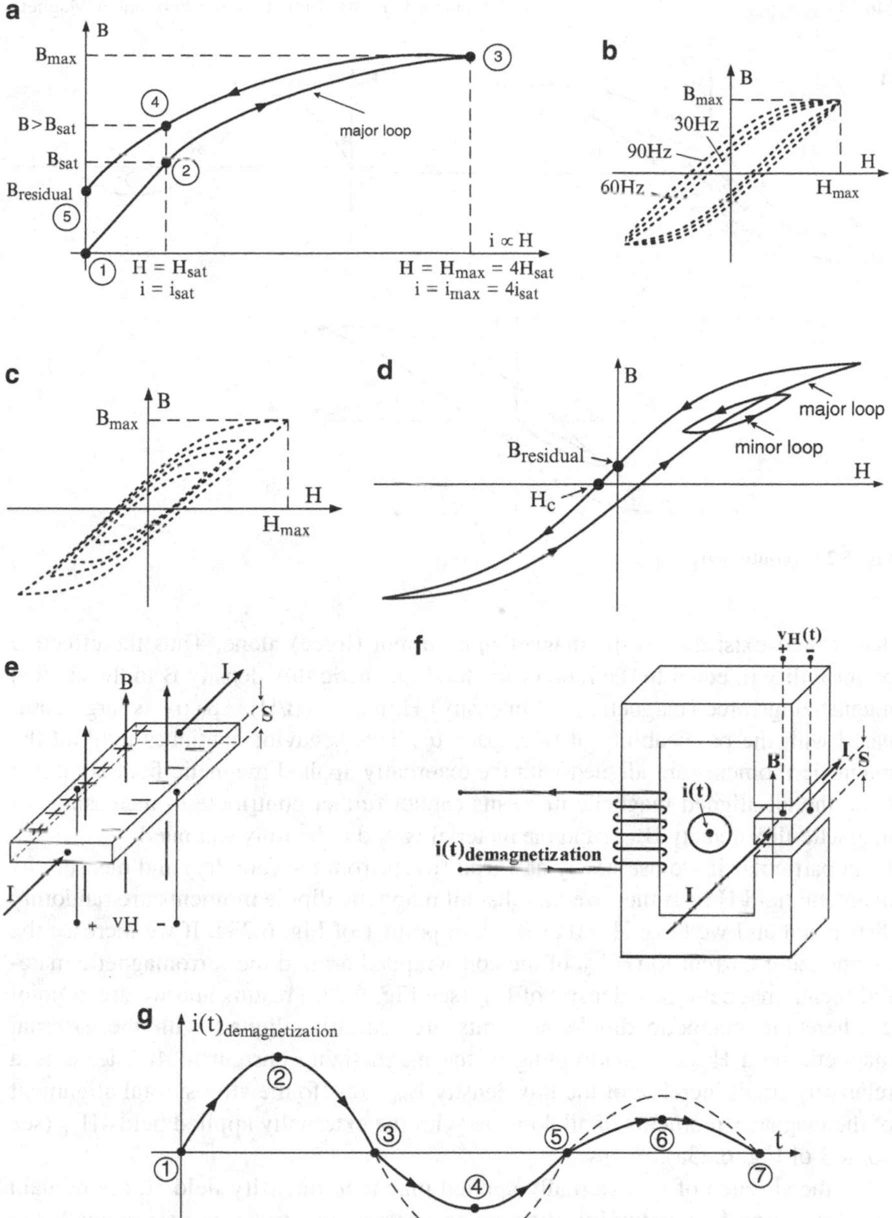

Fig. 6.23 (**a**) Nonlinear, double-valued (B-H) characteristic of ferromagnetic material. (**b and c**) Measured nonlinear multi-valued (B-H) characteristics of ferromagnetic materials; (**b**) Variation of B-H loop as a function of frequency with B_{max} and H_{max} constant, (**c**) Variation of B-H loops as a function of voltage with constant frequency. (**d**) Major and minor (B-H) characteristics. (**e**) Hall sensor. (**f**) Current probe consisting of iron core, demagnetization winding and Hall sensor. (**g**) Method #1 of demagnetization relying on decreasing current amplitudes of the demagnetization current. (**h**) Method #1 of demagnetization relying on decreasing major (B-H) characteristics. (**i**) Method #2 of demagnetization relying on controlled decreasing current amplitudes of the demagnetization current. (**j**) Method #2 of demagnetization relying on minor and major B-H loops (characteristics)

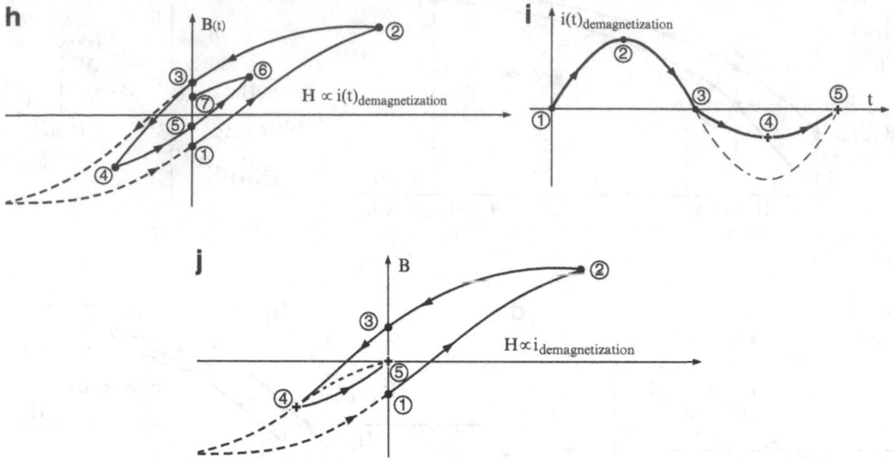

Fig. 6.23 (continued)

than would exist due to the magnetizing current (force) alone. Thus the effective permeability μ, equal to the ratio of the total magnetic flux density B to the applied magnetizing force (magnetic-field intensity) H, that is $(B/H) = \mu_r \mu_o$, is large compared with the permeability of free space μ_o. This behavior continues until all the magnetic moments are aligned with the externally applied magnetic field H; at this point the all-aligned magnetic moments cannot further contribute to increasing the magnetic-flux density B_{max}, and the material is said to be fully saturated.

In particular if we use newly cast iron (fresh from the foundry) and measure its magnetic field H or B then we find that all magnetic dipole moments are randomly distributed and we have $H = 0$ or $B = 0$ at point 1 of Fig. 6.23a. If we increase the magnetizing current to $i = i_{sat}$ of the coil wrapped around the ferromagnetic material then a magnetic flux density of B_{sat} (see Fig. 6.23a) results and we are at point 2 where the magnetic dipole moments are partially aligned with the external magnetic field H_{sat}. A quadrupling of the magnetizing current to $4i_{sat}$ leads to a relatively small increase of the flux density B_{max} due to the almost total alignment of the magnetic moments of all domains with the externally applied field $4H_{sat}$ (see point 3 of Fig. 6.23a).

In the absence of an externally applied magnetic intensity field H, the domain magnetic moments naturally align along certain directions associated with the crystal structure of the domains, known as axes of easy magnetization. These axes are not identical with those of the randomly oriented (Fig. 6.22c, point 1) domain magnetic moments. Thus if the applied magnetizing field is now reduced from $4H_{sat}$ to H_{sat}, the domain magnetic moments start to relax to the directions of easy magnetization nearest to that of the applied field. As a result, when the applied field is reduced to H_{sat} the magnetic dipole moments will no longer be aligned with the externally applied field, and a B larger than B_{sat} results. At $H = 0$, after having been exposed to $H \neq 0$, the dipole magnetic moments will no longer be totally random in their orientation; they will retain a net magnetization component along the applied-field direction (see Fig. 6.23a, point 5), and the residual flux density $B_{residual}$ is measured. It is this effect which is responsible for the phenomenon known as magnetic hysteresis. The relationship between B and H for a ferromagnetic material is both nonlinear and multi-valued due to the major loops of Figs. 6.23b, c [7]. The measured functions of Fig. 6.23b show that the enclosed area by the (B–H) characteristic is frequency-dependent: the higher the frequency the larger is the area due to the generation of eddy currents within the iron core. Figure 6.23d illustrates the occurrence of major and minor loops. The intercept of the B–H loop with the vertical axis is called residual flux density $B_{residual}$ and the intercept with the horizontal axis is called the coercivity H_c.

Demagnetization of Ferromagnetic Material: To demagnetize ferromagnetic material either the flux density B or the magnetic field intensity H must be measured. B can be measured via a Hall sensor as depicted in Fig. 6.23e where the Hall voltage is $V_H = R_H \frac{I \cdot B}{S}$. The coefficient R_H is depending on the semiconductor material used. For indium/antimony (InSb) this constant is $R_H = 240\ cm^3/(As)$. The current I is a calibration current, the thickness s is a design constant of the device and the flux density B is the quantity to be measured. Hall sensors have an error of about 1% and are not as accurate as compared with low inductive shunts for measuring currents i which are proportional to H and $B = \mu_o H$ in air or free space.

The demagnetization of current probes (see Fig. 6.23f) employing Hall sensors can be accomplished in several ways. Two methods highlight the procedure of demagnetization of current probes. This demagnetization of the iron core of a current probe is necessary in order to minimize the measuring error or such probes. Figure 6.23f depicts the iron core with demagnetization winding, Hall sensor, and conductor whose current i(t) must be measured via the flux density B developed in the iron core.

Method #1 of demagnetization relies on the decrease of the amplitude of the demagnetization current (Fig. 6.23g) resulting in the decrease of the major B–H loops as illustrated in Fig. 6.23h. *Method #2* of demagnetization relies on the controlled decrease of the amplitude of the demagnetization current (Fig. 6.23i) resulting in the employment of major and minor (B–H) characteristics as illustrated in Fig. 6.23j.

Application Example 6.7: Design of a Current Probe

A current probe is used to measure large currents of any waveshape and frequency (e.g., 100 A, triangular, 1,000 Hz). It consists of an iron core and an InSb Hall sensor (see Fig. 6.24).

What is the Hall voltage V_{Hall} for $B = 0.1$ T, $s = 1$ mm, $I_{Hall} = 1.0$ A and $R_{Hall} = 240\ cm^3/(As)$?

Solution

The Hall voltage is $V_{Hall} = \frac{R_{Hall} \cdot I_{Hall} \cdot B}{s} = \frac{240 \cdot 10^{-6} \cdot 1 \cdot 0.1}{10^{-3}} = 24\,mV.$

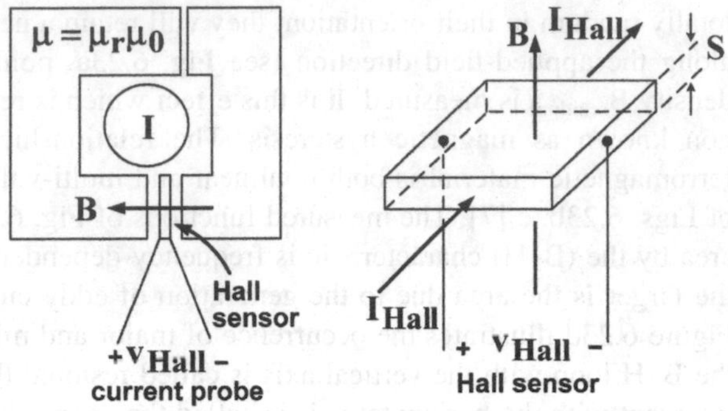

Fig. 6.24 Construction of current probe with Hall sensor (demagnetization coil not shown)

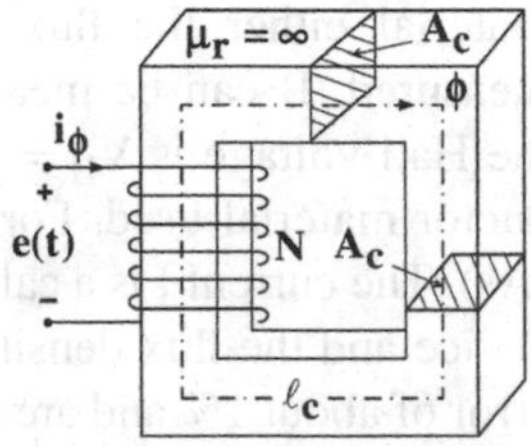

Fig. 6.25 Magnetic circuit with closed magnetic core and practically no air gap

Exciting Current at AC Excitation: In AC power systems the waveforms of voltage and flux (created within inductors and transformers by the applied terminal voltage) closely approximate sinusoidal functions of time. Even if the applied terminal voltage is not quite sinusoidal the fluxes generated by this somewhat nonsinusoidal voltage are still quite sinusoidal, due to the integration $\phi = (1/N)\int vdt$ where all higher harmonic amplitudes a_h of hth order are reduced with $a_h \propto (1/h)$.

This subsection describes the excitation characteristics and losses associated with steady-state AC operation of magnetic materials as used in inductors. We shall use as our model a closed-core magnetic circuit with practically no air gap (Fig. 6.25).

The magnetic path length is ℓ_c, and the cross-sectional area is A_c throughout the length of the core. Assuming a sinusoidal variation of the core flux

$$\phi(t) = \phi_{max} \sin \omega t = A_c B_{max} \sin \omega t, \tag{6.115}$$

where ϕ_{max} is the amplitude of the core flux, ω the angular velocity (frequency), and B_{max} the amplitude of the flux density within the core.

From Faraday's law $e(t) = N(d\phi/dt)$ follows

$$e(t) = \omega N \cos \omega t = \omega N A_c B_{max} \cos \omega t = E_{max} \cos \omega t, \tag{6.116}$$

where

$$E_{max} = 2\pi f N A_c B_{max} \tag{6.117}$$

is the amplitude of the induced (terminal voltage, if winding resistance and flux leakage are neglected) voltage e(t).

At steady-state AC operation we are usually interested in the rms value of voltages and currents

$$E_{rms} = \frac{E_{max}}{\sqrt{2}} = \frac{2\pi}{\sqrt{2}} fNA_cB_{max} = 4.44fNA_cB_{max}. \qquad (6.118)$$

This is an important relation for any electromagnetic device (e.g., inductors, transformers discussed in Chaps. 7 and 8, machines discussed in Chap. 10). It relates the rms value of the induced voltage (approximately terminal voltage) to the frequency of voltage (and current for linear conditions) applied. Note that for inductors the rated maximum flux density B_{max} is about 0.4 T, for transformers B_{max} it is about 2 T, and for rotating machines B_{max} is about from 0.6 T to 1.0 T. It is of utmost importance that rated flux densities are about maintained, independent of the frequency of the applied voltage. Therefore,

$$\left(\frac{E_{rms}}{f}\right) = 4.44\,NA_cB_{max} = \text{constant} \qquad (6.119)$$

must be maintained for all operating conditions; this is especially important for variable-speed drives.

The production of a magnetic field in non-ideal iron cores requires current in the excitation winding known as exciting current i_Φ. The nonlinear, multi-valued (e.g., double-valued) magnetic properties of the core mean that the waveform of the exciting current differs from the sinusoidal (impressed) waveform of the flux or voltage. The waveform of the exciting current as a function of time can be found graphically from the magnetic B–H characteristic, as illustrated below. B and H are related to Φ and i_Φ, respectively, by known geometric constants: from the continuity of flux condition $\phi = B_cA_c$ and from Ampere's law follows $i_\Phi = (H_c\ell_c)/N$. The AC (B–H) loop has been drawn in Fig. 6.26b in terms of ϕ and i_Φ, and minor loops as well as the trajectory through the origin have been neglected. Cosine and sine waves of induced voltage e(t) and flux Φ(t), respectively, are shown in Fig. 6.26a. At any given time t′ the flux Φ' can be identified in Fig. 6.26a which corresponds to Φ' of the (Φ–i_Φ) loop of Fig.6.26b, which corresponds to the value of i_Φ'. The i_Φ' will be then plotted in Fig. 6.26a which represents one point of the i_Φ'(t) function. This procedure can be repeated for many more points. Notice that because the B–H loop is multi-valued (e.g., double-valued), it is necessary to associate the rising-flux values of the time function with the rising-flux portion of the (B–H) loop; the same applies to the falling portions. Also, the amplitude of the flux-time function must have the same amplitude value as the maximum value of the (Φ–i_Φ) loop.

Notice that the graphically constructed exciting current (graphical construction can be replaced by numerical computer solution) is nonsinusoidal: it is about triangular for a sinusoidal flux. The rms value of the exciting current is defined in the standard way as

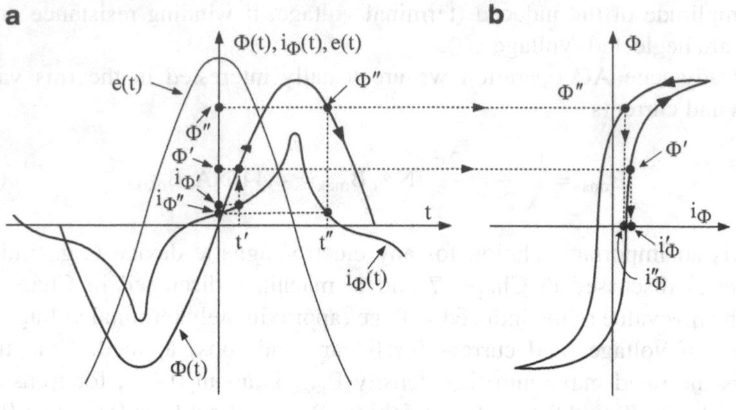

Fig. 6.26 (**a** and **b**) Graphical construction [2] of the exciting current i_ϕ based on given time-dependent flux function and the associated Φ–i_Φ characteristic; (**a**) induced voltage e, flux Φ, and exciting current i_Φ, (**b**) corresponding (Φ–i_Φ) loop

$$I_{\phi rms}=\left[\frac{1}{T}\int_0^T i_\phi^2 dt\right]^{1/2}. \tag{6.120}$$

The corresponding rms value of the magnetic field intensity within the core is $H_{crms}=\frac{NI_{\phi frms}}{\ell_c}$ or

$$I_{\phi rms}=\frac{\ell_c H_{crms}}{N}. \tag{6.121}$$

Application Example 6.8: Construction of Exiting Current $i_\phi(t)$ of a Magnetic Circuit

Construct the values of the exciting current $i_\phi(t)$ for the time instants t_1, t_2, t_3, t_4, t_5, t_6 (point by point) and plot these values $\{i_\phi(t_1), i_\phi(t_2), i_\phi(t_3), i_\phi(t_4), i_\phi(t_5), i_\phi(t_6)\}$ as a function of time in Fig. 6.27.

Solution

Consider the increasing branches of B(t) and [B(t) versus $i_\phi(t)$] characteristic (Fig. 6.27):

At $t=t_1=0$ the flux density has the average value of $B(t)=0$ T resulting in $i_\phi(t_1)=1.0$ A,
At $t=t_2$ the flux density has a value of $B(t)=1.0$ T resulting in $i_\phi(t_2)=3.5$ A,
At $t=t_3$ the flux density has a value of $B(t)=0.5$ T resulting in $i_\phi(t_3)=0$ A,
At $t=t_4$ the flux density has an average value of $B(t)=0$ T resulting in $i_\phi(t_4)=-1$ A,
At $t=t_5$ the flux density has a value of $B(t)=-1.0$ T resulting in $i_\phi(t_5)=-3.5$ A,

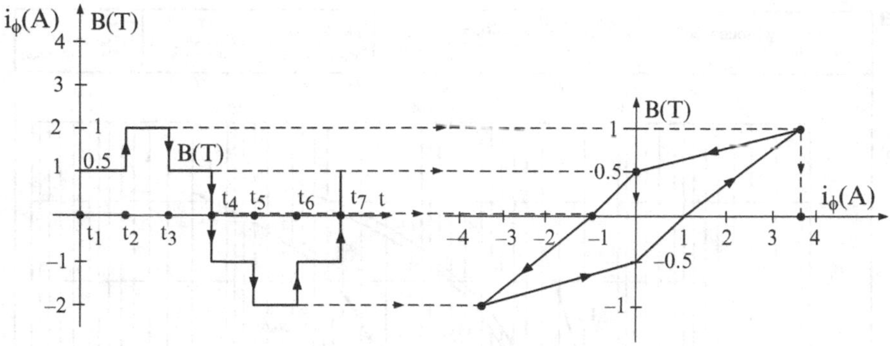

Fig. 6.27 Pointwise construction of the exciting current $i_\phi(t)$ of a magnetic circuit

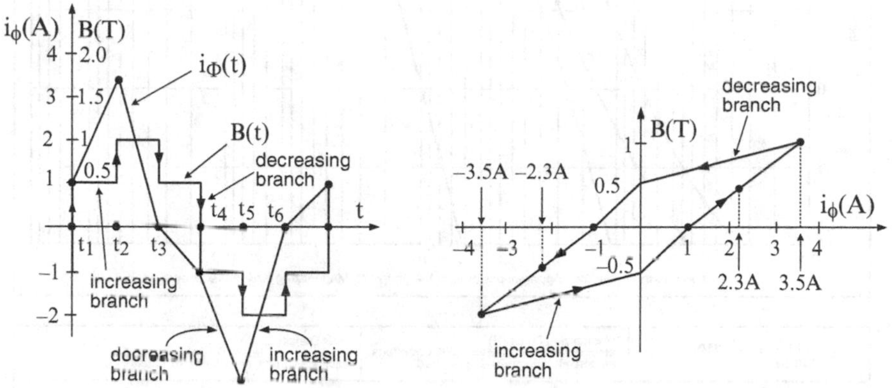

Fig. 6.28 Pointwise construction of the exciting current $i_\phi(t)$ of a magnetic circuit

At $t = t_6$ the flux density has a value of $B(t) = -0.5$ T resulting in $i_\phi(t_6) = 0$ A,
At $t = t_7 = t_1 = 0$ the flux density has an average value of $B(t) = 0$ T resulting in $i_\phi(t_1) = 1.0$ A.

Apparent Power Requirements for AC Excitation of Cores: The AC excitation characteristics of core materials are usually expressed in terms of apparent power (volt-amperes) S_c rather than a magnetization curve relating B and H, see Fig. 6.29a. The apparent power S_c required to excite the core to a specified flux density is

$$S_c = E_{rms} I_{\phi rms} = 4.44 f N A_c B_{cmax} \left(\frac{\ell_c H_{crms}}{N} \right). \tag{6.122}$$

For a magnetic material of density $\gamma_c = 7.86$ (kg-force)/dm^3 = 7,860 (kg-force)/m^3 the weight of the core is

$$\text{weight}_{core} = A_c \ell_c \gamma_c \tag{6.123}$$

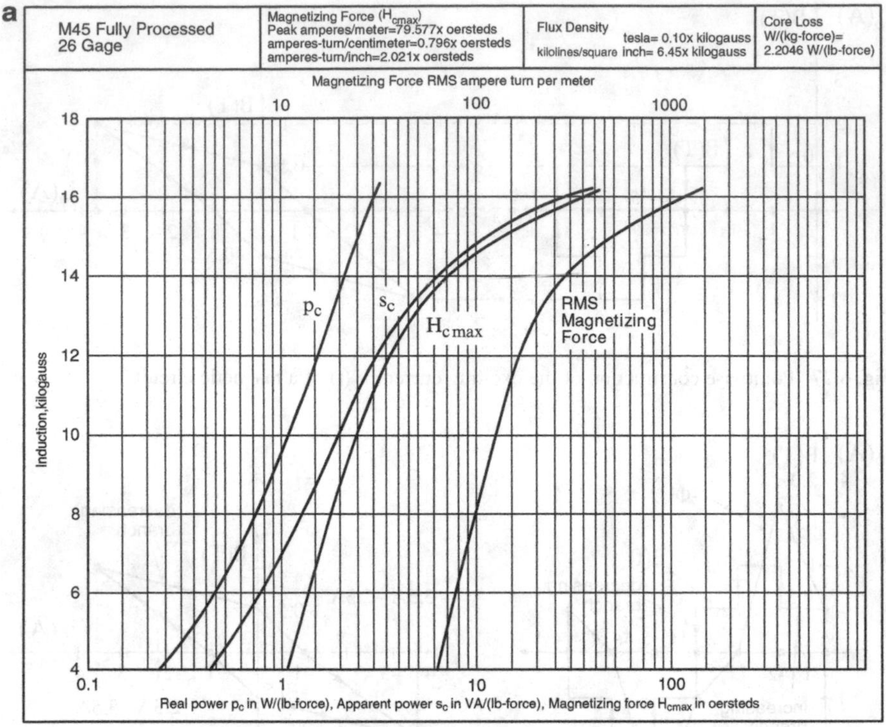

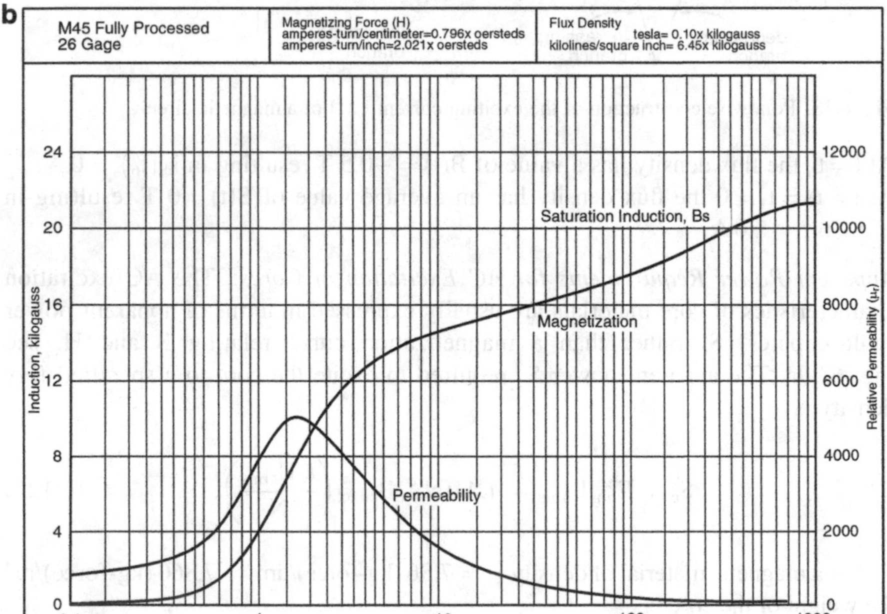

Fig. 6.29 (**a**) Apparent power per unit weight s_c, real power per unit weight p_c, magnetizing force H_{cmax} in oersteds, and rms magnetizing force (see upper horizontal scale) for M45 fully processed

and the apparent power per unit weight is (see Fig. 6.29a)

$$s_c = \frac{S_c}{\text{weight}_{core}} = \frac{4.44f}{\gamma_c} B_{cmax} H_{crms}. \quad (6.124)$$

The excitation apparent power measured in volt-amperes at a given frequency f (e.g., f = 60 Hz) is dependent only on B_{cmax} within the core because H_{cmax} is a unique function of B_{cmax} and is independent of turns and geometry. As a result the AC excitation requirements for a magnetic material are often given in terms of apparent power s_c expressed in volt-amperes per unit weight, as is illustrated in Fig. 6.29a for M45 Fully Processed 26 Gauge electrical steel. This figure also shows the (real) power loss per unit weight p_c within the iron core, and the magnetizing force H_{cmax} required. Figure 6.29b depicts the DC single-valued B–H characteristic for the same electrical steel [8, 9].

Core Losses due to AC Excitation of Cores: The apparent power S_c required by cores can be separated into real power (P_c) and reactive power (Q_c) at a certain frequency and a given B_{cmax}:

$$S = S_c = P_c + jQ_c. \quad (6.125)$$

The reactive power Q_c is cyclically supplied and absorbed by the excitation source and contributes to the excitation current i_Φ delivered by the source. In Chap. 7 we will call the reactive current causing the reactive power the "magnetization" current i_m. The real power P_c absorbed by the core is due to two loss mechanisms which are associated with time-varying fluxes in magnetic materials.

- I^2R heating associated with eddy currents in cores.
- Second loss mechanism is due to the hysteretic nature of magnetic materials.

To reduce eddy current losses the iron cores must be laminated and consist of electrical steel sheets as indicated in Fig. 6.30b.

Due to the core flux

$$\phi(t) = \phi_{max} \sin \omega t, \quad (6.126)$$

and Faraday's law in integral form

$e(t) = \oint_C \vec{E}(t) \bullet d\vec{l} = Nd(\int_S \vec{B}(t) \bullet d\vec{S})/dt = N\,\phi_{max}\omega\cos\omega t$ the induced voltage becomes

$$e(t) = E_{max} \cos \omega t = \sqrt{2} E_{rms} \cos \omega t, \quad (6.127)$$

Fig. 6.29 (continued) 26 Gauge electrical steel (From *Non-oriented Sheet Steel for Magnetic Applications*, United States Steel, 600 Grant Street, Pittsburgh, PA, 15219, May 1978, [8]). (**b**) DC single-valued flux density (induction) B versus magnetic-field strength H characteristic, and relative permeability μ_r for M45 fully processed 26 Gauge electrical steel (From *Non-oriented Sheet Steel for Magnetic Applications*, United States Steel , 600 Grant Street, Pittsburgh, PA, 15219, May 1978 [8])

where $\vec{E}(t)$is the electric-field strength within/outside the core due to the changing flux $\phi(t)$ as depicted in Figs. 6.31a, b. Note e(t) is the induced voltage with its amplitude E_{max} and its rms value E_{rms}.

Figures 6.31c, d illustrate in detail the generation of eddy currents in solid (Fig. 6.31c) and laminated (Fig. 6.31d) iron cores. It is assumed that the solid core has the horizontal dimension of 1 m and one lamination of the laminated core has the horizontal dimension of 1 mm. In the solid core (Fig. 6.31c) the eddy

Fig. 6.30 (**a**) Solid iron core, and (**b**) laminated iron core to suppress eddy current losses within core

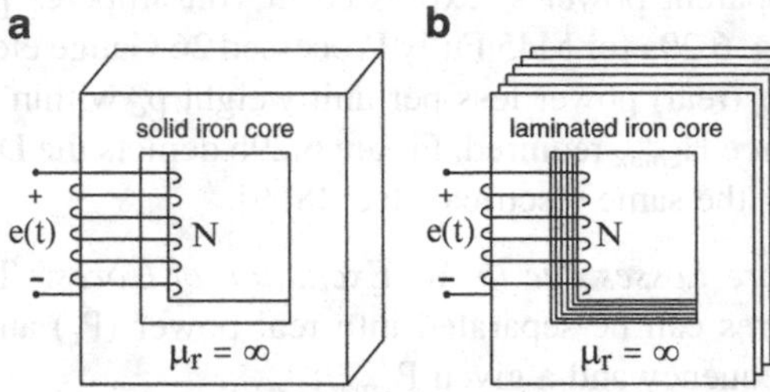

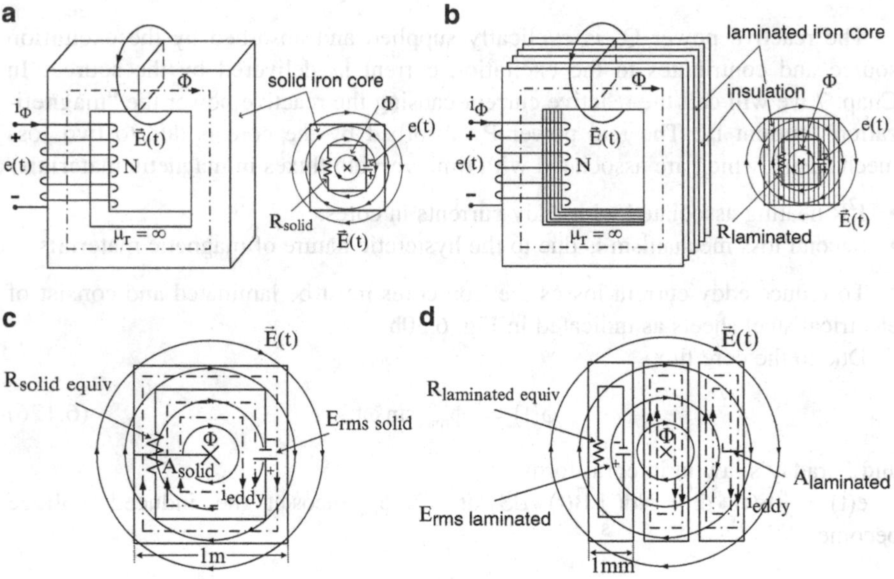

Fig. 6.31 (**a** and **b**) Cross-section of core with time-varying flux and associated electric-field strength $\vec{E}(t)$, and e(t) is the induced voltage; (**a**) solid iron core, (**b**) laminated iron core. (**c**) Cross-section of solid core with horizontal width of 1 m, where $\vec{E}(t)$ is the electric-field strength and E_{rms} is the rms value of the induced voltage e(t). (**d**) Cross-section of laminated core with (total) horizontal width of 1 m, where one lamination is 1 mm wide

currents flow within the cross-section A_{solid}, and within one lamination (Fig. 6.31d) the eddy currents flow within the cross-section $A_{laminated}$. This means the ratio $R_{laminated}/R_{solid} \approx 1{,}000$. If the induced voltage in the solid core is $E_{rms\ solid} = 1$ V then the induced voltage in one lamination is $E_{rms\ laminated} \approx E_{rms\ solid}$. As a result the loss within the core laminations is much less than that of the solid core because $R_{laminated} \approx 1{,}000\ R_{solid}$.

Application Example 6.9: Estimation of Eddy Current Losses in Solid and Laminated Cores

If in Fig. 6.31a and (6.127) $E_{rms\ solid} = 1$ V and the equivalent resistance of the solid core is $R_{solid\ equiv} = 1\ \Omega$, then $I_{rms} = (E_{rms\ solid}/R_{solid\ equi}) = 1$ A and the loss is $P_{eddy\ solid} = (I_{rms})^2\ R_{solid\ equiv} = 1$ W. Calculate eddy-current loss $P_{eddy\ laminated}$ for Fig. 6.31b.

Solution

For the laminated core $E_{rms\_laminated} \approx E_{rms\_solid} = 1$ V and $R_{laminated\ equiv} \approx 1\ k\Omega$ one obtains the current $I_{rms\_laminated} = 1$ mA or the eddy-current loss $P_{eddy\ laminated} = (10^{-3})^2\ (10^{+3}) = 10^{-3}$ W. This demonstrates the effectiveness of laminated cores with respect to the suppression of eddy current losses.

Hysteresis losses of cores come about when each time the magnetic material undergoes one magnetic-field intensity (H) cycle, there is a net energy input into the material. This energy is required to move around the magnetic dipoles in the material as discussed earlier in this subsection and this input energy is dissipated in heat. Since there is an energy loss per cycle, hysteresis loss is proportional to the frequency of the applied excitation. In many older textbooks and publications the hysteresis losses are related to the area enclosed by the (B–H) characteristic. Reference [7] shows that the area of the (B–H) characteristic (or loop) represents the total core losses: that is, eddy current and hysteresis losses are evident from the measured (B–H) characteristics/loops of Figs. 6.23b, c. The total core losses can be formulated based on Faraday's law $e(t) = d\lambda/dt$ and the relation between dissipated energy W and power

$$W = \int p dt = \int e i dt = \int \frac{dl}{dt} i dt = \int i d\lambda. \tag{6.128}$$

If (6.128) is applied to the excitation current one obtains with

$$\lambda = N\phi = NBA \tag{6.129}$$

and

$$d\lambda = NA_c dB_c,\ i_\phi = \frac{H_c \ell_c}{N}. \tag{6.130}$$

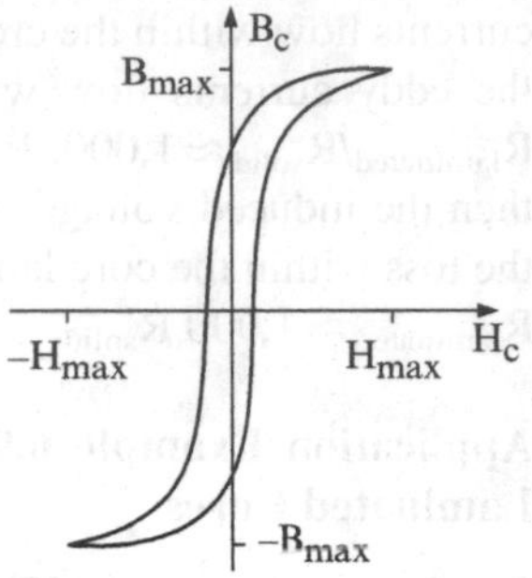

Fig. 6.32 (B_c–H_c) characteristic

$$W_{absorbed} = \oint_{(B-H)loop} i_\phi d\lambda = \oint_{(B-H)loop} \frac{H_c \ell_c}{N} A_c N dB_c$$
$$= A_c \ell_c \oint_{(B-H)loop} H_c dB_c [Ws]. \quad (6.131)$$

Recognizing that $A_c \ell_c$ is the volume of the core and that the line integral is the area enclosed by the (B_c–H_c) characteristic of Fig. 6.32, we can rewrite (6.131) as follows:

$$W_{absorbed} = A_c \ell_c \oint_{(B-H)loop} \frac{B_c}{\mu_c} dB_c. \quad (6.132)$$

The real power absorbed is

$$P_c = W_{absorbed} f = A_c \ell_c \oint_{(B-H)loop} H_c dB_c f, \quad (6.133)$$

or the specific absorbed total core losses per volume are

$$p_c = \frac{P_c}{volume_{core}} = f \cdot \oint H_c dB_c$$
$$= \{\text{area enclosed by } (B - H) \text{ loop}\} \cdot f \ [W/m^3], \quad (6.134)$$

or the total power is

$$P_c = \{\text{area enclosed by } (B - H) \text{ loop}\} \cdot f \cdot (\text{volume of core}) \ [W]. \quad (6.135)$$

In general, the losses depend upon the metallurgy of the material as well as on the maximum flux density within the core B_{cmax} and frequency f. Information on core loss (p_c) is typically presented in graphical form. It is plotted in terms of watts per unit weight (1 kg-force = 2.2 lbs-force) as a function of flux density, as illustrated in Fig. 6.29a.

Application Example 6.10: Determination of Iron-Core Losses of Inductor

The data for the top half of a symmetric (B–H) characteristic for the core of Fig. 6.33 are given in Table 6.4.

The mean length of the flux paths in the core is $\ell_c = 0.30\,\text{m}$. Find graphically the hysteresis and eddy-current loss in watts for a maximum core flux density of $B_{cmax} = 1$ T at a frequency of f = 60 Hz, (number of turns N = 1,000 is not needed for calculation), and core cross-section $A_c = 1.25 \cdot 10^{-3}$ m^2.

Solution

Figure 6.34 shows the upper half of the (B–H) characteristic, obtained by plotting the data of Table 6.4. The area of a trapezoid is $\text{area}_{\text{trapezoid}} = (1/2)\{a + b\}h$, where a is the base length, b is the top length, and h is the height.

Total area of (B_c–H_c) characteristic = 2{(1/2) [96 + 97]0.2 + (1/2)[97 + 98] 0.2 + (1/2)[98 + 102]0.2 + (1/2)[102 + 101]0.2 + (1/2)[101 + 60]0.1 + (75·0.1)/2} = 38.6 + 39 + 40 + 40.6 + 16.1 + 7.5

$$\frac{\text{At}}{\text{m}}\frac{\text{Wb}}{\text{m}^2}\text{per cycle} = 181.8\frac{\text{At}}{\text{m}}\frac{\text{Wb}}{\text{m}^2}\text{per cycle} = 181.1\frac{\text{joules}}{\text{m}^3}\text{ per 60 Hz cycle.}$$

The iron-core losses are now (6.135) $P_c = 181.8\frac{\text{joules}}{\text{m}^3\text{cycle}} \cdot 0.3\,\text{m} \cdot 1.25 \cdot 10^{-3}\text{m}^2 \cdot 60\frac{\text{cycles}}{\text{s}} = 4.09\,\text{W}.$

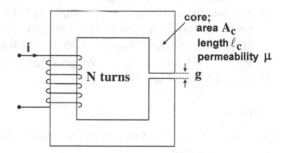

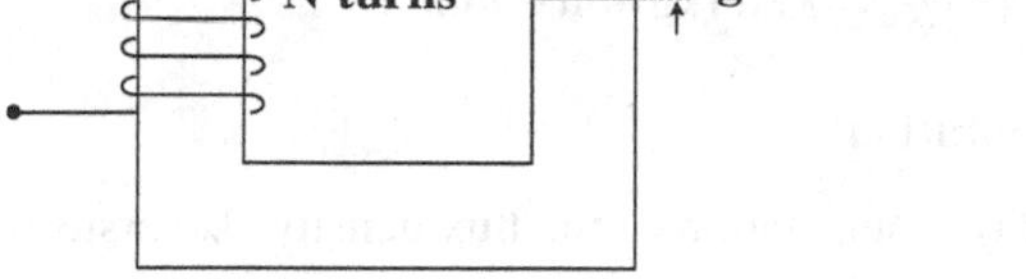

Fig. 6.33 "C" core inductor with uniform cross-sectional area

Table 6.4 (B_c–H_c) characteristic for the core of Fig. 6.33

B_c [T]	0	0.2	0.4	0.6	0.7	0.8	0.9	1.0	0.95
H_c [At/m]	48	52	58	73	85	103	135	193	80

B_c [T]	0.9	0.8	0.7	0.6	0.4	0.2	0
H_c [At/m]	42	2	−18	−29	−40	−45	−48

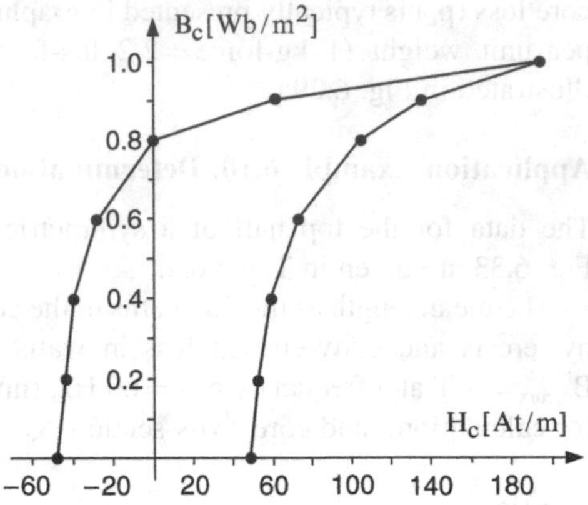

Fig. 6.34 Upper half of (B_c–H_c) characteristic/loop

Application Example 6.11: Iron-Core Loss of an Inductor as Used in Power Electronic Applications

The (λ–i) characteristic of an iron-core was recorded in the laboratory and is shown in Fig. 6.35.

(a) For $N = 680$ turns, a core cross-section of $A_c = 2.5\ \text{cm}^2$ and a core length of $\ell_c = 12$ cm compute the flux density B_c versus magnetic field intensity H_c function of the iron core: $B_c = f(H_c)$, that is, express B_c as a function of λ: $B_c = f(\lambda)$ and H_c as a function of current i : $H_c = f(i)$, and introduce a new coordinate system (B_c versus H_c) in Fig. 6.35.

(b) Find the energy input per cycle $W_{absorbed}$ measured in Ws.

(c) If the frequency of λ and i is $f = 60$ Hz compute the core-loss power per unit volume p_c measured in W/m^3, or W/dm^3.

(d) The weight of the iron-core sample is 0.12 (kg-force). What is the input power of this iron-core sample per unit weight measured in W/(kg-force) @ a flux density of $B_{max\_c} = 1.36$ T, and the input power of this iron-core $P_c^{0.12(kg-force)}$ measured in W provided the iron-core weight density is $\gamma_{Fe} = 7.86$ (kg-force)/dm^3?

Solution

(a) Computation of the flux density B_c versus magnetic field intensity H_c function of the iron core:

From Ampere's law

$$H_c = (N \cdot i)/\ell_c \tag{6.136}$$

one obtains the values of Table 6.5 for $N = 680$ turns, $A_c = 2.5 \cdot 10^{-4}\ \text{m}^2$, and $\ell_c = 0.12$ m.

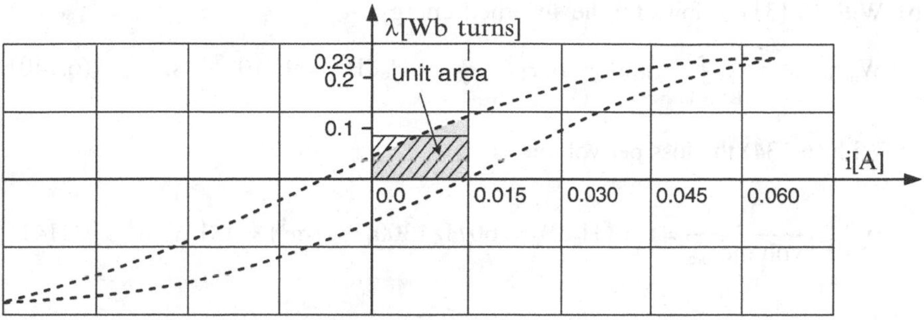

Fig. 6.35 Measured (λ–i) characteristic of iron core

Table 6.5 (i-H_c) relationship

i [A]	0.015	0.03	0.045	0.06	0.075
H_c [A/m]	85	170	255	340	425

Table 6.6 (λ-B_c) relationship

λ [Wb-turns]	0.1	0.2	0.23
B_c [T]	0.588	1.18	1.36

From the flux-linkage relation one gets

$$B_c = \lambda/(N \cdot A_c), \tag{6.137}$$

B_c and λ values are listed in Table 6.6.

The core loss depends upon the maximum core flux density $B_{max\_c} = 1.36$ T as listed in Table 6.6, fourth column.

In Fig. 6.35 one can define a unit area = 85(A/m)·0.8·0.588(T) ≈ 40 (AT/m). The total area of the (B_c–H_c) characteristic/loop of Fig. 6.35 can be estimated to be 7.5 unit areas. From this follows

$$\oint_{(B_c-H_c)\,loop} H_c dB_c = 7.5(\text{unit areas}) \cdot 40(\text{AT/m})/(\text{unit area})$$

$$= 300\,(\text{AT/m}) = 300\,(\text{AVs/m}^3) \tag{6.138}$$

or

$$\oint_{(B_c-H_c)\,loop} H_c dB_c = 300\,(\text{Ws/m}^3). \tag{6.139}$$

(b) With (6.131) follows for the absorbed energy

$$W_{absorbed} = \oint_{(B-H)loop} i_{\phi} d\lambda = A_c \ell_c \oint_{(B-H)loop} H_c dB_c = 9 \cdot 10^{-3}\,Ws. \qquad (6.140)$$

(c) With (6.134) the loss per volume is

$$p_c^{(m^3)} = \frac{P_c}{volume_{core}} = f \times \oint H_c dB_c = 60\,Hz \cdot 300(Ws/m^3) = 18\,kW/m^3 \qquad (6.141)$$

or

$$p_c^{(dm^3)} = \frac{18kW/m^3}{10^3 dm^3/m^3} = 18\,W/dm^3. \qquad (6.142)$$

(d) The input power in watts to the iron core sample is per (kg-force) weight

$$p_c^{(kg-force)} = \frac{p_c^{(dm^3)}}{\gamma_{Fe}} = (18\,W/dm^3)/(7.86\,(kg-force)/dm^3)$$
$$= 2.3\,W/(kg-force). \qquad (6.143)$$

The loss of the iron core with weight = 0.12 (kg-force) is at the maximum flux density of $B_{max\_c} = 1.36$ T (see Table 6.6)

$$P_c^{(0.12\,kg-force)} = |B_{max\_c} = 1.36\,T = 0.12(kg - force) \cdot 2.3\,W/(kg - force)$$
$$= 0.276\,W. \qquad (6.144)$$

Note, iron-core losses depend on mechanical stress induced by cutting and mounting (within a frame), and it may increase by a factor of 1.6 [10].

6.4 Permanent-Magnet Materials

In Subsection 6.3 we have investigated the properties of so-called “soft” magnetic electrical steel which is used for iron cores of inductors, transformers and electric machines. We have concluded that in order to suppress eddy-current losses within the cores, the core material cannot be of solid structure but must consist of sheets or laminations. We have also discovered that the total losses generated in the core are proportional to the area enclosed by the (B–H) characteristic: for low-loss devices this area has to be as small as possible and low-loss electrical steel has indeed low iron-core (eddy current and hysteresis) losses in the range of 2 W/kg-force at a maximum flux density B_{max} of 1 T. Figure 6.36 illustrates such a soft magnetic

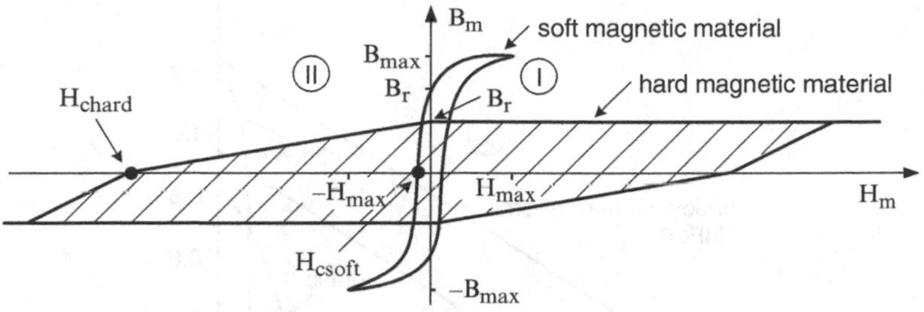

Fig. 6.36 Soft and hard magnetic material (B_m–H_m) characteristics, where subscript "m" stands for magnet, subscript "c" stands for coercivity, where for example $H_{c\ soft} = -5$ A/m, and $H_{c\ hard} = -10^6$ A/m, H_{max} and B_{max} relate to the soft magnetic material

material (e.g., M-45 electrical steel), and a hard magnetic material (permanent-magnet material).

The (B_m–H_m) characteristics can be identified by the residual flux density or remanent magnetization B_r, which is for M-45 in the range of $B_r = 1.4$ T and the value of coercivity H_c is for M-45 in the range of $H_c = -5$ A/m.

One obtains an excellent permanent magnet by making H_c very large, for example changing H_c from -5 A/m (for soft magnetic material M-45) to $-1{,}000$ kA/m (for hard magnetic material) makes the latter an excellent permanent magnet. The value of B_r is not so important. Neodymium-iron-boron (NdFeB) is such a hard magnetic material, the characteristic of which is shown in Fig. 6.37 [2].

Derivation of Load Line: In Fig. 6.38a a permanent magnet is embedded into a "C" core consisting of soft-magnetic material. Because there is no excitation current Ampere's law reduces to

$$F = 0 = H_g \cdot g + H_m \cdot \ell_m, \tag{6.145}$$

where H_g and H_m are the magnetic-field intensities in the air gap and within the permanent magnet, respectively; g and ℓ_m are the gap and magnet lengths, respectively. From (6.145) follows

$$H_g = -\left(\frac{\ell_m}{g}\right) H_m. \tag{6.146}$$

The flux must be continuous through the entire magnetic circuit (divB = 0)

$$\phi = A_g B_g = A_m B_m, \tag{6.147a}$$

where A_g and A_m are the cross-sectional areas of the air gap and the magnet, respectively (note in general $A_g \neq A_m$). B_g and B_m are the magnetic flux densities in the air gap and the permanent-magnet material, respectively.

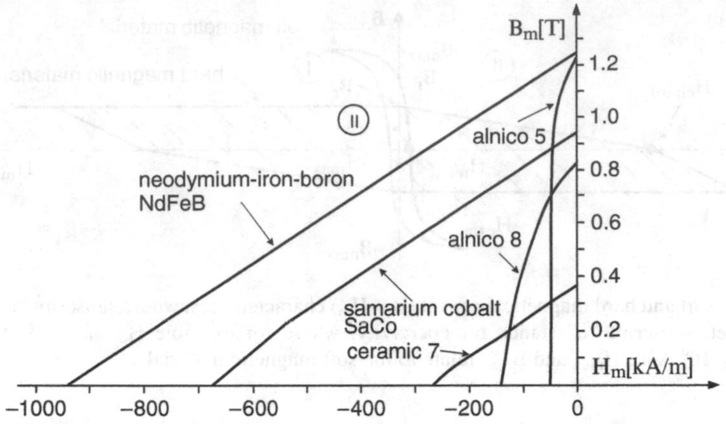

Fig. 6.37 (B_m–H_m) characteristics of neodymium-iron-boron, samarium-cobalt, alnico 5, alnico 8, and ceramic 7 permanent-magnet materials (From Fitzgerald, A.E.; Kingsley, Jr., Ch.; Umans, S. D.: *Electric Machinery*, 5th Edition, McGraw-Hill Publishing Company, 1990, copyright date is April 21, 2010, [2])

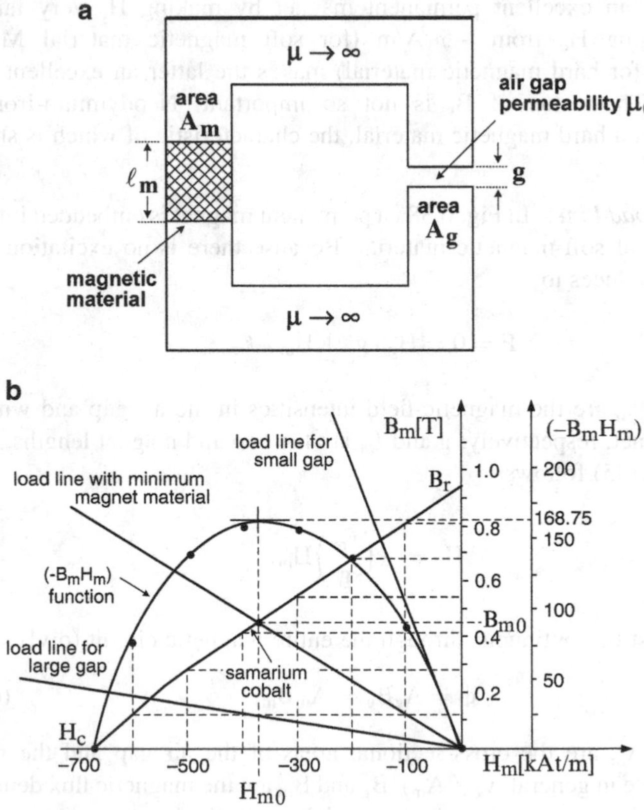

Fig. 6.38 (**a**) Permanent magnet residing within "C" core. (**b**) Permanent-magnet characteristic and various load lines

From Eq. (6.147a) follows

$$B_m = \left(\frac{A_g}{A_m}\right) B_g, \tag{6.147b}$$

with $B_g = \mu_0 H_g$, (6.146), and Eq. (6.147b) follows the load-line equation

$$B_m = -\left(\frac{A_g}{A_m}\right)\left(\frac{\ell_m}{g}\right)\mu_o H_m. \tag{6.148}$$

This equation is called the load line of a permanent-magnet circuit and is represented in Fig. 6.38b. The minus sign in the above equation originates from Ampere's law (6.146), and this is the reason why the load line exists in the second quadrant only and not in the first quadrant. Depending upon the application, the load line can be placed so that either a high, medium or low air-gap flux density results (see Fig. 6.38b). The load line resulting from a small air gap will be chosen for permanent-magnet machines; the load line where the least amount of permanent-magnet for a given air-gap flux density B_g is used is chosen when the cost of a device is important (e.g., loud speakers).

Construction of Energy-Product ($-B_mH_m$) Function: The cost of neodymium-iron-boron magnet material is about \$40/(lb-force). Significant cost savings can be achieved if the load line of a permanent magnet is placed at or near the operating point where the product of B_m and $(-H_m)$, that is $(-B_{m0}H_{m0}$, where $\{-H_{m0}\}$ is positive) is maximum. To find this point of the maximum energy product (which is a figure of merit for permanent-magnet materials), proceed as follows (see Fig. 6.39).

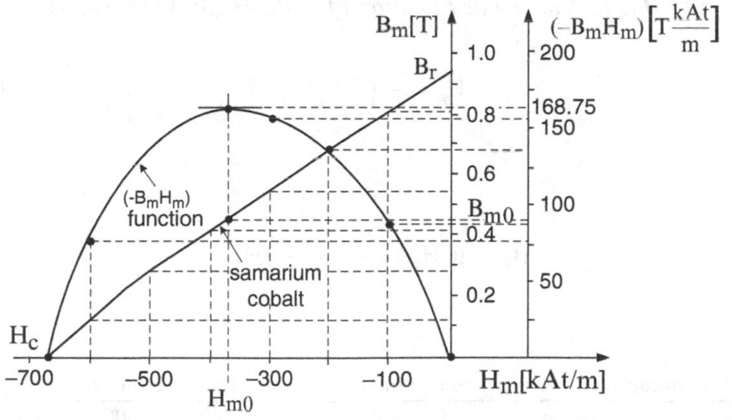

Fig. 6.39 Construction of energy product function $(-B_mH_m)$ and identification of maximum energy product $(-B_{m0}H_{m0})$ for samarium-cobalt permanent magnet material, where H_{m0} is negative

The operating point of a permanent magnet must lie on the (B_m–H_m) characteristic – which can be either linear or nonlinear – and on the load line (6.148). From (6.148) one notes that the load line exists in the second (and possibly fourth) quadrants only due to its negative slope (e.g., minus sign of load-line equation). For this reason the hard magnetic or permanent magnet material characteristic can be utilized only within the second quadrant, and for this quadrant we are going to construct (see Example 6.12) pointwise the energy-product function which leads to the maximum energy product point ($-B_{m0}H_{m0}$, where $\{-H_{m0}\}$ is positive). Note it is necessary to introduce a second vertical axis for the energy product ($-B_mH_m$) with the unit [T·kAt/m], where "t" stands for turns and is not really a unit and, therefore, can be deleted.

Application Example 6.12: Construction of Energy-Product Function for Samarium Cobalt Permanent-Magnet Material

For samarium cobalt permanent-magnet material find pointwise the energy-product function ($-B_mH_m$) as plotted in Fig. 6.39, and determine its maximum value ($-B_{mo}H_{mo}$).

Solution

For $H_m = 0$ and $B_m = B_r = 0.94$ T the energy product is $(-B_mH_m) = 0$; correspondingly, for $H_m = H_c = -675$ kA/m and $B_m = 0$ the product becomes $(-B_mH_m) = 0$; additional data points are tabulated in Table 6.7.

Plotting of the ($-B_mH_m$) values of Table 6.7 as a function of H_m leads to the ($-B_mH_m$) function of Fig. 6.39 and the identification of its maximum value $(-B_{m0}H_{m0}) = 168.75$ kAT/m, see Figs. 6.38b and 6.39: this number is a figure of merit for the permanent-magnet material samarium cobalt. From these two Figures one obtains $B_{mo} = 0.45$ T and $H_{m0} = -375$ kA/m

Derivation of Magnet Volume as Function of ($-B_{mo}H_{mo}$): From (6.146)

$$H_g = -\left(\frac{\ell_m}{g}\right)H_m \tag{6.149}$$

follows

$$B_g = \mu_o H_g = -\mu_o\left(\frac{\ell_m}{g}\right)H_m, \tag{6.150}$$

Table 6.7 Construction of energy-product function ($-B_mH_m$) for samarium cobalt

H_m [kA/m]	0	H_c	−600	−500	−400	−300	−200	−100
B_m [T]	B_r	0	0.13	0.29	0.41	0.54	0.68	0.82
($-B_mH_m$) [kAT/m] or [Ws/m³]	0	0	78	145	164	162	136	82

with (6.147b)

$$B_g = \left(\frac{A_m}{A_g}\right) B_m \tag{6.151}$$

one obtains, by multiplying the left-hand side of (6.151) with B_g and the right-hand side of (6.151) with $-\mu_o\left(\frac{\ell_m}{g}\right)H_m$, the relation

$$B_g^2 = \mu_o\left(\frac{\ell_m}{g}\right)\left(\frac{A_m}{A_g}\right)(-B_m H_m), \tag{6.152}$$

where $(\ell_m A_m)$ and $(g A_g)$ are the magnet volume V_{magnet} and air-gap volume $V_{air\ gap}$, respectively:

$$V_{magnet} = \ell_m A_m, \tag{6.153}$$

$$V_{air\ gap} = gA_g, \tag{6.154}$$

or

$$V_{magnet} = \frac{B_g^2}{\mu_0(-B_m H_m)} V_{air\ gap}. \tag{6.155}$$

The magnet volume V_{magnet} is a minimum if $(-B_m H_m)$ is a maximum, that is $(-B_m H_m) = (-B_{m0} H_{m0})$. This indicates that to achieve a desired flux density in the air gap B_g, the required volume of the magnet can be minimized by operating the magnet at the point of the maximum energy product, i.e., the point $(-B_{m0} H_{m0})$. Because the maximum energy product is a measure of the magnet volume required for a given application, it is often found on data sheets for permanent-magnet materials as a tabulated "figure of merit".

Application Example 6.13: Design of Circuit with a Samarium-Cobalt Permanent Magnet

It is desired to achieve a constant magnetic flux density $B_g = 1.0$ T in the air gap of the magnetic circuit of Fig. 6.40. The field is to be created by a samarium-cobalt magnet, see Fig. 6.39. For the air-gap dimensions of Fig. 6.40 find the magnet length ℓ_m and the magnet area A_m that will achieve the desired air-gap flux density and minimize the magnet volume.

Solution

From the samarium-cobalt characteristic (Fig. 6.39): $B_{m0} = 0.45$ T, $(-H_{m0}) =$ 375 kA/m.

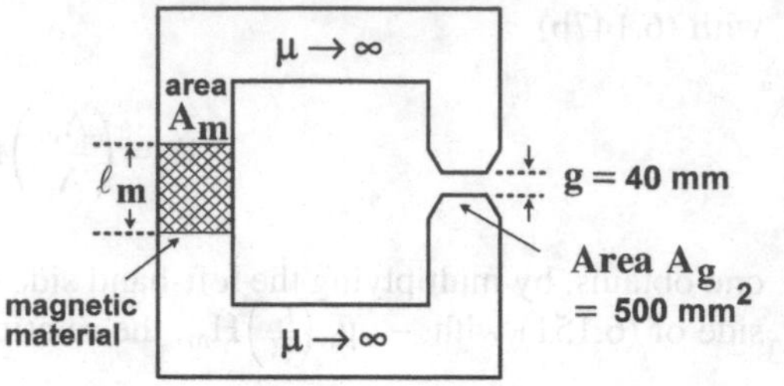

Fig. 6.40 C-core with permanent magnet

Using Ampere's law

$$H_m \ell_m + H_g \cdot g = 0 \tag{6.156}$$

follows

$$\ell_m = g\{H_g/(-H_{m0})\} = g\{B_g/(-H_{m0}\ \mu_o)\} = 0.0849\,\text{m} = 84.9\,\text{mm}. \tag{6.157}$$

With div(B) = 0 or $B_m A_m = B_g A_g$ follows

$$A_m = A_g(B_g/B_{m0}) = 11.11 \cdot 10^{-4}\text{m}^2. \tag{6.158}$$

Application Example 6.14: Maximum-Energy Product Applied to Permanent Magnets and Design of a Permanent-Magnet Circuit

It is desired to achieve a constant magnetic flux density $B_g = B_o = 1.80$ T in the air gap of the magnetic circuit of Fig. 6.41. The field is to be created by a samarium-cobalt magnet (see Fig. 6.39). For the air gap dimensions of Fig. 6.41 find the magnet length ℓ_m and the magnet area A_m that will achieve the desired air-gap flux density and minimize the magnet volume $V_{magnet} = A_m \cdot \ell_m$.

Solution

From Table 6.7, Figs. 6.38b and 6.39 one obtains for SaCo $B_{m0} = 0.45$ T and $H_{m0} = -375$ kA/m. With (6.149) follows for the magnet length

$$\ell_m = g \cdot \left(\frac{H_g}{-H_{m0}}\right) = g \cdot \left(\frac{B_g}{\mu_o(-H_{m0})}\right) = 4 \cdot 10^{-3} \frac{1.8}{4\pi \cdot 10^{-7} \cdot 375\,\text{k}}$$
$$= 0.0153\,\text{m} = 15.3\,\text{mm}. \tag{6.159}$$

Using (6.151) results in

$$A_m = \frac{B_g \cdot A_g}{B_{m0}} = \frac{1.8 \cdot 5 \cdot 10^{-4}}{0.45} = 20 \cdot 10^{-4}\text{m}^2 = 2{,}000\,\text{mm}^2. \tag{6.160}$$

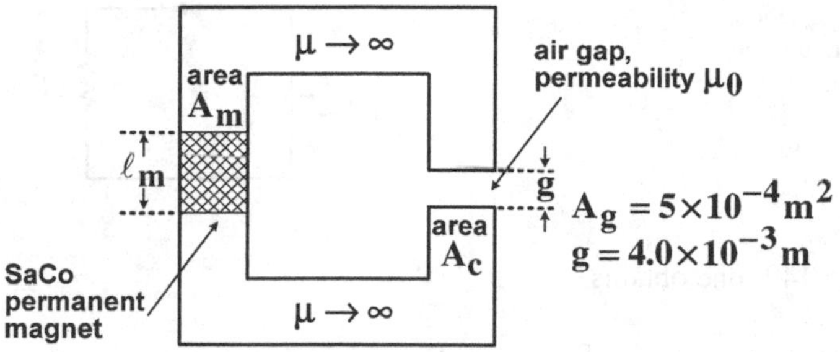

Fig. 6.41 C-core with SaCo permanent magnet

Fig. 6.42 Magnetic circuit of a loudspeaker (voice coil not shown)

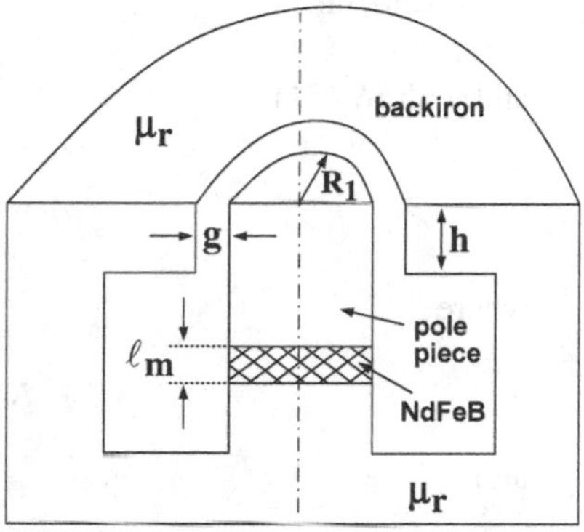

Application Example 6.15: Design of the Permanent Magnet for a Loudspeaker

Figure 6.42 shows the magnetic circuit of a loudspeaker. The voice coil (not shown) is in the form of a circular cylindrical winding which fits in the air gap. A neodymium-iron-boron (NdFeB) magnet (see Fig. 6.37) is used to create the air gap DC magnetic field which interacts with the voice coil currents to produce a magnetic force and therefore a motion of the voice coil. The designer has determined that the air gap must have a length of $g = 0.1$ cm and a height of $h = 1.5$ cm.

(a) Assuming that the finite relative permeability μ_r of the yoke and pole piece result in an increased equivalent (due to saturation) air gap of $g_{equiv} = 1.2$ g, draw the magnetic equivalent circuit, find the magnet height (length) ℓ_m and the magnet radius R_1 that will result in an air gap magnetic flux density of $B_g = 0.8$ T requiring the smallest magnet volume V_{magnet}.
(b) Compute the magnet volume V_{magnet}.

Solution

(a) From Fig. 6.37 one obtains for the maximum power product of NdFeB the values $B_{m0} = 0.6$ T and $H_{m0} = -490$ kA/m. The equivalent air gap length is $g_{equiv} = 1.2\ g = 1.2 \cdot 10^{-3}$ m. Figure 6.43 shows the magnetic equivalent circuit.

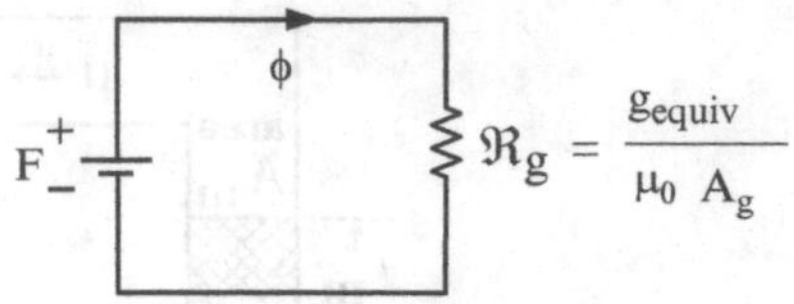

Fig. 6.43 Magnetic equivalent circuit

With (6.149) one obtains

$$\ell_m = \frac{g_{equiv.}H_g}{(-H_{m0})} = \frac{g_{equiv.}B_g}{\mu_o(-H_{m0})} = \frac{1.2 \cdot 10^{-3} \cdot 0.8}{4\,\pi \cdot 10^{-7} \cdot 490\,k} = 1.559 \cdot 10^{-3}\,m$$
$$= 1.56\,mm \qquad (6.161)$$

and with (6.151)

$$A_m = A_g \frac{B_g}{B_{m0}} \qquad (6.162)$$

where

$$A_m = \pi R_1^2 \qquad (6.163)$$

and

$$A_g = 2\ \pi R \cdot h = 2\ \pi (R_1 + g_{equiv}/2)\,h. \qquad (6.164)$$

Introducing (6.163) and (6.164) into (6.162) yields the second-order equation

$$R_1^2 - 2\left(R_1 + \frac{g_{equiv}}{2}\right) h \frac{B_g}{B_{mo}} = 0 \qquad (6.165)$$

or for the given values $R_1 = 8 \cdot 10^{-2}$ m = 80 mm.

(b) The smallest magnet volume is

$$V_{magnet} = \ell_m A_m = \ell_m\, \pi R_1^2 = 1.56\,mm \cdot \pi \cdot 6,400\,mm^2$$
$$= 31.36 \cdot 10^3\,mm^3 = 31.36\,cm^3. \qquad (6.166)$$

6.5 Measurement of Low Losses of High-Efficiency Inductors

The losses of high-efficiency inductors are difficult to measure accurately with regular wattmeters, because these losses are in the neighborhood of a few watts. A computer-aided measurement circuit [11] and the three-voltmeter method [12] will be described to measure accurately the losses of such inductors. The latter method can be replaced by the three-ampere meter method [12].

Computer-Aided Measurement Circuit: A power amplifier along with a signal generator supplies sinusoidal voltages at different frequencies to an inductor being tested, as shown in Fig. 6.44. Voltage and current waveforms of the inductor are sensed by low-inductance voltage dividers and shunts, respectively, and are sampled by a 12-bit A/D converter [13]. The sampling program sequentially samples 15,000 points for two channels at 900 kHz during one 60 Hz period (16.67 ms), resulting in 7,500 points for each channel. Similar computer-aided methods are described in [14] for transformers.

The input power, AC resistance and inductance at any frequency of harmonic order h are given by

$$P_{lossh}=\frac{1}{T}\int_0^T v_1(t)i_1(t)dt, \tag{6.167}$$

$$R_{ACh}=\frac{P_{lossh}}{I^2_{rmsh}}, \tag{6.168}$$

and

$$L_h=\frac{\sqrt{(Z_h^2-R^2_{ACh})}}{2\pi f}, \tag{6.169}$$

where $v_1(t)$ and $i_1(t)$ are the voltage and current of the inductor, respectively, $Z_h = V_{rmsh}/I_{rmsh}$, and f is the frequency.

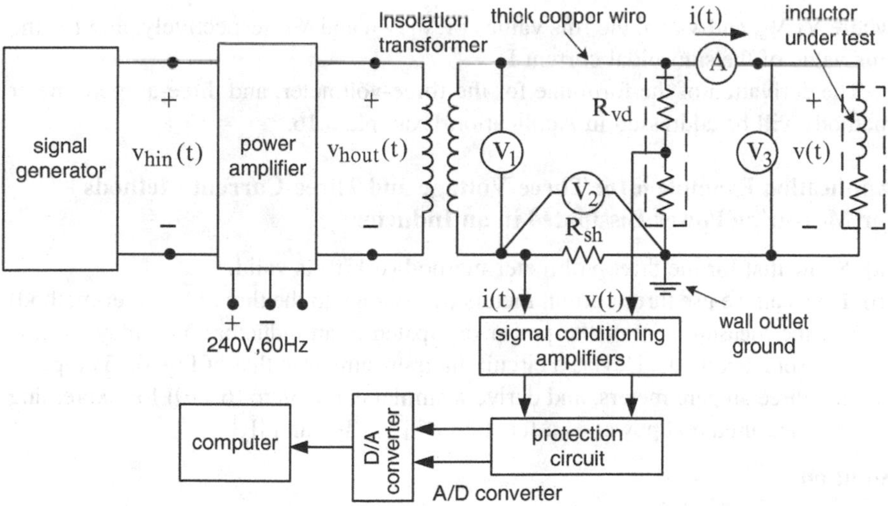

Fig. 6.44 Circuit for measuring losses of inductors with D/A converter, where h stands for harmonic in $v_{hin}(t)$ and $v_{hout}(t)$

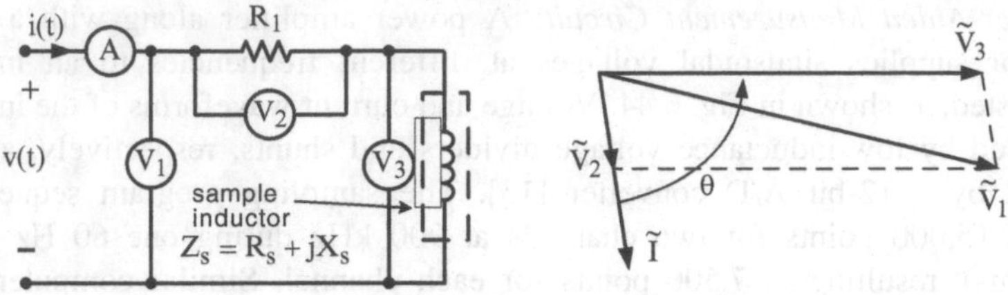

Fig. 6.45 Three-voltmeter method for measuring the loss of an inductor and corresponding phasor diagram

Three-Voltmeter Method for Power Measurement: The losses for some inductors are very small; nevertheless, one has to raise the question about the accuracy of the measured results obtained from the computer-aided measurement circuit of Fig. 6.44. For this reason an alternative approach is chosen and the results of both measurement approaches are compared. The three-voltmeter method is based on (sinusoidal) voltage measurements, as shown in the circuit of Fig. 6.45. The maximum value of the phase angle θ is 90°, which corresponds to an ideal lossless inductor. The value of resistor R_1 must be known in order to compute the loss P^{v}_{loss} and AC resistance R_s of the inductor. The three-voltmeter method can be implemented together with the computer-aided measurement circuit of Fig. 6.44 if the shunt resistance R_{sh} of Fig. 6.44 is used as resistance R_1 of Fig. 6.45. The loss is then computed from

$$P^{v}_{loss} = I^2 R_s = \frac{(V_1^2 - V_2^2 - V_3^2)}{2R_1}, \tag{6.170}$$

where V_1, V_2, and V_3 are the rms values of $\tilde{V}_1, \tilde{V}_2$, and $\tilde{V}_3$, respectively, and I is the rms value of the sinusoidal current $\tilde{I}$.

The derivation of the formulae for the three-voltmeter, and three-ampere meter methods will be addressed in Application Example 6.16.

Application Example 6.16: Three-Voltage and Three-Current Methods for Measuring Power Dissipated in an Inductor

(a) Show that for the three-voltmeter method (6.170) is valid.
(b) How can we use three current meters (in analogy to the three-voltmeter method) for the measurement of the power dissipated in an inductor? You may assume sinusoidal currents. Devise a circuit diagram similar to that of Fig. 6.45 employing three ampere meters, and derive a similar equation to (6.170) for expressing the measured real power as a function of $|\tilde{I}_1|$, $|\tilde{I}_2|$, and $|\tilde{I}_3|$

Solution

(a) There are two methods to show that (6.170) is true: method #1 is based on the phasor diagram and method #2 on Ohm's law.

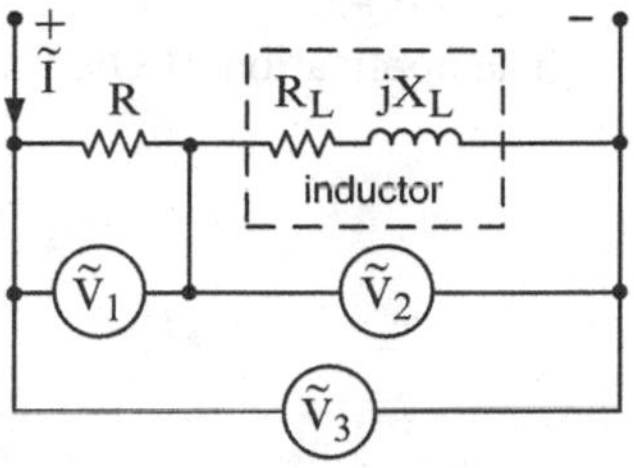

Fig. 6.46 Circuit of three-voltmeter method

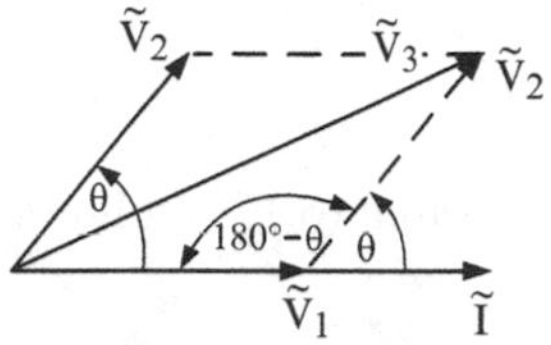

Fig. 6.47 Phasor diagram for three-voltmeter method

Method #1:
Figure 6.46 shows the definitions of the 3 voltmeters, inductor resistance R_L and reactance X_L, and calibrated resistor with resistance R.
The phasor diagram for circuit Fig. 6.46 is illustrated in Fig. 6.47.
Use of the cosine law yields

$$\left|\tilde{V}_3\right|^2 = \left|\tilde{V}_1\right|^2 + \left|\tilde{V}_2\right|^2 - 2\left|\tilde{V}_1\right|\left|\tilde{V}_2\right|\cos(180^\circ - \theta) \tag{6.171}$$

with

$$\cos(180^\circ - \theta) = -\cos\theta, \tag{6.172}$$

the inductor loss

$$P_{loss\_inductor} = \left|\tilde{V}_1\right|\left|\tilde{I}\right|\cos\theta \tag{6.173}$$

and Ohm's law

$$\left|\tilde{V}_1\right| = \left|\tilde{I}\right|R \tag{6.174}$$

follows

$$\left|\tilde{V}_3\right|^2 = \left|\tilde{V}_1\right|^2 + \left|\tilde{V}_2\right|^2 + 2R \cdot P_{loss\_inductor} \tag{6.175}$$

or

$$P_{loss\_inductor} = \frac{\left|\tilde{V}_3\right|^2 - \left|\tilde{V}_2\right|^2 - \left|\tilde{V}_1\right|^2}{2R}. \tag{6.176}$$

Method #2:
The application of Ohm's law yields for the current $|\tilde{I}|$

$$|\tilde{I}|^2 = \frac{|\tilde{V}_2|^2}{R_L^2 + X_L^2} \tag{6.177}$$

$$|\tilde{I}|^2 = \frac{|\tilde{V}_1|^2}{R^2} \tag{6.178}$$

$$|\tilde{I}|^2 = \frac{|\tilde{V}_3|^2}{(R + R_L)^2 + X_L^2}. \tag{6.179}$$

From (6.179) follows

$$|\tilde{I}|^2 = \frac{|\tilde{V}_3|^2}{R^2 + 2RR_L + R_L^2 + X_L^2} \tag{6.180}$$

from (6.177)

$$R_L^2 + X_L^2 = \frac{|\tilde{V}_2|^2}{|\tilde{I}|^2} \tag{6.181}$$

and from (6.178)

$$R^2 = \frac{|\tilde{V}_1|^2}{|\tilde{I}|^2}. \tag{6.182}$$

Introducing (6.182) and (6.181) into (6.180) results in

$$|\tilde{I}|^2 = \frac{|\tilde{V}_3|^2}{\frac{|\tilde{V}_1|^2}{|\tilde{I}|^2} + 2RR_L + \frac{|\tilde{V}_2|^2}{|\tilde{I}|^2}} \tag{6.183}$$

or

$$|\tilde{V}_3|^2 = |\tilde{V}_1|^2 + |\tilde{V}_2|^2 + 2R \cdot R_L|\tilde{I}|^2 = |\tilde{V}_1|^2 + |\tilde{V}_2|^2 + 2R \cdot P_{loss\_inductor} \tag{6.184}$$

and solved for the inductor loss

$$P_{loss\_inductor} = \frac{|\tilde{V}_3|^2 - |\tilde{V}_2|^2 - |\tilde{V}_1|^2}{2R}. \tag{6.185}$$

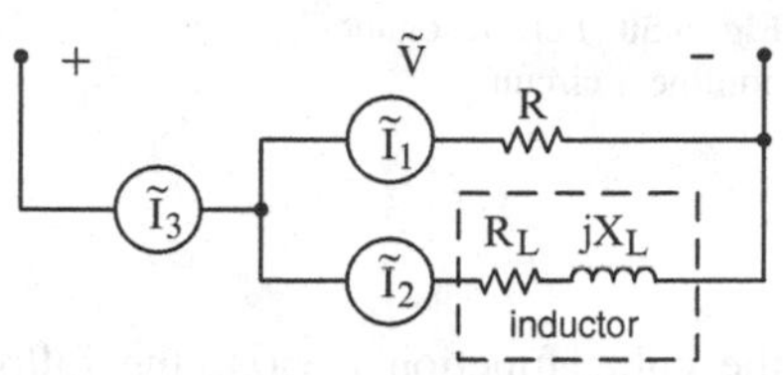

Fig. 6.48 Circuit of three-current meter method

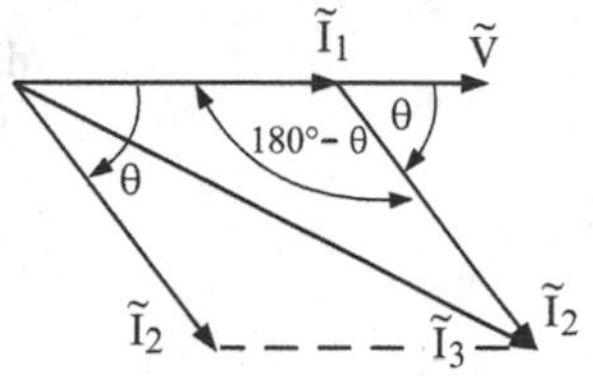

Fig. 6.49 Phasor diagram for three-current meter method

(b) The circuit for the three-current meter method is shown in Fig. 6.48 The phasor diagram for circuit Fig. 6.48 is illustrated in Fig. 6.49. Use of the cosine law yields

$$\left|\tilde{I}_3\right|^2 = \left|\tilde{I}_1\right|^2 + \left|\tilde{I}_2\right|^2 - 2\left|\tilde{I}_1\right|\left|\tilde{I}_2\right|\cos(180° - \theta) \qquad (6.186)$$

with

$$\cos(180° - \theta) = -\cos\theta, \qquad (6.187)$$

the inductor loss

$$P_{\text{loss\_inductor}} = \left|\tilde{V}\right|\left|\tilde{I}_2\right|\cos\theta. \qquad (6.188)$$

Introducing (6.187) and (6.188) into (6.186) yields

$$P_{\text{loss\_inductor}} = \left(\frac{\left|\tilde{I}_3\right|^2 - \left|\tilde{I}_1\right|^2 - \left|\tilde{I}_2\right|^2}{2}\right) \cdot R. \qquad (6.189)$$

Application Example 6.17: Ferroresonance in Nonlinear Inductors and Transformers

A ferroresonant circuit consists of a sinusoidal voltage source v(t), a (cable) capacitance C, and a nonlinear (magnetizing) impedance – consisting of a resistance R and an inductance L – connected in series, as shown in Fig. 6.50. If the saturation curve (flux linkages λ as a function of current i) is represented by

Fig. 6.50 Ferroresonant nonlinear circuit

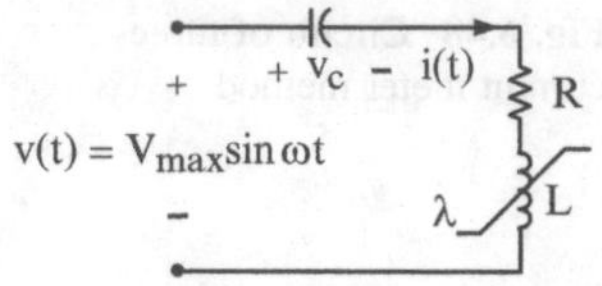

the cubic function $i = a\lambda^3$, the following set of first-order nonlinear differential equations can be formulated:

$$\frac{d\lambda}{dt} = V_{max}\sin\omega t - v_c - aR\lambda^3 \qquad (6.190)$$

$$\frac{dv_c}{dt} = \frac{a\lambda^3}{C} \qquad (6.191)$$

or in an equivalent manner, one can formulate the second-order nonlinear differential equation

$$\frac{d^2\lambda}{dt^2} + k\lambda^2\frac{d\lambda}{dt} + \lambda^3 = b\cos\omega t, \qquad (6.192)$$

where $k = 3aR$, $b = V_{max}\omega$, and $C = a$.

(a) Show that (6.190) to (6.192) are true for the circuit of Fig. 6.50.
(b) The system of Fig. 6.50 results in chaotic behavior for certain values of the parameters $b = V_{max}\omega$ and $k = 3aR$ ($\omega = 2\pi f$ where $f = 60$ Hz, $R = 0.1\ \Omega$, $C = 100\ \mu F$, $a = C$, low voltage $V_{max\_low} = \sqrt{2}\cdot 1{,}000$ V, and high voltage $V_{max\_high} = \sqrt{2}\cdot 20{,}000$ V), which characterize the source voltage and the resistive losses of the circuit, respectively. For zero initial conditions – using either Mathematica or Matlab – compute λ, dλ/dt, and i as a function of time from $t_{start} = 0$ to $t_{end} = 0.5$ s.
(c) Plot λ versus dλ/dt for the values as given in (b) for the low terminal voltage.
(d) Plot λ as a function of time for the values as given in (b) for the low terminal voltage.
(e) Plot dλ/dt as a function of time for the values as given in (b) for the low terminal voltage.
(f) Plot i as a function of time for the values as given in (b) for the low terminal voltage.
(g) Plot dλ/dt versus i for the values as given in (b) for the low terminal voltage.
(h) Plot λ versus dλ/dt for the values as given in (b) for the high terminal voltage.
(i) Plot λ as a function of time for the values as given in (b) for the high terminal voltage.
(j) Plot dλ/dt as a function of time for the values as given in (b) for the high terminal voltage.
(k) Plot i as a function of time for the values as given in (b) for the high terminal voltage.
(l) Plot dλ/dt versus i for the values as given in (b) for the high terminal voltage.

Solution

(a) Kirchhoff's law yields

$$V_{max} \sin \omega t = v_c + i \cdot R + \frac{d\lambda}{dt} = v_c + R \cdot a \cdot \lambda^3 + \frac{d\lambda}{dt} \tag{6.193}$$

solving this equation results in (6.190) With

$$i = C\frac{dv_c}{dt} \tag{6.194}$$

and

$$i = a\lambda^3 \tag{6.195}$$

leads to (6.191).

Introducing (6.194) and (6.195) into the time differential of (6.193) one obtains

$$V_{max}\omega \cos \omega t = \frac{dv_c}{dt} + 3a \cdot R \cdot \lambda^2 \frac{d\lambda}{dt} + \frac{d^2\lambda}{dt^2} \tag{6.196}$$

or with $k = 3aR$, $b = \omega V_{max}$, $C = a$, and (6.194) and (6.195)

$$\frac{d^2\lambda}{dt^2} + k\,\lambda^2 \frac{d\lambda}{dt} + \lambda^3 = b \cos \omega t. \tag{6.197}$$

(b) Use of Mathematica program is described in Application Example 2.8. The program for the low voltage with $V_{max\_low} = \sqrt{2} \cdot 1{,}000$ V is listed below.

```
f=60;
w=2*Pi*f;
Ca=0.0001;
U=1000*Sqrt[2];
R=0.1;
eqn={x''[t]+a *x[t]^2 *x'[t] +x[t] ^3= =d*Cos[w *t], x[0]= =0,x'[0]= =0};
sol=NDSolve[eqn/.{a->3*Ca*R, d->U*w}, x, {t, 0, 0.5}, MaxSteps -> 50000];
i[t_] :=Ca*x[t]^3/.sol;
ParametricPlot[{x'[t],x[t]}/. sol[[1]], {t, 0, 0.5}, PlotRange ->{{-1500, 1500}, {-10, 10}},
AxesLabel -> {"induced voltage", "lambda"}, AspectRatio->2.0]
Plot[x[t]/.sol[[1]], {t, 0, 0.5}, PlotRange ->{{0, 0.5}, {-10, 10}}, AxesLabel -> {"t" , "lambda"}]
Plot[x'[t]/.sol[[1]], {t, 0, 0.5}, PlotRange ->{{0, 0.5}, {-2000, 2000}}, AxesLabel -> {"t" ,
"induced voltage"}]
Plot[i[t], {t, 0, 0.5}, PlotRange ->{{0, 0.5}, {-0.06, 0.06}}, AxesLabel -> {"t" , "current"}]
ParametricPlot[{Ca*x[t]^3, x'[t]}/. sol[[1]], {t, 0, 0.5}, PlotRange ->{{-0.06, 0.06}, {-1500,
1500}}, AxesLabel -> {"i(t)", "induced voltage"}, AspectRatio->0.5]
```

Note that some of the characters employed by Mathematica, e.g., prime (') and := relying on the notebook file extension .nb are not compatible with Microsoft word files with the extension .doc and may have to be retyped in the notebook file.

The corresponding plots are shown in Figs. 6.51a–e. For the high voltage replace in the above listed program $V_{max\_low} = 1{,}000$*Sqrt [2] by $V_{max\_high} = 20{,}000$*Sqrt [2], the corresponding plots are depicted in Figs. 6.51f–j.

Application Example 6.18: Analysis of an Electromechanical Transducer

Position of the movable plunger in Fig. 6.52 is limited to $0 \le x \le w$ while maintaining a constant air gap of length g. Both yoke and plunger are considered to have infinite permeability. Neglect fringing and lamination effects. The first coil (N_1 turns) carries a constant DC current I_0 and the second coil (N_2 turns) is open-circuited.

(a) The mutual inductance between coils 1 and 2 as a function of x is: $L_{12} = (A + B\ x)$, compute A and B.

(b) The plunger is driven by an external mechanical source so that its motion is given by $x(t) = 0.0125 + 0.002 \cdot \sin(377t)$. Find the expression for the sinusoidal voltage $v_2(t)$ which is generated as a result of this motion. Note, $N_1 = 1{,}000$ turns, $N_2 = 500$ turns, $0 \le x \le w$, $w = 2.5$ cm, $g = 0.2$ cm, $D = 4$ cm, and $I_o = 5$ A.

Solution

(a)

$$L_{12} = N_1 N_2 \frac{\mu_o A_c}{2g} = \frac{\mu_o N_1 N_2}{2g}[D(w - x)] \tag{6.198}$$

$$\begin{aligned} L_{12} &= \frac{4\pi \times 10^{-7}(1{,}000)(500)}{2(0.2 \times 10^{-2})}[4 \times 10^{-2}(2.5 \times 10^{-2} - x)] \\ &= 0.157 - 6.283\,x. \end{aligned} \tag{6.199}$$

Therefore, $A = 0.157$ H, and $B = -6.283$ H/m.

(b)

$$v_2(t) = \frac{d\lambda}{dt} = I_o \frac{dL_{12}}{dt} = -6.283 I_o \frac{dx}{dt} = -31.415\,(0.002) \frac{d(\sin 377t)}{dt} \tag{6.200}$$

Therefore, $v_2(t) = -23.68 \cos 377t\ [V]$.

6.6 Summary

Maxwell's equations are the starting point for this chapter, in particular Ampere's and Faraday's laws are relied on for low-frequency applications where displacement currents can be neglected. These laws are applied to inductors and

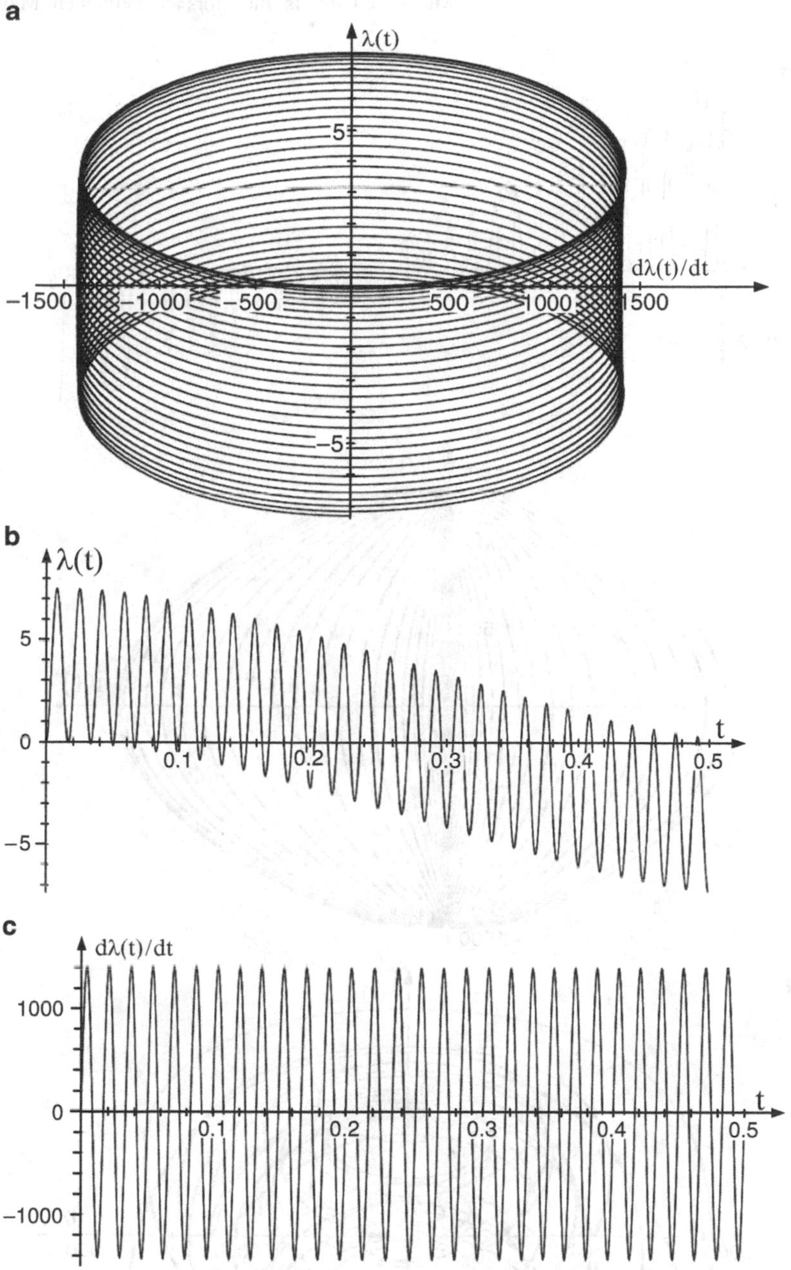

Fig. 6.51 (**a**) Flux linkages λ versus induced voltage $d\lambda/dt$ at low terminal voltage $V_{max\_low} = \sqrt{2} \cdot 1{,}000$ V. (**b**) Flux linkages λ as a function of time at low terminal voltage $V_{max\_low} = \sqrt{2} \cdot 1{,}000$ V. (**c**) Induced voltage $d\lambda/dt$ as a function of time at low terminal voltage $V_{max\_low} = \sqrt{2} \cdot 1{,}000$ V. (**d**) Current i as a function of time at low terminal voltage $V_{max\_low} = \sqrt{2} \cdot 1{,}000$ V. (**e**) Induced voltage $d\lambda/dt$ versus current i at low terminal voltage $V_{max\_low} = \sqrt{2} \cdot 1{,}000$ V. (**f**) Flux linkages λ versus induced voltage $d\lambda/dt$ at high terminal voltage $V_{max\_high} = \sqrt{2} \cdot 20{,}000$ V. (**g**) Flux linkages λ as a function of time at high terminal voltage $V_{max\_high} = \sqrt{2} \cdot 20{,}000$ V. (**h**) Induced voltage $d\lambda/dt$ as a function of time at high terminal voltage $V_{max\_high} = \sqrt{2} \cdot 20{,}000$ V. (**i**) Current i as a function of time at high terminal voltage $V_{max\_high} = \sqrt{2} \cdot 20{,}000$ V. (**j**) Induced voltage $d\lambda/dt$ versus current i at high terminal voltage $V_{max\_high} = \sqrt{2} \cdot 20{,}000$ V

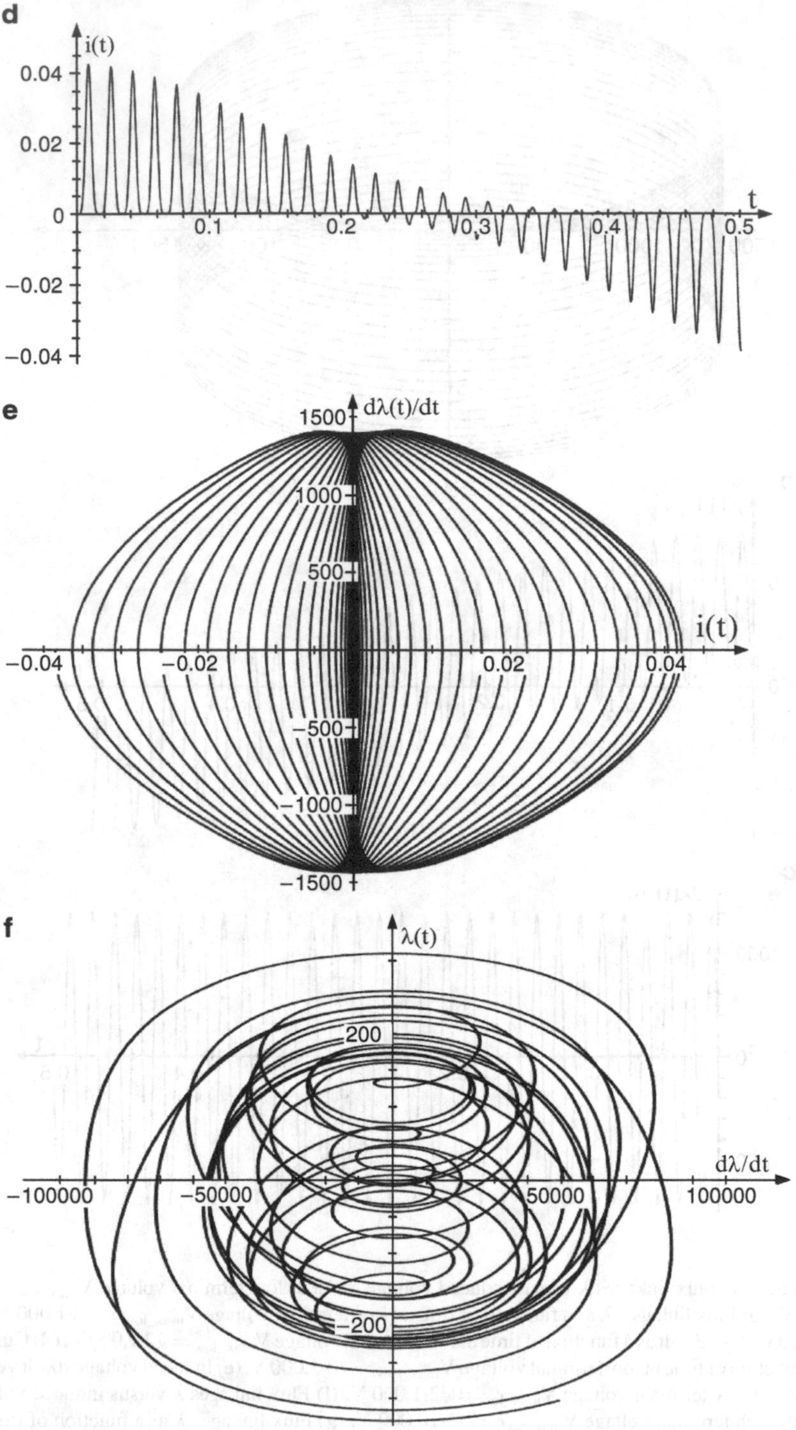

Fig. 6.51 (continued)

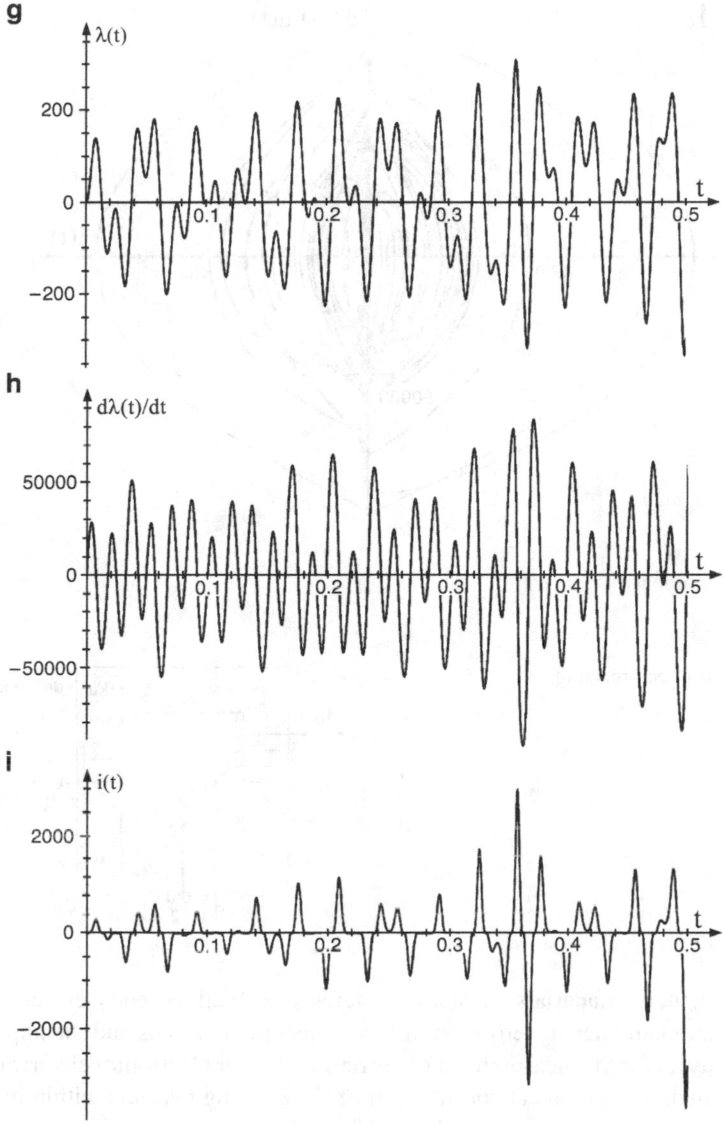

Fig. 6.51 (continued)

permanent magnets. After the definition of various approximations for (B–H) characteristics simple inductor circuits are analyzed based on self- and mutual inductances and their representation via Ohm's law for a magnetic circuit in analogy to Ohm's law for an electric circuit. American wire gauge (AWG) sizes are introduced in accordance with the National Electric Code®. Properties of magnetic materials are discussed next and the fact that the (B–H) characteristic

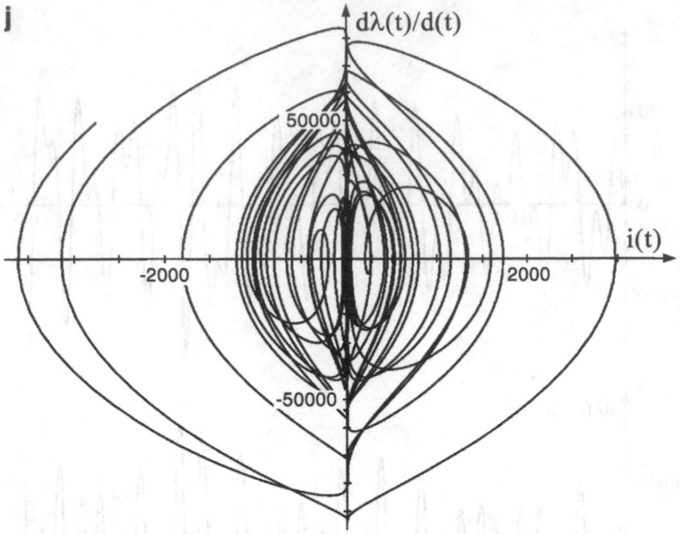

Fig. 6.51 (continued)

Fig. 6.52 The reciprocating generator

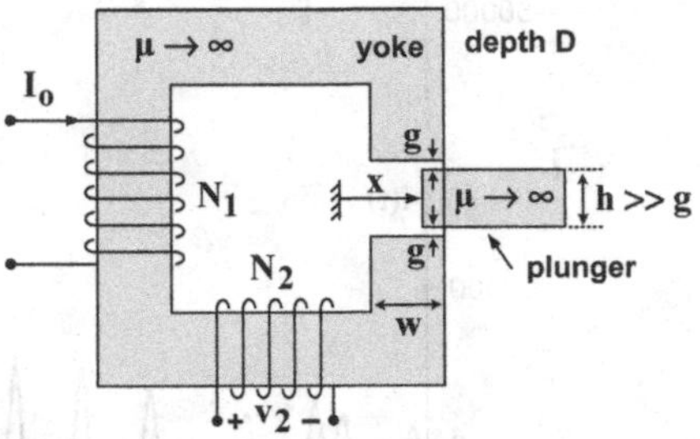

of soft magnetic materials includes hysteresis as well as eddy-current losses. Magnetization and demagnetization of ferromagnetic materials and the application of Hall sensors for the measurement of current in (power electronic) electric circuits are explained. The graphical construction of the exciting currents within magnetic circuits (e.g., soft magnetic materials used for inductors, transformers), the apparent power requirements of iron cores, their core losses due to hysteresis and eddy currents are introduced. The transition of soft to hard magnetic materials leads to permanent magnets. After the introduction of residual flux density and coercivity the energy-product function of permanent-magnet materials is derived. From this function the maximum energy product is defined which represents a figure of merit for permanent-magnet materials. Towards the end of the chapter measuring techniques for the determination of the low losses of highly efficient inductors are introduced. Application examples highlight the topics explained in this chapter. The concept of ferroresonance as it occurs in nonlinear inductors and transformers is introduced, and magnetic circuit theory is applied to electromechanical transducers.

6.7 Problems

Problem 6.1: *Calculation of reluctance, inductance, flux linkages and stored magnetic energy in a magnetic circuit using the MKSAK system of units*

An inductor for an electronic circuit is shown in Fig. 6.53: it has the iron-core stack height D = 1 cm, and a uniform core width of a = 1 cm. Assume that the iron core has an infinitely large relative permeability ($\mu_r = \mu_c = \infty$) and neglect the effects of magnetic leakage and flux fringing.

(a) Calculate the reluctance of the core $\mathfrak{R}_c$, and that of the air gap $\mathfrak{R}_g$ for a gap length of g = 1 cm.
(b) For N = 100 turns, calculate the self-inductance L.
(c) Determine the current I_{DC} required to operate the inductor at an air-gap flux density of $B_g = 1.0$ T.
(d) Calculate the flux linkages λ of the coil.
(e) Find the stored energy within the iron core W_{core}, and that within the air gap W_{gap}.

Problem 6.2: *Design of inductor for DC-to-DC converter*

The inductor of Fig. 6.54 is to be designed. Its iron core is of uniform cross-sectional area $A_c = A_g = 0.5 \cdot 10^{-3}$ m^2 and of mean length $\ell_c = 0.2$ m. It has an adjustable air gap of length g and will be wound with a coil of N turns.

(a) Draw its equivalent magnetic circuit. Calculate g and N such that the self-inductance is L = 15 mH and so that the inductor can operate at maximum currents of $i = I_{maxDC} = 5$ A without saturating the core; assume that saturation occurs when the maximum flux density in the core exceeds $B_{cmax} = B_{gmax} = 1.7$ T and that the core has a constant relative permeability of $\mu_r = \mu_c = 3{,}000$.
(b) For an inductor current of $I_{max} = 5$ A, calculate the magnetic stored energy in the air gap W_{gap} and the magnetic stored energy in the core W_{core}. Compare both stored energy values.

Problem 6.3: *Use of the wire table*

Figure 6.13 of Application Example 6.4 shows the geometric design of an inductor. Modify this design for a self-inductance of L = 0.1 henries wound on a high-permeability ($\mu_c \to \infty$) laminated core of rectangular cross-section. Neglect

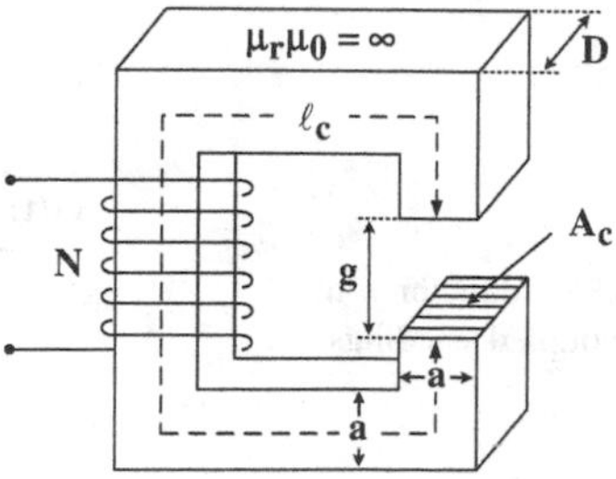

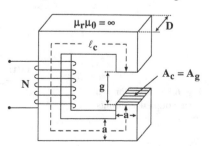

Fig. 6.53 Geometric design of an inductor with infinitely large permeability

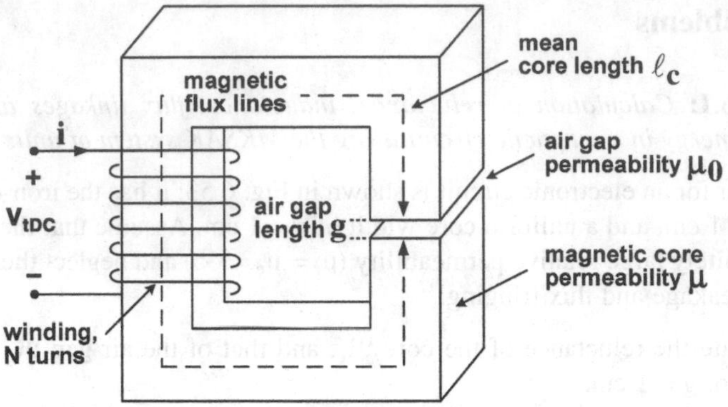

Fig. 6.54 Inductor with linear (finite or constant) core permeability

magnetic leakage and fringing. The winding consists of insulated copper wire with a specific resistivity of $\rho = 1.73 \cdot 10^{-6}\ \Omega$ cm. The coil is to be operated with a constant terminal voltage of $V_{tDC} = 24$ V and an air-gap flux density of $B_g = 1.00$ T. Find

(a) The average length of one turn of the coil ℓ_{avg},
(b) The number of turns N required,
(c) The coil current I_{DC}, the resistance of the coil R_{coil},
(d) The power dissipated in coil, the time constant τ,
(e) The wire cross-section a_{wire} including the wire size, and
(f) The copper fill factor f_w.

Problem 6.4: *Design of an inductor with 3 coupled windings*

(a) Draw the magnetic equivalent circuit for the magnetic circuit of Fig. 6.55.
(b) Determine the analytical expressions for the self-inductances L_{11}, L_{22}, L_{33}, and the mutual inductances L_{12}, L_{13}, L_{23} as a function of the parameters of Fig. 6.55 in terms of N_1, N_2, N_3, $\Re_{g1}$, $\Re_{g2}$, and $\Re_{g3}$. No numerical values are required.
(c) Draw the electrical equivalent circuit of Fig. 6.55.

Problem 6.5: *Computation of self- and mutual inductances*

The magnetic circuit of Fig. 6.56 has two windings and two air gaps (g_1 and g_2). The iron core can be assumed to be of infinite permeability ($\mu = \mu_r\mu_0 = \mu_c\mu_0 \to \infty$) and

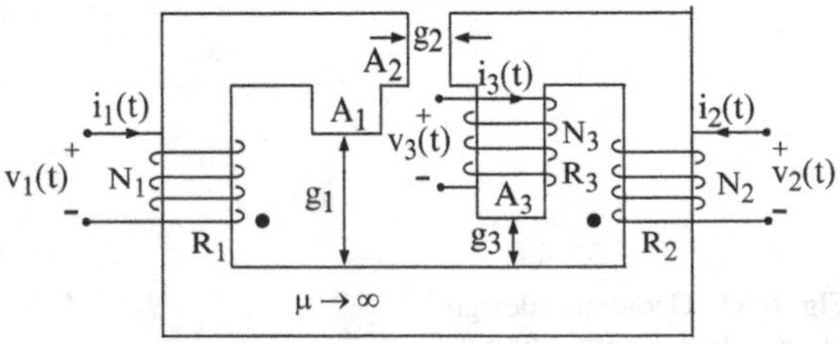

Fig. 6.55 Inductor with three-coupled windings (coils)

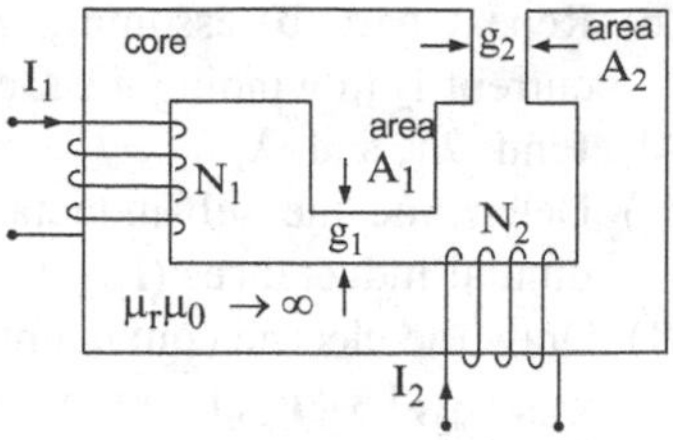

Fig. 6.56 Inductor with two-coupled windings (coils)

the core parameters (A_1, g_1 and A_2, g_2) are indicated in Fig. 6.56. The calculation of numerical values is not required.

(a) Introduce the dot markings in Fig. 6.56 and draw its magnetic equivalent circuit.
(b) Assuming coil #1 to be carrying a current I_1 and the current in coil #2 to be zero $I_2 = 0$ draw the magnetic equivalent circuit. Determine:
   1. The magnetic flux density in each of the air gaps ,
   2. The flux linkage of winding #1, and
   3. The flux linkage of winding #2.
(c) Repeat part (b) assuming zero current $I_1 = 0$ in winding #1 and a current I_2 in winding #2.
(d) Repeat part (b) assuming the current in winding #1 to be I_1 and the current in winding #2 to be I_2 (apply the principle of superposition).
(e) Find the self-inductances of windings #1 (L_{11}) and #2 (L_{22}) and the mutual inductance ($L_{12} = L_{21}$) between the two windings.

Problem 6.6: *Design of an inductor with two windings for power electronics applications*

The inductor core of Fig. 6.57 carries two windings and has two air gaps (g_1 and g_2). The core can be assumed to be of infinite permeability ($\mu_r \to \infty$), and the core dimensions (g_1, A_1, g_2, and A_2) are indicated in Fig. 6.57.

(a) Draw the magnetic equivalent circuit of Fig. 6.57.
(b) Assuming coil #1 to be carrying a non-zero current i_1 and the current in coil #2 to be zero $i_2 = 0$, calculate:
   1. the flux linkage of winding #1, $\lambda_1^{(b)}$, and
   2. the flux linkage of winding #2, $\lambda_2^{(b)}$.

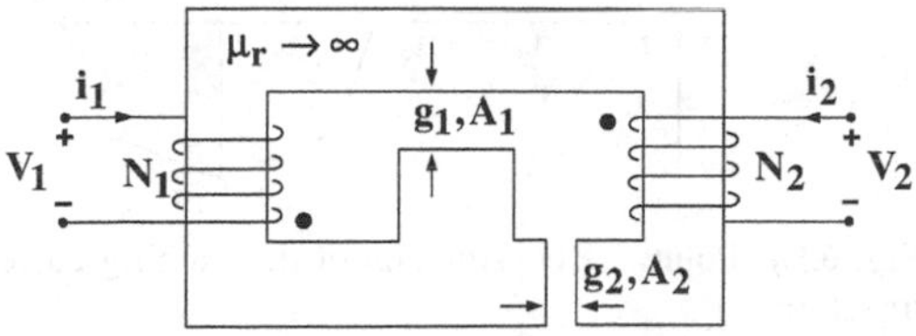

Fig. 6.57 Inductor circuit consisting of iron core and two windings

(c) Repeat part (b) assuming zero current $i_1 = 0$ in winding #1 and a non-zero current i_2 in winding #2. Determine $\lambda_1^{(c)}$, and $\lambda_2^{(c)}$.
(d) Find λ_1, and λ_2.
(e) Determine the self-inductances of windings #1 (L_{11}) and #2 (L_{22}) and the mutual inductances (L_{12}, L_{21}) between the two windings.
(f) Draw the electric equivalent circuit of Fig. 6.57 as a function of v_1, i_1, R_1, L_{11}, L_{12}, L_{21}, L_{22}, R_2, i_2, and v_2. Note R_1 is the ohmic resistance of winding #1, and R_2 that of winding #2.

Problem 6.7: *Graphical construction of exciting current $i_\phi(t)$ of a magnetic circuit (inductor or transformer)*

Construct the values of the exciting current $i_\phi(t)$ for the time instants t_1, t_2, t_3, t_4, t_5, t_6 and plot these values $\{i_\phi(t_1), i_\phi(t_2), i_\phi(t_3), i_\phi(t_4), i_\phi(t_5), i_\phi(t_6)\}$as a function of time in Fig. 6.58.

Problem 6.8: *Determination of the iron-core loss of an inductor from measurements as used in power electronics applications*

In Fig. 6.35 of Application Example 6.11 the (λ–i) characteristic of an iron-core sample has been recorded, where the current and flux linkage units are given on the abscissa (x-axis) and ordinate (y-axis) of the plot, respectively.

(a) For $N = 680$ turns, a core cross-section of $A_c = 5\ cm^2$ and a core length of $\ell_c = 12\,cm$ compute the flux density B_c versus magnetic field intensity H_c function of the iron core: $B_c = f(H_c)$, that is, express B_c as a function of λ ($B_c = \lambda/(N \cdot A_c)$) and H_c as a function of i ($H_c = N \cdot i/\ell_c$) and introduce a new coordinate system (B_c versus H_c) in Fig. 6.35. Find the energy input per cycle $W_{absorbed}$ (6.131) measured in W·s or joules.
(b) If the frequency of λ and i is f = 1,000 Hz compute the input power per unit volume p_c (6.134) measured in W/m^3 or W/dm^3.
(c) The weight of the iron-core sample is 0.12 (kg-force). What is the input power of this iron-core sample per unit weight measured in W/(kg-force) @ a flux density of 0.676 T, and the input power of this iron-core sample in watts @ 0.676 T provided the iron-core density is $\gamma_{Fe} = 7.86$ (kg-force)/dm^3? Note, that

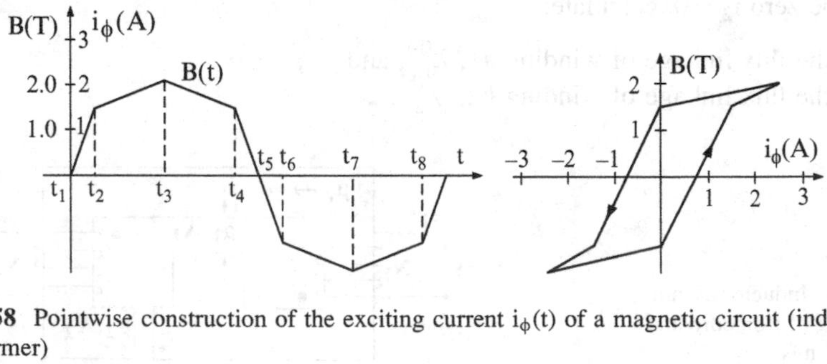

Fig. 6.58 Pointwise construction of the exciting current i_ϕ(t) of a magnetic circuit (inductor or transformer)

the losses of an iron core depend upon the mechanical stress [10] induced by cutting, compression, and mounting the core within a frame, this increase can be 1.6 times the measured loss value obtained from the iron-core sample performed with an Epstein frame [7].

Problem 6.9: *Use of inductor for other than rated frequency*

An inductor designed for $V_{rated} = 120\ V_{rms}$ @ $f_{rated} = 60$ Hz must be used for $f_{new} = 30$ Hz. Can a voltage of $V_{30Hz} = 120\ V_{rms}$ be applied without exceeding the rated flux density? What is $V_{30Hz\ max}$ that can be applied so that the rated flux density will not be exceeded?

Problem 6.10: *Design of a permanent-magnet circuit based on the maximum energy product*

It is desired to achieve a constant magnetic flux density $B_g = 2.0$ T in the air gap of the magnetic circuit of Fig. 6.59 through flux concentration where A_g is less than A_m. The field is to be created by a neodymium-iron-boron (NdFeB) permanent magnet. For the air gap dimensions of Fig. 6.59 find the magnet length ℓ_m and the magnet area A_m that will achieve the desired air-gap flux density and minimize the magnet volume $V_{magnet\_minimum} = A_m \cdot \ell_m$. Calculate $V_{magnet\_minimum}$.

Problem 6.11: *Design of a permanent magnet for a loudspeaker*

Figure 6.60 shows the magnetic circuit of the cross-section of a loudspeaker where the backiron, pole piece, air gap g, and the samarium-cobalt (SaCo) permanent magnet are identified. The voice coil (not shown) resides in the air gap g. The SaCo magnet is used to create the air gap DC magnetic field B_g which interacts with the voice coil current. The voice coil is attached to the paper cone which emits sound waves due to the vibrations of the voice coil. The designer has determined that the air gap must have a radial length of g = 2 mm and a height of h = 10 mm.

(a) Assuming that the relative permeability of the backiron and that of the pole piece is very large ($\mu_r \to \infty$) draw the magnetic equivalent circuit of the loudspeaker.
(b) Find the magnet height (length) ℓ_m and the magnet radius R_1 that will result in the air-gap flux density of $B_g = 0.5$ T and requires the smallest magnet volume $V_{magnet\_minimum}$.
(c) Compute the magnet volume $V_{magnet\_minimum}$.

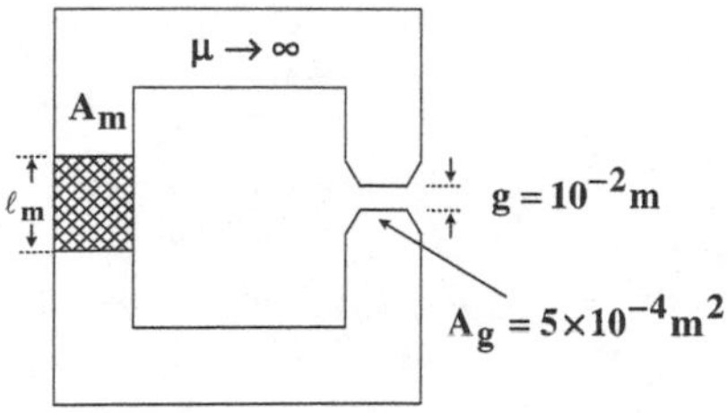

Fig. 6.59 Design of a permanent-magnet circuit with a C-core and flux concentration in the air gap

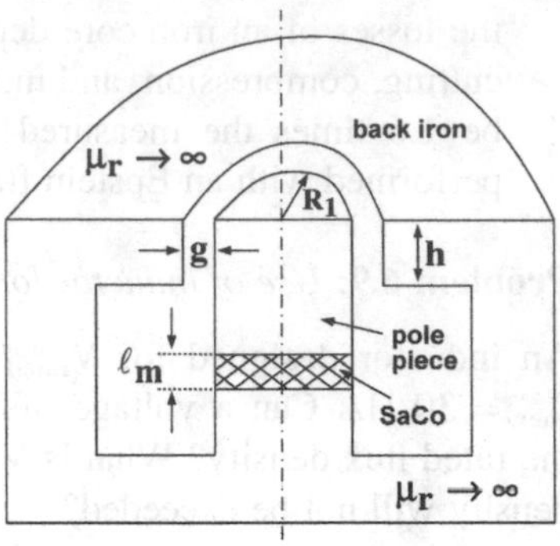

Fig. 6.60 Cross-section of a loudspeaker with backiron, pole piece, air gap and permanent magnet residing within pole piece

Problem 6.12: *Flux concentration within air gap of a C-core and application of the maximum energy product for permanent magnets*

It is desired to achieve a magnetic flux density $B_g = 2.0$ T in the air gap of the magnetic circuit of Fig. 6.59. The field is to be created by a samarium-cobalt (SaCo) permanent magnet. For the air-gap dimensions of Fig. 6.59 find the magnet length ℓ_m and the magnet area A_m that will achieve the desired air gap flux density and minimize the magnet volume $V_{magnet\_minimum}$. Calculate $V_{magnet\_minimum}$.

Problem 6.13: *Analysis of an inductor with given geometrical dimensions*

An inductor is designed using the magnetic core of Fig. 6.61. The core and gap have uniform cross-section areas $A_c = A_g = 5\ \text{cm}^2$ (neglect fringing and stacking lamination effects), the core average length is $\ell_c = 40$ cm, the excitation current is $i = 4.5$ A, the air-gap length is $g = 0.17$ mm, the constant permeability of the core is $\mu_r = 3{,}500$ and the number of turns is $N = 100$ turns. Calculate:

(a) The flux density B and the inductance L,
(b) The magnetic energy stored in air gap W_g,
(c) The magnetic energy stored in the core W_c, and
(d) The total magnetic energy stored in the inductor W_T.

Hint for parts (b) and (c): $W = A\ell \int_0^B H\,dB = A\ell \int_0^B (B/\mu)\,dB$.

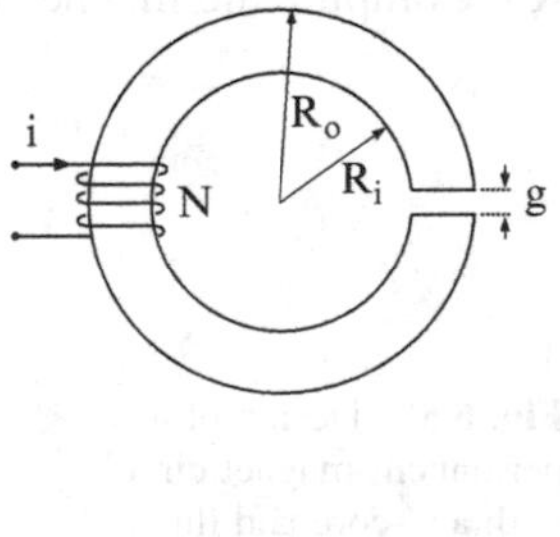

Fig. 6.61 Ring-type inductor

References

1. Johnk, C. T. A.: *Engineering Electromagnetic Fields and Waves*, John Wiley & Sons, New York, NY, 1975.
2. Fitzgerald, A. E.; Kingsley, Jr., Ch.; Umans, S. D.: *Electric Machinery*, 5th Edition, McGraw-Hill Publishing Company, 1990.
3. *Electrical Engineering Pocket Handbook*, Electrical Apparatus Service Association, Inc., St. Louis, MO.
4. *IEEE Standard Dictionary of Electrical and Electronics Terms*, 2nd Edition., IEEE Std 100-1977, Published by the Institute of Electrical and Electronics Engineers, Inc., New York, NY, December 1, 1977,
5. National Electric Code®
6. Gerthsen, Ch.; Kneser, H. O.: *Physik*, 6th Edition, Springer-Verlag, Berlin, 1960
7. Fuchs, E. F.; Fei, R.: "A new computer-aided method for the efficiency measurement of low-loss transformers and inductors under nonsinusoidal operation," *IEEE Transactions on Power Delivery*, Jan. 1996, Vol. 11, No. 1, pp. 292.
8. *Non-oriented Sheet Steel for Magnetic Applications*, United States Steel (USS), Pittsburgh, PA, May 1978.
9. *Armco Oriented and TRAN-COR H Electrical Steels*, ARMCO, Middletown, OH.
10. Fuchs, E. F.; Appelbaum, J.; I. Khan, A.; Höll, J.; Frank, U. V.: *Optimization of Induction Motor Efficiency, Volume 1: Three-Phase Induction Motors*, Report # EL-4152-CCM, Research Project 1944-1, Electric Power Research Institute, Palo Alto, CA.
11. Yildirim, D.; Fuchs, E. F.: "Computer-aided measurement of inductor losses at high frequencies (0 to 6 kHz)," *Proceedings of the 14th Annual Applied Power Electronics Conference and Exposition*, March 1999, Dallas, TX.
12. Stöckl, M.: *Elektrische Messtechnik*, B. G. Teubner Verlagsgesellschaft, Stuttgart, 1961.
13. *DAS-50 User's Guide*, Keithley MetraByte Corporation, 440 Myles Standish Boulevard, Taunton, MA, 1988.
14. Fuchs, E. F.; Masoum, M. A. S.: *Power Quality in Power Systems and Electrical Machines*, Elsevier/Academic Press, 2008, 638 pages, ISBN: 978-0-12-369536-9.

Chapter 7
Single-Phase, Two-Winding Transformers and Autotransformers/Variacs

Inductors discussed in Chap. 6 mostly have one winding only and a relatively large air gap to make it a linear device, where current and voltage are proportional to each other: if one defines in the time domain

$$v_L(t) = L\frac{di_L(t)}{dt}, \tag{7.1a}$$

then one can write in the phasor domain

$$\tilde{v}_L(t) = j\omega L\tilde{i}_L(t), \tag{7.1b}$$

where the inductance L is constant.

Transformers have the task to change voltages and currents while the power is invariant, if losses are neglected. Figures 7.1a, b illustrate the construction of a single-phase pole transformer (apparent power $S = 25$ kVA) with a wound iron core, and Figs. 7.1c, d depict its flux distribution [1–8]. Note there is no distinct air gap, however, there are many small (butt-to-butt and interlamination) air gaps due to the wound core: each wound-core layer has a butt-to-butt air gap and interlamination air gaps, and these gaps are made as small as possible.

7.1 Transformers at No-Load [9]

Figure 7.2a shows a two-winding transformer with infinitely large relative permeability ($\mu_r = \infty$) of its iron core with its secondary circuit open and an alternating voltage $v_1(t)$ applied to its primary terminals. In order to simplify the drawings it is common practice on schematic diagrams of transformers to show the primary and secondary windings as if they reside on separate legs/limbs of the core, even though the windings are actually interleaved in practice.

A schematic flux distribution of a two-winding transformer with finite relative permeability of its iron core at no-load is depicted in Fig. 7.2b. The flux at no-load,

E.F. Fuchs and M.A.S. Masoum, *Power Conversion of Renewable Energy Systems*,
DOI 10.1007/978-1-4419-7979-7_7,

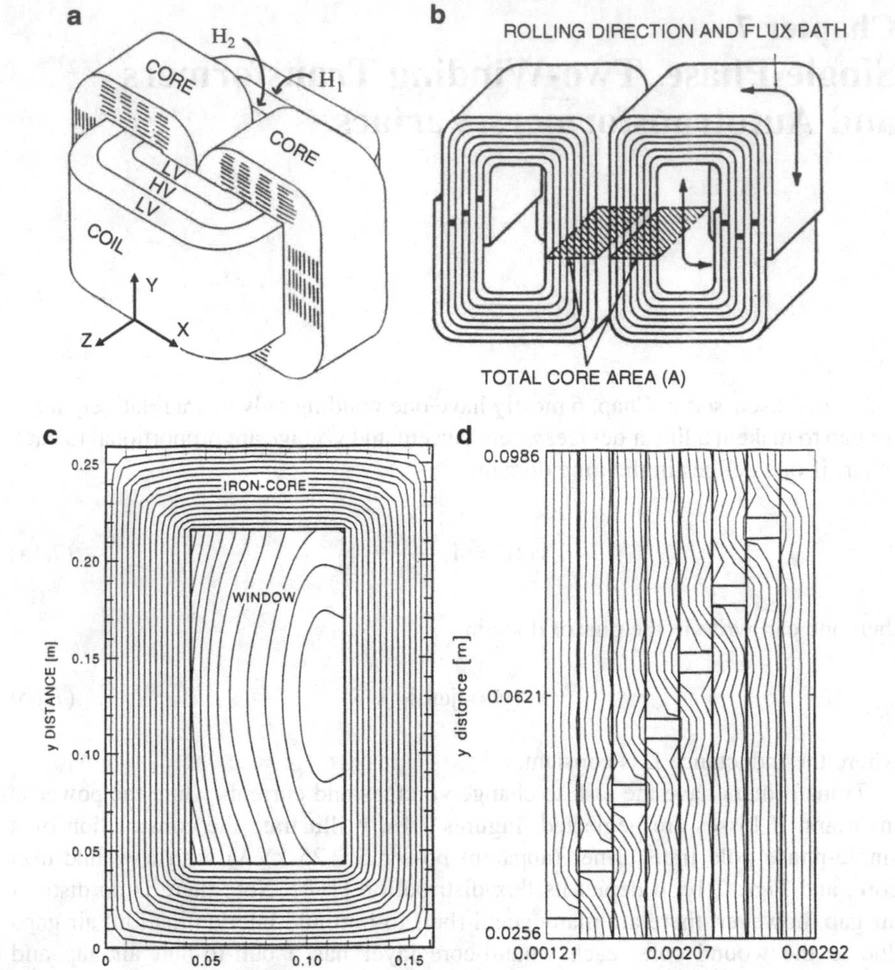

Fig. 7.1 Construction of S = 25 kVA single-phase (pole) transformer with a wound iron core and two (copper or/and aluminum) windings: (**a**) outside view of transformer, (**b**) wound iron core. Flux distribution of a S = 25 kVA single-phase (pole) transformer: (**c**) overall flux distribution in xy plane, F = magneto-motive force = 920 At within window of Fig. 7.1c, one flux tube inside window contains 0.000024 Wb/m, and one flux tube in iron core contains 0.00502 Wb/m, (**d**) flux distribution near butt-to-butt and interlamination air gaps

$i_2(t) = 0$, is $\phi_{11} = \phi_{21} + \phi_{l1}$, where ϕ_{21} is the flux linked with coil 2 (first subscript) and produced by coil 1 (second subscript), and $\phi_{\ell 1}$ is the leakage flux of coil 1.

A small steady-state current $i_{\phi}(t)$, called the exciting current, exists in the primary and establishes an alternating flux in the magnetic circuit (see Figs. 6.26a, b). The flux $\phi(t)$ (neglecting leakage and fringing fluxes) induces an electromagnetic force (emf) in the primary equal to

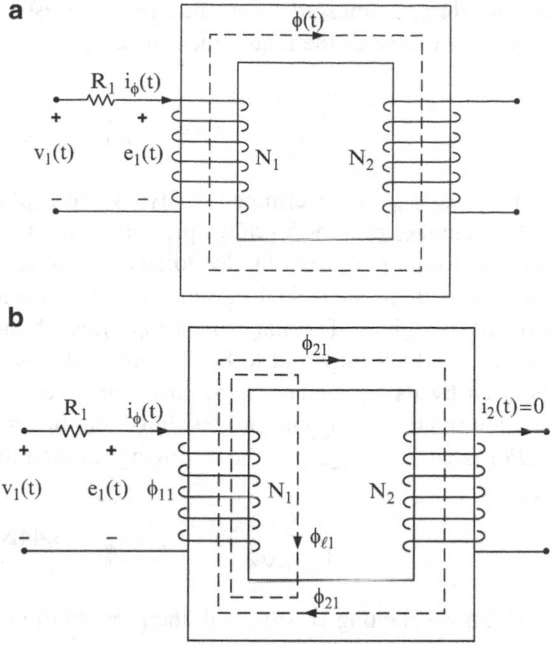

Fig. 7.2 (**a**) Single-phase transformer with infinitely large relative permeability ($\mu_r = \infty$) of its iron core with secondary open (no-load). (**b**) Detailed schematic flux distribution of a two-winding transformer with finite relative permeability of its iron core at no-load

$$e_1(t) = \frac{d\lambda_1}{dt} = N_1 \frac{d\phi_{11}}{dt}, \tag{7.2}$$

where $\lambda_1 = N_1\phi_{11}$ is flux linkage with primary, ϕ_{11} is primary flux, and N_1 is the number of turns of primary winding.

The electromagnetic force (emf) $e_1(t)$ together with the voltage drop (due to R_1) in the primary resistance R_1 (any leakage fluxes and fringing fluxes are neglected) must be balanced by the applied voltage $v_1(t)$; thus,

$$v_1(t) = R_1 i_\phi(t) + e_1(t). \tag{7.3}$$

In most power apparatus (e.g., transformers) the no-load resistance drop is very small indeed, and the induced emf $e_1(t)$ nearly equals the applied voltage $v_1(t)$. Furthermore, the waveforms of voltage and flux are very nearly sinusoidal. The analysis can then be greatly simplified: if the instantaneous flux is sinusoidal (see 6.115) then the rms value of the induced voltage is (6.118)

$$E_1 = \frac{2\pi}{\sqrt{2}} f N_1 \phi_{max} = 4.44 f N_1 \phi_{max}. \tag{7.4}$$

If the resistive voltage drop $R_1 \cdot i_\phi(t)$ is negligible, the emf $e_1(t)$ equals the applied voltage $v_1(t)$. Under these conditions, if a cosinusoidal voltage is applied to a winding, a sinusoidal core flux must be established whose maximum value (amplitude) satisfies the requirement that $E_{1\_max} = V_{1\_max}$ and, therefore,

$$\phi_{max} = \frac{V_1}{4.44 \cdot f \cdot N_1}. \tag{7.5a}$$

The flux ϕ_{max} is determined solely by the applied voltage V_1, its frequency f, and the number of turns N_1 of the primary winding. This important relation applies not only to transformers but also to any device (e.g., inductor, electric machines) operated with co-sinusoidal or sinusoidal impressed voltage V_1, if the resistance drop is negligible. The magnetic properties of the core determine the exciting current i_Φ which turns out to be nonsinusoidal (see Figs. 6.26a, b), but has been replaced by its equivalent (same rms value, same frequency, same losses as the nonsinusoidal exciting current) co-sinusoidal or sinusoidal components in Fig. 7.3.

For rated flux Φ_{max_rated} the following relation must be maintained

$$\frac{E_1}{f} = \frac{2\pi}{\sqrt{2}} N_1\ \phi_{max\_rated} = 4.44 N_1 \phi_{max\_rated}. \tag{7.5b}$$

If flux weakening is permitted, then the relation

$$\frac{E_1}{f} \approx \frac{V_1}{f} \langle \frac{2\pi}{\sqrt{2}} N_1\ \phi_{max\_rated} = 4.44\, N_1 \phi_{max\_rated} \tag{7.5c}$$

can be used.

If the exciting current is analyzed by Fourier-series methods, it will be found to comprise of a fundamental and a family of odd harmonics. The exciting current $i_\phi(t)$ can then be represented by its equivalent sinusoid, which has the same rms value and frequency and produces the same (average) losses as the actual wave

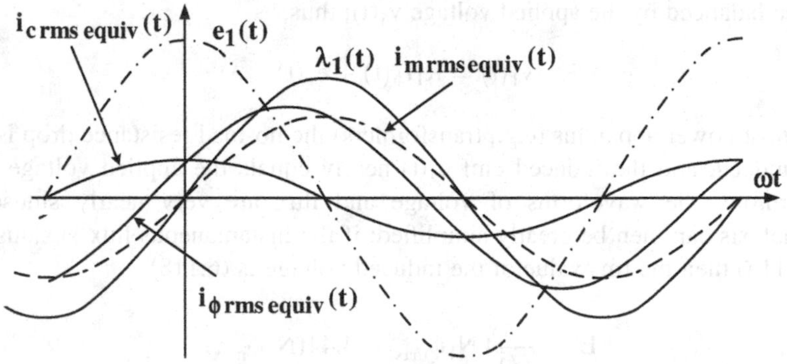

Fig. 7.3 Time domain: Sinusoidal flux linkage function $\lambda_1(t)$, cosine voltage $e_1(t)$, excitation current $i_\phi(t)$ (equivalent sinusoid), cosine core-loss current $i_{c\ rms\ equiv}(t)$, and sinusoidal magnetizing current $i_{m\ rms\ equiv}(t)$

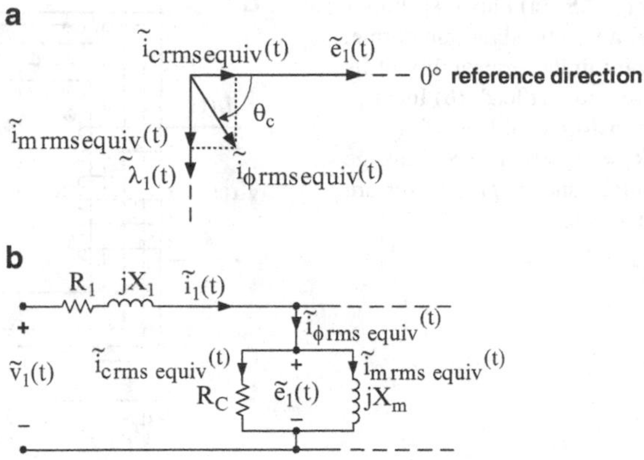

Fig. 7.4 **(a)** Phasor diagram representing $\tilde{\lambda}_1(t)$, $\tilde{i}_{\phi\,rms\,equiv}(t)$, $\tilde{i}_{c\,rms\,equiv}(t)$, $\tilde{i}_{m\,rms\,equiv}(t)$. **(b)** Equivalent circuit with $\tilde{i}_{\phi\,rms\,equiv}(t)$, $\tilde{i}_{c\,rms\,equiv}(t)$, $\tilde{i}_{m\,rms\,equiv}(t)$

(see Fig. 7.3). Such an equivalent sine/cosine representation will be essential to the construction of phasor diagrams.

The equivalent fundamental excitation current $i_{\phi\ rms\ equiv}(t)$ can, in turn, be resolved into two components, one in phase with the emf $e_1(t)$ and the other lagging the emf by 90°. The fundamental in-phase component accounts for the power absorbed by hysteresis and eddy-current losses in the core (see 6.133 and 6.134). It is called the core-loss component $i_{c\ rms\ equiv}$ (t), and the remainder is called the magnetizing current $i_{m\ rms\ equiv}$ (t). It comprises of a fundamental component lagging the emf $e_1(t)$ by 90°. A phasor diagram (Fig. 7.4a) and an equivalent circuit (Fig. 7.4b) can now be constructed representing the flux linkage $\tilde{\lambda}_1(t)$, the exciting current $\tilde{i}_{\phi\,rms\,equiv}(t)$, core-loss current $\tilde{i}_{c\,rms\,equiv}(t)$, and magnetizing current $\tilde{i}_{m\,rms\,equiv}(t)$ in the phasor domain (see Figs 7.4a, b).

The core losses can now be defined as

$$P_c = E_1 \left|\tilde{i}_{\phi\,rms\,equiv}(t)\right| \cos\,\theta_c, \tag{7.6}$$

where E_1 is the rms value of $\tilde{e}_1(t)$

7.2 Ideal Transformer at Load

The flux distribution of a two-winding transformer with finite permeability of its iron core at load is depicted in Fig. 7.5a. The fluxes at load are $\phi_{11} = \phi_{21} + \phi_{\ell 1}$ and $\phi_{22} = \phi_{12} + \phi_{\ell 2}$, where ϕ_{21} is the flux linked with coil 2 (first subscript) and produced by coil 1 (second subscript); ϕ_{12} is the flux linked with coil 1 (first subscript) and produced by coil 2 (second subscript). Fluxes $\phi_{\ell 1}$ and $\phi_{\ell 2}$ are the leakage fluxes of coil 1 and coil 2, respectively. The total flux of coil 1 is $\phi_1 = \phi_{11} - \phi_{12}$ and that of coil 2 is $\phi_2 = \phi_{22} - \phi_{21}$. The mutual flux is $\phi_m = \phi_{21} - \phi_{12}$.

As a first approximation to a quantitative theory of a loaded transformer, consider an ideal transformer with a primary of N_1 turns and a secondary winding of N_2 turns, as shown in Fig. 7.5b.

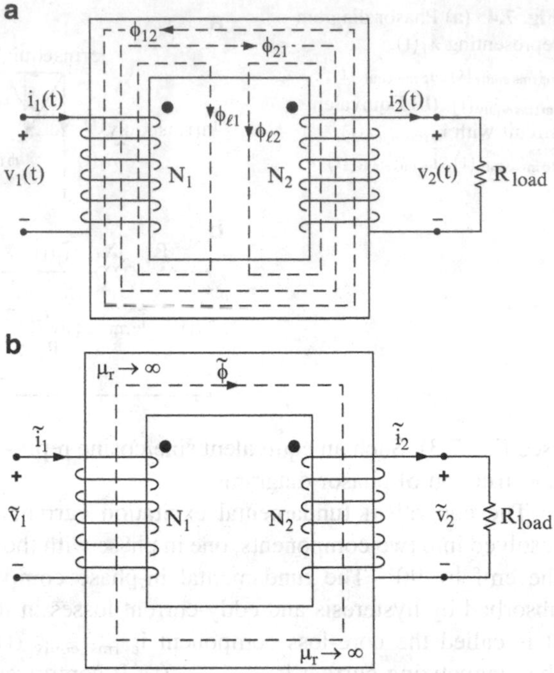

Fig. 7.5 (**a**) Flux distribution of a two-winding transformer with finite permeability of its iron core at load. (**b**) Ideal transformer at load with $R_1 = R_2 = 0$, $\mu_r = \infty$, leakage fluxes and fringing fluxes are neglected

Notice that the secondary current is defined as positive flowing out of the winding; thus the positive secondary current creates an mmf $(-N_2 i_2)$ in the opposite direction from that created by the positive primary current $(+N_1 i_1)$: the total mmf associated with the transformer (window) is (left-hand side of Ampere's law, see (6.2)

$$F = \text{mmf} = N_1 i_1 - N_2 i_2. \tag{7.7}$$

Let the properties of this transformer be idealized in that the winding resistances R_1, R_2 are negligibly small. Assume that all the flux is confined to the core (no leakage flux and no fringing flux) and links both windings. The core losses are negligible, and the permeability of the core is so high $(\mu_c = \mu_r \mu_0 = \infty)$ that only a negligibly small exciting current $(i_\phi = 0,\ H_c = 0)$ is required (see Fig. 6.3) to establish the flux $\phi(t)$: such a hypothetical transformer having these properties is often called an ideal transformer.

Ideal Transformer Relations: When a time-varying voltage $v_1(t)$ is impressed on the primary terminals, a core flux $\phi(t)$ will be established such that the emf $e_1(t)$ equals the impressed voltage when winding resistance R_1, leakage and fringing fluxes are negligible. Thus from Faraday's law (6.34) follows:

$$v_1(t) = e_1(t) = N_1 \frac{d\phi}{dt}. \tag{7.8}$$

The core flux also links the secondary winding and produces an induced emf $e_2(t)$ and an equal secondary terminal voltage $v_2(t)$ (note flux $\phi(t)$ is jointly produced by primary and secondary currents $i_1(t)$ and $i_2(t)$, respectively) given by

$$v_2(t) = e_2(t) = N_2 \frac{d\phi}{dt}. \tag{7.9}$$

Dividing (7.8) by (7.9) yields

$$\frac{v_1(t)}{v_2(t)} = \frac{N_1}{N_2}. \tag{7.10}$$

Now let a load (e.g., R_{load}) be connected to the secondary winding: a current i_2 and an mmf $(-N_2 i_2)$ are then present in the secondary and this secondary mmf must be counteracted by that of the primary. Hence a compensating primary mmf $(+N_1 i_1)$ and current i_1 must be called into being such that Ampere's law is satisfied: $F = H_c \ell_c = 0$ (due to $H_c = 0$ and $\mu_c = \mu_r \mu_0 = \infty$) or with (7.7):

$F = N_1 i_1 - N_2 i_2 = 0$

or

$$\frac{i_1(t)}{i_2(t)} = \frac{N_2}{N_1}. \tag{7.11}$$

The powers on primary (input) and secondary (output) must be conserved because all losses (ohmic and iron core) are neglected

$$p(t) = v_1(t) i_1(t) = v_2(t) i_2(t), \tag{7.12}$$

and the load resistance $\left(\text{e.g., } R_{load} = \frac{v_2(t)}{i_2(t)} = \frac{\tilde{v}_2(t)}{\tilde{i}_2(t)}\right)$ connected to the secondary can be expressed in terms of primary quantities $R'_{load} = \frac{v_1(t)}{i_1(t)} = \frac{\tilde{v}_1(t)}{\tilde{i}_1(t)}$ as follows (see Fig. 7.6):

$$R_{load} = \left(\frac{v_2(t)}{i_2(t)} = \frac{\tilde{v}_2(t)}{\tilde{i}_2(t)}\right), \tag{7.13}$$

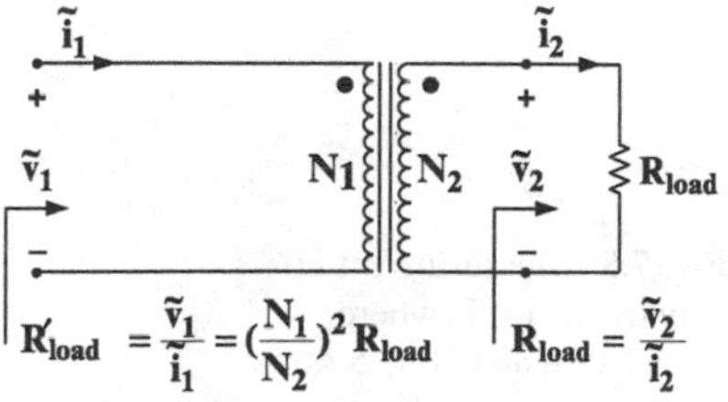

Fig. 7.6 Load resistance R_{load} as seen through ideal transformer becomes $R'_{load} = \frac{v_1(t)}{i_1(t)} = \frac{\tilde{v}_1(t)}{\tilde{i}_1(t)}$

with (7.10), $v_2(t) = (N_2/N_1)v_1(t)$, and from (7.11), $i_2(t) = (N_1/N_2)\,i_1(t)$, introduced into (7.13) yields

$$R_{load} = \frac{(N_2/N_1)}{(N_1/N_2)}\frac{v_1(t)}{i_1(t)} = \left(\frac{N_2}{N_1}\right)^2 \frac{v_1(t)}{i_1(t)} = \left(\frac{N_2}{N_1}\right)^2 R'_{load}$$

or the load resistance as seen through the ideal transformer is

$$R'_{load} = \left(\frac{N_1}{N_2}\right)^2 R_{load}. \tag{7.14}$$

The above analysis relates to load resistances. The same relations are valid for load impedances.

Application Example 7.1: Errors in a Measuring Circuit due to Current Transformer (CT)

Frequently specially designed single-phase transformers are used as current transformers (CT), where the current through an ampere meter (called the burden of CT) is reduced according to the turns ratio N_1/N_2, as shown in Figs. 7.7 and 7.8. A CT is essentially operated with the secondary winding short-circuited because the burden resistance (resistance of amperemeter) is always very small. A potential transformer (PT), however is essentially operated with the secondary winding open-circuited because the burden resistance (resistance of voltmeter) is always very large.

(a) Assuming an ideal transformer compute $\tilde{I}_1$, $\tilde{I}_2$, and the core flux ϕ_{max} for the values of Fig. 7.7.

(b) How would these values be affected if the secondary were opened ($R_{burden} = \infty$)?

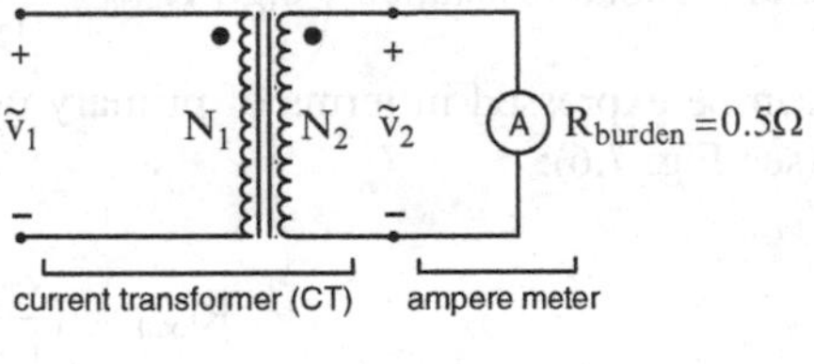

Fig. 7.7 Current transformer loaded with burden R_{burden} with $(N_1/N_2) = (1/5)$

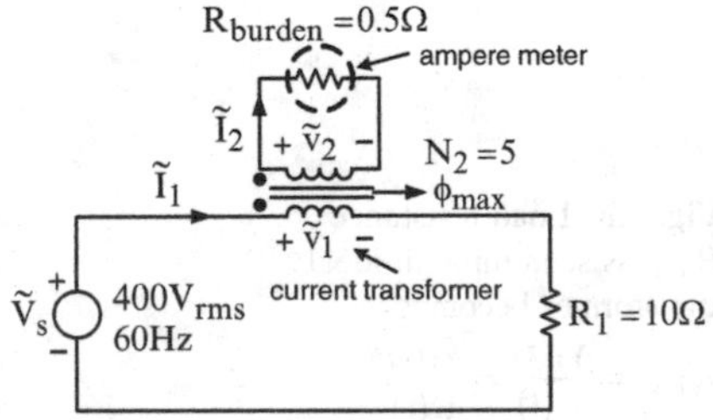

Fig. 7.8 Measuring set up employing a CT, where $N_1 = 1$ turn and $N_2 = 5$ turns

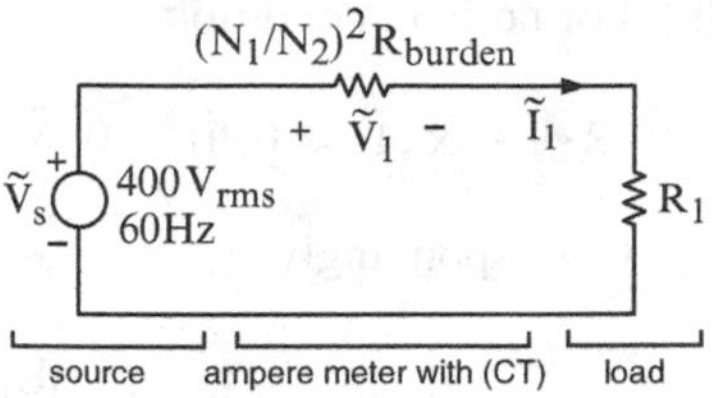

Fig. 7.9 Equivalent circuit with $R_1 = 10\ \Omega$ and $R_{burden} = R_2 = 0.5\ \Omega$

Solution

(a) With the ideal transformer relations and relying on the equivalent circuit of Fig. 7.9 one obtains the relations:

$$\frac{\tilde{v}_1}{\tilde{v}_2} = \frac{N_1}{N_2}, \ \frac{\tilde{i}_2}{\tilde{i}_1} = \frac{N_1}{N_2}, \tag{7.15}$$

and

$$R'_{burden} = R_{burden}\left(\frac{N_1}{N_2}\right)^2. \tag{7.16}$$

The primary current is

$$\tilde{I}_1 = \frac{\tilde{V}_S}{R_{burden}\left(\frac{N_1}{N_2}\right)^2 + R_1} = \frac{400\angle 0°\,\text{V}}{\{0.5\left(\frac{1}{5}\right)^2 + 10\}\ \Omega} = 39.92\angle 0°\,\text{A}, \tag{7.17}$$

and the secondary current is

$$I_2 = \tilde{I}_1/5 = 7.98\angle 0°\,\text{A}. \tag{7.18}$$

Ideally the load currents I_1 and the measured current I_2 should be 40 A and 8 A, respectively. That is the errors in the current measurement due to the current transformer are

$$\begin{aligned} \varepsilon_{i1} &= (40\,\text{A} - 39.92\,\text{A})/40\,\text{A} \times 100\% = 0.2\% \quad \text{and} \\ \varepsilon_{i2} &= (8\,\text{A} - 7.98\,\text{A})/8\,\text{A} \times 100\% = 0.2\%. \end{aligned} \tag{7.19}$$

The induced voltage in the primary is

$$\tilde{V}_1 = \tilde{V}_s - R_1\tilde{I}_1 = 400\,\text{V}\angle 0° - 399.2\,\text{V}\angle 0° = 0.8\angle 0°\text{V}, \tag{7.20}$$

and the flux becomes under load condition

$$\phi_{max} = \frac{|\tilde{V}_1|}{4.44 \cdot f \cdot N_1} = \frac{0.8\angle 0°}{4.44 \cdot 60 \cdot 1} = 3 \cdot 10^{-3}\,\text{Wb} = 3\,\text{mWb}. \tag{7.21}$$

(b) For no-load condition

$$R_2 \rightarrow \infty, \tilde{I}_2 = 0, \tilde{I}_1 = 0, \tilde{V}_1 = \tilde{V}_S \text{ and } \tilde{V}_2 = (N_2/N_1)\tilde{V}_1 = 2\,\text{kV}. \quad (7.22)$$

Correspondingly,

$$\phi_{max} = \frac{|\tilde{E}_1|}{4.44fN_1} = \frac{400\angle 0°}{4.44 \cdot 60 \cdot 1} = 1.5\,\text{Wb}. \quad (7.23)$$

Application Example 7.2: Ideal Transformer with Switch in the Load Circuit

For the ideal transformer of Fig. 7.10, calculate the following quantities for $\tilde{V}_1 = 440\,V\angle 0°$.

(a) $\tilde{I}_1$ when switch S is closed, and the current carried by switch when closed.
(b) $\tilde{I}_1$ when switch S is open, and the voltage across switch when open.

Solution

(a) When switch is closed one obtains the voltage and current definitions of Fig. 7.11.

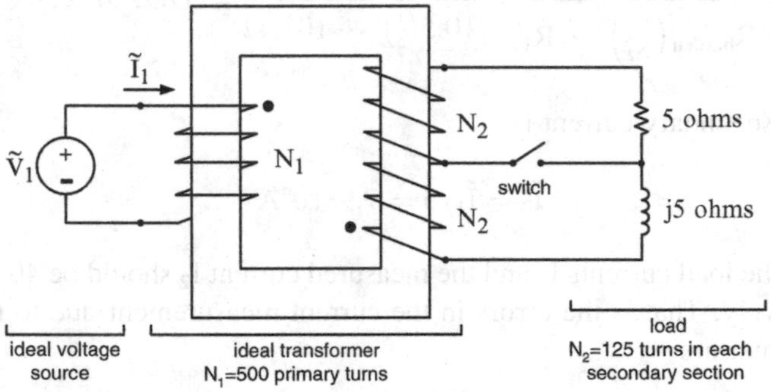

Fig. 7.10 Ideal transformer feeding load with switch S

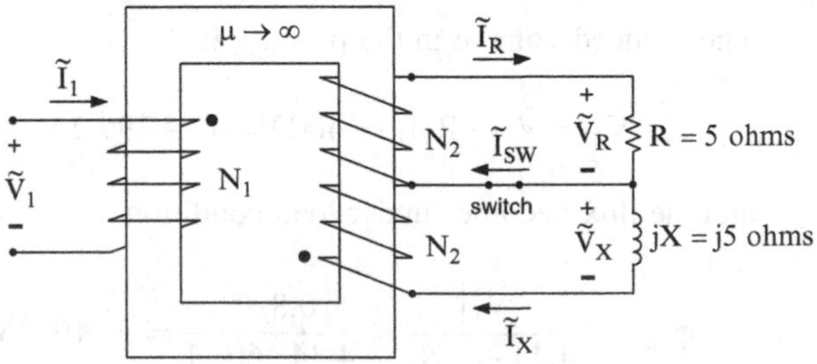

Fig. 7.11 Voltage and current definitions when switch is closed

Applying ideal transformer conditions for the voltages one gets

$$\tilde{V}_R = -\frac{N_2}{N_1}\tilde{V}_1, \tag{7.24}$$

$$\tilde{V}_X = -\frac{N_2}{N_1}\tilde{V}_1. \tag{7.25}$$

The application of Ohm's law yields

$$\tilde{V}_R = R \cdot \tilde{I}_R. \tag{7.26}$$

$$\tilde{V}_X = jX \cdot \tilde{I}_X \tag{7.27}$$

Based on Kirchhoff's current law one can write

$$\tilde{I}_{SW} = \tilde{I}_R - \tilde{I}_X. \tag{7.28}$$

From (7.24)

$$\tilde{V}_R = -440\,\angle 0^\circ \frac{125}{500} = -110\,\angle 0^\circ \mathrm{V}, \tag{7.29}$$

from (7.25)

$$\tilde{V}_X = -440\,\angle 0^\circ \frac{125}{500} = -110\,\angle 0^\circ \mathrm{V}, \tag{7.30}$$

from (7.26)

$$\tilde{I}_R = \frac{\tilde{V}_R}{R} = -22\,\angle 0^\circ \mathrm{A}, \tag{7.31}$$

from (7.27)

$$\tilde{I}_X = \frac{\tilde{V}_X}{jX} = j22\,\mathrm{A}, \tag{7.32}$$

from (7.28) the current carried by switch when closed is

$$\tilde{I}_{SW} = \tilde{I}_R - \tilde{I}_X = 31.108\,\angle -135^\circ \mathrm{A}. \tag{7.33}$$

Application of Ampere's law yields

$$N_1\tilde{I}_1 + N_2\tilde{I}_X + N_2\tilde{I}_R = 0. \tag{7.34}$$

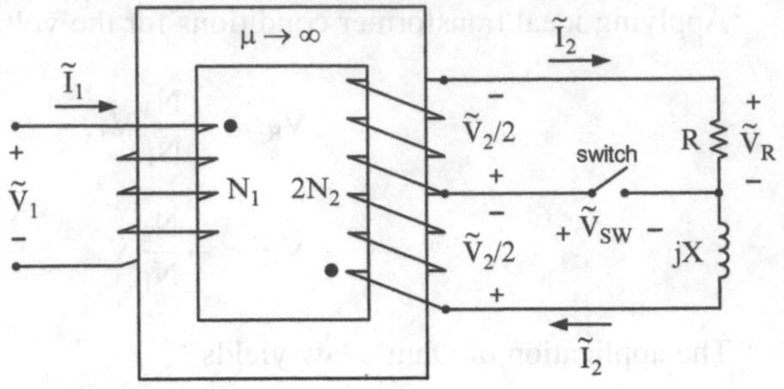

Fig. 7.12 Voltage and current definitions when switch is open

With (7.31) and (7.32) and (7.34) one obtains

$$\tilde{I}_1 = 7.77\angle -45°\mathrm{A}.$$

(b) When switch is open one obtains the voltage and current definitions of Fig. 7.12.
With the ideal transformer condition for the voltage one can write

$$\tilde{V}_2 = \left(-\frac{2N_2}{N_1}\right)\tilde{V}_1, \tag{7.35}$$

and with Ohm's law

$$\tilde{I}_2 = -\frac{\tilde{V}_2}{R + jX} = \left(-\frac{2N_2}{N_1}\right)\frac{\tilde{V}_1}{R + jX} = 31.117\angle -225°\mathrm{A}. \tag{7.36}$$

The application of Ampere's law leads to

$$\tilde{I}_1 = -\frac{2N_2}{N_1}\tilde{I}_2 = 15.558\angle -45°\mathrm{A}. \tag{7.37}$$

This result can be directly obtained from the load impedance as seen through the ideal transformer

$$\tilde{I}_1 = \frac{\tilde{V}_1}{Z'_2} = \frac{\tilde{V}_1}{\left(\frac{N_1}{2N_2}\right)^2 Z_2} = \frac{\tilde{V}_1}{\left(\frac{N_1}{2N_2}\right)^2 (R + jX)} = 15.558\angle -45°\mathrm{A}. \tag{7.38}$$

The voltage across the switch when open is obtained from Kirchhoff's voltage law:

$$\tilde{V}_{SW} = \frac{\tilde{V}_2}{2} + \tilde{V}_R = (-220 + j110)\,\mathrm{V}. \tag{7.39}$$

Application Example 7.3: Impedance Matching or Maximum Power Transfer

A voltage source (e.g., amplifier) can be represented by a $\tilde{V}_s = 5\angle 0° V_{rms}$ in series with an internal ohmic resistance of $R_S = 3{,}000\ \Omega$ (Thevenin equivalent circuit) and is connected to a $R_L = 150\ \Omega$ load resistance through an ideal transformer. Note subscript "s" stands for source.

(a) Draw the electrical equivalent circuit
(b) Calculate the power in milliwatts supplied to the load as a function of the transformer ratio $a = (N_1/N_2)$, covering ratios from $a = 1.0$ to 10.0. Note subscripts "1" and "2" stand for primary and secondary, respectively.
(c) For which turns ratio a_{max} is the power transferred from the source to the load a maximum?

Solution

(a) The electrical equivalent circuit is depicted in Fig. 7.13 where
The load resistance as seen through the ideal transformer is

$$R'_L = a^2 R_L = a^2 150\ \Omega, \tag{7.40}$$

and the current through the circuit is

$$\tilde{I}_s = \frac{\tilde{V}_s}{R_s + R'_L} = \frac{5\angle 0°}{3{,}000 + 150a^2}. \tag{7.41}$$

The power dissipated in R'_L is

$$P = |\tilde{I}_s|^2 R'_L = \frac{25}{(3{,}000 + 150a^2)^2} \cdot 150\,a^2. \tag{7.42}$$

The power P is listed in Table 7.1 as a function of the parameter $a = (N_1/N_2)$. The maximum power transfer occurs if R_S and R'_L are matched (principle of matching of impedances), that is $R'_L = R_S$ or

$$\left(\frac{N_1}{N_2}\right)^2 R_L = R_s \tag{7.43}$$

or

$$a = \frac{N_1}{N_2} = \sqrt{\frac{R_s}{R_L}} = \sqrt{20} = 4.4723 \tag{7.44}$$

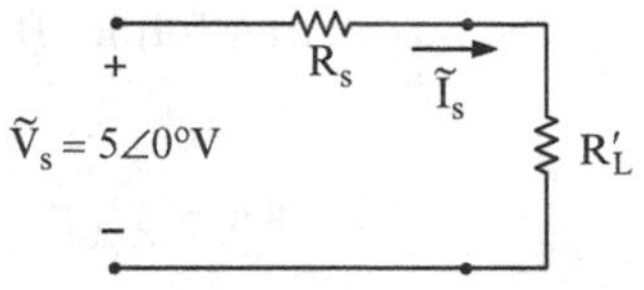

Fig. 7.13 Electrical equivalent circuit

Table 7.1 Power P in mW as a function of the turns ratio a

a [−]	0.1	1.0	2.0	3.0	4.0	4.4723
P [mW]	0.00416	0.3779	1.157	1.783	2.057	2.083
a [−]	5.0	6.0	7.0	8.0	9.0	10.0
P [mW]	2.057	1.913	1.745	1.512	1.323	1.157

resulting in the maximum power transferred from source to load

$$P_{max}\big|_{a=4.4723} = 2.083\,\text{mW}. \tag{7.45}$$

This result is confirmed by Table 7.1.

Application Example 7.4: Calculation of Branch Currents if a Resistive Load is Fed by Two Ideal Transformers

Two ideal single-phase transformers A and B are connected in series across a voltage source of 120 V_{rms} $\angle 0°$, @ f = 60 Hz.

(a) Calculate $\tilde{I}_1$, $\tilde{V}_{2B}$, and the transformer fluxes ϕ_A and ϕ_B if $R_A = 50\ \Omega$ and the transformer B has an open (unloaded) secondary winding, as indicated in Fig. 7.14. Draw the electrical equivalent circuit.

(b) Repeat these calculations, if a load resistance of $R_B = 25\ \Omega$ is added to the secondary winding of transformer B. Draw the electrical equivalent circuit.

Solution

(a) Ampere's law applied to transformer B yields

$$\left|\tilde{I}_{1B}\right| N_{1B} = \left|\tilde{I}_{2B}\right| N_{2B} \tag{7.46}$$

or with $\left|\tilde{I}_{2B}\right| = 0$ follows

$$\left|\tilde{I}_{1B}\right| = \left|\tilde{I}_{1A}\right| = \left|\tilde{I}_1\right| = 0 \tag{7.47}$$

and $\tilde{V}_{2A} = \tilde{V}_{1A} = 0$ resulting in the flux $\Phi_A = 0$. The entire source voltage drops off at the input of transformer B as is illustrated in Fig. 7.15, that is,

$$\tilde{V}_{1B} = \tilde{V}_1,\ \text{and}\ \tilde{V}_{2B} = \tilde{V}_{1B}\left(\frac{N_{2B}}{N_{1B}}\right) = 120\left(\frac{150}{50}\right) = 360\ \angle 0°\ \text{V}. \tag{7.48}$$

The flux in transformer B is

$$\phi_B = \frac{\left|\tilde{V}_{1B}\right| \angle 0°}{4.44 \cdot f \cdot N_{1B}} = \frac{120\ \angle 0°}{4.44 \cdot 60 \cdot 50} = 9.01\,\text{mWb}. \tag{7.49}$$

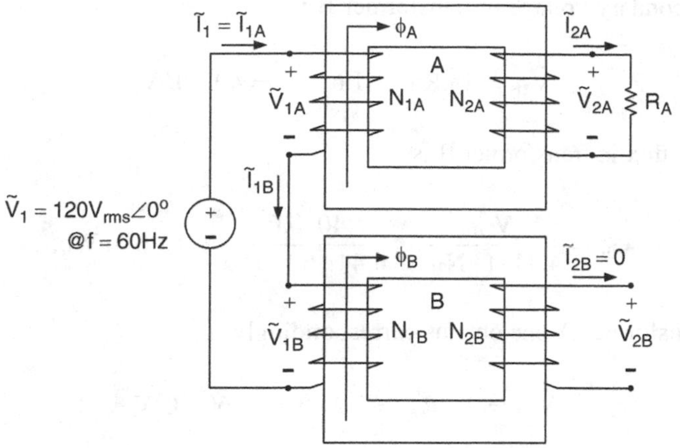

Fig. 7.14 Two ideal single-phase transformers A and B connected in series where $N_{1A} = 100$ turns, $N_{1B} = 50$ turns, and $N_{2A} = N_{2B} = 150$ turns

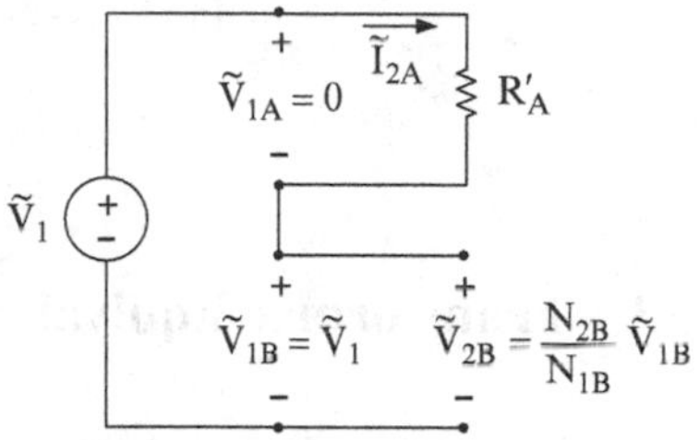

Fig. 7.15 Two ideal single-phase transformers connected in series with $R_A = 50\ \Omega$, and $R_B = \infty$, where $\tilde{I}_{2A} = 0$

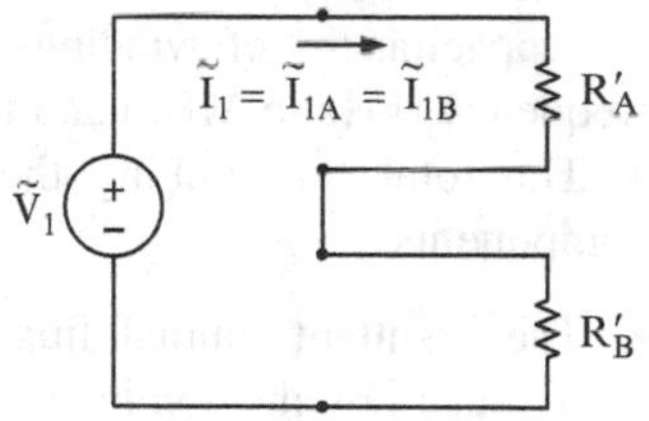

Fig. 7.16 Two ideal single-phase transformers connected in series with $R_A = 50\ \Omega$, and $R_B = 25\ \Omega$

(b) From Fig. 7.16 one obtains from Kirchhoff's current law

$$\tilde{I}_1 = \tilde{I}_{1A} = \tilde{I}_{1B} = \frac{120\ \angle 0°}{R'_A + R'_B} = 4.8\ \angle 0°\text{A}, \tag{7.50}$$

and from Ampere's law

$$\tilde{I}_{2B} = \tilde{I}_{1B}\frac{N_{1B}}{N_{2B}} = 4.8\frac{50}{150} = 1.6\ \angle 0°\ \text{A}. \tag{7.51}$$

The secondary voltage of transformer B is

$$\tilde{V}_{2B} = \tilde{I}_{2B} R'_B = 1.6 \cdot 25 = 40\,\angle 0^\circ\,\mathrm{V} \tag{7.52}$$

and the flux in transformer B is

$$\phi_B = \frac{\tilde{V}_{2B}}{4.44 \cdot f \cdot N_{2B}} = \frac{40\,\angle 0^\circ}{4.44 \cdot 60 \cdot 150} = 1\,\angle 0^\circ\,\mathrm{mWb}. \tag{7.53}$$

For transformer A one obtains correspondingly

$$\tilde{V}_{2A} = \tilde{I}_{2A} R'_A = 3.2 \cdot 50 = 160\,\angle 0^\circ \mathrm{V} \tag{7.54}$$

and the flux in transformer A is

$$\phi_A = \frac{\tilde{V}_{2A}}{4.44 \cdot f \cdot N_{2A}} = \frac{160\,\angle 0^\circ}{4.44 \cdot 60 \cdot 150} = 4\,\angle 0^\circ\,\mathrm{mWb}. \tag{7.55}$$

7.3 Transformer Equivalent Circuits [9]

A more complete theory of transformer performance than the ideal transformer approach must take into account the effects of winding resistances (R_1, R_2), magnetic leakage and fringing fluxes, as well as excitation current $i_\phi(t)$. Sometimes the capacitances of windings must be taken into account, in particular at high frequency (kHz to MHz) and transient operations.

The total flux linking the primary winding $\phi(t)$ can be divided into two components:

- The resultant mutual flux $\phi_m(t)$, confined essentially to the iron core and produced by the combined effect of the primary and secondary currents
- Primary (secondary) leakage flux $\phi_{\ell 1}(t)$ which links only the primary and $\phi_{\ell 2}(t)$ which links only the secondary.

Because the leakage path is largely in the air, the leakage flux and the voltage induced by it vary linearly with the primary and secondary current $\tilde{i}_1(t)$ and $\tilde{i}_2(t)$, respectively: the transformer primary leakage flux can be simulated by assigning to the primary a leakage inductance equal to the leakage-flux linkages divided by the primary current resulting in the primary leakage reactance

$$X_{\ell 1} = \omega L_{\ell 1} = \omega \frac{\lambda_{\ell 1}}{|\tilde{i}_1(t)|} = \omega \frac{N_1 \phi_{\ell 1}}{|\tilde{i}_1(t)|}, \tag{7.56}$$

and correspondingly for the secondary leakage reactance

$$X_{\ell 2} = \omega L_{\ell 2} = \omega \frac{\lambda_{\ell 2}}{\left|\tilde{i}_2(t)\right|} = \omega \frac{N_2 \phi_{\ell 2}}{\left|\tilde{i}_2(t)\right|}. \tag{7.57}$$

In addition there will be a voltage drop in the primary and secondary winding effective resistances R_1 and R_2, respectively. The impressed voltage $\tilde{v}_1(t)$ is then opposed by three phasor voltages: the voltage $\tilde{i}_1(t) \cdot R_1$ drop across the primary resistance, the $\tilde{i}_1(t) \cdot jX_{\ell 1}$ drop arising from the primary leakage flux $\phi_{\ell 1}(t)$, and the emf $\tilde{e}_1(t) = 4.44 \cdot f \cdot N_1 \tilde{\phi}_m$ induced in the primary by resultant mutual flux $\phi_m(t)$. This phasor addition is represented in Fig. 7.17 by an equivalent circuit.

The resultant mutual flux $\phi_m(t)$ links both the primary and secondary windings and is created by their combined mmfs: $(N_1 \cdot i_1 - N_2 \cdot i_2)$. It is convenient to treat these mmfs by considering that the primary mmf $(+N_1 \cdot i_1)$ must meet two requirements of the magnetic circuit: it must

1. Counteract the demagnetizing effect $(-N_2 \cdot i_2)$ of the secondary circuit current i_2
2. Produce sufficient mmf $(|N_1 \cdot i_1| > |N_2\ i_2|)$ or $(N_1 \cdot i_1 - N_2 \cdot i_2) \neq 0$ to create the resultant mutual flux $\phi_m(t)$.

According to this physical picture, it is convenient to resolve the primary current $\tilde{i}_1 = \tilde{i}_1(t)$ into two components:

- Load component $\tilde{i}_2^{\prime}(t)$, and
- Excitation component $\tilde{i}_{\phi rmsequiv}(t)$.

The load component $\tilde{i}_2^{\prime}(t)$ is defined as the component current in the primary whose mmf $N_1 \cdot i_2^{\prime}(t)$ exactly counteracts the mmf of the secondary current $\tilde{i}_2(t)$, that is,

$$N_1 \cdot i_2^{\prime}(t) - N_2 \cdot i_2(t) = 0.$$

Thus for opposing currents

$$\tilde{i}_2^{\prime}(t) = \left(\frac{N_2}{N_1}\right) \tilde{i}_2(t). \tag{7.58}$$

It equals the secondary current referred to the primary as in an ideal transformer.

The exciting component $i_\phi(t)$ is defined as the additional primary current required to produce the resultant mutual flux $\phi_m(t)$. It is a nonsinusoidal current

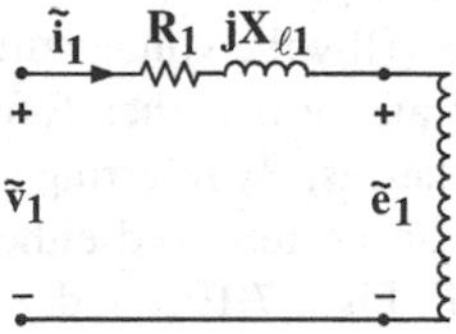

Fig. 7.17 Equivalent circuit of transformer primary winding

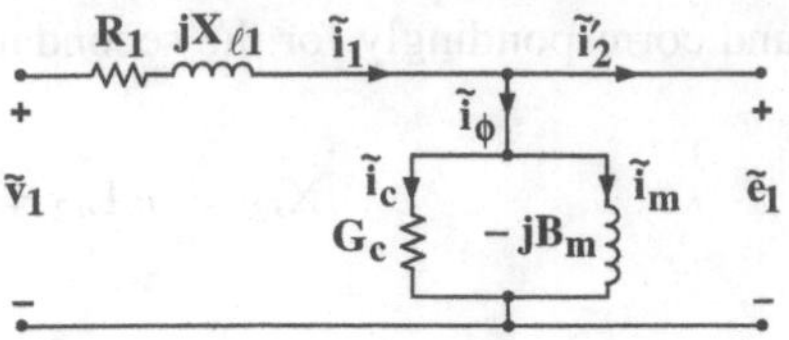

Fig. 7.18 Partial T-equivalent circuit of single-phase, two-winding transformer, where $\tilde{i}_\phi = \tilde{i}_{\phi \text{ rms equiv}}(t)$

(see Figs. 6.26a, b), as discussed in Chap. 6. The exciting current $\tilde{i}_\phi(t)$ can be represented by an equivalent sinusoidal current with an rms value $|\tilde{i}_{\phi \text{ rms equiv}}(t)|$ and it can be resolved into a core-loss component $\tilde{i}_c(t) = \tilde{i}_{c \text{ rms equiv}}(t)$ in phase with the emf $\tilde{e}_1(t)$and a magnetizing component $\tilde{i}_m(t) = \tilde{i}_{m \text{ rms equiv}}(t)$ lagging $\tilde{e}_1(t)$ by 90° (see Figs. 7.4a, b and 7.18).

T-Equivalent Circuit: In the partial T-equivalent circuit of Fig. 7.18 the equivalent sinusoidal exciting current $\tilde{i}_{\phi \text{rms equiv}}(t)$ is accounted for by means of a shunt branch connected across $\tilde{e}_1(t)$ comprising of a resistance ($R_c = 1/G_c$) whose conductance G_c is in parallel with a lossless reactance ($X_m = 1/B_m$) whose susceptance is B_m.

The power $|\tilde{e}_1|^2 G_c = E_1^2 \cdot G_c$ accounts for the core loss due to the resultant mutual flux. The magnetizing susceptance B_m varies with the saturation of the iron core. Both G_c and B_m are usually determined at rated voltage and frequency; they are then assumed to remain constant for small perturbations from rated values associated with normal operation. The resultant mutual flux $\tilde{\phi}_m(t)$ induces an emf $\tilde{e}_2(t)$ in the secondary, and since this flux links both windings, the induced-voltage ratio is

$$\frac{\tilde{e}_1(t)}{\tilde{e}_2(t)} = \frac{N_1}{N_2}, \tag{7.59}$$

where $E_i = 4.44 \cdot f \cdot N_i \cdot \phi_{m\,max}$ with $i = 1, 2$ just as in an ideal transformer: this voltage transformation and the current transformation

$$\tilde{i}_2'(t) = \left(\frac{N_2}{N_1}\right)\tilde{i}_2(t), \tag{7.60}$$

can be accounted for by introducing an ideal transformer in the complete T-equivalent circuit, as shown in Fig. 7.19.

The emf $\tilde{e}_2(t)$ is not the secondary terminal voltage, however, because of the secondary resistance R_2 and because the secondary current $\tilde{i}_2(t)$ creates secondary leakage flux. The secondary terminal voltage $\tilde{v}_2(t)$ differs from the induced voltage $\tilde{e}_2(t)$by the voltage drop due to R_2 and secondary leakage reactance $X_{\ell 2}$. The actual transformer, therefore, is equivalent to an ideal transformer plus external impedances. By referring all quantities to the primary or secondary the ideal transformer can be removed either to the right or left, respectively, of the equivalent circuit. In Figs. 7.19c, and 7.20a the secondary values are provided with primes ($'$), e.g.,

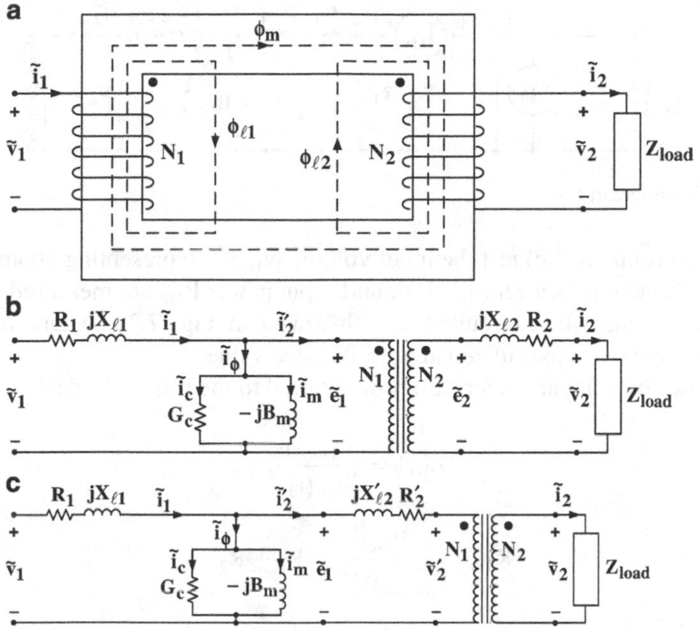

Fig. 7.19 Development stages of equivalent circuit of transformer

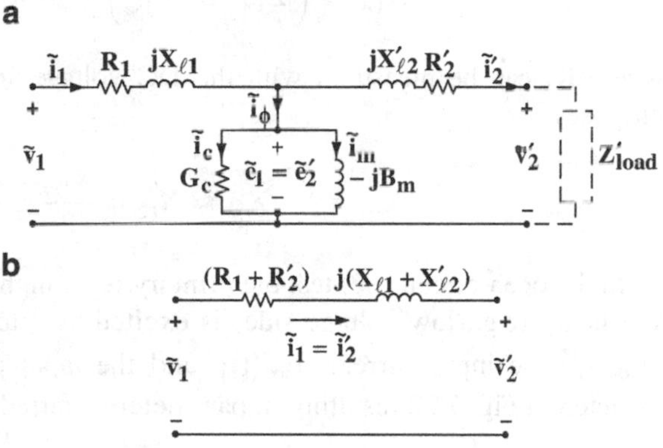

Fig. 7.20 **(a)** Complete transformer T-equivalent circuit. **(b)** Simplified transformer equivalent circuit

$X'_{\ell 2}$, R'_2, (Figs. 7.19c and 7.20a) Z'_{load} (Fig. 7.20a) to distinguish them from the actual values.

Simplified Equivalent Circuit: A great simplification of the T-equivalent circuit results from neglecting the exciting current, as shown in Fig. 7.20b in which the transformer is represented as an equivalent series impedance.

Experimental Determination of Transformer Equivalent Circuit Parameters: Two very simple tests serve to determine the approximate parameters of the equivalent circuit: the short-circuit test and the open-circuit test.

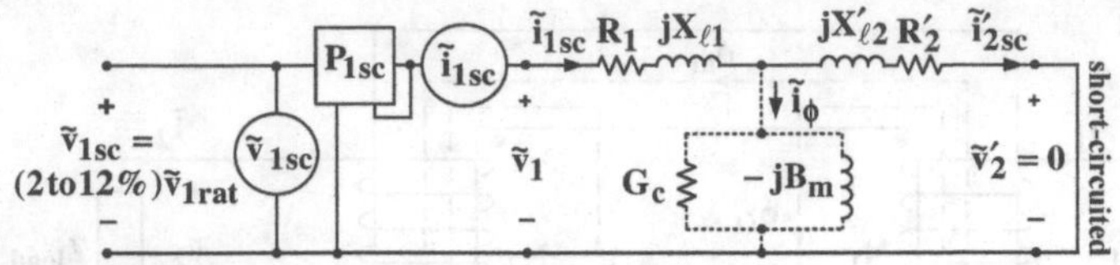

Fig. 7.21 Short-circuit test

In the short-circuit (sc) test the input voltage $|\tilde{v}_{1sc}(t)|$ representing about 2–12% of rated voltage, input current $|\tilde{i}_{1sc}(t)|$, and input power P_{1sc} are measured with the secondary winding short-circuited, as illustrated in Fig. 7.21, where the input current rms value is about identical with its rated value.

The series-branch parameters are now referred to the primary side:

$$|Z_{1sc}| = \frac{|\tilde{v}_{1sc}(t)|}{|\tilde{i}_{1sc}(t)|}, \tag{7.61}$$

$$R_{1sc} = \frac{P_{1sc}}{|\tilde{i}_{1sc}(t)|^2} = R_1 + R_2^{'}, \tag{7.62}$$

and

$$X_{1sc} = \left(|Z_{1sc}|^2 - R_{1sc}^2\right)^{1/2} = X_{\ell 1} + X_{\ell 2}^{'}, \tag{7.63}$$

where R_1 can be measured with the DC voltage-drop method. For all practical purposes

$$X_{\ell 1} \approx X_{\ell 2}^{'} \approx \frac{X_{1sc}}{2}. \tag{7.64}$$

In the open-circuit (oc) test the primary (e.g., high-voltage side) is open and the secondary (e.g., low-voltage side) is excited by rated voltage. The input voltage $|\tilde{v}_{2oc}(t)|$, the input current $|\tilde{i}_{2oc}(t)|$, and the input power P_{2oc} are measured, as depicted in Fig. 7.22 resulting in parameters referred to the secondary side:

$$|Y_{2oc}| = \frac{|\tilde{i}_{2oc}(t)|}{|\tilde{v}_{2oc}(t)|}, \tag{7.65}$$

$$G_{2oc} = \frac{P_{2oc}}{|\tilde{v}_{2oc}(t)|^2}, \tag{7.66}$$

and

$$B_{2oc} = \left(|Y_{2oc}|^2 - G_{2oc}^2\right)^{1/2}. \tag{7.67}$$

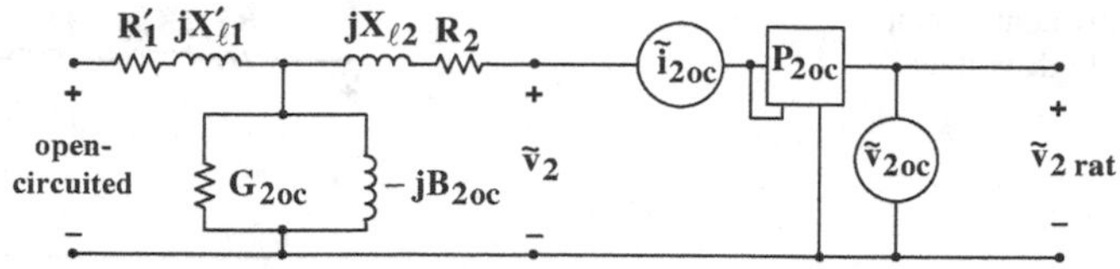

Fig. 7.22 Open-circuit test

Referring these parameters to the primary side one obtains

$$G_c = \left(\frac{N_2}{N_1}\right)^2 G_{2oc}, \tag{7.68}$$

$$B_m = \left(\frac{N_2}{N_1}\right)^2 B_{2oc}. \tag{7.69}$$

Equations (7.62)–(7.64), (7.68), (7.69) represent the complete set of parameters of a transformer referred to the primary side.

Application Example 7.5: Equivalent Circuit Parameters of Transformer Referred the to High- and Low-Voltage Sides

The resistances and leakage reactances in ohms of a $S = 25$ kVA, $f = 60$ Hz, $V_1 = 2{,}400$ V, $V_2 = 240$ V distribution transformer are $R_1 = 0.650\ \Omega$, $R_2 = 0.0065\ \Omega$, $X_{1\ell} = 8.40\,\Omega$, and $X_{2\ell} = 0.084\,\Omega$, where subscript "1" denotes the 2,400 V winding (high-voltage winding) and subscript "2" denotes the 240 V winding (low-voltage winding). Each quantity is referred to its own side of the transformer.

(a) Draw an equivalent circuit where all parameters ($R_1, R'_2, X_{1\ell}, X'_{2\ell}$) are referred to the high-voltage side. Subscript "ℓ" stands for leakage.
(b) Draw an equivalent circuit where all parameters (R_2, R'_1, $X_{2\ell}$, $X'_{1\ell}$) are referred to the low-voltage side.
(c) Consider the transformer to deliver its rated apparent power S (also called "voltampere" power) at $\cos\phi = 0.8$ lagging (consumer notation) to a load on the low-voltage side with 240 $V\angle 0°$across the load. Find the high-voltage (or high-tension, or primary) terminal voltage and draw the phasor diagram of this circuit.

Solution

(a) The equivalent circuit referred to the high-voltage side is shown in Fig. 7.23, where $N_1/N_2 = 2{,}400/240 = 10$, $R_1 = 0.65\Omega$, $R'_2 = R_2(N_1/N_2)^2 = 0.65\,\Omega$, $X_{1\ell} = 8.40\,\Omega$, and $X'_{2\ell} = X_{2\ell}(N_1/N_2)^2 = 8.40\,\Omega$.
(b) The equivalent circuit referred to the low-voltage side is shown in Fig. 7.24, where $N_1/N_2 = 2{,}400/240 = 10$, $R_2 = 0.0065\Omega$, $R'_1 = R_1(N_2/N_1)^2 = 0.0065\,\Omega$, $X_{2\ell} = 0.084\,\Omega$, and $X'_{1\ell} = X_{1\ell}(N_2/N_1)^2 = 0.084\,\Omega$.

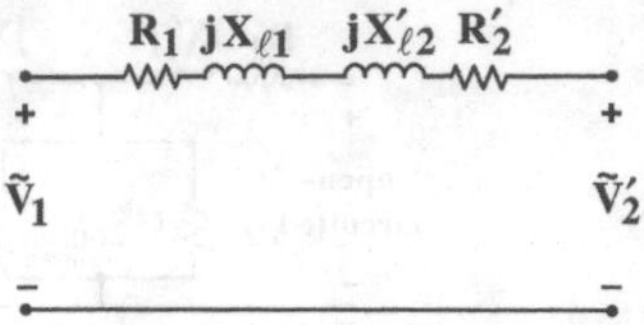

Fig. 7.23 Equivalent circuit referred to the high-voltage side

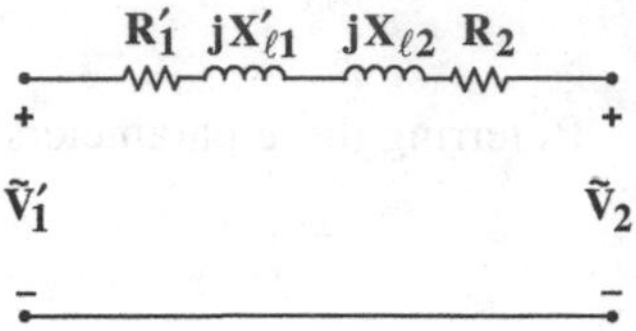

Fig. 7.24 Equivalent circuit referred to the low-voltage side

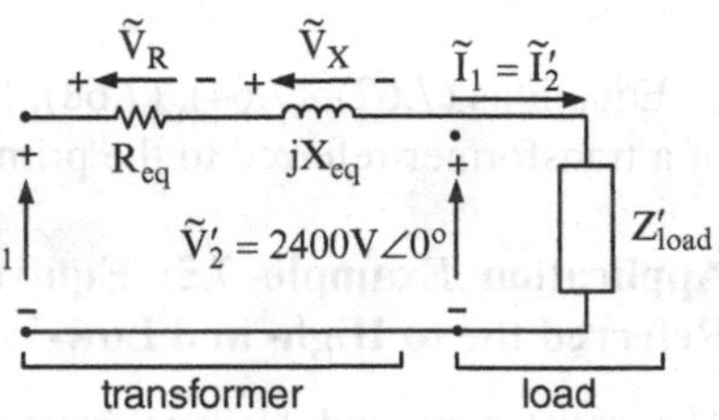

Fig. 7.25 Reduced equivalent circuit referred to the high-voltage side with impedance load

(c) Reduced equivalent circuit referred to high-voltage side with impedance load is given in Fig. 7.25 with $R_{eq} = R_1 + R_2(N_1/N_2)^2 = 1.3\,\Omega$, and $X_{eq} = X_{1\ell} + X_{2\ell}(N_1/N_2)^2 = 16.8\,\Omega$.

The magnitude of the load current is

$$|\tilde{I}_1| = |\tilde{I}'_2| = \frac{S}{|\tilde{V}_1|} = \frac{25{,}000\,\text{VA}}{2{,}400\,\text{V}} = 10.42\,\text{A}, \tag{7.70}$$

the lagging phase angle is

$$\phi = \cos^{-1}(0.8) = 36.87°. \tag{7.71}$$

The load current lagging (consumer notation) the load voltage (taken as reference, that is, 0°) is now

$$\tilde{I}_1 = |\tilde{I}_1| \angle - \phi = 10.42\angle - 36.87°\,\text{A} = (8.336 - j6.252)\,\text{A} = \tilde{I}'_2. \tag{7.72}$$

With Kirchhoff's voltage law and $\tilde{V}'_2 = 2{,}400\angle 0°\,\text{V}$

$$\tilde{V}_1 = \tilde{V}'_2 + \tilde{I}_1 R_{eq} + \tilde{I}_1 jX_{eq} = (2{,}515.8 + j131.87)\,\text{V} = 2{,}519\angle 2.987°\,\text{V}. \tag{7.73}$$

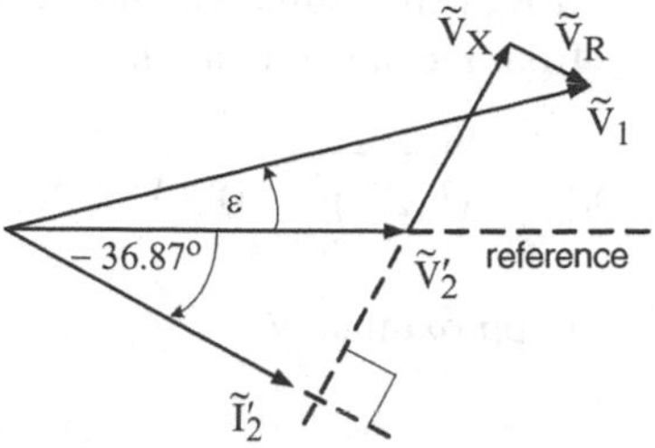

Fig. 7.26 Phasor diagram of the reduced equivalent circuit referred to the high-voltage side with impedance load at lagging power factor (consumer notation)

Figure 7.26 depicts the (schematic, not to scale) phasor diagram, where $\tilde{V}_X = \tilde{I}_1 jX_{eq}$, and $\tilde{V}_R = \tilde{I}_1 R_{eq}$.

Application Example 7.6: Short-Circuit (sc) and Open-Circuit (oc) Tests of a Transformer

A single-phase S = 1,500 VA, 240 V/120 V (=N_p/N_s) transformer is subjected to a short-circuit test and an open-circuit test for the determination of its equivalent circuit parameters. The short-circuit test (secondary short-circuited) yields the measured data: $V^p_{sc} = 10.0\,V$, $I^p_{sc} = 6.5\,A$, and $P^p_{sc} = 30.0\,W$. The open-circuit (primary open-circuited) test yields the data: $V^s_{oc} - 120.0\,V$, $I^s_{oc} = 0.5\,A$, and

$$P^S_{OC} = 25.0\,W.$$

(a) Calculate the series-branch and cross-branch equivalent circuit parameters and draw the equivalent circuit diagram where all parameters are referred to the primary (high-voltage side). You may assume that $R_p \approx R'_s$ and $X_{p\ell} \approx X'_{s\ell}$ where subscript "ℓ" stands for leakage.
(b) Draw the phasor diagram (schematically) for cosϕ = 0.90 lagging power factor (consumer notation) neglecting G_{cp} and B_{mp}.
(c) Draw the phasor diagram (schematically) for cosϕ = 0.90 lagging power factor (consumer notation) including G_{cp} and B_{mp}.

Solution

(a) The turns ratio is $N_p/N_s = 240/120 = 2$. From the short-circuit test one obtains the parameters referred to the high-voltage side

$$Z^p_{sc} = \frac{|\tilde{V}^p_{sc}|}{|\tilde{I}^p_{sc}|} = \frac{10}{6.5} = 1.538\,\Omega, \tag{7.74}$$

$$R^p_{sc} = \frac{|P^p_{sc}|}{|\tilde{I}^p_{sc}|^2} = R_p + R'_s = \frac{30}{(6.5)^2} = 0.710\,\Omega. \tag{7.75}$$

If R_p is measured via the DC voltage-drop method then $R_s^{'}$ can be determined from the above relation.

$$X_{sc}^{p} = \sqrt{(Z_{sc}^{p})^2 - (R_{sc}^{p})^2} = X_{p\ell} + X'_{s\ell} = \sqrt{2.365 - 0.504} = 1.364\ \Omega, \quad (7.76)$$

or approximately

$$X_{p\ell} \approx X'_{s\ell} \approx \frac{X_{sc}^{p}}{2} = 0.682\,\Omega. \quad (7.77)$$

From the open-circuit test one obtains the parameters referred to the low-voltage (secondary) side

$$Y_{oc}^{s} = \frac{|\tilde{I}_{oc}^{s}|}{|\tilde{V}_{oc}^{s}|} = \frac{0.5}{120} = 0.00417\,1/\Omega, \quad (7.78)$$

$$G_{oc}^{s} = \frac{|P_{oc}^{s}|}{|\tilde{V}_{oc}^{s}|^2} = \frac{25}{(120)^2} = 0.001740\,1/\Omega, \quad (7.79)$$

$$B_{oc}^{s} = \sqrt{(Y_{oc}^{s})^2 - (G_{oc}^{s})^2} = 0.003789\,1/\Omega, \quad (7.80)$$

Refer G_{oc}^{s} and B_{oc}^{s} to high-voltage (primary) side:

$$G_{c}^{p} = \left(\frac{N_2}{N_1}\right)^2 G_{oc}^{s} = 0.000435\,1/\Omega \quad (7.81)$$

$$B_{m}^{p} = \left(\frac{N_2}{N_1}\right)^2 B_{oc}^{s} = 0.000947\,1/\Omega. \quad (7.82)$$

Figure 7.27 shows the equivalent circuit referred to the high-voltage side including G_c^p and B_m^p.

(b) Figure 7.28 depicts the equivalent circuit including a load impedance neglecting G_c^p and B_m^p, and Fig. 7.29 shows the schematic (not to scale) phasor diagram. Note that $R_{eq} = R_p + R'_s$, $X_{eq} = X_{p\ell} + X'_{s\ell}$, and $\phi = \cos^{-1}(0.9) = -25.84°$ (lagging, consumer notation).

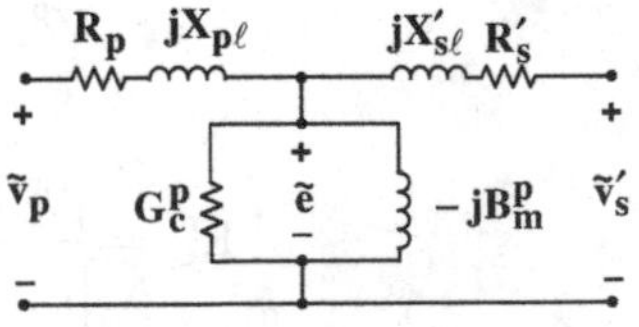

Fig. 7.27 Equivalent circuit referred to the high-voltage side including G_c^p and B_m^p

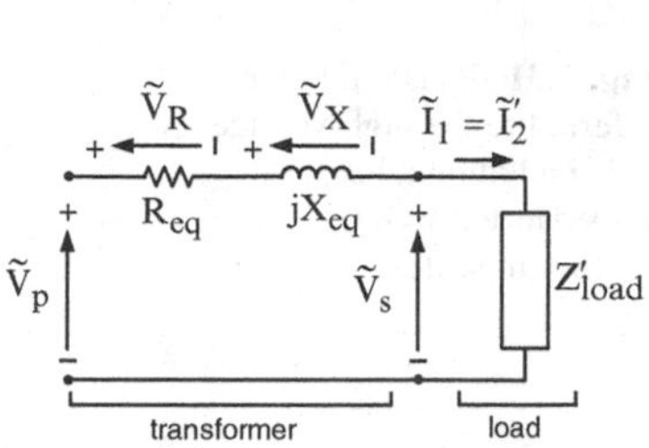

Fig. 7.28 Equivalent circuit referred to the high-voltage side including a load impedance neglecting G_c^p and B_m^p

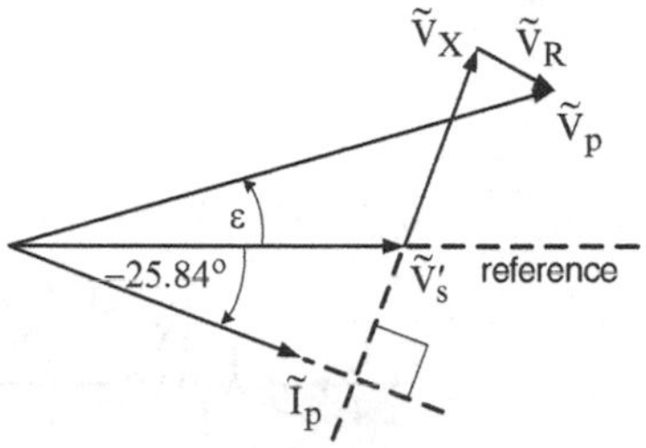

Fig. 7.29 Phasor diagram referred to the high-voltage side including a load impedance neglecting G_c^p and B_m^p (not to scale)

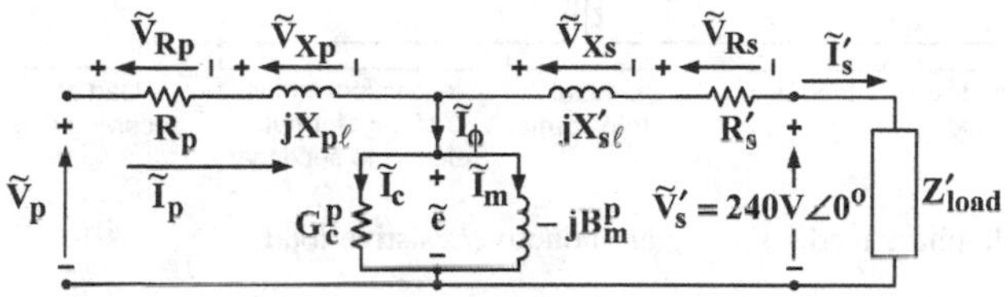

Fig. 7.30 Equivalent circuit referred to the high voltage side including a load impedance Z'_{load}, G_c^p and B_m^p, where $\tilde{i}_s' = \tilde{I}_s'$

(c) Figure 7.30 depicts the equivalent circuit including a load impedance Z'_{load} as well as G_c^p, and B_m^p. Figure 7.31 shows the schematic (not to scale) phasor diagram. Note, $\phi = \cos^{-1}(0.9) = -25.84°$(lagging, consumer notation).

Application Example 7.7: Single-Phase Feeder Serving a Load, and Principle of Power Flow

A single-phase load (see Fig. 7.32) is supplied through a 35 kV feeder whose impedance is $Z_f = (R_f + jX_f) = (100 + j300)\ \Omega$, and an apparent power $S = 500\,\text{kVA}$, (35 kV:2,400 V) transformer whose impedance is $Z_{eq}^s = R_{eq}^s + jX_{eq}^s = (0.2 + j1.0)\,\Omega$ referred to its low-voltage (secondary) side. The rated secondary rms voltage of the transformer is $|\tilde{V}_H'| = |\tilde{V}_{L\,rat}| = 2,400\,\text{V}$. The load absorbs the rated transformer current at 0.8 lagging power factor (consumer notation) at a sending end rms voltage

Fig. 7.31 Phasor diagram referred to the high-voltage side including a load impedance Z'_{load}, G_c^p and B_m^p (not to scale)

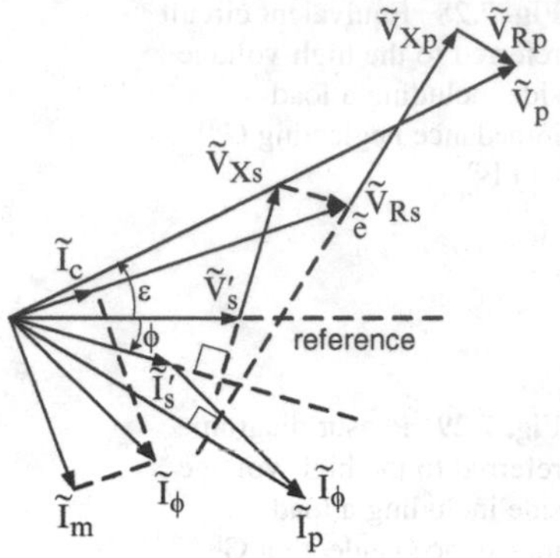

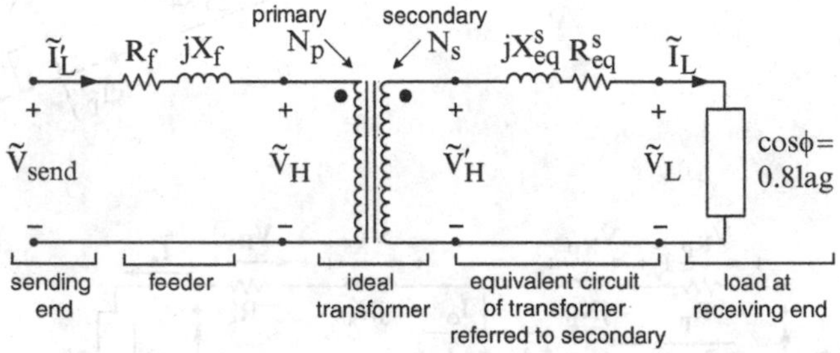

Fig. 7.32 Single-phase feeder serving an inductive/resistive load

of 35 kV. Note that the utility controls the voltage at the sending end, but not at the receiving end.

(a) Compute the voltage at the receiving (load) end of the feeder.
Hints: Compute rated current taking the secondary voltage $\tilde{V}_L = |\tilde{V}_L| \angle 0°$ as reference, refer all quantities to primary (high-voltage) side, and draw a phasor diagram referred to primary side.

(b) Compute the voltage regulation VR of the load voltage $\tilde{V}_L$. Is VR within acceptable limits?

Solution

(a) The rms value (magnitude) of the load (L) current is

$$|\tilde{I}_L| = \frac{S}{|\tilde{V}_L|} = \frac{500\,\text{kVA}}{2.4\,\text{kV}} = 208.33\,\text{A}, \tag{7.83}$$

at a power factor of $\cos\phi = 0.8$ lagging (consumer notation), or $\phi = -36.87°$ the load current is

$$\tilde{I}_L = 208.33 \angle -36.87° \text{A} = (166.67 - j124.99)\text{A}. \tag{7.84}$$

Refer $\tilde{V}_L$, $\tilde{I}_L$, X^s_{eq}, and R^s_{eq} to primary with $N_p/N_s = (35\ \text{kV}/2.4\ \text{kV})$:

$$\tilde{V}'_L = \tilde{V}_L\left(\frac{N_p}{N_s}\right) = \left(\frac{35}{2.4}\right)\tilde{V}_L, \tag{7.85}$$

$$|\tilde{I}'_L| = |\tilde{I}_L|\left(\frac{N_s}{N_p}\right) = \left(\frac{2.4}{35}\right)|\tilde{I}_L| = 14.285\ \angle -36.87°\text{A}, \tag{7.86}$$

$$R^P_{eq} = R^s_{eq}\left(\frac{N_p}{N_s}\right)^2 = 0.2(14.58)^2 = 42.53\ \Omega, \tag{7.87}$$

$$X^P_{eq} = X^s_{eq}\left(\frac{N_p}{N_s}\right)^2 = 1.0(14.58)^2 = 212.57\,\Omega. \tag{7.88}$$

The equivalent circuit referred to the primary is shown in Fig. 7.33.
In the above equivalent circuit $R = R_f + R^p_{eq} = 142.53\,\Omega$, and $X = X_f + X^p_{eq} = 512.57\,\Omega$. The schematic (not to scale) phasor diagram is depicted in Fig. 7.34 According to the Pythagorean theorem one obtains

$$\begin{aligned}|\tilde{V}_{send}|^2 &= \{|\tilde{V}'_L| + |\tilde{V}_R|\cos\phi + |\tilde{V}_X|\sin\phi\}^2 \\ &\quad + \{|\tilde{V}_X|\cos\phi - |\tilde{V}_R|\sin\phi\}^2,\end{aligned} \tag{7.89}$$

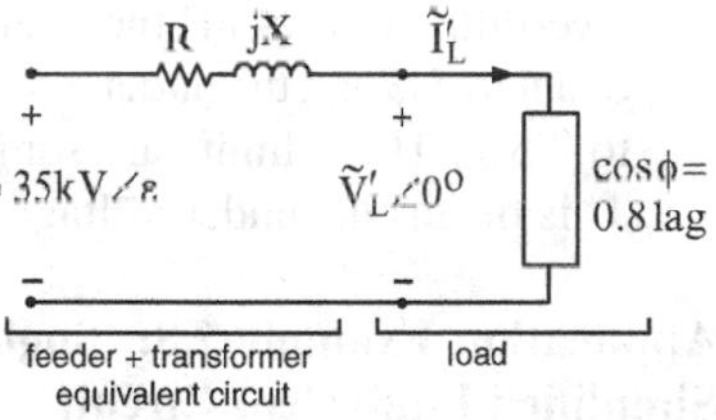

Fig. 7.33 Equivalent circuit referred to primary

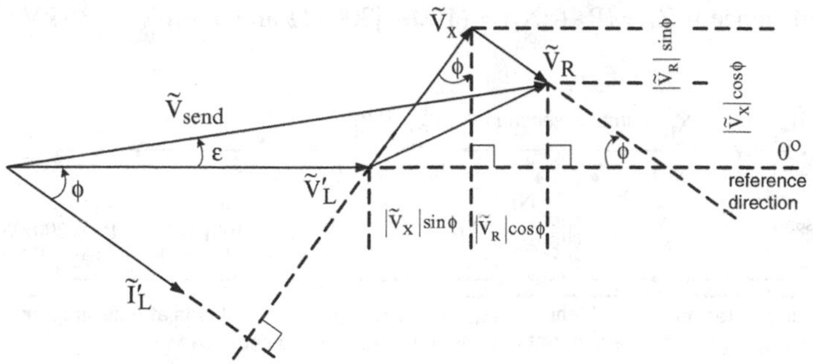

Fig. 7.34 Phasor diagram at lagging power factor (inductive/resistive load)

or

$$\left|\tilde{V}_L'\right| = \left[\left|\tilde{V}_{send}\right|^2 - \{\left|\tilde{V}_X\right|\cos\phi - \left|\tilde{V}_R\right|\sin\phi\}^2\right]^{1/2} - \left|\tilde{V}_R\right|\cos\phi$$
$$- \left|\tilde{V}_X\right|\sin\phi. \tag{7.90}$$

$$\left|\tilde{V}_R\right| = \left|\tilde{I}_L'\right|R = 14.285 \cdot 142.53 = 2,036\,\text{V}, \tag{7.91}$$

$$\left|\tilde{V}_X\right| = \left|\tilde{I}_L'\right|X = 7,322\,\text{V}, \tag{7.92}$$

$$\left|\tilde{V}_L'\right| = 283,359\,\text{V}, \tag{7.93}$$

or referred to secondary

$$\tilde{V}_L = 1.944\,\text{kV}. \tag{7.94}$$

(b) Voltage regulation VR is defined [10] as

$$\text{VR} = \frac{\{(\text{no} - \text{loadvoltage}) - (\text{full} - \text{loadvoltage})\} \cdot 100\%}{(\text{no} - \text{load voltage})}$$
$$= \left(\frac{2,400 - 1,944}{2,400}\right) \cdot 100\% = 19\%. \tag{7.95}$$

According to [10–14] the over-voltage voltage regulation VR should be less or equal to 5% and the under-voltage voltage regulation VR should be less or equal to 2.5%. These limits are somewhat device-(e.g., lighting, motors) dependent. This means the under-voltage regulation of VR = 19% is not acceptable.

Application Example 7.8: Single-Phase Feeder with Two Loads, and Use of Simplified Equivalent Circuit

Two single-phase loads (see Fig. 7.35) are supplied through a 35 kV feeder whose impedance is $Z_f = (R_f + jX_f) = (100 + j300)\ \Omega$ and a $V_H/V = 35\,\text{kV}/2,400\,\text{V}$

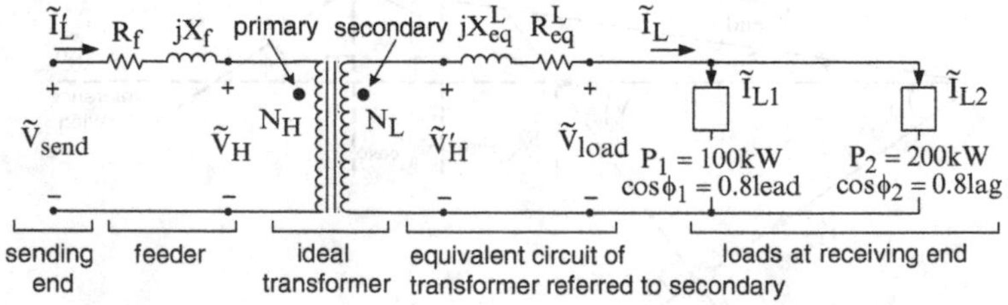

Fig. 7.35 Single-phase feeder, single-phase transformer and two single-phase loads

single-phase transformer whose equivalent (series-branch) impedance is $Z_{eq}^{L} = (R_{eq}^{L} + jX_{eq}^{L}) = (0.2 + j1.0)\,\Omega$ referred to its low-voltage (secondary) side. Note H and L stand for high side and low side, respectively. The two loads are $P_1 = 100$ kW at $\cos\phi_1 = 0.8$ leading power factor (consumer notation) at $V_{load} =$ 2,400 V, and $P_2 = 200$ kW at $\cos\phi_2 = 0.8$ lagging power factor (consumer notation) at the same voltage of $V_{load} = 2{,}400$ V. You may take $\tilde{V}_{load}$ as reference, that is, $\tilde{V}_{load} = 2{,}400\,\angle 0°$ V.

(a) Compute the total load current $\tilde{I}_L = \tilde{I}_{L1} + \tilde{I}_{L2}$ due to these two single-phase loads.
(b) Compute the voltage at the sending end of the feeder if the voltage at the load (receiving) end is maintained at $V_{load} = 2{,}400$ V.
(c) What is the real power P_{send} and reactive power Q_{send} at the sending end of the feeder?
(d) Calculate the voltage regulation VR (see (7.95)).

Solution

(a) The total load current is $\tilde{I}_L = \tilde{I}_{L2} + \tilde{I}_{L2}$ where the rms values of the two currents are

$$\left|\tilde{I}_{L1}\right| = \frac{P_1}{\left|\tilde{V}_{load}\right|\cos\phi_1} = \frac{100\,\text{k}}{2{,}400\cdot 0.8} = 52.08\text{ A}, \tag{7.96}$$

$$\left|\tilde{I}_{L2}\right| = \frac{P_2}{\left|\tilde{V}_{load}\right|\cos\phi_2} = \frac{200\,\text{k}}{2{,}400\cdot 0.8} = 104.17.\text{ A}, \tag{7.97}$$

or the two load currents become with $\phi_1 = +36.87°$ (leading power factor) and $\phi_2 = -36.87°$ (lagging power factor)

$$\tilde{I}_{L1} = \left|\tilde{I}_{L1}\right|\angle\,\phi_1 = 52.08\,\angle 36.87°\text{ A}, \tag{7.98}$$

$$\tilde{I}_{L2} = \left|\tilde{I}_{L2}\right|\angle\,\phi_2 = 104.17\,\angle -36.87°\text{A}, \tag{7.99}$$

$$\tilde{I}_L = \tilde{I}_{L1} + \tilde{I}_{L2} = 128.84\,\angle -14.03°\text{A} = (125 - j31.25)\text{A}. \tag{7.100}$$

(b) Applying Kirchhoff's voltage law one gets

$$\begin{aligned}\tilde{V}_H' &= \tilde{V}_{load} + \left(R_{eq}^{L} + jX_{eq}^{L}\right)\tilde{I}L \\ &= 2{,}400\,\angle 0° + (0.2 + j1.0)(125 - j31.25)\text{ V} \\ &= (2{,}456.25 + j118.75)\text{V}.\end{aligned} \tag{7.101}$$

With

$$\tilde{V}_H = \left(\frac{N_H}{N_L}\right)\tilde{V}'_H = 14.58\,\tilde{V}'_H = (35{,}820.3 + j1731.37)\,V \quad (7.102)$$

$$\tilde{I}_L{}' = \left(\frac{N_L}{N_H}\right)\tilde{I}_L = (1/14.58)\tilde{I}_L = (8.57 - j2.16)\,A, \quad (7.103)$$

the sending voltage is

$$\tilde{V}_{send} = \tilde{V}_H + (R_f + jX_f)\tilde{I}_L{}' = (37{,}321.7 + j4088.19)\,V. \quad (7.104)$$

(c) The apparent power is defined as

$$S = \tilde{V}_{send}(\tilde{I}_L{}')^* = P_{send} + jQ_{send} \quad (7.105)$$

where

$$P_{send} = \mathrm{Re}\{\tilde{V}_{send}(\tilde{I}_L{}')^*\} \approx 311\,kW \quad (7.106)$$

$$Q_{send} = \mathrm{Im}\{\tilde{V}_{send}(\tilde{I}_L{}')^*\} \approx -116\,kW. \quad (7.107)$$

At the sending end the power factor is lagging (consumer notation) and therefore $Q_{send} < 0$.

(d) The voltage regulation at the sending end is

$$VR = \frac{\{(no - load\ voltage) - (full - load\ voltage)\} \cdot 100\%}{(no - load\ voltage)}$$
$$= \left(\frac{37.55\,k - 35\,k}{37.55\,k}\right) \cdot 100\% = 6.8\%. \quad (7.108)$$

Application Example 7.9: Voltage regulation VR and Power Efficiency η_{power}

The equivalent circuit parameters of a S = 1,500 VA, 240 V/120 V single-phase transformer are $R_p \approx R'_s = 0.36\,\Omega$, $X_{p\ell} \approx X'_{s\ell} = 0.685\,\Omega$, and $G_{cp} = 0.00105$ S, $B_{mp} = 0.000955$ S. Note the unit siemens [S] is the inverse of ohms [1/Ω], and "ℓ" stands for leakage.

(a) Draw the equivalent circuit neglecting G_{cp} and B_{mp}.
(b) Calculate the voltage regulation VR (see (7.95)) at rated current and voltage ($I_{srat} = 12.5\,A$, $\tilde{V}_{srat} = 120\,V$) at a power factor of $\cos\phi = 0.9$ lagging (consumer notation).
(c) Assume that the load power factor $\cos\phi$ is varied while the load current and secondary terminal voltage V'_{srat} are held constant. Use a phasor diagram to determine the load power factor for which the voltage regulation is greatest. What is this voltage regulation?

(d) Use a phasor diagram to determine the load power factor for which the voltage regulation is smallest (zero).
(e) Compute the power efficiency $\eta_{power} = \frac{P_{out}}{P_{in}} = \frac{P_{out}}{(P_{out}+P_{loss})}$ at rated operation by using information given in Application Example 7.6 ($I^{p}_{sc} = 6.5$ A, $P^{p}_{sc} = 30.0$ W, $V^{s}_{oc} = 120.0$ V, $P^{s}_{oc} = 25.0$ W).

Solution

(a) The equivalent circuit referred to the primary (high-voltage side) neglecting G_{cp} and B_{mp} is shown in Fig. 7.36.
(b) Calculation of voltage regulation VR based on reduced equivalent circuit of Fig. 7.37 with
$R_{eq} = R_p + R'_s = 0.72\ \Omega$, and $X_{eq} = X_{p\ell} + X'_{s\ell} = 1.37\ \Omega$.
The rms value of the primary current is

$$|\tilde{I}_p| = \frac{S}{|\tilde{V}_p|} = \frac{1,500\ \text{VA}}{240\ \text{V}} = 6.25\ \text{A}, \tag{7.109}$$

and the primary current at a power factor of 0.9 lagging becomes

$$\tilde{I}_p = \tilde{I}'_s = 6.25\angle -25.84°\text{A} = (5.626 - j2.718)\ \text{A}. \tag{7.110}$$

Applying Kirchhoff's voltage law to the reduced equivalent circuit one can write with $N_p/N_s = 24/120$

$$\tilde{V}'_s = |\tilde{V}_{s\_rat}| \frac{N_p}{N_s} \angle 0° = 240\angle 0°\text{V}, \tag{7.111}$$

$$\tilde{V}_p = \tilde{V}'_s + \tilde{I}_p(R_{eq} + jX_{eq}) = (247.78 + j\,5.75) = 247.85\angle 1.33°\text{V}, \tag{7.112}$$

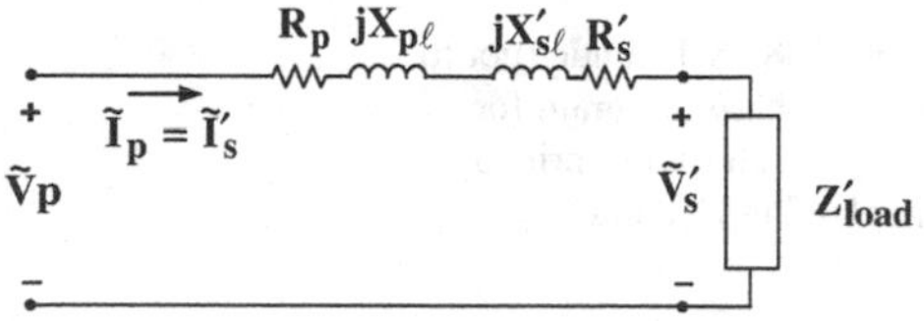

Fig. 7.36 Equivalent circuit referred to primary neglecting G_{cp} and B_{mp}

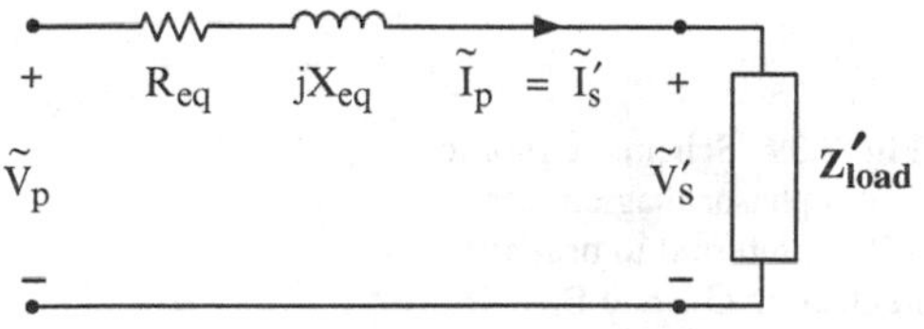

Fig. 7.37 Reduced equivalent circuit referred to primary neglecting G_{cp} and B_{mp}

and the voltage regulation is

$$VR = \left(\frac{247.85 - 240}{247.85}\right)100\% = 3.17\%. \tag{7.113}$$

(c) The greatest voltage regulation VR_{max} can be obtained from the phasor diagram Fig. 7.38 by inspection requiring that $\tilde{V}_p$and $\tilde{V}_s$ are in phase.
For $\phi = \theta$ one obtains . $\phi = \tan^{-1}(X_{eq}/R_{eq}) = 62.28°$ lagging power factor or $\phi = \theta = -62.28°$ resulting in

$$VR_{max} = \left(\frac{249.67 - 240}{249.67}\right)100\% = 3.87\%. \tag{7.114}$$

(d) The voltage regulation is zero if the rms values of $\tilde{V}_p$ and $\tilde{V}_s$ are the same as indicated in Fig. 7.39 in a schematic manner. Zero voltage regulation $VR_{min} = 0$ occurs at leading power factor which must be avoided in most cases because of stability problems associated with leading power factor operation within the distribution system.

e) The power efficiency can be determined from

$$\eta_{power} = \frac{P_{out}}{P_{in}} = \frac{P_{out}}{P_{out} + P_{loss}}. \tag{7.115}$$

The output power at a power factor of 0.9 lagging is

$$P_{out} = S_{rat}\cos\phi = 1.5\,kVA \cdot 0.9 = 1,350\,W, \tag{7.116}$$

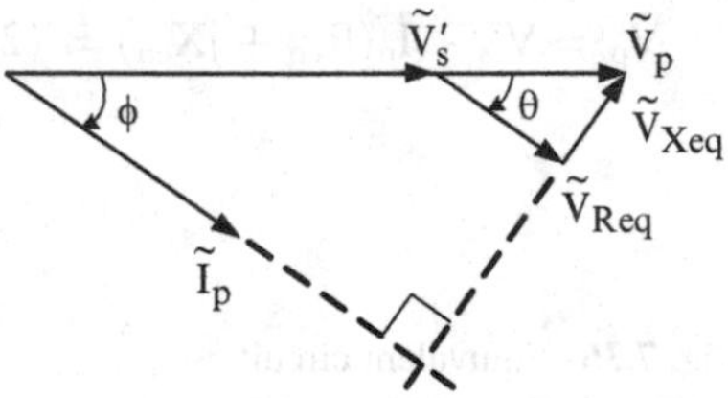

Fig. 7.38 Schematic (not to scale) phasor diagram for VR_{max}, referred to primary neglecting G_{cp} and B_{mp}

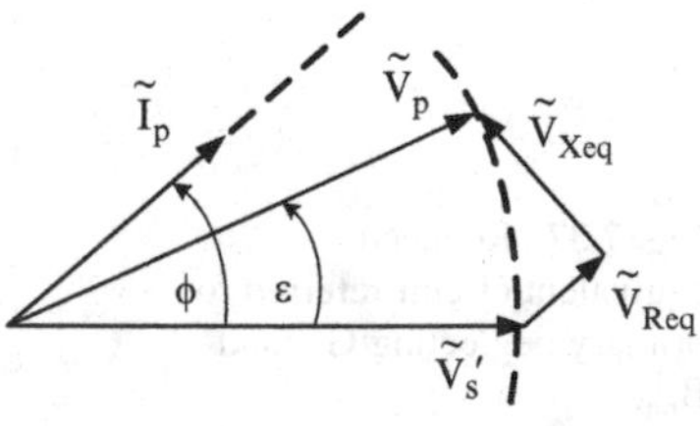

Fig. 7.39 Schematic (not to scale) phasor diagram for VR_{min}, referred to primary neglecting G_{cp} and B_{mp}

and the loss is

$$P_{loss} = P_{sc}^{p} + P_{oc}^{S} = 30\,W + 25\,W, \tag{7.117}$$

resulting in $\eta_{power} = 96\%$.

Application Example 7.10: Energy Efficiency of Transformer η_{energy}

A single-phase $S_{rat} = 100$ kVA, $V_{Hrat}/V_{Lrat} = 11{,}000$ V/550 V transformer has no-load loss $P_{no\ load} = P_{Fe\_rat} = 2$ kW at rated voltage and a copper loss of $P_{cu\_rat} =$ 4 kW at rated load current. It operates on a daily load cycle as indicated in Table 7.2.

Determine the energy efficiency for this load cycle.

Solution

The copper loss is proportional to the square of the load current $P_{cu} = I^2R$, and the rated current of the high-voltage winding is $I_{H\_rat} = S_{rat}/V_{H\_rat} = 100{,}000/11{,}000 = 9.091$ A. Thus, the copper loss when the high-voltage winding current is $I_H = 2.0$ A, as it is during the first period (8 h), becomes $P_{cu\_8hours} = (I_{H\_8h}/I_{H\_rat})^2 \cdot P_{cu\_rat} = (2.0/9.091)^2 \cdot 4$ kW $= 0.194$ kW. The output power during this period is $P_{out\_8} = V_H \cdot I_{H\_8} \cdot \cos\phi_{8h} = 11{,}000$ V$\cdot 2$ A$\cdot 0.9 = 19.8$ kW so that the input power during the same period is $P_{in\_8} = P_{out\_8} + P_{cu\_8h} + P_{Fe\_rat} = (19.8 + 0.194 + 2)$ kW $= 21.99$ kW. The output and input energies are obtained by multiplying these values for power by the length of the period, 8 h, to get the energy in kWh.

This problem is best done by tabulating the calculated values as shown in Table 7.3. It is convenient to number the lines/rows so that the calculation will proceed more quickly after the first column is completed. Note that whether the power factor is lagging or leading is irrelevant for this particular problem.

Table 7.2 Daily load cycle of transformer

Time period [h]	Current on high-voltage side [A]	Power factor $\cos\phi$ [−] (consumer notation)
8	2.0	0.9 lag
10	4.5	0.8 lag
6	9.0	0.9 lead

Table 7.3 Calculation of input and output energies

1. Time period [h]	8	10	6
2. Current [A]	2	4.5	9
3. Power factor $\cos\phi$[−]	0.9	0.8	0.9
4. Copper loss [kW]	0.194	0.98	3.92
5. Iron-core loss [kW]	2.0	2.0	2.0
6. Output power P_{out} [kW]	19.8	39.6	89.1
7. Input power P_{in} [kW]	21.99	42.58	95.02
8. Output energy E_{out} [kWh]	158.4	396	534.6
9. Input energy E_{in} [kWh]	175.92	425.8	570.1

The total output energy during the 24 h period is the sum of the values in line/row 8 of all three columns, $E_{out\ total} = 1{,}089$ kWh. The total input energy is the sum of the line/row 9 values of all three columns, $E_{in\ total} = 1{,}171.82$ kWh, so that the energy efficiency is $\eta_{energy} = (E_{out\ total}/E_{in\ total}) = 93\%$.

7.4 Autotransformers and Variacs

In two-winding transformers there is an electrical isolation between the primary and secondary windings and both sides can be assigned a different ground potential, as is shown in Fig. 7.40.

If the separation of the two grounds is not essential for the safe and proper operation of a network, autotransformers (fixed turns ratio N_1/N_2) or variacs (variable turns ratio N_1/N_2) can be used, where the primary and secondary are electrically (galvanically) connected, as is demonstrated in Fig. 7.41. Such galvanic connections can be made in a variety of ways. Autotransformers have an increased

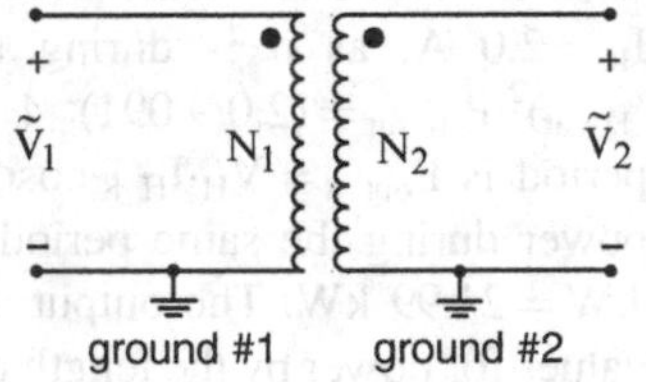

Fig. 7.40 Grounding of a two-winding transformer

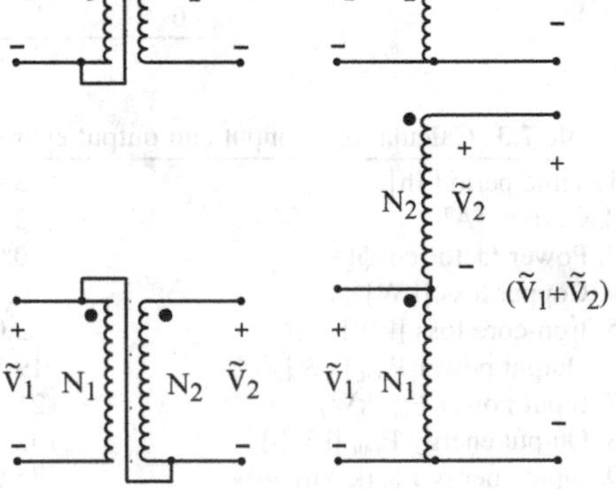

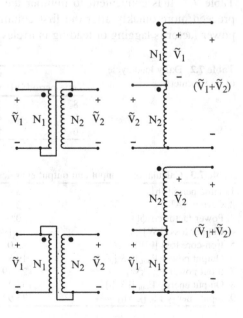

Fig. 7.41 Reconnecting a two-winding transformer in an autotransformer mode

power rating, a higher efficiency, no galvanic isolation between primary and secondary, and in some cases a high leakage (series) reactance, which makes them not suitable for feeding solid-state switching (e.g., thyristor) circuits where the di/dt is very large.

Application Example 7.11: Efficiency of Autotransformer as Compared with Two-Winding Transformer

The name plate on a $S = 50$ MVA, 60 Hz single-phase two-winding transformer indicates that it has a voltage rating of 8 kV:78 kV. An open-circuit test is performed from the low-voltage side, and corresponding instrument readings are $V_{Loc} = 8$ kV, $I_{Loc} = 61.9$ A, and $P_{Loc} = 136$ kW. Similarly, a short-circuit test (short-circuited at the low-voltage side) gives the readings $V_{Hsc} = 650$ V, $I_{Hsc} = 6.25$ A, and $P_{Hsc} = 103$ kW.

(a) Connect this transformer as an autotransformer so that the input voltage at the source is $V_{in} = 78$ kV and the output voltage at the load $V_{out} = 86$ kV. What is the apparent power rating of this autotransformer?
(b) Calculate the efficiency of the autotransformer in this connection when it is supplying rated load at unity ($\cos\phi = 1.0$) power factor. Compare the efficiency of the autotransformer with that of the two-winding transformer.

Solution

(a) Figure 7.42 represents the winding connection for a two-winding transformer while Fig. 7.43 is the configuration when this transformer is connected as an autotransformer. The primary line current of the two-winding configuration is $I_L = 50$ MVA/8 kV $= 6{,}250$ A.

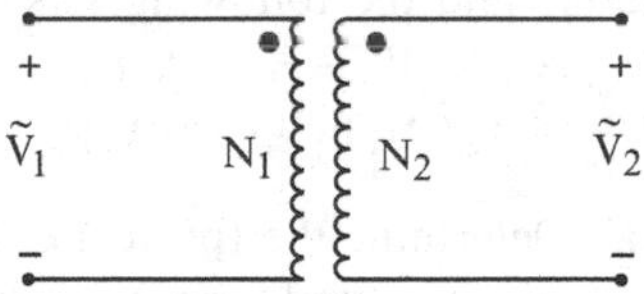

Fig. 7.42 Two-winding transformer connection

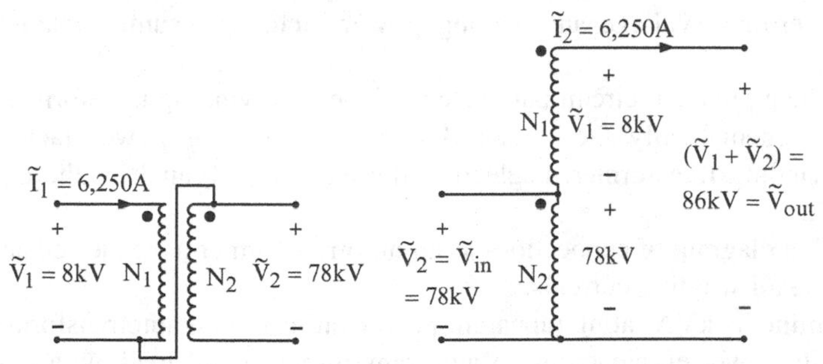

Fig. 7.43 Autotransformer connection

The input voltage of the autotransformer is $V_{in} = 78$ kV, and the output voltage of the autotransformer is $V_{out} = 78\text{ kV} + 8\text{ kV} = 86$ kV. Neglecting the losses, the rated apparent power of the autotransformer is $S_{out} = V_{out}\ I_{out} = 86\ \text{kV}\cdot 6{,}250\ \text{A} = 537.5\ \text{MVA} = S_{in}$.

(b) The rated losses are $P_{loss} = 136\text{ kW} + 103\text{ kW} = 239$ kW and at a power factor of $\cos\phi$ the output power is

$$P_{out} = S_{out}\cos\phi. \tag{7.118}$$

For $\cos\phi = 1.0$, therefore $P_{out} = S_{out}$, and the efficiency of the autotransformer becomes

$$\eta = \frac{P_{out}}{P_{in}} = \frac{P_{out}}{P_{out} + P_{loss}} = \frac{537.5}{537.5 + 0.239} = 0.9996 = 99.96\%. \tag{7.119}$$

Note the efficiency of the two-winding transformer at unity power factor is

$$\eta = \frac{P_{out}}{P_{in}} = \frac{P_{out}}{P_{out} + P_{loss}} = \frac{50}{50 + 0.239} = 0.995 = 99.5\%. \tag{7.120}$$

Application Example 7.12: Apparent Power and Efficiencies of Autotransformer and Two-Winding Transformer

A 240 V:120 V, $S_{rat} = 1.5$ kVA two-winding transformer is to be used as an autotransformer to supply a 360 V load circuit from a 120 V voltage source. When it is being tested as a two-winding transformer short-circuit and open-circuit tests yield the following sets of data: with low-voltage terminals short-circuited: $V_{sc}^{H} = 8\,\text{V}, I_{sc}^{H} = 6.25\,\text{A}, P_{sc} = 40\,\text{W}$; with high-voltage terminals open-circuited: $V_{oc}^{L} = 120\,\text{V}, I_{oc}^{L} = 1.0\,\text{A}, P_{oc} = 30\,\text{W}$.

(a) Determine the (power) efficiency of the two-winding transformer at full-load current, rated terminal voltage and unity power factor $\cos\phi = 1.0$.
(b) Determine the efficiency of the two-winding transformer at full-load current, rated terminal voltage and leading power factor (consumer notation) $\cos\phi = 0.8$.
(c) Find the equivalent circuit parameters of the two-winding transformer.
(d) Draw schematically the phasor diagram for leading power factor of the two-winding transformer, neglecting the excitation branch of the equivalent circuit.
(e) Sketch a diagram of connections as an autotransformer as requested above and indicate all winding currents.
(f) Determine its kVA rating (apparent power rating) as an autotransformer.
(g) Find its power efficiency as an autotransformer at full load, with $\cos\phi = 0.9$ power factor lagging (consumer notation).

Solution

The two-winding transformer of Fig. 7.44 is used as an autotransformer shown in Fig. 7.45.

(a) The power efficiency of the two-winding transformer at $\cos\phi = 1.0$ is

$$\eta_{\text{two-winding}}^{\cos\phi=1} = \frac{P_{out}}{P_{in}} = \frac{S\cos\phi}{S\cos\phi + P_{loss}} = \frac{S\cos\phi}{S\cos\phi + P_{sc} + P_{oc}} = \frac{1,500}{1,570} = 95.54\%. \tag{7.121}$$

(b) Power efficiency at $\cos\phi = 0.8$

$$\eta_{\text{two-winding}}^{\cos\phi=0.8} = \frac{P_{out}}{P_{in}} = \frac{S\cos\phi}{S\cos\phi + P_{loss}} = \frac{S\cos\phi}{S\cos\phi + P_{sc} + P_{oc}} = \frac{1,500 \cdot 0.8}{1,500 \cdot 0.8 + 70} = 94.5\%. \tag{7.122}$$

(c) Equivalent circuit parameters of two-winding transformer are obtained from the equivalent circuit of Fig. 7.46.
Short-circuit test yields if the low-voltage side is short-circuited:

$$Z_{sc}^{H} = \frac{V_{sc}^{H}}{I_{sc}^{H}} = \frac{8}{6.25} = 1.28\,\Omega \tag{7.123}$$

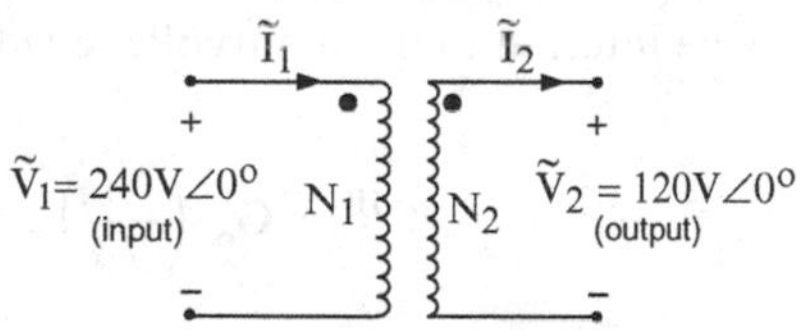

Fig. 7.44 Two-winding transformer

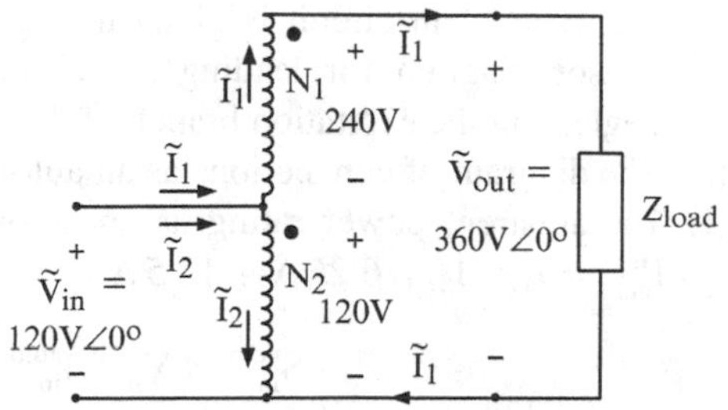

Fig. 7.45 Two-winding transformer reconnected as an autotransformer

Fig. 7.46 Equivalent circuit of two-winding transformer

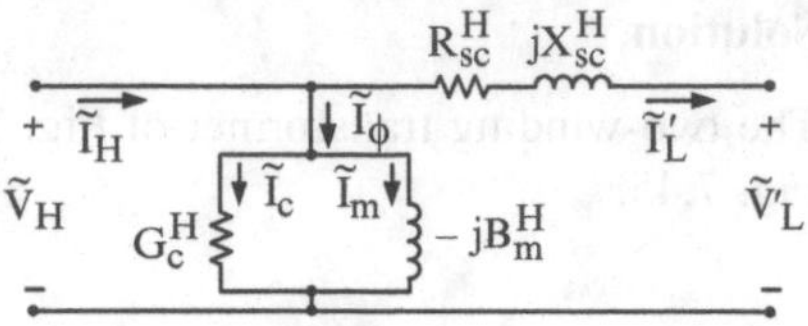

$$R_{sc}^H = \frac{P_{sc}^H}{\left(I_{sc}^H\right)^2} = \frac{40}{(6.25)^2} = 1.02\,\Omega \tag{7.124}$$

$$X_{sc}^H = \sqrt{\left(Z_{sc}^H\right)^2 - \left(R_{sc}^H\right)^2} = 0.78\,\Omega. \tag{7.125}$$

R_{sc}^H and X_{sc}^H are referred to the high(H)-voltage side of the transformer. Open-circuit test yields if the high-voltage side is open and the low-voltage side is energized:

$$Y_{oc}^L = \frac{I_{oc}^L}{V_{oc}^L} = \frac{1}{120} = 0.0083\,1/\Omega \tag{7.126}$$

$$G_{oc}^L = \frac{P_{oc}^L}{\left(V_{oc}^L\right)^2} = \frac{30}{(120)^2} = 0.00208\,1/\Omega \tag{7.127}$$

$$B_m^L = \sqrt{\left(Y_{oc}^L\right)^2 - \left(G_{oc}^L\right)^2} = 0.00804\,1/\Omega. \tag{7.128}$$

G_{oc}^L and B_m^L are referred to the low (L)-voltage side of the transformer and must be referred to the high-voltage side with $N_H = 240$ V and $N_L = 120$ V:

$$G_{oc}^H = G_{oc}^L\left(\frac{N_L}{N_H}\right)^2 = \frac{0.00208}{4} = 0.00052\,1/\Omega \tag{7.129}$$

$$B_m^H = B_m^L\left(\frac{N_L}{N_H}\right)^2 = \frac{0.00804}{4} = 0.00201\,1/\Omega. \tag{7.130}$$

(d) The equivalent circuit is given in Fig. 7.47 and the schematic (not to scale) phasor diagram for leading power factor of the two-winding transformer, neglecting the excitation branch of the equivalent circuit is shown in Fig. 7.48.
(e) The diagram of connections as an autotransformer is shown in Fig. 7.45.
(f) The apparent power rating as an autotransformer is with $V_{in}^{auto} = 120$ V and $I_{in}^{auto} = I_1 + I_2 = 6.25\text{ A} + 12.5\text{ A} = 18.75\text{ A}$ (as indicated in Fig. 7.45):

$$S_{in}^{auto} = V_{in}^{auto}\, I_{in}^{auto} = 2,250\text{ VA}. \tag{7.131}$$

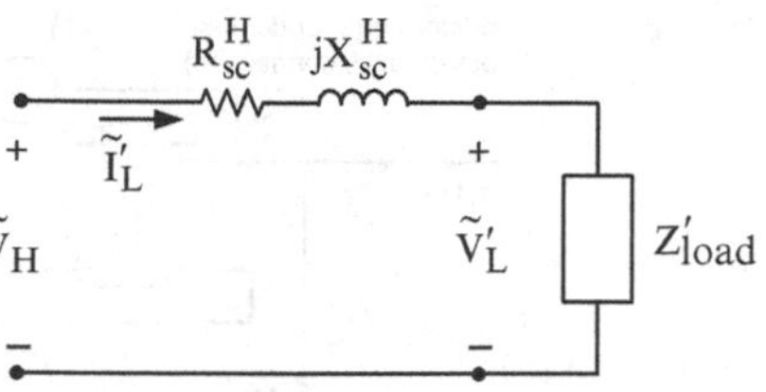

Fig. 7.47 Equivalent circuit of the two-winding transformer, neglecting the excitation branch of the equivalent circuit

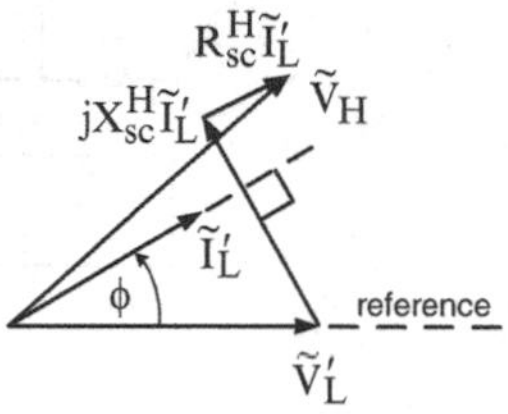

Fig. 7.48 Schematic phasor diagram for leading power factor of the two-winding transformer, neglecting the excitation branch of the equivalent circuit

(g) Power efficiency as an autotransformer at full load, with $\cos\phi = 0.9$ power factor lagging:

$$\eta_{auto}^{\cos\phi=0.9} = \frac{P_{out}}{P_{in}} = \frac{S\cos\phi}{S\cos\phi + P_{loss}} = \frac{S\cos\phi}{S\cos\phi + P_{sc} + P_{oc}}$$
$$= \frac{2,250 \cdot 0.9}{2,250 \cdot 0.9 + 70} = 96.7\%. \tag{7.132}$$

7.5 Parameters of Single-Phase Transformers from Computer-Aided Tests [15–23]

Consider the single-phase transformer connection of Fig. 7.49. In this case

$$P_{loss} = P_{in} - P_{out} = \frac{1}{T}\int_0^T (v_1 i_1 - v'_2 i'_2)dt$$
$$= \frac{1}{T}\int_0^T [(v_1 + v'_2)(i_1 - i'_2)/2 + (v_1 - v'_2)(i_1 + i'_2)/2]dt, \tag{7.133}$$

the first part of (7.133) is

$$\frac{1}{T}\int_0^T (v_1 + v'_2)(i_1 - i'_2)/2dt = \frac{1}{T}\int_0^T p_{fe}(t)dt$$

and corresponds to the iron-core loss. The second part of (7.133) is

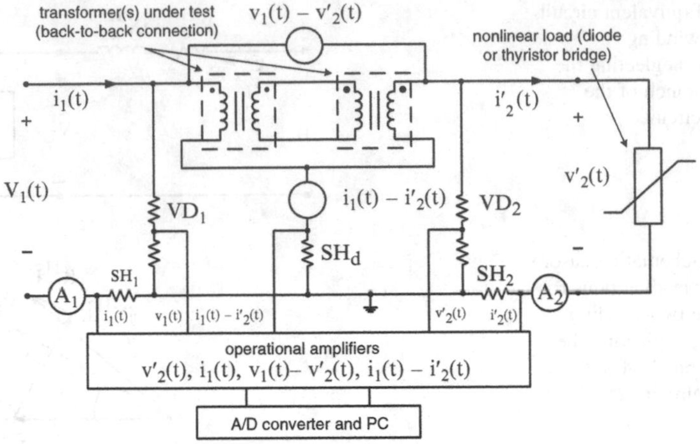

Fig. 7.49 Five-channel CAT (computer-aided testing) circuit and instrumentation for accurate single-phase transformer loss monitoring employing back-to-back method

$$\frac{1}{T}\int_0^T (v_1 - v'_2)(i_1 + i'_2)/2dt = \frac{1}{T}\int_0^T p_{cu}(t)dt$$

and corresponds to the copper loss.

From these operating-point dependent loss measurements one can now (see Sect. 7.3) compute the admittance of the magnetizing (cross) branch and the impedance of the series branch.

7.6 Summary

The application of Ampere's and Faraday's laws leads to the relations governing the ideal transformer. These relations can be applied to single-phase and poly-phase transformers with isolated primary and secondary windings including autotransformers and variacs. The ideal transformer model is augmented by resistance and reactance parameters which represents a more realistic transformer model. These parameters are derived from short-circuit and open-circuit tests of a transformer. An important application of transformers is their employment within power system feeders and to the study of power flow with multiple loads. Voltage regulation, power efficiency and energy efficiency are important concepts which maintain power quality [16] and conservation of energy. The application of autotransformers and variacs leads to highly efficient systems–albeit at the cost of electrical isolation and grounding issues, that is, the primary and secondary cannot have different ground potentials. Computer-aided tests permit the determination of operating point-dependent transformer losses which can be measured on-line.

7.7 Problems

Problem 7.1: *Ideal transformer with inductive/resistive load*

Calculate the primary current $\tilde{I}_p$ and the primary core flux Φ_{max_p} of the ideal transformer of Fig. 7.50.

Problem 7.2: *Impedance matching using an ideal transformer*

A certain power source will deliver maximum power when it is loaded with 980 Ω. The load is to be connected to the source through an ideal transformer having 350 primary turns. How many secondary turns should there be if the maximum power is to be dissipated in the load resistor of (a) 320 Ω, (b) 2,880 Ω?

Problem 7.3: *Resistive load fed by two ideal transformers*

Calculate the currents flowing in all branches if the two ideal transformers of Application Example 7.4 are loaded as illustrated in Fig. 7.51.

Problem 7.4: *Resistive load is fed by an ideal transformer*

Repeat the calculations of Problem 7.3 if the four coils of Fig. 7.51 are wound on a common core as shown in Fig. 7.52.

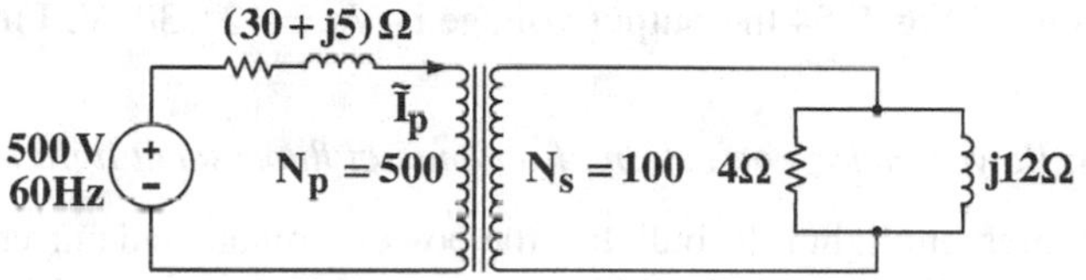

Fig. 7.50 Ideal transformer fed by non-ideal source and with inductive/resistive load

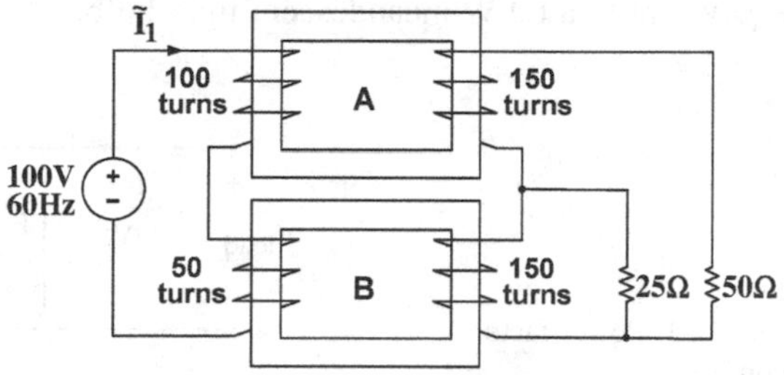

Fig. 7.51 Resistive load fed by two ideal transformers

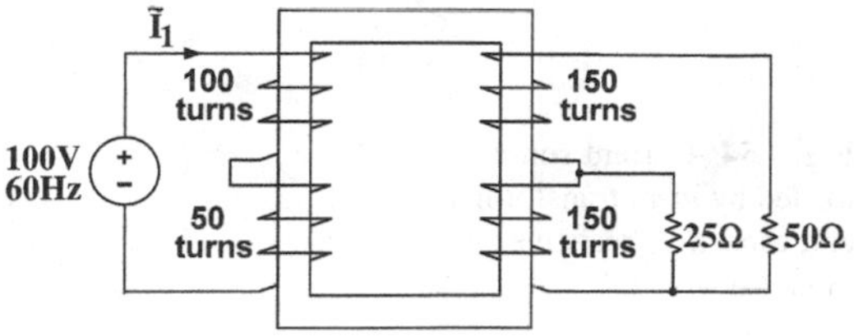

Fig. 7.52 Resistive load fed by an ideal transformer

Problem 7.5: *Use of a 60 Hz transformer at frequencies of f = 6 Hz and 600 Hz*

A f = 60 Hz single-phase transformer has the primary voltage rms rating of $V^{p}_{rat} = 240\,V$ with $a = (N_p/N_s) = 2$ at an apparent power rating of $S_{rat} = 10\,kVA$. It is available in the laboratory and the question arises whether it can be used for frequencies other than 60 Hz. Explain your answers to the following questions using basic transformer relations.

(a) What is the primary rated current I^{p}_{rat} of this 60 Hz transformer?
(b) Can this 60 Hz transformer be used at f = 6 Hz and a voltage rms value of $V^{p}_{6\,Hz} = 120\,V$?
(c) Can this 60 Hz transformer be used at f = 600 Hz and a voltage rms value of $V^{p}_{600\,Hz} = 120\,V$?

Problem 7.6: *Power-factor correction/compensation*

A single-phase load (see Fig. 7.53) is connected to a f = 60 Hz line and the load voltage is $\tilde{V}_{load} = 220\,V\angle 0°\,V$. The load consumes a (real) power of P = 25 kW at a power factor of $\cos\theta_{load} = 0.6$ lagging (consumer notation). Determine the capacitor value C needed to raise the load power factor to $\cos\theta_{load\_corrected} = 0.9$ lagging.

Problem 7.7: *Circuit with ideal transformer*

For the network of Fig. 7.54 the output voltage is $\tilde{V}_o = 12\angle 30°\,V$. Find the input current $\tilde{I}_s$.

Problem 7.8: *Power-factor correction of a compact fluorescent light bulb*

A compact fluorescent light (cfl) bulb has the power, voltage and current ratings of P = 14 W, $V_{rms} = 120\,V$, and $I_{rms} = 0.2\,A$, respectively, where the cfl-bulb terminal voltage is $v_{cfl}(t) = v_s(t) = \sqrt{2}\cdot V_{rms}\cos\omega t$, and the cfl-terminal current is $i_{cfl}(t) = \sqrt{2}\cdot I_{rms}\cos(\omega t - \Phi_{cfl})$. Note that from a light emission point of view the 14 W cfl-bulb is equivalent to a 60 W incandescent light bulb.

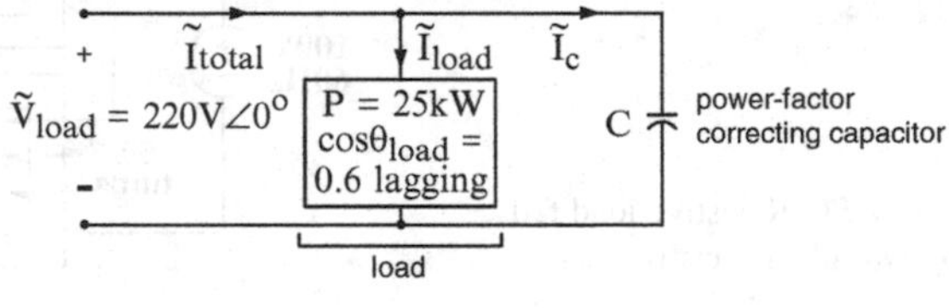

Fig. 7.53 Power-factor compensation

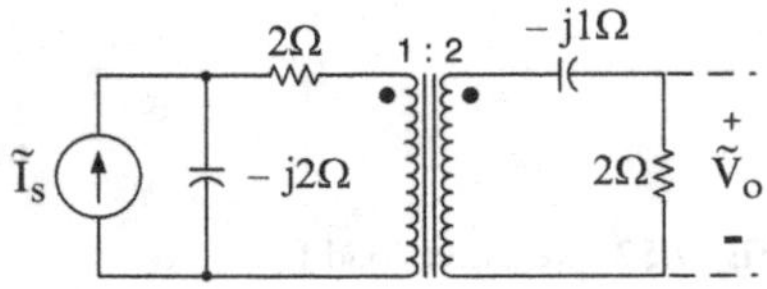

Fig. 7.54 Current source loaded by ideal transformer and capacitive/resistive impedance

(a) Compute the power factor of the cfl-bulb, that is, $\cos\Phi_{cfl}$, where $\tilde{i}_{cfl}(t)$ lags $\tilde{v}_{cfl}(t)$ by Φ_{cfl} degrees.
(b) The circuit of Fig. 7.55 can be relied on to improve the power factor as seen by the power system. Determine the value of the capacitance C such that the voltage $v_s(t)$ and the current $i_s(t)$ are in phase, that is, the power factor $\cos\Phi_s = 1.0$. The angle $\Phi_s = 0$ is the phase angle between $v_s(t)$ and $i_s(t)$. For your calculations you may assume a power system's frequency of f = 60 Hz.
(c) Draw the phasor diagram for the circuit of Fig. 7.55 showing in a qualitative manner (not to scale) $\tilde{v}_{cfl}(t)$, $\tilde{i}_{cfl}(t)$, $\tilde{i}_c(t)$, and $\tilde{i}_s(t)$.

Problem 7.9: *Residential power system, use of ideal transformer*

The AC power of a residence is fed from a single-phase transformer mounted on a pole in the alley ("pole transformer"). The pole transformer has the rated (nominal) data: apparent power S = 25 kVA, primary voltage $\tilde{V}_p = 7,200\angle 0°$V and secondary voltages $(2\tilde{V}_s) = 240\angle 0°$V and $\tilde{V}_s = 120\angle 0°$V as illustrated in Fig. 7.56.

Note that the secondary winding of the ideal transformer of Fig. 7.56 has a center tap. $(2\tilde{V}_s) = 240\angle 0°$ V supplies power to large appliances (e.g., dryer, washing machine) and the two $\tilde{V}_s = 120\angle 0°$ V circuits provide power for light bulbs and small appliances (e.g., electric clock, printer, computer).

(a) Calculate the primary current of the transformer $\tilde{I}_p$ for the resistive loads R = 25 Ω, $R_{top} = 50$ Ω and $R_{bottom} = 30$ Ω.
(b) Calculate the total power P consumed by the residence for the resistive loads of a).

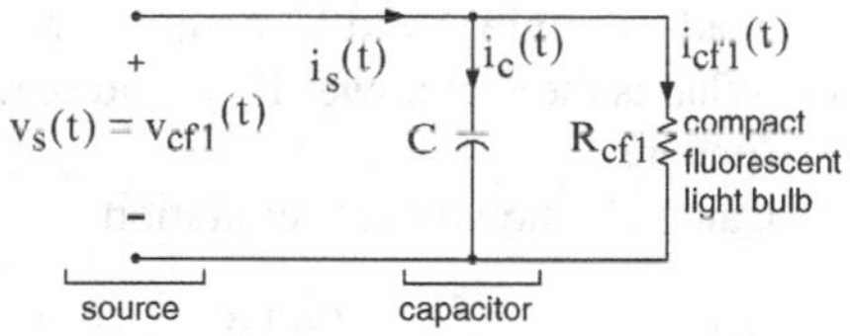

Fig. 7.55 Power-factor compensation of a compact fluorescent light bulb

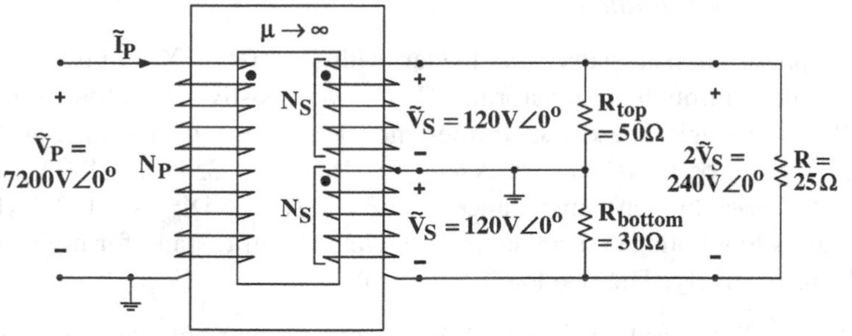

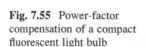

Fig. 7.56 Pole transformer approximated by ideal transformer model

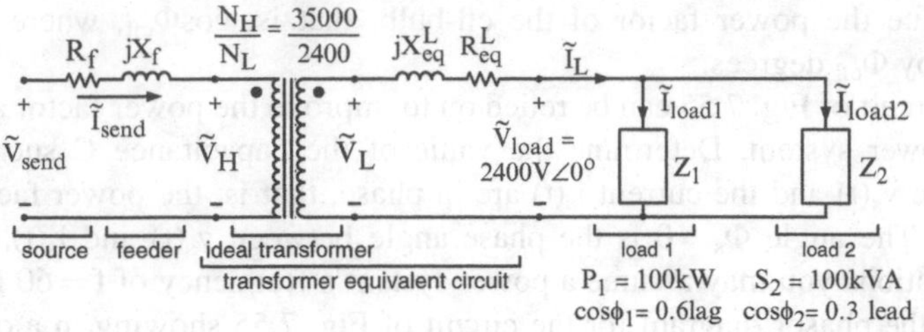

Fig. 7.57 Single-phase feeder with transformer and two single-phase loads (reference $\tilde{V}_{load} = 2,400\angle 0°V$)

Problem 7.10: *Power flow and voltage regulation (VR) of a single-phase feeder*

Two single-phase loads (see Fig. 7.57) are supplied through a 35 kV feeder whose impedance is $Z_f = (R_f + jX_f) = (100 + j300)\ \Omega$ and a $|\tilde{V}_H|/|\tilde{V}_L| = 35,000\ V/2,400\ V$ single-phase transformer whose equivalent (series-branch) impedance is $Z^L_{eq} = \left(R^L_{eq} + jX^L_{eq}\right) = (0.2 + j1.0)\ \Omega$ referred to its low-voltage (secondary) side. Note H and L stand for high side and low side, respectively. The two loads are $P_1 = 100$ kW at $\cos\phi_1 = 0.6$ lagging power factor (consumer notation) at $\tilde{V}_{load} = 2,400\ \angle 0°V$, and $S_2 = 100$ kVA at $\cos\phi_2 = 0.3$ leading power factor (consumer notation) at the same voltage of $\tilde{V}_{load} = 2,400\ \angle 0°V$.

(a) Compute $\tilde{I}_{load1}$.
(b) Compute $\tilde{I}_{load2}$.
(c) Determine the total load current $\tilde{I}_L = \tilde{I}_{load1} + \tilde{I}_{load2}$.
(d) Compute the voltage at the sending end of the feeder $\tilde{V}_{send}$ if the voltage at the load (receiving) end is maintained at $\tilde{V}_{load} = 2,400\ \angle 0°V$.
(e) What is the real power P_{send} and reactive power Q_{send} at the sending end of the feeder?
(f) Calculate the voltage regulation

$$VR = \frac{\{(\text{no}-\text{loadvoltage}) - (\text{full}-\text{loadvoltage})\} \cdot 100\%}{(\text{no–load voltage})}.$$

Problem 7.11: *Power flow along a distribution feeder supplying two loads with reactive power compensation*

A (single-phase) transmission line feeder with $Z_f = (R_f + jX_f) = (100 + j300)\ \Omega$ supplies power through a transformer T to two resistive/inductive loads (see Fig. 7.58). The single-phase transformer model consists of an ideal transformer with the voltage or turns ratio of $V_H/V_L = N_H/N_L = 35\,kV/2,400\,V$, and an equivalent (series-branch) impedance of $Z_{eq} = (R^L_{eq} + jX^L_{eq}) = (0.2 + j1.0)\ \Omega$ referred to its low-voltage (secondary) side. Note H and L stand for high side and low side, respectively. The two loads are as follows:

Load #1: $P_1 = 200$ kW at $\cos\Phi_1 = 0.60$ lagging (consumer notation) to a resistive-inductive load represented by the impedance Z_1.
Load #2: $S_2 = 300$ kVA at $\cos\Phi_2 = 0.80$ lagging (consumer notation) to a resistive-inductive load represented by the impedance Z_2.

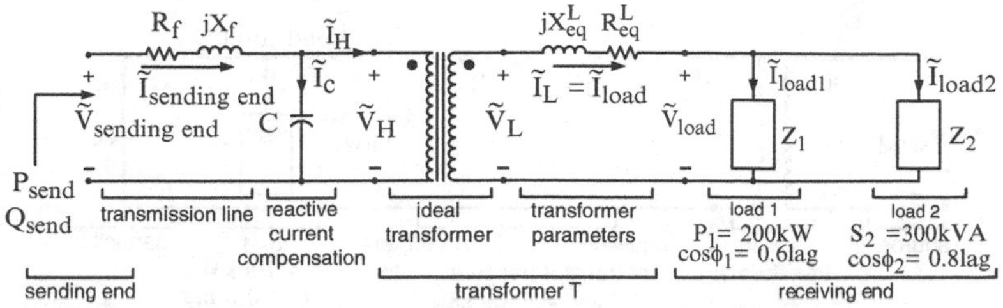

Fig. 7.58 Single-phase feeder supplying two resistive/inductive loads Z_1 and Z_2. To minimize the transmission line feeder current a reactive-current compensation at the primary of the transformer with capacitor C is employed

The resulting power factor angle Φ_H at the primary of the transformer T (angle between $\tilde{V}_H$ and $\tilde{I}_H$) is rather large and it is decided to connect a capacitor with capacitance $C = 1\ \mu F$ in parallel with the primary terminals of the transformer T (so that the total current at the sending end of the feeder $\tilde{I}_{sending\ end}$ lags the sending end voltage $\tilde{V}_{sending\ end}$ only by a few degrees).

(a) Assume that the load voltage is $\tilde{V}_{load} = 2,400\,V\,\angle 0°$ (reference voltage) compute the load currents $\tilde{I}_{load1}$, $\tilde{I}_{load2}$ and $\tilde{I}_{load}$.
(b) Compute the current at the primary side of the transformer $\tilde{I}_{II}$ and the voltage $\tilde{V}_{II}$.
(c) Compute the current through the capacitor C, that is $\tilde{I}_C$ (you may assume $\omega = 2\pi f$, $f = 60$ Hz).
(d) Compute the sending end current $\tilde{I}_{sending\ end} = \tilde{I}_H + \tilde{I}_C$.
(e) Compute the voltage $\tilde{V}_{sending\ end}$.
(f) What are the real power P_{send} and reactive power Q_{send} at the sending end?
(g) Draw a qualitative phasor diagram of load currents, capacitor current, sending-end current, load voltage, transformer voltage, and sending-end voltage for the given operating condition.

Problem 7.12: *Single-phase transformer supplying power to a resistive-inductive load Z with reactive power compensation using a capacitor C*

A single-phase transformer supplies power ($P = 100$ kW at $\cos\Phi = 0.60$ lagging, consumer notation) to a resistive-inductive load represented by the impedance Z as illustrated in Fig. 7.59. The power factor angle Φ (angle between $\tilde{V}_{load}$ and $\tilde{I}_{load}$) is rather large and it was decided to connect a capacitor with capacitance $C = 50\ \mu F$ in parallel with this resistive-inductive load so that the total current $\tilde{I}_{load\ total}$ lags the terminal voltage $\tilde{V}_{load}$ only by a few degrees. The single-phase transformer model consists of an ideal transformer with the voltage or turns ratio of $V_H/V_L = N_H/N_L = 35\,kV/2,400\,V$, and an equivalent (series-branch) impedance of $Z_{eq} = (R_{eq} + jX_{eq}) = (0.2 + j1.0)\,\Omega$ referred to its low-voltage (secondary) side. Note H and L stand for high side and low side, respectively.

(a) Assuming that the load voltage is $\tilde{V}_{load} = 2,400\,V\,\angle 0°$ (reference voltage), compute the load current $\tilde{I}_{load}$.

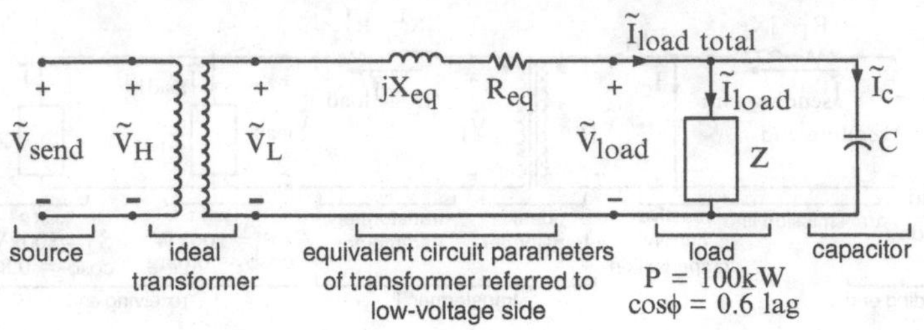

Fig. 7.59 Single-phase transformer supplying power to a resistive-inductive load Z with reactive power compensation using a capacitor C

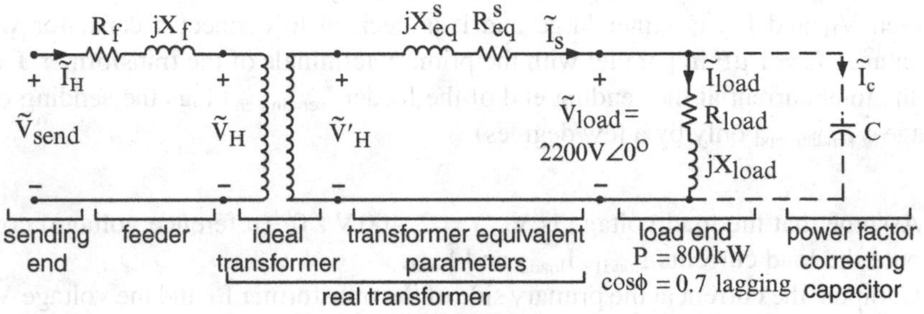

Fig. 7.60 Single-phase equivalent circuit of distribution feeder

(b) Compute the current through the capacitor C, that is $\tilde{I}_C$ (You may assume $\omega = 2\pi f$, $f = 60$ Hz).
(c) Compute the total load current $\tilde{I}_{loadtotal} = \tilde{I}_{load} + \tilde{I}_C$.
(d) Compute the voltage $\tilde{V}_{send}$, if the voltage at the load (receiving) end is maintained at $\tilde{V}_{load} = 2{,}400\,\text{V}\,\angle 0°$.
(e) Compute the real power P_{send} and reactive power Q_{send} at the source.
(f) Draw a qualitative phasor diagram for the given operating condition.

Problem 7.13: *Single-phase feeder supplying power via (real) transformer to load and power factor compensation at the load*

A balanced three-phase utility feeder can be represented by the single-phase equivalent circuit of Fig. 7.60.

In Fig. 7.60 the single-phase load is supplied through a (7.2 kV, 60 Hz) feeder whose impedance is $Z_f = (R_f + jX_f) = (1 + j6)\,\Omega$ and a $N_p/N_s = V_H/V'_H = 7.2\,\text{kV}/2{,}400\,\text{V}$ single-phase transformer whose series branch impedance, referred to its low-voltage side, is $Z^s_{eq} = (R^s_{eq} + jX^s_{eq}) = (0.4 + j1.0)\,\Omega$.

The load is $P = 800$ kW at $\cos\Phi = 0.7$ lagging power factor (consumer notation) at $\tilde{V}_{load} = \tilde{V}_L = 2{,}200\,\angle 0°$ V.

(a) Compute the load current $\tilde{I}_{load} (= \tilde{I}_s)$ ignoring the capacitor $C = 400\,\mu$F.
(b) Compute $\tilde{V}_H$.

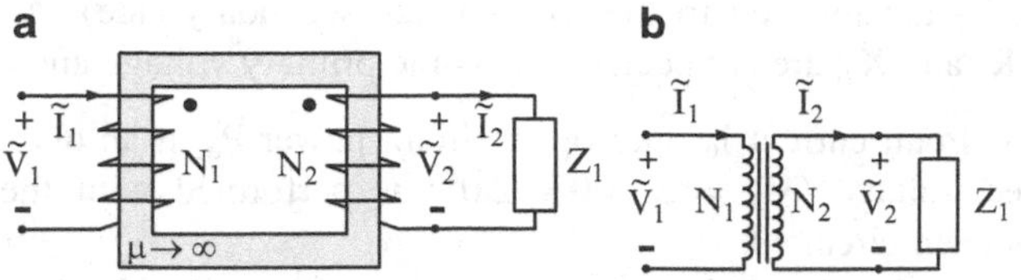

Fig. 7.61 Two-phase transformer; (**a**) actual connection, (**b**) schematic connection

Fig. 7.62 Auto-transformer; (**a**) actual connection, (**b**) schematic connection

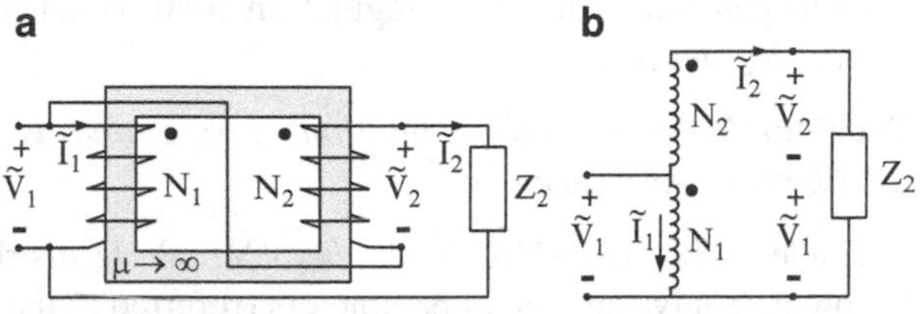

(c) Compute the sending end voltage $\tilde{V}_{send}$.
(d) Compute the real power P_{send} and reactive power Q_{send} at sending end of the feeder.
(e) Repeat your calculations with the capacitor $C = 400\ \mu F$ connected to the load.

Problem 7.14: *Ideal single-phase transformer used in auto-transformer connection*

Figure 7.61(a) shows the magnetic circuit of a single-phase pole transformer with $|\tilde{V}_1| = 7,200$ V and $|\tilde{V}_2| = 240$ V feeding a load with apparent power of $S_{2two\text{-}winding} = 25$ kVA at $\cos\Phi_2 = 0.8$ lagging (consumer notation). Fig. 7.61(b) illustrates the same circuit in a symbolic manner.

(a) Find $\tilde{I}_2$ and $\tilde{I}_1$.
(b) Figure 7.62(a) depicts the magnetic circuit of the same transformer connected as an auto-transformer; note that the primary and secondary windings are connected with two conducting wires. An auto-transformer can deliver much more power than the two-winding transformer of Fig. 7.61(a), where the two windings are electrically insulated.

Find the apparent output power $S_{2auto\text{-}transformer}$ of Figs. 7.62a, b for the condition where the currents flowing in the primary ($|\tilde{I}_1|$) and secondary ($|\tilde{I}_2|$) windings are the same as in Figs. 7.61a, b. This will ensure that the auto-transformer is not being overheated (rated losses exist) although it can deliver much more output apparent power to the load Z_2 than the two-winding transformer to Z_1 (Figs. 7.61a, b). This implies that the voltages applied to the windings are not changed either.

Problem 7.15: *Voltage regulation of a single-phase transformer*

A 15 kVA, 2,200 V/440 V, 60 Hz single-phase transformer has the following equivalent circuit parameters: $R_1 = 5\ \Omega$, $X_1 = 6\ \Omega$, $R_2 = 0.2\ \Omega$, $X_2 = 0.24\ \Omega$, $R_c = 15\ k\Omega$ and $X_m = 20\ k\Omega$ (R_1, X_1, R_c, X_m are referred to the high-voltage primary side and R_2, X_2 are referred to the low-voltage secondary side). Assume shunt components R_c and X_m are connected across the primary voltage and compute:

(a) The open-circuit current $\tilde{I}_{oc}$ and open-circuit power P_{oc} if an open-circuit test with rated voltage ($\tilde{V}_{oc} = 2,200\ \text{V} \angle 0°$) is performed with the secondary terminals open-circuited.
(b) The short-circuit voltage $\tilde{V}_{sc}$ and corresponding power P_{sc}, if a short-circuit test with rated current ($\tilde{I}_{sc} = 6.82 \angle 0°$) is performed with the secondary terminals short-circuited.
(c) The percentage of voltage regulation at 80% load at unity power factor neglecting R_c and X_m.

Problem 7.16: *Voltage regulation, losses and efficiency of a single-phase distribution transformer*

A single-phase, 150 kVA, 3.3 kV/415 V, 50 Hz distribution transformer has the following equivalent circuit parameters (referred to the high-voltage primary side): $R_{eq} = 1.57\ \Omega$, $X_{eq} = 3.68\ \Omega$, $R_c = 9.3\ \text{k}\Omega$, and $X_m = 1.5\ \text{k}\Omega$. The output (secondary) voltage is $\tilde{V}_s = 415\ \text{V} \angle 0°$ when a load of $Z_L = (1.0 + j\,0.75)\ \Omega$ is connected across the secondary terminals.

(a) Draw the approximate equivalent circuit (including the load) referred to the primary (assume shunt components are connected at the front end of the equivalent circuit).
(b) Find the load current $\tilde{I}_s'$ (referred to the primary) and load power factor $\cos\Phi$.
(c) Calculate the primary voltage $\tilde{V}_p$.
(d) Compute the percentage of voltage regulation %VR.
(e) Use the equivalent circuit to compute copper losses P_{copper}, core losses P_{core} and the efficiency η of transformer.

Problem 7.17: *Calculation of circuit parameters from open-circuit and short-circuit tests*

A single phase distribution transformer has the following specifications: Apparent power rating = 150 kVA, primary voltage $V_p = 3.3$ kV, secondary voltage $V_s = 415$ V, and frequency f = 50 Hz. Open-circuit and short-circuit test results are as follows. Open-circuit test: $V_{oc} = 3{,}300$ V, $I_{oc} = 2.27$ A, $P_{oc} = 1.17$ kW; short-circuit test: $V_{sc} = 182$ V, $I_{sc} = 45.45$, $P_{sc} = 3.24$ kW.

(a) Determine parameters R_c, X_m, R_{eq} and X_{eq}, all referred to the primary.
(b) Draw the approximate equivalent circuit referred to the primary. Assume the shunt components (R_c, X_m) are connected across the primary voltage.
(c) If the output (secondary) voltage is $\tilde{V}_s = 415\ \text{V} \angle 0°$ when a load of $Z_L = (1.0 + j\,0.75)\ \Omega$ is connected across the secondary terminals, calculate the primary voltage $\tilde{V}_p$.
(d) Use the equivalent circuit to compute losses and efficiency of transformer when supplying the load of part c).

Problem 7.18: *Efficiency of single-phase transformer as a function of power factor*

A 25 kVA, 3,300 V/440 V, 50 Hz, single phase transformer has high voltage winding resistance and leakage reactance of $R_1 = 4.53\ \Omega$ and $X_1 = 17.11\ \Omega$, respectively. The resistance and leakage reactance of the low voltage winding are $R_2 = 0.0183\ \Omega$ and $X_2 = 0.190\ \Omega$. The core-loss and magnetizing components $R_c = 53.4\ k\Omega$ and $X_m = 8.825\ k\Omega$ are referred to the high-voltage side.

(a) Draw the equivalent circuit and show values of all parameters referred to the low voltage side.
(b) Determine the voltage regulation at full load with a power factor of 0.85 lagging (consumer notation).
(c) Determine the voltage regulation at full load with a power factor of 0.85 leading (consumer notation).
(d) Compute the efficiency at full load with a power factor of 0.85 lagging.
(e) Compute the efficiency at full load with a power factor of 0.85 leading.

References

1. Stensland, T.; Fuchs, E. F.; Grady, W. M.; Doyle, M.: "Modeling of magnetizing and core-loss currents in single-phase transformers with voltage harmonics for use in power flow," *IEEE Trans. on Power Delivery*, Vol. 12, No. 2, April 1997, pp. 768–774
2. Stensland, T. D.: *Effects of Voltage Harmonics on Single-Phase Transformers and Induction Machines Including Pre-Processing for Power Flow*, M.S. Thesis, University of Colorado, Boulder, May 1995.
3. Fuchs, E. F., Masoum, M. A. S., Roesler, D. J.: "Large signal nonlinear model of anisotropic transformers for nonsinusoidal operation, Part I: (λ – i) characteristic," *IEEE Trans. on Power Delivery*, Jan. 1991, Vol. TPWRD-6, pp. 174–186.
4. Masoum, M. A. S.; Fuchs, E. F.; Roesler, D. J.: "Large signal nonlinear model of anisotropic transformers for nonsinusoidal operation, Part II: Magnetizing and core-loss currents," *IEEE Trans. on Power Delivery*, Oct. 1991, Vol. TPWRD-6, pp. 1509–1516.
5. Masoum, M. A. S.; Fuchs, E. F.; Roesler, D. J.: "Impact of nonlinear loads on anisotropic transformers," *IEEE Trans. on Power Delivery*, Oct.1991, Vol. TPWRD-6, pp. 1781–1788.
6. Masoum, M. A. S.: *Generation and Propagation of Harmonics in Power System Feeders Containing Nonlinear Transformers and Loads,* Ph.D. Dissertation, University of Colorado, Boulder, April 1991.
7. Chowdhury, A. H.; Grady, W. M.; Fuchs, E. F.: "An investigation of harmonic characteristics of transformer excitation current under nonsinusoidal supply voltage," *IEEE Trans. on Power Delivery*, Vol. 14, No. 2, April 1999, pp. 450–458
8. Fuchs, E. F.: "Transformers, liquid filled," *Encyclopedia of Electrical and Electronics Engineering online*, http://www.interscience.wiley.com:38/eeee/32/6132/w.6132-toc.html, Sept. 1, 2000, J. Webster (ed.), John Wiley & Sons, 605 Third Avenue, New York, NY 10158, paper No. 934C, 40 pages.
9. Fitzgerald, A. E.; Kingsley, Jr., C.; Umans, S.D.: *Electric Machinery,* 5th Edn, McGraw-Hill Publishing Company, New York, 1990
10. IEEE, *IEEE Recommended Practice for Electric Power Systems in Commercial Buildings (Gray Book)*, Published by the Institute of Electrical and Electronics Engineers, Inc., IEEE Std. 241–1974.

11. IEEE, *IEEE Recommended Practice for Electric Power Distribution for Industrial Plants (Red Book)*, Published by the Institute of Electrical and Electronics Engineers, Inc., IEEE Std. 141–1976.
12. IEEE, *IEEE Recommended Practice for Grounding of Industrial and Commercial Power systems (Green Book)*, Published by the Institute of Electrical and Electronics Engineers, Inc., IEEE Std. 142–1972.
13. IEEE, *IEEE Recommended Practice for Power System Analysis (Brown Book)*, Published by the Institute of Electrical and Electronics Engineers, Inc., IEEE Std. 399–1980.
14. IEEE, *IEEE Recommended Practice for Design of Reliable Industrial and Commercial Power Systems (Gold Book)*, Published by the Institute of Electrical and Electronics Engineers, Inc., IEEE Std. 493–1980.
15. Yildirim, D.; Fuchs, E. F.; Batan, T.: "Measurement of derating of distribution transformers at any (non) linear load," *Proceedings of the ICEM'98 International Conference on Electrical Machines*, Istanbul, Turkey, 2–4 Sept., 1998.
16. Fuchs, E. F.; Masoum, M. A. S.: *Power Quality in Power Systems and Electrical Machines*, Elsevier/Academic Press, Feb. 2008, 638 pages. ISBN: 978-0-12-369536-9.
17. Fuchs, E. F.; Fei, R.: "A new computer-aided method for the efficiency measurement of low-loss transformers and inductors under nonsinusoidal operation," *IEEE Trans. on Power Delivery*, January 1996, Vol. PWRD-11, No. 1, pp. 292–304.
18. Lin, D.; Fuchs, E. F.; Doyle, M.: "Computer-aided testing program of electrical apparatus supplying nonlinear loads," *IEEE Trans. on Power Systems*, Vol. 12, No. 1, Feb. 1997, pp. 11–21.
19. Fuchs, E.F.; Yildirim, D.; Batan, T.: "Innovative procedure for measurement of losses of transformers supplying nonsinusoidal loads," *IEE Proceedings - Generation, Transmission and Distribution*, Vol. 146, No. 6, Nov.1999, IEE Proceedings online no. 19990684.
20. Fuchs, E. F.; Yildirim, D.; Grady, W. M.: "Measurement of eddy-current loss coefficient P_{EC-R}, derating of single-phase transformers, and comparison with K-factor approach," *IEEE Trans. on Power Delivery*, Vol. 15, No. 1, Jan. 2000, pp. 148–154
21. Fuchs, E. F.; Yildirim, D.; Grady, W. M.: Corrections to "Measurement of eddy-current loss coefficient P_{EC-R}, derating of single-phase transformers, and comparison with K-factor approach," *IEEE Trans. on Power Delivery,* Vol. 15, Issue 4, Oct. 2000 Page(s): 1357–1357
22. Yildirim, D.; Fuchs, E. F.: "Measured transformer derating and comparison with harmonic loss factor F_{HL} approach," *IEEE Trans. on Power Delivery*, Vol. 15, No. 1, January 2000, pp. 186–191
23. Yildirim, D.; Fuchs, E. F.: Corrections to "Measured transformer derating and comparison with harmonic loss factor F_{HL} approach," *IEEE Trans. on Power Delivery,* Vol. 15, Issue 4, Oct. 2000 Page(s): 1357–1357

Chapter 8
Three-Phase Power System and Three-Phase Transformers

Single-phase systems are adequate for residential applications up to an apparent power S of 5 kVA (per residence), and 25 kVA single-phase pole transformers supply power to about 5–10 residences. In rural areas irrigation applications require single-phase motor ratings of up to 50 kVA. Above this rating three-phase power equipment must be employed for commercial and industrial applications.

8.1 Three-Phase AC Systems

Generation, transmission and heavy power utilization is based on three-phase systems, introduced more than 100 years ago [1] following the invention of the two-phase (or poly-phase) induction motor by Tesla [2]. Although Edison proposed DC transmission at that time, it was only 50 years later when the first DC transmission lines were commissioned. Due to stability problems associated with transmission of AC electrical power over distances of more than 1,000 miles, the utilization of high-voltage DC transmission tie lines became popular and cost-effective (see Chap. 3).

8.1.1 Generation of Three-Phase Voltages

Consider the elementary 3-phase, 2-pole generator of Fig. 8.1; on the armature (member, where voltages are induced) there are three coils (phase belts) aa′, bb′, and cc′ , whose axes are displaced 120° in space from each other.

Note that this text uses in this section clockwise (cw) notation for the phase (belt) sequence abc of a three-phase system. When the field winding residing on the rotor (moving member) of Fig. 8.1 is excited and rotated, voltages will be induced in the three phase belts abc in accordance with Faraday's law. If the field winding structure is so designed that the flux ϕ is distributed sinusoidally around the rotor periphery (θ), the flux linkages of phase belt aa' will vary sinusoidally with time

E.F. Fuchs and M.A.S. Masoum, *Power Conversion of Renewable Energy Systems*,
DOI 10.1007/978-1-4419-7979-7_8, © Springer Science+Business Media, LLC 2011

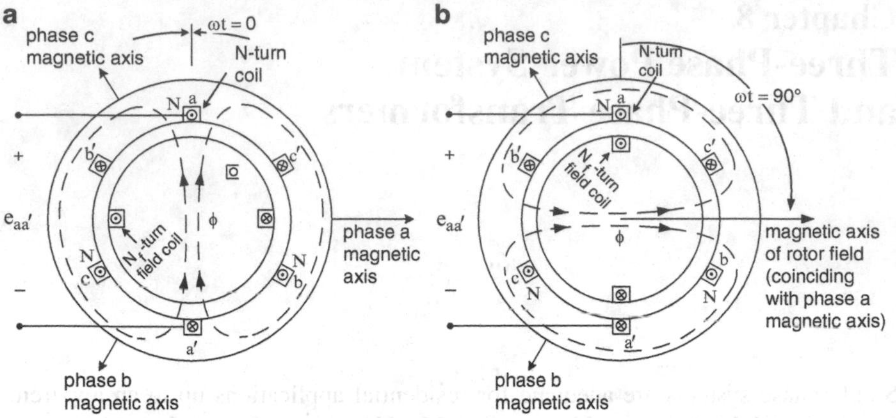

Fig. 8.1 Three-phase, two-pole generator with two rotor positions; (**a**) $\omega t = 0°$, (**b**) $\omega t = 90°$

Fig. 8.2 Three-phase voltage system $e_{aa'}$ (t), $e_{bb'}$ (t), and $e_{cc'}$ (t) in time domain

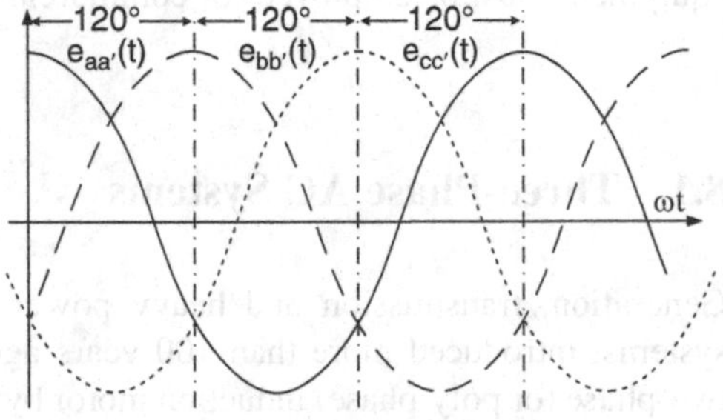

$$\lambda_{aa'} = N\phi(t)\sin\omega t \tag{8.1}$$

and according to Faraday's law the induced voltage in phase belt aa′ is

$$\begin{aligned} e_{aa'}(t) &= d\lambda_{aa'}/dt = \omega N\phi\cos\omega t + N(d\phi/dt)\sin\omega t \\ &= \text{speed voltage} + \text{transformer voltage}. \end{aligned} \tag{8.2}$$

The cosinusoidal speed voltage (flux ϕ is stationary with respect to rotating rotor, and does not change in time with respect to rotor)

$$e_{aa'}(t) = N\phi\omega\cos\omega t = E_{max}\cos\omega t \tag{8.3}$$

will be induced in phase belt aa′ (Fig. 8.2). Because phase belt bb′ is displaced by 120 mechanical degrees (for a 2-pole configuration) from phase belt aa′ one concludes that whatever happens to phase belt aa' at time $t = 0$ corresponding to a time angle of $\omega t = 0$ will happen to phase belt bb′ after a time of $t = 120°/\omega$ or when a time angle of $\omega t = 120°$ has elapsed; the same happens to the voltage induced in phase belt cc′. As shown below, these three voltage waves $e_{aa'}(t)$, $e_{bb'}(t)$ and $e_{cc'}(t)$ will be displaced 120° electrical degrees in time as a result of the phase belts being displaced by 120° mechanical degrees in space. Note, for a two-pole (p = 2) machine the mechanical degrees are identical with the electrical degrees. This is not so for a machine with pole numbers larger than two (e.g., p = 4, 6, 8...), as will be shown in Chap. 10.

The sinusoidal time-domain voltages $e_{aa'}(t)$, $e_{bb'}(t)$, and $e_{cc'}(t)$ can be represented in the phasor domain (see Fig. 8.3) as

$$\tilde{e}_{aa'}(t) = \sqrt{2}Ee^{j0} = \sqrt{2}E\angle 0°, \tag{8.4a}$$

$$\tilde{e}_{bb'}(t) = \sqrt{2}Ee^{-j120} = \sqrt{2}E\angle - 120°, \tag{8.4b}$$

$$\tilde{e}_{cc'}(t) = \sqrt{2}Ee^{-j240} = \sqrt{2}E\angle - 240° = \sqrt{2}E\angle + 120°, \tag{8.4c}$$

where $E = N\phi\omega/\sqrt{2}$ is the rms value of the induced phase voltage [see (8.3)].

A three-phase AC system employs voltage sources (see Fig. 8.4) which consist of three voltages equal in amplitude and displaced by electrical phase angles of

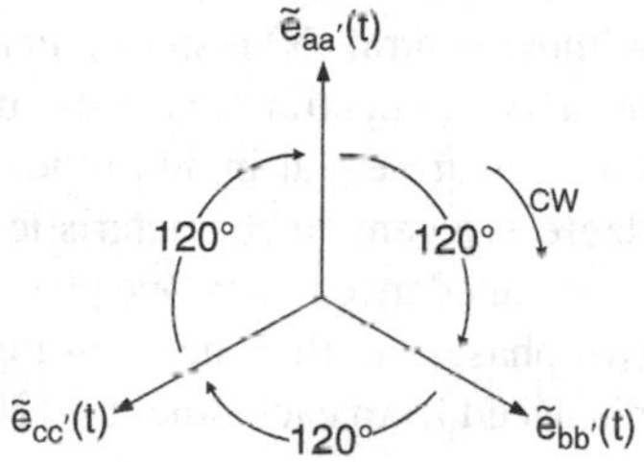

Fig. 8.3 Phasor diagram of induced three-phase voltage system

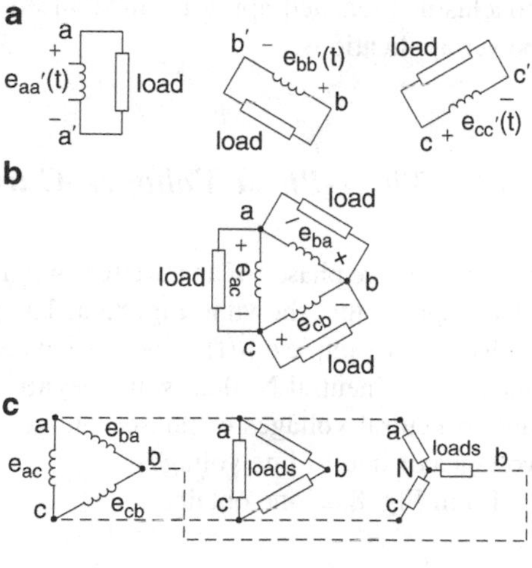

Fig. 8.4 **(a–c)** Three-phase voltage sources and loads; **(a)** Three separate systems, **(b)** three-phase system in Δ-connection, where a′ = b, b′ = c, c′ = a, and **(c)** Δ three-phase system with Δ and Y-connected loads, where N = a′ = b′ = c′

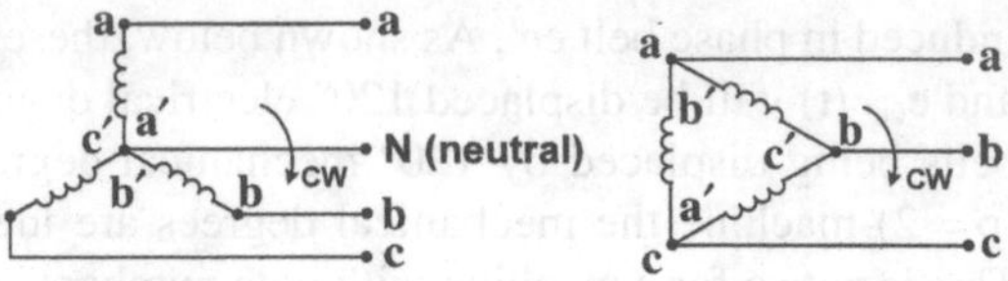

Fig. 8.5 Y and Δ configurations for armature windings representing voltage sources

360°/3 = 120° from one another. A three-phase system requires the least amount of copper/aluminum for the transmission lines to transmit a given amount of power from point A to point B.

The three individual voltages of a three-phase source may each be connected to its own independent load circuit. We would have then three separate single-phase systems (Fig. 8.4a). Alternatively, symmetrical electric connections can be made between the three voltages and the associated load circuitry to form a three-phase system (Figs. 8.4b, c). Note that the word "phase" now has two distinct meanings:

- It may refer to a portion of a three-phase system or circuit or
- In the familiar steady-state circuit theory, it may be used in reference to the angular displacement or phase shift between voltage and current phasors.

The three phases of the armature winding (representing voltage sources) may be interconnected in two possible ways, as shown in Fig. 8.5: in a Y-connection with or without neutral (N) resulting in a 4-wire or 3-wire three-phase system, respectively, or a Δ connection resulting in a 3-wire three-phase system. The three-phase voltages are equal in amplitude and are phase-displaced in time by 120 electrical degrees, a general characteristic of a balanced three-phase system. Furthermore, the load impedance in any one phase (see Fig. 8.4c) is equal to that in either of the other two phases, so that the resulting phase currents are equal in amplitude and phase displaced from each other by 120 electrical degrees. Likewise, equal real power and equal reactive power flow in each phase exist. It is important to note that only balanced systems are treated in this section and none of the methods developed or conclusions reached apply to unbalanced systems as they may occur in real-life power applications.

8.1.2 Three-Phase Voltages, Currents and Power

When the three phases of an armature winding are Y-connected, the phasor diagram of voltages is that shown in Fig. 8.6a. The polarity of the voltages is defined by the indices, for example $\tilde{e}_{aN}(t) = -\tilde{e}_{Na}(t)$, where "a" is the positive terminal of phase a, and N is the neutral N, that is, the negative terminal: $\tilde{e}_{aN}(t)$, $\tilde{e}_{bN}(t)$, $\tilde{e}_{cN}(t)$ are the line-to-neutral voltages or phase voltages, and $\tilde{e}_{ab}(t)$, $\tilde{e}_{bc}(t)$, $\tilde{e}_{ca}(t)$ are called line voltages or line-to-line voltages.

From Fig. 8.6a one obtains

$$|\tilde{e}_{ac}(t)| = 2|\tilde{e}_{cN}(t)|\sin 60^{\circ} = 2|\tilde{e}_{cN}(t)|\frac{\sqrt{3}}{2} = \sqrt{3}|\tilde{e}_{cN}(t)|, \quad (8.5)$$

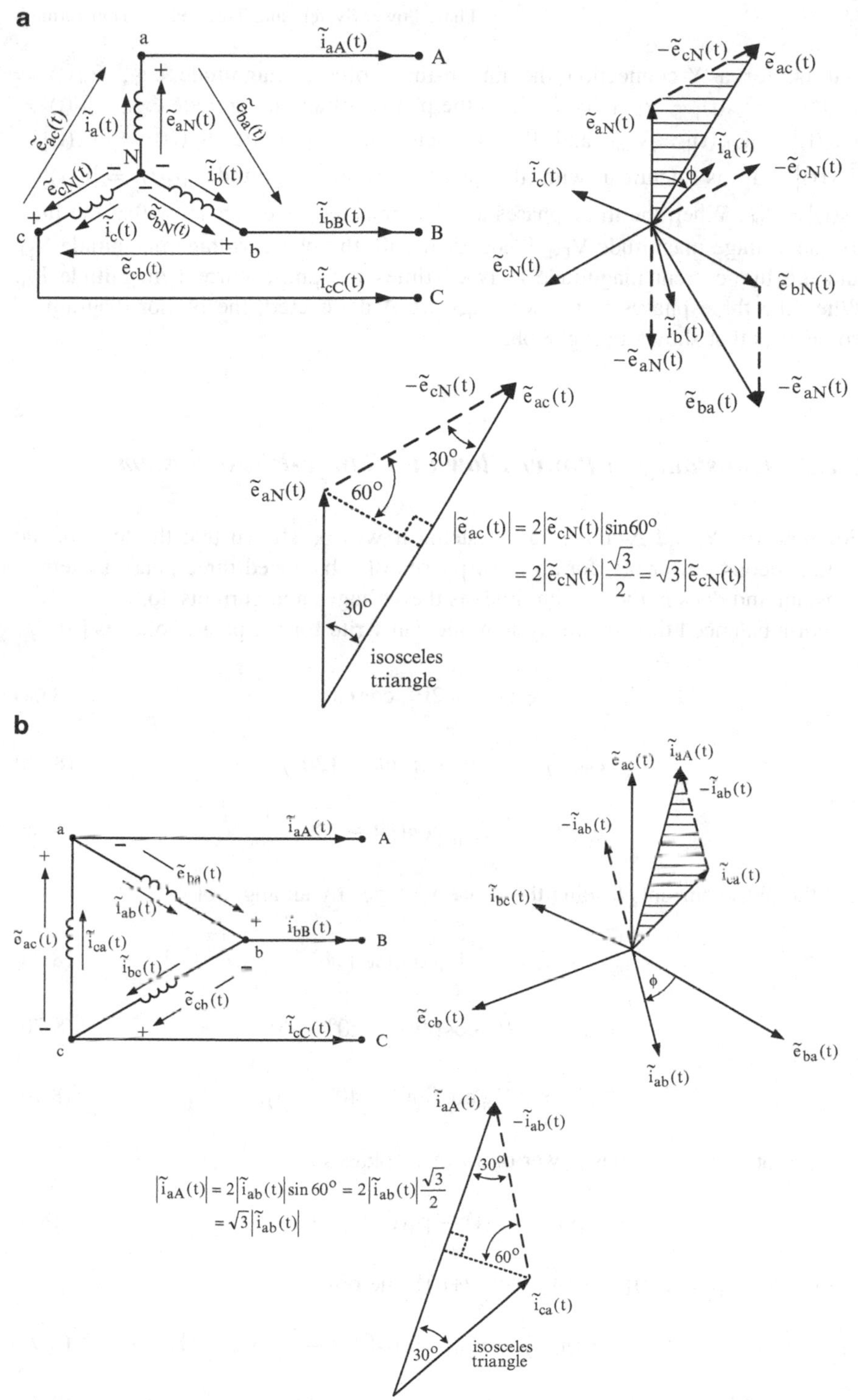

Fig. 8.6 (**a**) Phasor diagram of phase (e.g., $\tilde{e}_{aN}(t)$) voltages, line-to-line (e.g.,$\tilde{e}_{ba}(t)$) voltages, and phase (line) (e.g.,$\tilde{i}_a(t) = \tilde{i}_{aA}(t)$) currents for Y-connection (**b**) Phasor diagram of phase (line-to-line) (e.g.,$\tilde{e}_{ac}(t)$) voltages and phase (e.g., $\tilde{i}_{ab}(t)$) and line (e.g., $\tilde{i}_{aA}(t)$) currents for Δ-connection

that is, for a Y-connection the line-to-line voltage magnitude e.g., $|\tilde{e}_{ac}(t)| = |\tilde{e}_{ba}(t)| = |\tilde{e}_{cb}(t)| = V_{L-L}$, is $\sqrt{3}$ times the phase voltage magnitude e.g., $|\tilde{e}_{aN}(t)| = |\tilde{e}_{bN}(t)| = |\tilde{e}_{cN}(t)| = V_{ph}$, and the line current magnitude $|\tilde{i}_{aA}(t)| = |\tilde{i}_{bA}(t)| = |\tilde{i}_{cA}(t)| = I_L$ is identical with the phase current magnitude $|\tilde{i}_a(t)| = |\tilde{i}_b(t)| = |\tilde{i}_c(t)| = I_{ph}$. When the three phases are Δ-connected, one can show that the line-to-line voltage magnitude $V_{L\text{-}L}$ is identical with the phase voltage magnitude V_{ph} and the line-current magnitude I_L is $\sqrt{3}$ times the phase current magnitude I_{ph}. When the three phases of the windings are Δ-connected, the phasor diagram of voltages is that shown in Fig. 8.6b.

8.1.3 Constancy of Power Flow in a Three-Phase System

For both the Y and Δ- connected systems it will be shown that the total of the instantaneous power p(t) for all three phases of a balanced three-phase system is constant and does not vary with time, as the voltages and currents do.

For a balanced three-phase system one can write for the phase voltages [3]

$$e_{aN}(t) = \sqrt{2}E_{ph}\cos\omega t, \tag{8.6a}$$

$$e_{bN}(t) = \sqrt{2}E_{ph}\cos(\omega t - 120°), \tag{8.6b}$$

$$e_{cN}(t) = \sqrt{2}E_{ph}\cos(\omega t - 240°), \tag{8.6c}$$

and the phase currents leading the phase voltages by an angle of ϕ

$$i_a(t) = \sqrt{2}I_{ph}\cos(\omega t + \phi), \tag{8.7a}$$

$$i_b(t) = \sqrt{2}I_{ph}\cos(\omega t - 120° + \phi), \tag{8.7b}$$

$$i_c(t) = \sqrt{2}I_{ph}\cos(\omega t - 240° + \phi). \tag{8.7c}$$

The total instantaneous power of all three phases is

$$p(t) = p_a(t) + p_b(t) + p_c(t), \tag{8.8}$$

with $\cos\alpha\cos\beta = (1/2)\{\cos(\alpha\text{-}\beta)+\cos(\alpha+\beta)\}$ one obtains

$$p_a(t) = e_{aN}(t)i_a(t) = E_{ph}I_{ph}\{\cos(2\omega t + \phi) + \cos\phi\}, \tag{8.9a}$$

$$p_b(t) = e_{bN}(t)i_b(t) = E_{ph}I_{ph}\{\cos(2\omega t + \phi - 240°) + \cos\phi\}, \tag{8.9b}$$

$$p_c(t) = e_{cN}(t)i_c(t) = E_{ph}I_{ph}\{\cos(2\omega t + \phi - 480°) + \cos\phi\}, \tag{8.9c}$$

or

$$p(t) = 3E_{ph}I_{ph}\cos\phi + E_{ph}I_{ph}\{\cos(2\omega t + \phi) + \cos(2\omega t + \phi - 240°) + \cos(2\omega t + \phi - 480°)\}. \tag{8.10a}$$

It will be shown below that the second term of (8.10a) adds up to zero!

Proof:

$$\{\cos(2\omega t + \phi) + \cos(2\omega t + \phi - 240°) + \cos(2\omega t + \phi - 480°)\} = 0, \tag{8.10b}$$

with $\cos(\alpha + \beta) = \cos\alpha\cos\beta - \sin\alpha\sin\beta$ and
$\cos(\alpha + \beta + \gamma) = \cos\alpha\cos\beta\cos\gamma - \sin\alpha\sin\beta\cos\gamma - \sin\alpha\cos\beta\sin\gamma - \cos\alpha\sin\beta\sin\gamma$

Equation 8.10b becomes in an expanded manner:
$\cos2\omega t\cos\phi - \sin2\omega t\sin\phi + \cos2\omega t \ \cos\phi\cos120° - \sin2\omega t \ \sin\phi \ \cos120° - \sin2\omega t \ \cos\phi \ \sin120° - \cos2\omega t \ \sin\phi \ \sin120° + \cos2\omega t \ \cos\phi \ \cos120° - \sin2\omega t \ \sin\phi\cos120° + \sin2\omega t \ \cos\phi \ \sin120° + \cos2\omega t \ \sin\phi \ \sin120° = 0$,

or with $\cos120° = -(1/2)$:

$\cos2\omega t \ \cos\phi - \sin2\omega t \ \sin\phi - (1/2)\cos2\omega t \ \cos\phi + (1/2)\sin2\omega t \ \sin\phi - (1/2)\cos2\omega t\cos\phi + (1/2)\sin2\omega t\sin\phi = 0$.

This shows that the time-dependent powers of a balanced three-phase system add up to zero and the power flow within such a system is

$$P = p(t) = 3E_{ph}I_{ph}\cos\phi = \text{independent of time!} \tag{8.11}$$

The total instantaneous power for a balanced three-phase system is constant and is equal to 3 times the average power per phase. This is of particular advantage in the operation of three-phase (or poly-phase) motors, for example, because it means that the shaft-power output is constant and that no torque pulsations arise due to the time-dependent sinusoidal variations of AC currents and voltages.

In the following [4] the time-dependent power relations of single, two, and three-phase systems will be visualized. Figures 8.7a, b illustrate the voltage, current and power at unity-power factor, and lagging power factor (consumer notation), e.g., $\cos\Phi = 0.707$ corresponding to $\Phi = -45°$, of a single-phase system, respectively. Figure 8.8 shows the voltage, current and power at unity-power factor of a two-phase system. Figures 8.9a, b depict the voltage, current and power at unity-power factor, and lagging power factor (consumer notation), e.g., $\cos\Phi = 0.80$ corresponding to $\Phi = -36.87°$, of a three-phase system, respectively. The influence of a 5% imbalance is studied in Fig. 8.10 and that of loss of phase c in Fig. 8.11.

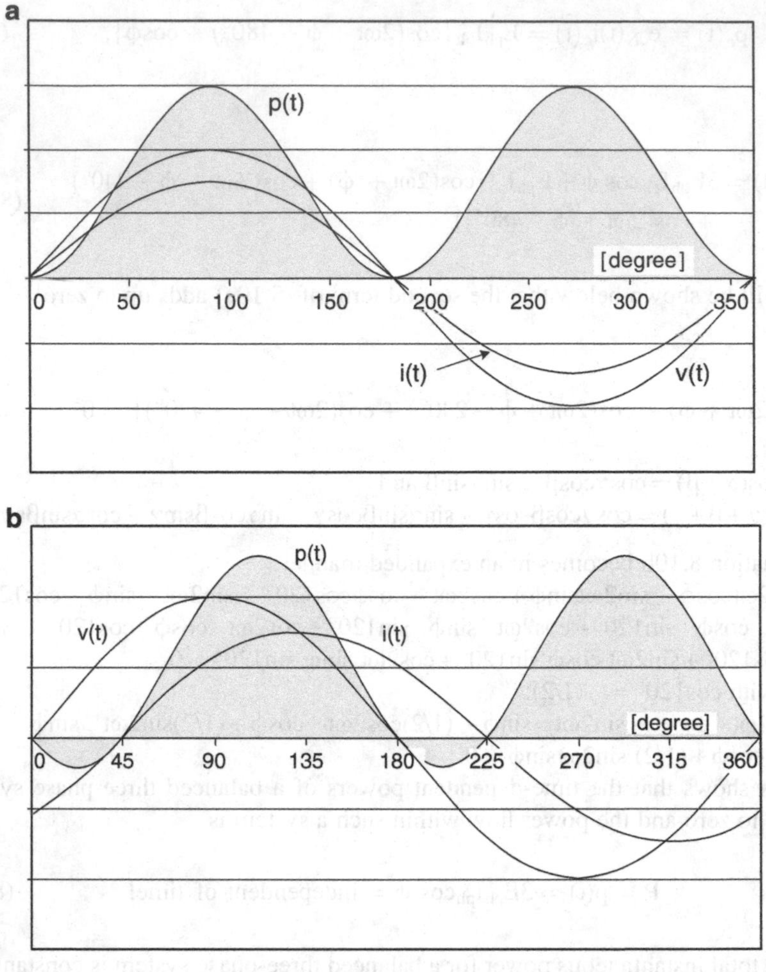

Fig. 8.7 (**a**) Voltage, current and power as a function of time for a single-phase system at unity-power factor of $\cos\Phi = 1.0$ (Courtesy of John M. Cowdrey, P.O. Box 847, Lyons, CO 80540, [4]) (**b**) Voltage, current and power as a function of time for a single-phase system at lagging-power factor (consumer notation) of $\cos\Phi = 0.707$ (Courtesy of John M. Cowdrey, P.O. Box 847, Lyons, CO 80540, [4])

8.2 Three-Phase Transformers [5–7]

8.2.1 Equivalence of Y and Δ-Connected Circuits

For Δ (Fig. 8.12a) and Y (Fig. 8.12b)-connected load circuits the line-to-line voltage ($V_{L\text{-}L}$), line current (I_L), power factor ($\cos\phi$), total real power (P_{tot}), total reactive power (Q_{tot}) and total apparent power or voltampere power (S_{tot}) are

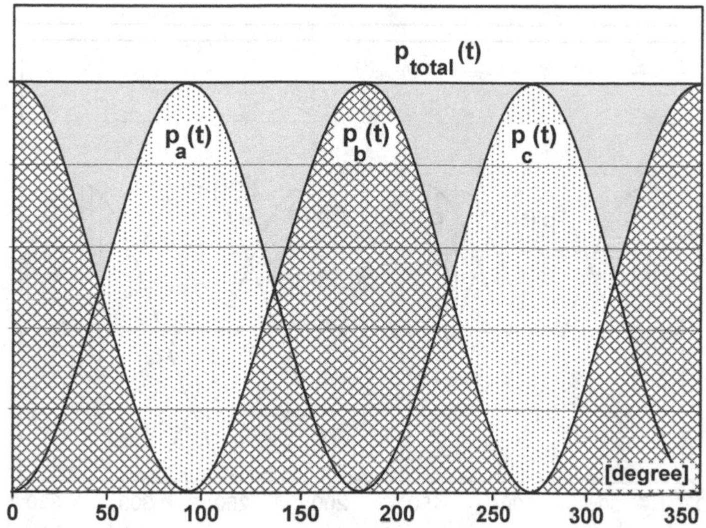

Fig. 8.8 Powers as a function of time for a two-phase system at unity-power factor of cosΦ = 1.0 (Courtesy of John M. Cowdrey, P.O. Box 847, Lyons, CO 80540, [4])

precisely equal for both cases. In other words, conditions viewed from the terminals A, B, and C of Figs. 8.12a, b are identical and one cannot distinguish between the two circuits from their terminal quantities. It will also be seen that the impedance (resistance and reactance) per phase of the balanced Y connection is exactly one-third of the corresponding values per phase of the balanced Δ connection. Consequently, a balanced Δ connection can be replaced by a balanced Y connection, providing that the circuit impedances per phase obey the relation

$$Z_Y = (1/3)Z_\Delta. \tag{8.12}$$

A general computational scheme for balanced circuits can be based entirely upon Y-connected circuits (or entirely Δ-connected circuits).

8.2.2 *Three-Phase Transformer Circuits*

Three single-phase transformers can be connected to form a three-phase bank in any of the four ways shown in Figs. 8.13a–d.

Application Example 8.1: Three-Phase (Ideal) Transformer Connections

Figure 8.13a represents the Y-Δ connection, that is, the primary is connected in Y and the secondary in Δ. Starting out with the primary line-to-line voltage magnitude V_{L-L}^p one obtains the primary line-to-neutral (phase) voltage magnitude

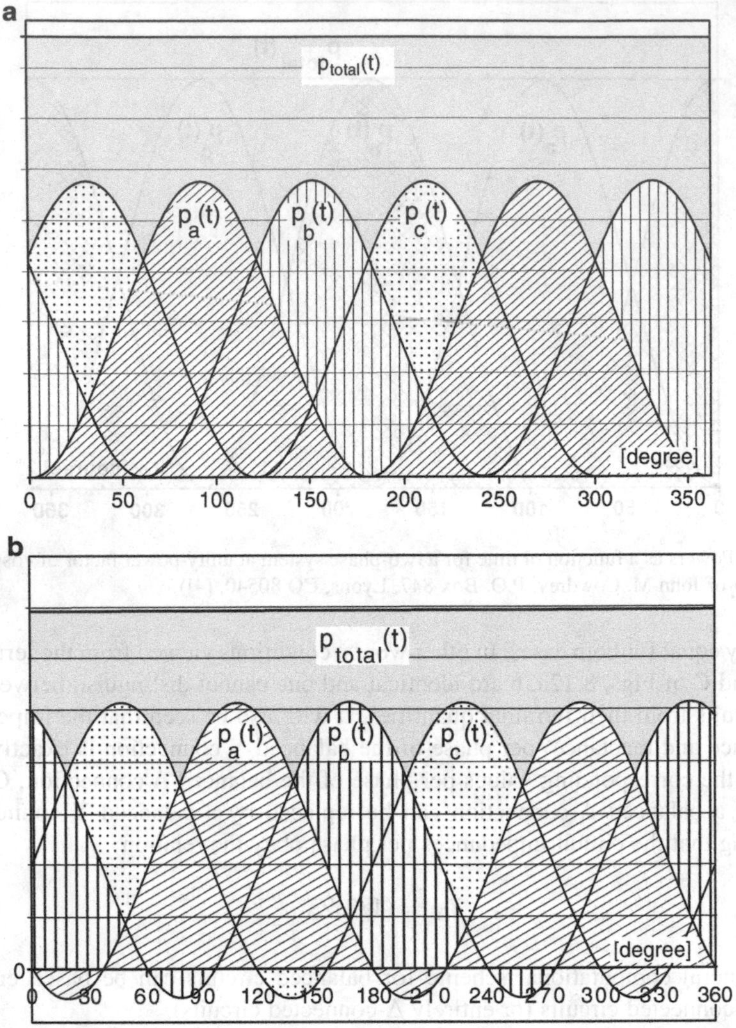

Fig. 8.9 (**a**) Powers as a function of time for a three-phase system at unity-power factor of $\cos\Phi = 1.0$ (Courtesy of John M. Cowdrey, P.O. Box 847, Lyons, CO 80540, [4]) (**b**) Powers as a function of time for a three-phase system at lagging-power factor (consumer notation) of $\cos\Phi = 0.80$ (Courtesy of John M. Cowdrey, P.O. Box 847, Lyons, CO 80540, [4])

$V^p_{ph} = V^p_{L-L}/\sqrt{3}$. The primary and secondary of phase "a" are wound on the same iron-core limb and, therefore, the primary ($\tilde{V}^p_{ph}$) and secondary ($\tilde{V}^s_{ph}$) are in phase for the given dot markings of Fig. 8.13a. If the turns ratio between primary (N_p) and secondary (N_s) phase windings is $a = (N_p/N_s)$ then

$$V^s_{ph} = \frac{V^p_{ph}}{a} = \frac{V^p_{L-L}/\sqrt{3}}{a}.$$

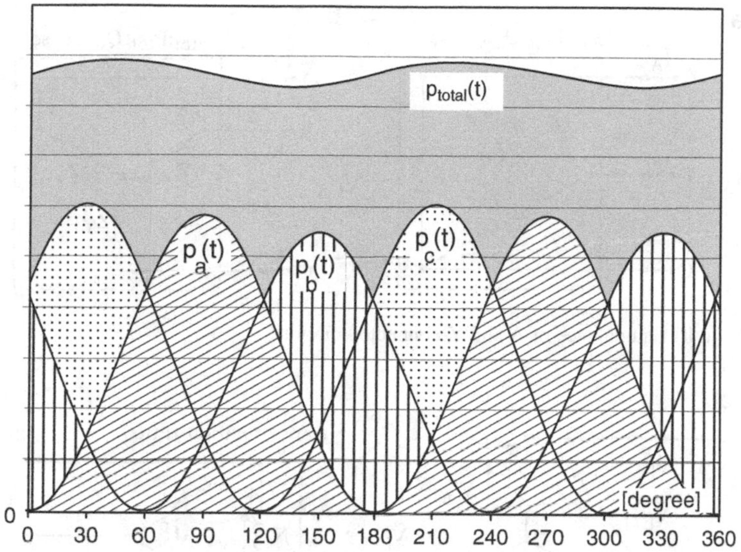

Fig. 8.10 Powers at an imbalance of 5% as applied to a three-phase system (Courtesy of John M. Cowdrey, P.O. Box 847, Lyons, CO 80540, [4])

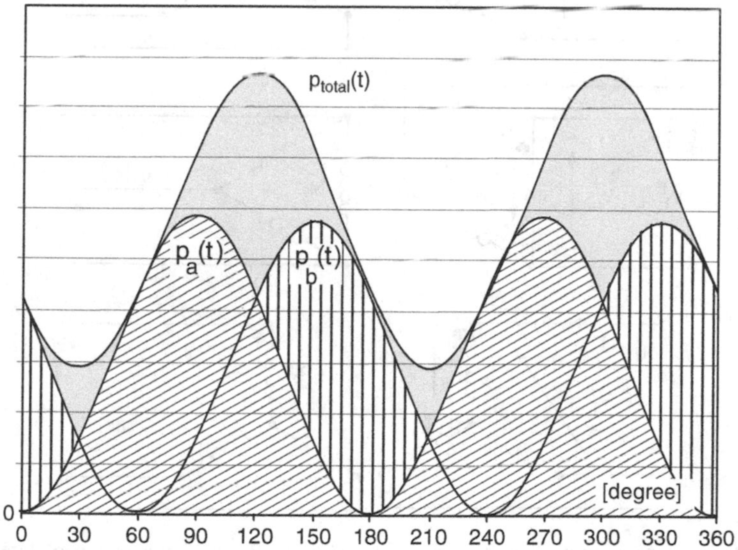

Fig. 8.11 Powers at a loss of phase c as applied to a three-phase system (Courtesy of John M. Cowdrey, P.O. Box 847, Lyons, CO 80540, [4])

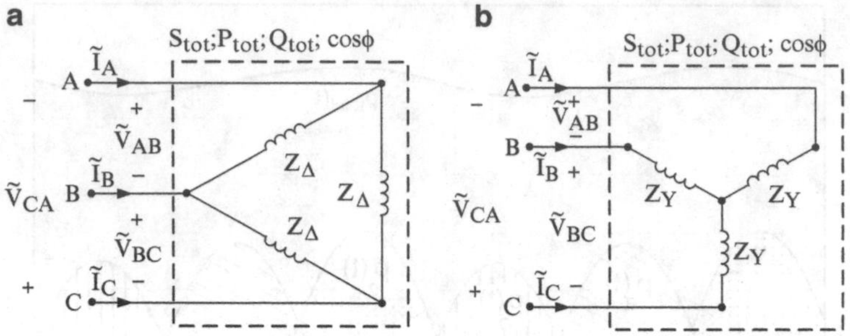

Fig. 8.12 (**a**, **b**) Equivalence of Δ and Y-connected circuits

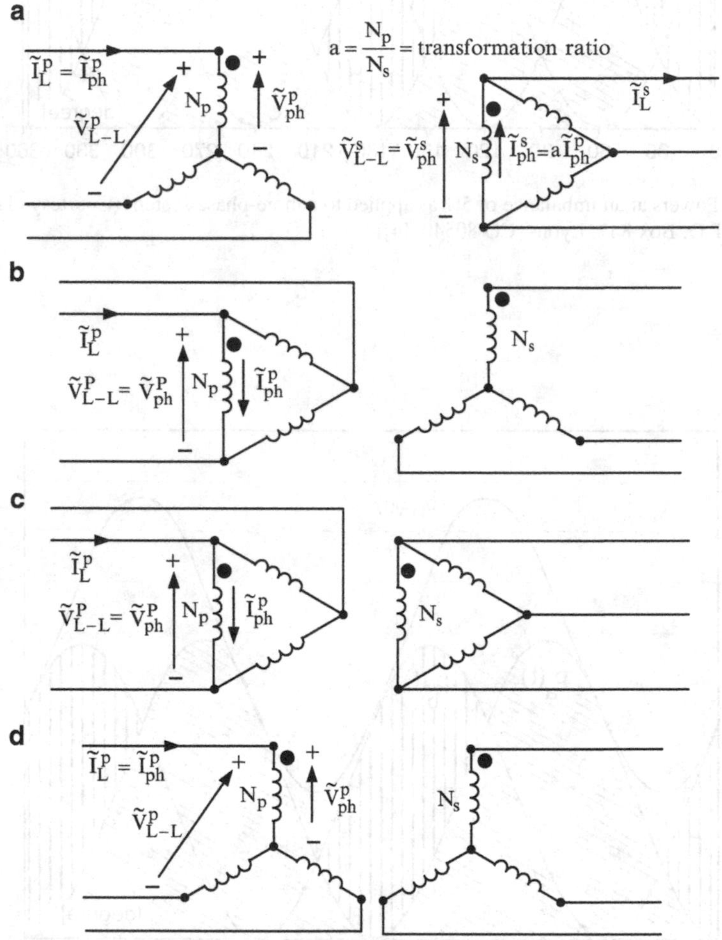

Fig. 8.13 (**a**) Y-Δ connection of three-phase transformer (**b**) Δ-Y connection of three-phase transformer (**c**) Δ-Δ connection of three-phase transformer (**d**) Y-Y connection of three-phase transformer

The secondary line-to-line voltage magnitude for a Δ-connection is identical with the phase voltage magnitude

$$V^{s}_{L-L} = V^{s}_{ph} = \frac{V^{p}_{L-L}/\sqrt{3}}{a}.$$

Starting out with the primary line current I^{p}_{L} which is identical with the primary phase current for a Y connection $\tilde{I}^{p}_{ph} = \tilde{I}^{p}_{L}$ one obtains for the secondary phase current $I^{s}_{ph} = aI^{p}_{ph} = aI^{p}_{L}$. The secondary line current is then $I^{s}_{L} = \sqrt{3}I^{s}_{ph} = \sqrt{3}aI^{p}_{ph} = \sqrt{3}aI^{p}_{L}$.

The primary (input) apparent power is

$$S^{p} = 3V^{p}_{ph}I^{p}_{ph} = 3\frac{V^{p}_{L-L}}{\sqrt{3}}I^{p}_{ph} = \sqrt{3}V^{p}_{L-L}I^{p}_{L}$$

and the secondary (output) apparent power is

$$S^{s} = 3V^{s}_{ph}I^{s}_{ph} = 3V^{s}_{L-L}I^{s}_{ph} = 3\frac{V^{p}_{L-L}/\sqrt{3}}{a}aI^{p}_{L} = \sqrt{3}V^{p}_{L-L}I^{p}_{L} = S^{p}.$$

This means the input apparent power is identical to the output apparent power because transformer losses have been neglected due to the use of the ideal transformer relations. The reader is encouraged to perform the corresponding analysis for Figs. 8.13b–d.

Application Example 8.2: Application of Short–Circuit and Open–Circuit Tests to 3-Phase Δ-Y Transformer and Power Flow in a 3-Phase Feeder at Balanced Load

A Δ-Y connected bank of three identical single–phase transformers, with one transformer having the ratings S = 100 kVA, V_p = 2,400 V, V_s = 120 V at 60 Hz, is supplied with power through a feeder whose impedance is $Z^{p}_{F} = R^{p}_{F} + jX^{p}_{F} = (0.30 + j0.80)\ \Omega$ per phase. The line-to-line voltage magnitude at the sending end of the feeder is held constant by the utility at $V_{send\ L\text{--}L}$ = 2,400 V. The results of a short-circuit test on one of the transformers with its low (secondary)-voltage terminals short-circuited are V_{p_sc} = 57.5 V, f = 60 Hz, I_{p_sc} = 41.6 A, and P_{p_sc} = 875 W.

(a) Determine the secondary line-to-line voltage when the bank delivers rated current to a balanced three-phase unity ($\cos\phi = 1.0$) power factor load.
(b) Compute the currents in the transformer primary and secondary windings and in the feeder wires if a solid three-phase short-circuit occurs at the secondary line terminals.

Solution

First find single-phase equivalent circuit parameters because test results are available for one transformer (see Fig. 8.14).

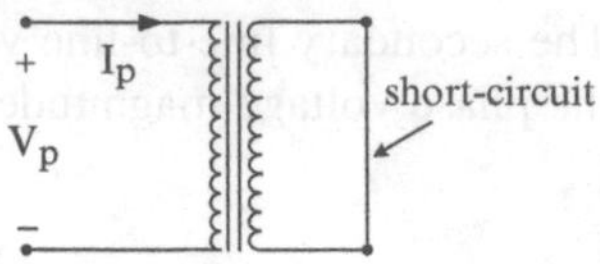

Fig. 8.14 Single-phase transformer under short-circuit condition

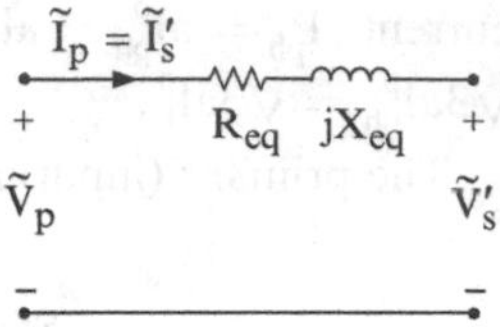

Fig. 8.15 Single-phase equivalent circuit referred to primary without short-circuit

The equivalent circuit of transformer of Fig. 8.14 referred to primary is shown in Fig. 8.15.

The equivalent circuit parameters are:

$$R^p_{eq\Delta} = \frac{P_{sc}}{(I^p_{sc})^2} = \frac{875}{(41.6)^2} = 0.5056\,\Omega, Z^p_{eq\Delta} = \frac{V^p_{sc}}{I^p_{sc}} = \frac{57.5}{41.6} = 1.382\,\Omega,$$

and

$$X^p_{eq\Delta} = \sqrt{(Z^p_{eq\Delta})^2 - (R^p_{eq\Delta})^2} = \sqrt{(1.382)^2 - (0.5056)^2} = 1.286\Omega.$$

The single-phase transformer tested is one phase of a Δ-connected primary of the three-phase transformer, therefore, the equivalent-circuit series impedance is that of a Δ connection referred to the primary $Z^p_{eq\Delta} = R^p_{eq\Delta} + jX^p_{eq\Delta} = (0.5056 + j1.286)\Omega$. The Δ-connected primary of the three-phase transformer (see Fig. 8.16) is now transformed to an equivalent Y-connected primary (see Fig. 8.17).

This above-mentioned transformation will be performed by the relation (well-known Δ-Y transformation for balanced system) $Z^p_{eqY} = (1/3)(R^p_{eq\Delta} + jX^p_{eq\Delta}) = (0.16853 + j0.42867)\Omega$. Now we can reduce the three-phase transformer problem (Fig. 8.17) to a single-phase transformer problem (see Fig. 8.18, with an ideal transformer, and Fig. 8.19 without an ideal transformer) by analyzing one phase only, and we can proceed as we have learned it in Chap. 7.

After such preliminary considerations we are ready to answer the questions posed.

(a) Determination of secondary line-to-line voltage for the load condition power factor of $\cos\phi = 1.0$, and $P_{transformer} = S_{transformer}\ \cos\phi = S_{transformer} = 100$ kVA = 100 kW:

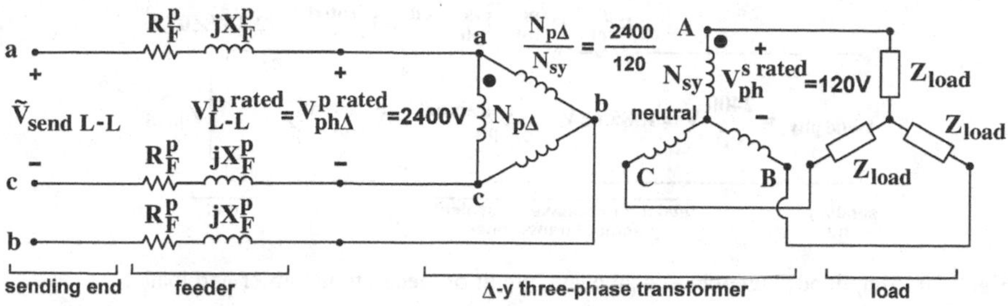

Fig. 8.16 Three-phase transformer with Δ-connected primary

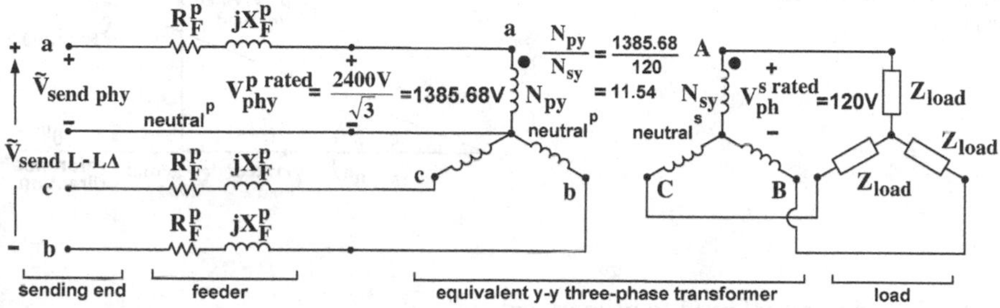

Fig. 8.17 Three-phase transformer with equivalent Y-connected primary

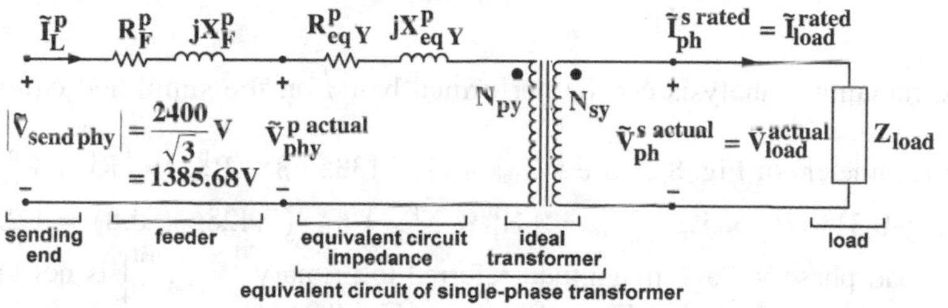

Fig. 8.18 Single-phase equivalent circuit for feeder, transformer (with ideal transformer) and load, where the sending-end voltage is maintained constant at $|V_{sendphy}| = 2400/\sqrt{3} = 1385.68\text{V}$

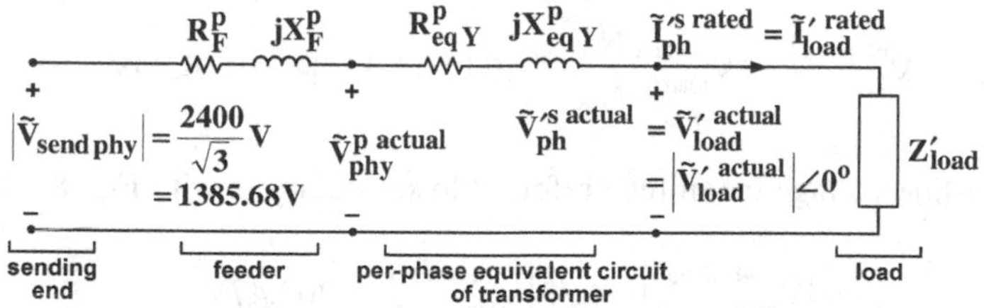

Fig. 8.19 Single-phase equivalent circuit for feeder, transformer (without ideal transformer) and load

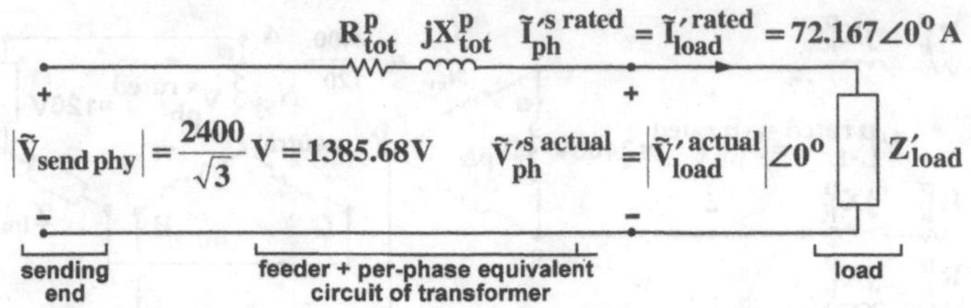

Fig. 8.20 Simplified single-phase equivalent circuit of feeder, transformer and load

Fig. 8.21 Phasor diagram for unity power factor, not to scale

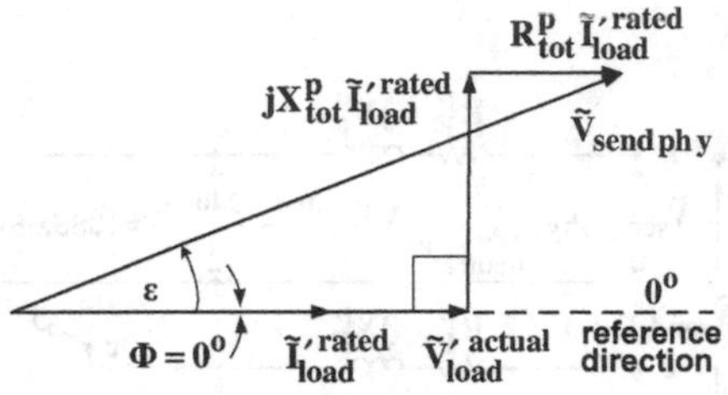

the load current referred to primary side is

$$\tilde{I}_{ph}^{'s\ rated} = \frac{S_{transformer}^{rated}}{\tilde{V}_{ph}^{'s\ rated}} = \frac{100kW}{1385.68\angle 0^0} = 72.167\angle 0^0 A,$$

and the remaining analysis can be performed based on the simplified equivalent circuit of Fig. 8.20:

The parameters of Fig. 8.20 are $\left|\tilde{V}_{send\ ph\ y}\right| = 1385.68V$, $R_{tot}^p = \left(R_F^p + R_{eqY}^p\right) = (0.1685 + 0.3) = 0.4685\Omega$, $X_{tot}^p = \left(X_F^p + X_{eqY}^p\right) = (0.4286 + 0.8) = 1.2286\Omega$, and the load phase voltage magnitude referred to primary $\left|\tilde{V}_{load}^{'s\ actual}\right|$ is not known, but can be obtained from the phasor diagram of Fig. 8.21.

$\left|\tilde{V}_{load}^{'actual}\right| = \sqrt{\left|\tilde{V}_{send\ ph\ y}\right|^2 - \left(X_{tot}^p\left|\tilde{I}_{load}^{'rated}\right|\right)^2} - R_{tot}^p\left|\tilde{I}_{load}^{'rated}\right| = 1349.07V = \tilde{V}_{load}^{'actual}$. The load phase voltage referred to secondary is

$$\tilde{V}_{load}^{ph\ actual} = \tilde{V}_{load}^{'\ actual}\frac{N_{sy}}{N_{py}} = 1349.07/11.54 = 116.90V.$$

The line-to-line voltage magnitude referred to secondary is with Fig. 8.22

$$\left|\tilde{V}_{load}^{L-L\ actual}\right| = \sqrt{3}\left|\tilde{V}_{load}^{ph\ actual}\right| = 202.47V.$$

(c) At short-circuit $\left|\tilde{V}_{load}^{L-L}\right| = \left|\tilde{V}_{load}^{'}\right| = 0$ as indicated in Fig. 8.23.

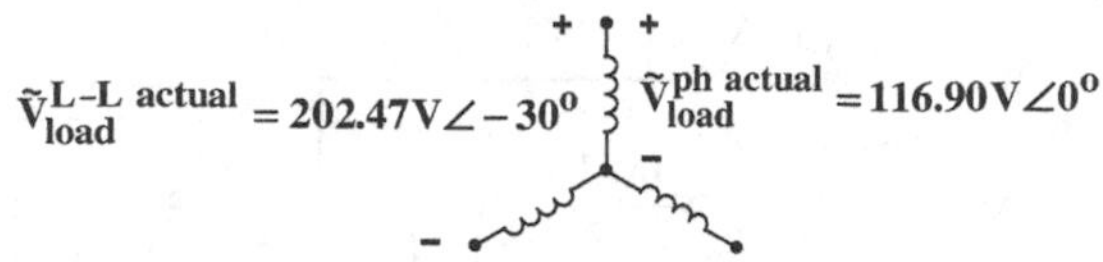

Fig. 8.22 Relation between phase and line-to-line voltages

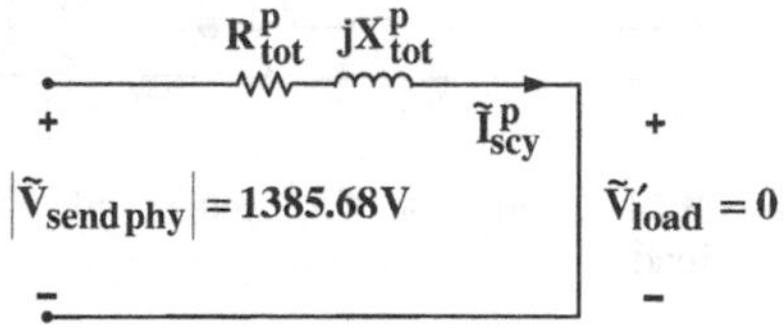

Fig. 8.23 Short-circuit on load side

The short-circuit current referred to the primary is $\left|\tilde{I}_{scy}^{P}\right| = \left|\tilde{I}_{L}^{P}\right| = \frac{1{,}385.68V}{\sqrt{(0.468)^2+(1.228)^2}} = 1{,}054.4\,A$. With Fig. 8.18 the load current referred to the secondary becomes $\left|\tilde{I}_{load}\right| = \left|\tilde{I}_{ph}^{S}\right| = \left|\tilde{I}_{scy}^{P}\right| \cdot \left(\frac{1385.68}{120}\right) = 1054.4 \cdot 11.54A = 12.176kA$. With Fig. 8.16 the phase current in the primary Δ is

$\left|\tilde{I}_{ph\Delta}^{P}\right| = \left|\tilde{I}_{L}^{P}\right|/\sqrt{3} = \left|\tilde{I}_{scy}^{P}\right|/\sqrt{3} = 1054.4/\sqrt{3} = 608.78A$

Application Example 8.3: Δ-(Ungrounded Y), Three-Phase Transformer with Diode Rectifier

(a) Perform a PSpice analysis [8–11] for the circuit of Fig. 8.24, where a three-phase diode rectifier with filter (e.g., capacitor C_f) serves a load R_{load}. You may assume ideal transformer conditions. For your convenience you may assume $(N_1/N_2) = 1$. The circuit parameters are $R_{syst} = 0.01\ \Omega$, $X_{syst} = 0.05\ \Omega$ @ f = 60 Hz, $v_{AB}(t) = \sqrt{2}600V\cos\omega t$, $v_{BC}(t) = \sqrt{2}600V\cos(\omega t - 120°)$, $v_{CA}(t) = \sqrt{2}600V\cos(\omega t - 240°)$, ideal diodes D_1 to D_6, $C_f = 500\ \mu F$, and $R_{load} = 10\ \Omega$. Plot one or two periods of both voltages and currents after steady-state has been reached as requested in parts b) to e).

(b) Plot the line-to-line voltages $v_{AB}(t)$ and line-to-neutral voltages $v_a(t)$. Why are they different? Subject $v_{AB}(t)$ and $v_{ab}(t)$ to Fourier analysis.

(c) Plot and subject the input line current $i_{AL}(t)$ to the Δ primary to a Fourier analysis. The input line currents of the primary Δ do not contain the 3rd, 6th, 9th, 12th, ...harmonics, that is, there are no harmonic zero-sequence current components [12–14]. This is one advantage of the Δ-(ungrounded Y) transformer connection.

(d) Plot and subject the phase current $i_{Aph}(t)$ of the Δ primary to a Fourier analysis. The phase currents of the primary Δ do not contain the 3rd, 6th, 9th, 12th, ... harmonics.

(e) Plot and subject the output current $i_{aph}(t)$ of the Y secondary to a Fourier analysis. The output currents of the secondary Y do not contain the 3rd, 6th, 9th, 12th, ...harmonics.

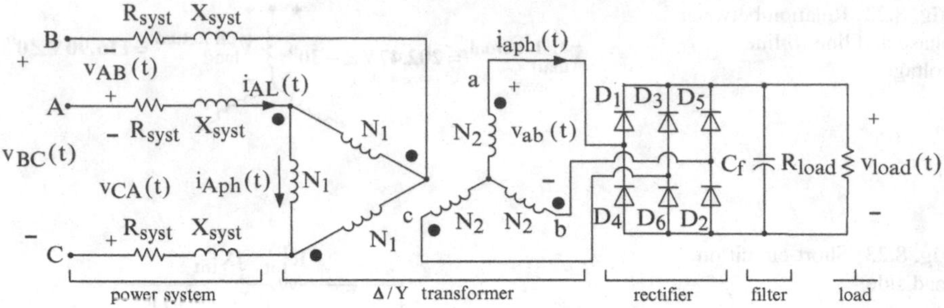

Fig. 8.24 Connection of a Δ/(ungrounded Y) three-phase transformer with a diode rectifier, filter, and load

(a) The input program for the PSpice analysis with extension .cir is listed below.

```
*delta/(ungrounded wye) transformer with diode rectifier, filter, and load
L1 8 7 10
L2 7 9 10
L3 40 8 10
Lxsysa 7 6 132.6u
Lxsysb 9 4 132.6u
Lxsysc 8 5 132.6u
*R8 8 0 10meg
Rxsysa 6 1 10m
Rxsysb 4 2 10m
Rxsysc 5 3 10m
L4 10 11 10
L6 10 13 10
L5 10 12 10
D1 11 15 Dideal
D3 13 15 Dideal
D5 12 15 Dideal
D4 14 11 Dideal
D6 14 13 Dideal
D2 14 12 Dideal
Rload 15 14 10
Cload 15 14 500u
Rn 10 0 10meg
VAN 1 0 sin(0 489.9 60 0 0 -30)
VBN 2 0 sin(0 489.9 60 0 0 -150)
VCN 3 0 sin(0 489.9 60 0 0 -270)
*V_BC 2 3 SIN(0 848.52 60 0 0 -120)
*V_AB 1 2 SIN(0 848.52 60 0 0 0)
*V_CA 3 1 SIN(0 848.52 60 0 0 -240)
R3 9 40 1u
.model Dideal d(is=1m)
.tran 100u 200m 160m 100u uic
.four 60 12 V(1,2) V(11,12) I(L1) I(L4) I(Lxsysa)
.options abstol=1m chgtol=0.1m reltol=0.1 vntol=1m
.options itl4=200 itl5=0
.probe
K14 L1 L4 0.9999
K25 L2 L5 0.9999
K36 L3 L6 0.9999
.end
```

(b) Figures 8.25 and 8.26 show the line-to-line voltages $v_{AB}(t)$, $v_{BC}(t)$, $v_{CA}(t)$, and the line-to-neutral voltages $v_a(t)$, $v_b(t)$, $v_c(t)$, respectively. The line-to-line voltages $v_{AB}(t)$, $v_{BC}(t)$, $v_{CA}(t)$ are sinusoidal and are imposed by the power system while the line-to-neutral voltages $v_a(t)$, $v_b(t)$, $v_c(t)$ are distorted because of the rectifier limiting the voltage amplitude and causing commutation spikes.

For $v_{AB}(t)$ one obtains the Fourier components of Table 8.1. FOURIER COMPONENTS OF TRANSIENT RESPONSE V(1,2).

For $v_{ab}(t)$ one obtains the Fourier components of Table 8.2. FOURIER COMPONENTS OF TRANSIENT RESPONSE V(11,12).

(c) Input line current $i_{AL}(t)$ of primary Δ is depicted in Fig. 8.27, and Fig. 8.28 illustrates its Fourier spectrum

For $i_{AL}(t)$ one obtains the Fourier components of Table 8.3. FOURIER COMPONENTS OF TRANSIENT RESPONSE I(Lxsysa).

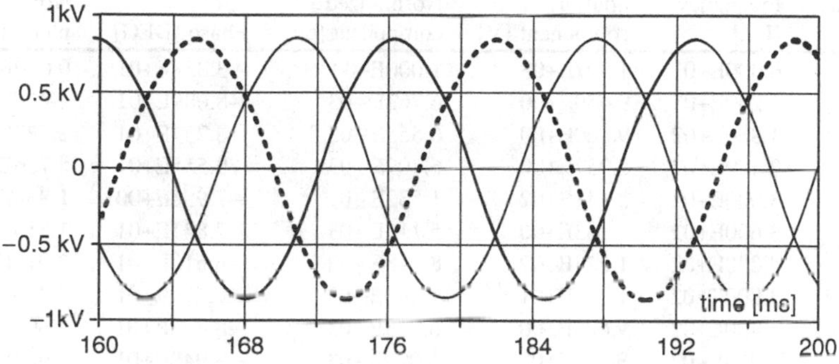

Fig. 8.25 Line-to-line voltages $v_{AB}(t)$, $v_{BC}(t)$, $v_{CA}(t)$

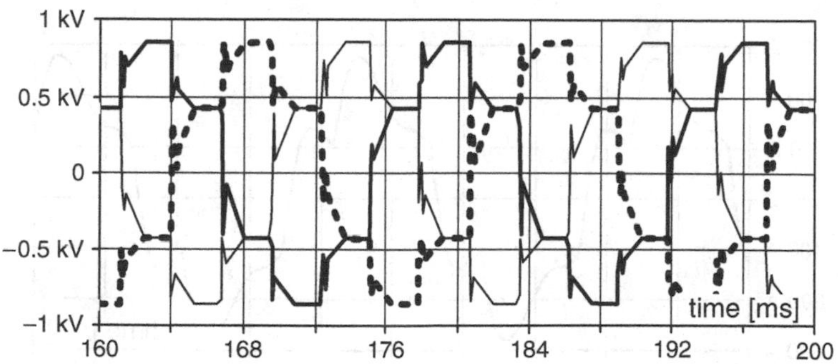

Fig. 8.26 Line-to-neutral voltages $v_a(t)$, $v_b(t)$, $v_c(t)$

Table 8.1 Fourier analysis for $v_{AB}(t)$ DC COMPONENT = −3.461398E−02 V

Harmonic no	Frequency [HZ]	Fourier component [V]	Normalized component [pu]	Phase [DEG]	Normalized phase [DEG]
1	6.000E+01	8.485E+02	1.000E+00	−1.296E−04	0.000E+00
2	1.200E+02	3.945E−02	4.649E−05	6.832E+01	6.832E+01
3	1.800E+02	1.187E−02	1.399E−05	1.402E+02	1.402E+02
4	2.400E+02	5.987E−03	7.057E−06	1.705E+02	1.705E+02
5	3.000E+02	7.246E−03	8.540E−06	−9.826E+01	−9.825E+01
6	3.600E+02	8.070E−03	9.512E−06	−9.327E+00	−9.326E+00
7	4.200E+02	6.069E−03	7.153E−06	4.424E+01	4.424E+01
8	4.800E+02	5.775E−03	6.807E−06	8.930E+01	8.930E+01
9	5.400E+02	5.857E−03	6.903E−06	1.723E+02	1.723E+02
10	6.000E+02	5.853E−03	6.898E−06	−1.006E+02	−1.006E+02
11	6.600E+02	4.978E−03	5.867E−06	−4.124E+01	−4.124E+01
12	7.200E+02	6.017E−06	6.017E−06	1.178E+01	1.178E+01

Total harmonic distortion = 5.323514E−03%

Table 8.2 Fourier analysis for $v_{ab}(t)$ DC COMPONENT = −3.499218E+00 V

Harmonic no	Frequency [HZ]	Fourier component [V]	Normalized component [pu]	Phase [DEG]	Normalized phase [DEG]
1	6.000E+01	1.397E+03	1.000E+00	−3.828E+01	0.000E+00
2	1.200E+02	9.450E+00	6.762E−03	−8.089E+01	−4.333E+00
3	1.800E+02	9.160E+00	6.555E−03	−8.757E+01	2.727E+01
4	2.400E+02	8.957E+00	6.409E−03	−9.555E+01	5.756E+01
5	3.000E+02	2.141E+02	1.532E−01	−7.956E+00	1.834E+02
6	3.600E+02	7.143E+00	5.111E−03	−7.841E+01	1.513E+02
7	4.200E+02	1.223E+02	8.748E−02	−6.612E+01	2.018E+02
8	4.800E+02	1.006E+01	7.198E−03	−7.774E+01	2.285E+02
9	5.400E+02	9.604E+00	6.873E−03	−8.433E+01	2.602E+02
10	6.000E+02	8.524E+00	6.099E−03	−8.948E+01	2.933E+02
11	6.600E+02	6.488E+01	4.642E−02	−1.859E−02	4.210E+02
12	7.200E+02	8.326E+00	5.958E−03	−6.751E+01	3.918E+02

Total harmonic distortion = 1.832974E+01%

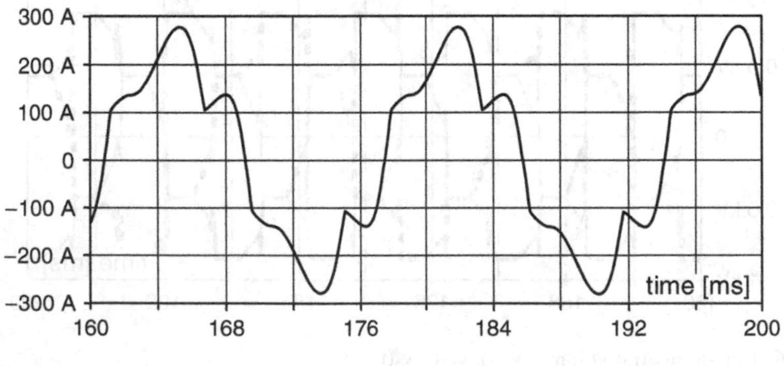

Fig. 8.27 Input line current $i_{AL}(t)$ of primary Δ winding of Δ-Y transformer

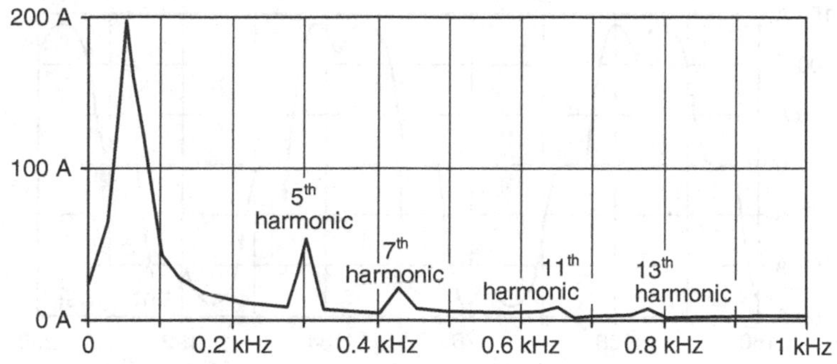

Fig. 8.28 Fourier spectrum of the input line current $i_{AL}(t)$ of primary Δ

Table 8.3 Fourier analysis for $i_{AL}(t)$ DC COMPONENT = −4.756973E−01 A

Harmonic no	Frequency [HZ]	Fourier component [A]	Normalized component [pu]	Phase [DEG]	Normalized phase [DEG]
1	6.000E+01	2.421E+02	1.000E+00	1.271E+02	0.000E+00
2	1.200E+02	3.075E−01	1.270E−03	−1.021E+02	−3.564E+02
3	1.800E+02	2.957E−01	1.222E−03	−8.977E+01	−4.712E+02
4	2.400E+02	3.080E−01	1.273E−03	−9.725E+01	−6.058E+02
5	3.000E+02	4.722E+01	1.951E−01	−9.658E+01	−7.322E+02
6	3.600E+02	3.053E−01	1.261E−03	−9.013E+01	−8.529E+02
7	4.200E+02	1.781E+01	7.357E−02	−1.563E+02	−1.046E+03
8	4.800E+02	3.256E−01	1.345E−03	−9.204E+01	−1.109E+03
9	5.400E+02	2.730E−01	1.128E−03	−9.022E+01	−1.234E+03
10	6.000E+02	3.231E−01	1.335E−03	−9.699E+01	1.368E+03
11	6.600E+02	6.285E+00	2.597E−02	−8.478E+01	−1.483E+03
12	7.200E+02	2.982E−01	1.232E−03	−9.513E+01	−1.621E+03

Total harmonic distortion = 2.101297E+01%

(d) The phase current $i_{Aph}(t)$ of primary Δ is depicted in Fig. 8.29, and Fig. 8.30 illustrates its Fourier spectrum

For $i_{Aph}(t)$ one obtains the Fourier components of Table 8.4. FOURIER COMPONENTS OF TRANSIENT RESPONSE I(L1).

(e) The phase current $i_{aph}(t)$ of the secondary Y is depicted in Fig. 8.31, and Fig. 8.32 illustrates its Fourier spectrum

For $i_{aph}(t)$ one obtains the Fourier components of Table 8.5. FOURIER COMPONENTS OF TRANSIENT RESPONSE I(L4).

Application Example 8.4: Δ/Zigzag, Three-Phase Transformer Configuration

Figure 8.33 depicts the so-called Δ/zigzag configuration [5] of a three-phase transformer which is used for supplying power to unbalanced loads and three-phase rectifiers. You may assume ideal transformer conditions.

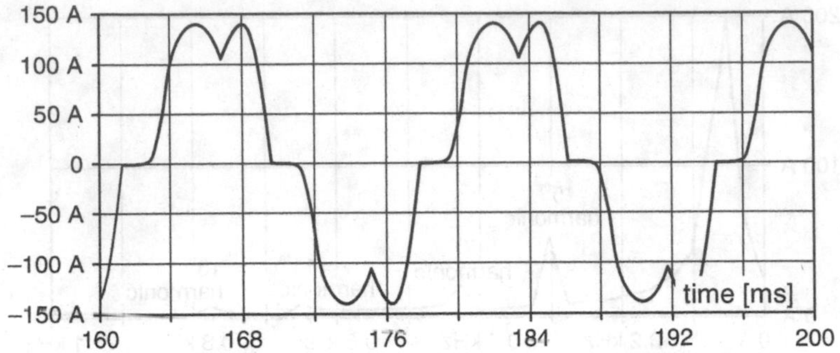

Fig. 8.29 Phase current $i_{Aph}(t)$ of primary Δ winding of Δ-Y transformer

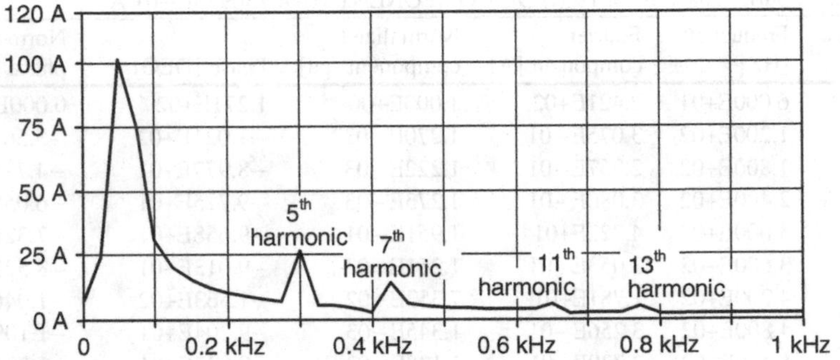

Fig. 8.30 Fourier spectrum of phase current $i_{Aph}(t)$ of primary Δ

Table 8.4 Fourier analysis for $i_{Aph}(t)$ DC COMPONENT = −1.562151E−01 A

Harmonic no	Frequency [HZ]	Fourier component [A]	Normalized component [pu]	Phase [DEG]	Normalized phase [DEG]
1	6.000E+01	1.398E+02	1.000E+00	9.710E+01	0.000E+00
2	1.200E+02	1.143E−01	8.174E−04	−1.098E+02	−3.040E+02
3	1.800E+02	9.257E−02	6.622E−04	−9.115E+01	−3.825E+02
4	2.400E+02	1.188E−01	8.500E−04	−9.380E+01	−4.822E+02
5	3.000E+02	2.719E+01	1.945E−01	−6.671E+01	−5.522E+02
6	3.600E+02	8.275E−02	5.919E−04	−8.000E+01	−6.626E+02
7	4.200E+02	1.021E+01	7.305E−02	1.734E+02	−5.063E+02
8	4.800E+02	1.141E−01	8.164E−04	−9.963E+01	−8.764E+02
9	5.400E+02	9.559E−02	6.838E−04	−9.236E+01	−9.663E+02
10	6.000E+02	1.255E−01	8.974E−04	−9.307E+01	−1.064E+03
11	6.600E+02	3.551E+00	2.540E−02	−5.538E+01	−1.123E+03
12	7.200E+02	7.731E−02	5.530E−04	−9.096E+01	−1.256E+03

Total harmonic distortion = 2.093031E+01%

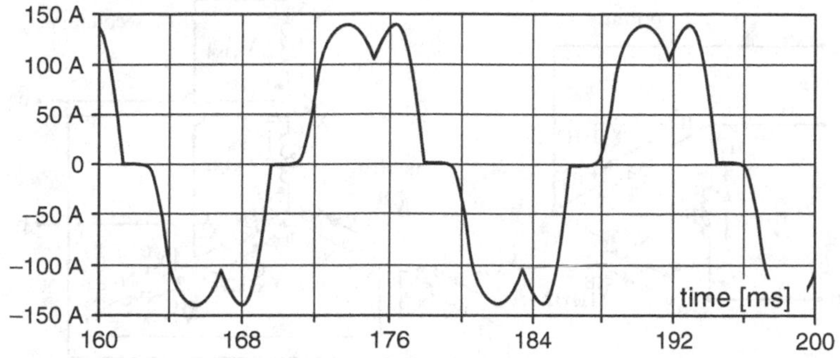

Fig. 8.31 Phase current $i_{aph}(t)$ of secondary Y winding of Δ-Y transformer

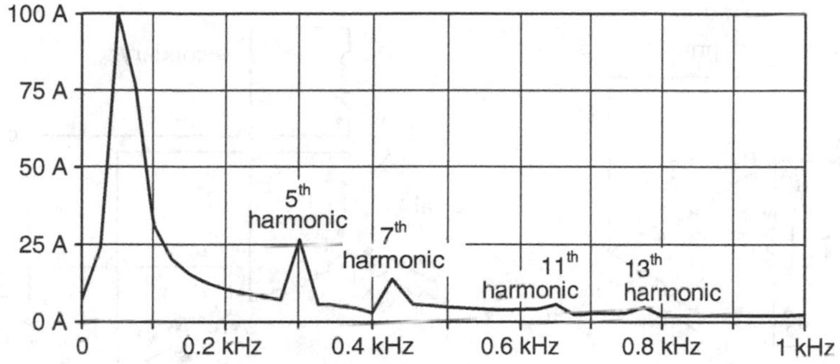

Fig. 8.32 Fourier spectrum of phase current $i_{aph}(t)$ of secondary Y

Table 8.5 Fourier analysis for $i_{aph}(t)$ DC COMPONENT = 4.182258E−02 A

Harmonic no	Frequency [HZ]	Fourier component [A]	Normalized component [pu]	Phase [DEG]	Normalized phase [DEG]
1	6.000E+01	1.397E+02	1.000E+00	−8.281E+01	0.000E+00
2	1.200E+02	1.143E−01	8.180E−04	7.020E+01	2.358E+02
3	1.800E+02	9.259E−02	6.626E−04	8.885E+01	3.373E+02
4	2.400E+02	1.189E−01	8.506E−04	8.620E+01	4.175E+02
5	3.000E+02	2.719E+01	1.946E−01	1.133E+02	5.274E+02
6	3.600E+02	8.276E−02	5.923E−04	1.000E+02	5.969E+02
7	4.200E+02	1.021E+01	7.310E−02	−6.598E+00	5.731E+02
8	4.800E+02	1.142E−01	8.170E−04	8.037E+01	7.429E+02
9	5.400E+02	9.561E−02	6.842E−04	8.765E+01	8.330E+02
10	6.000E+02	1.255E−01	8.980E−04	8.693E+01	9.151E+02
11	6.600E+02	3.551E+00	2.542E−02	1.246E+02	1.036E+03
12	7.200E+02	7.732E−02	5.533E−04	8.904E+01	1.083E+03

Total harmonic distortion = 2.094341E+01%

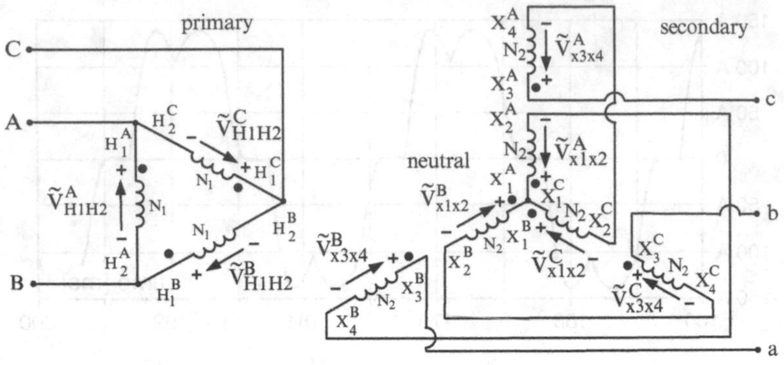

Fig. 8.33 Connection of a Δ /zigzag, three-phase transformer with the definition of primary and secondary voltages

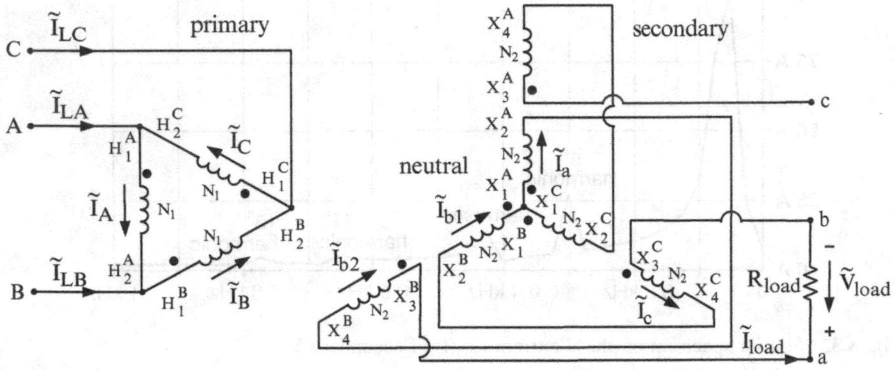

Fig. 8.34 Connection of a Δ/zigzag, three-phase transformer with the definition of current for line-to-line load

(a) Draw a phasor diagram of the primary and secondary voltages when there is no-load on the secondary side. For your convenience you may assume $(N_1/N_2) = 1$. The Δ/zigzag three-phase configuration is used for feeding unbalanced loads and three-phase rectifiers. Even when only one line-to-line load (e. g., R_{load}) of the secondary is present as indicated in Fig. 8.34 the primary line currents $\tilde{I}_{LA}$, $\tilde{I}_{LB}$, and $\tilde{I}_{LC}$ will be balanced because the line-to-line load is distributed to all three (single-phase) transformers. This is the advantage of a Δ/zigzag configuration.

(b) If there is a resistive line-to-line load on the secondary side $\left(\left|\tilde{I}_{load}\right| = 10A\right)$ present as illustrated in Fig. 8.34 draw a phasor diagram of the primary and secondary currents. For your convenience you may assume that the same voltage definitions apply, as in Fig. 8.33 and $(N_1/N_2) = 1$. You may assume ideal transformer conditions.

(c) Repeat the analysis of (b) if there is a resistive line-to-neutral load on the secondary side $\left(\left|\tilde{I}_{load}\right| = 10A\right)$ present, as illustrated in Fig. 8.35; that is, draw

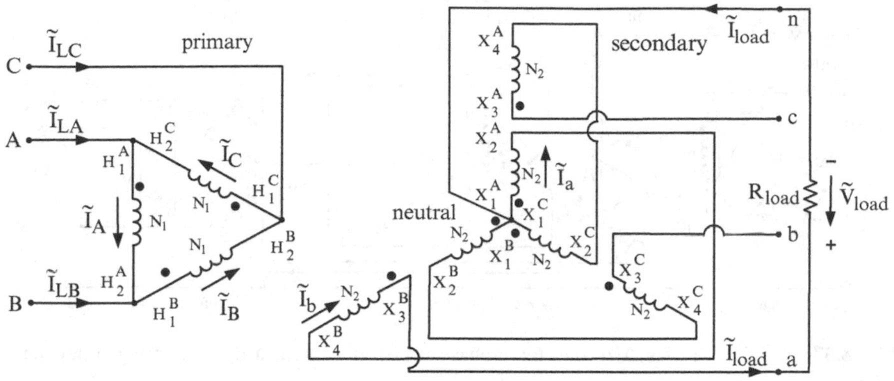

Fig. 8.35 Connection of a Δ/zigzag, three-phase transformer with the definition of current for line-to-neutral load

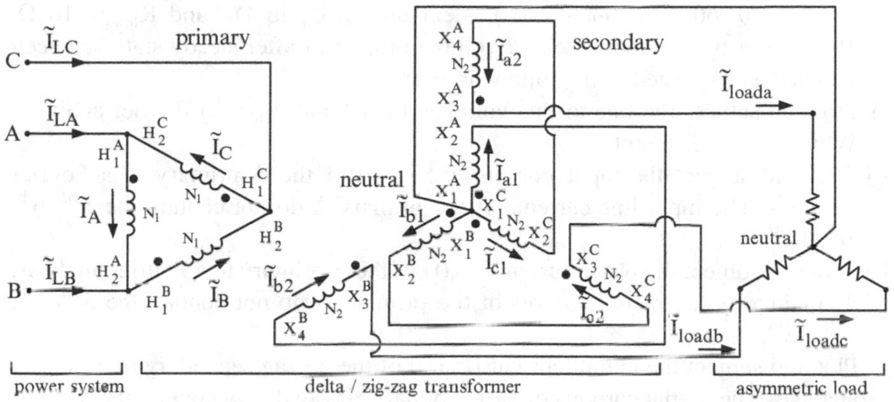

Fig. 8.36 Connection of a Δ/zigzag, three-phase transformer with the definition of currents for unbalanced line-to-neutral loads

a phasor diagram of the primary and secondary currents as defined in Fig. 8.35. For your convenience you may assume that the same voltage definitions apply as in Fig. 8.33 and $(N_1/N_2) = 1$. In this case the load is distributed to two (single-phase) transformers.

(d) Repeat the analysis of (b) if there is a resistive unbalanced load on the secondary side $\left(\left|\tilde{I}_{loada}\right| = 30\text{A}, \left|\tilde{I}_{loadb}\right| = 20\text{A}, \left|\tilde{I}_{loadc}\right| = 10\text{A}\right)$ present as illustrated in Fig. 8.36; that is, draw a phasor diagram of the primary and secondary currents as defined in Fig. 8.36. For your convenience you may assume that the same voltage definitions apply as in Fig. 8.33 and $(N_1/N_2) = 1$. What conclusion can you draw from the phasor diagrams?

(e) Perform a PSpice analysis for the circuit of Fig. 8.37 where a three-phase diode rectifier with filter (e.g., capacitance $C_f = 500\ \mu\text{F}$) serves the load R_{load}. You may assume ideal transformer conditions. For your convenience you may

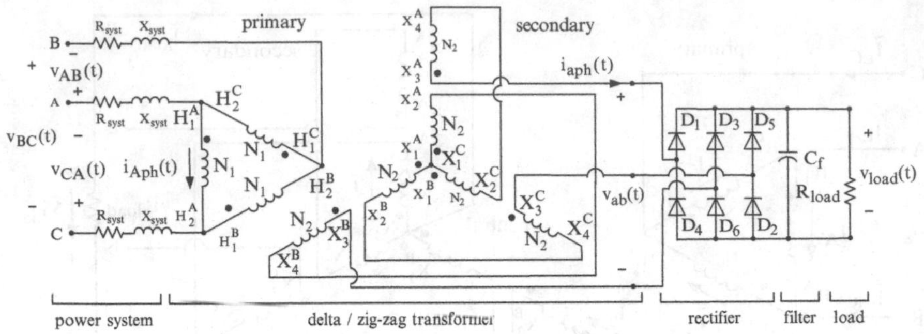

Fig. 8.37 Connection of a Δ/zigzag, three-phase transformer with a diode rectifier, filter, and load R_{load}

assume $(N_1/N_2) = 1$. The circuit parameters are $R_{syst} = 0.01\ \Omega$, $X_{syst} = 0.05\ \Omega$ @ f = 60 Hz, $v_{AB}(t) = \sqrt{2}600\text{V} \cos \omega t$, $v_{BC}(t) = \sqrt{2}600\text{V} \cos(\omega t - 120°)$, $v_{CA}(t) = \sqrt{2}600\text{V} \cos(\omega t - 240°)$, ideal diodes D_1 to D_6, and $R_{load} = 10\ \Omega$. Plot one or two periods of either voltage or current after steady state has been reached as requested in the following parts.

(f) Plot and subject the line-to-line voltages $v_{AB}(t)$ and $v_{ab}(t)$ to Fourier analysis. Why are they different?

(g) Plot and subject the input line current $i_{AL}(t)$ of the Δ primary to a Fourier analysis. The input line currents of the primary Δ do not contain the 3rd, 6th, 9th, 12th, ...

(h) Plot and subject the phase current $i_{Aph}(t)$ of the Δ primary to a Fourier analysis. Explain why the phase currents of the primary Δ do not contain the 3rd, 6th, 9th, 12th, ...

(i) Plot and subject the output current $i_{aph}(t)$) of the zigzag secondary to a Fourier analysis. The output currents of the secondary zigzag do not contain the 3rd, 6th, 9th, 12th, ...

(j) Plot and subject the output voltage $v_{load}(t)$ to a Fourier analysis.

Solution

(a) Phasor diagram of the primary and secondary voltages at no-load is shown in Fig. 8.38.

(b) Phasor diagram of the primary and secondary currents if there is a resistive line-to-line load on the secondary side $\left(\left|\tilde{I}_{load}\right| = 10\text{A}\right)$ is depicted in Fig. 8.39. Note, $\tilde{I}_{LA} = \tilde{I}_A - \tilde{I}_C$, $\tilde{I}_{LB} = \tilde{I}_B - \tilde{I}_A$, $\tilde{I}_{LC} = \tilde{I}_C - \tilde{I}_B$, $\tilde{I}_A = -\tilde{I}_a$, $\tilde{I}_B = \tilde{I}_{b1} + \tilde{I}_{b2}$, and $\tilde{I}_C = -\tilde{I}_c$.

(c) Phasor diagram of the primary and secondary currents if there is a resistive line-to-neutral load on the secondary side $\left(\left|\tilde{I}_{load}\right| = 10\text{A}\right)$ is depicted in Fig. 8.40. Note, $\tilde{I}_{LA} = \tilde{I}_A$, $\tilde{I}_{LB} = \tilde{I}_B - \tilde{I}_A$, $\tilde{I}_{LC} = -\tilde{I}_B$, $\tilde{I}_A = -\tilde{I}_a$, $\tilde{I}_B = \tilde{I}_b$, and $\tilde{I}_C = -\tilde{I}_c = 0$.

(d) Phasor diagram of the primary and secondary currents if there is a resistive unbalanced load on the secondary side $\left(\left|\tilde{I}_{loada}\right| = 30\text{A}, \left|\tilde{I}_{loadb}\right| = 20\text{A},\right.$

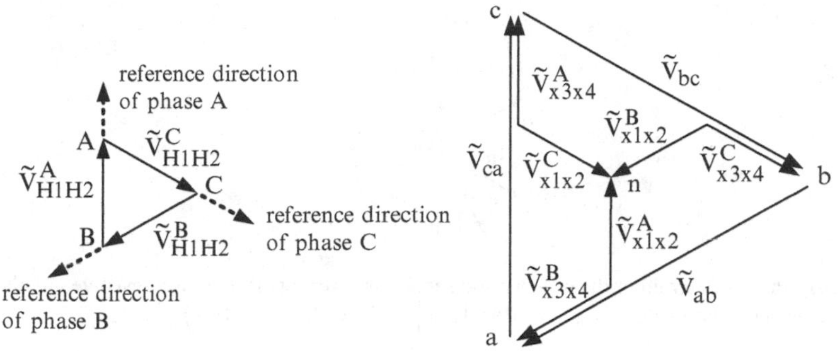

Fig. 8.38 Phasor diagram of the primary and secondary voltages at no-load. You may use hexagonal paper

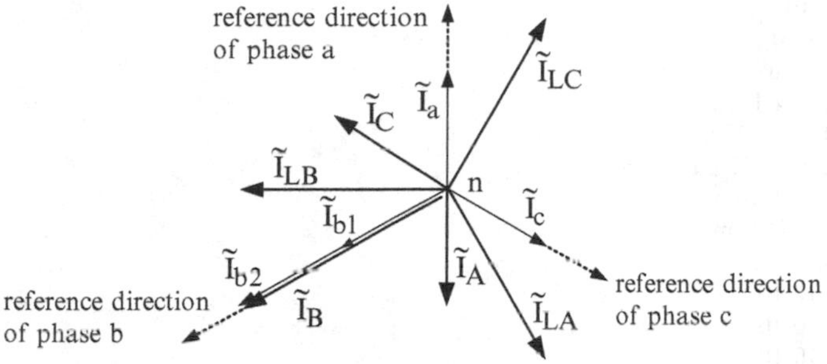

Fig. 8.39 Phasor diagram of the primary and secondary currents if there is a resistive line-to linc load on the secondary side $\left(\left|\tilde{I}_{load}\right| = \left|\tilde{V}_{ab}\right|/R_{load} = 10A\right)$

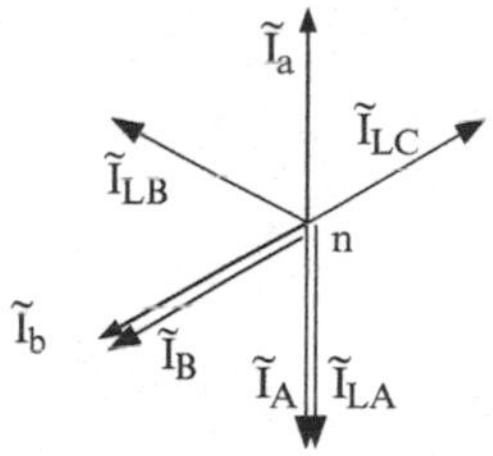

Fig. 8.40 Phasor diagram of the primary and secondary currents if there is a resistive line-to-neutral load on the secondary side $\left(\left|\tilde{I}_{load}\right| = \left|\tilde{V}_{an}\right|/R_{load} = 10A\right)$

$\left|\tilde{I}_{loadc}\right| = 10A$) is shown in Fig. 8.41. Note, $\tilde{I}_{a1} = \tilde{I}_{loadb} = 20A = 4$ units, $\tilde{I}_{a2} = \tilde{I}_{loada} = 30A = 6$ units, $\tilde{I}_{b1} = \tilde{I}_{loadc} = 10A = 2$ units, $\tilde{I}_{b2} = \tilde{I}_{loadb} = 20A = 4$ units, $\tilde{I}_{c1} = \tilde{I}_{loada} = 30A = 6$ units, $\tilde{I}_{c2} = \tilde{I}_{loadc} = 10A = 2$ units, $\tilde{I}_{LA} = \tilde{I}_{A} - \tilde{I}_{C} = -\tilde{I}_{a1} + \tilde{I}_{a2}$, $\tilde{I}_{LB} = \tilde{I}_{B} - \tilde{I}_{A} = -\tilde{I}_{b1} + \tilde{I}_{b2}$, and $\tilde{I}_{LC} = \tilde{I}_{C} - \tilde{I}_{B} = -\tilde{I}_{c1} + \tilde{I}_{c2}$.

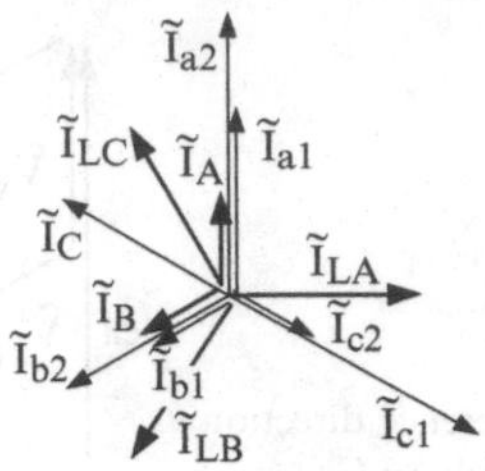

Fig. 8.41 Phasor diagram of the primary and secondary currents if there is a resistive unbalanced load on the secondary side $\left(\left|\tilde{I}_{loada}\right| = 30A, \left|\tilde{I}_{loadb}\right| = 20A, \left|\tilde{I}_{loadc}\right| = 10A\right)$

(e) The input program for the PSpice analysis with extension .cir is listed below.

```
*delta/zigag three-phase transformer configuration with diode rectifier, filter, and load
L1 8 7 10
L2 7 9 10
L3 40 8 10
R3 9 40 1u
Lxsysa 7 4 132.6u
Lxsysb 9 5 132.6u
Lxsysc 8 6 132.6u
Rxsysa 4 1 10m
Rxsysb 5 3 10m
Rxsysc 6 2 10m
L5 13 12 10
L4 11 10 10
L6 14 10 10
L7 11 16 10
L8 13 10 10
L9 14 15 10
D1 12 17 Dideal
D3 16 17 Dideal
D5 15 17 Dideal
D4 18 12 Dideal
D6 18 16 Dideal
D2 18 15 Dideal
Rload 17 18 10
Cload 17 18 500u
Rn 10 0 10meg
VAN 1 0 sin(0 489.9 60 0 0 -30)
VBN 2 0 sin(0 489.9 60 0 0 -150)
VCN 3 0 sin(0 489.9 60 0 0 -270)
*V_BC 2 3 SIN(0 848.52 60 0 0 -120)
*V_AB 1 2 SIN(0 848.52 60 0 0 0)
*V_CA 3 1 SIN(0 848.52 60 0 0 -240)
.model Dideal d(is=1u)
.tran 100u 200m 160m 100u uic
.four 60 12 V(1,2) V(12,16) I(L1) I(L5) I(Lxsysa)
.options abstol=100u chgtol=0.1m reltol=0.1 vntol=100u
.options itl4=200 itl5=0
.probe
K145 L1 L4 L5 0.9999
K389 L3 L8 L9 0.9999
K267 L2 L6 L7 0.9999
.end
```

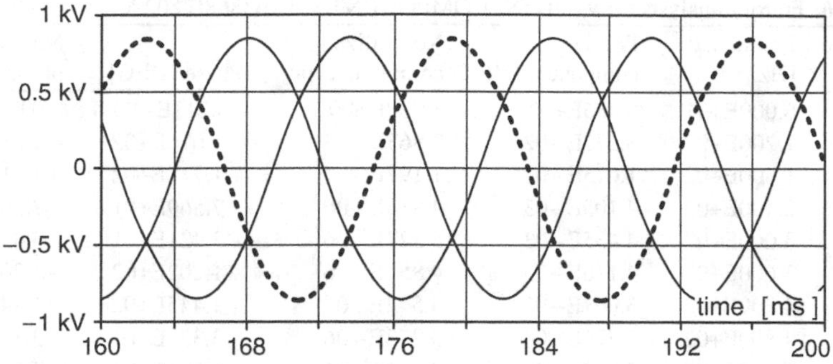

Fig. 8.42 Line-to-line voltages $v_{AB}(t)$, $v_{BC}(t)$, $v_{CA}(t)$

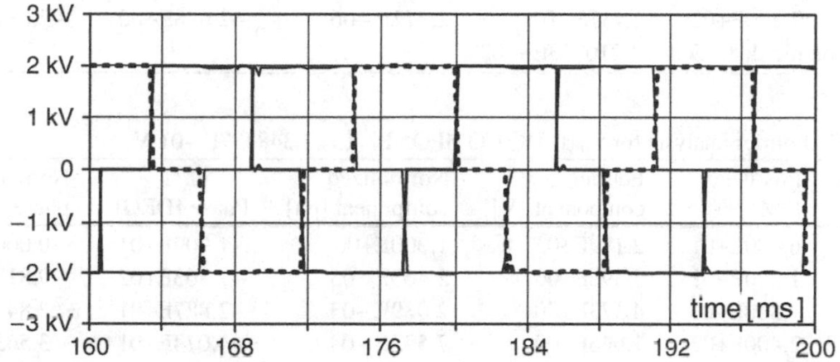

Fig. 8.43 Line-to-line voltages $v_{ab}(t)$, $v_{bc}(t)$, $v_{ca}(t)$

(f) Figures 8.42 and 8.43 show the line-to-line voltages $v_{AB}(t)$, $v_{BC}(t)$, $v_{CA}(t)$, and the line-to-line voltages $v_{ab}(t)$, $v_{bc}(t)$, $v_{ca}(t)$, respectively. The line-to-line voltages $v_{AB}(t)$, $v_{BC}(t)$, $v_{CA}(t)$ are sinusoidal and are imposed by the power system while the line-to-line voltages $v_{ab}(t)$, $v_{bc}(t)$, $v_{ca}(t)$ are distorted because of the rectifier limiting the voltage amplitude.

For $v_{AB}(t)$ one obtains the Fourier components of Table 8.6. FOURIER COMPONENTS OF TRANSIENT RESPONSE V(1,2).

For $v_{ab}(t)$ one obtains the Fourier components of Table 8.7. FOURIER COMPONENTS OF TRANSIENT RESPONSE V(12,16).

(g) Input line current $i_{AL}(t)$ of primary Δ is depicted in Fig. 8.44.

For $i_{AL}(t)$ one obtains the Fourier components of Table 8.8. FOURIER COMPONENTS OF TRANSIENT RESPONSE I(Lxsysa).

(h) The phase current $i_{Aph}(t)$ of primary Δ is depicted in Fig. 8.45.

For $i_{Aph}(t)$ one obtains the Fourier components of Table 8.9. FOURIER COMPONENTS OF TRANSIENT RESPONSE I(L1).

(i) The phase current $i_{aph}(t)$ of the secondary zigzag-Y winding is depicted in Fig. 8.46.

For $i_{aph}(t)$ one obtains the Fourier components of Table 8.10. FOURIER COMPONENTS OF TRANSIENT RESPONSE I(L5).

Table 8.6 Fourier analysis for $v_{AB}(t)$ DC COMPONENT = 3.263868E−02 V

Harmonic no	Frequency [HZ]	Fourier component [V]	Normalized component [pu]	Phase [DEG]	Normalized phase [DEG]
1	6.000E+01	8.485E+02	1.000E+00	3.213E−04	0.000E+00
2	1.200E+02	3.277E−02	3.863E−05	−1.016E+02	−1.016E+02
3	1.800E+02	1.015E−02	1.197E−05	−1.718E+02	−1.718E+02
4	2.400E+02	4.105E−03	4.838E−06	7.569E+01	7.569E+01
5	3.000E+02	4.435E−03	5.227E−06	−3.421E+01	−3.421E+01
6	3.600E+02	4.116E−03	4.852E−06	−1.202E+02	−1.202E+02
7	4.200E+02	3.007E−03	3.544E−06	1.445E+02	1.445E+02
8	4.800E+02	2.776E−03	3.272E−06	3.333E+01	3.333E+01
9	5.400E+02	3.023E−03	3.563E−06	−6.504E+01	−6.505E+01
10	6.000E+02	2.799E−03	3.299E−06	−1.574E+02	−1.574E+02
11	6.600E+02	2.389E−03	2.815E−06	1.003E+02	1.003E+02
12	7.200E+02	2.437E−03	2.872E−06	−2.628E+00	−2.632E+00

Total harmonic distortion = 4.210323E−-03%

Table 8.7 Fourier analysis for $v_{ab}(t)$ DC COMPONENT = 7.348437E−01 V

Harmonic no	Frequency [HZ]	Fourier component [V]	Normalized component [pu]	Phase [DEG]	Normalized phase [DEG]
1	6.000E+01	2.189E+03	1.000E+00	−1.909E+01	0.000E+00
2	1.200E+02	5.390E+00	2.463E−03	−1.363E+02	−9.812E+01
3	1.800E+02	4.573E+00	2.089E−03	−2.887E+01	2.841E+01
4	2.400E+02	1.646E+00	7.522E−04	−4.074E+01	3.563E+01
5	3.000E+02	4.328E+02	1.977E−01	8.410E+01	1.796E+02
6	3.600E+02	4.420E+00	2.020E−03	−8.943E+00	1.056E+02
7	4.200E+02	3.128E+02	1.429E−01	4.814E+01	1.818E+02
8	4.800E+02	4.568E+00	2.087E−03	−3.234E+01	1.204E+02
9	5.400E+02	4.258E+00	1.946E−03	2.487E+01	1.967E+02
10	6.000E+02	3.945E−01	1.803E−04	1.752E+02	3.661E+02
11	6.600E+02	1.926E+02	8.801E−02	1.511E+02	3.612E+02
12	7.200E+02	5.720E+00	2.614E−03	7.959E+01	3.087E+02

Total harmonic distortion = 2.594178E+01%

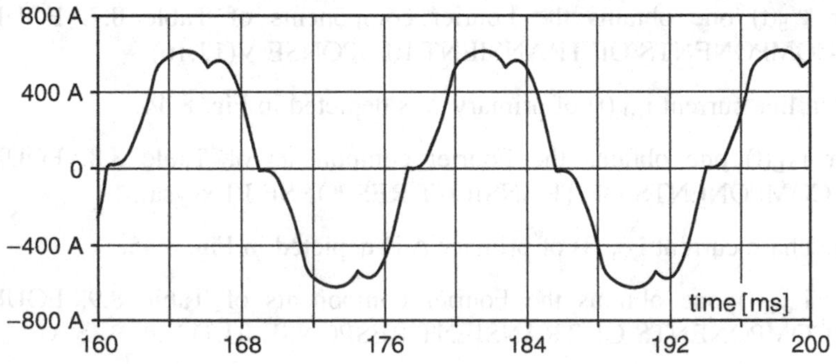

Fig. 8.44 Input line current $i_{AL}(t)$ of primary Δ winding of Δ-Y transformer

Table 8.8 Fourier analysis for $i_{AL}(t)$ DC COMPONENT = −3.180521E+00 A

Harmonic no	Frequency [HZ]	Fourier component [A]	Normalized component [pu]	Phase [DEG]	Normalized phase [DEG]
1	6.000E+01	6.410E+02	1.000E+00	1.186E+02	0.000E+00
2	1.200E+02	2.370E+00	3.698E−03	9.199E+01	−1.452E+02
3	1.800E+02	8.691E−01	1.356E−03	−8.694E+01	−4.427E+02
4	2.400E+02	6.599E−01	1.030E−03	−5.802E+01	−5.324E+02
5	3.000E+02	6.397E+01	9.980E−02	2.376E+01	−5.692E+02
6	3.600E+02	1.215E+00	1.896E−03	−1.002E+02	−8.117E+02
7	4.200E+02	3.264E+01	5.093E−02	−7.252E+01	−9.026E+02
8	4.800E+02	4.476E−01	6.984E−04	−1.692E+02	−1.118E+03
9	5.400E+02	3.560E−01	5.555E−04	−2.860E+01	−1.096E+03
10	6.000E+02	2.008E−01	3.132E−04	1.647E+01	−1.169E+03
11	6.600E+02	1.266E+01	1.976E−02	8.981E+01	−1.215E+03
12	7.200E+02	7.099E−01	1.108E−03	−2.756E+01	−1.451E+03

Total harmonic distortion = 1.138696E+01%

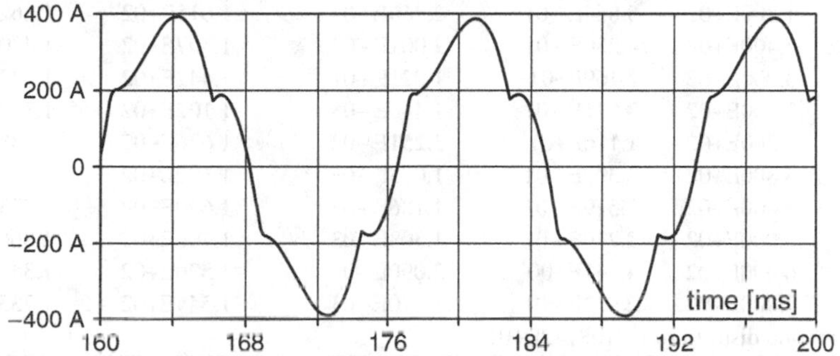

Fig. 8.45 Phase current $i_{Aph}(t)$ of primary Δ winding of Δ-Y transformer

Table 8.9 Fourier analysis for $i_{Aph}(t)$ DC COMPONENT = −2.267984E+00 A

Harmonic no	Frequency [HZ]	Fourier component [A]	Normalized component [pu]	Phase [DEG]	Normalized phase [DEG]
1	6.000E+01	3.695E+02	1.000E+00	1.484E+02	0.000E+00
2	1.200E+02	1.329E+00	3.597E−03	1.034E+02	−1.934E+02
3	1.800E+02	7.434E−01	2.012E−03	−3.978E+01	−4.850E+02
4	2.400E+02	6.372E−01	1.725E−03	−4.569E+01	−6.393E+02
5	3.000E+02	3.734E+01	1.011E−01	−6.291E+00	−7.483E+02
6	3.600E+02	6.879E−01	1.862E−03	−8.714E+01	−9.775E+02
7	4.200E+02	1.907E+01	5.162E−02	−4.205E+01	−1.081E+03
8	4.800E+02	2.314E−01	6.263E−04	−9.855E+01	−1.286E+03
9	5.400E+02	3.663E−01	9.916E−04	−1.596E+01	−1.351E+03
10	6.000E+02	3.393E−01	9.184E−04	−4.445E+00	−1.488E+03
11	6.600E+02	7.434E+00	2.012E−02	5.805E+01	−1.574E+03
12	7.200E+02	5.277E−01	1.428E−03	−2.650E+01	−1.807E+03

Total harmonic distortion = 1.153665E+01%

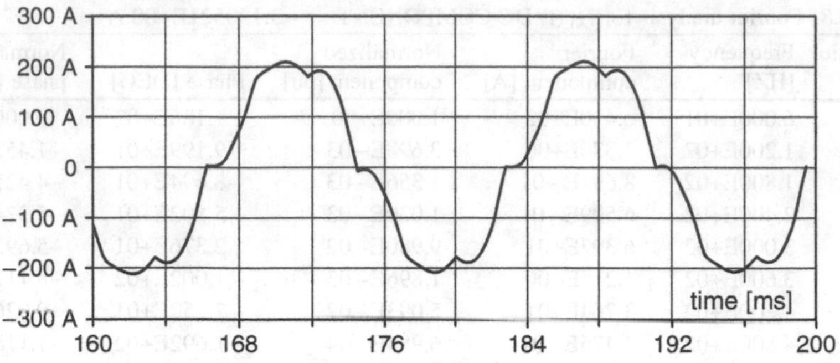

Fig. 8.46 Phase current $i_{aph}(t)$ of the secondary zigzag-Y winding

Table 8.10 Fourier analysis for $i_{aph}(t)$ DC COMPONENT = 1.100945E+00 A

Harmonic no	Frequency [HZ]	Fourier component [A]	Normalized component [pu]	Phase [DEG]	Normalized phase [DEG]
1	6.000E+01	2.125E+02	1.000E+00	−1.597E+00	0.000E+00
2	1.200E+02	5.758E−01	2.710E−03	−6.091E+01	−5.772E+01
3	1.800E+02	5.864E−01	2.760E−03	1.615E+02	1.662E+02
4	2.400E+02	4.250E−01	2.001E−03	1.407E+02	1.470E+02
5	3.000E+02	2.169E+01	1.021E−01	1.442E+02	1.522E+02
6	3.600E+02	3.071E−01	1.446E−03	1.102E+02	1.197E+02
7	4.200E+02	1.116E+01	5.251E−02	1.676E+02	1.788E+02
8	4.800E+02	2.302E−01	1.083E−03	1.192E+02	1.319E+02
9	5.400E+02	2.519E−01	1.186E−03	1.699E+02	1.843E+02
10	6.000E+02	2.779E−01	1.308E−03	1.706E+02	1.866E+02
11	6.600E+02	4.441E+00	2.090E−02	−1.520E+02	−1.344E+02
12	7.200E+02	2.912E−01	1.370E−03	1.544E+02	1.735E+02

Total harmonic distortion = 1.168142E+01%

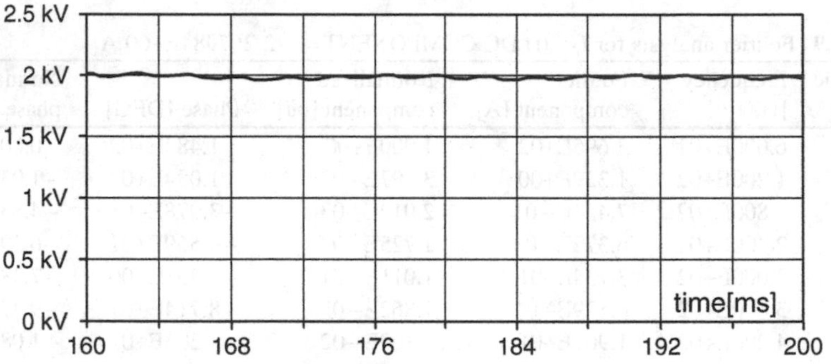

Fig. 8.47 Output DC voltage $v_{out}(t)$ of rectifier

(j) Output DC voltage $v_{out}(t)$ of rectifier is depicted in Fig. 8.47.

For $v_{out}(t)$ one obtains the Fourier components of Table 8.11. FOURIER COMPONENTS OF TRANSIENT RESPONSE V(17,18).

Table 8.11 Fourier analysis for $v_{out}(t)$ DC COMPONENT = 1.982841E+03 V

Harmonic no	Frequency [HZ]	Fourier component [V]	Normalized component [pu]	Phase [DEG]	Normalized phase [DEG]
1	6.000E+01	7.534E+00	1.000E+00	−7.072E+01	0.000E+00
2	1.200E+02	3.150E+00	4.181E−01	1.377E+02	2.791E+02
3	1.800E+02	1.214E+00	1.611E−01	1.598E+02	3.720E+02
4	2.400E+02	8.743E−01	1.160E−01	1.695E+02	4.524E+02
5	3.000E+02	4.622E−01	6.136E−02	1.480E+02	5.016E+02
6	3.600E+02	1.087E+01	1.443E+00	−1.270E+02	2.974E+02
7	4.200E+02	3.508E−01	4.656E−02	−1.790E+02	3.160E+02
8	4.800E+02	3.932E−01	5.218E−02	1.647E+02	7.305E+02
9	5.400E+02	3.058E−01	4.060E−02	1.763E+02	8.128E+02
10	6.000E+02	2.658E−01	3.528E−02	1.800E+02	8.872E+02
11	6.600E+02	2.617E−01	3.473E−02	1.666E+02	9.445E+02
12	7.200E+02	1.045E+00	1.387E−01	−4.537E+01	8.033E+02

Total harmonic distortion = 1.526312E+02%

Application Example 8.5: Decomposition of Unbalanced Load into Positive/Negative/Zero (+/−/0) Sequence Components

Determine (+/−/0) sequence components of the current set $\tilde{I}_a = 5A\angle 0°$, $\tilde{I}_b = 3A\angle 150°$, $\tilde{I}_c = 3A\angle 210°$. Can this set of currents exist in a three-wire system? View positive/negative/zero sequence components at Internet addresses [18,19]

Solution

According to [13] the zero-sequence component is defined as

$$\tilde{I}_a^{(0)} = \frac{1}{3}(\tilde{I}_a + \tilde{I}_b + \tilde{I}_c) = \frac{1}{3}(5\angle 0° + 3\angle 150° + 3\angle 210°)A$$
$$= -0.067\angle 0°A = 0.067\angle 180°A. \quad (8.13)$$

The zero-sequence component is not zero and therefore there must be a connection to ground and this set of currents cannot exist in a 3-wire system.

The positive-sequence system is defined as

$$\tilde{I}_a^{(1)} = \frac{1}{3}(\tilde{I}_a + a\cdot\tilde{I}_b + a^2\cdot\tilde{I}_c)$$
$$= \frac{1}{3}(5\angle 0° + 3\angle(150° + 120°) + 3\angle(210° + 240°) = 1.67\angle 0°A. \quad (8.14)$$

The negative-sequence system is defined as

$$\tilde{I}_a^{(2)} = \frac{1}{3}(\tilde{I}_a + a^2\cdot\tilde{I}_b + a\cdot\tilde{I}_c)$$
$$= \frac{1}{3}(5\angle 0° + 3\angle (150° + 240°) + 3\angle (210° + 120°)A = 3.4\angle 0°A. \quad (8.15)$$

These results can be checked because the following relation must hold for $\tilde{I}_a$

$$\tilde{I}_a = (\tilde{I}_a^{(0)} + \tilde{I}_a^{(1)} + \tilde{I}_a^{(2)}) = (-0.067\angle 0^\circ + 1.67\angle 0^\circ + 3.4\angle 0^\circ)\,A = 5\angle 0^\circ\,A.$$

Now we can express the b- and c-phase components in terms of the a-phase component as follows:

$$\tilde{I}_b^{(0)} = \tilde{I}_a^{(0)} = -0.067\angle 0^\circ A,\ \tilde{I}_b^{(1)} = a^2 \cdot \tilde{I}_a^{(1)} = 1.67\angle -120^\circ A,\ \tilde{I}_b^{(2)} = a \cdot \tilde{I}_a^{(2)} = 3.4\angle 120^\circ A, \tag{8.16}$$

and

$$\tilde{I}_c^{(0)} = \tilde{I}_a^{(0)} = -0.067\angle 0^\circ A,\ \tilde{I}_c^{(1)} = a \cdot \tilde{I}_a^{(1)} = 1.67\angle 120^\circ A,\ \tilde{I}_c^{(2)} = a^2 \cdot \tilde{I}_a^{(2)} = 3.4\angle -120^\circ A. \tag{8.17}$$

Recalculation of the given currents $\tilde{I}_b$, and $\tilde{I}_c$ from their components

$$\tilde{I}_b = \tilde{I}_a^{(0)} + a^2 \cdot \tilde{I}_a^{(1)} + a \cdot \tilde{I}_a^{(2)} = -0.067\angle 0^\circ A + 1.67\angle 240^\circ A + 3.4\angle 120^\circ A = 3\angle 150^\circ A,$$

$$\tilde{I}_c = \tilde{I}_a^{(0)} + a \cdot \tilde{I}_a^{(1)} + a^2 \cdot \tilde{I}_a^{(2)} = -0.067\angle 0^\circ A + 1.67\angle 120^\circ A + 3.4\angle 240^\circ A = 3\angle 210^\circ A.$$

Application Example 8.6: Per Unit (pu) Analysis of Unbalanced Three-Phase Load

It is easy to measure the line-to-line voltage magnitudes of the voltages $\tilde{V}_{ab}$, $\tilde{V}_{bc}$, and $\tilde{V}_{ca}$ with a voltmeter, that is $|\tilde{V}_{ab}|$, $|\tilde{V}_{bc}|$, and $|\tilde{V}_{ca}|$ (see Fig. 8.48). The measurement of the phase angles of a three-phase system requires an oscilloscope. Alternatively, one can determine the phase angles of the voltages $\tilde{V}_{ab}$, $\tilde{V}_{bc}$, and $\tilde{V}_{ca}$ by using the cosine law. The apparent base power is $S_{base}^{total} = 500$ kVA, and the base voltage is $|\tilde{V}_{base}^{L-L}| = 2,300$ V.

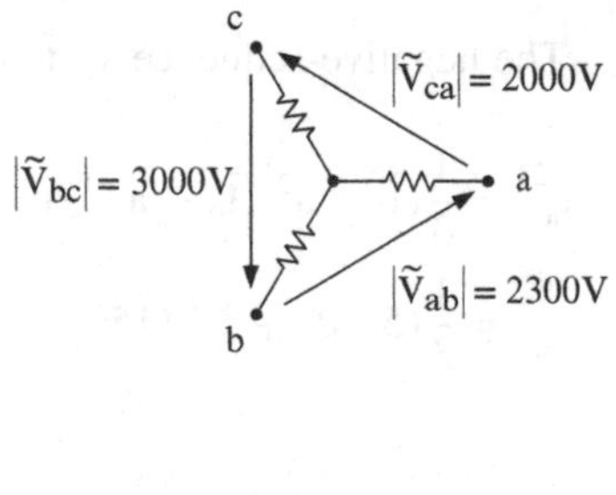

Fig. 8.48 Unbalanced three-phase load voltages, not to scale, clockwise phase rotation a, b, c

(a) Express all voltages of Fig. 8.48 in pu.
(b) Determine the phase angles of $\tilde{V}_{ab}$, $\tilde{V}_{bc}$, and $\tilde{V}_{ca}$ using pu values for the magnitudes. This avoids the use of large values (e.g., 2,300 V).
(c) Find the symmetrical components of $\tilde{V}_{ab}$, $\tilde{V}_{bc}$, $\tilde{V}_{ca}$.
(d) Determine the symmetrical components of the phase voltages $\tilde{V}_{an}$, $\tilde{V}_{bn}$, and $\tilde{V}_{cn}$.
(e) Determine the symmetrical components of the phase currents $\tilde{I}_a$, $\tilde{I}_b$, and $\tilde{I}_c$, if the 3 loads consist of the base impedances/resistances R_{base} or R^{pu}.

Solution

(a) There are several ways how the per-unit values can be defined. According to [13] the line-to-line voltage is taken as the base as well as the total 3-phase apparent power.

Therefore,

$$\left|\tilde{I}_{base}\right| = \frac{S_{base}^{total}}{\left|\tilde{V}_{base}^{L-L}\right|} = \frac{500\,\text{kVA}}{2,300\,\text{V}} = 217.4\,\text{A}, \qquad (8.18)$$

$$Z_{base} = R_{base} = \frac{\left|\tilde{V}_{base}^{L-L}\right|}{\left|\tilde{I}_{base}\right|} = \frac{2,300\,\text{V}}{217.4\,\text{A}} = 10.6\,\Omega, \qquad (8.19)$$

and the line-to-line voltage magnitudes are in per unit

$$\left|\tilde{V}_{ab}^{pu}\right| = \frac{2,300\,\text{V}}{2,300\,\text{V}} = 1.0\text{ pu}, \left|\tilde{V}_{bc}^{pu}\right| = \frac{3,000\,\text{V}}{2,300\,\text{V}} = 1.3\text{ pu, and } \left|\tilde{V}_{ca}^{pu}\right| = \frac{2,000\,\text{V}}{2,300\,\text{V}} = 0.87\text{ pu.} \qquad (8.20)$$

(b) Figure 8.48 shows the line-to-line voltage diagram where the angles A, B, and C are unknown (see Fig. 8.49). One can obtain these angles by applying the cosine law. If the voltage $\tilde{V}_{ab}^{pu} = 1\angle 0°$pu is taken as reference then one gets

$$\cos A = \frac{\left|\tilde{V}_{ab}^{pu}\right|^2 + \left|\tilde{V}_{ca}^{pu}\right|^2 - \left|\tilde{V}_{bc}^{pu}\right|^2}{2\cdot\left|\tilde{V}_{ab}^{pu}\right|\left|\tilde{V}_{ca}^{pu}\right|} = \frac{1+0.87^2-1.3^2}{2\cdot 1\cdot 0.87} = 0.038 \text{ or } A = 87.8°. \quad (8.21)$$

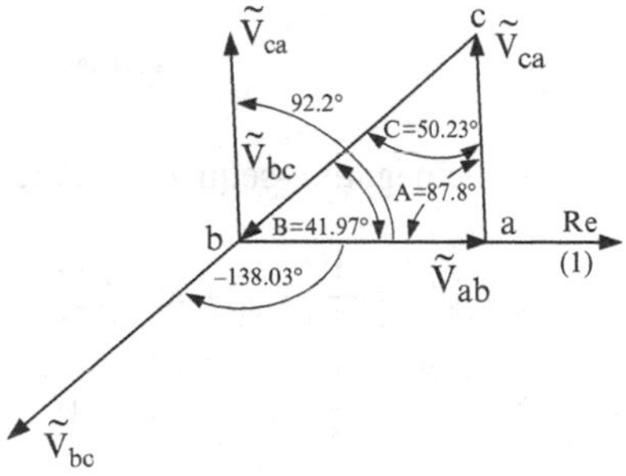

Fig. 8.49 Phasor diagram for the determination of the line-to-line voltages, magnitudes and phase angles, not to scale, clockwise phase rotation a, b, c

$$\cos B = \frac{\left|\tilde{V}_{bc}^{pu}\right|^2 + \left|\tilde{V}_{ab}^{pu}\right|^2 - \left|\tilde{V}_{ca}^{pu}\right|^2}{2 \cdot \left|\tilde{V}_{bc}^{pu}\right| \left|\tilde{V}_{ab}^{pu}\right|} = \frac{1.3^2 + 1^2 - 0.87^2}{2 \cdot 1.3 \cdot 1} = 0.744 \text{ or } B = 41.97^\circ. \quad (8.22)$$

$$\cos C = \frac{\left|\tilde{V}_{bc}^{pu}\right|^2 + \left|\tilde{V}_{ca}^{pu}\right|^2 - \left|\tilde{V}_{ab}^{pu}\right|^2}{2 \cdot \left|\tilde{V}_{bc}^{pu}\right| \left|\tilde{V}_{ca}^{pu}\right|} = \frac{1.3^2 + 0.87^2 - 1^2}{2 \cdot 1.3 \cdot 0.87} = 0.64 \text{ or } C = 50.23^\circ. \quad (8.23)$$

Knowing these angles one can determine the line-to-line voltages from phasor diagram Fig. 8.49.

From Figure E8.6.2 follows $\tilde{V}_{ab}^{pu} = 1\angle 0^\circ$ pu,

$$\tilde{V}_{bc}^{pu} = 1.3\angle -138.03^\circ \text{ pu}, \quad (8.24)$$

$$\tilde{V}_{ca}^{pu} = 0.87\angle 92.2^\circ \text{ pu}$$

(c) Calculation of the symmetrical components of $\tilde{V}_{ab}^{pu}$, $\tilde{V}_{bc}^{pu}$, and $\tilde{V}_{ca}^{pu}$. For the positive-sequence components one obtains:

$$\tilde{V}_{ab}^{(1)pu} = \frac{1}{3}\left[\tilde{V}_{ab}^{pu} + a \cdot \tilde{V}_{bc}^{pu} + a^2 \cdot \tilde{V}_{ca}^{pu}\right]$$
$$= \frac{1}{3}[1\angle 0^\circ + 1.3\angle(-138.03 + 120)^\circ + 0.87\angle(92.2 + 240)^\circ]$$

$$\tilde{V}_{ab}^{(1)pu} = 1.037\angle -15.03^\circ \text{pu}, \quad (8.25)$$

$$\tilde{V}_{bc}^{(1)pu} = \frac{1}{3}\left[\tilde{V}_{bc}^{pu} + a \cdot \tilde{V}_{ca}^{pu} + a^2 \cdot \tilde{V}_{ab}^{pu}\right]$$
$$= \frac{1}{3}[1.3\angle -138.03^\circ + 0.87\angle(92.2 + 120)^\circ + 1\angle 240^\circ]$$

$$\tilde{V}_{bc}^{(1)pu} = 1.037\angle 224.96^\circ \text{pu}, \quad (8.26)$$

$$\tilde{V}_{ca}^{(1)pu} = \frac{1}{3}\left[\tilde{V}_{ca}^{pu} + a \cdot \tilde{V}_{ab}^{pu} + a^2 \cdot \tilde{V}_{bc}^{pu}\right]$$
$$= \frac{1}{3}[0.87\angle 92.2^\circ + 1\angle 120^\circ + 1.3\angle(-138.03 + 240)^\circ]$$

$$\tilde{V}_{ca}^{(1)pu} = 1.037\angle 104.97^\circ \text{pu}. \quad (8.27)$$

For the negative-sequence components one obtains:

$$\tilde{V}_{ab}^{(2)pu} = \frac{1}{3}\left[\tilde{V}_{ab}^{pu} + a^2 \cdot \tilde{V}_{bc}^{pu} + a \cdot \tilde{V}_{ca}^{pu}\right]$$
$$= \frac{1}{3}[1\angle 0^\circ + 1.3\angle(-138.03 + 240)^\circ + 0.87\angle(92.2 + 120)^\circ]$$

$$\tilde{V}_{ab}^{(2)pu} = 0.269\angle 90.43^\circ \text{pu}, \quad (8.28)$$

$$\tilde{V}_{bc}^{(2)pu} = \frac{1}{3}\left[\tilde{V}_{bc}^{pu} + a^2 \cdot \tilde{V}_{ca}^{pu} + a \cdot \tilde{V}_{ab}^{pu}\right]$$
$$= \frac{1}{3}[1.3\angle -138.03^\circ + 0.87\angle(92.2+240)^\circ + 1\angle 120^\circ]$$

$$\tilde{V}_{bc}^{(2)pu} = 0.269\angle 210.38^\circ \text{pu}, \tag{8.29}$$

$$\tilde{V}_{ca}^{(2)pu} = \frac{1}{3}\left[\tilde{V}_{ca}^{pu} + a^2 \cdot \tilde{V}_{ab}^{pu} + a \cdot \tilde{V}_{bc}^{pu}\right]$$
$$= \frac{1}{3}[0.87\angle 92.2^\circ + 1\angle 240^\circ + 1.3\angle(-138.03+120)^\circ]$$

$$\tilde{V}_{ca}^{(2)pu} = 0.269\angle -29.61^\circ \text{pu}. \tag{8.30}$$

For the zero-sequence components one obtains:

$$\tilde{V}_{ab}^{(0)pu} = \frac{1}{3}\left[\tilde{V}_{ab}^{pu} + \tilde{V}_{bc}^{pu} + \tilde{V}_{ca}^{pu}\right] = 0. \tag{8.31}$$

Note there are no zero-sequence components because of the 3-wire system.

(d) Determination of the symmetrical components of the phase voltages $\tilde{V}_{an}^{pu}$, $\tilde{V}_{bn}^{pu}$, and $\tilde{V}_{cn}^{pu}$ from the phasor diagrams.

By inspection one gets from phasor diagram Fig. 8.50 the positive-sequence components of the line-to-neutral voltages as follows:

$$\tilde{V}_{an}^{(1)pu} = \tilde{V}_{ab}^{(1)pu}\angle -30^\circ \text{ pu} = 1.037\angle -45.03^\circ \text{ pu}, \tag{8.32}$$

with the base voltage of $\left|\tilde{V}_{base}^{L-N}\right| = \frac{\left|\tilde{V}_{base}^{L-L}\right|}{\sqrt{3}} = 1328\text{V}$.
Correspondingly,

$$\tilde{V}_{bn}^{(1)pu} = \tilde{V}_{bc}^{(1)pu}\angle -30^\circ \text{ pu} = 1.037\angle -165.04^\circ \text{ pu}, \tag{8.33}$$

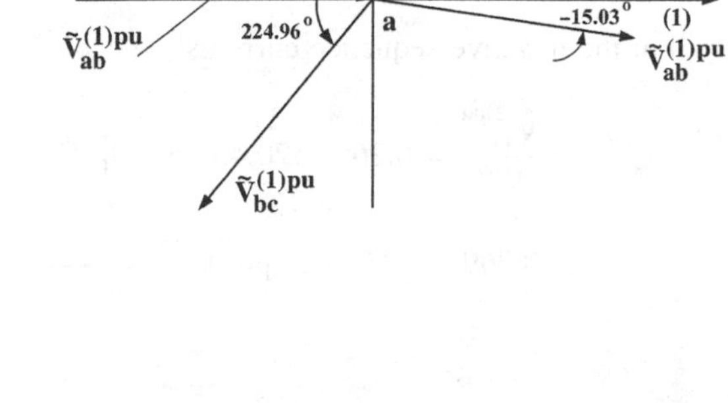

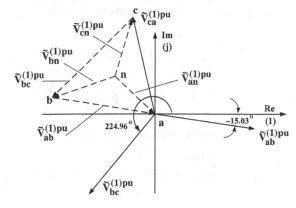

Fig. 8.50 Phasor diagram for the positive-sequence components of the line-to-line and line-to-neutral voltages, not to scale, clockwise phase rotation a, b, c

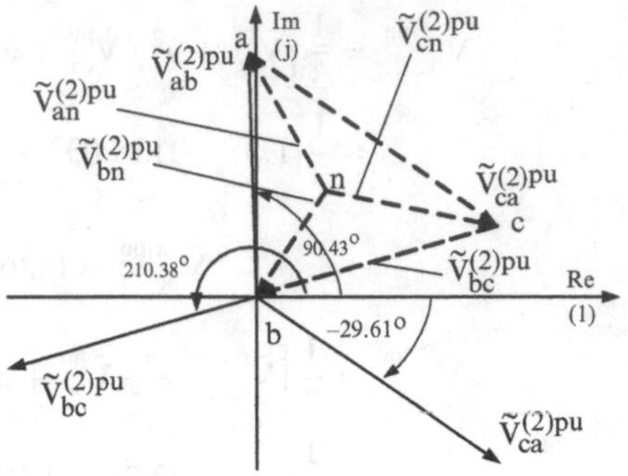

Fig. 8.51 Phasor diagram for the negative-sequence components of the line-to-line and line-to-neutral voltages, not to scale, counter-clockwise phase rotation a, b, c

$$\tilde{V}_{cn}^{(1)pu} = \tilde{V}_{ca}^{(1)pu}\angle -30^\circ \text{ pu} = 1.037\angle 74.97^\circ \text{ pu}. \tag{8.34}$$

By inspection one gets from phasor diagram Fig. 8.51 the negative-sequence components of the line-to-neutral voltages as follows:

$$\tilde{V}_{an}^{(2)pu} = \tilde{V}_{ab}^{(2)pu}\angle 30^\circ \text{ pu} = 0.269\angle 120.43^\circ \text{ pu}, \tag{8.35}$$

with the base voltage of $\left|\tilde{V}_{base}^{L-N}\right| = \frac{\left|\tilde{V}_{base}^{L-L}\right|}{\sqrt{3}} = 1328\text{V}$. Correspondingly,

$$\tilde{V}_{bn}^{(2)pu} = \tilde{V}_{bc}^{(2)pu}\angle 30^\circ \text{ pu} = 0.269\angle 240.38^\circ \text{ pu}, \tag{8.36}$$

$$\tilde{V}_{cn}^{(2)pu} = \tilde{V}_{ca}^{(2)pu}\angle 30^\circ \text{ pu} = 0.269\angle 0.39^\circ \text{ pu}. \tag{8.37}$$

As mentioned before, there are no zero-sequence components because of the 3-wire system and the neutral is not grounded.

(e) Determination the symmetrical components of the phase currents $\tilde{I}_a^{pu}$, $\tilde{I}_b^{pu}$, and $\tilde{I}_c^{pu}$ from Ohm's law.

With $R_{base} = 10.6\ \Omega$ or $R^{pu} = R_{load}/R_{base} = 10.6/10.6 = 1$ pu one obtains for the positive-sequence currents in per unit

$$\tilde{I}_a^{(1)pu} = \frac{\tilde{V}_{an}^{(1)pu}}{R^{pu}} = 1.037\angle -45.03^\circ \text{ pu}, \quad \tilde{I}_b^{(1)pu} = \frac{\tilde{V}_{bn}^{(1)pu}}{R^{pu}}$$
$$= 1.037\angle -165.04^\circ \text{ pu } \tilde{I}_c^{(1)pu} = \frac{\tilde{V}_{cn}^{(1)pu}}{R^{pu}} = 1.037\angle 74.97^\circ \text{ pu}. \tag{8.38}$$

For the negative-sequence currents

$$\tilde{I}_a^{(2)pu} = \frac{\tilde{V}_{an}^{(2)pu}}{R^{pu}} = 0.269\angle 120.43^\circ \text{ pu}, \quad \tilde{I}_b^{(2)pu} = \frac{\tilde{V}_{bn}^{(2)pu}}{R^{pu}}$$
$$= 0.269\angle -119.62^\circ \text{ pu}, \ \tilde{I}_c^{(2)pu} = \frac{\tilde{V}_{cn}^{(2)pu}}{R^{pu}} = 0.269\angle 0.39^\circ \text{ pu}. \tag{8.39}$$

There are no zero-sequence currents $\tilde{I}_a^{(0)pu} = \tilde{I}_b^{(0)pu} = \tilde{I}_{ca}^{(0)pu} = 0$ because the neutral is not grounded.

The ratio of the negative-sequence current to the positive-sequence current is

$$\frac{\left|\tilde{I}^{(2)}\right|}{\left|\tilde{I}^{(1)}\right|} = \frac{0.269}{1.037} = 25.9\% \ . \tag{8.40}$$

A rule of thumb permits a ratio of less than 8%. That is, the above value of 25.9% is not permissible.

Application Example 8.7: Calculation of Powers Using Symmetrical Components

Using the symmetrical components of Application Example 8.6, calculate the power absorbed in the load of Fig. 8.48, and alternatively check the answer by not relying on the symmetrical components.

Solution

The total per unit power in terms of per unit symmetrical components is

$$S_{3\phi}^{pu} = \tilde{V}_{an}^{(0)}\tilde{I}_{an}^{(0)*} + \tilde{V}_{an}^{(1)}\tilde{I}_{an}^{(1)*} + \tilde{V}_{an}^{(2)}\tilde{I}_{an}^{(2)*}, \tag{8.41}$$

or

$$S_{3\phi} = S_{3\phi}^{pu} \cdot 500\text{kVA} = 1.149 \cdot 500\text{kVA} \approx 575\text{kVA}. \tag{8.42}$$

Not relying on symmetrical components one can write with $R_\Delta = 3R_{base} = 3R_Y = 31.8\ \Omega$

$$S_{3\phi} = \frac{\left|\tilde{V}_{ab}\right|^2}{R_\Delta} + \frac{\left|\tilde{V}_{bc}\right|^2}{R_\Delta} + \frac{\left|\tilde{V}_{ca}\right|^2}{R_\Delta} = \frac{(2300)^2 + (3000)^2 + (2000)^2}{31.8} \approx 575\text{kVA}, \tag{8.43}$$

and because of the purely resistive load $S_{3\Phi} = P_{3\Phi} = 575$ kW.

8.3 Summary

The introduction of the 3-phase system followed the invention of the 2-phase (or poly-phase) induction motor by Tesla. This chapter discusses the voltages, currents and powers of 3-phase transformers based on the ideal transformer. The phase relationship between phase and line-to-line voltages and phase and line currents is explained with phasor diagrams. After the discussion of the single-phase power flow circuit (see Chapter 7), it is shown that a balanced three-phase power flow problem can be reduced to a single-phase system. The advantages of the Δ-Y and Δ-zigzag 3-phase transformers are highlighted in application examples: the Δ-zigzag transformer balances on the primary side unsymmetrical fundamental load currents –which may exist on the secondary side– and eliminates zero-sequence harmonic components in a similar manner as a Δ-Y configuration, however, in a more effective manner. The Δ-zigzag transformer harmonic current distortion on the primary Δ is significantly less than that of the Δ -Y connected transformer. The concept of power factor and its compensation via capacitor banks is addressed, and the imbalance of 3-phase current and voltage systems is illustrated in application examples based on symmetrical components. Lastly the per-unit system is introduced.

8.4 Problems

Problem 8.1: *Constancy of power flow in a 2-phase system*

Although the phase voltages and currents are time-dependent, any balanced poly-phase power system generates a constant (time-independent) power flow. A poly-phase system consists of at least $m=2$ phases or a higher number of phases, that is $m=3, 4, 5, \ldots$. A single-phase system where $m=1$ does not maintain the power to be constant, that is, independent of time. Show that for a two-phase balanced system ($m=2$) the power flow is constant and independent of time. The phase voltages are $e_{aa'}(t)=\sqrt{2}E_{ph}\cos\omega t$, $e_{bb'}(t)=\sqrt{2}E_{ph}\cos(\omega t-90°)$, and the currents leading the voltages by an angle of ϕ are $i_a(t)=\sqrt{2}I_{ph}\cos(\omega t+\phi)$, and $i_b(t)=\sqrt{2}I_{ph}\cos(\omega t-90°+\phi)$.

Problem 8.2: *Ideal transformer principles applied to 3-phase transformers*

In Application Example 8.1 for the Y-Δ connection (Fig. 8.13a) the secondary line-to-line and phase voltages as well as the secondary line and phase currents are derived as a function of the primary line-to-line voltages as well as the primary line and phase currents, respectively. The transformation ratio "a" plays a role in this derivation. Repeat the same derivation for

(a) Δ-Y connection (see Fig. 8.13b).
(b) Δ-Δ connection (see Fig. 8.13c).
(c) Y-Y connection (see Fig. 8.13d).

Problem 8.3: *Neutral current i_o in a balanced three-phase system*

Figure 8.52 shows the secondary winding of a Δ-Y three-phase transformer. Compute the current $i_0=i_a+i_b+i_c$ provided $i_a=I_{max}\cos(\omega t)$, $i_b=I_{max}\cos(\omega t-120°)$ and $i_c=I_{max}\cos(\omega t-240°)$. Note that $\cos(\alpha-\beta)=\cos\alpha\cos\beta+\sin\alpha\sin\beta$.

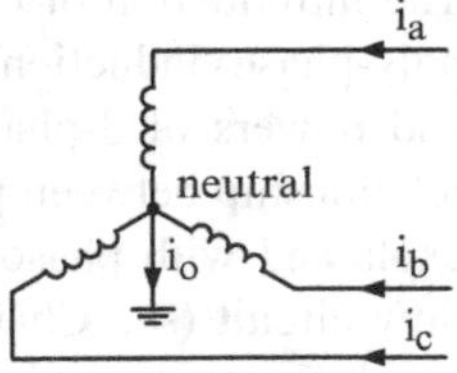

Fig. 8.52 Neutral current i_o in a three-phase Y transformer winding

Problem 8.4: *Ideal three-phase transformer feeding a three-phase load*

A three-phase Δ/Y-connected transformer bank consisting of three ideal single-phase transformers supplies a (total) three-phase apparent power of $S_{out} = 300$ kVA to a Y-connected load at a terminal phase voltage of $V_{load} = 2{,}400$ V (magnitude only) (see Fig. 8.53). One single-phase transformer has a rated secondary (low-voltage side) voltage of $V_{srat} = 2{,}400$ V and a rated primary (high-voltage side) voltage of $V_{prat} = 35{,}000$ V.

(a) Find the turns ratio N_p/N_s.
(b) Compute the load current I_{load} (magnitude only).
(c) Find V_s and I_s.
(d) Compute V_p and I_p.
(e) Compute I_L^P and the (total) three-phase apparent power S_{in} at the input terminals.

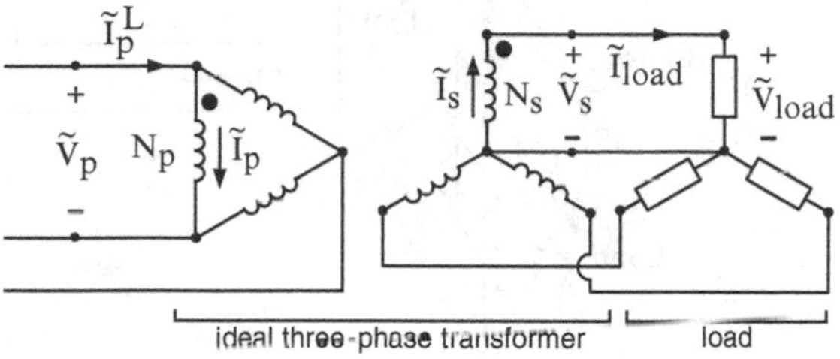

Fig. 8.53 Three-phase transformer bank supplying apparent power to Y-connected three-phase load

Problem 8.5: *Bank of three single-phase transformers*

Three single-phase transformers can be arranged to form a bank, as shown in Fig. 8.54. If the 3 single-phase transformers are excited by a balanced three-phase voltage system, one obtains for the fluxes $\Phi_A = \Phi_{max}\cos\omega t$, $\Phi_B = \Phi_{max}\cos(\omega t - 120°)$, and $\Phi_C = \Phi_{max}\cos(\omega t - 240°)$. Compute Φ_{total} within the common limb or leg of the three single-phase transformers.

Problem 8.6: *Modeling of three-phase transformer based on reluctances* $\mathfrak{R}$

The primary Y-connected winding of a three-limb, three-phase transformer shown in Fig. 8.55 is supplied with the currents $i_1 = I_{max}\sin\omega t$, $i_2 = I_{max}\sin(\omega t - 120°)$, and $i_3 = I_{max}\sin(\omega t - 240°)$. The Y-connected secondary winding is open-circuited. Assume that $N_1 = N_2 = N_3 = N$ and that the linear reluctance of each limb and each yoke is $\mathfrak{R}$ [20–22].

(a) Derive an expression in the time domain for the induced voltage in phase 1 of the three-limb transformer.
(b) Repeat the same analysis for a bank of three single-phase transformers. The assumptions about the number of turns, and the reluctance of limbs and yokes are the same.
(c) Compare the results of a) and b) in terms of the magnetizing current to obtain the same amplitude of the induced voltage in phase 1. Hint: Perform the comparison in the phasor domain.

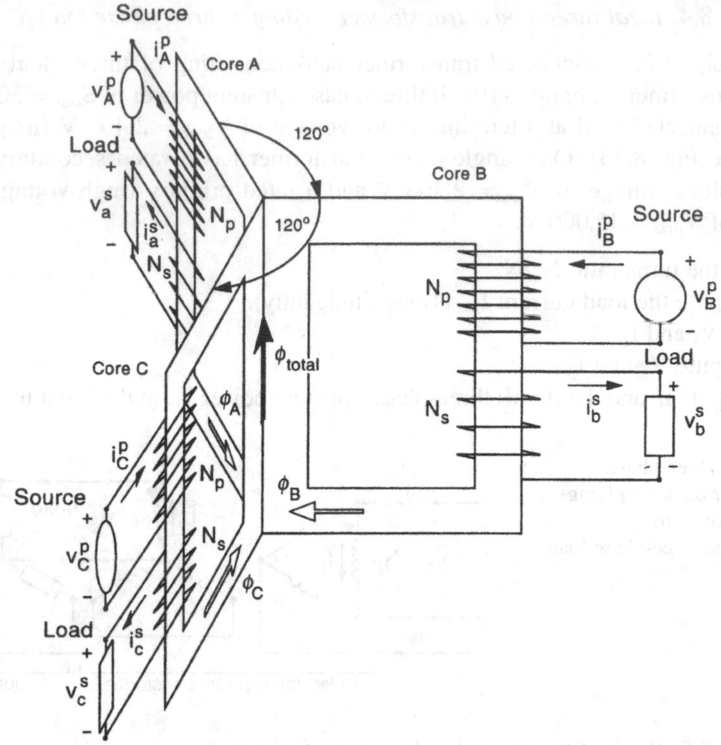

Fig. 8.54 Bank of three single-phase transformers

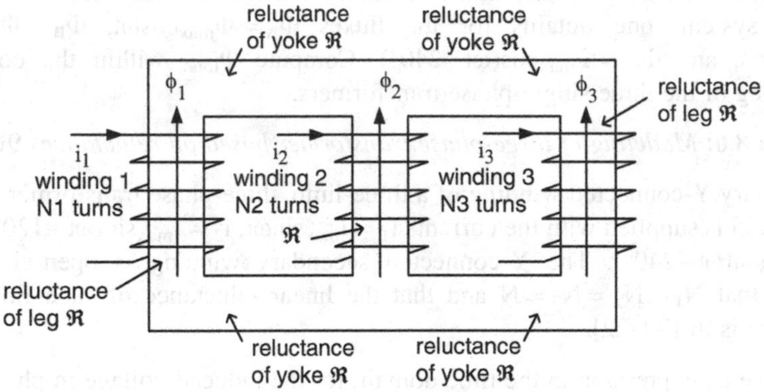

Fig. 8.55 Modeling of three-phase transformer with reluctances $\mathfrak{R}$

Problem 8.7: *Three-phase transformer in Δ-Y connection*

A three-phase transformer is connected on the primary in Δ and on the secondary in Y. Measurements at no-load reveal that for a primary line-to line voltage $V^{p}_{L-L} = 230\,\text{kV}$ the line-to-line voltage of the secondary is $V^{s}_{L-L} = 115\,\text{kV}$.

You may assume ideal transformer(s). What is the turns ratio $a = N_p/N_s$ of this Δ- Y transformer configuration?

Problem 8.8: *Three-phase system*

(a) Figure 8.56 represents a (power system) three-phase equivalent circuit, where the voltage sources are connected in Y and the loads in Δ. Compute the mesh currents $\tilde{i}_1$, $\tilde{i}_2$, $\tilde{i}_3$ and the phase currents $\tilde{i}_a$, $\tilde{i}_b$, $\tilde{i}_c$.

(b) Performing a Δ-Y transformation for the load circuit of Fig. 8.56 one obtains the circuit of Fig. 8.57 which represents a power system three-phase equivalent circuit, where the sources are connected in Y and the loads in Y. Compute $\tilde{i}_a$, $\tilde{i}_b$, $\tilde{i}_c$ and $\tilde{i}_o = \tilde{i}_a + \tilde{i}_b + \tilde{i}_c$.

(c) The three-phase equivalent circuit of Fig. 8.57 can be simplified to that of Fig. 8.58. Compute the phase current $\tilde{i}_{ph}$ and compare it with $\tilde{i}_a$ of Figs. 8.56 and 8.57.

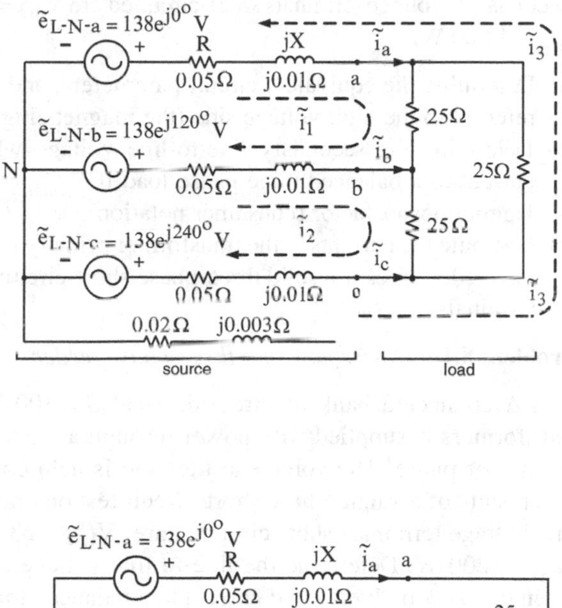

Fig. 8.56 Balanced three-phase circuit with load in Δ-connection

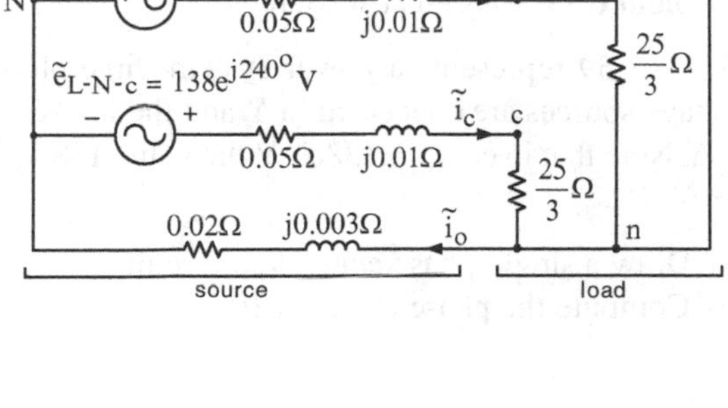

Fig. 8.57 Balanced three-phase circuit with load in Y-connection

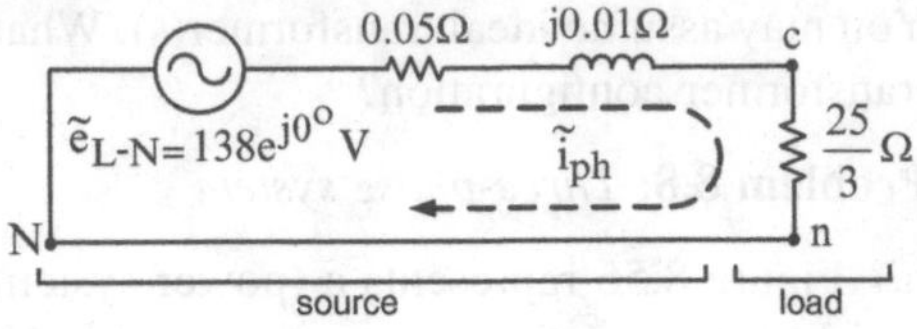

Fig. 8.58 Single-phase equivalent circuit for a balanced three-phase circuit

Problem 8.9: *Application of short-circuit test to a 3-phase transformer bank and power flow in a 3-phase feeder at any balanced load*

A Y-Y connected bank of three identical single–phase transformers, with one transformer having the ratings $S=200$ kVA, $V_p=2{,}400$ V, $V_s=120$ V at 60 Hz, is supplied with power through a feeder whose impedance is $Z_F^p = R_F^p + jX_F^p = (0.20 + j0.40)\Omega$ per phase. The line-to-line voltage at the sending end of the feeder is held constant by the utility at $V_{send}^{L-L} = \sqrt{3} \cdot 2400$ V.

The results of a short-circuit test on one of the transformers with its low (secondary)–voltage terminals short-circuited are $V_{psc}=40$ V, $f=60$ Hz, $I_{psc}=80$ A, $P_{psc}=1{,}000$ W.

(a) Determine the equivalent circuit parameters, and sketch the equivalent circuit referred to the high-voltage side (the magnetizing branch can be neglected).
(b) Determine the secondary line-to-line voltage when the bank delivers rated current to a balanced three-phase load if $V_{send}^{L-L} = \sqrt{3} \cdot 2400$ V at $\cos\phi=0.95$ lagging power factor (consumer notation).
(c) Compute the currents in the transformer primary and secondary windings and in the feeder wires if a solid three-phase short-circuit occurs at the secondary line terminals.

Problem 8.10: *Power flow in a three-phase feeder*

A Δ-Δ connected bank of three identical $S=300$ kVA, 2,400: 240 V, 60 Hz transformers is supplied with power through a feeder whose impedance is (1.0+ j1.0)Ω per phase. The voltage at the load is held constant at 240 V line-to-line. The results of a single-phase short-circuit test on one of the transformers with its low-voltage terminals short-circuited are: $V_{Hsc}=63.25$ V, f=60 Hz, $I_{Hsc}=20$ A, $P_{Hsc}=1200$ W. Determine the line-to-line voltage at the sending end $\tilde{V}_{send\_L-L}$ when the bank delivers rated current to a balanced three-phase unity power factor load.

Problem 8.11: *Three-phase system*

Figure 8.59 represents a power system three-phase equivalent circuit, where the voltage sources are connected in Y and the load (e.g., three-phase induction motor) in Δ. Note that in $\tilde{e}_{L-N} = 138e^{j0}$ V the value 138 V is its rms-value at a frequency of f = 60 Hz.

(a) Draw a single-phase equivalent circuit.
(b) Compute the phase current $\tilde{i}_a(t)$.

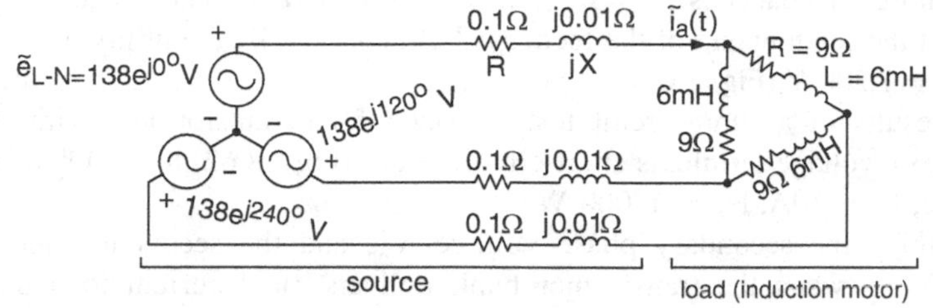

Fig. 8.59 Equivalent circuit of three-phase power system with load

Problem 8.12: *Power flow in a 3-phase feeder with a Δ-Y transformer*

A Δ-Y connected bank of three identical single–phase transformers, with one transformer having the ratings S = 300 kVA, V_p=2,400 V, V_s = 240 V at 60 Hz, is supplied with power through a feeder whose impedance is $Z_F^p = R_F^p + jX_F^p = (0.20 + j0.50)\Omega$ per phase. The line-to-line voltage (rms value) at the sending end of the feeder is held constant by the utility at $|\tilde{V}_{sendL-L}| = 2,400$ V. The results of a short-circuit test on one of the transformers with its low (secondary)–voltage terminals short-circuited are (see Fig. 8.60) $V_{sc}^p = 60$ V, f = 60Hz, $I_{sc}^p = 40$ A, $P_{sc}^p = 900$ W.

(a) Calculate $R_{eq\Delta}^p$, and $X_{eq\Delta}^p$. The (iron-core) conductance and the (magnetic) susceptance can be assumed to be zero.
(b) Draw the three-phase actual circuit consisting of three-phase feeder, Δ-Y transformer, and Y-connected three-phase load.
(c) Redraw the three-phase circuit of part b) but replace the Δ-Y transformer by an equivalent Y-Y transformer.
(d) Replace the three-phase circuit of part c) by a single-phase equivalent circuit (one-phase representation) including R_{eqy}^p, and X_{eqy}^p.
(e) Determine the secondary phase voltage $|\tilde{V}_{load}|$ when one phase supplies a load current of $|\tilde{I}_{ph}^s| = 30$ A at a power factor of $\cos\phi = 0.8$ lagging (consumer notation). Note, the utility controls/maintains a line-to-neutral sending end voltage of $|\tilde{V}_{sendL-N}| = 2400/\sqrt{3}\,V = 1386$ V.

Fig. 8.60 Single-phase transformer

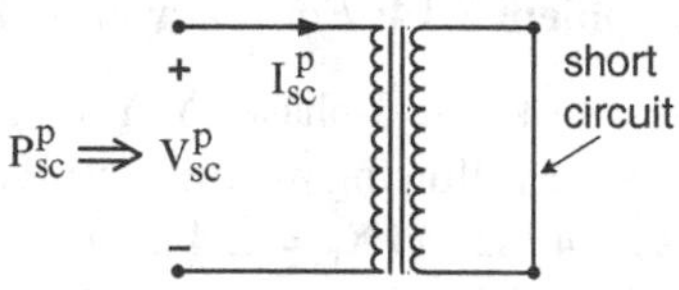

Problem 8.13: *Application of short-circuit and open-circuit tests to Δ-Y 3-phase transformer and power flow in a 3-phase feeder at balanced load*

A delta-wye Δ-Y connected bank of three identical single–phase transformers, with one transformer having the ratings (nominal values) S = 100 kVA, $V_{ph\Delta}^p = 7,200$ V, $V_{ph}^s = 240$ V at f = 60 Hz, is supplied with power through a feeder whose impedance is $Z_F^p = R_F^p + jX_F^p = (1 + j3)\Omega$ per phase. The line-to-line voltage at the sending end of the feeder is held constant by the utility at $V_{send\ L\text{-}L} = 7{,}200$ V (Fig. 8.61).

The results of a short-circuit test on one of the transformers with its low (secondary)–voltage terminals short-circuited are (Figs. 8.62, 8.63) $V_{sc}^p = 120$ V, f = 60 Hz, $I_{sc}^p = 50$A, $P_{sc}^p = 1{,}000$ W.

Determine the secondary phase voltage $\tilde{V}_{ph}^s$ and the secondary line-to-line voltage $\tilde{V}_{L-L}^s$ when the transformer bank delivers rated current to a balanced three-phase cosϕ = 0.8 lagging power-factor (consumer notation) load.

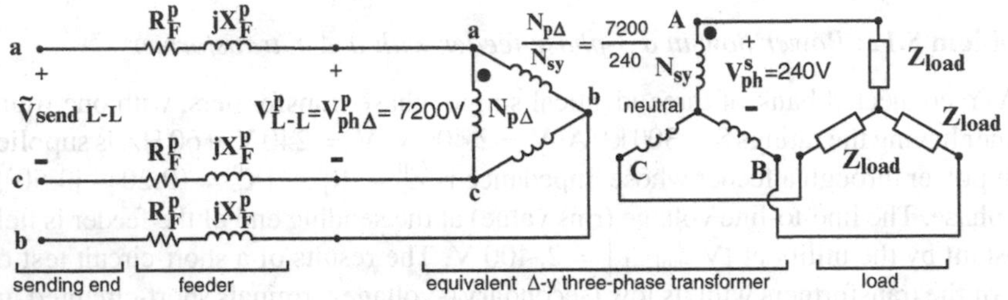

Fig. 8.61 Three-phase transformer with Δ-connected primary

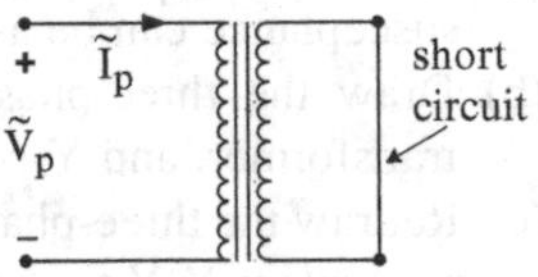

Fig. 8.62 Single-phase transformer under short-circuit condition

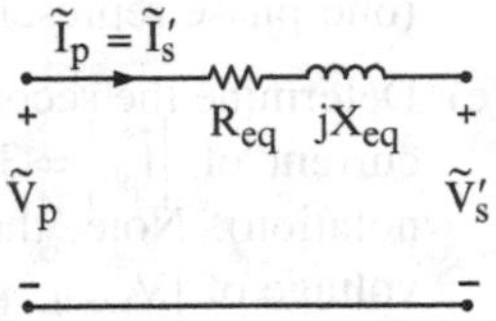

Fig. 8.63 Equivalent circuit of single-phase transformer referred to primary

Problem 8.14: *Efficiency of a (Y-Y) transformer*

A 180 kVA, 3-phase, Y-Y connected, $3800V_{L\text{-}L}/380V_{L\text{-}L}$ step-down transformer has the following parameters: $R_1 = 0.5\ \Omega$, $X_1 = 1.2\ \Omega$, $R_2 = 0.005\ \Omega$, $X_2 = 0.007\ \Omega$, $R_c = 4$ kΩ and $X_m = 2.4$ kΩ (R_1, X_1, R_c, X_m are referred to the high-voltage primary side and R_2, X_2 are referred to the low-voltage secondary side). The transformer supplies full load with a lagging power factor (consumer notation) of 0.8 at rated output voltage. Use the exact equivalent circuit of the transformer and calculate:

(a) The induced voltage $\tilde{E}_1$
(b) The core-loss current $\tilde{I}_c$, the magnetizing current $\tilde{I}_m$, and the primary current $\tilde{I}_1$ drawn from the supply
(c) The input power P_{in}, output power P_{out} and the transformer efficiency η

References

1. Neidhöfer, G.: "Early three-phase power," *IEEE Power and Energy Magazine*, Sept./Oct. 2007, pp. 88–100.
2. Tesla, N.: "A new system of alternate current motors and transformers," *American Institute of Electrical Engineers*, May 1888.
3. Fitzgerald, A. E.; Kingsley, Ch. Jr.; Umans, S. D.: *Electric Machinery*, 5th Edition, McGraw-Hill Publishing Company, New York, 1990.
4. Cowdrey, J. M.: Private communication.
5. *Electrical Transmission and Distribution Reference Book*, 4th Edition, 1950, Westinghouse Electric Corporation, East Pittsburgh, Pennsylvania.
6. Nilsson, J. W.: *Electric Circuits*, Addison-Wesley Publishing Company, Reading MA, 1983.
7. You, Y.; Fuchs, E. F.; Lin, D.; Barnes, P. R.: "Reactive power demand of transformers with DC bias," *IEEE Industry Applications Society Magazine*, Vol. 2, No. 4, July/Aug. 1996, pp.45–52.
8. Roberts, G. W.; Sedra, A. S.: *SPICE*, 2nd Edition, Oxford University Press, New York, 1997.
9. Tuinenga, P. W.: *SPICE, A Guide to Circuit Simulation & Analysis Using PSpice*, 2nd Edition, Prentice Hall, Englewood Cliffs, NJ 07632, 1988.
10. Rashid, M. H.: *SPICE for Power Electronics and Electric Power*, Prentice Hall, Englewood Cliffs, NJ 07632, 1993.
11. Banzhaf, W.: *Computer-Aided Circuit Analysis Using PSpice*, 2nd Edition, Regents/Prentice Hall, Englewood Cliffs, NJ, 1992.
12. Wagner, C. F.; Evans, R. D.: *Symmetrical Components*, McGraw-Hill Publishing Company, New York, 1933.
13. Grainger, J. J.; Stevenson, Jr., W. D.: *Power System Analysis*, McGraw-Hill Publishing Company, New York, 1994.
14. Fortescue, C. L.: Method of symmetrical co-ordinates applied to the solution of poly-phase networks, *Presented at the 34th annual convention of the AIEE (American Institute of Electrical Engineers) in Atlantic City, N.J.* on 28 July 1918. Published in: *AIEE Transactions*, Vol. 37, part II, pages 1027–1140 (1918).
15. Fuchs, E. F.; Hanna, W. J.: Measured efficiency improvements of induction motors with thyristor/triac controllers, *IEEE Transaction on Energy Conversion*, Vol. 17, No. 4, Dec. 2002, pp. 437–444.
16. Fuchs, E. F.; Masoum, M. A. S.: *Power Quality in Power Systems and Electrical Machines*, Elsevier/Academic Press, February 2008, 638 pages. ISBN: 978-0-12-369536-9.
17. Standard IEEE 519.
18. http://www.powerstandards.com/PQTeachingToyIndex.htm
19. http://www.powerstandards.com/
20. Fuchs, E. F. ; You, Y.; Roesler, D. J.: Modeling, simulation and their validation of three-phase transformers with three legs under DC bias, *IEEE Transaction on Power Delivery*, Vol. 14, No. 2, April 1999, pp. 443–449.
21. Fuchs, E. F.; You, Y.: "Measurement of (λ - i) characteristics of asymmetric three-phase transformers and their applications," *IEEE Transaction on Power Delivery*, Vol. 17, No. 4, Oct. 2002, pp. 983–990.
22. Fuchs, E. F.: Transformers, liquid filled, *Encyclopedia of Electrical and Electronics Engineering online*, Sept. 1, 2000, J. Webster (ed.), John Wiley & Sons, New York, NY 10158, paper No. 934 C, 40 pages, http://www.interscience.wiley.com:38/eeee/32/6132/w.6132-toc.html

Chapter 9
Force and Torque Production in Electromagnetic Circuits

In Chap. 6 inductors have been analyzed. In these devices there are no moving parts. In this chapter the force ($\vec{F}$) and torque ($\vec{T}$) production in electromagnetic circuits such as magnets, actuators, relays, linear and rotating machines will be discussed. These devices have a stationary part (stator) and a moving part (rotor). Figures 9.1a, b illustrate linear asynchronous actuator fields, Fig. 9.1c a linear synchronous actuator field [1, 2], and Fig. 9.1d shows a rotary permanent-magnet actuator [3].

There are basically three approaches for the calculation of the forces and torques developed in electromagnetic devices:

- The Lorentz force law [4]
- The principle of conservation of energy [4]
- The Maxwell stress [5–7].

The Lorentz force law can be applied to configurations without an iron core, while the second one can be applied to all types of configurations. Unfortunately, the more general conservation-of-energy approach is more difficult to apply because in the presence of an iron core the forces and torques act mainly on the iron-core material and not on the conductors. The Maxwell stress can be applied also in the presence of iron cores but requires numerical field solutions [8–12].

9.1 Lorentz Force Law Approach

The Lorentz force law for static (not moving) systems reads

$$\vec{F} = (\vec{I} \times \vec{B}) \cdot \ell[\mathrm{N}], \tag{9.1}$$

where $\vec{F}$ is the magnitude of the force in newtons, $\vec{I}$ is the magnitude of the current in amperes, $\vec{B}$ is the magnitude of the flux density in teslas, and ℓ is the axial length of the arrangement in meters. The direction of force, current and flux density can be

E.F. Fuchs and M.A.S. Masoum, *Power Conversion of Renewable Energy Systems*, DOI 10.1007/978-1-4419-7979-7_9, © Springer Science+Business Media, LLC 2011

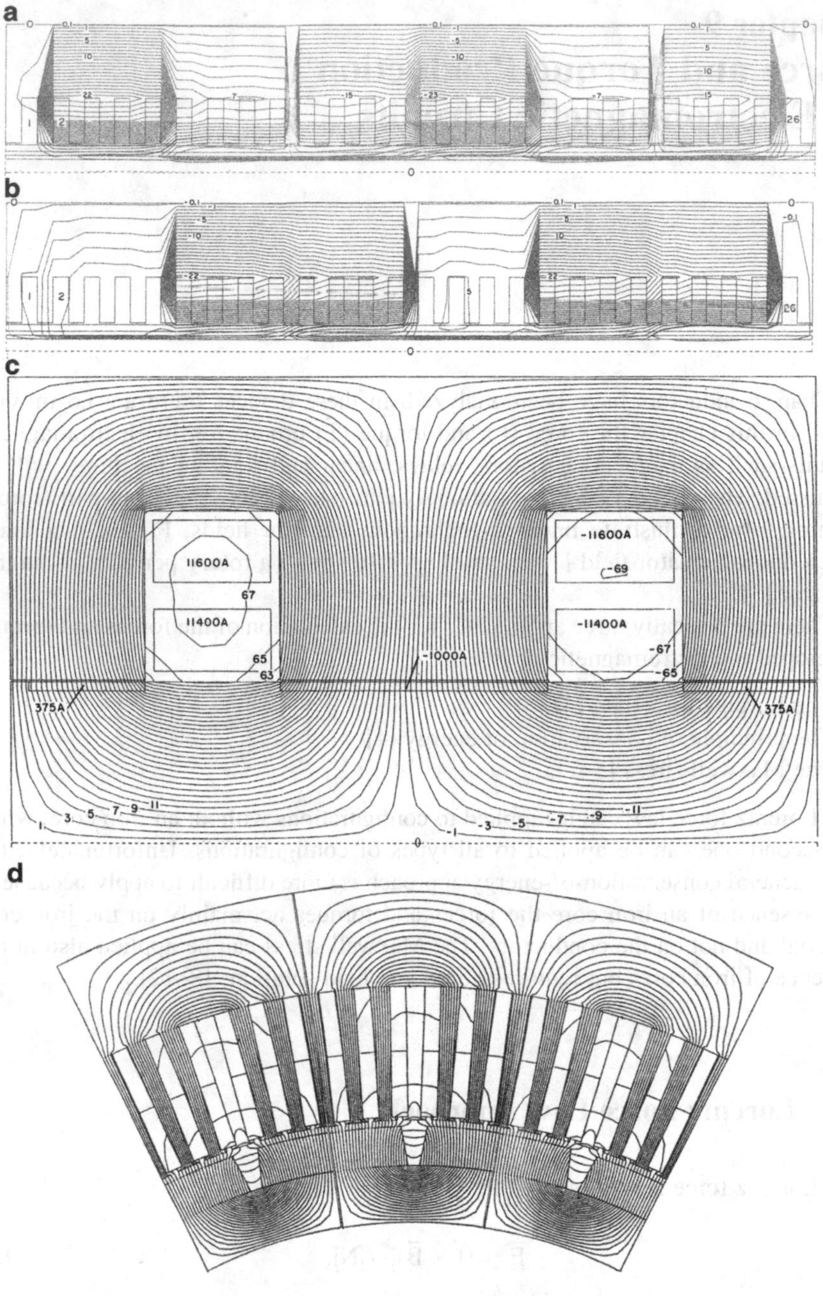

Fig. 9.1 (**a**) Real part of complex vector potential distribution of one half of a linear 2-pole asynchronous (induction motor) actuator with non-magnetic sleeve. All vector potentials must be multiplied by 10^{-3} Wb/m. (**b**) Imaginary part of complex vector potential distribution of one half of a linear 2-pole asynchronous (induction motor) actuator with non-magnetic sleeve. All vector potentials must be multiplied by 10^{-3} Wb/m. (**c**) Vector potential plot of one half of a linear synchronous actuator with 2 DC coils on the stator and 3 AC magnetic steel coils on moving part. All vector potentials must be multiplied by 10^{-3} Wb/m. (**d**) Permanent-magnet motor type rotary-type actuator with 18 poles

Fig. 9.2 Single-coil residing on nonmagnetic rotor. The current I_{1z} is positive when going into the paper x, y, z are rectangular and θ, r are polar coordinates

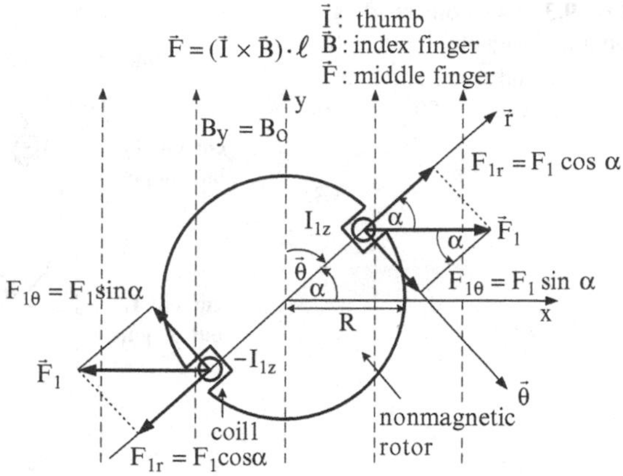

identified via the right-hand rule: the thumb corresponds to the direction of $\vec{I}$, the index finger to the direction of $\vec{B}$, and the middle finger gives then the direction of $\vec{F}$.

Application Example 9.1: Torque Calculation for Nonmagnetic Rotor with One Coil

A nonmagnetic rotor containing a single-turn coil is placed in a uniform magnetic field B_0, as shown in Fig. 9.2. The coil sides are residing at a radius R, and the wire carries a DC current I_{1z} as indicated. Find the z-directed torque as a function of rotor position when $I = 20$ A, $B_0 = 0.3$ T and $R = 0.05$ m. Assume that the rotor is $\ell = 0.3$ m long.

Solution

Apply the right-hand rule for the computation of the total force F_1 using the Lorentz-force equation

$$\vec{F} = (\vec{I} \times \vec{B}) \cdot \ell, \tag{9.2}$$

where $\vec{I}$ flows in the direction of the thumb, $\vec{B}$ in the direction of the index finger then $\vec{F}$ is in the direction of the middle finger. The force generated by coil side 1 through which current I_{1z} flows becomes $F_1 = I \cdot B \cdot \ell$ (note $\vec{I}$ and $\vec{B}$ are perpendicular and $\vec{I} \times \vec{B}$ becomes $I \cdot B$) with the tangential and radial components $F_{1\theta} = F_1 \sin\alpha$ and $F_{1r} = F_1 \cos\alpha$, respectively. The radial component does not contribute to rotational motion, while the tangential component produces the torque component

$$T_{I1z} = \vec{R} \times \vec{F}_{1\theta} = R \cdot F_{1\theta} = R \cdot F_1 \cdot \sin\alpha. \tag{9.3}$$

The current $(-I_{1z})$ contributes the torque component

$$T_{(-I1z)} = R \cdot F_{1\theta} = R \cdot F_1 \cdot \sin\alpha, \tag{9.4}$$

or the total torque developed by coil 1 is

$$T = T_{I1z} + T_{(-I1z)} = 2 \cdot I \cdot B \cdot \ell \cdot R \cdot \sin\alpha. \tag{9.5}$$

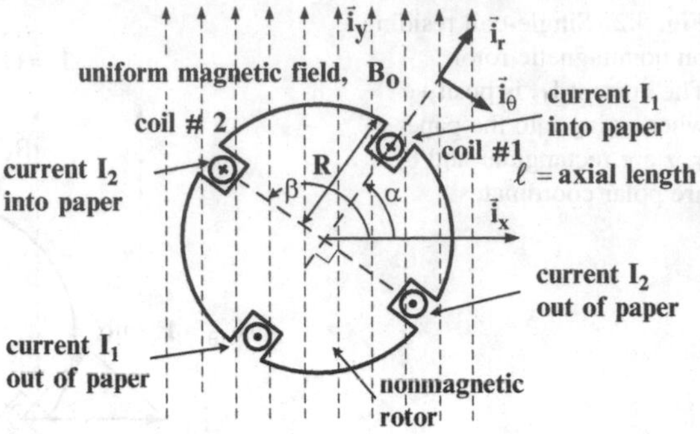

Fig. 9.3 Two-coils residing on a nonmagnetic rotor; note $\vec{i}_x$, $\vec{i}_y$, $\vec{i}_r$, and $\vec{i}_\theta$ are unit vectors; $\beta = \alpha + 90°$

For the given geometric dimensions, flux density and current values one obtains

$$T = \{2 \cdot 20 \cdot 0.3 \cdot 0.3 \cdot 0.05\} \sin\alpha = 0.18 \sin\alpha \ [\text{Nm}]. \tag{9.6}$$

Application Example 9.2: Application of the Lorentz Force Law to Two Coils

Consider the two-coil rotor of Fig. 9.3. Assume the two rotor windings to be carrying a constant current $I_1 = I_2 = I$, and the rotor to have a polar moment of inertia J.

(a) Find the force components F_r, F_θ and the resultant torque $\vec{T}$ due to the two coils.
(b) Find the equilibrium position α_0 of the rotor. Is it stable?
(c) Write the dynamic equation for this system.
(d) Find the natural frequency f_0 in hertz for incremental rotor motion α_1 around the equilibrium position α_0. Hint: linearize differential equation and use $\sin(\alpha_0 + \alpha_1) = (\sin\alpha_0 \ \cos\alpha_1 + \cos\alpha_0 \ \sin\alpha_1)$, $\cos(\alpha_0 + \alpha_1) = (\cos\alpha_0 \ \cos\alpha_1 - \sin\alpha_0 \ \sin\alpha_1)$.

Solution

(a) The Lorentz force law $\vec{F} = (\vec{I} \times \vec{B})\ell$ can be applied to circuits having components with permeability of free space (μ_o) only. The force component in $\vec{i}_r$ direction is

$$F_r = I_1 B_o \ell \cos\alpha, \tag{9.7}$$

and the force component in $\vec{i}_\theta$ direction is

$$F_\theta = I_1 B_o \ell \sin\alpha. \tag{9.8}$$

The torque due to coil #1 is

$$\vec{T}_{1\theta} = R\ I_1 B_o \ell \sin\alpha + R\ (-I_1 B_o \ell)\sin(\alpha + 180°) = R\ I_1 B_o \ell \sin\alpha + R\ I_1 B_o \ell \sin\alpha$$

$$\vec{T}_{1\theta} = 2R(I_1 B_o \ell)\ \sin\alpha. \tag{9.9}$$

The torque due to coil #2 is

$$\vec{T}_{2\theta} = 2R(I_1 B_o \ell)\ \sin\beta. \tag{9.10}$$

The total torque is with $\beta = \alpha + 90°$

$$\begin{aligned} T = \vec{T} = \vec{T}_{1\theta} + \vec{T}_{2\theta} &= 2R(I_1 B_o \ell)\sin\alpha + 2R(I_1 B_o \ell)\ \sin(\ \alpha + 90°) \\ &= 2RB_0\ell\{I_1 \sin\alpha + I_2 \cos\ \alpha\}. \end{aligned} \tag{9.11}$$

(b) At equilibrium the torque is $T = \vec{T} = 0$, that results for $I_1 = I_2$ in $I_1\sin\alpha_o + I_2\cos\alpha_o = 0$ or

$$\tan\alpha_o = \frac{\sin\alpha_o}{\cos\alpha_o} = -1 \text{ or } \alpha_o = \tan^{-1}(-1) = -0.785 \text{ rad or } \alpha_o = -45°. \tag{9.12}$$

That is, there is no net tangential component and the coils are at rest, the equilibrium position is stable.

(c) The dynamic equation is

$$J\frac{d^2\alpha}{dt^2} + T = 0 \tag{9.13}$$

or with (9.11) $2RB_oI\,\ell = K$

$$J\frac{d^2\alpha}{dt^2} = -K(\sin\alpha + \cos\alpha). \tag{9.14}$$

(d) Linearization around the equilibrium point α_o yields with $\sin\,\alpha_1 \approx \alpha_1$ the linearized differential equation

$$J\frac{d^2\alpha_1}{dt^2} = -K\alpha_1(\cos\alpha_0 - \sin\alpha_0). \tag{9.15}$$

Assume the solution of (9.15) to be of the form $\alpha_1 = \gamma_1\cos\,\omega_o t + \gamma_2\sin\,\omega_o t$, with $d\alpha_1/dt = -\gamma_1\omega_o\sin\,\omega_o t + \gamma_2\omega_o\cos\,\omega_o t$, and $d^2\alpha_1/dt^2 = -\gamma_1(\omega_o)^2\cos\,\omega_o t - \gamma_2(\omega_o)^2\sin\omega_o t$ one obtains with (9.15) the natural angular frequency $\omega_0 = 2\pi f_0$ or the natural frequency

$$f_0 = \frac{1}{2\,\pi}\sqrt{K(\cos\alpha_0 - \sin\alpha_0)/J)}. \tag{9.16}$$

Note that the rotors of the above two examples are nonmagnetic. The Lorentz force law can be applied only to nonmagnetic configurations, where no iron core is present. This is the disadvantage of this method. The next section will discuss a more general method for force and torque calculation based on the principle of conservation of energy. In this method forces and torques acting on windings and those acting on iron cores can be dealt with as well.

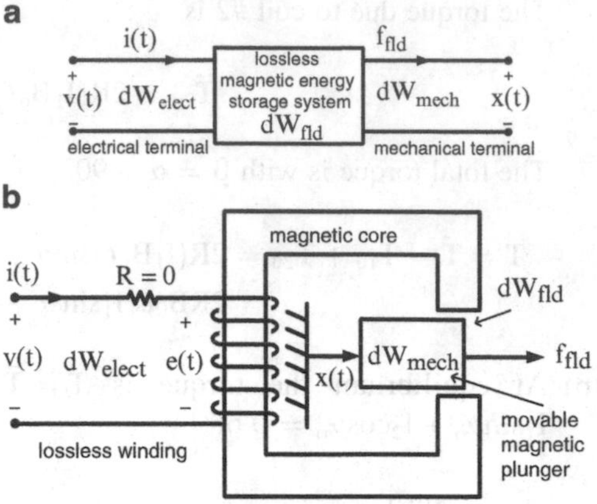

Fig. 9.4 (**a**) Lossless magnetic energy storage system with electric and mechanical terminals. (**b**) Actuator, an example for a magnetic energy storage system with electric and mechanical terminals

In Chap. 10 we will see that most electric machines use iron cores for guiding and increasing magnetic fluxes. As a consequence the magnetic forces and torques of electric machines act mainly on the iron cores because the windings are residing within slots and are therefore not greatly exposed to magnetic fields.

9.2 General Approach Based on Conservation of Energy

The principle of conservation of energy [4] states that energy can neither be created nor destroyed; it is merely changed in form. Figure 9.4a represents a lossless magnetic energy storage system with electric and mechanical terminals. Figure 9.4b is the magnetic circuit of an actuator which has an electrical input, either v(t) or i(t), and a mechanical output x(t).

The differential energy balance for such a system is given by the differential equation:

$$dW_{elec} = dW_{mech} + dW_{fld}, \tag{9.17}$$

where dW_{elec} is the differential electric energy input, dW_{mech} is the differential mechanical energy output, and dW_{fld} is the differential change in stored energy.

9.3 Magnetic Energy and Co-energy

9.3.1 *Magnetic Energy* $W_{fld}(\lambda, x)$ *and Force* $f_{fld}(\lambda, x)$ *as a Function of the Independent Variables* λ *and* x

The instantaneous power is $p(t) = e(t)i(t)$ and the differential energy input is

$$dW_{elec} = p(t)dt = e(t)i(t)dt, \tag{9.18}$$

with $e = d\lambda/dt$ follows

$$dW_{elec}(\lambda, x) = \frac{d\lambda}{dt} i(t)dt = i(\lambda, x)d\lambda. \tag{9.19}$$

The differential mechanical energy output is

$$dW_{mech} = f_{fld}dx, \tag{9.20}$$

where f_{fld} is the mechanical force due to the magnetic field within the system, and dx is the differential displacement. Thus with (9.17)

$$dW_{fld}(\lambda, x) = i(\lambda, x)d\lambda - f_{fld}(\lambda, x)dx, \tag{9.21}$$

note λ and x are independent variables, and $i(\lambda, x)$, $f_{fld}(\lambda, x)$ and $W_{fld}(\lambda, x)$ are dependent variables. Since our magnetic energy system is lossless, it is a conservative system and the value of W_{fld} is uniquely specified by the values of λ and x and does not depend on how λ and x are brought to their final values. Consider the λ–x plane of Fig. 9.5 in which two separate paths are shown over which (9.21) can be integrated.

Path #1 is the general case and is difficult to integrate unless both $i(\lambda, x)$ and $f_{fld}(\lambda, x)$ are known explicitly as a function of λ and x. Path #2 gives the same result and is much easier to integrate:

$$W_{fld}(\lambda_0, x_0) = \int_{\text{path \#2a}} i(\lambda, x)d\lambda - \int_{\text{path \#2a}} f_{fld}(\lambda, x)dx + \int_{\text{path \#2b}} i(\lambda, x)d\,\lambda - \int_{\text{path \#2b}} f_{fld}(\lambda, x)dx. \tag{9.22}$$

The first integral of the right-hand side of (9.22) is zero because for path #2a $\lambda = 0$, thus $d\lambda = 0$. The second integral is zero due to $\lambda = 0$ along path #2a: a zero field ($\lambda = 0$) produces a zero force $f_{fld} = 0$. The third integral is non-zero and the fourth

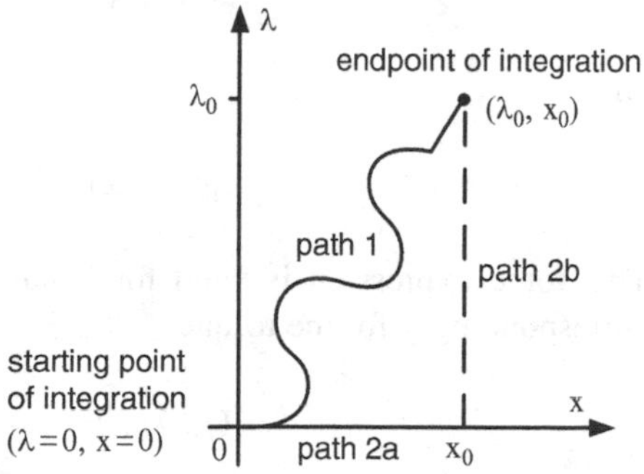

Fig. 9.5 Integration paths for $dW_{fld}(\lambda, x)$

Fig. 9.6 Graphical interpretation of magnetic energy $W_{fld}(\lambda, x)$ and of magnetic co-energy $W'_{fld}(i, x)$

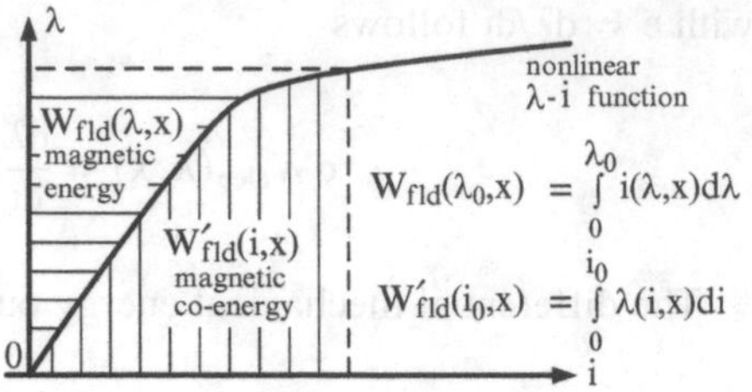

integral is zero, due to the fact that $x = x_0$ along path #2b and $dx = 0$. Therefore, the magnetic energy is defined as (Fig. 9.6)

$$W_{fld}(\lambda_0, x_0) = \int_0^{\lambda_0} i(\lambda, x_0)d\lambda, \tag{9.23}$$

where λ and x are the independent variables, or state variables. $W_{fld}(\lambda_0, x_0)$ is a dependent variable or a dependent state function.

It has been found in (9.21)

$$dW_{fld}(\lambda, x) = i(\lambda, x)d\lambda - f_{fld}(\lambda, x)dx, \tag{9.24}$$

which represents the total differential of the function $W_{fld}(\lambda, x)$.

For any function $F(x_1, x_2)$, the total differential of $F(x_1, x_2)$ with respect to the two independent state variables x_1 and x_2 can be written as

$$dF(x_1, x_2) = \frac{\partial F}{\partial x_1}dx_1 + \frac{\partial F}{\partial x_2}dx_2. \tag{9.25}$$

Application of the concept of (9.25) to $W_{fld}(\lambda, x)$ yields

$$dW_{fld}(\lambda, x) = \frac{\partial W_{fld}}{\partial \lambda}d\lambda + \frac{\partial W_{fld}}{\partial x}dx. \tag{9.26}$$

Comparing the coefficients of (9.24) and (9.26) yields

$$i(\lambda, x) = \frac{\partial W_{fld}(\lambda, x)}{\partial \lambda}, \tag{9.27}$$

and

$$f_{fld}(\lambda_0, x) = -\frac{\partial W_{fld}(\lambda_0, x)}{\partial x}. \tag{9.28a}$$

This force expression is valid for linear motion. For rotary motion one obtains correspondingly for the torque

$$T_{fld}(\lambda_0, \theta) = -\frac{\partial W_{fld}(\lambda_0, \theta)}{\partial \theta}. \tag{9.28b}$$

The direction of the force f_{fld} or of the torque T_{fld} can in some circumstances be obtained from the principle of minimum stored energy. As an example, consider the familiar case of a marble on the edge of a bowl. If we consider the marble and bowl to be an isolated system, then when the marble drops into the bowl, the potential energy will be converted to the kinetic energy of motion of the marble. In a similar manner a magnetic system reaches equilibrium, if the stored magnetic energy is a minimum. In Application Example 9.3 the magnetic energy is stored in the air gap, because of the assumed infinitely large permeability of the iron core: that is, the magnetic plunger comes to rest or finds its equilibrium point (where the stored energy is a minimum), if it symmetrically in x direction resides within the air gap of the C-core as demonstrated in Application Example 9.3.

Application Example 9.3: Polarities of Mechanical Forces Due to Magnetic Fields f_{fldx} and f_{fldy} Acting on a Plunger, Application of the Principle of Minimum Stored Energy in the Air Gap

In a C-core the flux Φ is produced by a permanent magnet PM. For the various plunger (piece of magnetic iron) positions as shown in Figs. 9.7–9.12 indicate the polarities of the forces f_{fldx} and f_{fldy} (e.g., $f_{fldx} < 0$).

Solution:

Figure 9.7: $f_{fldx} > 0$;
Figure 9.8: $f_{fldx} = 0$ (stored energy in air gap is smallest);
Figure 9.9: $f_{fldx} < 0$;

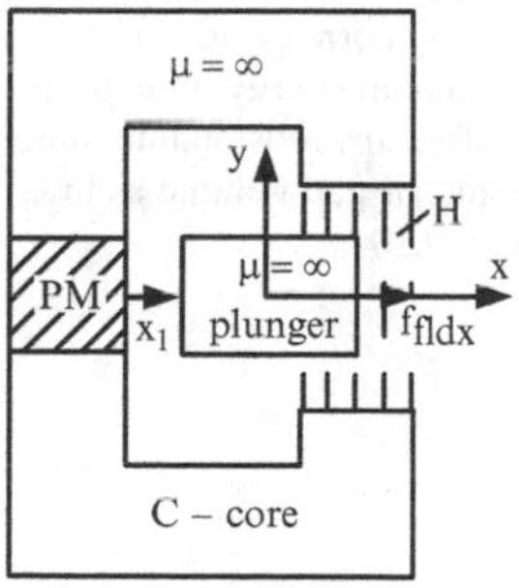

Fig. 9.7 Stored energy in air gap is larger than minimum energy

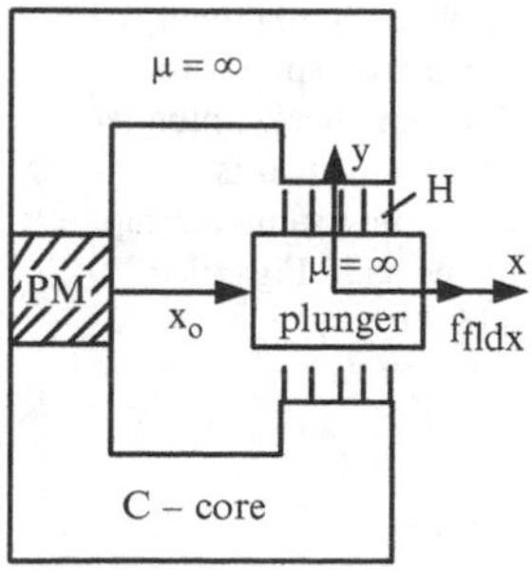

Fig. 9.8 Stored energy in air gap corresponds to minimum energy

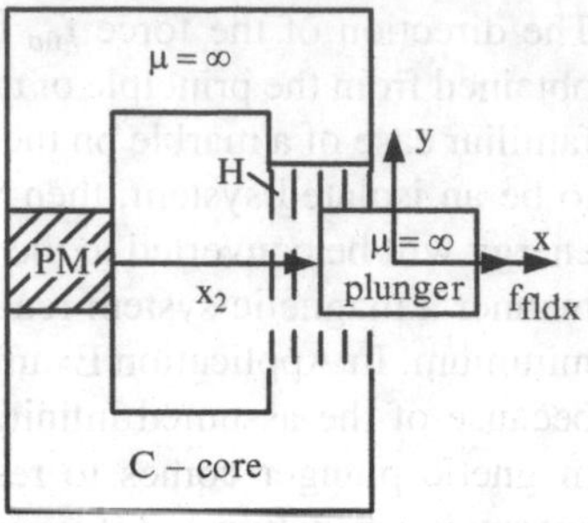

Fig. 9.9 Stored energy in air gap is larger than minimum energy

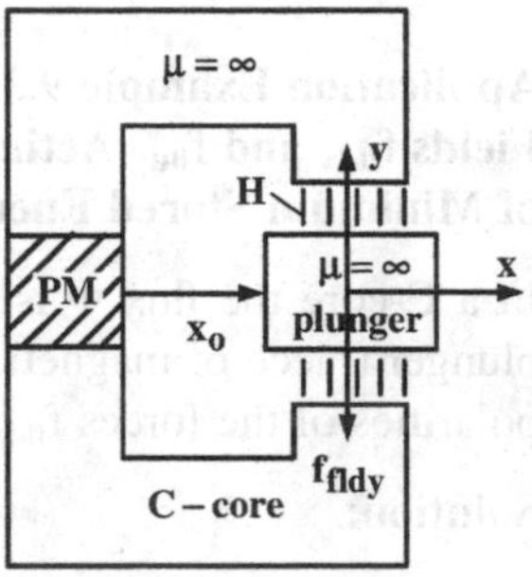

Fig. 9.10 Stored energy in air gap corresponds to minimum energy, plunger resides in a symmetrical manner in the air gap

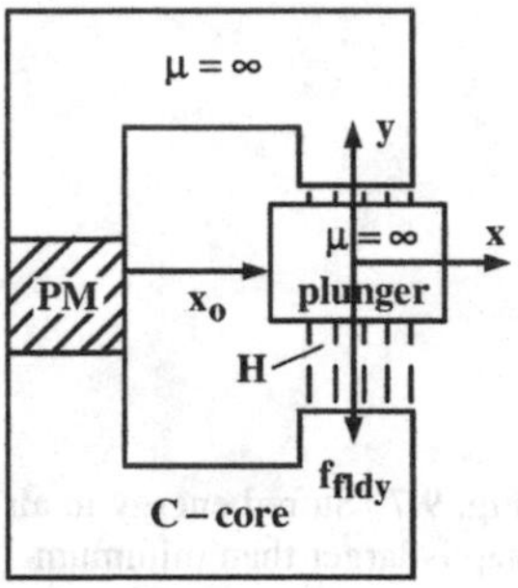

Fig. 9.11 Stored energy in air gap corresponds to minimum energy, plunger is shifted upwards, maintaining same air gap volume as in Fig. 9.10

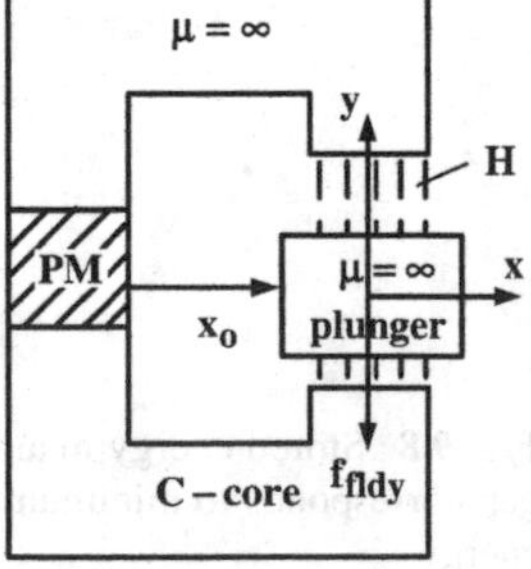

Fig. 9.12 Stored energy in air gap corresponds to minimum energy, plunger is shifted downwards, maintaining same air gap volume as in Figs. 9.10 and 9.11

Figure 9.10: $f_{fldy} = 0$ (stored energy in air gap is smallest);
Figure 9.11: $f_{fldy} = 0$ (stored energy in air gap is not changed and is smallest);
Figure 9.12: $f_{fldy} = 0$ (stored energy in air gap is not changed and is smallest).

Application Example 9.4: Computation of Magnetic Energy $W_{fld}(\lambda_o, x)$ and Force $f_{fld}(\lambda_o, x)$, if $i = f(\lambda)$ is Given

For $i = xA\lambda^3 + B\lambda^2 + C\lambda$ compute $W_{fld}(\lambda_o, x)$ and find $f_{fld}(\lambda_0, x)$.

Solution

$$W_{fld}(\lambda_0, x) = \int_0^{\lambda_0} i(\lambda, x)d\lambda, \tag{9.29}$$

$$\begin{aligned} W_{fld}(\lambda_0, x) &= \int_0^{\lambda_0} (xA\lambda^3 + B\lambda^2 + C\lambda)d\lambda = \left|\frac{xA\lambda^4}{4} + \frac{B\lambda^3}{3} + \frac{C\lambda^2}{2}\right|_0^{\lambda_0} \\ &= \frac{xA\lambda_0^4}{4} + \frac{B\lambda_0^3}{3} + \frac{C\lambda_0^2}{2}, \end{aligned} \tag{9.30}$$

$$f_{fld}(\lambda_0, x) = -\frac{\partial W_{fld}(\lambda_0, x)}{\partial x} = -\frac{A\lambda_0^4}{4}[N]. \tag{9.31}$$

Application Example 9.5: Computation of Magnetic Energy Compute $W_{fld}(\lambda, x)$ and Force $f_{fld}(\lambda, x)$, if $\lambda = f(i)$ is Given, Rationale for Introducing the Co-energy

For $\lambda = x{\cdot}a{\cdot}i^3 + bi^2 + ci$ compute $W_{fld}(\lambda_0, x)$ and find $f_{fld}(\lambda_0, x)$.

Solution

$$W_{fld}(\lambda_0, x) = \int_0^{\lambda_0} i(\lambda, x)d\lambda, \tag{9.32}$$

It is not easy to find $i = f(\lambda)$ from $\lambda = xai^3 + bi^2 + ci$ and, therefore, a second approach, that of the magnetic co-energy will have to be used, as demonstrated in the next Subsection.

9.3.2 *Magnetic Co-energy $W'_{fld}(i, x)$ and Force $f_{fld}(i, x)$ as a Function of the Independent Variables i and x*

The magnetic co-energy $W'_{fld}(i, x)$ is defined as a function of the independent variables or state variables i and x such that (Fig. 9.6)

$$W'_{fld}(i, x) = i\lambda - W_{fld}(\lambda, x), \tag{9.33}$$

this new dependent state function $W'_{fld}(i, x)$ is graphically identified in Fig. 9.6. The differential of (9.33) results in $dW'_{fld}(i, x) = d(i\lambda) - dW_{fld}(\lambda, x)$, where $d(i\lambda) = id\lambda + \lambda di$, thus with (9.21) one obtains $dW'_{fld}(i, x) = id\lambda + \lambda di - id\lambda + f_{fld}dx$, or

$$dW'_{fld}(i, x) = \lambda(i, x)di + f_{fld}(i, x)dx. \tag{9.34}$$

Comparing the coefficients of (9.34) with the total differential $dW'_{fld}(i, x) = \frac{\partial W'_{fld}}{\partial i}di + \frac{\partial W'_{fld}}{\partial x}dx$ generates the identities

$$\lambda(i, x) = \frac{\partial W'_{fld}(i, x)}{\partial i} \text{ or } W'_{fld}(i_0, x) = \int_0^{i_0}\lambda(i, x)di, \tag{9.35}$$

$$f_{fld}(i_0, x) = \frac{\partial W'_{fld}(i_0, x)}{\partial x}. \tag{9.36a}$$

This force expression is valid for linear motion. For rotary motion one obtains correspondingly for the torque

$$T_{fld}(i_0, \theta) = \frac{\partial W'_{fld}(i_0, \theta)}{\partial \theta}. \tag{9.36b}$$

Application Example 9.6: (Example 9.5 Revisited)

For $\lambda = xai^3 + bi^2 + ci$ compute $W'_{fld}(i_0, x)$ and find $f_{fld}(i_0, x)$

Solution

$$W'_{fld}(i_0, x) = \int_0^{i_0}\lambda(i, x)di = \int_0^{i_0}(xai^3 + bi^2 + ci)di = \left|\frac{xai^4}{4} + \frac{bi^3}{3} + \frac{ci^2}{2}\right|_0^{i_0}$$

$$= \frac{xai_0^4}{4} + \frac{bi_0^3}{3} + \frac{i_0^2}{2}, \tag{9.37}$$

$$f_{fld}(i_0, x) = \frac{\partial W'_{fld}(i_0, x)}{\partial x} = \frac{ai_0^4}{4}\ [\text{N}]. \tag{9.38}$$

Application Example 9.7: Calculation of Mechanical Force f_{fld} from Magnetic Energy $W_{fld}(\lambda, x)$ and Magnetic Co-energy $W'_{fld}(i, x)$ Generating the Same Results

For the magnetic field transducer of Fig. 9.13 it has been determined experimentally that the flux linkages λ relate to the displacement x as $\lambda = \frac{\sqrt{i}}{3(1 - ax^3)}$, where $a = 10^4\ (1/m^3)$, i is measured in amperes, and x is measured in meters.

(a) Use the magnetic co-energy $W'_{fld}(i, x)$ to calculate the force due to the magnetic field (fld), that is, $f_{fld}(i,x)$

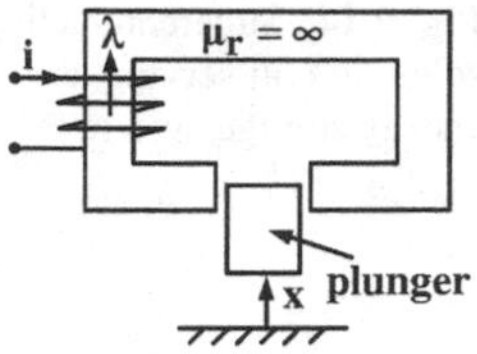

Fig. 9.13 Magnetic transducer translating the displacement x into either the electrical variable $\lambda = N\Phi$ or i

(b) Use the magnetic energy $W_{fld}(\lambda, x)$ to calculate the force due to the magnetic field, that is, $f_{fld}(\lambda, x)$

(c) Show that the calculated force in Part (a) $f_{fld}(i, x)$ is identical to that calculated in Part (b) $f_{fld}(\lambda, x)$.

Solution

(a) The magnetic co-energy is defined as

$$W'_{fld}(i, x) = \int_0^{\lambda} \lambda di = \int_0^{\lambda} \frac{i^{\left(\frac{1}{2}\right)}}{3(1-a\cdot x^3)} di = \frac{2\cdot i^{\left(\frac{3}{2}\right)}}{9(1-a\cdot x^3)}, \tag{9.39}$$

and the force due to the magnetic co-energy is

$$f_{fld}(i, x) = \frac{\partial W'_{fld}(i, x)}{\partial x} = \frac{2a\cdot i^{\left(\frac{3}{2}\right)} x^2}{3(1-a\cdot x^3)^2}. \tag{9.40}$$

(b) The magnetic energy is

$$W_{fld}(\lambda, x) = \int_0^{\lambda} i d\lambda = \int_0^{\lambda} \lambda^2 9(1-a\cdot x^3)^2 d\lambda = 3\lambda^3(1-a\cdot x^3)^2, \tag{9.41}$$

and the force due to the magnetic energy is

$$f_{fld}(\lambda, x) = -\frac{\partial W_{fld}(\lambda, x)}{\partial x} = 18\lambda^3(1-a\cdot x^3)a\cdot x^2. \tag{9.42}$$

(c) To show that (9.40) and (9.42) are identical introduce $\lambda^3 = \frac{i^{\left(\frac{3}{2}\right)}}{27(1-a\cdot x^3)^3}$ into (9.42) resulting in the identical relation as given by (9.40).

Application Example 9.8: Storage of Electrical Energy in a Superconducting Solenoid Coil

Electrical energy can be stored in the magnetic field of a superconducting coil without an iron core. Figure 9.14 illustrates such a superconducting solenoid with radius r_o and height h. The magnetic field inside this solenoid is axially directed

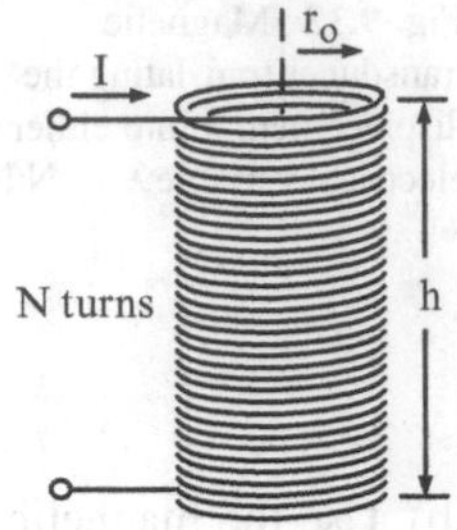

Fig. 9.14 Superconducting solenoid coil serving as energy storage system

and essentially uniform and equal to $H = (N \cdot I)/h$, where $N = 1{,}000$ turns is the number of turns and $I = 5$ kA is the DC current within the solenoid coil. The magnetic field outside the solenoid coil can be neglected.

(a) Calculate the magnetic field strength H and the flux density B for $h = 1$ m and the above-given number of turns N and DC excitation current I.
(b) Calculate the co-energy stored in the coil $W'_{fld}(i, r)$ for the above conditions and $r_o = 1$ m.
(c) How many hours can the co-energy stored in the solenoid coil supply a residence if 0.5 kW are used per hour?
(d) Calculate the radial pressure p_{radial} in newtons per square meter (N/m^2) for the above conditions.
(e) Calculate the pressure in $\frac{\text{lbs} - \text{force}}{\text{in.}^2}$ (or in colloquial terms pounds per inch squared = psi), if $\frac{1\text{N}}{\text{m}^2} = 1.45 \cdot 10^{-4} \frac{\text{lbs} - \text{force}}{\text{in.}^2}$. How does this calculated pressure value compare with that of a tire of a car?
(f) In which direction does the pressure act on the solenoid coil?
(g) How could the energy storage system be redesigned to fit in an automobile?
(h) Provided an internal combustion engine has an efficiency of 18%, how many liters of gasoline must be in the tank to provide the energy of 10 kWh at the wheels?
(i) How heavy (kg-force) will be a lithium-ion battery to store 10 kWh?

Solution

(a) The magnetic field strength is $H = (N \cdot I)/h = 5 \cdot 10^6$ A/m and the flux density is $B = \mu_o H = 6.28$ T.

(b) The magnetic co-energy is (neglecting the field outside the solenoid)

$$W'_{fld}(i, r) = \int_{\text{vol. of solenoid}} \frac{\mu_o}{2} H^2 d\text{Volume} = \frac{\mu_o}{2} \frac{N^2 I_o^2}{h^2} (\pi \cdot r_o^2 h) = \frac{\mu_o}{2} \frac{N^2 I_o^2}{h} (\pi \cdot r_o^2)$$
$$= 13.7\,\text{kWh}. \tag{9.43}$$

(c) The solenoid can supply for 13.7 kWh/(0.5 kW) = 27.38 h the house with power.

(d) The radial magnetic force developed is

$$f_{fld}(i_o, r_o) = \frac{\partial W'_{fld}}{\partial r}\bigg|_{r=r_o} = \mu_o H^2 \, \pi r_o h = \frac{\mu_o N^2 I_o^2 \pi r_o}{h}, \tag{9.44}$$

and the radial pressure – defined as force divided by surface area – is

$$\text{pressure} = \frac{\text{force}}{\text{cylinder area}} = \frac{f_{fld}(i_o, r_o)}{2\pi r_o h} = \frac{\frac{\mu_o N^2 I_o^2 \pi r_o}{h}}{2\pi r_o h} = \frac{\mu_o N^2 I_o^2}{2h^2}$$
$$= 15.71 \cdot 10^6 \frac{N}{m^2}. \tag{9.45}$$

(e) The pressure in psi is pressure = 2.28 kpsi. The pressure within the tire of a car is $\text{pressure}_{car} = 40$ psi, that is, the pressure applied to the solenoid cylindrical surface is 57 times larger than that applied to the wall of a tire.
(f) The pressure acts in r direction because the pressure value is positive or the pressure acts in outward direction.
(g) In order to fit the solenoid within the trunk of the car and to increase the magnetic co-energy of the solenoid according to (9.43) the height h can be decreased and the radius r can be increased, which will result in an increased pressure.
(h) At an efficiency of 18% the total required energy is 10 kWh/ 0.18 = 55.56 kWh. One liter of gasoline contains energy of 8.8 kWh, that is, 6.3 ℓ of gasoline are required to provide the energy of 10 kWh at the wheel.
(i) The energy density of lithium-ion battery is (100–160) Wh/(kg-force). The weight of a battery storing 10 kWh is therefore (100–62.5) kg-force. Note that a battery cannot be completely discharged. Assuming a depth of discharge of DoD = 50% the battery weight will be (200–125) kg-force.

Application Example 9.9: Computation of Magnetically Developed Forces from Stored Magnetic Energy = Stored Magnetic Co-energy (for linear Conditions) at DC and AC Excitations (see Application Example 6.2)

An electromagnet (see Fig. 9.15) is to be used in a foundry to lift ingots (iron bars). The surface roughness of the ingots is such that when they and the electromagnet are in contact, there is an average air gap of g = x in each leg. The coil resistance of the magnet is R = 0.1 Ω.

(a) Draw the magnetic equivalent circuit of this magnet configuration.
(b) Calculate the self-inductance L(x) of the magnet as a function of x, provided the relative permeability of the magnetic circuit is $\mu_r = \infty$. Compute inductance for $x = g = x_0 = 0.005$ m = 5 mm.
(c) Draw the electric equivalent circuit for the magnet-ingot configuration, if the magnet is powered by an AC voltage of $v(t) = \sqrt{2}V_{rms}\cos\omega t$, where $V_{rms} = 120$ V and $\omega = 2\pi f = 377$ rad/s, indicate ohmic resistance R and inductance L(x).

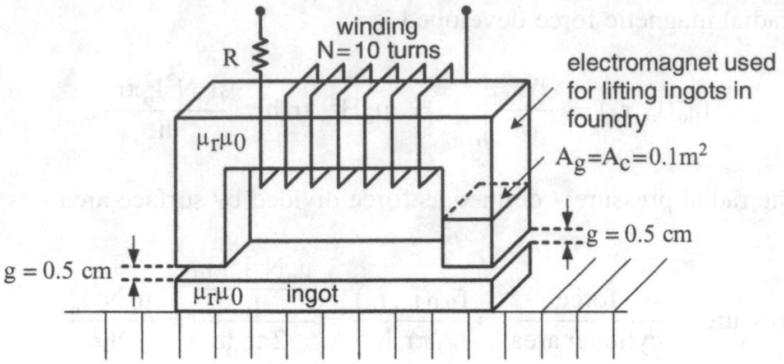

Fig. 9.15 Calculation of forces of a magnet when operating with either DC or AC excitation

(d) Compute the current $\tilde{i}(t)$ in the phasor domain and the current i(t) in the time domain as a function of V_{rms}, R, L(x), ω, and θ. You may assume the voltage as reference quantity. Compute the rms value of the current for $x_0 = 0.005$ m = 5 mm.
(e) Compute the time-averaged magnetic stored co-energy in the inductor $<W_{fld}>$. *Hint*: For the solution of the integral encountered you may remember the so-called orthogonality conditions of sine and cosine as listed in Application Example 6.2.
(f) Compute the force f_{fldAC}. Note L(x) occurs in numerator and denominator of $<W_{fld}>$. and therefore, you will have to use the formula $(u/v)' = (u'v–uv')/v^2$ for performing differentiation.
(g) Compute the force f_{fldDC} if the winding is excited by an equivalent DC current of the same magnitude as the rms current obtained in (d).
(h) Compare the forces f_{fldAC} and f_{fldDC} for $x_0 = 0.005$ m = 5 mm.

Solution

(a) The magnetic equivalent circuit consisting of mmf F and reluctance or magnetic resistance is $\Re_g$ depicted in Fig. 9.16.
(b) The self-inductance of the winding is

$$L = \frac{N^2}{\Re_g} = \frac{N^2 \mu_o A_g}{2x} = \frac{62.83 \cdot 10^{-7}}{x} \text{H}. \quad (9.46)$$

For $g = x_o = 0.005$ m the inductance is L = 1.257 mH.
(c) The electric equivalent circuit consisting of voltage source, resistance R, and reactance ωL(x) in the phasor domain is shown in Fig. 9.17.
(d) The current as defined in Fig. 9.17 is in the phasor domain

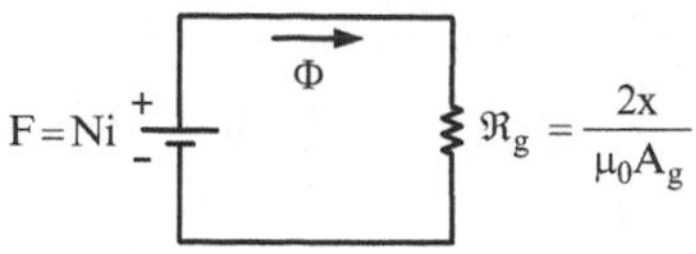

Fig. 9.16 Magnetic equivalent circuit

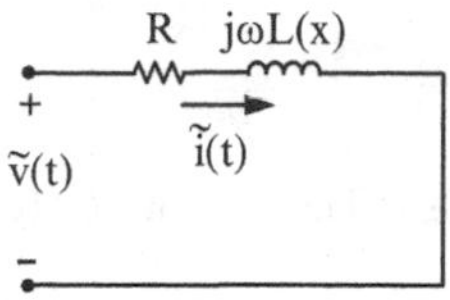

Fig. 9.17 Electric equivalent circuit

$$\tilde{i}(t) = \frac{\tilde{v}(t)}{(R + j\omega L(x))} = \frac{V_{rms}\angle -\theta}{\sqrt{R^2 + [\omega L(x)]^2}} \tag{9.47}$$

or in the time domain

$$i(t) = \frac{\sqrt{2}V_{rms}}{\sqrt{R^2 + [\omega L(x)]^2}} \cos(\omega t - \theta) \tag{9.48}$$

where $\theta = \tan^{-1}\left\{\frac{\omega L(x)}{R}\right\}$ and $I_{rms} = \frac{V_{rms}}{\sqrt{R^2 + +[\omega L(x)]^2}}$. For $g = x = 0.005$ m the rms value of the current is $I_{rms} = \frac{120}{\sqrt{0.01 + +0.225}} = 247.54$ A.

(e) The time-averaged magnetic stored co-energy = magnetic stored energy (because the system is linear, where both energy expressions are identical) due to AC excitation is

$$\left\langle W'_{fld\,AC} \right\rangle = \frac{1}{2} L(x) I_{rms}^2 = \frac{L(x)}{2} \frac{V_{rms}^2}{R^2 + [\omega L(x)]^2}. \tag{9.49}$$

(f) The magnetic force developed by the AC current is

$$f_{fldAC} = \frac{\partial W'_{fld\,AC}(i, x)}{\partial x} = \frac{V_{rms}^2}{2} \frac{\partial \left\{ \frac{L(x)}{R^2 + [\omega L(x)]^2} \right\}}{\partial x}, \tag{9.50}$$

$$f_{fldAC} = \frac{(120)^2}{2} 62.83 \cdot 10^{-7} \frac{\partial \left(\frac{1}{xR^2 + \frac{x\omega^2[62.83 \cdot 10^{-7}]^2}{x^2}} \right)}{\partial x}, \tag{9.51}$$

$$f_{fldAC} = 0.0452\left\{\frac{(-0.01x^2 + 5.61 \cdot 10^{-6})}{(0.01x^2 + 5.61 \cdot 10^{-6})^2}\right\}N. \tag{9.52}$$

For g = x = 0.005 m the force developed by the AC current becomes

$$f_{fldAC}\big|_{g=x=0.005\ m} = 7{,}059\ N. \tag{9.53}$$

(g) The magnetic stored co-energy due to DC excitation is

$$\left\langle W'_{fld\ DC}\right\rangle = \frac{1}{2}L(x)I_{DC}^2 = \frac{1}{2}\frac{N^2\mu_o A_g}{2x}I_{DC}^2. \tag{9.54}$$

For I_{DC} = 247.54 A the stored magnetic co-energy becomes

$$\left\langle W'_{fld\ DC}\right\rangle = \frac{0.1924}{x}, \tag{9.55}$$

and the force developed by the DC current is

$$\langle f_{fld\ DC}\rangle = (-)\frac{0.1924}{x^2} \tag{9.56}$$

or at g = x = 0.005 m

$$f_{fld\,DC}\big|_{g=x=0.005\,m} = (-)7{,}696\,N. \tag{9.57}$$

The force developed by AC and DC excitation is about the same. The AC excitation is preferred because of the zero crossings of the AC current: in case of a short-circuit the arc development will be minimized for AC excitation. The force developed by AC excitation is $f_{fld\ AC}$ = 7,059 N = 1,586 (lb-force) = 721 (kg-force) = 0.721 (ton-force). The minus sign in (9.57) will be addressed in the next application example.

Application Example 9.10: Computation of Magnetically Developed Force Acting on a Conductor Residing in a Slot

Figure 9.18 illustrates the general nature of the slot-leakage flux produced by the current i in a rectangular conductor embedded in a rectangular slot in iron. Assume that the iron reluctance is negligibly small ($\mu \to \infty$) and that the slot-leakage flux ϕ_{sx} goes straight across the slot in the region between the top of the conductor and the slot opening.

(a) Derive an expression for the flux density B_s in the region between the top of the conductor and the opening of the slot.

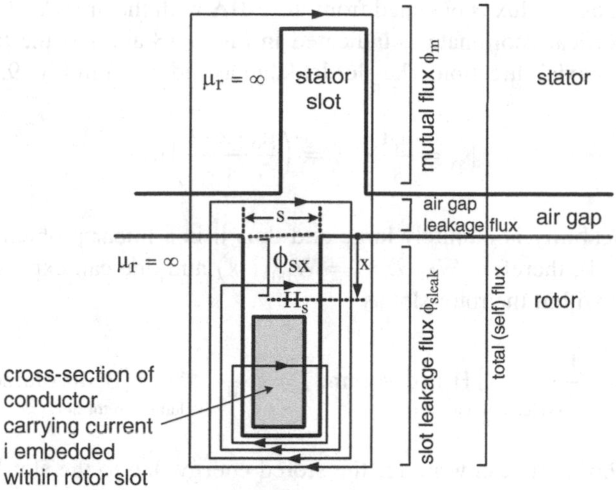

Fig. 9.18 Calculation of force on a conductor residing in a slot embedded in iron with infinite permeability

(b) Derive an expression for the slot-leakage flux ϕ_{sx} crossing the slot above the conductor, in terms of the variable height coordinate x of the slot above the conductor, the slot width s, and the length ℓ perpendicular to the cross-section.

(c) Derive an expression for the force f_{fld} created by this magnetic leakage field on a conductor of length ℓ. Use MKSAK units. In what direction does this force act on the conductor?

(d) Compute in pound-force (lbs-force) the force on a conductor $\ell = 1$ foot (ft) long in a slot s = 1 in. wide when the current in the conductor is i = 1 kA or 1kAt if the conductor consists of several ($N \neq 1$) turns.

Solution

(a) Ampere's law

$$\int_S \vec{J} \cdot d\vec{S} = \oint_C \vec{H} \cdot d\vec{\ell} \tag{9.58}$$

applied to the rotor slot cross-section one obtains

$$N \cdot i = H_s \cdot s. \tag{9.59}$$

With N = 1 turns the magnetic field intensity is $H_s \cdot s = i$, and with the constituent relation

$$H_s = B_s/\mu_o \tag{9.60}$$

the flux density within the rotor slot becomes

$$B_s = \frac{\mu_o i}{s}. \tag{9.61}$$

(b) The slot leakage flux is obtained from $\Phi = BA$ with the area $A = x \cdot \ell$, where x is the vertical coordinate as indicated in Fig. 9.18 and ℓ is the length of the machine in axial direction. The slot leakage as indicated in Fig. 9.18 is now

$$\phi_{sx} = \frac{\mu_o i}{s} x \cdot \ell = \left(\frac{\mu_o \cdot x \cdot \ell}{s} \right) i. \tag{9.62}$$

(c) The permeability is infinitely large and thus it is a linear problem (saturation is neglected), therefore $W_{fld}(\lambda, x) = W'_{fld}(i, x)$ and one can express the stored co-energy within the rotor slot as

$$W'_{fld}(i, x) = \frac{1}{2} \int_{\text{(volume of rotor slot)}} H_s B_s d(\text{volume}) = \frac{1}{2} \frac{1}{\mu_o} \int_{\text{(volume of rotor slot)}} B_s^2 d(\text{volume}) \tag{9.63}$$

or with (9.61) one can write for the stored energy due to the slot-leakage flux ϕ_{sx} crossing the slot above the conductor, in terms of the variable height coordinate x:

$$W'_{fld}(i, x) = \frac{1}{2} \frac{1}{\mu_o} \int_{\text{(volume of rotor slot)}} B_S^2 d(\text{volume}) = \frac{1}{2} \frac{1}{\mu_o} \left(\frac{\mu_o i}{s} \right)^2 (x \ell s) = \frac{1}{2} \frac{x \cdot \ell \cdot \mu_o}{s} i^2, \tag{9.64}$$

where $(x \cdot \ell \cdot s)$ represents the volume above the conductor with current i. The force due to the magnetic field (fld) is now

$$f_{fld}(i, x) = \frac{\partial W'_{fld}(i, x)}{\partial x} = \frac{\mu_o \cdot \ell \cdot i^2}{2s}. \tag{9.65}$$

The force f_{fld} acts to increase x, that is, f_{fld} is pushing the conductor into the slot, or in other words f_{fld} acts in the direction of x: as a consequence, the insulation at the slot bottom wears out and represents a failure mode of an electrical machine.

(d) For the given data one obtains the force due to the magnetic field

$$f_{fld}(i, x)|_{\ell=1\text{ft}, i=1\text{kAt}} = \frac{\mu_o \cdot \ell \cdot i^2}{2s} = \frac{\mu_o}{2} \left(\frac{12 \text{ in.}}{1 \text{ in.}} \right) 10^6 \text{N} = 7.54 \text{ N}, \tag{9.66}$$

$$f_{fld}(i, x)|_{\ell=1\text{ft}, i=1\text{kAt}} = 1.696(\text{lbs} - \text{force}).$$

Application Example 9.11: Plunger Supported by a C-Core Excited by Permanent Magnet

A magnetic block (plunger) with mass M is supported against the force of gravity by a C-core excited by a permanent magnet with the characteristic $B_m = \mu_R (H_m - H_c)$, where μ_R is the recoil permeability and H_c the coercivity of the

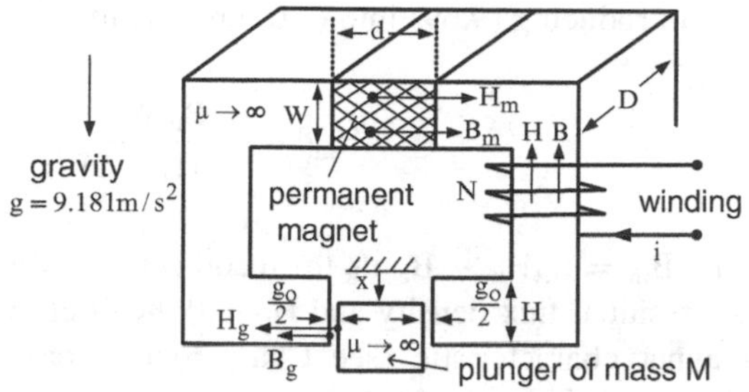

Fig. 9.19 Plunger supported by permanent-magnet system

permanent magnet (see Chap. 6). Figure 9.19 illustrates the system including a permanent magnet and N-turn winding. The function of this winding is to counteract the field produced by the magnet so that the plunger can be removed from the C-core. The air gaps at each side of the plunger remain constant at $g_o/2$. Assume that the winding direction is such that a positive winding current reduces the air-gap flux produced by the permanent magnet. Leakage and fringing effects can be neglected.

(a) Find the force f_{fld} acting on the plunger in the x-direction due to the permanent magnet for $0 \leq x \leq H$ as a function of i and x.
(b) Assume that the winding current i is zero, find the force f_{fld} acting on the plunger in the x-direction due to the permanent magnet for $0 \leq x \leq H$.
(c) Find the maximum mass M of the plunger that can be supported against gravity for $0 \leq x \leq H$.
(d) For $M = M_{max}/2$, find the minimum current required to ensure that the mass will fall out of the system when the current is applied.

Solution

The definition of the current in the winding of Fig. 9.19 is such that the magnetic field strength caused by the winding H opposes that of the permanent magnet H_m. This weakening of the permanent-magnet field makes it possible to reduce the field in the air gap (H_g) so that the plunger can be removed (fall out) of the air gap at a minimum current $i = I_{min}$ of the winding.

(a) The application of Ampere's law yields

$$H_m d + H_g g_o = -Ni, \tag{9.67}$$

with the continuity of flux condition (div B)

$$B_m WD = B_g(H - x)D = \mu_o H_g(H - x) \tag{9.68}$$

one obtains for the magnetic field strength

$$H_g = \frac{B_m W}{\mu_o(H - x)}. \tag{9.69}$$

Introducing (9.69) into (9.67) results in

$$H_m d + \frac{B_m W}{\mu_o (H - x)} g_o = -N \cdot i, \tag{9.70}$$

with $B_m = \mu_R H_m + B_r$, μ_R the recoil permeability, and $B_r = -\mu_R H_c$ with B_r being the residual flux density and $H_c < 0$ the coercivity characterizing the permanent-magnet characteristic (see Chap. 6) one can represent the permanent-magnet characteristic by the function

$$H_m = \frac{B_m}{\mu_R} + H_c. \tag{9.71}$$

Introducing (9.71) into (9.70) results in the flux density of the permanent magnet as a function of H_c, μ_R, and i

$$B_m = -\frac{N \cdot i + H_c d}{\left[\frac{d}{\mu_R} + \frac{W \cdot g_o}{\mu_o (H - x)}\right]}. \tag{9.72}$$

The flux linkages are

$$\lambda = -N\phi = -N[WDB_m] = \frac{NWD[N \cdot i + H_c d]}{\left[\frac{d}{\mu_R} + \frac{W g_o}{\mu_o (H - x)}\right]}. \tag{9.73}$$

The minus sign in (9.73) is required because the B_m is in opposite direction as the flux density B generated by the winding. The flux linkages are zero if

$$I_o = -\frac{H_c d}{N}. \tag{9.74}$$

The magnetic stored co-energy in the air gap is

$$W'_{fld}(i, x) = \int_{I_o}^{i} \lambda(i, x) di = \int_{I_o}^{i} \frac{NWD(Ni + H_c d)}{\left[\frac{d}{\mu_R} + \frac{W g_o}{\mu_o (H - x)}\right]} di$$

$$= \frac{NWD\left(\frac{Ni^2}{2} + H_c d \cdot i\right)}{\left[\frac{d}{\mu_R} + \frac{W g_o}{\mu_o (H - x)}\right]}\Bigg|_{I_o = -\frac{H_c d}{N}}^{i}, \tag{9.75}$$

or

$$W'_{fld}(i,x) = \frac{NWD}{\left[\dfrac{d}{\mu_R} + \dfrac{Wg_o}{\mu_o(H-x)}\right]}\left(\frac{Ni^2}{2} + H_c d \cdot i + \frac{H_c^2 d^2}{2N}\right). \quad (9.76)$$

The magnetic force is $f_{fld}(i,x) = \frac{\partial W'_{fld}(i,x)}{\partial x}$. Applying the quotient rule $\left(\frac{u}{v}\right)' = \frac{u'v - uv'}{v^2}$
with

$u = NWD\,\mu_R\,\mu_o(H-x)\left[\frac{Ni^2}{2} + H_c d \cdot i + \frac{H_c^2 d^2}{2N}\right]$, and $v = \left[\frac{\mu_o d(H-x)}{\mu_R} + Wg_o\right]$
one obtains for the force as a function of i, H_c, and μ_R

$$f_{fld}(i,x) = -\frac{\mu_o W^2 D N g_o\left(\dfrac{N \cdot i^2}{2} + H_c d \cdot i + \dfrac{H_c^2 d^2}{2N}\right)}{\left[Wg_o + \dfrac{\mu_o d}{\mu_R}(H-x)\right]^2}. \quad (9.77)$$

(b) For i = 0 (9.77) becomes

$$f_{fld}(i,x)|_{i=0} = -\frac{\mu_o W^2 D g_o H_c^2 d^2}{2\left[Wg_o + \dfrac{\mu_o d}{\mu_R}(H-x)\right]^2}. \quad (9.78)$$

The minus sign in the above force relation indicates that the force f_{fld} acts in opposite direction of the spatial coordinate x, that is, the plunger is drawn into the air gap. In other words, the plunger tends to move such that the stored energy becomes smaller or a minimum.

(c) The maximum force occurs for i = 0, and H = x. The summation of the forces yields the relation $\left(f_{fld}(i,x)|_{i=0}^{\max \text{ at } x=H}\right) + (M_{max} g) = \left(-\frac{\mu_o D H_c^2 d^2}{2g_o}\right) + (M_{max} g) = 0$, where g_o is the air gap length and $g = 9.81 m/s^2$. From this equation follows the maximum mass that can be supported by the permanent magnet

$$M_{max} = \frac{\mu_o D H_c^2 d^2}{2 g_o g}. \quad (9.79)$$

(d) The minimum current I_{min} required to ensure that the mass $M = M_{max}/2$ will fall out of the air gap follows from the relation

$$\left(f_{fld}(i,x)\Big|_{i=I_{min}}^{max\,at\,x=H}\right)+(M_{max}g/2)=\left(-\frac{\mu_o W^2 D N g_o\left(\frac{N\cdot I_{min}^2}{2}+H_c d\cdot I_{min}+\frac{H_c^2 d^2}{2N}\right)}{[Wg_o]^2}\right.$$
$$\left.+(M_{max}g/2)=0\right)$$

or

$$-\frac{\mu_o W^2 D N g_o\left(\frac{N\cdot I_{min}^2}{2}+H_c d\cdot I_{min}+\frac{H_c^2 d^2}{2N}\right)}{[Wg_o]^2}+\frac{\mu_o D H_c^2 d^2}{2g_o g}g/2=0$$ resulting in the second order equation

$$\frac{N^2}{2}I_{min}^2+H_c d N I_{min}+\frac{H_c^2 d^2}{4}=0 \quad (9.80)$$

with the two solutions $I_{min\_1}=1.707I_o$ and $I_{min\_2}=0.292I_o$. The first solution is not feasible because $I_{min\_1}>I_o$.

Application Example 9.12: Torque Production in a Single-Phase Motor

A single phase $p=2$, round-rotor machine (either motor or generator) has a single-phase (concentrated) winding on the stator (s, −s) and a single-phase (concentrated) winding on the rotor (r, −r). The stator and rotor self- inductances are $L_{ss}=2.5$ H, $L_{rr}=0.5$ H, respectively, and the stator-to-rotor mutual inductance is $L_{sr}=1\cos\theta$ H, where θ is the circumferential coordinate of the machine at the air gap. The resistances of the stator and rotor windings are R_s and R_r, respectively. Figures 9.20a–c illustrate the flux linkage definitions λ_s, λ_r, λ_{sr}, respectively. The currents i_s and i_r can be assumed to be known.

(a) Formulate λ_s and λ_r as a function of the currents and inductances.
(b) Find $v_s(t)$ and $v_r(t)$ as a function of the resistances, inductances, and currents.
(c) Derive an expression for the co-energy $W_{fld}(i_s, i_r, \theta)$ for a given $\theta=\theta_0$.
(d) Determine the torque T_{fld} as a function of p, L_{sr}, i_s, i_r and θ.

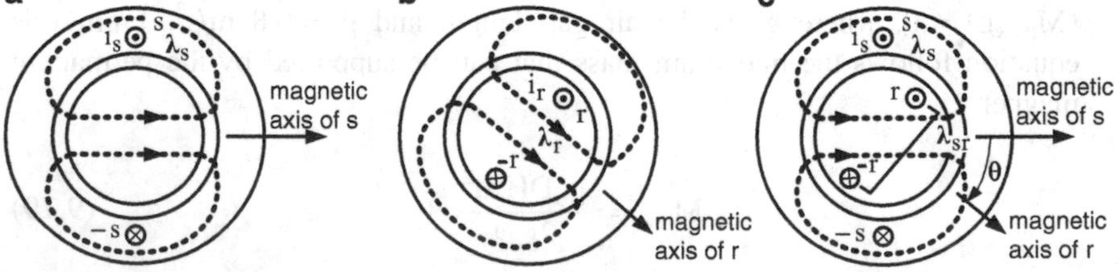

Fig. 9.20 (a–c) Definition of self-(L_{ss}, L_{rr}) and mutual ($L_{sr}\cos\theta$) inductances via flux linkages λ_s, λ_r, and λ_{sr}, respectively. s stands for stator and r stands for rotor

(e) Under what current (constraints) conditions (e.g., i_s and i_r are either time-dependent or independent of time) is T_{fld} of part (d) a standing wave, an alternating wave or a traveling wave?

(f) For $i_s = I_s$ and $i_r = I_{rmax}\sin\omega t$, determine the average torque by integrating over a spatial period, that is find $T_{avg} = \frac{1}{2\pi}\int_0^{2\pi} T_{fld} d\,\theta$.

Solution

(a) The stator and rotor flux linkages are

$$\lambda_s = L_{ss}i_s + 1 \cdot \cos\theta i_r, \tag{9.81}$$

$$\lambda_r = 1 \cdot \cos\theta i_s + L_{rr}i_r. \tag{9.82}$$

(b) The stator and rotor voltages are

$$v_s = R_s i_s + L_{ss}\frac{di_s}{dt} + 1 \cdot \cos\theta\frac{di_r}{dt} - 1 \cdot i_r \sin\theta\frac{d\theta}{dt}, \tag{9.83}$$

$$v_r = R_r i_r + L_{rr}\frac{di_r}{dt} + 1 \cdot \cos\theta\frac{di_s}{dt} - 1 \cdot i_s \sin\theta\frac{d\theta}{dt}. \tag{9.84}$$

(c) The co-energy is by integrating from the starting point (at origin) to the end point $(i_{so}, i_{ro}, \theta_o)$ of Fig. 9.21

$$W'_{fld}(i_{so}, i_{ro}, \theta_o) = \int_0^{i_{so}} L_{ss}i_s di_s + \int_0^{i_{so}} 1 \cdot \cos\theta_o i_r di_s + \int_0^{i_{ro}} 1 \cdot \cos\theta_o i_{so} di_r$$
$$+ \int_0^{I_{ro}} L_{rr}i_r di_r, \tag{9.85}$$

$$W'_{fld}(i_{so}, i_{ro}, \theta_o) = \frac{1}{2}L_{ss}i_{so}^2 + 0 + i_{so}i_{ro}\cos\theta_o + \frac{1}{2}L_{rr}i_{ro}^2. \tag{9.86}$$

(d) The torque is at any θ

$$T_{fld}(i_{so}, i_{ro}, \theta) = \frac{\partial W'_{fld}(i_{so}, i_{ro}, \theta)}{\partial\theta_m} = \frac{\partial W'_{fld}(i_{so}, i_{ro}, \theta)}{\partial\theta} \cdot \frac{d\theta}{d\theta_m} = \frac{\partial W'_{fld}(i_{so}, i_{ro}, \theta)}{\partial\theta} \cdot \left(\frac{p}{2}\right)$$

$$T_{fld}(i_{so}, i_{ro}, \theta) = -\left(\frac{p}{2}\right) i_{so}i_{ro}\sin\theta. \tag{9.87}$$

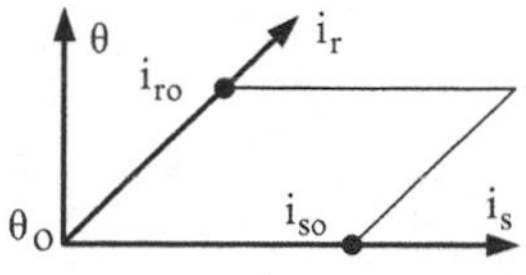

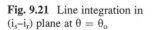

Fig. 9.21 Line integration in (i_s–i_r) plane at $\theta = \theta_o$

(e) Torque developed under various current constraints
   1. If $i_{so} = i_{ro}$ = constant a standing wave (e.g., $T_{fld}(\theta) = -(\frac{p}{2}) i_{so} i_{ro} \sin\theta$) is obtained.
   2. For either i_{so} or i_{ro} being a sine or cosine function and either i_{ro} or i_{so} being constant, an alternating wave (e.g., $T_{fld}(\omega, \theta) = -(\frac{p}{2}) I_{rmax} \sin\omega t \cdot \sin\theta$) is obtained.
   3. A traveling wave cannot be obtained for any i_{ro} or i_{so} combinations.

(f)

$$T_{avg} = \frac{1}{2\pi} \int_0^{2\pi} T_{fld} d\theta = -\frac{1}{2\pi} \int_0^{2\pi} \left(\frac{p}{2}\right) I_s \cdot I_{rmax} \sin\omega t \cdot \sin\theta \, d\theta = 0. \tag{9.88}$$

9.4 Radial Forces Acting on Iron Core of an Induction Machine

Figures 9.22a–e show computed radial forces acting on the stator core of an induction motor, resulting in core vibrations of various modes [13–15]. For the calculation of such forces (acting on iron cores) the Lorentz force law cannot be applied. The radial forces were obtained with the Maxwell-stress approach [5–7, 16–18]. The starting torque of machines [18, 19] can be obtained based on either the equivalent circuit or on the Maxwell stress.

Figure 9.23 demonstrates that the magnetic field exists mainly in the teeth and not in the slots [8, 9].

Application Example 9.13: Torque Production and Stored Energy in a Doubly Excited Machine

An electromagnetic system has two windings and a non-uniform air gap as shown in Fig. 9.24. The self- and mutual inductances are given by $L_{11} = 6 + 1.5 \cos 2\beta$ [H], $L_{22} = 4 + \cos 2\beta$ [H], and $L_{12} = L_{21} = 5 \cos\beta$ [H]. The windings are connected to DC voltage sources. The first winding requires a current $I_s = 10$ A and the second winding draws a current $I_r = 2$ A. The developed (d) torque and the stored energy = stored co-energy (for linear systems) may be written (as a function of β) in the following forms:

$T_d = T_{fld} = (A_1 + A_2 \sin\beta + A_3 \sin 2\beta)$ [Nm], $W_{fld} = (B_1 + B_2 \sin\beta + B_3 \cos\beta + B_4 \sin 2\beta + B_5 \cos 2\beta)$ [joules].

(a) Determine the values of A_1, A_2 and A_3.
(b) Determine the values of B_1, B_2, B_3, B_4 and B_5.
(c) For $\beta = 30^{\circ}$, find the magnitude and direction of torque on the rotor.

Solution

(a)

$$W'_{fld} = \frac{1}{2} L_{11} i_1^2 + L_{12} i_1 i_2 + \frac{1}{2} L_{22} i_2^2, \tag{9.89}$$

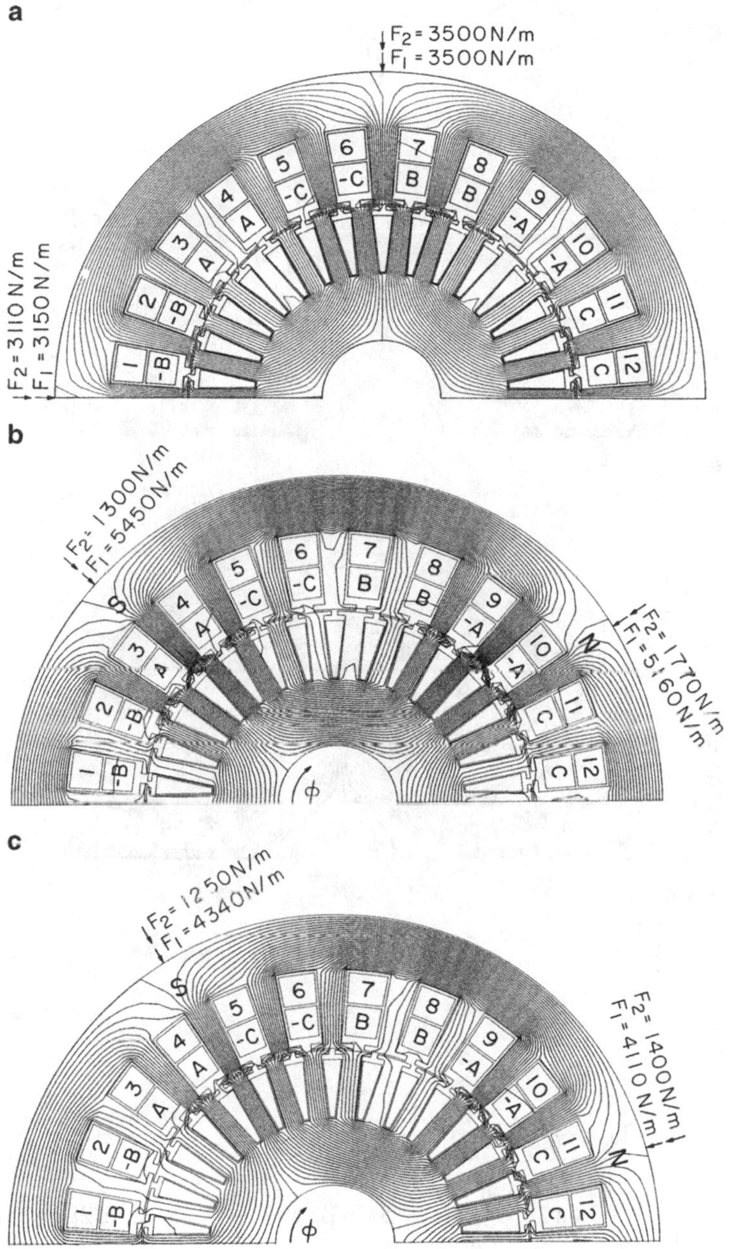

Fig. 9.22 (**a**–**e**) Radial magnetic forces acting on the iron core of an induction motor; (**a**) rotor position 1, (**b**) rotor position 2, (**c**) rotor position 3, (**d**) rotor position 4, (**e**) rotor position 5 (From Fuchs, E. F. ; Roesler, D. J.; Chang, L. H.: "Magnetizing current, iron losses and forces of three-phase induction machines at sinusoidal and nonsinusoidal terminal voltages, Part II: Results," *IEEE Transactions on Power Apparatus and Systems*, Nov. 1984, Vol. PAS-103, No. 11, pp. 3313–3326, [18])

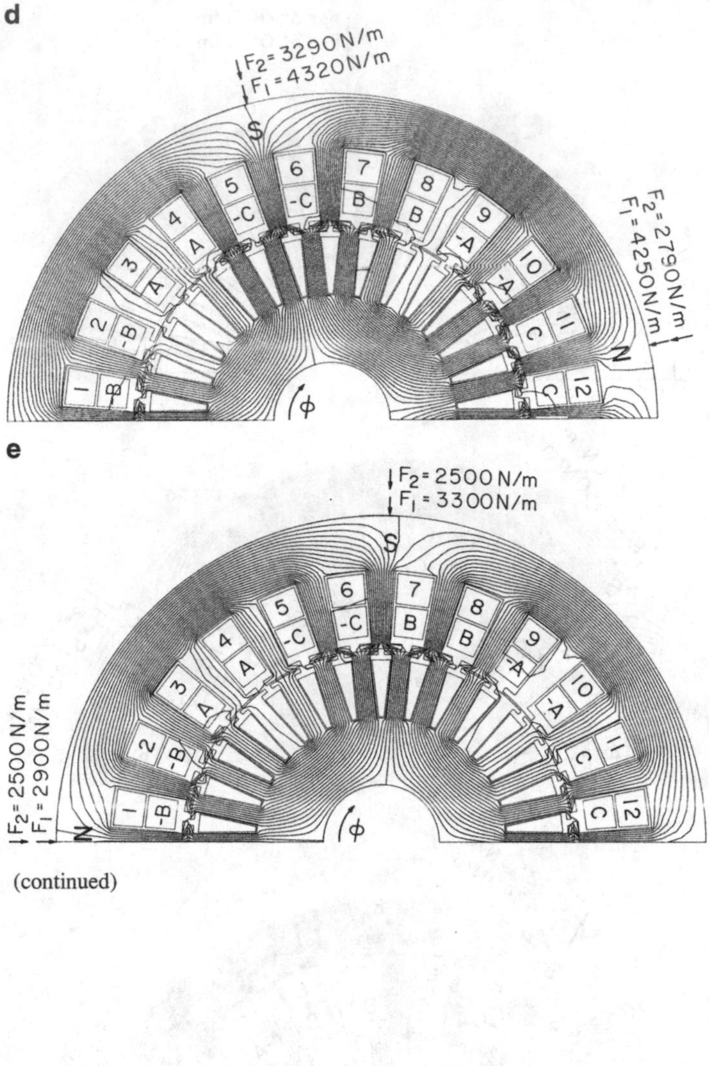

Fig. 9.22 (continued)

$$T_{fld} = \frac{\partial W'_{fld}}{\partial \beta} = \frac{1}{2} i_1^2 \frac{\partial L_{11}}{\partial \beta} + i_1 i_2 \frac{\partial L_{12}}{\partial \beta} + \frac{1}{2} i_2^2 \frac{\partial L_{22}}{\partial \beta}. \tag{9.90}$$

With $\frac{\partial L_{11}}{\partial \beta} = -3 \sin 2\beta$, $\frac{\partial L_{12}}{\partial \beta} = -5 \sin \beta$, and $\frac{\partial L_{22}}{\partial \beta} = -2 \sin 2\beta$, we have

$$T_{fld} = \frac{1}{2}(10)^2(-3 \sin 2\beta) + (10)(2)(-5 \sin \beta) + \frac{1}{2}(2)^2(-2 \sin 2\beta)$$
$$= -100 \sin \beta - 154 \sin 2\beta. \tag{9.91}$$

Therefore, $A_1 = 0\,\text{Nm}$, $A_2 = -100\,\text{Nm}$, $A_3 = -154\,\text{Nm}$.

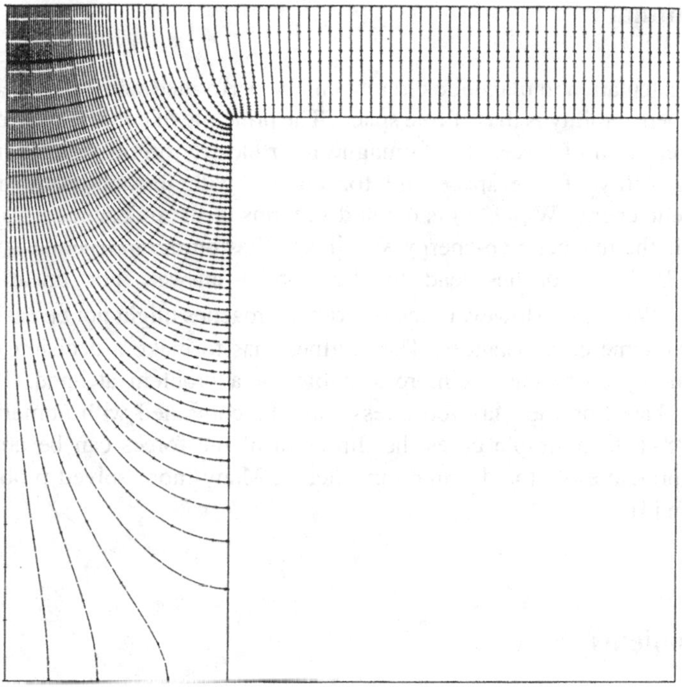

Fig. 9.23 Vector equipotential lines for a deep slot with finite depth. The lines enclose unit flux tubes

Fig. 9.24 Doubly excited machine

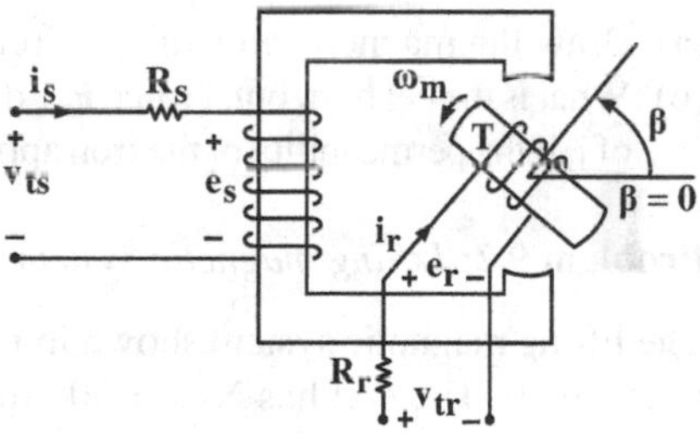

(b) With (9.89) one obtains

$$
\begin{aligned}
W'_{fld} &= \frac{1}{2}(6 + 1.5\cos 2\beta)(10)^2 + 5\cos\beta(10)(2) + \frac{1}{2}(4 + \cos 2\beta)(2)^2 \\
&= 308 + 100\cos\beta + 77\cos 2\beta.
\end{aligned} \tag{9.92}
$$

Therefore, $B_1 = 308$ Ws, $B_2 = 0$ Ws, $B_3 = 100$ Ws, $B_4 = 0$ Ws, $B_5 = 77$ Ws.

(c) Using (9.91) the torque is
$T_{fld} = -100\sin(30°) - 154\sin 2(30°) = -183.4$ Nm (opposite to the direction of rotation).

9.5 Summary

The force calculation in magnetic circuits can be performed with the Lorentz force law, if the permeability is that of free space. The principle of conservation of energy for the calculation of forces due to magnetic or electric origin works for circuits with permeability of free space and for circuits with ferromagnetic materials. The magnetic energy $W_{fld}(\lambda, x)$ is defined in terms of the independent variables λ and x while the magnetic co-energy $W'_{fld}(i, x)$ relies on the independent variables i and x. Both approaches lead to the correct answer. For linear systems $W'_{fld}(i, x) = W_{fld}(\lambda, x)$. However, one of both approaches might not be easily applicable under some circumstances. The engineer has to make a decision which of the two energy approaches is more suitable for a problem at hand. The force calculation based on the Maxwell stress must be combined with numerical field solutions [8–12]. In simple cases the direction of the forces can be determined using the principle of stored minimum energy. Many more solved problems are available in [4].

9.6 Problems

Problem 9.1: *Force of electromagnet exerted on plunger*

The total air-gap length of the electromagnet shown in Fig. 9.25 is $g = (x + y) =$ constant.

(a) Draw the magnetic equivalent circuit for Fig. 9.25.
(b) What is the net horizontal force F_{net} acting on the iron plunger in the air-gap in terms of i, if the permeability of the iron approaches infinity and fringing flux is neglected?

Problem 9.2: *Lifting magnetic system*

The lifting magnetic system shown in Fig. 9.26 has a cross section of $A_g = 0.01\ m^2$ (one-sided). The coil has $N = 1{,}000$ turns and a resistance $R = 10\ \Omega$. Neglect the reluctance of the magnetic core and field fringing in the air gap.

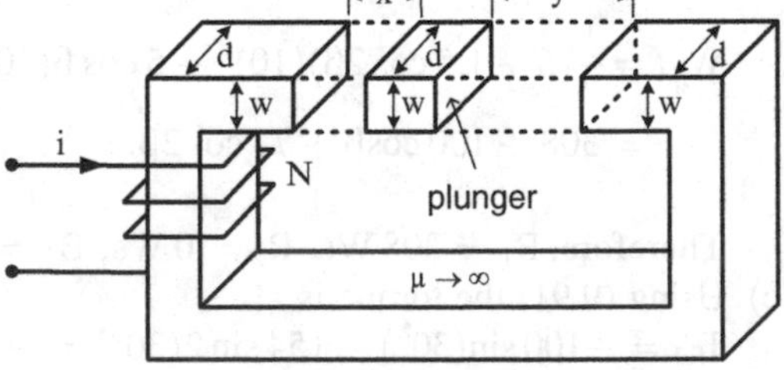

Fig. 9.25 Electromagnet with movable plunger

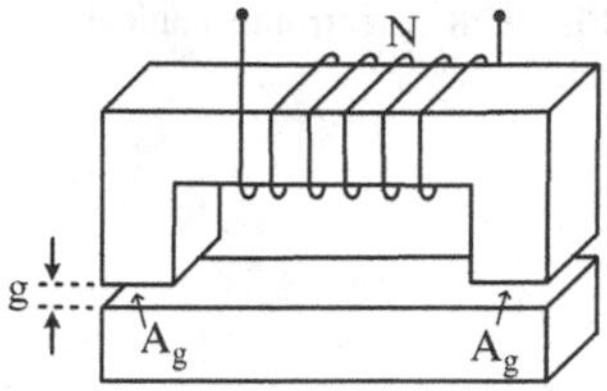

Fig. 9.26 Lifting magnetic system

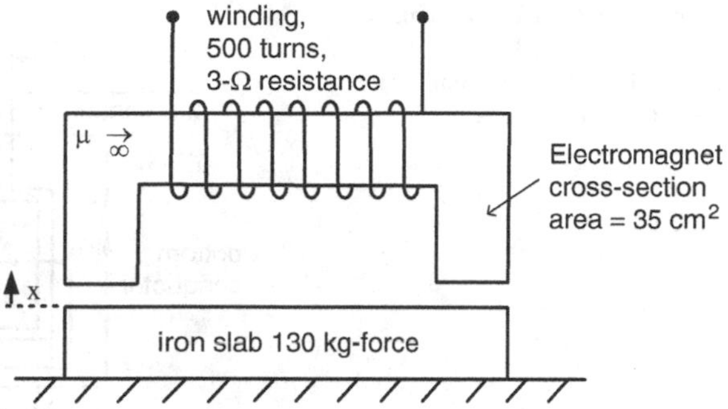

Fig. 9.27 Electromagnet lifting an iron slab

1) The air gap is initially held at g = 5 mm and a DC source of 120 V is connected to the coil.
   (a) Determine the stored magnetic co-energy $W'_{fld}(i, x)$.
   (b) Determine the lifting force f_{fld}.
2) The air gap is again held at g = 5 mm and an AC source of 120 V (rms) at 60 Hz is connected to the coil. Determine the average value of the lifting force.

Problem 9.3: *Design of a magnet*

An electromagnet (see Fig. 9.27) is to be used to lift a 130 (kg-force) slab of iron. The surface roughness of the iron is such that when the iron and the electromagnet are in contact, there is a minimum air gap of x = 0.1 cm in each leg. The coil resistance is 3 Ω. Calculate the minimum coil voltage which must be used to lift the slab against the force of gravity. The reluctance of the iron is negligible.

Problem 9.4: *Calculation of natural frequency for incremental rotor motion*

Consider the two-coil rotor of Fig. 9.3. Assume the two rotor windings to be carrying a constant current $I_1 = I_2 = 100$ A, the rotor to have a moment of inertia $J = 0.5$ kgm^2, $B_o = 1$ T, R = 0.1 m and $\ell = 1$ m.

(a) Find the forces (components F_r, F_θ) and the resultant torque T.
(b) Find the equilibrium position α_0 of the rotor. Is it stable?

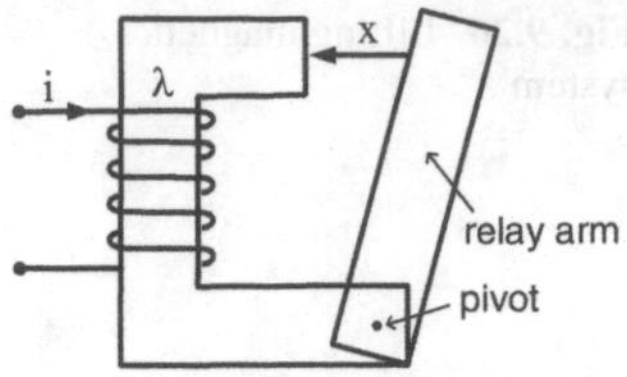

Fig. 9.28 Electromechanical relay

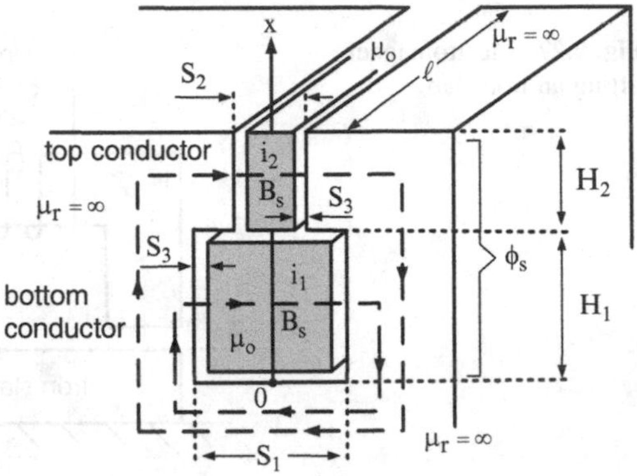

Fig. 9.29 Two conductors in a slot. $H_1 = 3$ cm, $H_2 = 2$ cm, $\ell = 10$ cm, $S_1 = 3$ cm, $S_2 = 1.5$ cm, $S_3 = 0.2$ cm, $i_1 = 100$ A, $i_2 = 20$ A

(c) Write the dynamic equations for the system.
(d) Find the natural frequency in hertz for incremental rotor motion α_1 around the equilibrium position α_0.

Problem 9.5: *Force calculation from magnetic energy and co-energy*

The force f_{fld} acting on the relay arm of Fig. 9.28 is to be calculated for the experimentally determined (λ–i) relation $\lambda = \sqrt{\frac{i}{(1-X)}}$

(a) Calculate $f_{fld}^{coenergy}$ from the co-energy.
(b) Calculate f_{fld}^{energy} from the energy.
(c) Show that $f_{fld}^{coenergy} = f_{fld}^{energy}$.

Problem 9.6: *Magnetic force within slot with two conductors*

Figure 9.29 shows the general nature of the slot-leakage flux produced by currents i_1 and i_2 in two rectangular conductors embedded in a slot in iron ($\mu_r \to \infty$). Assume that the leakage flux ϕ_s goes straight across the slot.

(a) Derive an expression for the flux density $B_s(x)$ within the entire slot height and plot $B_s(x)$ versus x.
(b) Derive an expression for the slot-leakage flux ϕ_s crossing the slot within the bottom conductor in terms of the height x, the slot width S_1, and the length ℓ which is perpendicular to the cross-section.

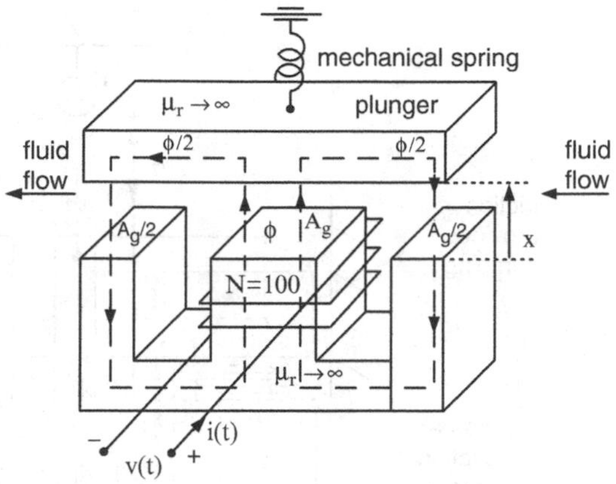

Fig. 9.30 Forces developed by a magnet of a linear pump operating with either DC or AC excitation

(c) Derive an expression for the force f_{fld} created by the magnetic field penetrating the bottom conductor. In what direction does this force act on the conductor?

Problem 9.7: *Force calculation for an actuator as used in a linear pump*

An electromagnet (Fig. 9.30) attracts an iron plunger suspended by a mechanical spring. There is an average air gap of x in each of the three legs (two outer legs and one center leg). The coil resistance of the magnet winding is $R = 1\ \Omega$.

(a) Draw the magnetic equivalent circuit of this magnet configuration.
(b) Calculate the self-inductance of the magnet L(x) as a function of x provided the relative permeability of the plunger and the magnet core is $\mu_r = \infty$. Compute inductance for $x = x_0 = 0.003\ \text{m} = 3\ \text{mm}$, $N = 100$ turns, $A_g = 4 \cdot 10^{-4}\ \text{m}^2$.
(c) Draw the electrical equivalent circuit for this magnet configuration provided the magnet winding is powered by an AC voltage of $v(t) = \sqrt{2}V_{rms}\cos\omega t$, where $V_{rms} = 10$ V and $\omega = 2\pi f = 377$ rad/s. Indicate ohmic resistance $R = 1\ \Omega$ and inductance L(x) in this electrical equivalent circuit.
(d) Find an expression for the current $\tilde{i}(t)$ in the phasor domain. Formulate the current i(t) in the time domain as a function of V_{rms}, R, L(x), ω, and θ. You may assume the voltage as reference quantity. Compute the rms value of the current I_{rms} for $x = x_0 = 0.003\ \text{m} = 3\ \text{mm}$ and $V_{rms} = 10$ V.
(e) Derive an expression for the time-averaged stored magnetic co-energy of the inductor $<W'_{fld}>$.
(f) Compute the force f_{fldAC} for $x = x_0 = 0.003\ \text{m} = 3\ \text{mm}$ and $V_{rms} = 10$ V.
(g) Compute the force f_{fldDC} if the winding is excited by an equivalent DC current I_{DC} of the same magnitude as the rms current I_{rms} obtained in (d) for $x = x_0 = 0.003\ \text{m} = 3\ \text{mm}$.
(h) Compare the forces f_{fldAC} and f_{fldDC} for $x = x_0 = 0.003\ \text{m} = 3\ \text{mm}$.

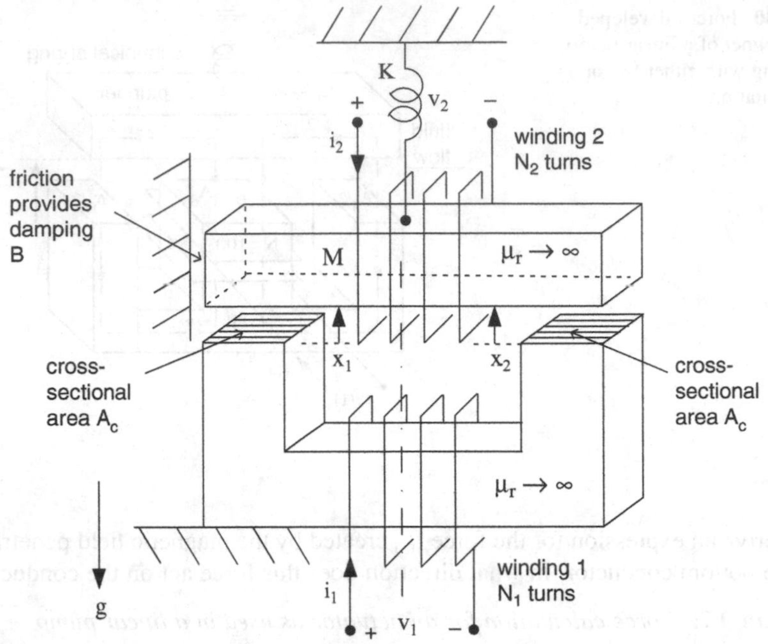

Fig. 9.31 Two-winding magnetic circuit

Problem 9.8: *Dynamic system equations*

The two-winding magnetic circuit of Fig. 9.31 consists of a stationary U-shaped part and a movable part which is suspended by a spring, and is constrained to motion such that the lengths of both air gaps remain equal $x_1 = x_2 = x$. The mass of the movable element is M and the direction of the force of gravity is indicated by g. The damping coefficient is B. The resistances of winding 1 and winding 2 are R_1 and R_2, respectively.

(a) Find the self-inductances of windings 1 and 2 using Ohm's law for magnetic circuits.
(b) Find the mutual inductance.
(c) Calculate the co-energy $W'_{fld}(i_1, i_2, x)$.
(d) Find an expression for the force acting on the movable element as a function of the winding currents.
(e) Find the three nonlinear differential equations of motion, taking into account the forces due to the spring, damper, mass, and gravity, in addition to the magnetic force. Note that at $x = x_0$ the two coils are not excited ($i_1 = i_2 = 0$).

Problem 9.9: *Force calculation of an actuator*

The cylindrical iron-clad solenoid magnet shown in Fig. 9.32 has a plunger which can move a relatively short distance x developing a large force f_{fld}. The plunger is guided so that it can move in vertical direction only. The radial air gap between the

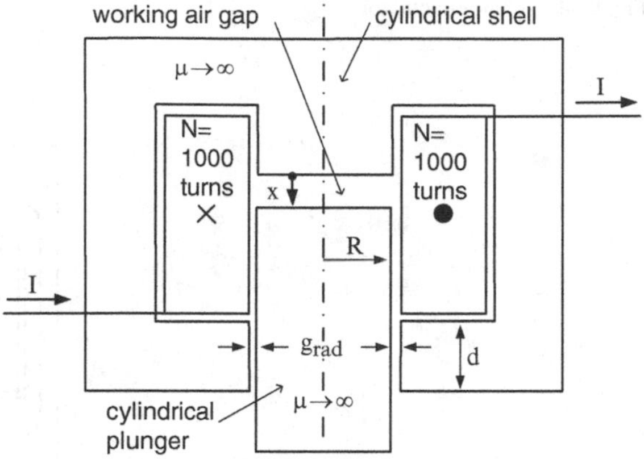

Fig. 9.32 Cylindrical iron-clad solenoid magnet

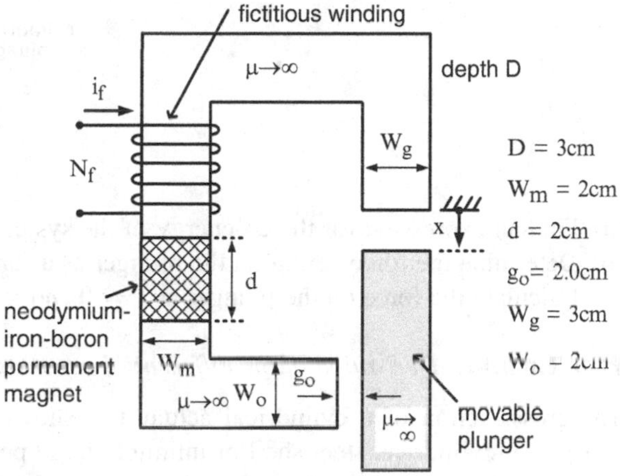

Fig. 9.33 Magnetic plunger suspended by permanent magnet

shell and the plunger is $g_{rad} = 1$ mm, R = 20 mm, and d = 10 mm. The exciting coil has N = 1,000 turns and carries a constant current of I = 10 A.

(a) Draw the magnetic equivalent circuit.
(b) Compute the flux density in the working air gap for x = 10 mm,
(c) The value of the energy stored in W_{fld}, and
(d) Value of the inductance L.
(e) For a force f_{fld} of −1,000 N determine for x = 10 mm the current $I = I_0$ required.

Problem 9.10: *Suspension of a plunger using permanent magnet*

The C-core of Fig. 9.33 is excited with a neodymium-iron-boron (NdFeB) permanent magnet and suspends a plunger with infinitely large permeability. For the purpose of the analysis a fictitious winding with N_f turns with a current i_f is residing on the C-core. Neglect fringing and leakage fluxes.

Fig. 9.34 Solenoid actuator

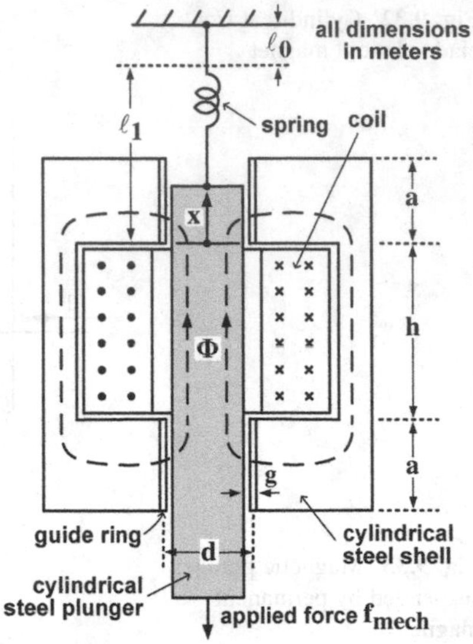

(a) Find an expression for the co-energy of the system as a function of x.
(b) Determine the force acting on the plunger as a function of x.
(c) Calculate the force on the plunger at $x = 0$ and $x = 0.5$ cm.

Problem 9.11: *Derivation of the differential equations of an actuator*

The cross-section of a cylindrical actuator is shown in Fig. 9.34. The actuator consists of cylindrical steel shell of infinitely large permeability, cylindrical magnetic plunger of mass M and infinitely large permeability, solenoid coil with N turns and resistance R, brass guide ring with permeability of free space, mechanical spring with spring constant K, and a mechanical force f_{mech} applied to the plunger. The plunger moves vertically and is supported by the spring. The un-stretched length of the plunger suspension is ℓ_o. The frictional force between plunger and brass ring is proportional to the velocity of the plunger and the coefficient of friction is B. The terminal voltage of the solenoid winding is v_t and its current is i. Magnetic leakage and fringing fluxes can be neglected.

(a) Determine the reluctances of the upper ($\Re_u$) and lower ($\Re_\ell$) air gaps.
(b) Determine the total reluctance ($\Re$).
(c) Find an expression for the inductance L of the solenoid.
(d) Derive the dynamic equations of motion of the electromechanical system, that is, the differential equations expressing the dependent variables v_t, f_{mech} in terms of the independent variables i and x and the given constants.

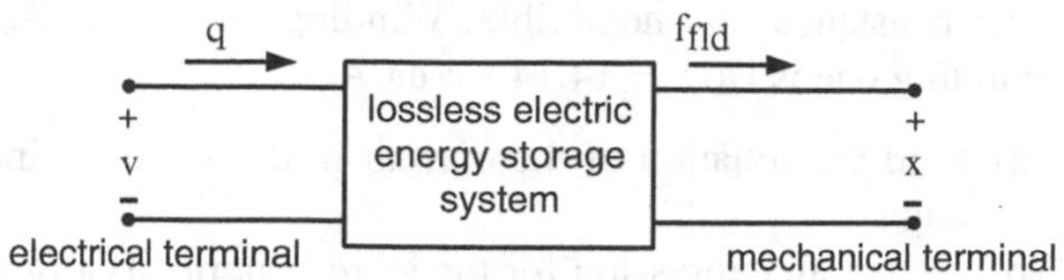

Fig. 9.35 Lossless electric energy storage system

Problem 9.12: *Energy storage in electric fields*

An electromechanical system in which energy storage is in electric fields is shown in Fig. 9.35. For the single electric terminal the differential energy is $dW_{elect} = vidt = v\,dq$, where v is the electric terminal voltage and q is the net charge associated with electric energy storage (e.g., capacitor). The stored energy is $W_{fld} = vdq - f_{fld}dx$.

(a) Derive an expression for the electric stored energy $W_{fld}(q, x)$.
(b) Determine an expression for the force $f_{fld}(q, x)$.
(c) Derive an expression for the electric stored co-energy $W'_{fld}(v, x)$.
(d) Find an expression for the force $f_{fld}(v, x)$.

Problem 9.13: *Instantaneous and average torques*

An electromagnetic system has two windings and a uniform air gap. The self- and mutual inductances are given by (θ is the angle between the two axes of windings): $L_{11} = 5$ H, $L_{22} = 2$ H, and $L_{12} = L_{21} = M = 3\cos(\theta)$ H. The winding currents are $i_1 = 10\sin(\omega t)$ A and $i_2 = -5\sin(\omega t)$ A, where ω is the angular frequency of voltage source applied to the first coil. While winding two is short-circuited the instantaneous and average developed torques (as a function of θ) may be written in the following forms:

$$T_{instant} = A_1 \sin(\theta) + A_2 \cos(2\,\omega t)[\text{Nm}],\, T_{avg} = B_1 \sin(\theta)[\text{Nm}].$$

(a) Determine the values of A_1 and A_2. Hint: $\sin^2(\omega t) = 0.5\text{–}0.5\cos(2\omega t)$.
(b) Determine the value of B_1.
(c) Compute θ for an average torque of 50 Nm.

Problem 9.14: *Energy stored and average torque in a singly excited system*

An electromagnetic system consists of a stator winding and a rotor with the following self- and mutual inductances: $L_{11} = 1.0$ H, $L_{22} = 0.5$ H, and $L_{12} = L_{21} = 1.414\cos\theta$ H, where θ is the angle between the two axes of the windings.

The resistances are negligible. Winding two is short-circuited and the current in winding one is $i_1(t) = 14.14 \sin \omega t$ A.

(a) Find the equation of the current in the second winding, $i_2(t)$, as a function of angle θ.
(b) Derive an expression for the stored energy as a function of θ.
(c) Derive an expression for instantaneous torque on the rotor as a function of angle θ.
(d) Compute the time averaged torque when $\theta = 45°$.

References

1. Fuchs, E. F.; Sasaki, K.: "Design of a resonant electromagnetic pile driver," *Proceedings of the International Conference on Electrical Machines,* Lausanne, Switzerland, September 17–20, 1984.
2. Sasaki, K.: *Magnetic Field Analysis of a Linear Motion Electromagnetic Pile Driver*, M.S. Thesis, University of Colorado, Boulder, CO, 1980.
3. Fardoun, A. A.; Fuchs, E. F.; Huang, H.: "Modeling and simulation of an electronically commutated permanent-magnet machine drive system using PSpice," *IEEE Transactions on Industry Applications*, Vol. 30, No. 4, July/August 1994, pp. 927–937.
4. Fitzgerald, A. E.; Kingsley, Ch., Jr.; Umans, S. D.: *Electric Machinery*, 5th Edition, McGraw-Hill Publishing Co., New York, NY. 1990.
5. Fuchs, E. F.; Masoum, M. A. S.: *Power Quality in Power Systems and Electrical Machines*, Elsevier/Academic Press, Feb. 2008, 638 pages. ISBN: 978-0-12-369536-9.
6. Carpenter, C. J.: "Surface-integral methods of calculating forces on magnetized iron parts," *Proceedings of IEEE*, 106C, 1960, pp. 19–28.
7. Reichert, K.; Freundl, H.; and Vogt, W.: The calculation of forces and torques within numerical magnetic field calculation methods, *Proceedings of Compumag Conference*, London, 1976.
8. Fuchs, E. F.; McNaughton, G. A.: "Comparison of first-order finite difference and finite element algorithms for the analysis of magnetic fields, Part I: Theoretical analysis," *IEEE Transactions on Power Apparatus and Systems*, May 1982, Vol. PAS-101, No. 5, pp. 1170–1180.
9. McNaughton, G. A.; Fuchs, E. F.: "Comparison of first-order finite difference and finite element algorithms for the analysis of magnetic fields, Part II: Numerical results," *IEEE Transactions on Power Apparatus and Systems*, May 1982, Vol. PAS-101, No. 5, pp. 1181–1201.
10. Fuchs, E. F.; McNaughton, G. A.: "Properties of orthogonal and triangular grids for the analysis of magnetic fields based on the finite difference and finite element methods, Part I: Theoretical analysis," *Acta Technica*, March/April 1982, No. 2, pp. 168–205.
11. McNaughton, G. A.; Fuchs, E. F.; Siegl, M.: "Properties of orthogonal and triangular grids for the analysis of magnetic fields based on the finite difference and finite element methods, Part II: Numerical results based on Gaussian elimination," *Acta Technica,* March/April 1982, No. 2, pp. 206–238.
12. Fuchs, E. F.; Siegl, M.: "Properties of orthogonal and triangular grids for the analysis of magnetic fields based on the finite difference and finite element methods, Part III: Comparison of iterative solutions," *Acta Technica*, May/June 1982, No. 3, pp. 261–290.
13. Verma, S. P.; Li, W.: "Measurement techniques of vibrations and radiated acoustic noise of electrical machines," *Proceedings of the Sixth International Conference on Electrical Machines and Systems,* Vol. 2, 9–11 Nov., 2003, pp. 861–866.

14. Verma, S. P.; Balan, A.: "Determination of radial forces in relation to noise and vibration problems of squirrel-cage induction motors," *IEEE Transactions on Energy Conversion*, Vol. 9, Issue 2, June 1994, pp. 404–412.
15. Erdelyi, E. A.: *Predetermination of the Sound Pressure Levels of Magnetic Noise in Medium Induction Motors*, Doctoral Dissertation, University of Michigan, Ann Arbor, January 1955.
16. Fuchs, E. F.; Chang, L. H.; Appelbaum, J.; Moghadamnia, S.: "Sensitivity of transformer and induction motor operation to power system's harmonics, *Topical Report*," Prepared for the US Department of Energy DOE-RA-50150-18, April 1983.
17. Fuchs, E. F.; Chang, L. H.; Appelbaum, J.: "Magnetizing current, iron losses and forces of three-phase induction machines at sinusoidal and nonsinusoidal terminal voltages, Part I: Analysis," *IEEE Transactions on Power Apparatus and Systems*, Nov. 1984, Vol. PAS-103, No. 11, pp. 3303–3312.
18. Fuchs, E. F.; Roesler, D. J.; Chang, L. H.: "Magnetizing current, iron losses and forces of three-phase induction machines at sinusoidal and nonsinusoidal terminal voltages, Part II: Results," *IEEE Transactions on Power Apparatus and Systems*, Nov. 1984, Vol. PAS-103, No. 11, pp. 3313–3326.
19. Fuchs, E. F.; Poloujadoff, M.; Neal, G. W.: "Starting performance of saturable three-phase induction motors," *IEEE Transactions on Energy Conversion*, Sept. 1988, Vol. EC-3, No. 3, pp. 624–635.

Chapter 10
Rotating and Linear Motion Electric Machines

The objective of this chapter is to analyze/present the different types and flux patterns of electric machines. These flux distributions are based on numerical solutions [1–5]. Moreover, to discuss the techniques and approximations involved in reducing the physical machine to a simple mathematical model and to give some simple concepts relating to the basic machine types.

There are two types of electric machines with respect to the interaction of currents ($\vec{I}$) with a given flux density ($\vec{B}$) to produce electromagnetic forces:

- The longitudinal types (flux density is uniform in z-direction, see Figs. 10.1a, b)
- The transversal types (flux density is uniform in θ-direction, see Fig. 10.1c)

Note that the word "machine" can be interpreted as a device that works either as a motor or a generator. Any motor can be used as a generator, and vice versa any generator can be used as a motor. All rotating or linearly moving machines are based on the same principles:

- Faraday's law (see Chap. 6)
- Ampere's law (see Chap. 6)
- Constraints (e.g., constant speed, frequency)

Electromechanical energy conversion (production of force or torque) takes place when a change in flux Φ, that is, change in co-energy $dW'_{fld}(i, x)$ is associated with linear or angular mechanical motion dx or dθ, respectively:

$$f_{fld}(i_o, x) = \frac{\partial W'_{fld}(i_o, x)}{\partial x}, \tag{10.1a}$$

or

$$T_{fld}(i_o, \theta) = \frac{\partial W'_{fld}(i_o, \theta)}{\partial \theta} \tag{10.1b}$$

as discussed in Chap. 9.

E.F. Fuchs and M.A.S. Masoum, *Power Conversion of Renewable Energy Systems*,
DOI 10.1007/978-1-4419-7979-7_10,

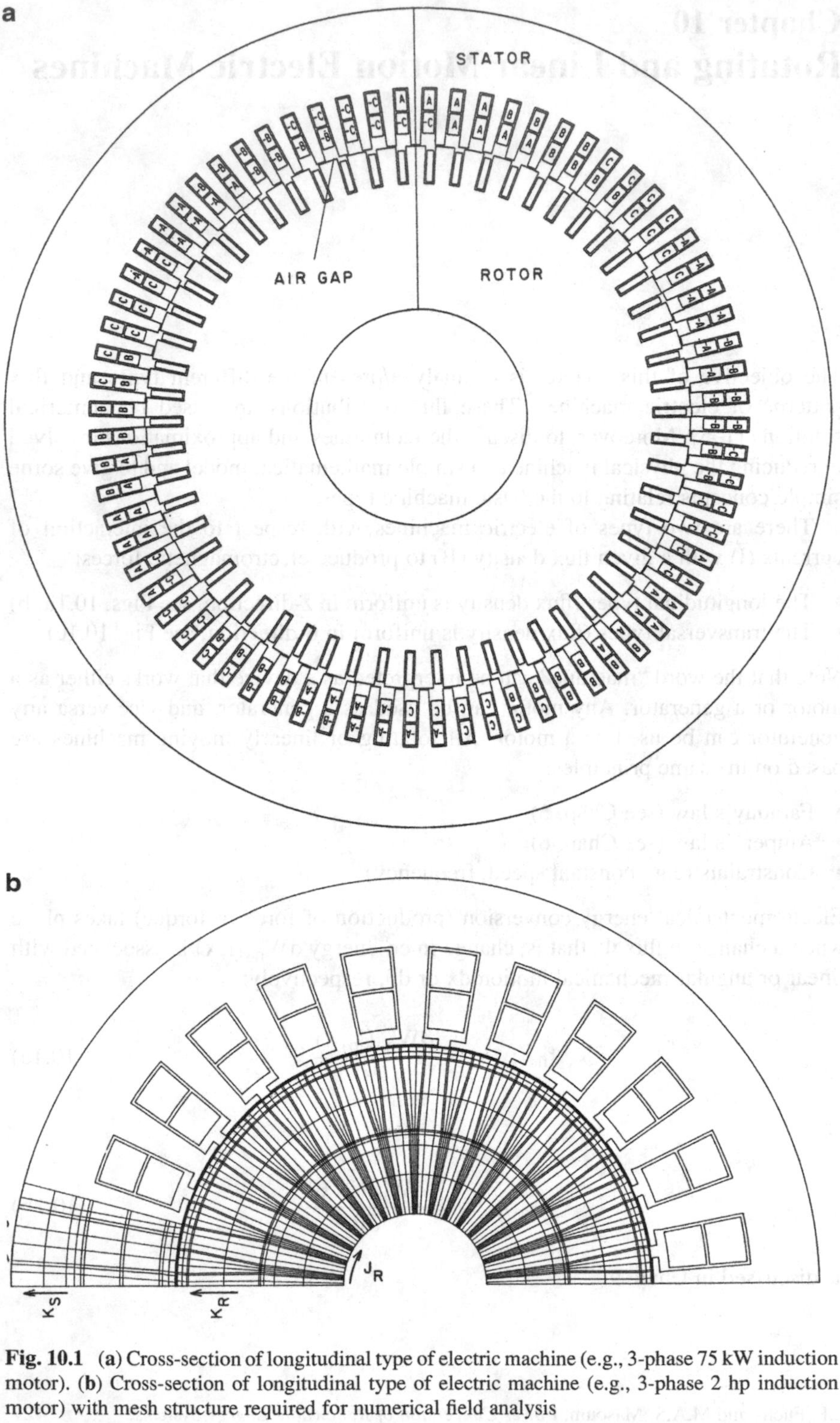

Fig. 10.1 (**a**) Cross-section of longitudinal type of electric machine (e.g., 3-phase 75 kW induction motor). (**b**) Cross-section of longitudinal type of electric machine (e.g., 3-phase 2 hp induction motor) with mesh structure required for numerical field analysis

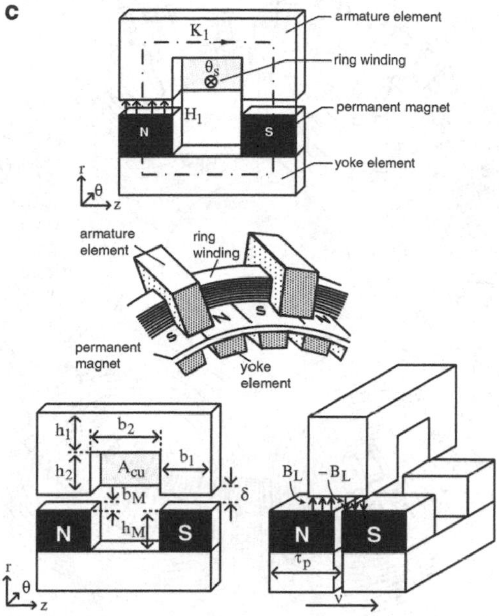

Fig. 10.1 (continued) (**c**) Various views of the transversal type of electric permanent magnet machine with a large pole number

10.1 Flux Patterns of Longitudinal and Transversal Rotating and Linear Machines

All machines are operating on the above mentioned principles (10.1a, b); any violation of these principles leads to zero torque and force production. Figures 10.1a, b illustrate the cross-sections of two longitudinal induction machines illustrating stator and rotor members as well as the air gap [6, 7]. Figure 10.1c shows the cross-sections and assembly of a transversal permanent-magnet machine [8, 9].

10.1.1 Longitudinal Types of Machines

Induction Machines: The most commonly used machine is the three-phase induction machine. About 63% of the total installed power capacity (1,000 GW) within the US is being consumed by such machines (motors). Typical flux patterns for a six-pole and a four-pole machine are shown in Figs. 10.2a, b, c [6], respectively: this type of machine consists of a stationary member (stator) which carries three distributed phase windings (A, B, C) displaced in space around the stator periphery

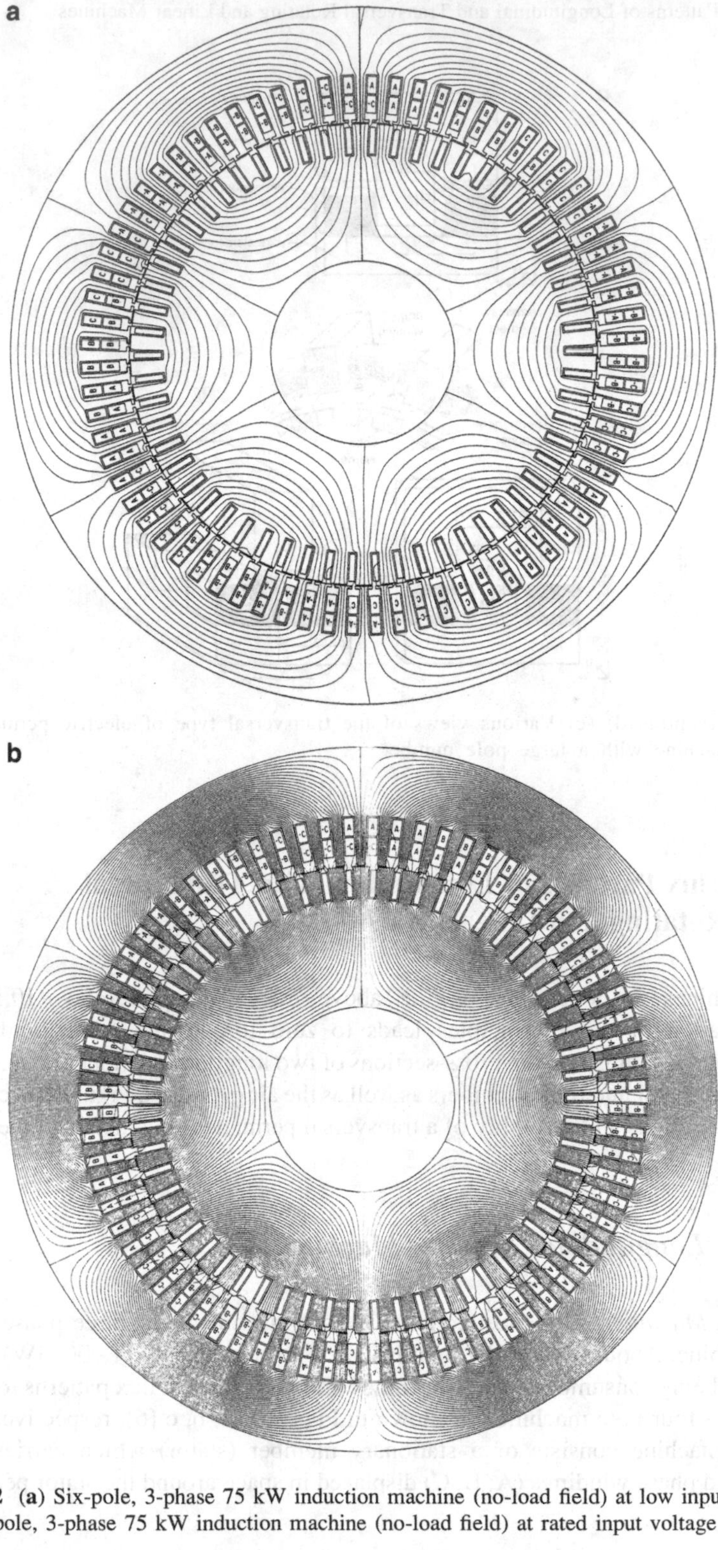

Fig. 10.2 (**a**) Six-pole, 3-phase 75 kW induction machine (no-load field) at low input voltage. (**b**) Six-pole, 3-phase 75 kW induction machine (no-load field) at rated input voltage

c

d

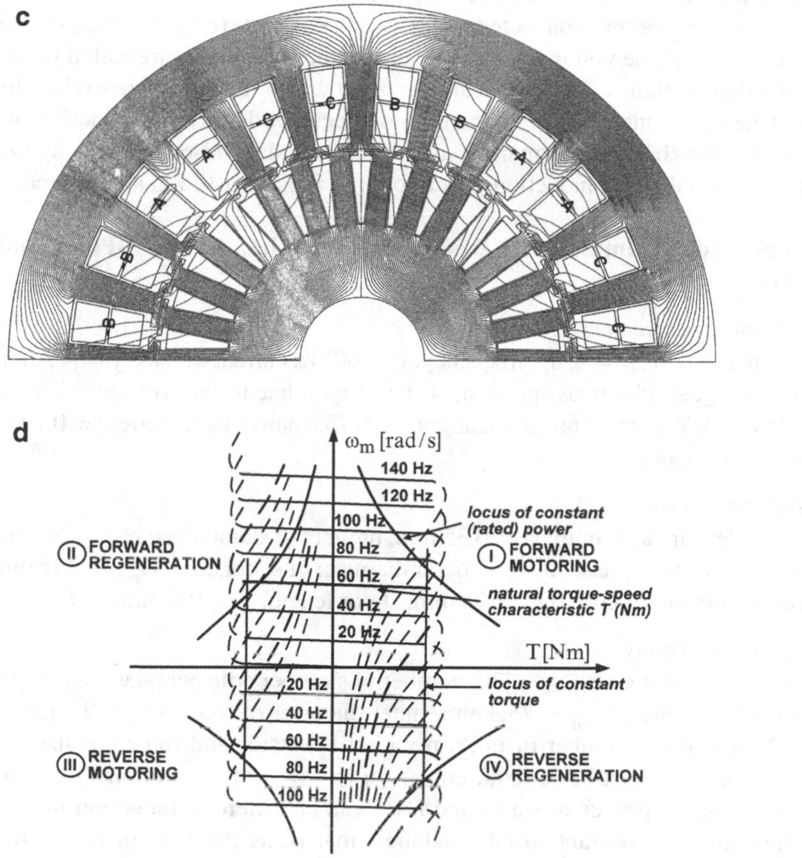

Fig. 10.2 (continued) (**c**) Four-pole, 3-phase 2 hp induction machine (no-load field) at rated input voltage. (**d**) Torque-speed characteristics of 3-phase induction machine for various frequencies of the induced voltage E (e.g., E/f control)

of the machine by 120 electrical degrees from each other. These three-phase windings are fed either from a balanced three-phase voltage or current source, where the (phase) voltages or currents are displaced in time by 120 electrical degrees. The moving part (rotor) carries (in most cases) a short-circuited three-phase winding, that is, the rotor is not excited by an external source, and there is no need to make provisions for electrical connections from the stationary reference frame of the stator to the moving reference frame of the rotor. This type of rotor winding is called a "squirrel-cage" rotor winding, and the machine is called a "squirrel-cage" type of three-phase induction motor/generator. The excitation currents within the short-circuited rotor windings are induced through the stator field across the air gap of the machine: this is the reason why the air gap of such machines must be as small as possible from a mechanical point of view: 0.1 mm for small (kW range) machines and up to 1.0 mm for large (MW range) machines [6, 7]. Some induction machines carry a 3-phase winding on the rotor similar to that of the stator, and this rotor winding can be either connected to passive components (e.g., resistors) or excited by a separate 3-phase voltage or current source: these machines are called wound-rotor induction machines. If these wound-rotor induction machines are excited by either voltage or current 3-phase source then we speak of doubly fed induction machines (DFIM) which are used mainly in variable-speed wind-power and hydro-power plants as doubly fed induction generators (DFIG) discussed in a later section.

Application Example 10.1: Design Data of 2 hp, 4-Pole, 3-Phase Induction Motor

Electrical parameters
Output power: $P_{out} = 2$ hp, frequency: $f = 60$ Hz, current density: $j = 4$ A/mm^2 with air cooling, synchronous speed: $n_s = 1{,}800$ rpm, line-to-line voltage: $V_{L\text{-}L} = 220$V/380 V in Δ/Y-connection, efficiency: $\eta = 0.735$, power factor: $\cos\phi = 0.705$ lagging (consumer notation).

Stator dimensions
One-sided air gap length: $g = 0.35$ mm, number of stator slots: $N_1 = 24$, number of slots per pole per phase: $q_1 = 2$, outer diameter of stator core: $D^s_{out} = 125$ mm, inner diameter of stator core: $D^s_{in} = 74$ mm, stator length: $\ell = 100$ mm.

Rotor dimensions
Number of rotor slots: $N_2 = 30$, number of slots per pole per phase: $q_2 = 2.5$, outer diameter of rotor: $D^r_{out} = 73.3$ mm, inner diameter of rotor: $D^r_{in} = 22$ mm.

Note that the number of poles p within the stator and rotor members must be the same – this is a basic requirement for any type of electric machine. However, the number of phases of stator and rotor can be different. Induction machines are approximately constant-speed machines, that is, as the load increases the speed decreases in a small measure only and the torque-speed characteristic of Fig. 10.2d is obtained: this type of machine can develop a torque at any speed. Induction generators are mainly used for variable-speed and constant-speed hydro-power and wind-power plants.

Synchronous Machines: Synchronous machines are mostly used in the generation of electric power via coal, gas, and nuclear power plants. The stator of such machines carries a three-phase winding similar to those of induction machines. The rotor carries a field winding which is excited by DC current. To feed the DC current from the stationary reference frame into the field winding which is moving, slip-rings will be employed. Figures 10.3a, b, c illustrate a two-pole [10] and a four-pole [11] synchronous machine, respectively. Figures 10.3d, e depict the magnetic fields of a six-pole synchronous aircraft alternator [12]. The name "synchronous" stems from the fact that the (mechanical) rotor moves in synchronism with the rotating stator field, that is, this machine is truly a constant-speed machine and has the torque-speed characteristic of Fig. 10.3f: this type of machine can develop torque at a single speed (the so-called synchronous speed) only [10–14] and is used mainly in large power generation plants up to an output power of 1,500 MVA.

a

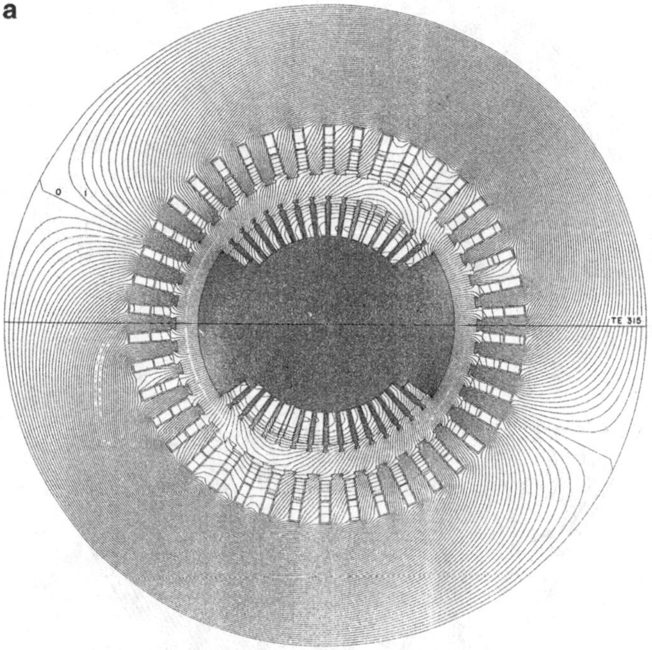

b

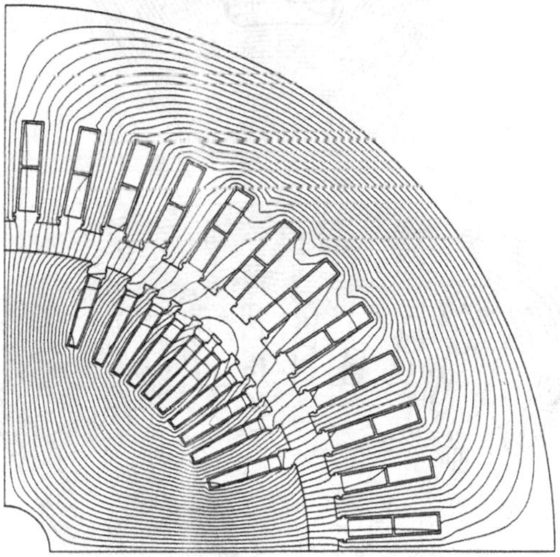

Fig. 10.3 (**a**) Two-pole, 3-phase synchronous machine, full-load field at rated operation (From Fuchs E. F.; Pohl, G.: "Computer generated polycentric grid design and novel dynamic acceleration of Convergence for the iterative solution of magnetic fields based on the finite difference method," *IEEE Transaction on Power Apparatus and Systems*, August 1981, Vol. PAS-100, No. 8, pp. 3911–3920, [10]). (**b**) Four-pole, 3-phase synchronous machine (no-load field) within one pole pitch (half a period)

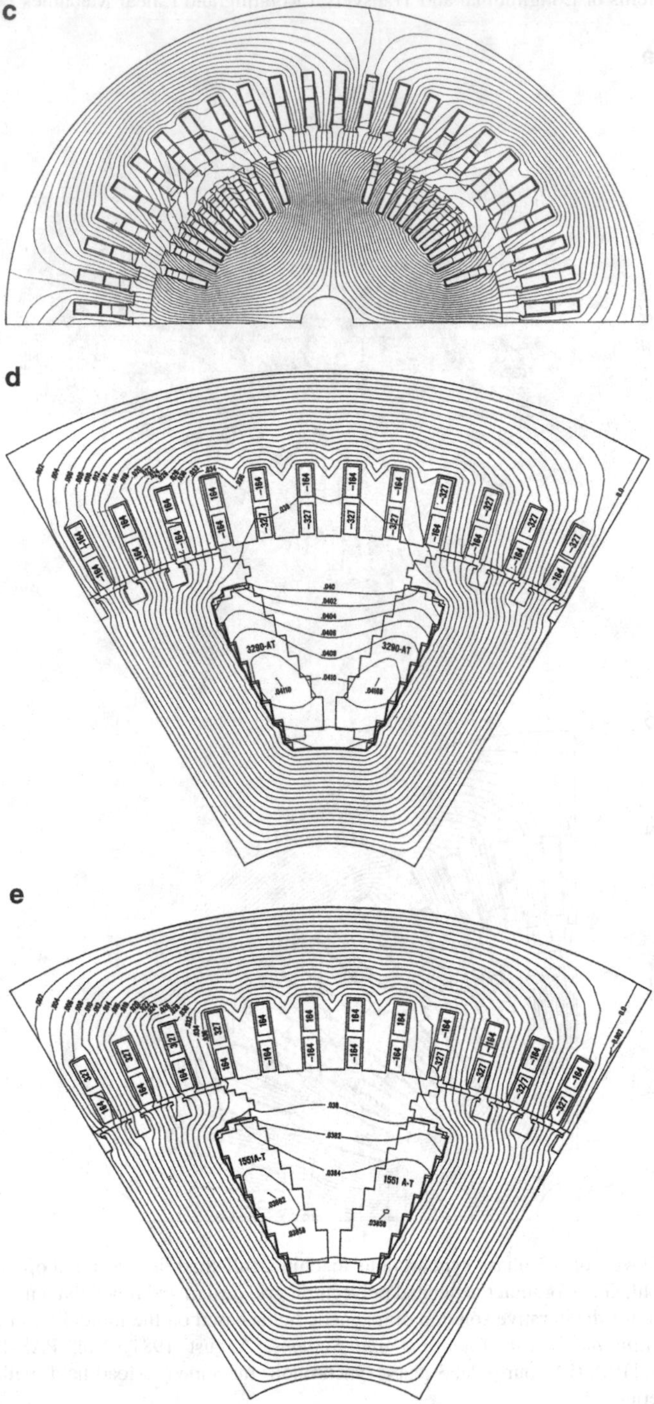

Fig. 10.3 (continued) (**c**) Four-pole, 3-phase synchronous machine (full-load field at rated operation) within 2 pole pitches or one period. (**d**) Six-pole, 3-phase 250 kVA, 120/240 V, 400 Hz, and 8,000 rpm aircraft synchronous machine (full-load field at rated operation, when stator and rotor mmfs are in phase). (**e**) Six-pole, 3-phase 250 kVA, 120/240 V, 400 Hz, and 8,000 rpm aircraft synchronous machine (full-load field at rated operation, when stator and rotor mmfs are not in phase)

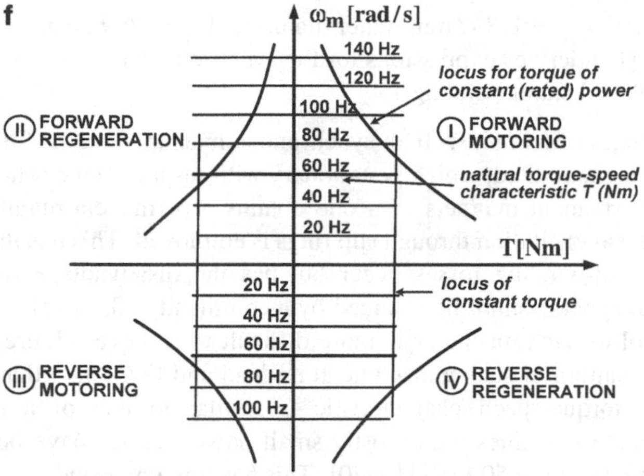

Fig. 10.3 (**f**) Torque-speed characteristics of 3-phase synchronous machine (e.g., E/f control)

Application Example 10.2: Design Data [12] of a 6-Pole Aircraft Synchronous Alternator

Electrical parameters
Output power: $S_{out} = 250$ kVA, efficiency: $\eta = 0.9$, power factor: $\cos\phi = 0.8$ lagging (generator notation), line-to-line voltage: $V_{L-L} = 120/240$ V, Y-connected, forced-air cooling, frequency: $f = 400$ Hz, synchronous speed: $n_s = 8{,}000$ rpm.

Stator dimensions
Stator skew: 1 slot pitch, number of stator poles: $p = 6$, housing outer diameter: 335.28 mm, overall housing length: 571.5 mm, total weight: 132.73 kg-force, number of stator slots: $N_1 = 72$, air gap radial length (one sided); $g = 1.143$ mm, stack length of core: $\ell = 152.4\,\text{mm}$, stator-bore diameter: $D^s_{in} = 226.06\,\text{mm}$.

Rotor dimensions
Damper bars per pole: 5.

Application Example 10.3: Design Data of a 4-Pole Turboalternator

Electrical parameters
Output power: $S_{out} = 200$ MVA, efficiency: $\eta = 0.96$, indirect hydrogen cooling, power factor: $\cos\phi = 0.8$ lagging (generator notation), line-to line voltage: $V_{L-1} = 15$ kV, frequency: $f = 60$ Hz, direct-axis synchronous reactance [15] $X_d = 1.7$–1.38 pu, quadrature-axis synchronous reactance [16] $X_q = 1.63$–1.36 pu, field/rotor excitation current $I_f = 2{,}000$ A, synchronous rotor speed: $n_s = 1{,}800$ rpm.

Stator dimensions
Number of stator poles: $p = 4$, outer diameter: $D^s_{out} = 2,540\,\text{mm}$, overall housing length: 5,000 mm, number of stator slots: $N_1 = 48$, stack length of iron core: $\ell = 4,000\,\text{mm}$, stator-bore diameter: $D^s_{in} = 1533.5\,\text{mm}$.

Rotor dimensions
Outer diameter: $D^r_{out} = 1,397\,mm$, inner diameter: $D^r_{in} = 209.55\,mm$.

Reference [17] derives expressions for the synchronous reactances of machines with two displaced stator windings.

Permanent-Magnet Machines: If in synchronous machines the DC field winding producing a magnetic field, which is stationary with respect to the rotating rotor, is replaced by permanent magnets then one obtains a permanent-magnet machine, where no external excitation through slip rings is employed. This has the advantage that no I^2R field-winding losses occur but has the disadvantage that the field produced by magnets cannot be changed by a command and, therefore, speed and voltage control of such machines is more difficult to achieve. Figure 10.4a illustrates the flux pattern of such a machine at no-load and Fig. 10.4b that at full-load [18–20]. The torque-speed characteristic is similar to that of a synchronous machine. To date machines with mostly small power ratings have been built for drive applications (up to 50 kW) [18–20]. This has a few reasons:

- Cost of magnet material (NdFeB: $40/lb-force)
- Difficulty in mounting the rotor: mounting gear is required
- Speed control is more difficult as compared with induction and conventional synchronous machines

However, variable-speed permanent-magnet wind-power generators with a high pole number and with a rating up to 5 MW are in service.

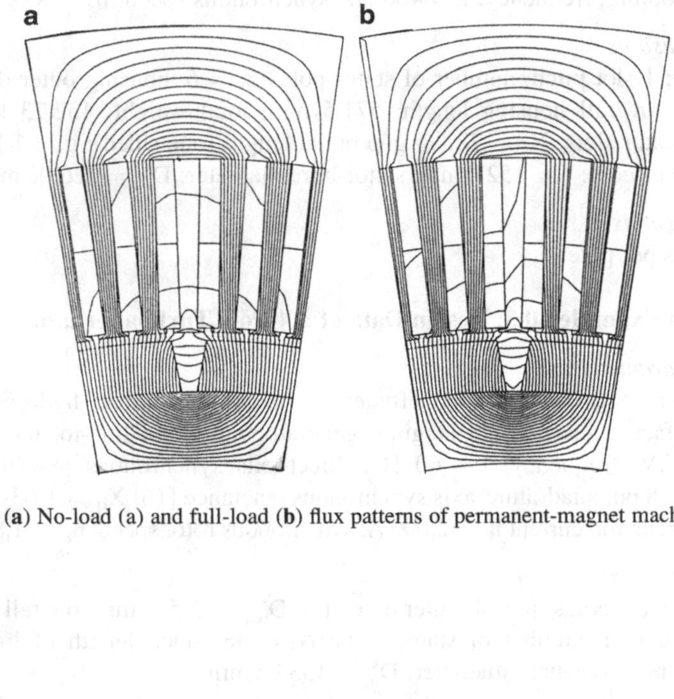

Fig. 10.4 **(a)** No-load (a) and full-load (**b**) flux patterns of permanent-magnet machine

Application Example 10.4: Design Data of a Permanent-Magnet Machine

Electrical parameters
Output power: $P = 50$ kW, efficiency: $\eta = 0.95$, indirect water cooling, line-to-line voltage: $V_{L\text{-}L} = 240$ V, frequency: $f = 400$ Hz, power factor: $\cos\phi = 0.9$ lagging (generator notation).

Stator dimensions
Core length $\ell = 300$ mm, outer diameter: $D^s_{out} = 244$ mm, inner diameter: $D^s_{in} = 170.4$ mm, number of slots in stator: $N_1 = 120$, number of poles: $p = 20$.

Rotor dimensions
Outer diameter: $D^r_{out} = 169.4$ mm, inner diameter: $D^r_{in} = 140.2$ mm.

Brushless DC Machines: These machines are mainly used in electric/hybrid automobile drives [18]. They consist of a permanent-magnet machine fed by either a six-step or PWM inverter. In Application Example 5.14 such a drive is analyzed with PSpice. To synchronize the switching of the inverter with the speed an encoder must be employed. The flux weakening/reduction at high speed is performed with the stator mmf which may oppose -depending on the phase angle between stator and rotor mmfs- the constant rotor mmf due to the permanent magnets residing on the rotor.

DC Machines [11, 21]: DC machines (motors/generators) were invented around 1870 (Gramme 1871, see Chap. 2) prior to the invention of any other type of electrical machine (see Fig. 10.5). When the machine is motoring the mechanical commutator (acting like a mechanical inverter) converts the applied DC current to an AC current within the rotor/armature of such a machine; when operated as a generator the mechanical commutator rectifies the induced AC currents within the rotor/armature and behaves like a mechanical rectifier; that is, the terminal voltage and current are of the DC type (see Chap. 2). In the case of the brushless DC machine (motor/generator) the mechanical commutator is replaced by an electronic commutator (called an electronic inverter for motoring or an electronic rectifier for regeneration). DC machines are very complicated machines from a mechanical and an electrical point of view, due to the mechanical commutator.

Reluctance Machines: Such machines have a relatively simple stator structure housing concentrated windings (Fig. 10.6a) which are excited by current pulses: these machines rely on the generation of current pulses by power electronic circuits (Fig. 10.6b). The rotor member is simple as well and does not carry any winding. These machines are ideal for very high-speed applications ($n_m > 20{,}000$ rpm). The only drawbacks are low torque production and generated acoustic noise. For these reasons such machines have not yet gained a wide application. Figures 10.6c–f illustrate flux patterns for different rotor positions [22].

Longitudinal types of machines are built with a relatively small pole number ($p < 60$, e.g., the hydro-generators installed at the Hoover Dam have $p = 40$ [14]) and produce at a relatively low torque and a high speed (e.g., $n_s = 3{,}600$ rpm @ $p = 2$)

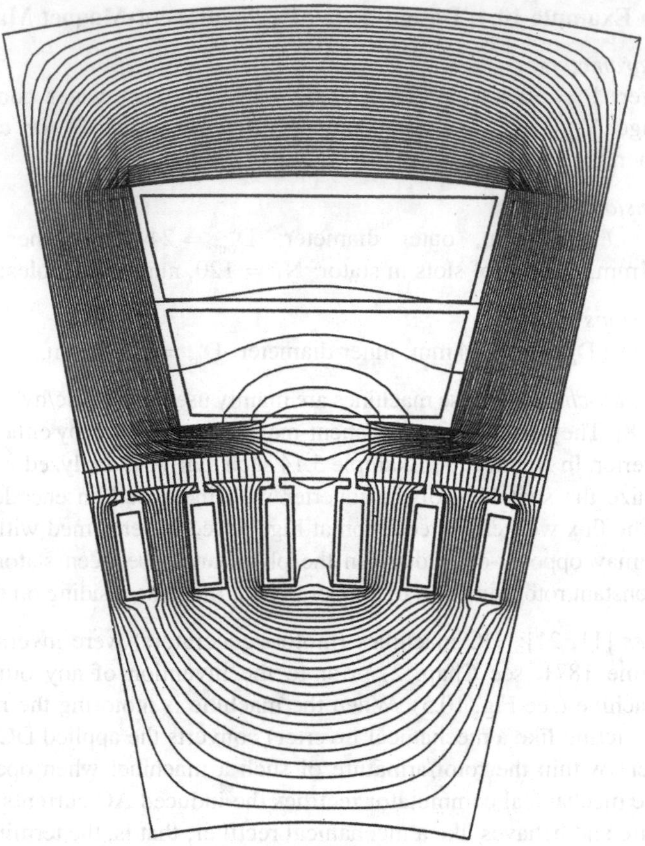

Fig. 10.5 No-load flux pattern of one pole pitch of a 12-pole DC machine

a relatively high output power. Unfortunately, the utilization of such machines at low-speed requires (e.g., electric car) a mechanical gear. Such gears are heavy and reduce the overall efficiency of the drive by about 2–4%. It would be best not to rely on any mechanical gears in such applications [23–25, 30]. One possibility is to employ transversal types of electric machines producing high torques at low speeds but they might have the disadvantage of a low power factor [8, 9, 22], due to the large inductances involved.

10.1.2 Transversal Types of Machines

While all longitudinal machines are of the poly-phase type (number of phases m $\geq$2) some of the transversal machines are of the single-phase type. Their main advantage is that they can be operated at relatively large air-gap flux densities (e.g., $B_{max} = 2$ T) as compared to longitudinal machines which have air-gap

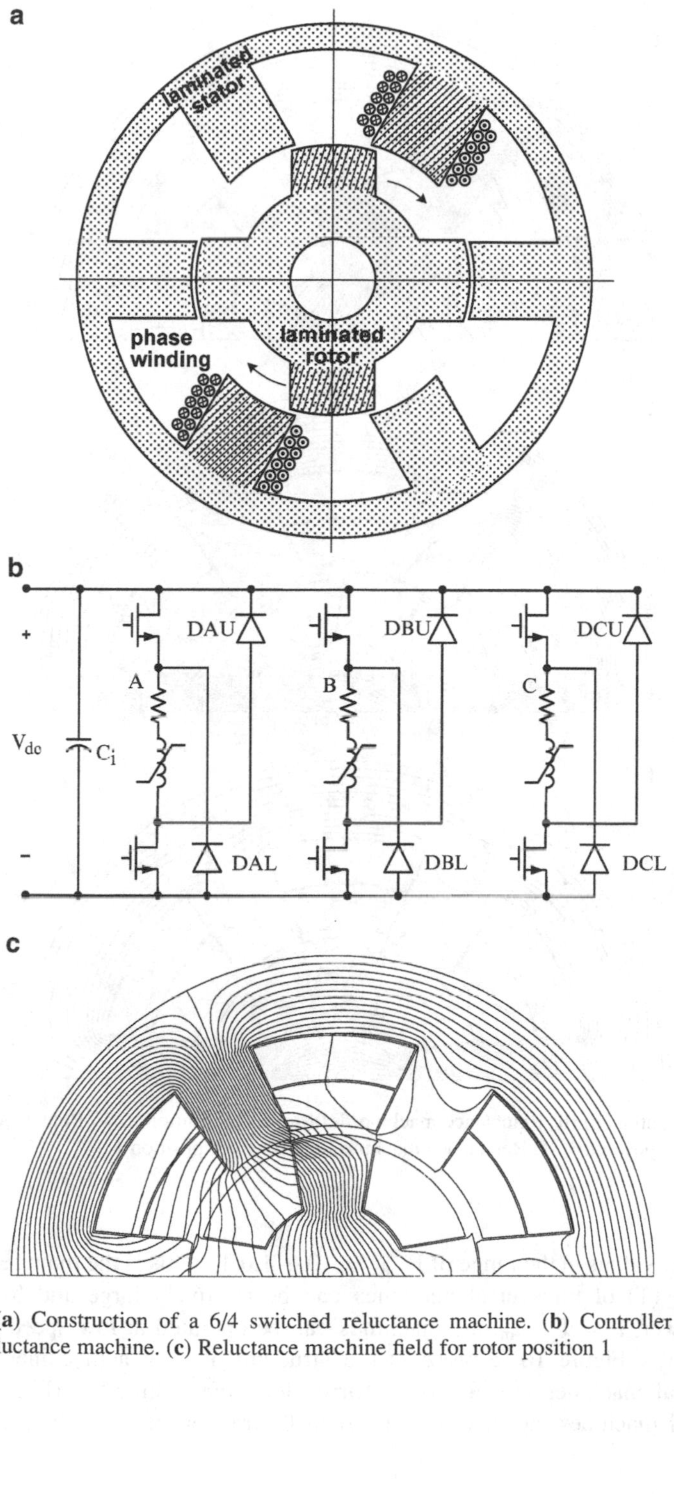

Fig. 10.6 (**a**) Construction of a 6/4 switched reluctance machine. (**b**) Controller circuit for switched reluctance machine. (**c**) Reluctance machine field for rotor position 1

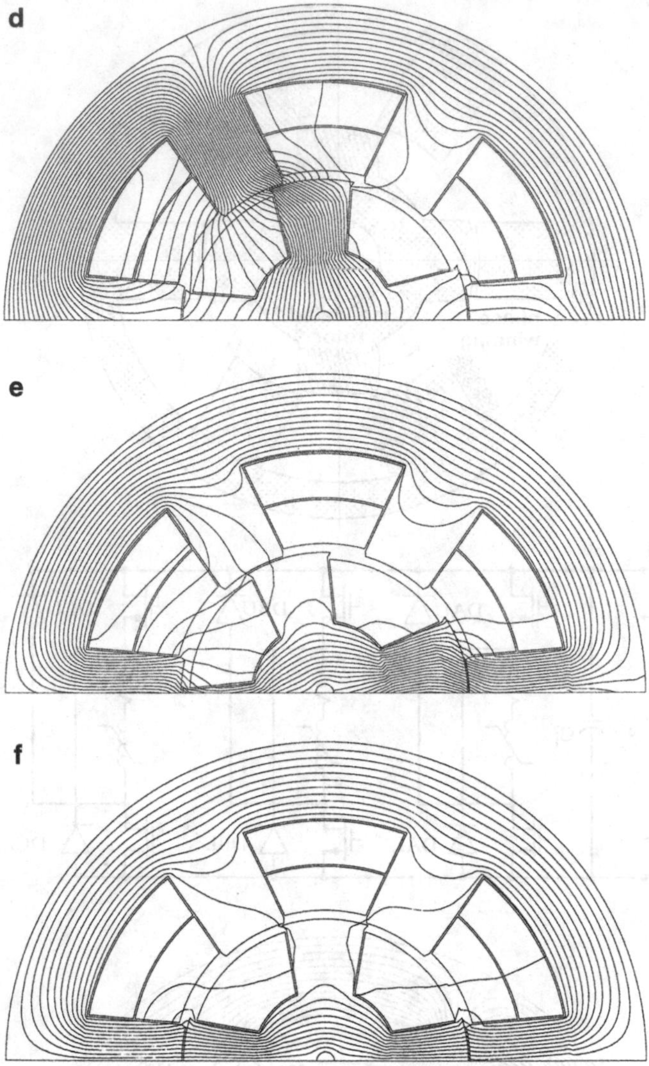

Fig. 10.6 (continued) (**d**) Reluctance machine field for rotor position 2. (**e**) Reluctance machine field for rotor position 3. (**f**) Reluctance machine field for rotor position 4

flux density values in the range of $B_{max} = (0.6–1.0\ T)$. This is the main reason why the torque (T) of transversal machines can be relatively large and for a given output power, $P = T \cdot \omega_m$, the machines can be operated at low speeds (angular velocity ω_m). Figure 10.1c presents the structure of such a machine. While in longitudinal machines the flux is uniform along the z-direction (Fig. 10.1a) in transversal machines the flux is uniform in θ-direction of the r, θ, z cylindrical

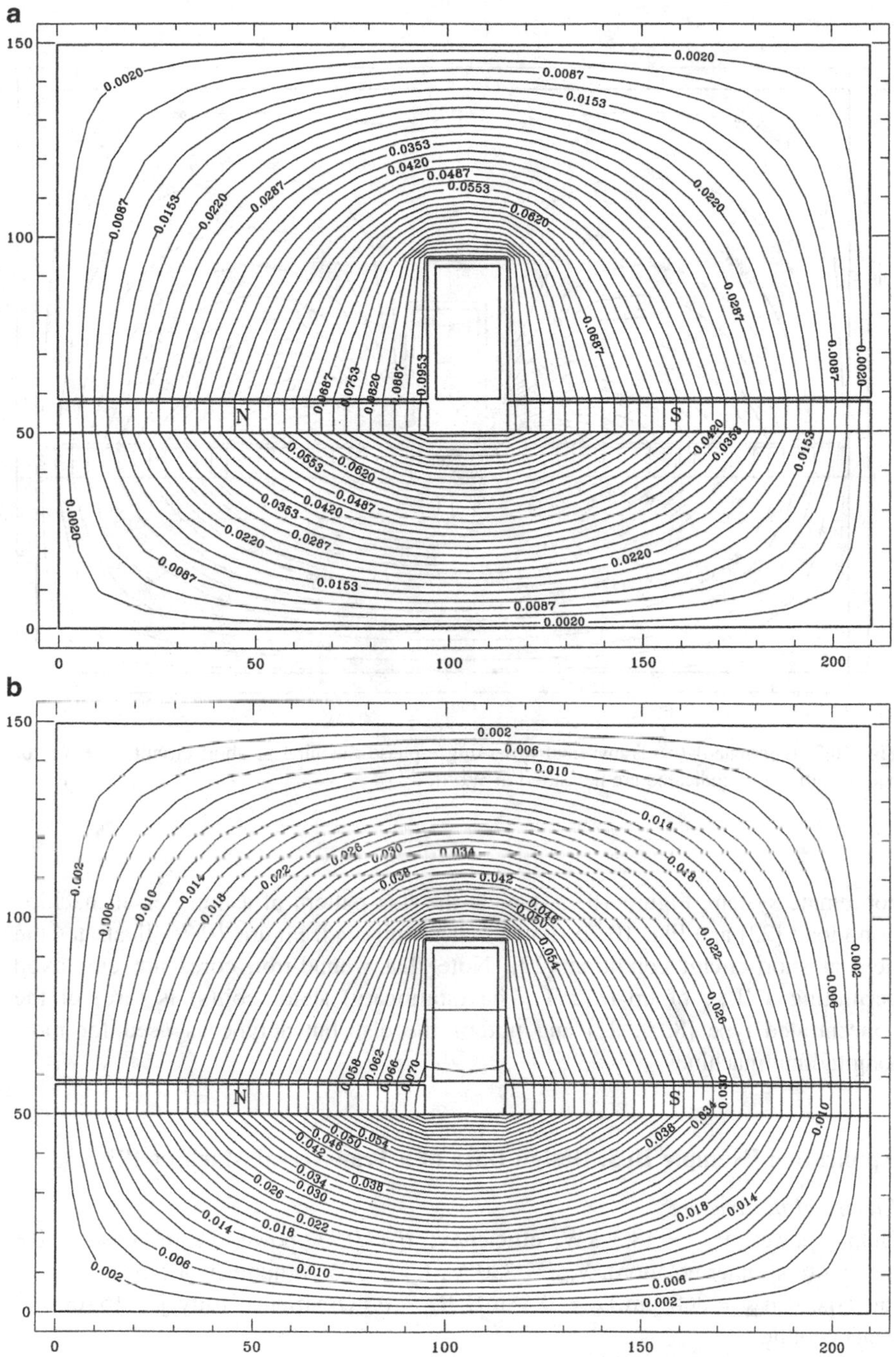

Fig. 10.7 (**a**) Transversal type, single-phase machine at no-load. The vector equipotential lines indicated are measured in Wb/m. (**b**) Transversal type single-phase machine at full-load. The vector equipotential lines indicated are measured in Wb/m

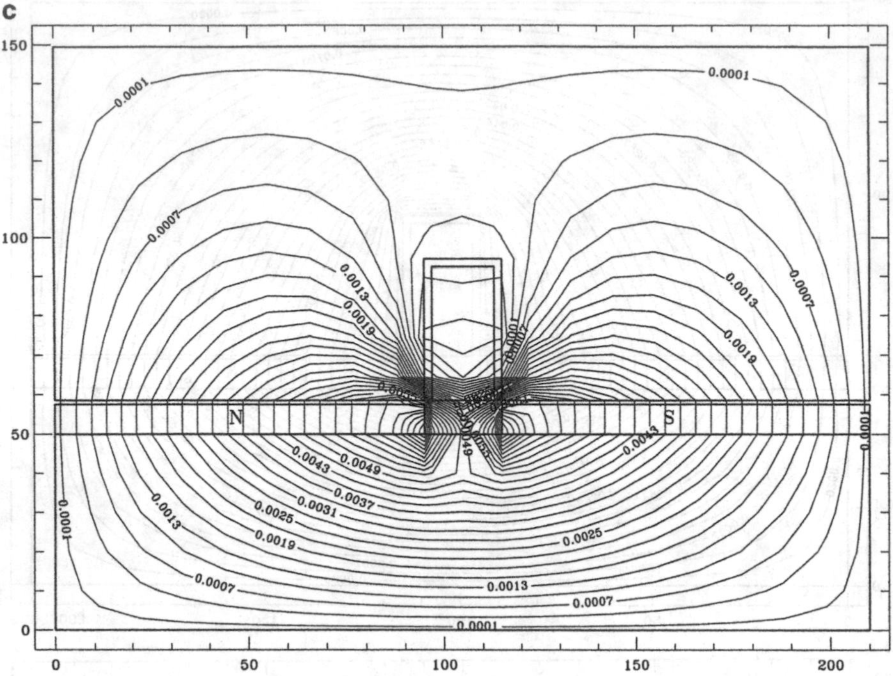

Fig. 10.7 (continued) (**c**) Transversal type single-phase machine at short-circuit. The vector equipotential lines indicated are measured in Wb/m

coordinate system. Figure 10.7a depicts the flux pattern of a transversal machine at no-load [22] and Fig. 10.7b shows that at full-load. Figure 10.7c illustrates the field at short-circuit condition [22]. Note that wound-iron cores are employed permitting a B_{max} of up to 2 T. The (stationary) stator coil(s) is (are) of the concentrated type [8, 9, 22] and not of the distributed type as used for most longitudinal machines.

Application Example 10.5: Design Data of a Transversal Machine [9]

Electrical parameters
Output power: $P_{out} = 18.3$ kW, efficiency: $\eta = 0.93$, power factor: $\cos\phi = 0.8$ lagging (consumer notation), $V_{rat} = 230$ V, $I_{rat} = 105$ A, number of poles: $p = 60$, rated speed: $n_{rat} = 360$ rpm, $L_s = 0.46$ mH, $R_s = 0.1\,\Omega$, current density: $j = 15$ A/mm^2, water cooling.

Stator dimensions (see Fig. 10.1c)
Air gap (one sided): $g = 1$ mm, magnet height: $h_M = 8$ mm, magnet depth: $b_M = 10$ mm, $b_1 = 36.6$ mm, $h_1 = 20.3$ mm, $b_2 = 21.6$ mm, $h_2 = 32.4$ mm.

10.2 Elementary Concepts [26]

Iterative Solution of Nonlinear Equations: In Chap. 6 the nonlinear B-H characteristics of ferromagnetic materials are discussed. The presence of any nonlinear dependency (e.g., nonlinear iron characteristic, product of two independent variables) within a design process necessarily leads to iterative solutions. In the case of an electric machine the sequence of steps of the solution process is shown in Fig. 10.8: one starts out with the assumption of the current distribution around the circumference with the coordinate θ of a machine: i(θ), and computes the magnetomotive force F = mmf(θ) by using Fourier series techniques. The application of Ampere's law leads to the magnetic field intensity H(θ), and via the constituent relation B = μH one arrives at the flux density B(θ). The continuity of flux condition (div B = 0) leads to the flux ϕ(θ), the application of Faraday's law yields the induced voltage e(t), Kirchhoff's voltage and current laws lead to the impedance Z, and Ohm's law permits us to re-compute i(θ), thus completing one iteration.

MMF of a Concentrated Coil (Winding): The most elementary winding consists of a concentrated N-turn coil carrying a current i, that is, the entire coil resides in two slots: one coil side resides in one slot and the span of the coil is 180 electrical degrees, and is, therefore, called a full-pitch coil. All following considerations refer to a p = 2 pole winding: for such a winding the electrical degrees are identical with the mechanical degrees, and there is no need to differentiate between both. Figure 10.9a depicts how a concentrated coil resides within the stator member of an electric machine.

The field lines of H are indicated in this figure and the axis of the field lines is identical with the axis of the concentrated winding aa′. The cylindrical coordinate system is assumed to be r, θ, and z. If the circumference at the stator inner surface is expanded (developed) then one obtains Fig. 10.9b, where a current coming out (•)

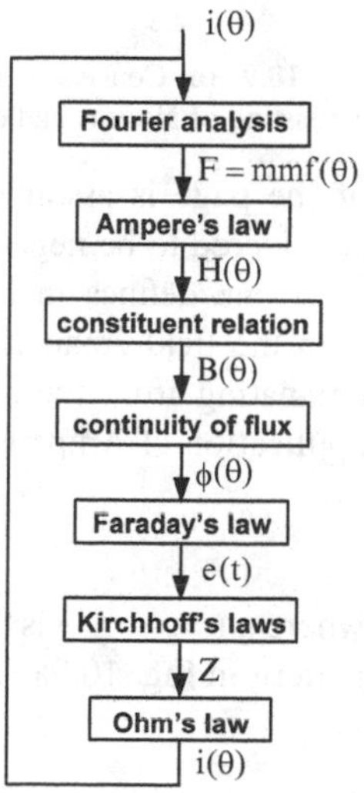

Fig. 10.8 Iterative solution of the nonlinear equations of a magnetic device

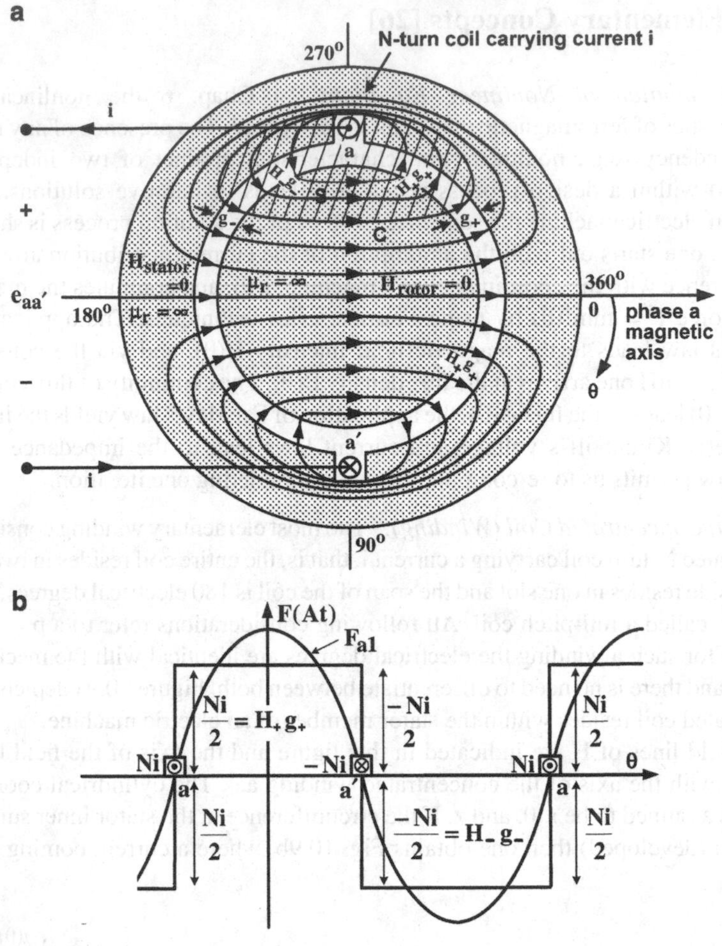

Fig. 10.9 (a) Concentrated winding (coil aa′) residing in stator. (b) MMF of a concentrated coil consisting of N-turns and conducting a current of i amperes per turn

of the page is assumed to be positive, and a current going into (x) the page is considered to be negative.

If one defines in Fig. 10.9a the magnetic field intensity in the air gap as H then the field emanating from the rotor and going into the stator is H_+, and that emanating from the stator and going into the rotor is H_- (see Fig. 10.9a). The application of Ampere's law along path C enclosing the surface S leads to

$$N \cdot i = H_+ \cdot g_+ + H_- \cdot g_- \tag{10.2a}$$

where g ($=g_+=g_-$) is the one-sided air gap length. Because of symmetry of the flux pattern in Fig. 10.9a $H_+=H_-=H$

$$|H_+ \cdot g_+| = |H_- \cdot g_-|, \tag{10.2b}$$

therefore,

$$N \cdot i = 2|H_+ \cdot g_+| \text{ or } |H_+ \cdot g_+| = \frac{N \cdot i}{2}. \tag{10.3}$$

Now the mmf can be plotted by identifying

$H_+ \cdot g_+ = \frac{Ni}{2}$, and $H_- \cdot g_- = -\frac{Ni}{2}$, as shown in Fig. 10.9b. A Fourier series of the rectangular mmf space wave mmf(θ) –for the 2 pole (p = 2) machine of Fig. 10.9a– leads to the fundamental

$$F_{a1} = \frac{4}{\pi}\left(\frac{Ni}{2}\right)\cos\theta. \tag{10.4}$$

This is the equation of a sinusoidal standing wave in space. All higher harmonics are neglected for the time being. In general, for a p-pole machine one obtains

$$F_{a1} = \frac{4}{\pi}\left(\frac{N_{ph}}{p}\right)i_a\cos\theta, \tag{10.5a}$$

where i_a is the current through the coil of phase "a", and N_{ph} is the number of turns of one phase(belt) such as aa′.

MMF of a Distributed Winding: In order to approximately achieve sinusoidal waveshapes one employs distributed windings by displacing several concentrated coils in space around the circumference. Figure 10.10a illustrates how the N_{ph}-turns of one phase of a winding (remember we are concerned with three-phase windings a, b, and c) are divided up in equal parts and distributed in eight slots that means (N_{ph}/4)-turns (one coil side) within one slot of a 60° phase(belt). Note that one phase occupies a sector of 360°/3 = 120° or four coil sides of phase "a" 120°/2 = 60°. This means there is a mechanical angle of 60°/4 = 15° between adjacent slots.

The mmf of coil 1-1′ is in phasor representation:

$$\tilde{F}_{a1}\Big|_{1-1'} = \frac{4}{\pi}\frac{(N_{ph}/4)}{p}i_a\cos\theta\angle 0°, \tag{10.6a}$$

those of coils 2-2′, 3-3′ and 4-4′ are

$$\tilde{F}_{a1}\Big|_{2-2'} = \frac{4}{\pi}\frac{(N_{ph}/4)}{p}i_a\cos\theta\angle -15°, \tag{10.6b}$$

$$\tilde{F}_{a1}\Big|_{3-3'} = \frac{4}{\pi}\frac{(N_{ph}/4)}{p}i_a\cos\theta\angle -30°, \tag{10.6c}$$

$$\tilde{F}_{a1}\Big|_{4-4'} = \frac{4}{\pi}\frac{(N_{ph}/4)}{p}i_a\cos\theta\angle -45°. \tag{10.6d}$$

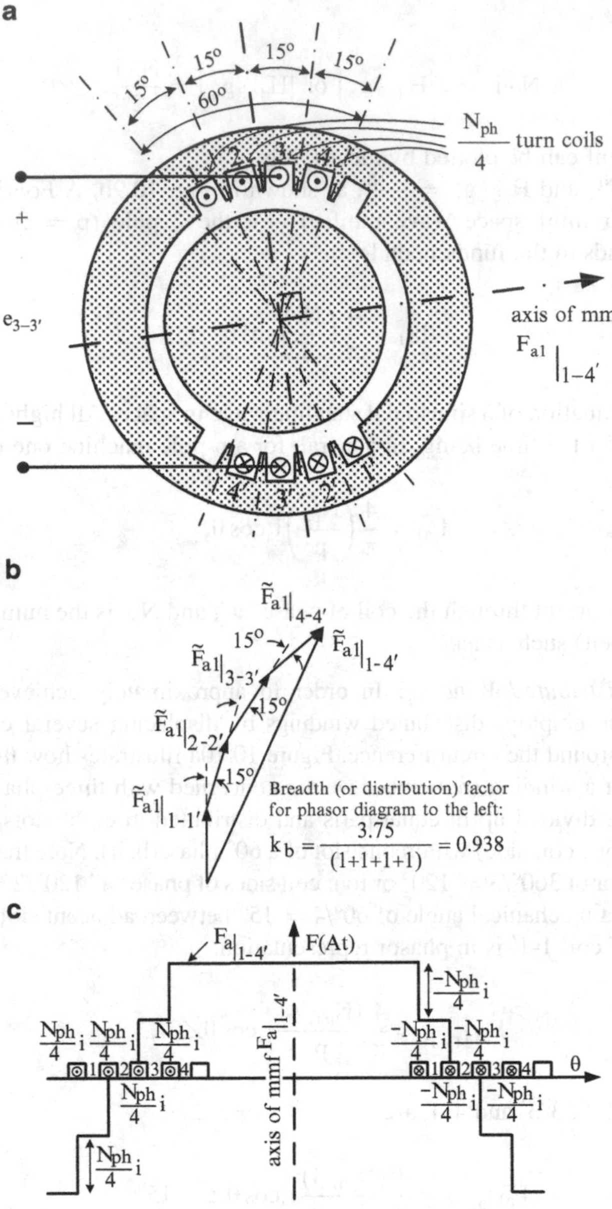

Fig. 10.10 (**a**) Distribution of turns within a 60° phase(belt). (**b**) Resultant (geometric sum) of mmf of a 60° phase(belt) in the phasor domain. (**c**) Resultant (geometric sum) of mmf of a 60° phase(belt) where $i = i_a$

One concludes that the mmf of coil 2-2′ lags that of coil 1-1′ and the mmf of coil 3-3′ lags that of coil 2-2′, and so on. Figure 10.10b shows the geometric addition of these phasors and leads to the resultant sum $\tilde{F}_{resultant}$=geometric sum $\Sigma\tilde{F}_{a1}|_{i-i'}$. This leads to the definition of the breadth factor

$$k_b = \frac{\text{geometric sum}\Sigma\tilde{F}_{a1}\big|_{i-i'}}{\text{algebraic sum } \Sigma\tilde{F}_{a1}\big|_{i-i'}}, \tag{10.7}$$

k_b is part of the winding factor $k_w = k_b \cdot k_p$. This breadth factor is also called distribution factor $k_d = k_b$. One concludes the resultant (geometric) fundamental mmf wave $|\tilde{F}_{resultant}|$ of a distributed winding is less than the algebraic sum of the fundamental components of the individual coils, because the magnetic axes of the individual coils are not aligned with the resultant, therefore, $0 \le k_b \le 1.0$.

The employment of distributed windings cancels harmonics but reduces the fundamental by the breadth factor k_b. Figure 10.10c illustrates the mmf as a function of θ.

A further improvement of the sinusoidal waveshape is possible through short pitching of the coils as demonstrated in Example 10.8, resulting in the pitch factor $k_p \le 1.0$.

The total winding factor is then $k_w = k_b \cdot k_p$, where $0.8 \le k_w \le 1.0$, and therefore, (10.5a) can be augmented by k_w

$$F_{a1} = \frac{4}{\pi}\left(\frac{N_{ph}}{p}\right)k_w i_a \cos\theta. \tag{10.5b}$$

For plotting mmfs it is convenient to adhere to some simple rules: Examples 10.6 and 10.7 convey these rules. Depending upon the number of current-carrying slots in either stator or rotor the starting point for plotting the mmfs is different. If these rules are not followed then one obtains a DC component within the mmfs even though there should not exist one.

Application Example 10.6: Construction of mmf for an Even Number of Distributed Current-Carrying Coils Residing Within a Pole Pitch (One Half of Period)

A convenient starting point for plotting the mmf of a current distribution with an even number of current-carrying slots per pole pitch (half a period) is a horizontal line along the abscissa (θ-axis) at the line of symmetry of the current distribution within one pole pitch, as illustrated in Fig. 10.11 for four current-carrying slots per pole pitch.

Application Example 10.7: Construction of mmf for an Odd Number of Distributed Current-Carrying Coils

A convenient starting point for plotting the mmf of a current distribution with an odd number of current-carrying slots per pole pitch (half a period) is a vertical line intersecting with the abscissa (θ-axis) at the line of symmetry of the current distribution within one pole pitch, as illustrated in Fig. 10.12 for three current-carrying slots per pole pitch.

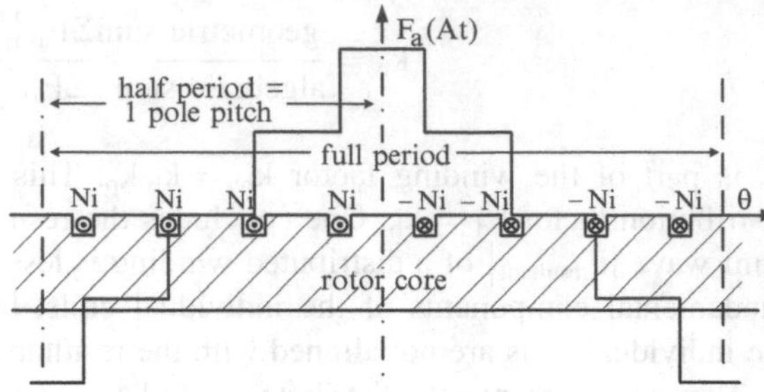

Fig. 10.11 Even number of distributed current-carrying coils within a pole pitch

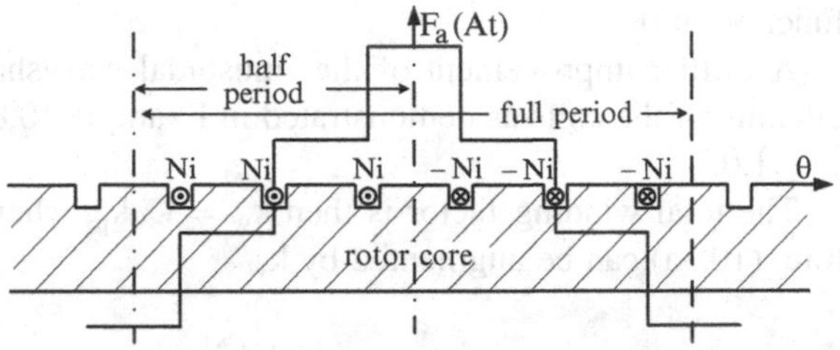

Fig. 10.12 Odd number of distributed current-carrying coils within a pole pitch

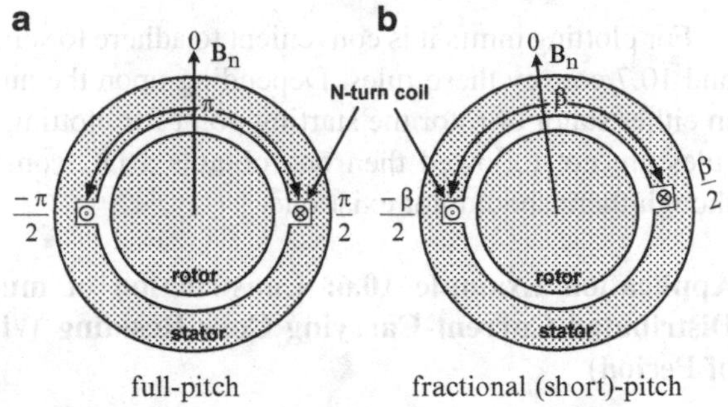

Fig. 10.13 Computation of the pitch factor for a short-pitch winding

Application Example 10.8: Computation of Pitch Factor k_p of a Winding

This example investigates the advantages of short-pitching the stator coils of an AC machine. Figure 10.13a shows a single full-pitch coil in a two-pole machine. Figure 10.13b shows a fractional-pitch coil; the coil sides are β radians apart rather than π radians (180°), as in the full-pitch case. For an air-gap radial flux distribution of the form

$$B_{radial} = \sum_{n=odd} B_n \cos n\theta, \tag{10.8}$$

where n = 1 corresponds to the fundamental space harmonic, n = 3 the third spatial harmonic, etc., the flux linkage of each coil is the integral of B_{radial}, over the surface spanned by that coil. Thus for the *n*th space harmonic, the ratio of the flux linkage of the fractional-pitch coil to that of the full-pitch coil is given by the so-called pitch factor defined as

$$k_p(n) = \frac{\int_{-\beta/2}^{\beta/2} B_n \cos n\theta d\theta}{\int_{-\pi/2}^{\pi/2} B_n \cos n\theta d\theta} = \frac{\int_{-\beta/2}^{\beta/2} \cos n\theta d\theta}{\int_{-\pi/2}^{\pi/2} \cos n\theta d\theta}. \quad (10.9)$$

It is common to (fractional) pitch the coils of an AC machine by 30° ($\beta = 5\pi/6$ radians = 150°). For n = 1, 3, and 5 calculate the reduction in flux linkage due to short-pitching.

Solution

It can be shown that the integration of (10.9) will lead to the following expression

$$k_p(n) = \frac{\sin\left(\frac{n\beta}{2}\right)}{\sin\left(\frac{n\pi}{2}\right)} = \left|\sin\left(\frac{n\beta}{2}\right)\right|. \quad (10.10)$$

For $\beta = \frac{5\pi}{6}$ radians corresponding to 150° one obtains the pitch factor as a function of the harmonic order n

$$k_p(n) = \left|\sin\left(n\frac{5\pi}{6 \cdot 2}\right)\right|. \quad (10.11)$$

$n = 1$ leads to $k_p(n = 1) = 0.9659$, $n = 3$ leads to $k_p(n = 3) = 0.707$, $n = 5$ leads to $k_p(n = 5) = 0.2588$. One concludes that the fundamental ($n = 1$) is hardly reduced, while the higher harmonics ($n = 3, 5, \ldots$) are greatly reduced, making the wave shape more sinusoidal.

Magnetic Fields in Rotating Machinery: Once the mmf(θ) is known, the magnetic field intensity H can be obtained from

$$H(\theta) = \frac{mmf(\theta)}{g}, \quad (10.12)$$

or in particular for the fundamental (1) field strength of coil a H_{a1}

$$H_{a1} = \frac{F_{a1}}{g} = \frac{4}{\pi}\left(\frac{N_{ph}}{p \cdot g}\right) k_w i_a \cos\theta, \quad (10.13)$$

N_{ph} is number of series turns per phase, θ is electrical angle measured with respect to magnetic axis of winding, $k_w = k_b \cdot k_p$ is winding factor, k_b is breadth factor or distribution factor k_d, k_p is pitch factor, and g is the one-sided air gap length.

10.3 Alternating and Rotating (Circular, Elliptic) Magnetic Fields, Induced Voltage, and Torques

Rotating mmf Waves in AC Machines: To understand the theory and operation of poly-(three) phase AC machines it is necessary to study the nature of the mmf wave produced by a three-phase winding generating a rotating circular wave. Attention will be focused on a p = 2 pole machine or one pair of a p-pole winding. To develop insight into the three-phase arrangement it is helpful to begin with an analysis of a single-phase winding.

(a) Single-Phase Winding
Figure 10.14 shows the space-fundamental mmf distribution of a single-phase winding

$$F_{a1} = \frac{4}{\pi}\left(\frac{N_{ph}}{p}\right)k_w i_a \cos\theta, \qquad (10.14)$$

for i_a = constant this relation represents a standing wave.
When this winding "a" is excited by a sinusoidally varying current in time

$$i_a = I_{max}\cos\omega t \qquad (10.15)$$

the mmf distribution is given by

$$F_{a1} = (F_{max}\cos\theta)\cos\omega t, \qquad (10.16)$$

and $F_{max} = \frac{4}{\pi}\left(\frac{N_{ph}}{p}\right)k_w I_{max}$. This represents an alternating wave, as is depicted in Fig. 10.15 for times corresponding to $\omega t = 0$, $\omega t = \omega t_1$, $\omega t = \omega t_2$, and $\omega t = \pi$, respectively.

With the trigonometric relation

$$\cos\theta\ \cos\omega t = (1/2)\{\cos(\theta - \omega t) + \cos(\theta + \omega t)\} \qquad (10.17a)$$

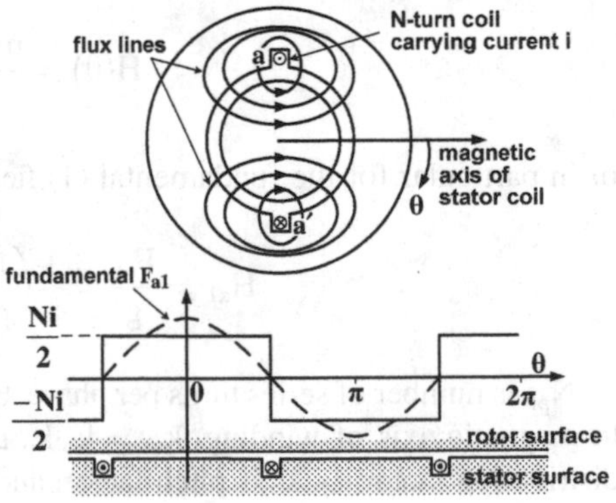

Fig. 10.14 Standing wave: the mmf of a concentrated full-pitch coil aa′

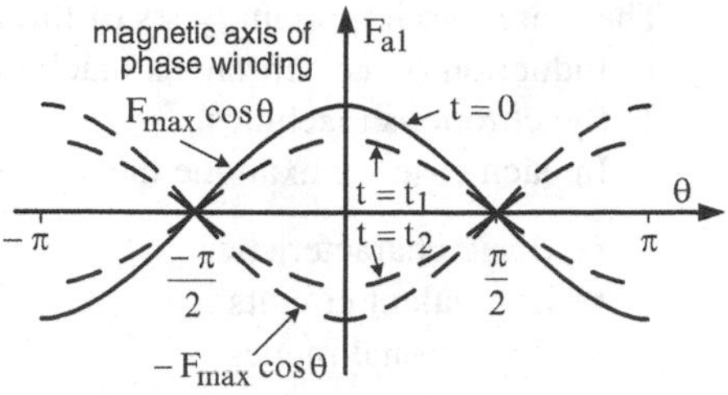

Fig. 10.15 Alternating wave: mmf distribution of a single-phase winding at various times

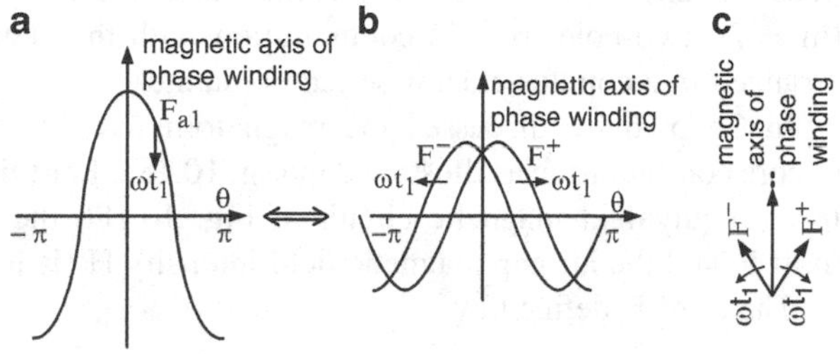

Fig. 10.16 (**a**) Decomposition of the alternating wave F_{a1} (**b**) into two oppositely rotating waves F^+ and F^- of half amplitude of the alternating wave F_{a1}, and (**c**) its phasor decomposition

one learns that any alternating wave of amplitude F_{max} (10.16) can be decomposed into two oppositely rotating circular waves of amplitude $F_{max}/2$ each. In other words,

$$F_{a1} = F^+ + F^-, \tag{10.17b}$$

where

$$F^+ = \frac{F_{max}}{2}\cos(\theta - \omega t), \tag{10.17c}$$

$$F^- = \frac{F_{max}}{2}\cos(\theta + \omega t). \tag{10.17d}$$

Equations 10.17b, c, d are represented in Fig. 10.16 at the time t_1 corresponding to ωt_1. The fact that the mmf of a single-phase winding excited by a source of alternating current can be resolved into two oppositely rotating (traveling) waves is an important conceptual step in understanding AC machinery. As shown in the next subsection, in three-phase AC rotating machines the three windings are equally (120 electrical degrees) displaced in space phase, and the three-winding currents are similarly displaced in time phase (120 electrical degrees). As a consequence, the negative-traveling (rotating) flux (mmf) waves of the three windings add up to zero while the positive-traveling mmf waves reinforce (add up), giving a single positive-traveling (rotating) circular mmf (flux) wave.

(b) Three-Phase (Poly-Phase) Windings of an Electric Machine and Their Rotating (Revolving) Magnetic Fields

There are two important types of three-phase machines

1. Induction or asynchronous machines
2. Synchronous machines

   In each case we examine the

   - Basic characteristics
   - Equivalent circuits
   - Operational modes

The Rotating (Revolving) Magnetic Field: Tesla invented the two-phase revolving field [27, 28]. Figure 10.17 depicts the stator, rotor and the air gap of a two-phase (m = 2), two-pole (p = 2) configuration with the stator phase belts aa' and bb' arranged in a counter-clockwise (ccw) manner.

In Chap. 6 we discussed the magnetomotive force (mmf) of an inductor in C-core configuration as illustrated in Fig. 10.18a. For this simple inductor we depict in the equivalent magnetic circuit of Fig. 10.18b the air gap reluctance $\Re_g$, the mmf F, and the air gap magnetic field intensity H_g is indicated in Fig. 10.18a.

The mmf is defined by

$$F = N \cdot I = \Phi \cdot \Re_g \tag{10.18}$$

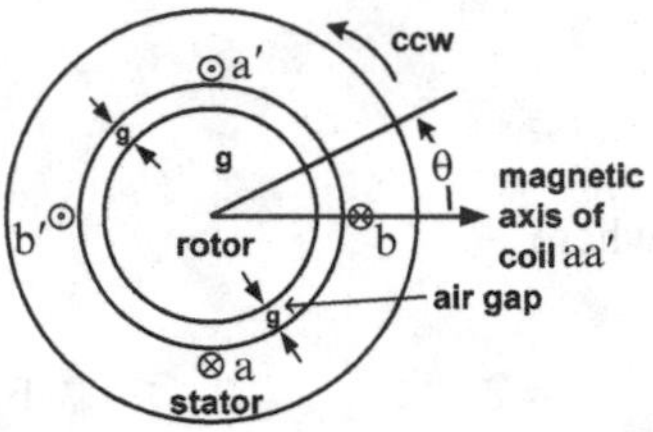

Fig. 10.17 Two-phase (m = 2), two-pole (p = 2) stator winding arranged in a counter-clockwise (ccw) manner

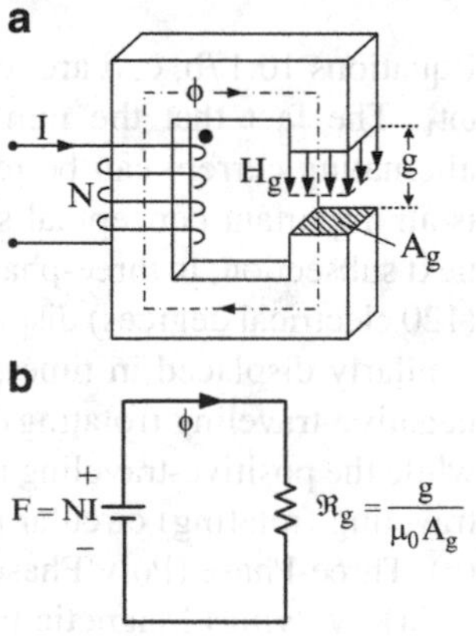

Fig. 10.18 (**a**) Inductor with C-core. (**b**) Equivalent magnetic circuit of C-core inductor of Fig. 10.18a

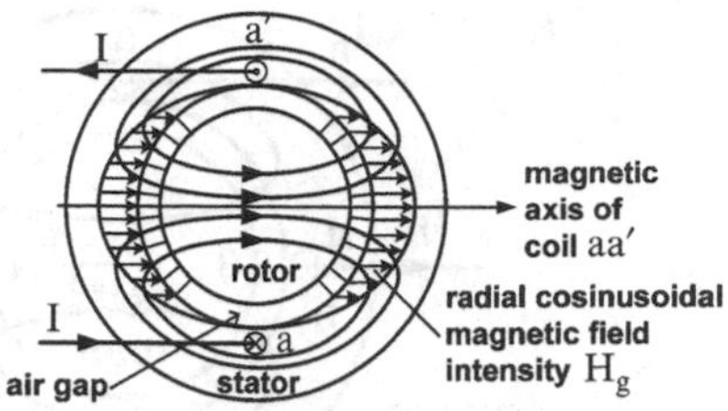

Fig. 10.19 Phase belt aa′ excited by DC current I residing on stationary stator, fundamental of H_g only

with $B_g = \mu_o \cdot H_g$ one obtains the flux $\Phi = B_g \cdot A_g = \mu_o \cdot H_g \cdot A_g$ or the mmf $F = N \cdot I = g \cdot H_g$ or

$$H_g = F/g. \tag{10.19}$$

If for linear conditions the mmf F is known, then the magnetic field intensity in the air gap is known as well. Figures 10.9a, b show that the magnetic field intensity of a single phase belt is co-sinusoidal, if the space harmonics are neglected. Figure 10.19 demonstrates this condition where for a DC current I the magnetic field intensity H_g is plotted around the circumference of the air gap. For phase belt aa′ one gets

$$F_a = NI\cos\theta, \tag{10.20}$$

correspondingly phase belt bb' displaced from phase belt aa' by 90° in counter-clockwise (ccw) direction

$$F_b = NI\sin\theta. \tag{10.21}$$

If the excitation currents of phase belts a-a' and b-b' are

$$i_a = I_{max}\cos\omega t, \tag{10.22}$$

$$i_b = I_{max}\sin\omega t \tag{10.23}$$

where $\omega = 2\pi f$ and $f = 60$ Hz then we can study the influence of the time-dependent currents i_a and i_b on the motion of the generated (rotating, revolving) magnetic field. Figures 10.20a, b, c illustrate the rotation of the mmf F_i for $i = 1, 2$, and 3 if the currents i_a and i_b assume values corresponding to $\omega t = 0, \pi/2$, and π, respectively.

The above graphical interpretation of the rotating field is now supported by an analysis of a two-phase ($m = 2$) current system, where the phase belts are arranged in a two-pole configuration.

The mmfs for time-dependent cosinusoidal/sinusoidal currents i_a and i_b are

$$F_a(\theta, t) = N \cdot i_a \cdot \cos\theta = N \cdot I_{max} \cdot \cos\omega t \cdot \cos\theta, \tag{10.24}$$

$$F_b(\theta, t) = N \cdot i_b \cdot \sin\theta = N \cdot I_{max} \cdot \sin\omega t \cdot \sin\theta. \tag{10.25}$$

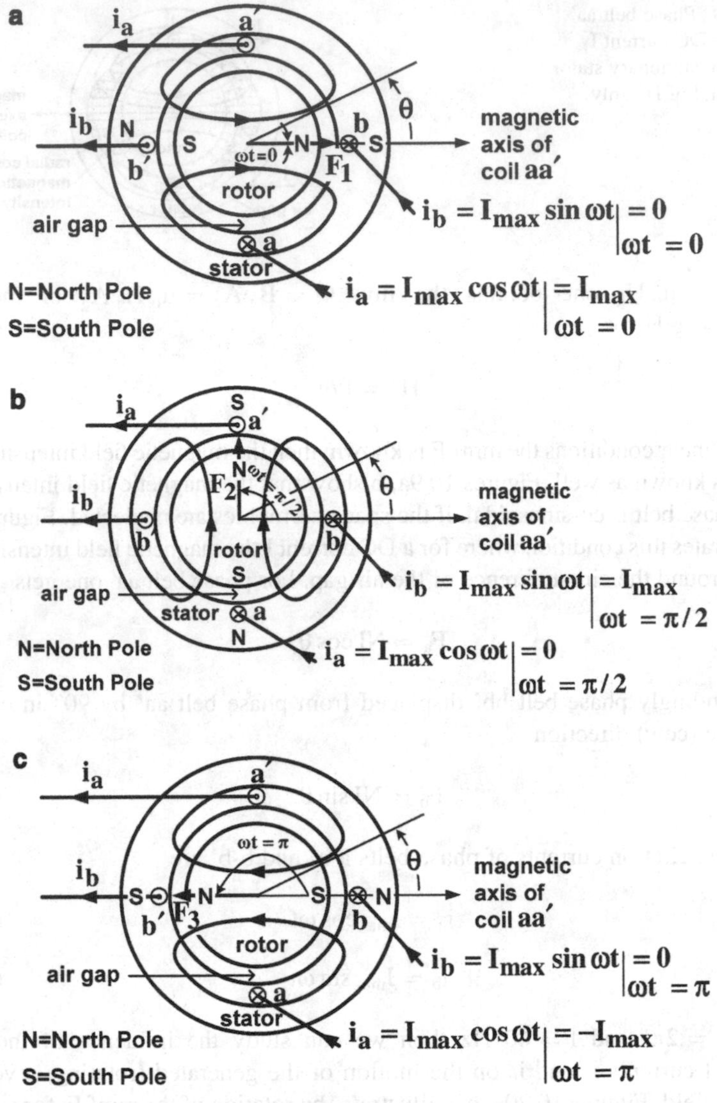

Fig. 10.20 (**a**) Location of mmf F_1 for $\omega t = 0$. (**b**) Location of mmf F_2 for $\omega t = \pi/2$. (**c**) Location of mmf F_3 for $\omega t = \pi$

The sum of both is

$$F_{m=2}(\theta, t) = N \cdot I_{max} \cdot \cos(\omega t - \theta). \tag{10.26}$$

This function is the analytical expression for a rotating (revolving) wave. We can observe the maximum of this rotating wave if (ωt-θ) = 0 or the angular displacement is

$$\theta = \omega t. \tag{10.27}$$

From this follows the angular (mechanical) velocity of the rotating wave as

$$\frac{d\theta}{dt} = \omega = \omega_m. \tag{10.28}$$

For an m-phase system one obtains the relation

$$F_m(\theta, t) = \frac{m \cdot N \cdot I_{max}}{2} \cos(\omega t - \theta). \tag{10.29}$$

In particular one obtains for a three-phase system

$$F_3(\theta, t) = \frac{3 \cdot N \cdot I_{max}}{2} \cos(\omega t - \theta). \tag{10.30}$$

Figure 10.3a represents the magnetic field of a two-pole (synchronous) machine.

A rotating electric machine can have $p = 2, 4, 6, 8, 10, \ldots$ (even) poles. For this reason we are going to derive the angular (mechanical) velocities for a $p = 4$ pole machine with two phases ($m = 2$) as depicted in Fig. 10.21. Note that $i_a = i_{a1} = i_{a2}$ and $i_b = i_{b1} = i_{b2}$ with $i_a = I_{max}\cos\omega t = I_{max}$, and $i_b = I_{max}\sin\omega t = 0$ for $\omega t = 0$.

Provided one measures the magnetic field around the air gap circumference of Fig. 10.20a with a Hall sensor as discussed in Chap. 6 then one obtains the (fundamental) magnetic field H_g variation of Fig. 10.22 where one north (N) pole and one south (S) pole per mechanical period T_m can be identified. In this case the electrical (e) period T_e is identical with the mechanical (m) period T_m, that is, $T_e = T_m$. If the same measurement of the magnetic field intensity is applied to the four-pole configuration of Fig. 10.21 one obtains the (fundamental) magnetic field variation of Fig. 10.23 where two north (N) poles and two south (S) poles per mechanical period T_m can be identified. In this case the electrical period T_e is identical with one half of the mechanical period T_m, that is, $T_e = T_m/2$.

One concludes that for a two-pole machine $T_e = T_m$, and with $f = 1/T$ follows $T_e = 1/f = 1/f_m = T_m$ resulting in $f = f_m$ or with $\omega = 2\pi f$ yields

$$\omega = \omega_m = 2\pi f. \tag{10.31}$$

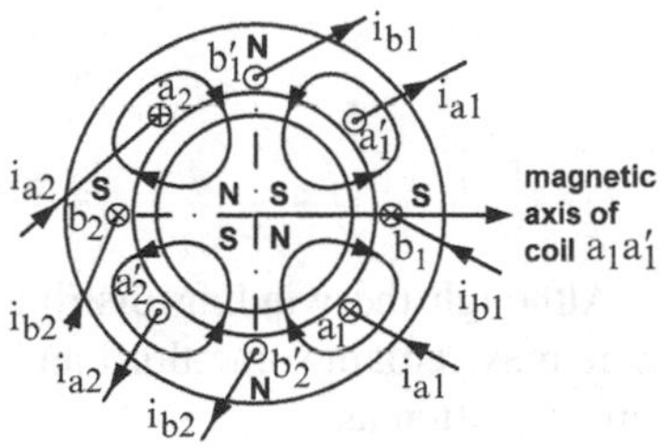

Fig. 10.21 Schematic field plot of a two-phase machine in four-pole configuration

Fig. 10.22 Magnetic field-intensity variation H_g around the air-gap circumference of a two-pole machine

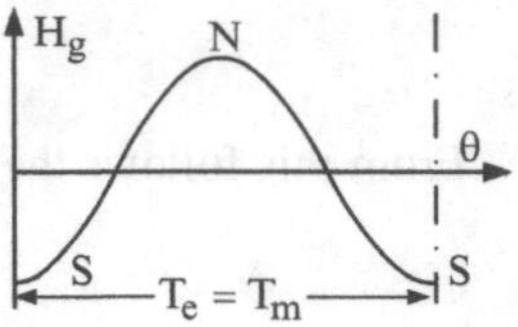

Fig. 10.23 Magnetic field-intensity H_g variation around the air-gap circumference of a four-pole machine

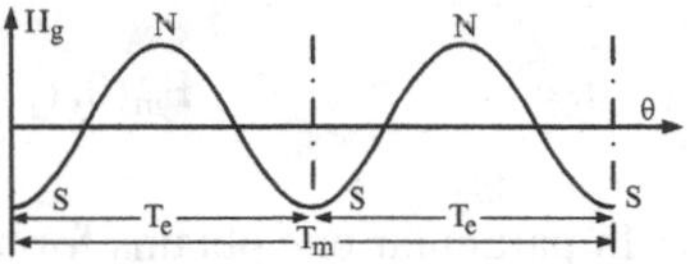

For a four-pole machine (see Figs. 10.2c and 10.21) $2T_e = T_m$ or $2/f = 1/f_m$ resulting in $f = 2f_m$ or

$$\omega = 2\omega_m = 2\pi f. \tag{10.32}$$

For a p-pole machine one can generalize

$$\omega = \left(\frac{p}{2}\right)\omega_m = 2\pi f. \tag{10.33}$$

Introducing the (mechanical) synchronous (s) angular velocity of the rotating stator field as $\omega_s = \omega_m$ one obtains

$$\omega_s = \left(\frac{2}{p}\right)\omega = \left(\frac{2}{p}\right)2\pi f. \tag{10.34}$$

Application Example 10.9: Three-Phase Rotating mmf with Spatial Harmonics

A three-phase, two-pole winding is excited by balanced three-phase 60 Hz currents as described below

$$i_a = I_{max}\cos\omega t, \tag{10.35}$$

$$i_b = I_{max}\cos(\omega t - 120°), \tag{10.36}$$

$$i_c = I_{max}\cos(\omega t + 120°). \tag{10.37}$$

Although the winding distribution has been designed to minimize harmonics, there may remain some third and fifth spatial harmonics. Thus the phase "a" mmf can be written as

$$F_a = (A_1\cos\theta + A_3\cos 3\theta + A_5\cos 5\theta)i_a. \tag{10.38}$$

Similar expressions can be written for phases "b" (replace θ by $\theta-120°$) and "c" (replace θ by $\theta+120°$ or $\theta-240°$). Calculate the total three-phase mmf. What are the angular velocity and the rotational direction of each component of the mmf?

Solution

Magnetomotive force (mmf) of phase "a" for number of poles of $p = 2$, and number of phases $m = 3$:

$$F_a = (A_1\cos\theta + A_3\cos 3\theta + A_5\cos 5\,\theta)i_a, \tag{10.39}$$

Correspondingly,

$$F_b = (A_1\cos(\theta - 120°) + A_3\cos 3(\theta - 120°) + A_5\cos 5(\theta - 120°))i_b, \tag{10.40}$$

$$F_c = (A_1\cos(\theta + 120°) + A_3\cos 3(\theta + 120°) + A_5\cos 5(\theta + 120°))i_c. \tag{10.41}$$

The total mmf is the sum of the contributions from each of the three phases

$$F_{tot}(\theta, t) = F_a + F_b + F_c. \tag{10.42}$$

Specifically,

$$F_a = (A_1\cos\theta + A_3\cos 3\theta + A_5\cos 5\,\theta)I_{max}\cos\omega t, \tag{10.43}$$

$$\begin{aligned} F_b = {} & (A_1\cos(\theta - 120°) + A_3\cos 3(\theta - 120°) \\ & + A_5\cos 5(\theta - 120°))I_{max}\cos(\omega t - 120°), \end{aligned} \tag{10.44}$$

$$\begin{aligned} F_c = {} & (A_1\cos(\theta + 120°) + A_3\cos 3(\theta + 120°) \\ & + A_5\cos 5(\theta + 120°))I_{max}\cos(\omega t + 120°). \end{aligned} \tag{10.45}$$

Expanded:

$$\begin{aligned} F_a = {} & (A_1/2)I_{max}\{\cos(\theta - \omega t) + \cos(\theta + \omega t)\} \\ & + (A_3/2)I_{max}\{\cos(3\theta - \omega t) + \cos(3\theta + \omega t)\} \\ & + (A_5/2)I_{max}\{\cos(5\theta - \omega t) + \cos(5\theta + \omega t)\}, \end{aligned} \tag{10.46}$$

$$\begin{aligned} F_b = {} & (A_1/2)I_{max}\{\cos(\theta - \omega t) + \cos(\theta + \omega t - 240°)\} \\ & + (A_3/2)I_{max}\{\cos(3\theta - \omega t - 240°) + \cos(3\theta + \omega t - 480°)\} \\ & + (A_5/2)I_{max}\{\cos(5\theta - \omega t - 480°) + \cos(5\theta + \omega t - 720°)\}, \end{aligned} \tag{10.47}$$

$$
\begin{aligned}
F_c = {} & (A_1/2)I_{max}\{\cos(\theta-\omega t)+\cos(\theta+\omega t+240°)\} \\
& +(A_3/2)I_{max}\{\cos(3\theta-\omega t+240°)+\cos(3\theta+\omega t+480°)\} \\
& +(A_5/2)I_{max}\{\cos(5\theta-\omega t+480°)+\cos(5\theta+\omega t+720°)\}. \quad (10.48)
\end{aligned}
$$

The total mmf is now

$$F_{tot}(\theta,t)=3(A_1/2)I_{max}\cos(\theta-\omega t)+3(A_5/2)I_{max}\cos(5\theta+\omega t). \quad (10.49)$$

$$\text{Fundamental spatial angular velocity: } \frac{d\theta^{(1)}}{dt}=\omega. \quad (10.50)$$

$$\text{Fifth spatial harmonic angular velocity: } \frac{d\theta^{(5)}}{dt}=-\frac{\omega}{5}. \quad (10.51)$$

Application Example 10.10: Three-Phase Rotating mmf with Time Harmonics

A three-phase, two-pole winding is excited by balanced three-phase 60 Hz currents containing a fifth time harmonic

$$i_a=\{I_{max1}\cos\omega t+I_{max5}\cos 5\omega t\}, \quad (10.52)$$

$$i_b=\{I_{max1}\cos(\omega t-120°)+I_{max5}\cos(5\omega t-120°)\}, \quad (10.53)$$

$$i_c=\{I_{max1}\cos(\omega t+120°)+I_{max5}\cos(5\omega t+120°)\}. \quad (10.54)$$

The winding has been designed to eliminate all spatial harmonics. Thus for phase "a" the mmf is

$$F_a=A_1\cos\theta i_a \quad (10.55)$$

(a) Write similar expressions for F_b and F_c
(b) Calculate the total three-phase mmf F_{tot}
(c) What are the angular velocities and the rotational directions of the components of the mmf?

Solution

$$F_a=A_1\cos\theta\{I_{max1}\cos\omega t+I_{max5}\cos 5\omega t\}, \quad (10.56)$$

$$F_b=A_1\cos(\theta-120°)\{I_{max1}\cos(\omega t-120°)+I_{max5}\cos(5\omega t-120°)\}, \quad (10.57)$$

$$F_c=A_1\cos(\theta+120°)\{I_{max1}\cos(\omega t+120°)+I_{max5}\cos(5\omega t+120°)\}, \quad (10.58)$$

$$F_{tot}=F_a+F_b+F_c. \quad (10.59)$$

Expanded:

$$F_a = (A_1/2)I_{max1}\cos(\theta - \omega t) + (A_1/2)I_{max1}\cos(\theta + \omega t) + (A_1/2)I_{max5}\cos(\theta - 5\omega t) + (A_1/2)I_{max5}\cos(\theta + 5\omega t), \quad (10.60)$$

$$F_b = (A_1/2)I_{max1}\cos(\theta - \omega t) + (A_1/2)I_{max1}\cos(\theta + \omega t - 240°) + (A_1/2)I_{max5}\cos(\theta - 5\omega t) + (A_1/2)I_{max5}\cos(\theta + 5\,\omega t - 240°), \quad (10.61)$$

$$F_c = (A_1/2)I_{max1}\cos(\theta\omega t) + (A_1/2)I_{max1}\cos(\theta + \omega t + 240°) + (A_1/2)I_{max5}\cos(\theta 5\omega t) + (A_1/2)I_{max5}\cos(\theta + 5\,\omega t + 240°). \quad (10.62)$$

or

$$F_{tot} = (A_1/2)I_{max1}\cos(\theta - \omega t) + (A_1/2)I_{max5}\cos(\theta - 5\,\omega t). \quad (10.63)$$

$$\text{Fundamental time - dependent angular velocity: } \frac{d\theta^{(1)}}{dt} = \omega. \quad (10.64)$$

$$\text{Fifth time harmonic angular velocity: } \frac{d\theta^{(5)}}{dt} = 5\,\omega. \quad (10.65)$$

Reference [29] analyses torques for time and spatial harmonics of induction machines.

Generated (Induced) Voltage in AC Machines: An elementary two-pole AC machine is shown in Fig. 10.24. The windings on both the rotor (ff′) and the stator (aa′, bb′, cc′) are depicted as single multiple-turn, full-pitch concentrated coils. The field winding ff′ on the rotor is assumed to produce a cosinusoidal spatial wave of flux density B(θ) on the rotor surface. The rotor is spinning at a constant angular velocity of $\omega_m = \omega$ electrical radians per second. Although a two-pole machine is shown, the derivations presented here are for the general case of a p-pole machine.

When the rotor pole axis is in line with the magnetic axis of the "a" stator coil, the flux linkage with the stator N-turn coil is Nϕ, where ϕ is the air-gap flux per pole. For the assumed cosinusoidal flux density wave we can write

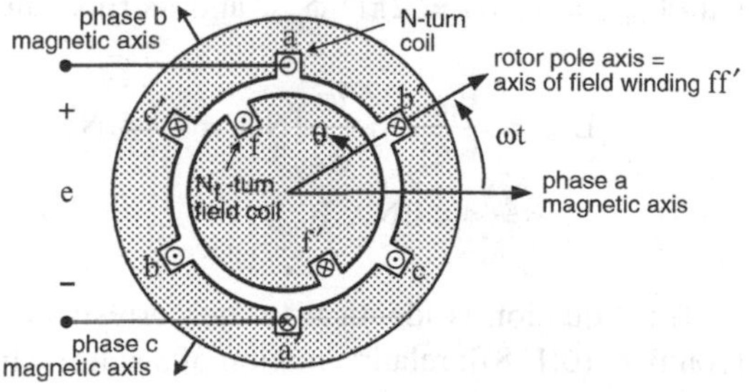

Fig. 10.24 Induced voltage in a (synchronous) AC machine

$$B(\theta) = B_{max} \cos\theta, \tag{10.66}$$

B_{max} is its amplitude at the rotor pole center and θ is measured in electrical radians from the rotor pole axis. The air-gap flux per pole is the integral of the flux density over the pole area; thus for a two-pole machine

$$\phi = \int_{-\pi/2}^{\pi/2} B_{max} \cos\theta dS = \int_{-\pi/2}^{\pi/2} B_{max} \cos\theta \cdot \ell \cdot r \cdot d\,\theta = 2B_{max} \cdot \ell \cdot r. \tag{10.67}$$

Note, $dS = r\, d\theta \cdot \ell$ is the differential surface, where ℓ is the axial length of the stator and r is its radius at the air gap. For a p-pole machine

$$\phi = \left(\frac{2}{p}\right) \cdot 2B_{max} \cdot \ell \cdot r, \tag{10.68}$$

note, the pole area of a p-pole machine is (2/p) times that of a 2-pole machine of the same length and diameter.

As the rotor turns, the flux linkage with stator coil a-a′ varies as the cosine of the angle ωt between the magnetic axes of the stator winding aa′ and the rotor axis (axis of winding ff′). With the rotor spinning at constant angular velocity $\omega_m = \omega$, the flux linkage with the N-turn stator coil aa’ is

$$\lambda_{aa'}(\omega t) = N\phi \cos\omega t, \tag{10.69}$$

where time t is arbitrarily chosen as zero when the amplitude of the flux-density wave coincides with the magnetic axis of the stator winding aa′. By Faraday’s law the voltage induced in the stator winding aa′ is

$$e_{aa'}(t) = \frac{d\lambda_{aa'}}{dt} = N\frac{d\phi}{dt}\cos\omega t - \omega N\phi \sin\omega t. \tag{10.70}$$

The first term on the right-hand side of (10.70) is the transformer voltage, the second term is the speed voltage. In normal steady-state operations of most rotating machines φ = constant, and the so-called electromotive force (emf) becomes

$$e_{aa'}(t) = -\omega N\phi \sin\omega t, \tag{10.71}$$

with $E_{max} = \omega N\phi = 2\pi f N\phi_{max}$ and its rms value follows

$$\begin{aligned} E_{rms} &= \frac{E_{max}}{\sqrt{2}} = \frac{2\pi}{\sqrt{2}} fN\phi = 4.44 fN\phi_{max} = \frac{2\pi}{\sqrt{2}} fNA_c B_{max} \\ &= 4.44 fNA_c B_{max}. \end{aligned} \tag{10.72}$$

This equation is identical to the corresponding emf equation of a transformer [compare (6.118)]: relative motion of a stator winding aa′ and a constant-amplitude spatial flux-density wave $B(\theta) = B_{max}\cos\theta$ attached to a rotating rotor induces in stator winding aa′ the same rms voltage E_{rms} as a time-varying flux (in association with stationary primary and secondary coils) does in a transformer. For distributed windings,

$$E_{rms} = \frac{E_{max}}{\sqrt{2}} = \frac{2\pi}{\sqrt{2}} f k_w N_{ph} \phi = 4.44 f k_w N_{ph} \phi_{max}. \quad (10.73)$$

This relation can be employed to achieve (E/f) control in electric machines resulting in constant (rated) flux operation

$$\left(\frac{E_{rms}}{f}\right) = 4.44 k_w N_{ph} \phi_{max\ rated}. \quad (10.74)$$

Flux-weakening operation occurs at above rated speed if one maintains rated voltage $V_{rms\ rated}$ with increasing frequency f

$$\left(\frac{V_{rms\ rated}}{f}\right) < 4.44 k_w N_{ph} \phi_{max\ rated} \text{ or } \left(\frac{V_{rms\ rated}}{f}\right) = 4.44 k_w N_{ph}\ \phi_{max}, \quad (10.75)$$

where $\phi_{max} \le \phi_{max\ rated}$.

A compensation of flux weakening at above rated speed is possible based on (V/f·N) control [23, 30, 31], and a torque increase at low speed is achievable by (V·p/f) control [23, 30, 31]. The induced voltage E of an alternating current electric machine (either motor or generator) is related to the rated maximum flux density B_{max}, the rated number of series turns per phase N_{rated}, the radius of the location of the stator winding R, the active (core) machine length ℓ, the frequency of the voltages/currents f, and the pole number p by

$$E_{rms} = 4.44 \cdot f \cdot B_{max} \cdot N_{rated} \cdot 4 \cdot R \cdot \ell/p. \quad (10.76)$$

For rated induced voltage $E_{rms\ rated}$ the frequency f and the maximum flux density assume rated values. If $E_{rms} < E_{rms\ rated}$ then the flux density is less than its rated value, and for $E_{rms} > E_{rms\ rated}$ the flux density will be above its rated value. The latter case cannot be permitted during steady-state conditions and the relation

$$\frac{E_{rms}}{f} = 4.44 \cdot B_{max} \cdot N_{rated} \cdot 4 \cdot R \cdot \ell/p = \text{constant} \quad (10.77)$$

must be maintained for all operating conditions where $E_{rms} < E_{rms\ rated}$ and $f < f_{rated}$.

For $f > f_{rated}$ the induced voltage can be at the most $E_{rms} = E_{rms\ rated}$ and the flux density will be less than its rated value, called the flux-weakening operation. In some

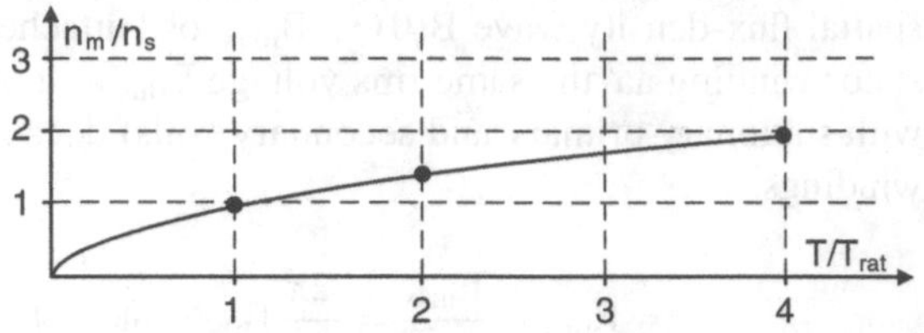

Fig. 10.25 Torque-speed characteristic of a wind turbine $T \propto n_m^2$

cases the induced voltage E_{rms} can be replaced in the above formulas by the terminal voltage V_{rms}. By changing the number of series turns per phase N and the pole number p of the machine the flux weakening can be compensated [31] and the machine can be operated at rated flux density under most operating conditions. This results in superior performance (e.g., increased torque and increased output power) above and below rated speed of variable-speed drives. For example, it is well known that the torque-speed characteristic of wind turbines (Fig. 10.25 [23]) and commonly available variable-speed generators employing field (flux) weakening (Fig. 10.26a) do not match because the torque of a wind turbine is about proportional to the square of the speed, and that of a variable-speed generator is inversely proportional to the speed in the field-weakening region [23]. The same applies to variable-speed drives (motors/generators) of hybrid/electric cars where the critical (maximum) speed is limited by the reduction of the developed torque, resulting in less than the rated output power. One possibility to mitigate this mismatch is an electronic change in the number of poles p and/or a change in the number of series turns per phase N of the stator winding. The change in the number of poles p is governed by the relation

$$\frac{E_{rms} \cdot p}{f} = 4.44 \cdot B_{max} \cdot N_{rated} \cdot 4 \cdot R \cdot \ell, \tag{10.78}$$

where B_{max} is the rated maximum flux density at the radius R of the machine, N_{rated} is the rated number of series turns of a stator phase winding, and ℓ is the axial iron-core length of the machine.

The change of the number of series turns per phase N is governed by

$$\frac{E_{rms}}{f \cdot N} = 4.44 \cdot B_{max} \cdot 4 \cdot R \cdot \ell / p. \tag{10.79}$$

The introduction of an additional degree of freedom – either by a change of p or N – permits an extension of the constant (rated) flux density (B_{max}) region up to a multiple of the rated speed [31]. For example, adding this degree of freedom will permit wind turbines to operate under stalled conditions at all speeds, generating the maximum possible power at a given speed with no danger of a runaway. This will make the blade-pitch control less important – however, a furling (feathering) of the blades at excessive wind velocities must be provided – and wind turbines become more reliable and less expensive due to the absence of mechanical control. This might be important for off-shore wind-power plants. Figure 10.26b illustrates the speed and torque increase due to the change of the number of turns from N_{rated} to

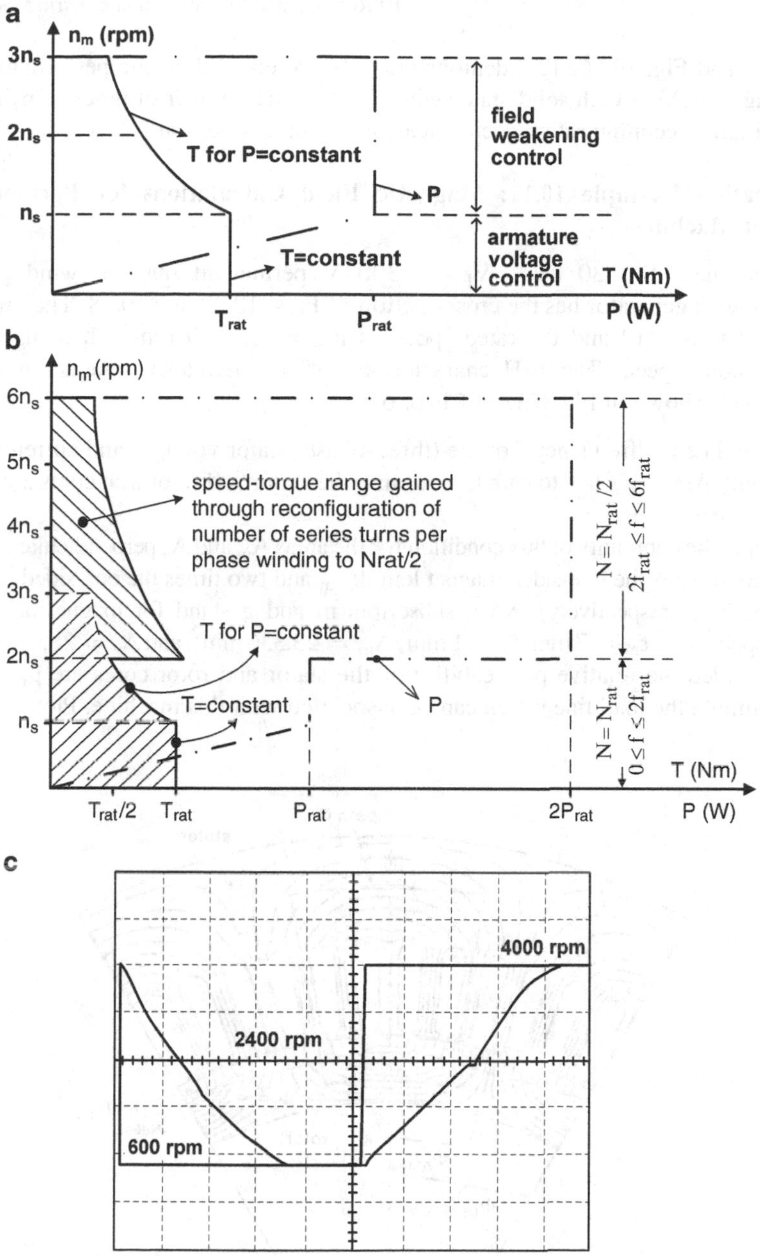

Fig. 10.26 (**a**) Variable-speed drive torque-speed characteristic without winding switching $T \propto 1/n_m$. (**b**) Variable-speed drive torque-speed characteristic by changing the number of series turns of the stator winding from N_{rated} to $N_{rated}/2$ (From Fuchs, E. F.; Schraud, J.; Fuchs, F. S.: "Analysis of critical-speed increase of induction machines via winding reconfiguration with solid-state switches," *IEEE Transaction on Energy Conversion*, Vol. 23, No. 3, Sept. 2008, pp. 774–780, [24]). (**c**): Measured variable-speed drive response with winding switching from pole number p_1 to p_2 resulting in a speed range from 600 to 4,000 rpm. One horizontal division corresponds to 200 ms. One vertical division corresponds to 800 rpm (From Schraud, J.; Fuchs, E. F.; Fuchs, H. A.: "Experimental verification of critical-speed increase of induction machines via winding reconfiguration with solid-state switches," *IEEE Transaction on Energy Conversion*, Vol. 23, No. 2, June 2008, pp. 460–465, [25])

$N_{rated}/2$, and Fig. 10.26c [25] demonstrates the excellent dynamic performance of winding switching with solid-state switches from the number of poles of p_1 to p_2. The winding reconfiguration occurs near the horizontal axis at 2,400 rpm.

Application Example 10.11: Magnetic Field Calculations for Permanent-Magnet Machine

A three-phase $S = 30$ kVA, $V_{L\text{-}L} = 300$ V permanent magnet, wind power synchronous generator has the cross-section of Figs. 10.27 and 10.28. The number of poles is $p = 12$ and the rated speed is $n_{rat} = n_s = 60$ rpm, where n_s is the synchronous speed. The B-H characteristic of the permanent magnet material (NdFeB) is shown in Fig. 6.37 of Chap. 6.

(a) Calculate the frequency f of the (three-phase) stator voltages and currents
(b) Apply Ampere's law to path C assuming the stator and rotor iron cores are ideal ($\mu_r \to \infty$)
(c) Apply the continuity of flux condition for the areas A_m and A_g perpendicular to path C (two times the one-sided magnet length ℓ_m and two times the one-sided air gap length ℓ_g, respectively). Note, subscripts m and g stand for magnet and gap, respectively; $\ell_m = 7$ mm, $\ell_g = 1$ mm, $A_m = 23{,}336\ \text{mm}^2$, and $A_g = 28{,}938\ \text{mm}^2$.
(d) Provided the relative permeability of the stator and rotor cores are $\mu_r \to \infty$, compute the load line which can be associated with this machine. Plot the load

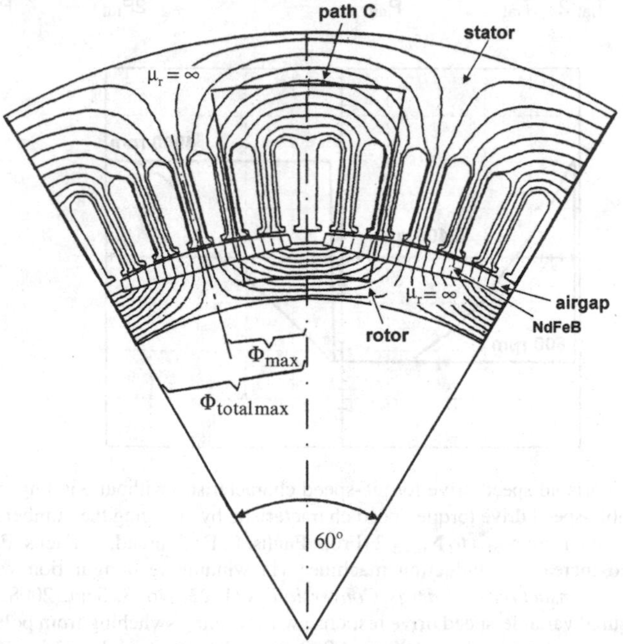

Fig. 10.27 Cross-section of one period (2 pole pitches) of permanent-magnet machine with no-load field

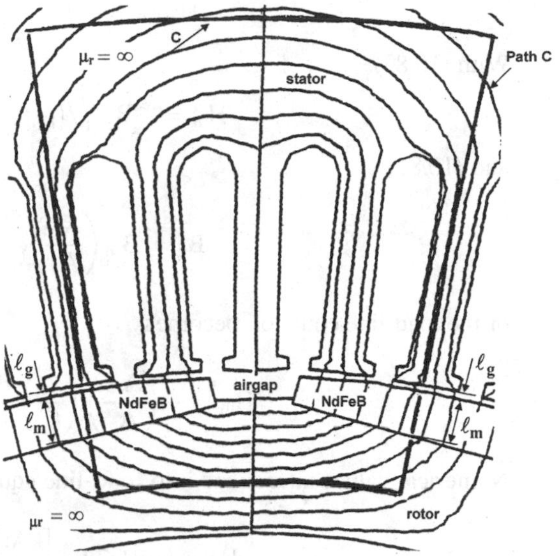

Fig. 10.28 Magnified cross-section of area enclosed by path C of a pole pitch of permanent-magnet machine (see Fig. 10.27)

line in a figure similar to that of Fig. 6.38b. Is this magnetic material operated at the point of the maximum energy product?

(e) What is the recoil permeability μ_R of the NdFeB material?

(f) Compute the fluxes ϕ_{max}, $\phi_{totalmax} = 2\phi_{max}$, and the induced voltage provided there are N = 360 turns per stator phase, the winding factor is $k_w = 0.8$, and the iron-core stacking factor (taking into account the insulation between the core laminations) is $k_s = 0.94$.

Solution

(a)

$$n_s = \frac{120f}{p}, \text{ or } f = \frac{n_s p}{120} = \frac{60 \cdot 12}{120} = 6\,\text{Hz}. \tag{10.80}$$

(b) Ampere's law:

$$H_m \ell_m + H_g \ell_g = 0. \tag{10.81}$$

(c) Continuity of flux:

$$\phi = B_m A_m = B_g A_g. \tag{10.82}$$

(d) Constituent relation:

$$B_g = \mu_0 H_g, \tag{10.83}$$

from (10.81)

$$H_m \ell_m = -H_g \ell_g \tag{10.84}$$

With (10.83)

$$H_m = -B_g \ell_g / (\mu_0 \ell_m) \tag{10.85}$$

and (10.82)

$$B_g = B_m \left(\frac{A_m}{A_g} \right), \tag{10.86}$$

or the load-line equation becomes

$$H_m = -\left(\frac{A_m}{A_g} \right) \left(\frac{\ell_g}{\ell_m} \right) \frac{B_m}{\mu_0}. \tag{10.87}$$

Numerical values introduced into load-line equation:

$$H_m = -91.7\, B_m \left[\frac{kAt}{m} \right]. \tag{10.88}$$

This equation is plotted in Fig. 10.29. Note the magnet material is not operated in the point of maximum energy product !

(e) Recoil permeability of NdFeB:

$$\mu_R = \frac{(y_2 - y_1)}{(x_2 - x_1)} = \frac{(B_r - 0)}{(0 - (-H_c))} = \frac{1.25}{950 \cdot 10^3} = 1.316 \cdot 10^{-6} [H/m] \tag{10.89}$$

or

$$\mu_R = 1.047\, \mu_0. \tag{10.90}$$

(f) Induced voltage:

$$\phi_{max} = B_m A_m / 2 \tag{10.91}$$

$$\phi_{total\ max} = 2\phi_{max} = B_m A_m. \tag{10.92}$$

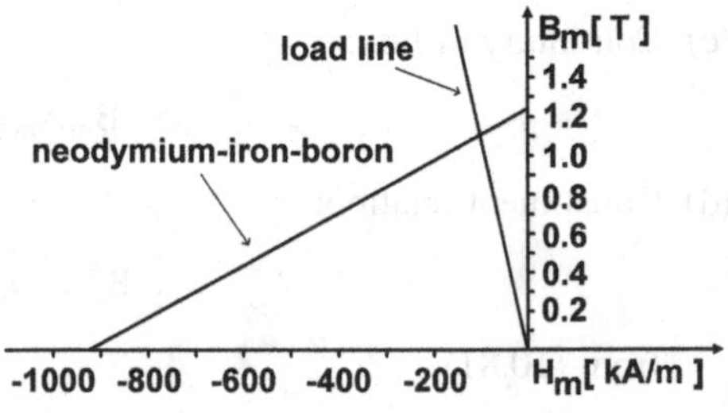

Fig. 10.29 Magnet (NdFeB) characteristic with load-line

Flux:

$$\phi(t) = \phi_{total\ max} \cos \omega t. \tag{10.93}$$

Induced voltage:

$$E_{max} = N(d\ \phi/dt) = \omega N_{ph} k_w k_s \phi_{total\ max}, \tag{10.94}$$

from load line and magnet characteristic

$$B_{m0} = 1.1\,T, \tag{10.95}$$

$$H_{m0} = -100\,kAt/m. \tag{10.96}$$

Therefore,

$$\phi = \phi_{total\ max} = 1.1 \cdot 23336 \cdot 10^{-6}\,Wb = 0.0257\,Wb, \tag{10.97}$$

the induced voltage is now

$$e(t) = 360 \cdot 0.0257 \cdot 2\pi \cdot 6 \cdot 0.8 \cdot 0.94 \sin \omega t = E_{max} \sin \omega t, \tag{10.98}$$

or $E_{max} = 262.3$ V, $E_{rms} = E_{max}/\sqrt{2} = 185.48$ V, and the line-to-line terminal voltage is approximately
$V_{L-L} \approx \sqrt{3}\,E_{rms} = 321.3$ V.

Torque Production in Non-Salient-Pole (Round Rotor) Rotating Machines [26]: The behavior of any electromagnetic device as a component in an electromechanical system can be described in terms of its flow diagram of Fig. 10.8 and its electromagnetic torque, (10.1b). The purpose of this section is to derive the voltage and torque equations for an idealized elementary machine, results which can readily be extended later to real-life machines.

We can derive these equations from two viewpoints:

1. Coupled-circuit viewpoint
2. Magnetic-field viewpoint

Both viewpoints are equivalent and produce the same results. Employing the coupled-circuit viewpoint, the machine will be regarded as a circuit element [32] whose inductances depend on the angular position (θ) of the rotor with respect to the stator. The flux linkages λ and the magnetic field co-energy (currents and angular rotor position are chosen to be independent variables) will be expressed in terms of the currents, angular position and inductances. The torque can then be found from the partial derivative of the magnetic field co-energy with respect to θ, and the terminal voltages from the sum of the resistance drops $R \cdot i$ and the Faraday-law (induced) voltages $d\lambda/dt$. The result will be a set of nonlinear (due to products of independent variables) differential equations describing the dynamic performance of the machine.

Consider the elementary single-phase machine shown below with one winding on the stator and one on the rotor. Note that such a single-phase configuration does not really work in real-life due to the development of alternating fields, and for any constant, continuous (time-independent, steady-state) torque development we need to rely on poly-phase [27] winding configurations. However, for the sake of simplicity in the following derivation, we use this "impractical" single-phase configuration. The two phase(belts) ss′ and rr′ are residing on stator and rotor, respectively. The stator and rotor are concentric (round, non-salient) cylinders, and slot openings are neglected.

Consequently, our elementary model does not include the effects of salient poles [33]. We shall also assume that the reluctances of the stator and rotor iron are negligible ($\mu_r = \infty$). Based on these assumptions the stator and rotor self-inductances L_{ss} and L_{rr}, respectively, are constant, as illustrated in Figs. 10.30a, b. However, the stator-rotor mutual inductance $L_{sr}(\theta)$ depends upon the position angle θ between the magnetic axes of the stator and rotor coils (see Fig. 10.30c).

The mutual inductance L_{sr} has a positive maximum when $\theta = 0$ or 2π, it is zero when $\theta = \mp \pi/2$, and has a negative maximum when $\theta = \mp \pi$. On the assumption of sinusoidal mmf waves (neglecting all higher harmonics) and a uniform air gap, the space distribution of the air-gap flux wave is sinusoidal, and the mutual inductance is

$$L_{sr}(\theta) = L_{sr} \cos \theta, \tag{10.99}$$

where the bold character $\mathbf{L}_{sr}$ is used to denote an inductance which is a function of the mechanical position angle θ. The character L_{sr} denotes a constant value. Thus L_{rs} is the value of the mutual inductance when the magnetic axes of the stator

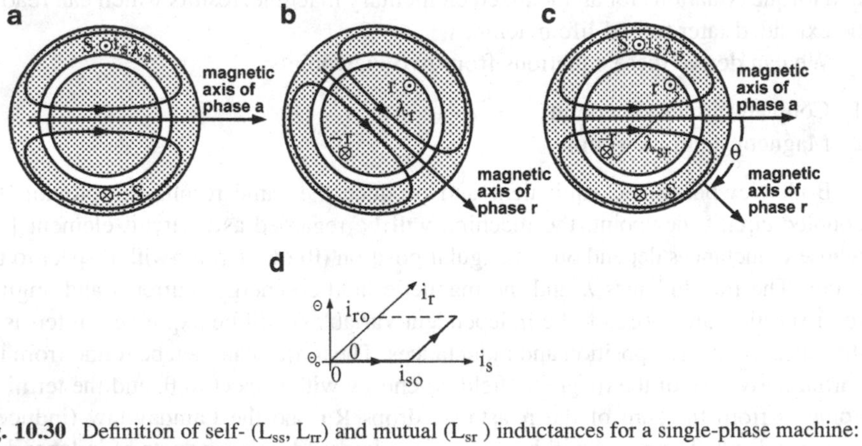

Fig. 10.30 Definition of self- (L_{ss}, L_{rr}) and mutual (L_{sr}) inductances for a single-phase machine: **(a)** L_{ss}, **(b)** L_{rr}, **(c)** $L_{sr}(\theta)$. **(d)** Line integration

and rotor are aligned. In terms of the inductances, the stator and rotor flux linkages λ_s and λ_r are

$$\lambda_s = L_{ss} i_s + L_{sr}(\theta) i_r = L_{ss} i_s + L_{sr} \cos\theta\, i_r, \tag{10.100}$$

$$\lambda_r = L_{sr}(\theta) i_s + L_{rr} i_r = L_{sr} \cos\theta\, i_s + L_{rr} i_r. \tag{10.101}$$

The terminal voltages are, using Kirchhoff's and Faraday's laws

$$v_s = R_s i_s + p\,\lambda_s, \tag{10.102}$$

$$v_r = R_r i_r + p\,\lambda_r. \tag{10.103}$$

Introducing (10.100), (10.101) into (10.102) and (10.103) and replacing the operator p by (d.../dt) yields after applying the product rule of differentiation

$$v_s = R_s i_s + L_{ss}\frac{di_s}{dt} + L_{sr}\cos\theta\frac{di_r}{dt} - L_{sr} i_r \sin\theta\frac{d\theta}{dt}, \tag{10.104}$$

$$v_r = R_r i_r + L_{rr}\frac{di_r}{dt} + L_{sr}\cos\theta\frac{di_s}{dt} - L_{sr} i_s \sin\theta\frac{d\theta}{dt}, \tag{10.105}$$

where $p\theta = d\theta/dt = \omega$ is the instantaneous angular velocity due to the relation $\theta = \int\omega dt$. In a two-pole machine $\theta = \theta_m$ and $\omega = \omega_m$. However, in a p-pole machine $\theta = (p/2)\theta_m$ and $\omega = (p/2)\omega_m$. The electromagnetic torque can be found from the co-energy of the magnetic field in the air gap where i_s, i_r, and θ are the independent variables. Integrating from the starting point ($i_s = 0$ and $i_r = 0$ at constant $\theta = \theta_o$) to the end point ($i_s = i_{so}$ and $i_r = i_{ro}$ at constant $\theta = \theta_o$) as illustrated in Fig. 10.30d, one obtains for the co-energy

$$W'_{fld}(i_s, i_r, \theta_o) = \int_0^{i_{so}} \lambda_s(i_s, i_r, \theta_o)\, di_s + \int_0^{i_{ro}} \lambda_r(i_{so}, i_r, \theta_o)\, di_r, \tag{10.106}$$

or with (10.100) and (10.101)

$$W'_{fld}(i_s, i_r, \theta_o) = \int_0^{i_{so}} L_{ss} i_s di_s + \int_0^{i_{so}} L_{sr}\cos\theta_o i_r di_s + \int_0^{i_{ro}} L_{rr} i_r di_r + \int_0^{i_{ro}} L_{sr}\cos\theta_o i_{so} di_r, \tag{10.107}$$

$$W'_{fld}(i_{so}, i_{ro}, \theta_o) = \frac{1}{2} L_{ss} i_{so}^2 + \frac{1}{2} L_{rr} i_{ro}^2 + i_{so} i_{ro} L_{sr}\cos\theta_o. \tag{10.108}$$

If any endpoint i_s, i_r, θ is considered one obtains for a 2-pole machine the torque due to the magnetic field (fld)

$$T_{fld} = \frac{\partial W'_{fld}(i_s, i_r, \theta)}{\partial \theta} = -L_{sr} i_s i_r \sin\theta \tag{10.109}$$

or the torque for a p-pole machine is

$$T_{fld} = \frac{\partial W'_{fld}(i_s, i_r, \theta_m)}{\partial \theta_m} = \frac{\partial W'_{fld}(i_s, i_r, \theta)}{\partial \theta} \cdot \frac{d\theta}{d\theta_m}, \text{ where } \frac{d\theta}{d\theta_m} = \frac{p}{2} \text{ or } \theta = \frac{p}{2}\theta_m \tag{10.110}$$

T_{fld} is the electromagnetic torque acting in the positive direction of θ_m; note the derivative must be taken with respect to the actual mechanical angle θ_m. Differentiation of the above equation gives for a p-pole single-phase machine

$$T_{fld} = -\frac{p}{2} L_{sr} i_s i_r \sin\theta = -\frac{p}{2} L_{sr} i_s i_r \sin\left(\theta_m \frac{p}{2}\right), \tag{10.111}$$

T_{fld} is measured in Nm. The negative sign in the above equation means that the electromagnetic torque acts in such a direction to bring the magnetic fields of stator and rotor into alignment. Equations (10.104), (10.105), and (10.111) are a set of three (differential) equations relating the electrical variables v_s, i_s, v_r, i_r, and the mechanical variables T_{fld}, θ_m. These equations are a function of the independent variables i_s, i_r, θ_m. The constraints imposed on the electrical variables by networks connected to the terminals (sources or loads and external impedances) and the constraints imposed on the mechanical variables (applied torques and inertial, frictional and spring torques) determine the performance of the device as a coupling element between electrical and mechanical networks.

They are nonlinear differential equations which are difficult to solve, and the Laplace approach does not work except under special circumstances. We are not concerned with their solution here, as we are using them merely as steps in the development of the theory of rotating machines.

The magnetic field viewpoint generates the relations [26] for the electromagnetic torque of a three-phase machine

$$T_{fld} = -\frac{p}{2}\frac{\pi D \ell}{2} B_{sr} F_r \sin\delta_r = -\frac{\pi}{2}\left(\frac{p}{2}\right)^2 \Phi_{sr} F_r \sin\delta_r, \tag{10.112}$$

B_{sr} is the resultant flux density wave between stator and rotor, F_r is the rotor mmf, δ_r is the angle between B_{sr} and F_r and ϕ_{sr} is the resultant flux produced by the combined effect of the stator and rotor mmfs F_s and F_r, respectively.

All rotating machines are based on the same principles:

1. Faraday's law
2. Ampere's law
3. Constraints (e.g., frequency, speed)

In general, electromechanical energy conversion takes place when a change in flux or stored co-energy ($\partial W'_{fld}$) (or stored energy ∂W_{fld}) is associated with mechanical motion ($\partial \theta_m$). In rotating machines voltages are generated in windings or groups of coils by

1. Rotating these windings mechanically through a magnetic field
2. By mechanically rotating a magnetic field past the winding
3. Designing the magnetic circuit so that the reluctance varies with the rotation of the rotor

While we had ample examples for the first two voltage generation mechanisms, Fig. 10.31 shows a so-called homopolar alternator, where the voltage is induced on the basis of varying the reluctance with the rotation of the rotor [34, 35].

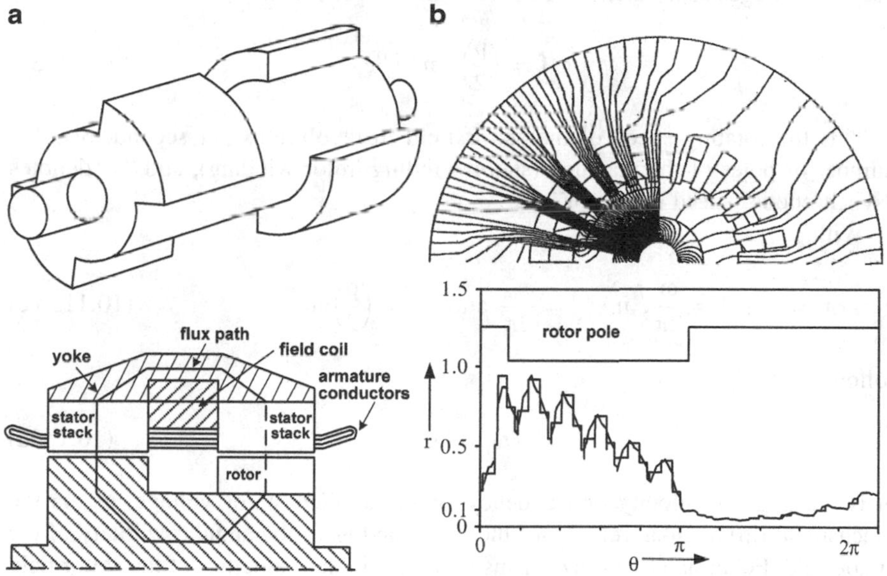

Fig. 10.31 Structure of a homopolar alternator: (**a**) solid rotor and coaxial cross-section, (**b**) cross-section in rθ-plane and magnetic field (From Fuchs, E. F.; Frank, U.V.: "High-speed motors with reduced windage and eddy-current losses, Parts I and II," *etzArchiv*, Vol. 5, Feb. 1983, H. 2, pp. 17–23 and 55–62, [35])

10.4 Non-Salient-Pole (Round-Rotor) Synchronous Machines [26]

Figure 10.3a shows the magnetic field of a two-pole, non-salient-pole synchronous generator, Figs. 10.3b, c those of a four-pole, non-salient-pole synchronous generator, and Figs. 10.3d, e those of a six-pole salient-pole synchronous generator or alternator. The configuration of Fig. 10.3a generates within one mechanical revolution one voltage cycle measured in Hz; its frequency in cycles per second (hertz) is the same as the speed of the rotor in revolutions per second, i.e., the electrical frequency f is synchronized with the mechanical speed measured in revolutions per second (rps) and this is the reason for the designation synchronous machine. Thus a 2-pole synchronous machine must revolve at 3,600 rpm to produce a 60 Hz voltage. A great many of synchronous machines have more than 2 poles, however, when a machine has more than 2 poles, it is convenient to concentrate on a single pair of poles and to recognize that the electric, magnetic, and mechanical conditions associated with every other pole pair are repetitions of those of the pair under consideration.

Concept of Stationary Stator and Rotor Fields for Steady-State Torque Production: The configuration of Figs. 10.3b, c generates within one mechanical revolution two voltage cycles, and the fields of Figs. 10.3d, e generate within one mechanical revolution three voltage cycles. In general one can write for the frequency f of the induced voltages as a function of the revolutions per second

$$f = \left(\frac{p}{2}\right) \cdot n_s^{(rps)}, \qquad (10.113a)$$

$n_s^{(rps)}$ is the rotating speed of the stator field in revolutions per second, p is the number of poles of the machine (stator winding, rotor winding), and "s" denotes "synchronous" speed or "stator".

With

$$f = \frac{\omega}{2\pi}, \quad n_s^{(rps)} = \frac{\omega_m}{2\pi} \quad \text{and} \quad \omega = \left(\frac{p}{2}\right)\omega_s \qquad (10.113b,c)$$

follows

$$\omega_s = \omega_m, \qquad (10.113d)$$

ω is the angular velocity or frequency of the applied stator voltage, ω_m is the mechanical (m) angular velocity of the rotor, and ω_s is the angular velocity of the stator field. Equation (10.113d) tells us that the angular velocities of the stator field and that of the mechanical rotor are identical.

Integrating (10.113c) leads with (10.113d) to

$$\omega t = \left(\frac{p}{2}\right)\omega_m t, \text{ or } \theta = \left(\frac{p}{2}\right)\theta_m. \qquad (10.114)$$

Equation (10.113a) can be rewritten in terms of $n_s^{(rpm)}$

$$f = \left(\frac{p}{2}\right) \cdot \frac{n_s^{(rpm)}}{60}. \tag{10.115}$$

Equivalent Circuit and Phasor Diagram of a Three-Phase (Round-Rotor) Synchronous Machine: With very few exceptions, synchronous machines are of the three-phase type because of the advantages of three-phase systems with respect to generation, transmission and heavy-power utilization [28]. For the production of a set of three induced voltages ($\tilde{E}_a$ per phase), phase-displaced by 120 electrical degrees in time, a minimum of three coils phase-displaced 120 electrical degrees in space must be used. When a synchronous generator supplies electrical power to a load, the armature current ($\tilde{I}_a$ within one phase) creates a component flux wave in the machine which rotates at the same (synchronous) speed as the rotor field stationary with respect to the mechanical rotor. This stator flux can be characterized by stator flux linkages and if these linkages are referred to the armature (stator) current $\tilde{I}_a$ they yield the so-called synchronous reactance X_s [33]. This stator (armature) flux interacts with the flux created by the field current I_f, and electromagnetic torque T_{fld} results from the tendency of the two magnetic fields to align themselves. In order to produce a steady-state electromagnetic torque T_{fld}, the magnetic fields of stator and rotor must be constant in amplitude and stationary with respect to each other. Figures 10.32a, b show the per-phase equivalent circuits of a non-salient-pole synchronous machine identifying induced voltage $\tilde{E}_a$which is a function of the field current I_f and the electrical angular velocity ω. The armature current $\tilde{I}_a$ is assumed to flow into the machine (consumer notation, Fig. 10.32a) through the synchronous reactance X_s and the induced voltage source, where the stator winding resistance is R_a, and the terminal voltage is denoted as $\tilde{V}_a$.

The phasor diagram corresponding to the equivalent circuit of Fig. 10.32b (generator notation) is shown in Fig. 10.33 for overexcited (lagging power factor) operation.

Operating characteristics of a three-phase, non-salient-pole synchronous machine: The most important operating characteristics of a synchronous machine are the so-called V-curves shown in Fig. 10.34a, and the compounding curves of Fig. 10.34b.

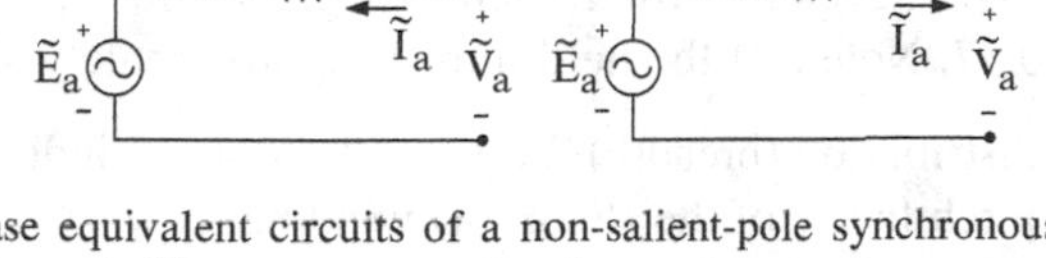

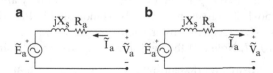

Fig. 10.32 Per-phase equivalent circuits of a non-salient-pole synchronous machine; (**a**) consumer notation for current, (**b**) generator notation for current

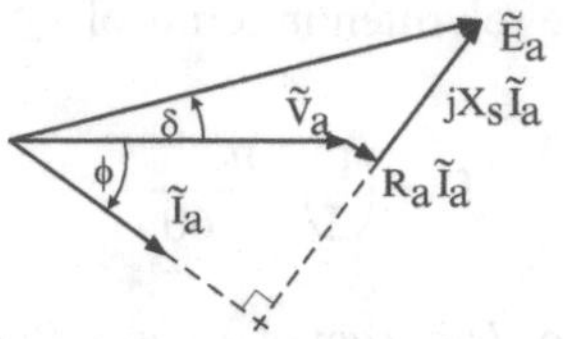

Fig. 10.33 Phasor diagram of non-salient-pole (cylindrical, round rotor) synchronous machine at lagging current operation for generator notation (overexcited): $\tilde{E}_a$ leads $\tilde{V}_a$; overexcited $|\tilde{E}_a|>|\tilde{V}_a|$; for corresponding leading current operation or underexcited operation (not shown) $|\tilde{E}_a|<|\tilde{V}_a|$

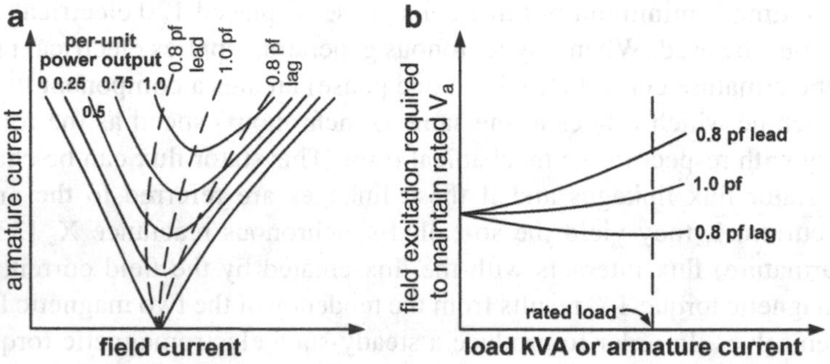

Fig. 10.34 V-curves (**a**) of a synchronous machine for generator notation (Fig. 10.32b), and compounding curves (**b**) of a synchronous machine for consumer notation (Fig. 10.32a)

Application Example 10.12: Calculation of the Synchronous Reactance X_S of a Cylindrical-Rotor (or Non-Salient-Pole) Synchronous Machine

A p = 2 pole, f = 60 Hz synchronous machine (generator or alternator, motor) is rated S_{rat} = 16 MVA at a lagging (generator notation) power factor $\cos\Phi = 0.8$, a line-to-line terminal voltage of $V_{L\text{-}L}$ = 13,800 V, and a (one-sided) air gap length of g = 0.0127 m (0.5 inch). Fig. 10.3a illustrates the magnetic field distribution of a similar machine at full-load. The stator of the machine of this example has 48 slots and has a three-phase double-layer, 60° phase-belt winding with 48 armature coils with a short pitch of 20/24 corresponding to (20/24)180°= 150° (see Fig. 10.35). Each coil consists of one turn $N_{coil} = 1$, and 16 coils are connected in series, that is, the number of series turns per phase of the stator (st) are $N_{st\ phase} = 16$ (see Fig. 10.36). The maximum air-gap flux density at no-load (stator current $I_{st\ phase} = 0$) is $B_{st\ max} = 1$ T.

There are eight field coils per pole (see Fig. 10.36) and they are pitched 44-60-76-92-124-140-156-172 degrees, respectively as indicated in Fig. 10.36. Each field coil has 18 turns. The developed rotor (field) magneto-motive force (mmf) F_r is depicted in Fig. 10.37. Note that the field mmf is approximately sinusoidal.

(a) Calculate the distribution (breadth) factor of the stator winding $k_{st\ d} = k_{st\ b}$
(b) Calculate the pitch factor of the stator (st) winding $k_{st\ p}$
(c) Determine the total winding factor of the stator winding k_{st}

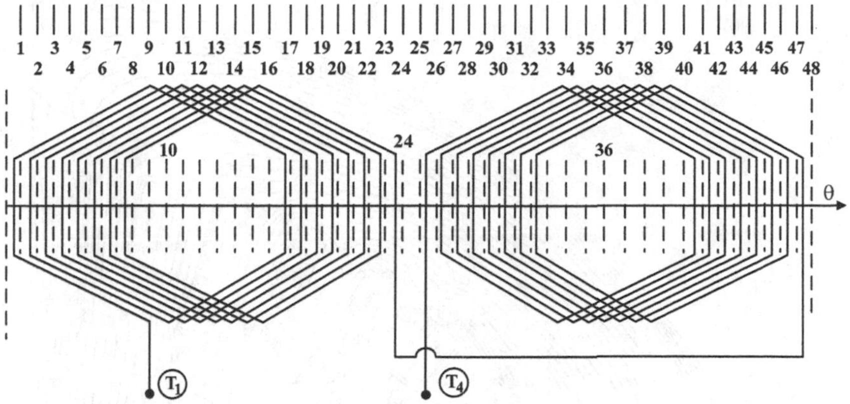

Fig. 10.35 Stator winding of 2-pole, three-phase synchronous machine with a short pitch (20/24) of 150°

Fig. 10.36 Rotor winding of 2-pole, three-phase synchronous machine

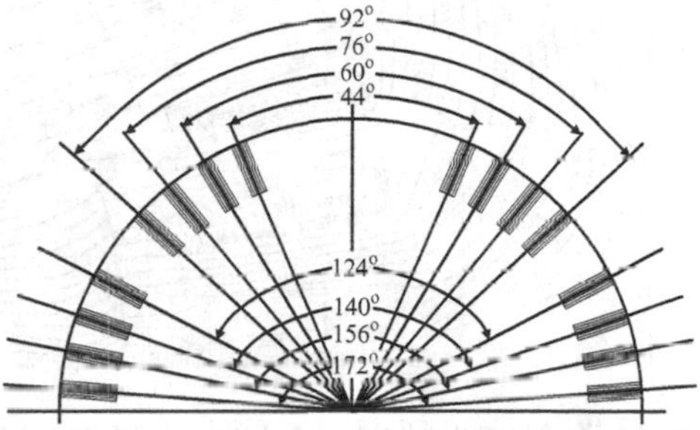

Fig. 10.37 Rotor mmf F_r of 2-pole, three-phase synchronous machine

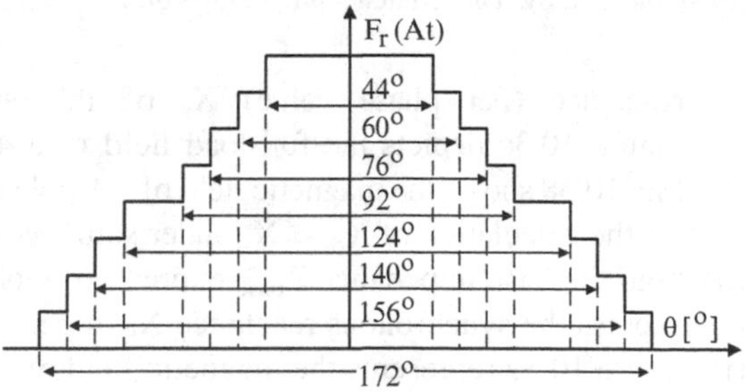

(d) Calculate the pitch factor of the rotor (r) winding $k_{r\ p}$

(e) Determine the stator flux $\Phi_{st\ max}$

(f) Compute the area $area_p$ of one pole

(g) Compute the field (f) current I_{fo} required at no-load. Figure 10.3b illustrates the no-load field of a 4-pole synchronous machine. Determine the synchronous (s)

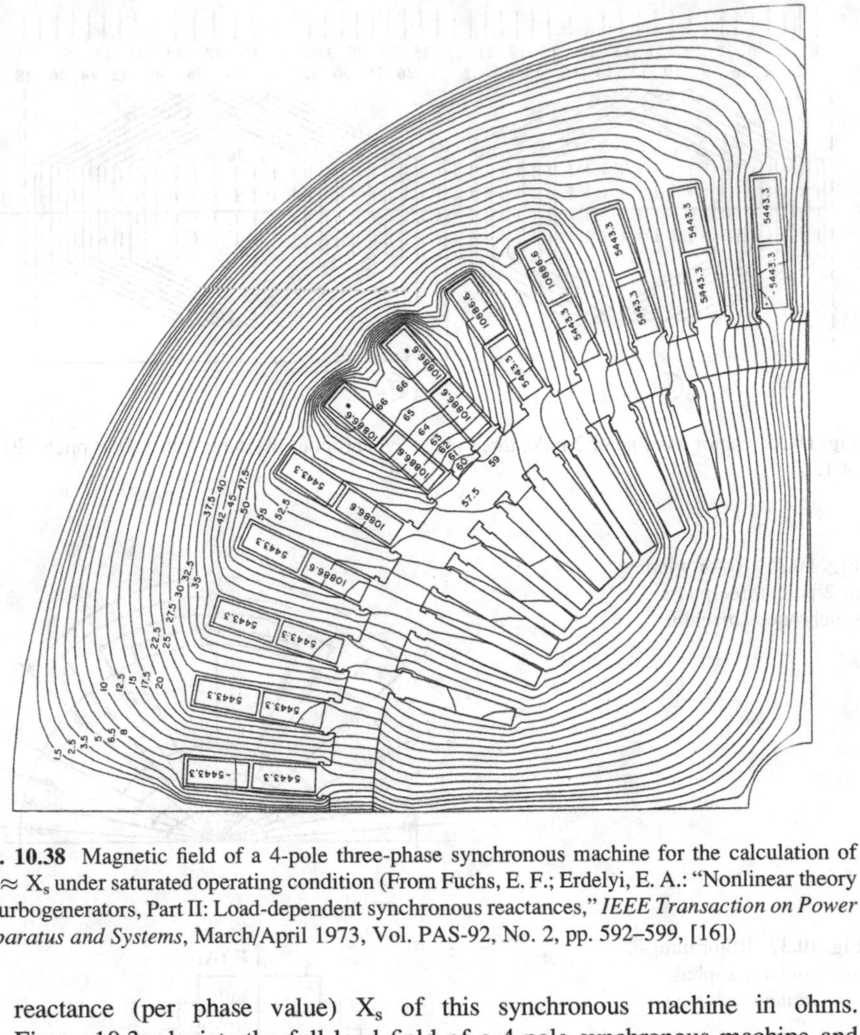

Fig. 10.38 Magnetic field of a 4-pole three-phase synchronous machine for the calculation of $X_d \approx X_s$ under saturated operating condition (From Fuchs, E. F.; Erdelyi, E. A.: "Nonlinear theory of turbogenerators, Part II: Load-dependent synchronous reactances," *IEEE Transaction on Power Apparatus and Systems*, March/April 1973, Vol. PAS-92, No. 2, pp. 592–599, [16])

reactance (per phase value) X_s of this synchronous machine in ohms. Figure 10.3c depicts the full-load field of a 4-pole synchronous machine and Fig. 10.38 shows the magnetic field of a 4-pole three-phase synchronous machine for the calculation of $X_d \approx X_s$ under saturated operating condition [16].

(h) Find the base impedance Z_{base} expressed in ohms

(i) Express the synchronous reactance X_s in per unit (pu)

(j) Figure 10.39 represents the magnetic field of a 4-pole three-phase synchronous machine for the calculation of X_q under saturated operating condition [16]

Solution

(a) Calculation of the distribution (breadth) factor of the stator winding $k_{st\,d} = k_{st\,b}$. The distribution or breadth factor can be obtained graphically (see Fig. 10.10b or from the formula [26] of (10.116). The winding of Fig. 10.35 has eight coils per phase belt, separated by an electrical angle of 7.5°, therefore

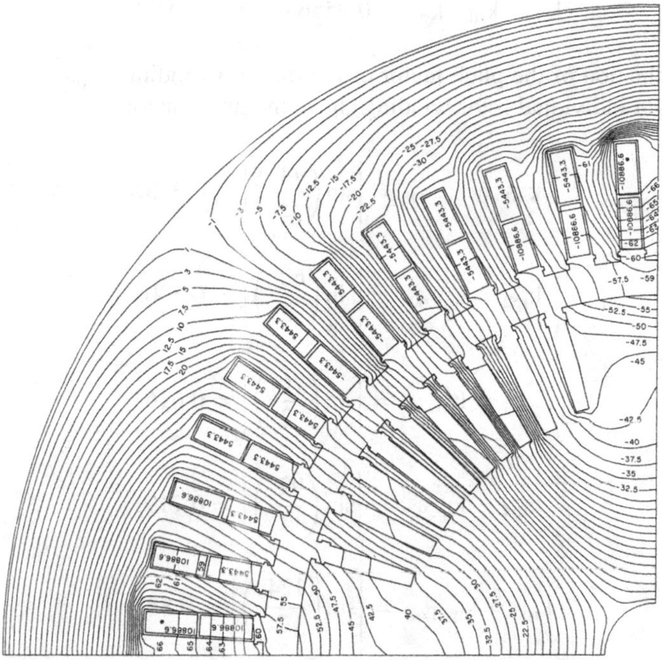

Fig. 10.39 Magnetic field of a 4-pole three-phase synchronous machine for the calculation of X_q under saturated operating condition (From Fuchs, E. F.; Erdelyi, E. A.: "Nonlinear theory of turbogenerators, Part II: Load-dependent synchronous reactances," *IEEE Transaction on Power Apparatus and Systems*, March/April 1973, Vol. PAS-92, No. 2, pp. 592–599, [16])

$$k_{stb} = k_{std} = \frac{\sin(n\gamma/2)}{n\sin(\gamma/2)} = \frac{\sin(8 \cdot 7.5^\circ/2)}{8\sin(7.5^\circ/2)} = 0.9556. \tag{10.116}$$

(b) Computation of the pitch factor of the stator winding $k_{st\,p}$ according to (10.10), where n = 1, 5, 7, 11,... are the orders of the harmonic components that can occur in a three-phase field and $\beta = (20/24)\cdot 180^\circ = 150^\circ$.

$$k_{stp}(n) = \frac{\sin\left(\frac{n\beta}{2}\right)}{\sin\left(\frac{n\pi}{2}\right)} = \left|\sin\left(\frac{n\beta}{2}\right)\right| \text{ yields for } n = 1 \text{ (fundamental)}$$

$$k_{stp}(n = 1) = 0.966, \text{ for } n = 5 \tag{10.117}$$

$$k_{stp}(n = 5) = 0.259, \text{ for } n = 7 k_{stp}(n = 7) = 0.259, \text{ and for } n = 11$$

$$k_{stp}(n = 11) = 0.966.$$

Note that the pitch factor for the 11th harmonic is large and the winding configuration (e.g., pitch) must be changed if the stator field contains a significant magnitude of the 11th harmonic.

(c) The total winding factor for the fundamental (n = 1) is

$$k_{st} = k_{stb} \cdot k_{stp} = 0.9556 \cdot 0.966 = 0.923. \tag{10.118}$$

(d) Determination of the pitch factor of the rotor (r) winding $k_{r\,p}$.
The eight individual rotor windings have the pitch factors

$$k_{rp1} = k_{rp\_44^\circ} = \left|\sin\left(\frac{44^\circ}{2}\right)\right| = 0.375, \tag{10.119}$$

$$k_{rp2} = k_{rp\_60^\circ} = \left|\sin\left(\frac{60^\circ}{2}\right)\right| = 0.500, \tag{10.120}$$

$$k_{rp3} = k_{rp\_76^\circ} = \left|\sin\left(\frac{76^\circ}{2}\right)\right| = 0.616, \tag{10.121}$$

$$k_{rp4} = k_{rp\_92^\circ} = \left|\sin\left(\frac{92^\circ}{2}\right)\right| = 0.719, \tag{10.122}$$

$$k_{rp5} = k_{rp\_124^\circ} = \left|\sin\left(\frac{124^\circ}{2}\right)\right| = 0.883, \tag{10.123}$$

$$k_{rp6} = k_{rp\_140^\circ} = \left|\sin\left(\frac{140^\circ}{2}\right)\right| = 0.937, \tag{10.124}$$

$$k_{rp7} = k_{rp\_156^\circ} = \left|\sin\left(\frac{156^\circ}{2}\right)\right| = 0.978, \tag{10.125}$$

$$k_{rp8} = k_{rp\_172^\circ} = \left|\sin\left(\frac{172^\circ}{2}\right)\right| = 0.998. \tag{10.126}$$

The resultant rotor-pitch factor is

$$k_{rp} = \frac{\sum_{i=1}^{8} k_{rpi}}{8} = 0.751. \tag{10.127}$$

(e) Calculation of the stator flux $\Phi_{st\,max}$.
The induced voltage in one phase-taking into account the stator winding factor $k_w = k_{st}$ is according to (10.73)

$$\begin{aligned} E_{rms\text{-}ph} &= \frac{E_{max}}{\sqrt{2}} = \frac{2\pi}{\sqrt{2}} f \cdot k_w N_{ph}\,\phi = 4.44 f \cdot k_w N_{ph}\,\phi_{max} \\ &= 4.44 f \cdot k_{st} N_{stph} B_{stmax}\,\text{area}_p \end{aligned} \tag{10.128}$$

with $\phi_{stmax} = (B_{stmax} \cdot area_p)$ the line-to-line induced voltage is

$$E_{L-L} = \sqrt{3} \cdot 4.44f \cdot k_{st} N_{st\,ph}\ \phi_{stmax} \tag{10.129}$$

or the stator flux is with (at no-load) $E_{L-L} \approx V_{L-L} = 13{,}800$ V

$$\phi_{stmax}^{no-load} \frac{E_{L-L}}{\sqrt{3} \cdot 4.44f \cdot N_{st\,ph} k_{st}} = \frac{13,800}{\sqrt{3} \cdot 4.44 \cdot 60 \cdot 16 \cdot 0.923}$$
$$= 2.025\ \text{Wb/pole.} \tag{10.130}$$

(f) Computation of the area per pole
The area per pole is

$$area_p = \frac{\phi_{stmax}^{no-load}}{B_{stmax}} = \frac{2.025}{1} = 2.025\,m^2. \tag{10.131}$$

(g) Calculation of the field (f) current I_{fo} required at no-load.
The maximum rotor mmf is according to (10.14)

$$F_r = \frac{4}{\pi} k_r \frac{N_r}{p} I_r \tag{10.132}$$

with $H_g = F_r/g$ and $B_g = \mu_o(F_r/g)$one obtains for the maximum stator flux

$$B_{stmax} = \frac{\mu_o \frac{4}{\pi} k_{rp} N_{rtotal} I_{fo}}{p \cdot g}, \tag{10.133}$$

or for the field current at no-load operation

$$I_{fo} = \frac{B_{stmax} p \cdot g \cdot \pi}{4\mu_o k_{rp} N_{rtotal}} = \frac{1 \cdot 2 \cdot 0.0127 \cdot \pi}{4 \cdot 4\ \pi \cdot 10^{-7} 0.751 \cdot 18 \cdot 16} = 73.4\text{A}. \tag{10.134}$$

Determination of the synchronous (s) reactance (per phase value) X_s of this synchronous machine in ohms.
The self-inductance of phase a is

$$L_{aa} = \frac{\lambda_{aa}}{i_a} = \frac{N_a \phi_{aa}}{i_a} = \frac{N_{st\,ph} \phi_{stmax}}{I_{st}}. \tag{10.135}$$

with $F_{st} = \frac{4}{\pi} k_{st} \frac{N_{st\,ph}}{p} I_{st}$, $H_{st} = \frac{F_{st}}{g}$, and $B_{stmax} = \mu_o \frac{F_{st}}{g}$, and $\phi_{stmax} = (B_{stmax} \cdot area_p)$ one obtains for the maximum stator flux

$$\phi_{stmax} = \mu_o \frac{4}{\pi} k_{st} \frac{N_{st\,ph} I_{st}}{p \cdot g}\, area_p, \tag{10.136}$$

and for the self-inductance of phase "a"

$$L_{aa} = \mu_o \frac{4}{\pi} k_{st} \frac{(N_{st\,ph})^2}{p \cdot g} \text{area}_p = 4\,\pi \cdot 10^{-7} \frac{4}{\pi} 0.92 \frac{3 \cdot (16)^2 2.025}{2 \cdot 0.0127}$$
$$= 30.04\,\text{mH}. \tag{10.137}$$

According to Concordia [33] the synchronous reactance X_s for a round rotor machine is about identical with the direct axis reactance X_d and relates to the self- and mutual reactances X_{aa} and X_{ab}, respectively, as follows

$$X_s \approx X_d = X_{aa} + X_{ab} \approx 2X_{aa}. \tag{10.138}$$

The synchronous reactance in ohms is now

$$X_s^{\Omega} = 2\omega L_{aa} = 2 \cdot 377 \cdot 30.04\,\text{m}\,\Omega = 22.65\ \Omega. \tag{10.139}$$

(h) The base impedance Z_{base} expressed in ohms is with

$$V_{base\_phase} = \frac{V_{L-L}}{\sqrt{3}} = V_{phase} = 7{,}967.7\,\text{V}, \quad S_{base\_phase} = \frac{16\,\text{MVA}}{3} = 5.33\,\text{MVA}$$

$$I_{base\_phase} = \frac{5.33\,\text{MVA}}{7967.7\,\text{V}} = 669.37\,\text{A},$$

$$Z_{base} = \frac{V_{base\_phase}}{I_{base\_phase}} = \frac{7{,}967.7}{669.37} = 11.90\ \Omega. \tag{10.140}$$

(i) The synchronous reactance X_s in per unit (pu) is now

$$X_s^{pu} = \frac{X_s^{\Omega}}{Z_{base}} = \frac{22.65}{11.9} = 1.90\,\text{pu}. \tag{10.141}$$

From a synchronous machine stability point of view this value is satisfactory because $1\,\text{pu} \leqslant X_s \leqslant 2\,\text{pu}$ [13, 14, 23].

10.5 Induction Machines [26]

The induction machine mechanical structure is similar to a synchronous machine except the rotor winding is not necessarily excited by an external source. Its rotor winding is in most applications short-circuited (e.g., squirrel-cage winding) and the rotor currents are being induced by the rotating stator field. Due to this induction process this machine is called induction machine or asynchronous machine because there is not a fixed relationship between stator frequency and rotor speed. As a consequence of this induction process the mechanical rotor and the stator field cannot rotate in synchronism (asynchronous operation) and the mechanical rotor must have a slip with respect to the rotating stator field. Due to this induction process the air gap of an induction machine must be as small as possible from a mechanical point of view.

Basic law of energy conversion: One must remember that for steady-state torque production the interacting stator and rotor fields must be stationary with respect to one another.

Basic principles of operation: Figure 10.40 shows the cross-section of an induction machine for a three-phase, two-pole configuration.

Induction-machine principle: Figure 10.41 relates to the induction machine (motor or generator) principle depicting the stationary stator, the rotating rotor and the stator and rotor magnetic fields. The mechanical rotor rotates with respect to the stator with the mechanical angular velocity ω_m. The stator field rotates with respect to the stator with the mechanical synchronous angular ω_s, and the rotor field rotates with respect to the rotating rotor with the mechanical angular velocity ω_r. Satisfying the law that the interacting fields must be stationary, one can write

$$\omega_s = \omega_m + \omega_r. \tag{10.142}$$

Now we can perform two experiments:

1. Rotor is locked (Fig. 10.42) and does not rotate ($\omega_m = 0$), that is, it is stationary with respect to the stator and the machine behaves like a three-phase transformer. As a consequence the angular velocity of the rotor field is the same as that of the stator field, that is,

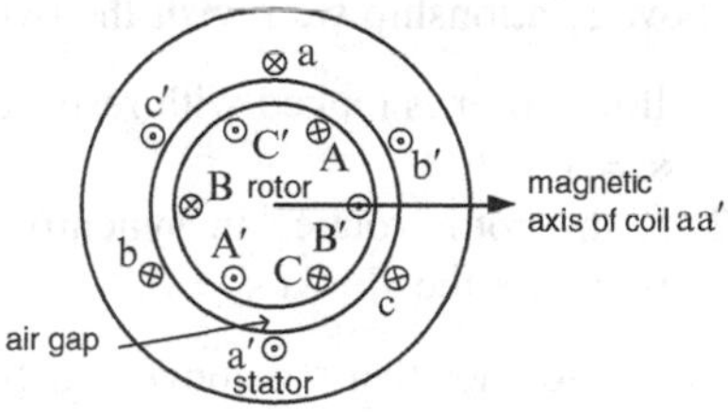

Fig. 10.40 Three-phase, two-pole induction machine cross-section

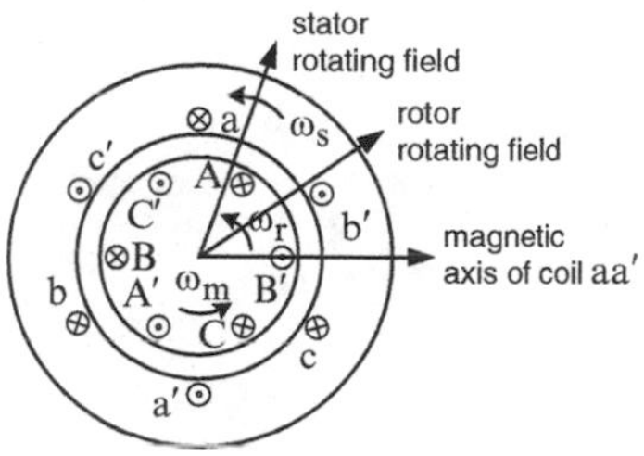

Fig. 10.41 Definition of stator and rotor fields. Note the stator field rotates with respect to the stator with ω_s, and the rotor field rotates with respect to the (rotating) rotor with ω_r

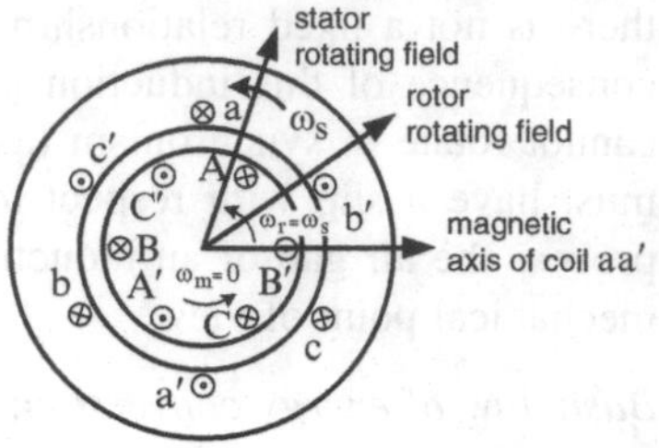

Fig. 10.42 Locked or blocked rotor ($\omega_m = 0$)

$$\omega_r = \omega_s = \frac{2}{p}(2\pi f). \tag{10.143}$$

2. Rotor rotates at the same angular mechanical velocity at the stator field, that is, $\omega_m = \omega_s$; because the rotor does not rotate relative to the stator field there are no rotor voltages/currents induced and their frequency is $f_r = 0$ or

$$\omega_s = \omega_m = \frac{2}{p}(2\pi f). \tag{10.144}$$

Generalizing the results of these two experiments we can state that $\omega_s = \omega_m + \omega_r$ is satisfied.

Definition of slip s: One defines the slip as

$$s = \frac{\omega_s - \omega_m}{\omega_s} \tag{10.145}$$

or $s \cdot \omega_s = \omega_s - \omega_m = \omega_r$, where $\omega_r = s \cdot \omega_s$ is the mechanical angular velocity of the rotating rotor field with respect to the rotating rotor. To check the validity of the above relationship we revisit the two experiments performed before:

1. If the rotor is locked with respect to the stator $\omega_m = 0$, and therefore the slip is $s = 1$.
2. If the rotor rotates in synchronism with the stator field then $\omega_m = \omega_s$ and therefore the slip is $s = 0$.

One concludes that the above equation for the slip s is satisfied for all slip values.

With

$$\omega_s = \frac{2}{p}(2\pi f) \tag{10.146}$$

and

$$\omega_r = \frac{2}{p}(2\pi f_r) \tag{10.147}$$

one obtains

$$\omega_m = \omega_s - \omega_r = \omega_s - s \cdot \omega_s = \frac{2}{p}(2\pi f) - s\left(\frac{2}{p}(2\pi f)\right) \tag{10.148}$$

or

$$\frac{\omega_s}{\omega_r} = \frac{f}{f_r} = \frac{f}{sf} \tag{10.149}$$

or

$$f_r = s \cdot f. \tag{10.150}$$

The induction machine may be regarded as a generalized transformer in which electric power is transferred between stator and rotor together with a change of frequency, voltage amplitude, current amplitude and a flow of mechanical power. *Relation between speed measured in revolutions per minute* ($n_s^{(rpm)}$) *and frequency f measured in hertz*: The synchronous speed measured in revolutions per second (rps) is

$$n_s^{(rps)} = \frac{n_s^{(rpm)}}{60} \tag{10.151}$$

and the synchronous angular velocity is

$$\omega_s = 2\pi n_s^{(rps)} = \frac{2}{p}(2\pi f), \tag{10.152}$$

therefore

$$n_s^{(rpm)} = \frac{120}{p} f, \tag{10.153}$$

$n_s^{(rpm)}$ is the synchronous (denoted by subscript s) speed of the rotating stator field measured in revolutions per minute (rpm). Table 10.1 lists the synchronous speeds and angular velocities at 60 Hz as a function of the pole number, and Table 10.2

Table 10.1 Synchronous speeds $n_s^{(rpm)}$ and angular velocities ω_s for f = 60 Hz

p	$n_s^{(rpm)}$	ω_s [rad/s]
2	3,600	376.98
4	1,800	188.49
6	1,200	125.66
8	900	75.39
...	...	...

Table 10.2 Synchronous speeds $n_s^{(rpm)}$ and angular velocities ω_s for f = 50 Hz

p	$n_s^{(rpm)}$	ω_s [rad/s]
2	3,000	314.15
4	1,500	157.08
6	1,000	104.72
8	750	78.54
...	...	...

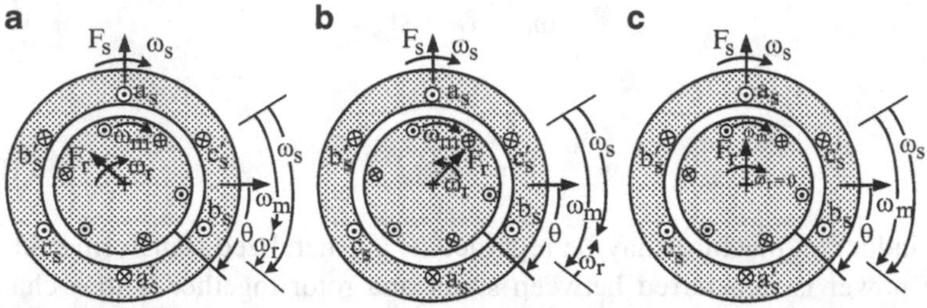

Fig. 10.43 Angular frequency (velocity) relations for motoring, regeneration, and synchronous operation at synchronous speed; (**a**) motoring, (**b**) regeneration, (**c**) synchronous operation

lists the synchronous speeds and angular velocities at 50 Hz as a function of the pole number.

For motor and generator operation the slip is in the range $-\infty \le s \ge \infty$. For motor operation $0 \le s \le 1$, and for generator operation $0 \ge s \ge -\infty$. Under normal steady-state operating conditions these ranges are much smaller: $0 \le s_{rated} \le 0.1$ for motoring, and $0 \ge s_{rated} \ge -0.1$ for generation. From this follows that for motoring $\omega_m < \omega_s$ and for generation $\omega_m > \omega_s$ (see Fig. 10.45). The slip s of an induction machine can be expressed in terms of the speeds as follows

$$s = \frac{n_s^{(rpm)} - n_m^{(rpm)}}{n_s^{(rpm)}}. \tag{10.154}$$

Figures 10.43a, b, c illustrate the operation of an induction machine as a motor, generator (regeneration), and at synchronous (no-load) operation, respectively.

Equivalent circuit of an induction machine: Balanced, constant-speed operation of the induction machine (either motor or generator) may be modeled by the AC per-phase Y-equivalent circuits of Figs. 10.44a, b. The parameters of these circuits are defined as follows.

$\tilde{V}_s$ = line-to-neutral stator phase voltage applied at the terminals (a, n), usually used as phase reference, where a is the terminal of phase a and n is the neutral which is identical with the terminal a′ of phase a, if the machine is connected in Y-configuration.

$\tilde{I}_s$ = stator phase current,

R_s = phase stator winding resistance

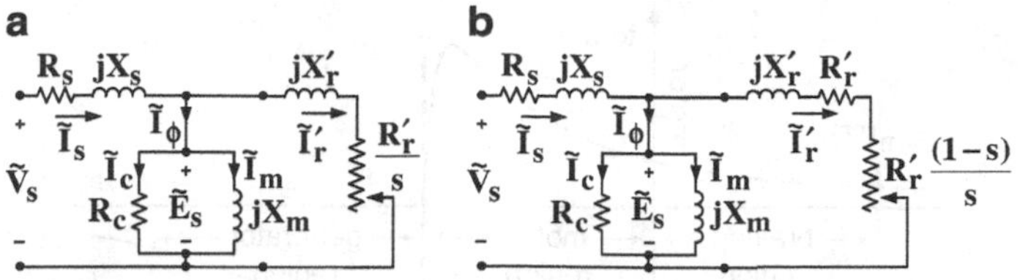

Fig. 10.44 Per-phase equivalent circuits of a three-phase induction machine: (**a**) R'_r and $R'_r(1-s)/s$ combined, (**b**) R'_r and $R'_r(1-s)/s$ not combined

X_s = stator phase leakage reactance associated with the stator flux that does not link the rotor windings
X_m = magnetizing reactance due to mutual flux between stator and rotor
$\tilde{I}'_r$ = rotor phase current, referred (reflected) to the stator
X'_r = rotor phase leakage reactance associated with the rotor flux that does not link the stator windings, referred to the stator
R'_r = phase rotor winding resistance, referred to the stator
s = (rotor) slip which is a measure of the rotor speed $n_m^{(rpm)} = n_s^{(rpm)}(1-s)$ or $s = (\omega_s - \omega_m)/\omega_s$

The mechanical output power of an induction machine is

$$P_m = 3\left|\tilde{I}'_r\right|^2 R'_r \frac{(1-s)}{s}, \tag{10.155}$$

and the torque developed is

$$T = \frac{P_m}{\omega_m} = 3\frac{\left|\tilde{I}'_r\right|^2 \frac{R'_r}{s}}{\omega_s}. \tag{10.156}$$

Note that 1 hp = 746 W in the US and 1 PS (Pferdestärke) = 736 W, that is 1 hp = 1.014 PS

Natural torque-speed characteristic of a three-phase induction machine [26]: Figure 10.45 shows the natural (rated frequency) torque-speed characteristic of an induction machine at rated voltage. Note the different regions for motoring, regeneration and braking. There are two abscissa axes: one in terms of the mechanical speed $n_m^{(rpm)}$ and one in terms of the slip s.

Power flow diagram for an induction motor [36]: In the following the various power and losses are defined as follows:

Input power is

$$P_{in} = 3\left|\tilde{V}_s\right|\left|\tilde{I}_s\right|\cos\phi, \tag{10.157}$$

ϕ is the phase angle between $\tilde{V}_s$ and $\tilde{I}_s$.

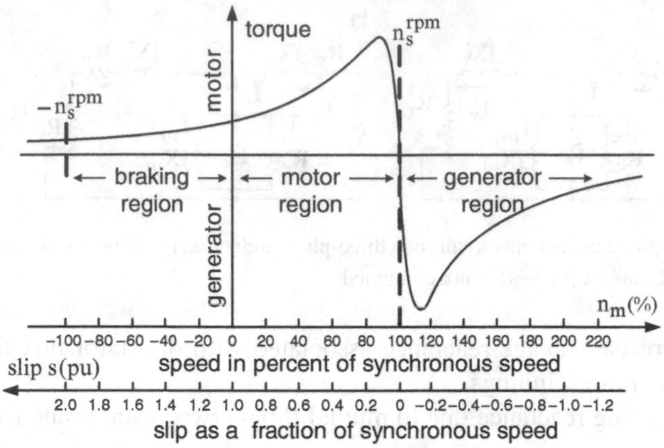

Fig. 10.45 Induction-machine torque-speed or torque-slip curve showing braking (counter torque or reverse voltage braking), motor, and generator (regeneration) regions

Fig. 10.46 Loss distribution within induction motor

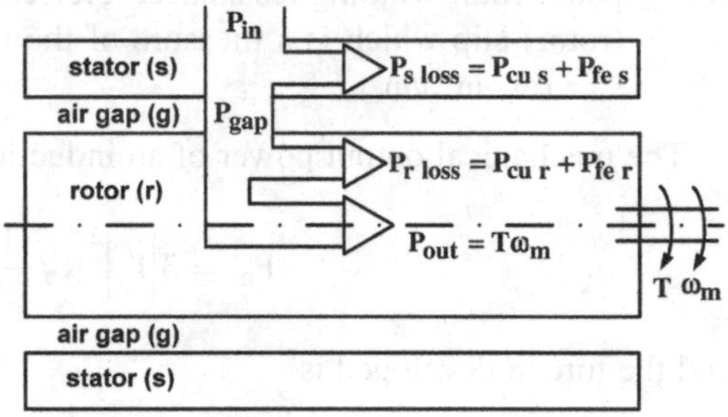

Stator copper losses are

$$P_{cu\,s} = 3\left|\tilde{I}_s\right|^2 R_s. \tag{10.158}$$

Rotor copper losses are

$$P_{cu\,r} = 3\left|\tilde{I}'_r\right|^2 R'_r. \tag{10.159}$$

In Fig. 10.46 the stator and rotor iron-core losses $P_{fe\ s}$ and $P_{fe\ r}$, respectively, are indicated. The friction and windage losses $P_{friction/windage}$ as well as stray losses P_{stray} are not shown. These losses are relatively small and can be mostly neglected.

The efficiency η is defined as output ($P_{out} = P_m$) over input (P_{in}) powers

$$\eta = \frac{P_m}{P_{in}}. \tag{10.160}$$

Efficiency optimization of three-phase induction machines: Reference [37] discusses the efficiency improvements of induction motors with thyristor and triac controllers, and reference [38–43] address the efficiency optimization of

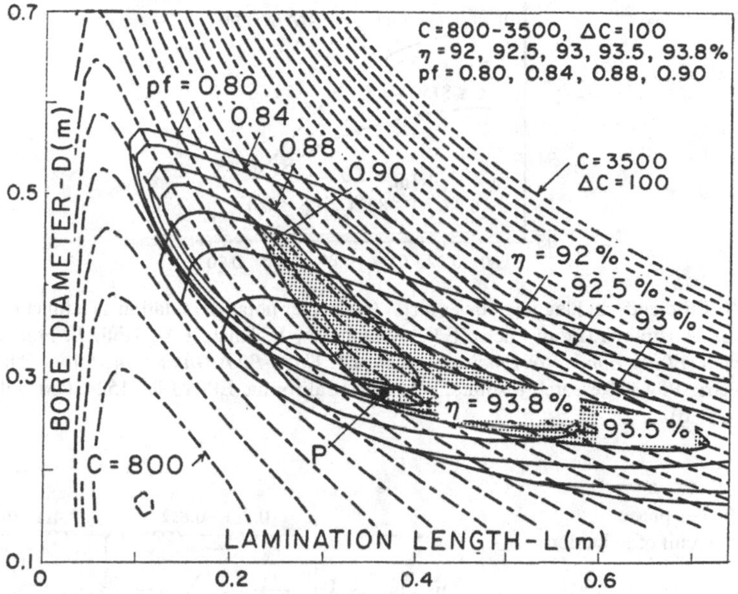

Fig. 10.47 Efficiency, lagging power factor (consumer notation), and production cost C in US$ in the (D, L) plane (From Fuchs, E.F.; Appelbaum, J.; Khan, I.A.; Höll, J.; Frank, U.V.: *Optimization of Induction Motor Efficiency, Volume 1: Three-Phase Induction Motors*, Publication of the Electric Power Research Institute, Palo Alto, California, EPRI EL-4152-CCM, July 1985, 336 pages, [43])

three-phase induction machines under given performance constraints. The technique developed in the above-mentioned references, which optimizes three-phase motor designs for efficiency, marks a departure from the traditional emphasis on cost reduction. Since motors use about two-thirds of the electric power generated in the United States (about 1,000 GW), an increase in their efficiency could make a major contribution to energy conservation. Figure 10.47 illustrates the efficiency η, the lagging power factor pf (consumer notation), and production cost C in US$ as a function of the bore-diameter (D) and lamination-length (L) [43].

An important finding is the efficiency-power factor relation shown in Fig. 10.48: as the efficiency increases the power factor reduces and vice versa [43].

Application Example 10.13: Three-Phase Induction Machine in Motoring and Generation Modes

Figures 10.2a, b, c illustrate the no-load magnetic fields of $p = 6$ pole and $p = 4$ pole three-phase induction machines, respectively. In this problem we will analyze the behavior of such machines in the motoring and generation mode. A $P_{out\_rat} = 27.3$ kW, $V_{L\text{-}L\_rat} = 460$ V, $f = 60$ Hz, $p = 6$ pole, $n_{m\_rat} = 1{,}180$ rpm, and Y-connected squirrel-cage induction machine has the following parameters per phase referred to the stator: $R_s = 0.20\ \Omega$, $X_s = 0.80\ \Omega$, $X_m = 20\ \Omega$, $R_r' = 0.10\ \Omega$, and $X_r' = 0.40\ \Omega$. Calculate for motor operation:

(a) The rated current, torque, power factor, and efficiency
(b) The starting torque and starting current as a ratio of their full-load values

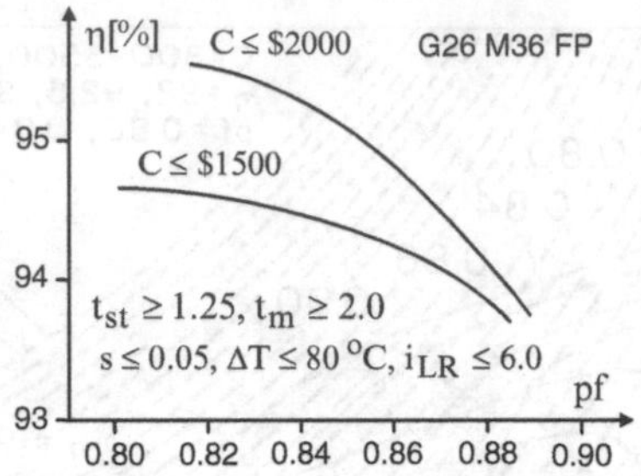

Fig. 10.48 Efficiency and lagging power factor (consumer notation) relation as a function of cost C for given constraints (From Fuchs, E.F.; Appelbaum, J.; Khan, I.A.; Höll, J.; Frank, U. V.: *Optimization of Induction Motor Efficiency, Volume 1: Three-Phase Induction Motors,* Publication of the Electric Power Research Institute, Palo Alto, California, EPRI EL-4152-CCM, July 1985, 336 pages, [43])

Fig. 10.49 Per-phase equivalent circuit of induction machine

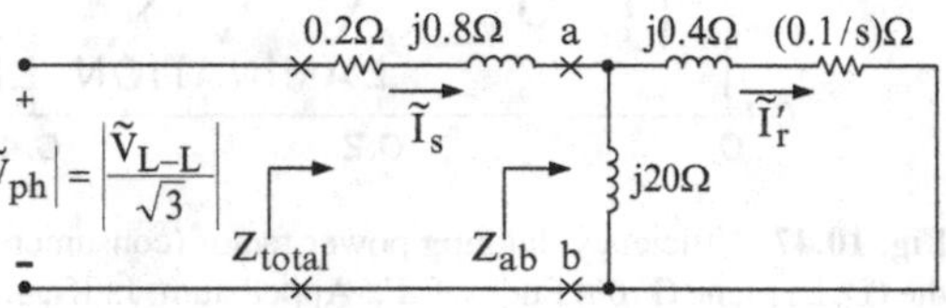

(c) The breakdown torque as a ratio of the full-load torque
(d) The sum of the core- and friction losses at full-load

Calculate for generator operation:

(e) The range of active load torque it can hold and corresponding speed range
(f) The maximum power it can generate
(g) The speed at the developed torque of −300 Nm.

Solution

(a) The rated current, torque, power factor, and efficiency can be obtained from the per-phase equivalent circuit Fig. 10.49.

$$\text{The rated phase voltage is } \left|\tilde{V}_{ph\_rated}\right| = 460\,\text{V}/\sqrt{3} = 265.6\,\text{V}, \qquad (10.161)$$

$$\text{the synchronous speed is } n_s = \frac{120f}{p} = \frac{120 \cdot 60}{6} = 1{,}200\,\text{rpm}, \qquad (10.162)$$

$$\text{the synchronous angular velocity is } \omega_s = \frac{2\pi \cdot n_s}{60} = \frac{2\pi \cdot 1{,}200}{60} = 125.66\,\text{rad/s}, \qquad (10.163)$$

$$\text{the rated slip is } s_{rated} = \frac{n_s - n_{m\_rated}}{n_s} = \frac{1,200 - 1,180}{1,200} = 0.0166. \quad (10.164)$$

With $\dfrac{R'_r}{s_{rated}} = \dfrac{0.1}{0.0166}\ \Omega = 6.024\ \Omega$ one obtains the impedance

$$Z_{ab} = \frac{(R'_r/s + jX'_r)jX_m}{(R'_r/s + jX'_r + jX_m)} = 5.67\angle 20.25^\circ\ \Omega, \quad (10.165)$$

or the input impedance $Z_{total} = R_s + jX_s + Z_{ab} = 6.16\angle 26.60^\circ\ \Omega.$ (10.166)

The rated stator current is now

$$\tilde{I}_{s\_rated} = \frac{|\tilde{V}_{L-L}|\angle 0^\circ}{\sqrt{3}\cdot Z_{total}} = \frac{460\angle 0^\circ}{\sqrt{3}\cdot 6.16\angle 26.62^\circ} = 43.11\angle -26.62^\circ\text{A}. \quad (10.167)$$

The rated rotor current is

$$\tilde{I}'_{r\_rated} = \left(\frac{jX_m}{R'_r/s_{rated} + jX'_r + jX_m}\right)\tilde{I}_{s\_rated} = 40.54\angle -10.17^\circ\text{A}. \quad (10.168)$$

The rated motor torque is

$$T_{rated} = \frac{3}{\omega_s}\left|\tilde{I}'_{r\_rated}\right|^2 \frac{R'_r}{s_{rated}} = 236.36\,\text{Nm}. \quad (10.169)$$

The rated power factor of the motor is obtained from $\tilde{I}_s = 43.11\angle -26.62^\circ$A as $\phi = -26.62^\circ$ or

$$\cos\phi = 0.894 \text{ lagging (consumer notation)}. \quad (10.170)$$

The rated efficiency is obtained from $P_{in\_rated} = \sqrt{3}\left|\tilde{V}_{L-L\_rated}\right|\cdot\left|\tilde{I}_{s\_rated}\right|\cos\phi = 30.71\,\text{kW}$ and $P_{out\_rated} = 27.3\,\text{kW}$ as

$$\eta_{rated} = \frac{P_{out\_rated}}{P_{in\_rated}} = \frac{27.3}{30.71} = 0.889. \quad (10.171)$$

(b) The starting torque and starting current as a ratio of their full-load values can be obtained for a slip of s=1 as follows:

$$Z_{ab} = \frac{(0.1 + j0.4)j20}{0.1 + j0.4 + j20} = 0.404\angle 76.24^\circ\ \Omega, \quad (10.172)$$

$$\begin{aligned} Z_{total} &= (0.2 + j0.8 + 0.09611 + j0.3924)\ \Omega \\ &= (0.29611 + j1.1924)\ \Omega \end{aligned} \quad (10.173)$$

resulting in the stator starting current

$$\tilde{I}_{s\_starting} = \frac{460\angle 0^\circ}{\sqrt{3}(0.29611 + j1.1924)} = 216.155\angle - 76.06^\circ \text{ A}, \quad (10.174)$$

or the current ratio

$$\frac{\left|\tilde{I}_{s\_starting}\right|}{\left|\tilde{I}_{s\_rated}\right|} = \frac{216.155}{43.11} = 5.01. \quad (10.175)$$

The rotor starting current is

$$\tilde{I}'_{r\_starting} = \left(\frac{jX_m}{R'_r + jX'_r + jX_m}\right)\tilde{I}_{s\_starting} = 211.92\angle - 75.78^\circ \text{ A}, \quad (10.176)$$

and the starting torque becomes

$$T_{starting} = \frac{3}{\omega_s}\left|\tilde{I}'_{r\_starting}\right|^2 R'_r = 107.22\,\text{Nm} \quad (10.177)$$

or the ratio

$$\frac{T_{starting}}{T_{rated}} = \frac{107.22}{228.15} = 0.47. \quad (10.178)$$

(c) The breakdown or maximum torque as a ratio of the full-load torque can be obtained from equivalent circuits Figs. 10.50a,b.
The Thevenin impedance Z_{TH} is obtained from Fig. 10.50a as

$$Z_{TH} = \frac{jX_m(R_s + jX_s)}{R_s + j(X_s + X_m)} = R_{TH} + jX_{TH}, \quad (10.179)$$

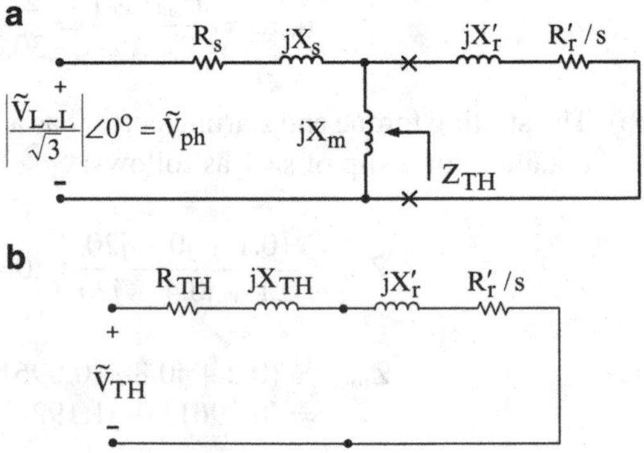

Fig. 10.50 (**a**) Definition of the Thevenin impedance Z_{TH} with $\tilde{V}_{ph} = 0$. (**b**) Thevenin equivalent circuit

where

$$R_{TH} = \frac{R_s X_m^2}{R_s^2 + (X_s + X_m)^2}, \tag{10.180a}$$

$$X_{TH} = \frac{X_m[R_s^2 + X_s(X_s + X_m)]}{R_s^2 + (X_s + X_m)^2}, \tag{10.180b}$$

and the Thevenin voltage magnitude

$$\left|\tilde{V}_{TH}\right| = \frac{(\left|\tilde{V}_{L-L\_rated}\right|/\sqrt{3})X_m}{\sqrt{R_s^2 + (X_s + X_m)^2}}. \tag{10.181}$$

The Thevenin parameters (10.180a,b) and (10.181) are represented in Fig. 10.50b, where $\tilde{V}_{TH} = \left|\tilde{V}_{TH}\right|\angle 0°$.

The breakdown torque can be derived from Fig. 10.51 as follows:

Maximum power transfer between stator and rotor occurs at slip $s = s_m$ where the maximum torque T_{max} occurs, if the load impedance Z_{load} and the air-gap impedance $Z_{air\ gap}$=Z_{out} are matched, that is, identical $Z_{load} = Z_{out} = Z_{air\ gap}$ resulting in the relation

$$\frac{R_r'}{s_m} = \pm\sqrt{R_{TH}^2 + (X_r' + X_{TH})^2}. \tag{10.182}$$

The air-gap real power is

$$P_g = 3\left|\tilde{I}_r'\right|^2 \frac{R_r'}{s} = T\omega_s, \tag{10.183}$$

where the torque T is

$$T = \frac{3}{\omega_s}\left|\tilde{I}_r'\right|^2 \frac{R_r'}{s}, \tag{10.184}$$

with the rotor current

$$\tilde{I}_r' = \frac{\tilde{V}_{TH}}{(R_{TH} + \frac{R_r'}{s}) + j(X_{TH} + X_r')} \tag{10.185}$$

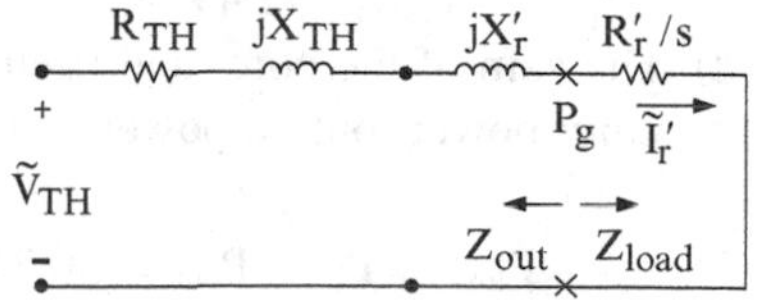

Fig. 10.51 Thevenin circuit for the derivation of breakdown torques in motoring and generation regions

one obtains for the torque

$$T = \frac{3}{\omega_s}\left[\frac{|\tilde{V}_{TH}|^2 \frac{R'_r}{s}}{(R_{TH} + \frac{R'_r}{s})^2 + (X_{TH} + X'_r)^2}\right]. \tag{10.186}$$

The maximum torque T_{max} occurs at the slip $s = s_m$. With (10.182) the torque becomes

$$T_{max} = \frac{3}{\omega_s}\left[\frac{|\tilde{V}_{TH}|^2 \frac{R'_r}{s_m}}{(R_{TH} + \frac{R'_r}{s_m})^2 + (X_{TH} + X'_r)^2}\right] \tag{10.187}$$

$$T_{max} = \frac{3}{\omega_s}\left[\frac{|\tilde{V}_{TH}|^2 \times \{\pm\sqrt{R_{TH}^2 + (X_{TH} + X'_r)^2}\}}{(R_{TH} \pm \sqrt{R_{TH}^2 + (X_{TH} + X'_r)^2})^2 + (X_{TH} + X'_r)^2}\right]$$

$$= \frac{3}{2\omega_s}\left[\frac{|\tilde{V}_{TH}|^2}{R_{TH} \pm \sqrt{R_{TH}^2 + (X_{TH} + X'_r)^2}}\right]. \tag{10.188}$$

The maximum torque in the motoring region is

$$T_{max}^{motoring} = \frac{3}{2\omega_s}\left[\frac{|\tilde{V}_{TH}|^2}{R_{TH} + \sqrt{R_{TH}^2 + (X_{TH} + X'_r)^2}}\right], \tag{10.189}$$

and in the (re-)generation region

$$T_{max}^{generation} = \frac{3}{2\omega_s}\left[\frac{|\tilde{V}_{TH}|^2}{R_{TH} - \sqrt{R_{TH}^2 + (X_{TH} + X'_r)^2}}\right]. \tag{10.190}$$

For the given induction machine parameters one obtains $R_{TH} = 0.185\ \Omega$, $X_{TH} = 0.771\ \Omega$, and $|\tilde{V}_{TH}| = 255.36\,V$, $T_{max}^{motoring} = 568\,Nm$, and $T_{max}^{motoring}/T_{rated} = 2.49$.

(d) The sum of the core- and friction losses at full-load can be obtained from the input power, output power and the ohmic losses.

$$P_{loss} = P_{in} - P_{out} = 30.71\,kW - 27.3\,kW = 3.41\,kW, \tag{10.191}$$

with the stator and rotor copper losses the friction/windage and core losses become

$$P_{\text{friction/windage,core}} = P_{\text{loss}} - P_{\text{cu s}} - P_{\text{cu r}} = P_{\text{loss}} - 3\left|\tilde{I}_s\right|^2 R_s - 3\left|\tilde{I}'_r\right|^2 R'_r. \\ = (3.41 - 1.12 - 0.493)\,\text{kW} = 1.797\,\text{kW}. \tag{10.192}$$

(e) The range of active load torque it can hold and corresponding speed range: with (10.182) the maximum slip in the generation region is

$$s_m^{\text{generation}} = \frac{R'_r}{-\sqrt{R_{TH}^2 + (X_{TH} + X'_r)^2}} = -0.084 \tag{10.193a}$$

and using (10.190) the maximum torque in the generation region is $T_{\max}^{\text{generation}} = -778.07\,\text{Nm}$. Note that the maximum torque in the generation region is (much) larger than that in the motoring region. The range of active load is from 0 to -778.07 Nm and the speed range is from 1,200 rpm to $n_{m\_\max}^{\text{generation}} = n_s(1 - s_m^{\text{generation}}) = 1{,}200(1 + 0.084) = 1{,}300.8\,\text{rpm}$. The maximum angular velocity is $\omega_{m\_\max}^{\text{generation}} = \frac{2\pi \cdot n_{m\_\max}^{\text{generation}}}{60} = 136.22\,\text{rad/s}$.

(f) The maximum power it can generate is

$$P_{\max}^{\text{generation}} = T_{\max}^{\text{generation}}\,\omega_{m\_\max}^{\text{generation}} + P_{\text{loss}}^{\text{generation}}. \tag{10.193b}$$

From Fig. 10.52 one gets for the indicated impedances

$$Z_{ab}^{\text{generation}} = \frac{\left(\frac{R'_r}{s_m^{\text{generation}}} + jX'_r\right) jX_m}{\frac{R'_r}{s_m^{\text{generation}}} + jX'_r + jX_m} = 1.23\angle 158.1^\circ\ \Omega, \tag{10.194}$$

$$Z_{\text{total}} = Z_{ab}^{\text{generation}} + R_s + jX_s = 1.572\angle - 233.25^\circ\ \Omega. \tag{10.195}$$

The stator current is for generation

$$\tilde{I}_s^{\text{generation}} = \frac{\tilde{V}_{ph}}{Z_{\text{total}}} = \frac{460/\sqrt{3}\angle 0^\circ}{1.572\angle - 233.25^\circ} = 168.95\angle - 126.8^\circ\text{A}, \tag{10.196}$$

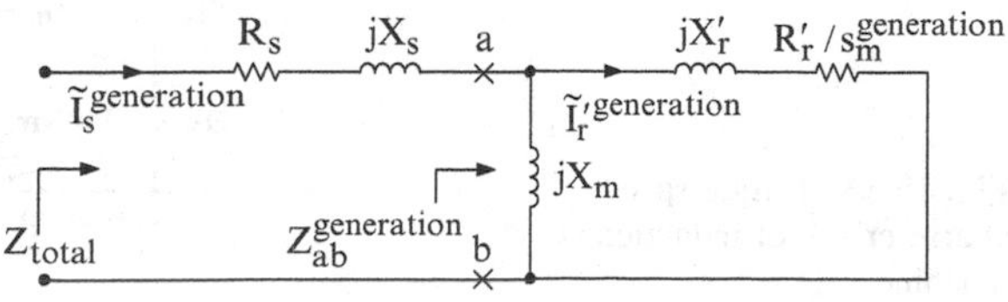

Fig. 10.52 Impedance definitions for maximum power generation

the corresponding rotor current is

$$\tilde{I}_r'^{\,generation} = \left(\frac{jX_m}{\frac{R'_r}{s_m} + jX'_r + jX_m} \right) \tilde{I}_s^{generation} = 165.36\angle -130.14^\circ A. \qquad (10.197)$$

The losses during (re)generation are

$$P_{loss}^{generation} = 3\left(\left|\tilde{I}_s\right|^2\right)R_s + 3\left(\left|\tilde{I}'_r\right|^2\right)R'_r + P_{friction/windage\ core}$$

$$= 17.13\,kW + 8.2\,kW + 1.797\,kW = 27.13\,kW, \qquad (10.198)$$

and the maximum power during generation operation is

$$P_{max}^{generation} = (-778.07 \cdot 136.22)\,W + 27.13\,kW = -78.86\,kW. \qquad (10.199)$$

(g) The speed at the developed load torque of $T_L = -300$ Nm is given by the induction machine torque (neglecting frictional and windage torques)

$$T_L = -300\,Nm = T = \frac{3}{\omega_s}\left[\frac{\left|\tilde{V}_{TH}\right|^2 \frac{R'_r}{s}}{\left(R_{TH} + \frac{R'_r}{s}\right)^2 + (X_{TH} + X'_r)^2}\right]. \qquad (10.200)$$

Let $R'_r/s = x$ from this follows $-300 = \frac{3}{125.66}\left[\frac{(255.36)^2 x}{(0.185+X)^2+(0.771+0.4)^2}\right]$ or the second-order equation $x^2 + 5.56x + 1.405 = 0$ leading to the two solutions $x_1 = -0.265$ and $x_2 = -5.295$ or the two slips $s_1 = -0.377$ and $s_2 = -0.019 = s_L$, slip s_1 leads to an unstable operation, as can be seen from Chap. 11 and Fig. 10.53. The speed of the induction generator at $T_L = -300$ Nm is $n_m = 1{,}200\,(1{+}0.019) = 1{,}222.8$ rpm.

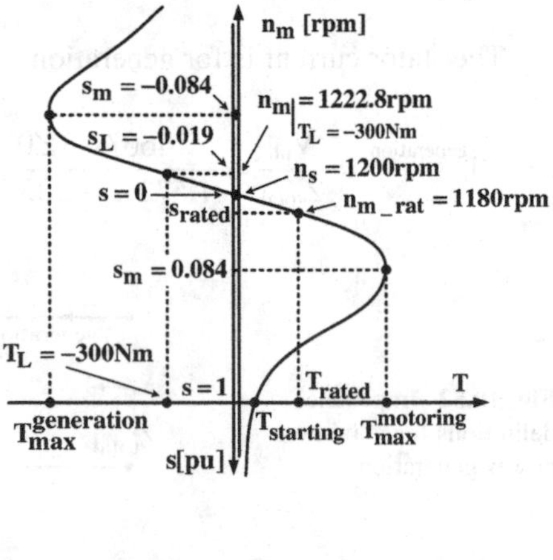

Fig. 10.53 Torque-speed characteristic of induction machine

Figure 10.54 illustrates the magnetic field during starting (slip s = 1) of a three-phase induction motor and Fig. 10.55 shows the corresponding starting currents and torques as a function of the starting voltage. The latter graph teaches that the skin effects for the calculation of the rotor resistance and leakage reactance cannot be neglected due to the deep-bar, squirrel-cage rotor winding [7].

Construction elements of a three-phase induction machine with squirrel-cage winding: Three-phase induction motors consume about 63% of the total installed power capacity of 1,000 GW within the US and, therefore are very important load

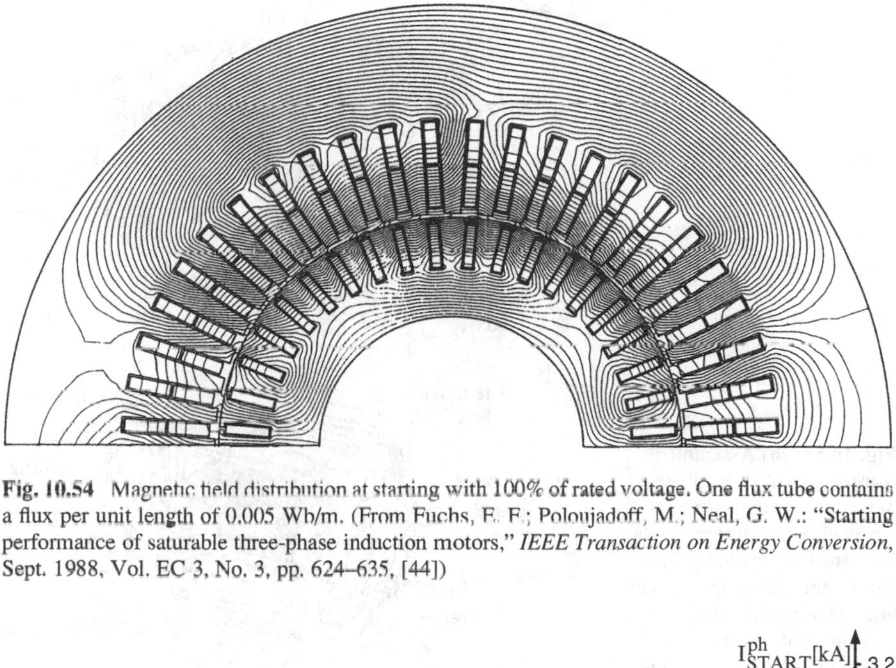

Fig. 10.54 Magnetic field distribution at starting with 100% of rated voltage. One flux tube contains a flux per unit length of 0.005 Wb/m. (From Fuchs, E. F.; Poloujadoff, M.; Neal, G. W.: "Starting performance of saturable three-phase induction motors," *IEEE Transaction on Energy Conversion*, Sept. 1988, Vol. EC 3, No. 3, pp. 624–635, [44])

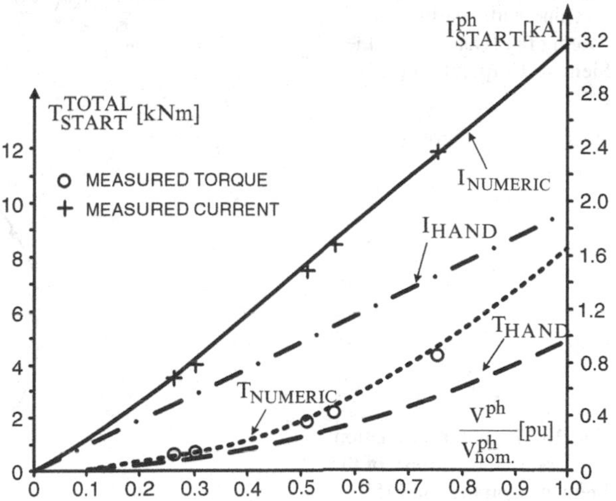

Fig. 10.55 Starting currents and torques as a function of the terminal voltage. (From Fuchs, E. F.; Poloujadoff, M.; Neal, G. W.: "Starting performance of saturable three-phase induction motors," *IEEE Transaction on Energy Conversion*, Sept. 1988, Vol. EC-3, No. 3, pp. 624–635, [44])

components within the power system. This type of machine consists of stator frame, stator core and stator winding, and the rotor core with the rotor winding. The stator and rotor core consist of laminations to suppress eddy currents. Figures 10.56a, b show the assembly of the entire machine, which can be used either as a motor or as a generator [45].

The stator is the stationary part of the machine's electromagnetic circuit, and is made up of thin metal sheets (so-called laminations) with semi-closed stator slots as depicted in Fig. 10.57 [45]. Laminations are used to reduce eddy currents (see Chap. 6) within the core thus increasing the efficiency of the machine.

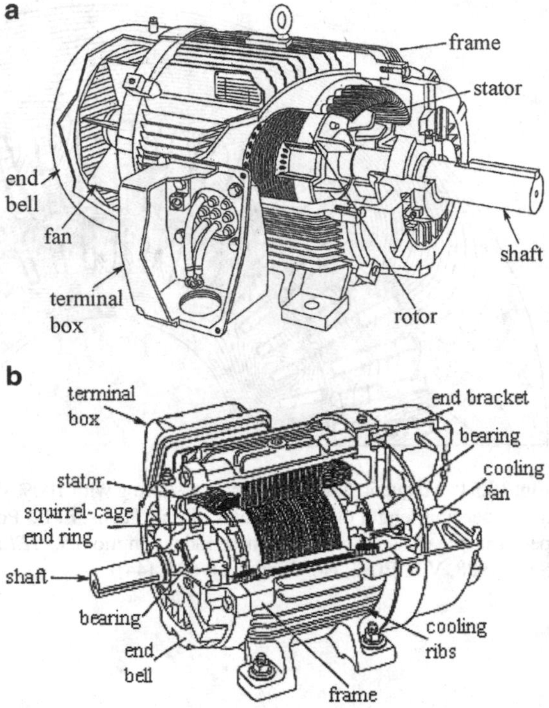

Fig. 10.56 (**a**) Assembled components of three-phase induction machine with terminal box (Courtesy of the Siemens Corporation [45]). (**b**) Assembled components of three-phase induction machine with bearing and cooling fan (Courtesy of the Siemens Corporation [45])

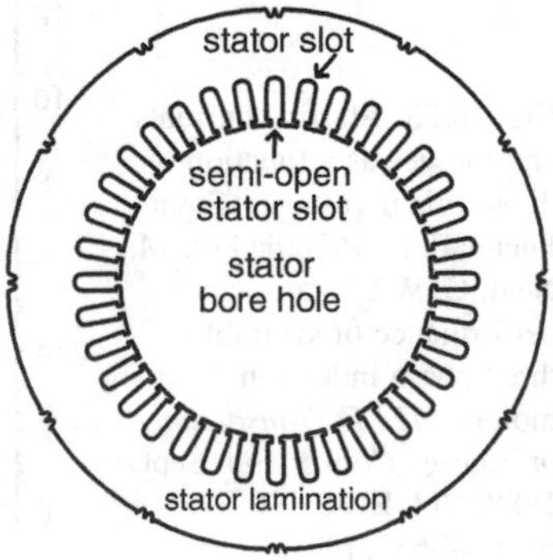

Fig. 10.57 Stator lamination cross-section (Courtesy of the Siemens Corporation [45])

The stator laminations are stacked together to form a hollow cylinder as illustrated in Figs. 10.58a, b, and coils of insulated wire are inserted into the slots of the stator core. Figure 10.58a shows the location of one coil and Fig. 10.58b illustrates the completed stator winding [45]. When the assembled machine is in operation, the

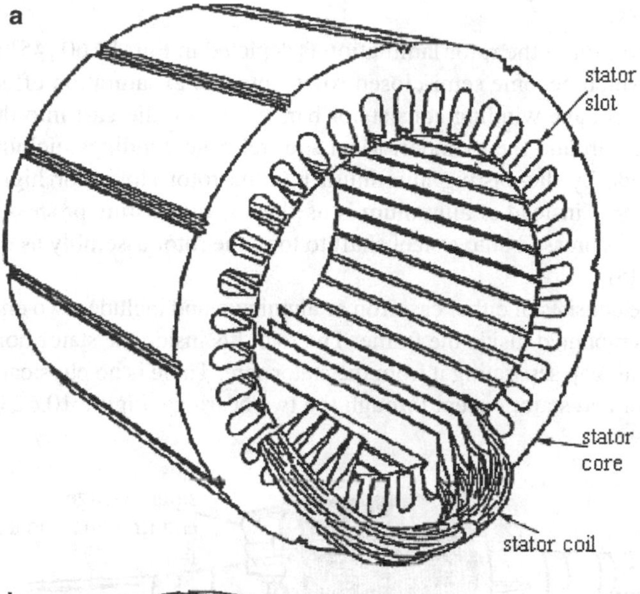

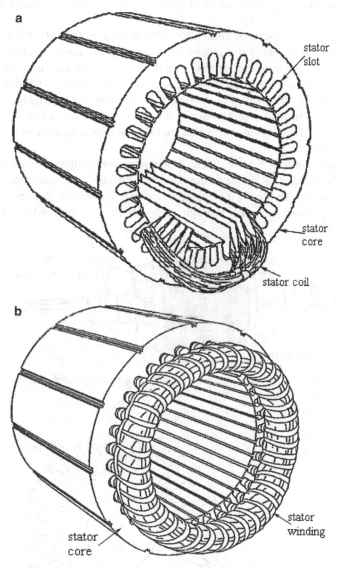

Fig. 10.58 (**a**) Location of one stator coil within stator slots (Courtesy of the Siemens Corporation [45]). (**b**) Assembled three-phase winding within stator slots (Courtesy of the Siemens Corporation [45])

stator windings are connected directly to the three-phase voltage source or to an inverter generating a rotating field within the machine.

The rotor is the rotating member of the machine's electromagnetic circuit. The most common type of rotor used in a three-phase machine is the squirrel-cage winding consisting of short-circuited conducting bars which can be skewed. The rotor core is made by stacking steel laminations to form a cylinder shown in Fig. 10.59 [45].

A cross-section of the rotor lamination is depicted in Fig. 10.60 [45] with closed rotor slots, which become semi-closed rotor slots due to saturation effects.

The squirrel-cage winding consists of bars which are die-cast into the slots and evenly spaced around the cylinder. Most squirrel-cage windings including the end rings are made by die-casting aluminum into the rotor slots. For high efficiency machines copper instead of aluminum is used in the die-casting process. Thereafter the rotor core is pressed onto a steel shaft to form the rotor assembly as illustrated in Fig. 10.61 [45]

The frame consists of either cast iron or aluminum and includes two end bells. The stator core is mounted inside the frame. The rotor fits inside the stator bore hole with a very small air gap separating it from the stator core. There is no physical connection between rotor and stator except through the two bearings. Figure 10.62 [45] depicts

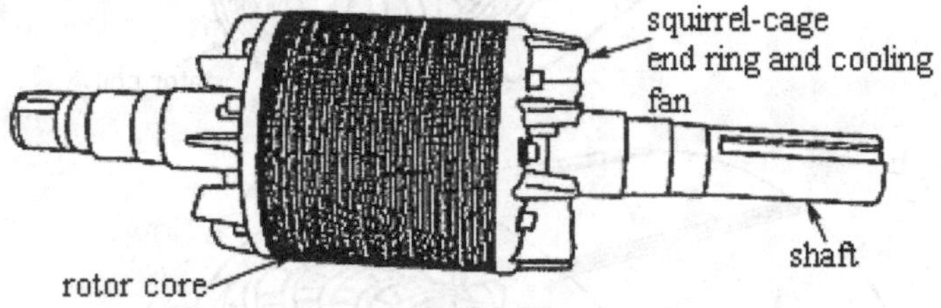

Fig. 10.59 Rotor assembly consisting of core, squirrel-cage winding, and shaft (Courtesy of the Siemens Corporation [45])

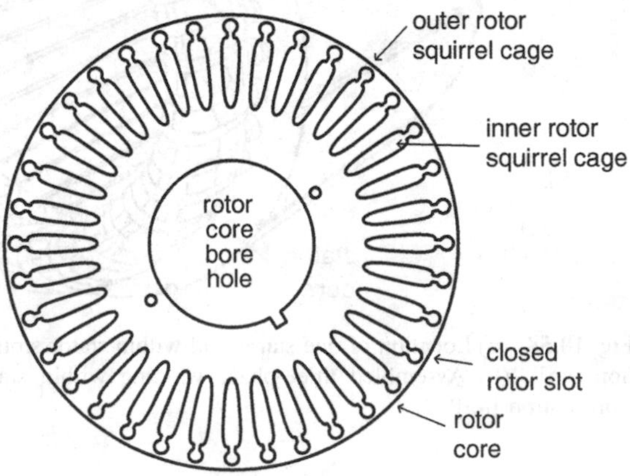

Fig. 10.60 Rotor lamination with closed rotor slots (Courtesy of the Siemens Corporation [45])

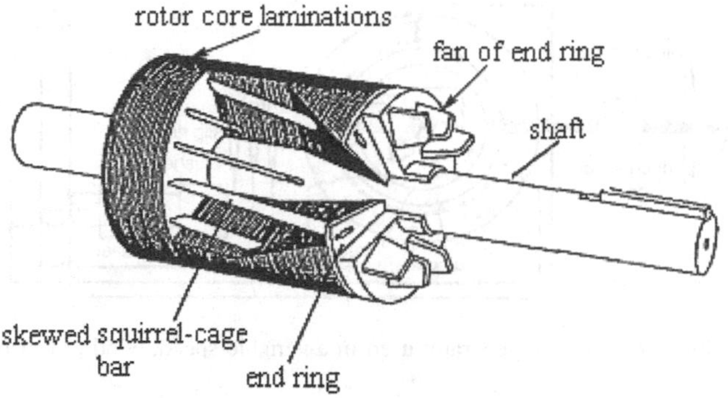

Fig. 10.61 Assembled rotor (Courtesy of the Siemens Corporation [45])

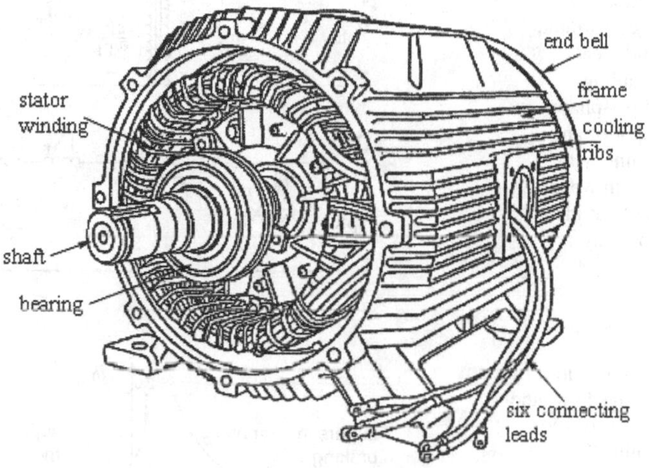

Fig. 10.62 Machine frame without left end bell (Courtesy of the Siemens Corporation [45])

the frame with the left end bell removed. Bearings mounted on the shaft support the rotor and allow it to rotate. Some motors use a fan mounted on the rotor shaft to cool the motor when the shaft rotates. The outside of the frame has ribs to increase the surface area of the frame, enhancing the cooling of the machine [23].

The doubly fed induction machine: The doubly fed induction machine has found an important application in variable-speed wind- and hydro-power generation in the form of the doubly fed induction generator (DFIG) as shown in Fig. 10.63 [23]. To minimize the rating of the solid-state rectifier and the PWM inverter the three-phase stator winding is directly connected to the power system and the rectifier/inverter combination is supplying/extracting power into/from the rotor circuit of the induction generator [46–57]. Figures 10.64a, b depict the per-phase equivalent circuit of an induction machine with an injected voltage in the wound-rotor circuit. Figure 10.65 [36] illustrates the speed control by injecting a voltage in the rotor winding of the induction machine.

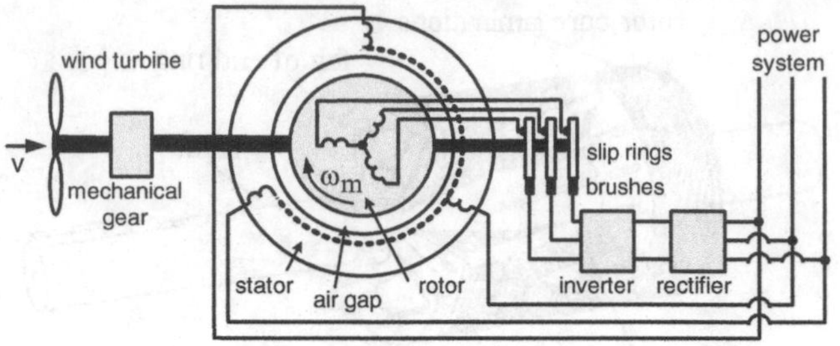

Fig. 10.63 Doubly fed induction generator used in a variable-speed, wind-power plant

Fig. 10.64 (**a**) Per-phase equivalent circuit of an induction machine with injected voltage in wound-rotor circuit and ideal transformer coupling. (**b**) Simplified per-phase equivalent circuit of an induction machine with injected voltage in wound-rotor circuit without ideal transformer coupling

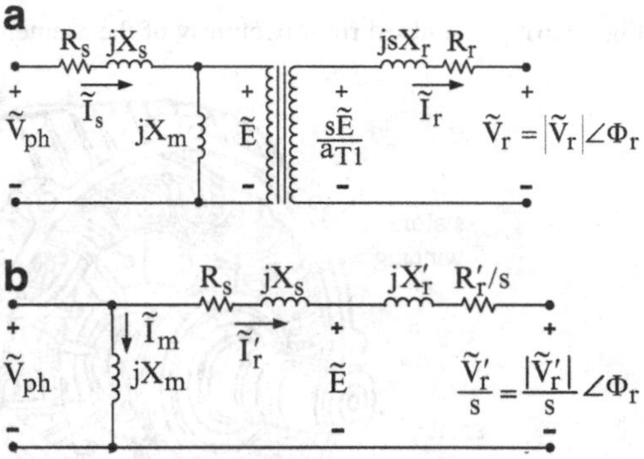

Fig. 10.65 Speed control by injection of voltage in wound-rotor winding of a 10 kW induction machine

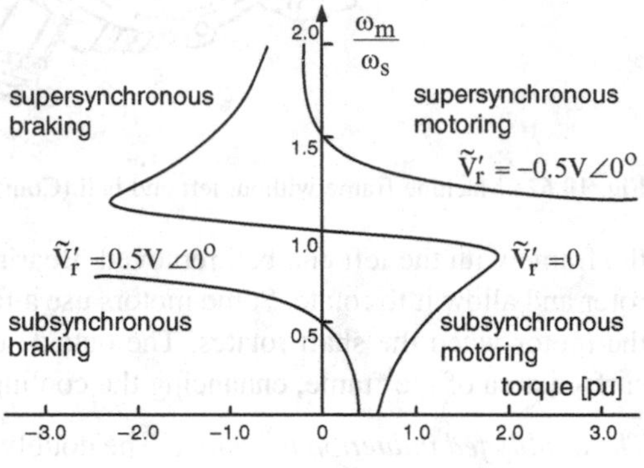

Variable-speed induction generators are applied in wind-power plants and in hydro (pump-storage) plants [58, 59]. In contrast to constant-speed machines these plants can control the output power within a few cycles due to the change in the stored energy within the rotating rotor. The required rotor excitation apparent power S_{rotor} via an inverter is about 8% of the stator apparent power S_{stator}, that is, for a hydro power plant rated $S_{stator} = 600$ MVA, the rotor power is $S_{rotor} = 48$ MVA for a variable-speed range of $\pm 10\%$ from the rated operating point.

Application Example 10.14: Operation of Induction Machine (Motor or Generator) with Injected Voltage in Wound-Rotor Circuit as Used for State-of-the-Art, Variable-Speed, Wind-Power (and Hydro-Power) Plants

A three-phase $P_{out} = 5.7$ MW, $V_{L\text{-}L} = 4$ kV, f = 60 Hz, 8-pole Y-connected wound-rotor induction machine has the parameters: $R_s = 0.02\ \Omega$, $X_s = 0.3\ \Omega$, $R_r' = 0.05\ \Omega$, $X_r' = 0.4\ \Omega$, $X_m \to \infty$, and a stator-to-rotor turns ratio $a_{T1} = 2$. The rated slip of this machine when operation as a motor is $s_{rat} = 0.02$.

When the above induction machine is operated as a motor with unexcited (short-circuited) rotor, that is, $\tilde{V}_r' = 0$ determine:

(a) Phase voltage $\left|\tilde{V}_{s\_ph\_rat}\right|$,
(b) Synchronous mechanical speed n_{s_rat},
(c) Synchronous angular velocity ω_{s_rat}
(d) Speed of the shaft n_{m_rat}
(e) Rated stator current and rated torque
(f) When the above machine is operated with injected voltage $\tilde{V}_r' = \left|\tilde{V}_r'\right| \angle \Phi_r$ in the rotor circuit compute the rotor current $\tilde{I}_r' = \left|\tilde{I}_r'\right| \angle \theta_r$ and the torque T for the rotor voltage magnitudes $\left|\tilde{V}_r'\right| = 350$ V, and 700 V with the phase angles $\phi_r = 0°$, $\pm 15°$, and $\pm 30°$ provided the induction machine operates at $n_{m\_new} = 600$ rpm. What is the required frequency of the rotor voltages/currents f_r if the stator frequency is f = 60 Hz?
(g) Determine the ratio of the mechanical gear if the rated speed of the wind turbine is $n_{m\_wind\ turbine} = 20$ rpm.

Solution

(a) The rated phase voltage is

$$V_{s\_ph\_rat} = \frac{V_{L-L}}{\sqrt{3}} = \frac{4,000\ \text{V}}{\sqrt{3}} = 2,309.5\ \text{V}. \tag{10.201}$$

(b) The rated synchronous speed is

$$n_{s\_rat} = \frac{120}{p} f = \frac{120}{8} 60 = 900\ \text{rpm}. \tag{10.202}$$

(c) The synchronous rated angular velocity is

$$\omega_{s\_rat} = \frac{2\pi}{60} n_{s\_rat} = \frac{2\pi}{60} 900 = 94.25\ \text{rad/s}. \tag{10.203}$$

(d) The rated speed is

$$n_{m\_rat} = n_{s\_rat}(1 - s_{rat}) = 900(1 - 0.02) = 882\ \text{rpm}. \tag{10.204}$$

(e) The rated motor current and torque can be calculated as follows:

$$\tilde{I}_s = \frac{\tilde{V}_{s\_ph}}{(R_s + \frac{R'_r}{s}) + j(X_s + X'_r)}, \tag{10.205}$$

and the induction machine torque becomes

$$T = \frac{3}{\omega_s}\left[\frac{(V_{s\_ph})^2\left(\frac{R'_r}{s}\right)}{\left(R_s + \frac{R'_r}{s}\right)^2 + (X_s + X'_r)^2}\right]. \tag{10.206}$$

For rated motor operation one obtains the stator current

$$\tilde{I}_{s\_rat} = \frac{2,309.5\angle 0^\circ}{(0.02 + \frac{0.05}{0.02}) + j(0.3 + 0.4)} = \frac{2309.5\angle - 0.27095\,\text{rad}}{\sqrt{(2.52)^2 + (0.7)^2}}$$

$$= 883.03\,\text{A}\angle - 0.27095\,\text{rad} \tag{10.207a}$$

and the rated induction machine (motor operation) torque

$$T_{rat} = \frac{3}{94.25}\left[\frac{(2309.5)^2\left(\frac{0.05}{0.02}\right)}{\left(0.02 + \frac{0.05}{0.02}\right)^2 + (0.3 + 0.4)^2}\right] = 62,048.94\,\text{Nm}. \tag{10.207b}$$

(f) If the new speed at subsynchronous operation is given as $n_{m\_new} = 600$ rpm then the new slip is

$$s = s_{new} = \frac{900 - 600}{900} = 0.3333. \tag{10.208}$$

Since the speed is to be reduced, $\tilde{V}'_r$ must be positive (see Fig. 10.65). From the equivalent circuit of Fig. 10.64b one obtains for the rotor current which is identical to the stator current for $X_m \to \infty$

$$\tilde{I}'_r = \tilde{I}_s = \frac{\left|\tilde{V}_{s\_ph\_rat}\right|\angle 0^\circ - \frac{|\tilde{V}'_r|\angle\phi_r}{s}}{(R_s + \frac{R'_r}{s}) + j(X_s + X'_r)} = \frac{2,309.5 - 3\left|\tilde{V}'_r\right|\angle\phi_r}{(0.02 + 0.15) + j0.7}$$

$$= \frac{(2,309.5 - 3\left|\tilde{V}'_r\right|\angle\phi_r)\angle - \theta_z}{0.7204} \tag{10.209}$$

where $\theta_z = \tan^{-1}\left(\frac{0.7}{0.17}\right) = 1.333$ rad
or

$$\tilde{I}'_r = \tilde{I}_s = (3205.86 - 4.1644\left|\tilde{V}'_r\right|\angle\,\phi_r)\angle - 1.333\,\text{rad}. \tag{10.210}$$

This equation indicates that by controlling ϕ_r, the induction machine power factor (consumer notation) can be changed.

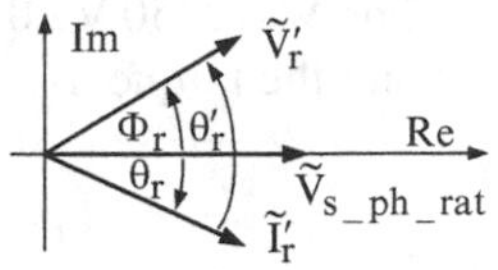

Fig. 10.66 Phasor diagram for doubly fed induction machine (either motor or generator) where θ_r is negative)

Taking the stator phase voltage as reference the phasor diagram of Fig. 10.66 is obtained.
The torque of an induction machine is defined by the air-gap power P_g and the synchronous angular velocity ω_s

$$T = \frac{P_g}{\omega_s} \tag{10.211}$$

The mechanical output power is

$$P_m = T\omega_m = \frac{P_g}{\omega_s}(1 - s)\omega_s = (1 - s)P_g. \tag{10.212}$$

The electrical dissipated or absorbed power P_e of the rotor circuit is

$$P_e = P_g - P_m = P_g - (1 - s)P_g = sP_g, \tag{10.213}$$

it is the sum of the power absorbed by $\tilde{V}_r'$ that is P_r, and the rotor copper loss P_{loss_r}. For motor operation both power components are positive. Therefore, $sP_g = (P_{loss\_r} + P_r)$ or the air-gap power is

$$P_g = \frac{(P_{loss\_r} + P_r)}{s} = 3\left[\left|\tilde{I}'_r\right|^2\left(\frac{R_r'}{s}\right) + \left(\frac{\left|\tilde{V}'_r\right|}{s}\right)\left|\tilde{I}'_r\right|\cos\theta_r'\right], \tag{10.214}$$

θ_r' is the phase angle between $\tilde{V}'_r$and $\tilde{I}'_r$. From the phasor diagram of Fig. 10.66 one obtains

$$\theta_r' = (\phi_r - \theta_r). \tag{10.215}$$

The machine torque is

$$\begin{aligned} T_{subsynchronous} &= T = \frac{P_g}{\omega_s} = \frac{3}{s\omega_s}\left[\left|\tilde{I}'_r\right|^2 R'_r + \left|\tilde{V}'_r\right|\left|\tilde{I}'_r\right|\cos(\phi_r - \theta_r)\right], \\ &= \frac{3}{0.3333 \cdot 94.25}\left[\left|\tilde{I}'_r\right|^2 0.05 + \left|\tilde{V}'_r\right|\left|\tilde{I}'_r\right|\cos(\phi_r - \theta_r)\right] \\ &= 0.0955\left[\left|\tilde{I}'_r\right|^2 0.05 + \left|\tilde{V}'_r\right|\left|\tilde{I}'_r\right|\cos(\phi_r - \theta_r)\right], \end{aligned} \tag{10.216}$$

where θ_r is negative.

For $\tilde{V}'_r = 350\,\text{V}\angle 0°$ and the slip $s = s_{new} = 0.3333$ one obtains the current $\tilde{I}'_r$ and the torque T:

$$\begin{aligned}\tilde{I}'_r &= (3205.86 - 4.1644 \cdot 350)\angle -1.333\ \text{rad}\\ &= 1,748.32\text{A}\angle -1.333\,\text{rad}\end{aligned} \tag{10.217}$$

or in pu (per unit)

$$\frac{\left|\tilde{I}'_r\right|}{\left|\tilde{I}_{s\_rat}\right|} = 1.98\,\text{pu}, \tag{10.218}$$

$$\begin{aligned}T &= 0.0955[(1748.32)^2 0.05 + 350 \cdot 1,748.32 \cdot \cos(+1.333)]\\ &= 28,212.53\,\text{Nm}\end{aligned} \tag{10.219}$$

or referred to the rated torque

$$\frac{T}{T_{rat}} = 0.4547\,\text{pu}. \tag{10.220}$$

For $\tilde{V}'_r = 350\,\text{V}\angle 30°$ and the slip $s = s_{new} = 0.3333$ one obtains the current $\tilde{I}'_r$ and the torque T:

$$\begin{aligned}\tilde{I}'_r &= [3,205.86 - 4.1644 \cdot 350\{\cos(30°) + j\sin(30°)\}]\angle -1.333\ \text{rad}\\ &= [3,205.86 - 1,262.23 - j728.77]\angle -1.333\ \text{rad}\\ &= [1943.63 - j728.77]\angle -1.333\ \text{rad}\\ &= 2,075.77\angle -1.708\ \text{rad}\end{aligned} \tag{10.221}$$

or in pu

$$\frac{\left|\tilde{I}'_r\right|}{\left|\tilde{I}_{s\_rat}\right|} = 2.35\,\text{pu}, \tag{10.222}$$

$$\begin{aligned}T &= 0.0955[(2075.77)^2 0.05 + 350 \cdot 2,075.77 \cdot \{\cos(0.52358\,\text{rad} + 1.708\,\text{rad})]\\ &= -22,008.15\,\text{Nm}\end{aligned} \tag{10.223}$$

or referred to the rated torque

$$\frac{T}{T_{rat}} = -0.35\,\text{pu}. \tag{10.224}$$

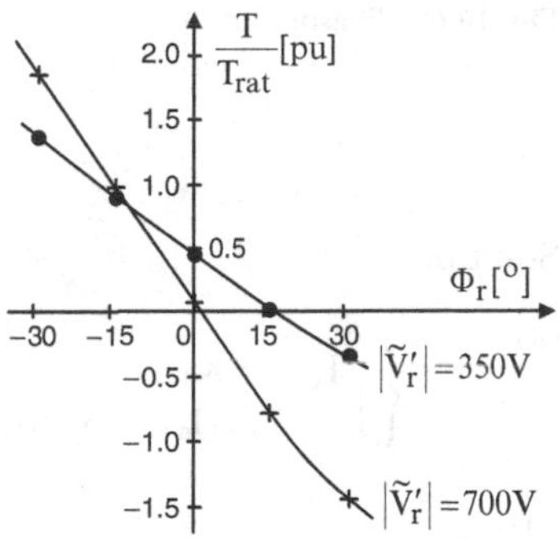

Fig. 10.67 Per-unit subsynchronous torque $T/T_{rat} = T_{subsynchronous}/T_{rat}$ as a function of ϕ_r for the slip $s_{new} = 0.3333$ corresponding to a subsynchronous speed of $n_{m\_new} = 600$ rpm where $|\tilde{V}'_r|$ is parameter

Figure 10.67 shows the per-unit subsynchronous torque $T/T_{rat} = T_{subsynchronous}/T_{rat}$. as a function of ϕ_r for the slip $s_{new} = 0.3333$ corresponding to a subsynchronous speed of $n_{m\_new} = 600$ rpm where $|\tilde{V}'_r|$ is parameter.

Note that for any slip $s = s_{new}$ similar torque relations as depicted in Fig. 10.67 can be obtained as a function of the phase angle ϕ_r of the rotor voltage $\tilde{V}'_r$, where for example $|\tilde{V}'_r| = 350$ V is parameter. With the stator-to-rotor turns ration $a_{T1} = 2$ one obtains the rotor voltage $\tilde{V}_r = \tilde{V}'_r/a_{T1}$ or for example for $|\tilde{V}'_r| = 350$ V the voltages $\tilde{V}_r = 175V\angle 0°$, and $\tilde{V}_r = 175\,V\angle 30°$ as demonstrated in the above example calculations. The required frequency of the rotor voltages/currents is $f_r = s\cdot f = s_{new}\cdot f = 0.3333\cdot 60$ Hz $= 19.998$ Hz.

(g) The ratio of the mechanical gear RG is

$$RG = \frac{n_{s\_rat}}{n_{m\_windturbine}} = \frac{900}{20} = 45. \quad (10.225)$$

Application Example 10.15: Speed-Control of Brushless DC Motor Drive

For a 2-pole, 3-phase brushless DC motor (or permanent-magnet AC motor fed by an inverter) drive, the torque and the voltage constants are 0.6 and 0.5 (in MKSAK system of units), respectively. The synchronous inductance is $L_s = 10$ mH and the winding resistance is neglected. The motor is supplying a load under balanced sinusoidal steady-state conditions. If the magnitude of per-phase output voltage across the inverter supplying the motor is $|\tilde{V}_{a\_ph}| = 180$ V and the phase current is $|\tilde{I}_a| = 6$ A, compute:

(a) The torque supplied by the motor.
(b) The speed (in rpm) of the drive, assuming $\theta_m(0) = -90°$ where $\theta_m(0)$ is the phase angle of the rotor-produced flux-density space vector ($\vec{B}_r(t) = \vec{B}_r\angle\theta_m(t)$) at time $t = 0$ (Eq. 10-1 of [60]).
(c) Draw the phasor diagram, showing $\tilde{V}_{a\_ph}$ and $\tilde{I}_a$.

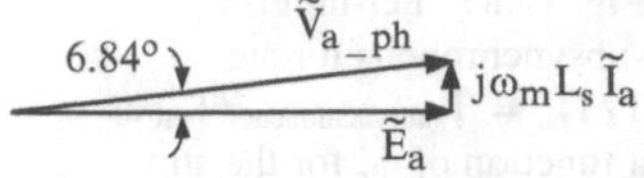

Fig. 10.68 Phasor diagram

Solution

(a)

$$\begin{cases} T_{em} = K_T I_s \\ I_s = 3/2(I_a) = 3/2(6) = 9\text{A} \end{cases} \Rightarrow T_{em} = 0.6(9) = 5.4\,\text{Nm}. \quad (10.226)$$

(b)

$$E_a = K_E \omega_m = 0.5\omega_m. \quad (10.227)$$

If $\theta_m(0) = -90°$, then $\Phi_{i_s}|_{t=0} = \theta_m(t) + (\pi/2) = 0$ (Eq. 10-10 of [60]) $\Rightarrow I_a = 6\angle 0°$ A and,

$$\begin{aligned} V_{a\_ph}\angle\,\Phi &= \tilde{E}_a + j\omega_m L_s \tilde{I}_a \Rightarrow 180\cos\Phi + j180\sin\Phi \\ &= 0.5\omega_m + j\omega_m(0.01)(6\angle 0°) \end{aligned} \quad (10.228)$$

$$\Rightarrow \begin{cases} 180\cos\Phi = 0.5\omega_m \\ 180\sin\Phi = 0.06\omega_m \end{cases} \Rightarrow \tan\Phi = 0.06/0.5 = 0.12 \Rightarrow \Phi = 6.84°.$$

$$\omega_m = 180\cos\Phi/0.5 = 180\cos 6.84/0.5 = 357.43\,\text{rad/s} \Rightarrow n_m = \frac{357.43(60)}{2\pi}$$
$$= 3{,}413\,\text{rpm}.$$

(c) The phasor diagram is shown in Fig. 10.68.

Application Example 10.16: Torque Developed by a 3-Phase Induction Motor as a Function of Supply Voltage Variation

A 230 V, 3-phase, Δ-connected, 4-pole, 60 Hz induction motor operates at a full-load speed of $n_m = 1{,}710$ rpm. The output power at this speed is $P_{out} = 2$ hp and the rotor current is $|I'_r| = 4.5$ A.

(a) Determine the slip s, and the developed torque T.
(b) If the supply voltage increases by 10%, determine torque T_1 and stator current $\tilde{I}_{s1}$.
(c) If the supply voltage decreases by 10%, determine torque T_2 and stator current $\tilde{I}_{s2}$.

Solution

(a)

$$n = \frac{120f}{P} = \frac{120(60)}{4} = 1{,}800\,\text{rpm} \Rightarrow s = \frac{1{,}800 - 1{,}710}{1{,}800} = 0.05. \quad (10.229)$$

$$T_d = \frac{P_d}{\omega} = \frac{2(746)}{1,710(2\pi/60)} = 8.33\,\text{Nm}. \qquad (10.230)$$

(b) Since $P_d \propto V^2/R \propto V^2$, for $P_d = 8.33$ Nm at one obtains

$$T_1 = \frac{[230(1+0.1)]^2}{230^2}(8.33) = 10.08\,\text{Nm}. \qquad (10.231)$$

Since $I_r' \propto V_t/Z \propto V_t$, for $I_r' = 4.5$ A at one gets

$$I_{s1} = \frac{[230(1+0.1)]}{230^2}(4.5) = 4.95\,\text{A}. \qquad (10.232)$$

(c) Using the same approach:

$$T_2 = \frac{[230(1-0.1)]^2}{230^2}(8.33) = 6.75\,\text{Nm}, \quad \text{and} \qquad (10.233)$$

$$I_{s2} = \frac{[230(1-0.1)]}{230^2}(4.5) = 4.05\,\text{A}. \qquad (10.234)$$

Application Example 10.17: Efficiency and Starting Performance of a 3-Phase Induction Motor

A 3-phase, four-pole, 60 Hz induction motor is rated at $P_{out} = 10$ hp, 208 V and 1,755 rpm. The parameters of the approximate equivalent circuit (neglecting core-loss resistance R_c) referred to the stator are as follows: $R_1 = 0.15\ \Omega$, $R'_2 = 0.15\ \Omega$, $X_1 = 0.4\ \Omega$, $X'_2 = 0.25\ \Omega$, and $X_m = 30\ \Omega$. The rotational losses (including frictional, windage, and core losses) are 500 W. The motor operates at rated speed when connected to a 208 V and 60 Hz source. Assume X_m is connected at the front end of the equivalent circuit. Calculate:

(a) Synchronous speed and slip, draw the equivalent circuit.
(b) Line current and input power factor (consumer notation).
(c) Developed internal power (P_g) and torque (T_g).
(d) Output power in hp, output torque.
(e) Efficiency.

Solution

(a)

$$n_{syn} = \frac{120f}{P} = \frac{120(60)}{4} = 1,800\,\text{rpm} \Rightarrow s = \frac{n_{syn} - n_m}{n_{syn}}$$

$$= \frac{1,800 - 1,755}{1,800} = 0.025. \qquad (10.235)$$

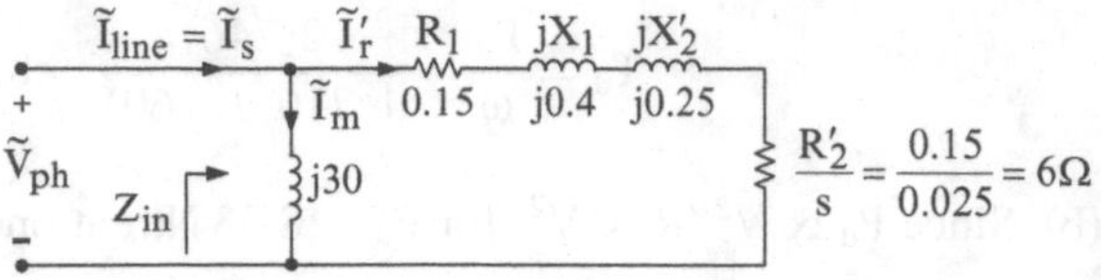

Fig. 10.69 Equivalent circuit

The equivalent circuit is shown in Fig. 10.69.

(b)

$$Z_{in} = \frac{1}{\frac{1}{j30} + \frac{1}{(0.15+6)+j(0.4+0.25)}} = 5.93\ \Omega\angle 17.38^\circ. \quad (10.236)$$

$$\tilde{I}_{line} = \frac{\tilde{V}_{ph}}{Z_{in}} = \frac{(208/\sqrt{3})\angle 0^\circ}{5.93\angle 17.38} = 20.25\ \text{A}\angle -17.30^\circ \Rightarrow I_{line} = 20.25\ \text{A},$$
$$\text{pf} = \cos(-17.30) = 0.95\ \text{lag}.$$

(c)

$$\tilde{I}'_r = \frac{208/\sqrt{3}}{(R_1 + R'_2/s) + j(X_1 + X'_2)} = \frac{120}{6.15 + j0.65} = 19.418\ \text{A}\angle -6.03^\circ. \quad (10.237)$$

$$P_g = 3I_r'^2 R'_2/s = 3(19.418)^2(0.15/0.025) = 6,787\ \text{W} \quad (10.238)$$

$$\left\{\begin{array}{l} T_g = P_g/\omega_{syn} \\ \omega_s = [(1,800)(2\pi)]/60 = 188.5\ \text{rad/s} \end{array}\right. \Rightarrow T_g = \frac{6,787}{188.5} = 36\ \text{Nm}. \quad (10.239)$$

(d)

$$P_d = P_g(1-s) = 6,787(1-0.025) = 6,617\ \text{W}. \quad (10.240)$$

$$P_{out} = P_d - P_{rot} = 6,617 - 550 = 6,117\ \text{W} \Rightarrow P_{out} = 6,117/746 = 8.2\ \text{hp}. \quad (10.241)$$

$$\left\{\begin{array}{l} T_{out} = P_{out}/\omega_m \\ \omega_m = (1,755)(2\pi)/60 = 183.78\ \text{rad/s} \end{array}\right. \Rightarrow T_{out} = 6,117/183.78 = 33.28\ \text{Nm}. \quad (10.242)$$

(e)

$$\eta = \frac{P_{out}}{P_{in}} \times 100 = \frac{6,117}{\sqrt{3}(20.25)(208)(0.95)} \times 100 = 88\%. \quad (10.243)$$

Application Example 10.18: Circle-Phasor Diagram of a 3-Phase Synchronous Motor

A 415 V, Y-connected, 3-phase synchronous machine has synchronous reactance of $X_s = 5\ \Omega$ and negligible armature winding resistance. At certain load, the motor has an output power of 11.5 kW at 0.8 lagging (consumer notation) power factor.

If the magnitude of the excitation voltage remains the same while the load power is decreased by 40%, determine:

(a) Excitation voltage $\tilde{E}$ and the new (power/torque) angle δ_{new}.
(b) The new armature current $\tilde{I}_{a\_new}$ and the new power factor $\cos\Phi_{new}$.
(c) Draw the phasor diagram for constant excitation voltage $\tilde{E}_a$. Show that for a constant magnitude of $\tilde{E}_a$ the locus of the head of $\tilde{E}_a$ describes a circle.

Solution

(a)

$$P = 11,500\ W = 3VI_a \cos\theta = 3(415/\sqrt{3})I_a(0.8) \Rightarrow \tilde{I}_a = 20\angle - 36.87°\ A. \tag{10.244}$$

$$\tilde{E}_a = \tilde{V} + jX_s\tilde{I}_a = 415/\sqrt{3} + j5(20\angle - 36.87°) = 197\angle - 24° \Rightarrow |\tilde{E}| = 197 \text{ and } \delta = -24°. \tag{10.245}$$

Therefore, $|\tilde{E}_a| = |\tilde{E}_{a-new}| = 197\ V$, and

$$P_{new} = 3\frac{VE_a}{X}\sin\delta_{new} \Rightarrow 11500(0.6) = 3\frac{197(415/\sqrt{3})}{5}\sin\delta_{new} \Rightarrow \delta_{new} = -14°. \tag{10.246}$$

(b)

$$\tilde{E}_{a-new} = \tilde{V} + jX_s\tilde{I}_{a\_new} \Rightarrow 197\angle - 14° = 415/\sqrt{3} + j5(\tilde{I}_{a\_new}) \tag{10.247}$$

$\Rightarrow \tilde{I}_{a\_new} = 13.65\angle - 44°$ A and $\cos\phi_{new} = \cos(44°) = 0.72$ lagging (consumer notation).

(c) The phasor diagram and the locus of the head of $\tilde{E}_a$(a circle) are shown in Fig. 10.70.

Application Example 10.19: Synchronous Generator Connected to Infinite Bus

A 3-phase synchronous generator with the synchronous impedance Z_s = (0+j0.85) pu is connected to an infinite (impedance is zero) bus. The prime mover torque is kept constant at a value corresponding to the output power P = 0.8 pu. The infinite bus voltage and the generated voltage have rated 1 pu magnitudes.

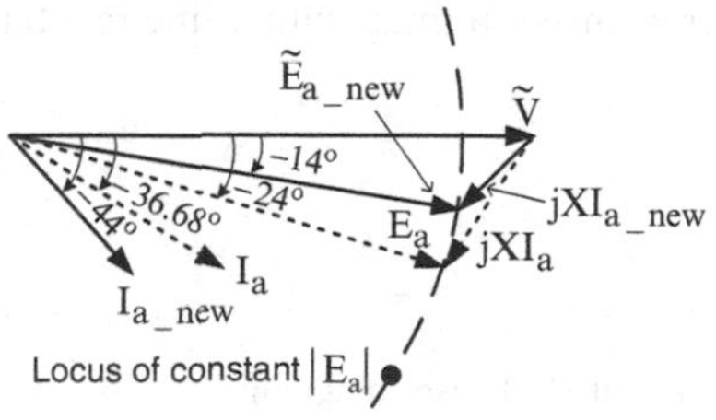

Fig. 10.70 Circle-phasor diagram of synchronous motor

(a) Compute (power/torque) angle δ, armature current $\tilde{I}_a$, power factor pf, and reactive power Q. Draw the phasor diagram.
(b) Compute the required value of the generated voltage $\tilde{E}_{a\_new}$ for unity power factor condition at constant power P = 0.8 pu. What is the new value of (power/torque) angle δ_{new}?

Solution:
(a)

$$P = \frac{VE_a}{X_s}\sin\delta \Rightarrow \sin\delta = \frac{0.8(0.85)}{1^2} \Rightarrow \delta = 42.84°. \quad (10.248)$$

$$\begin{aligned}\tilde{E}_a &= \tilde{V} + jX_s\tilde{I}_a \Rightarrow jX_s\tilde{I}_a = \tilde{E}_a - \tilde{V} \Rightarrow 0.85I_a\angle(90° + \theta)\\ &= (1\angle 42.84° - 1\angle 0°)\,\text{pu}\end{aligned} \quad (10.249)$$

$\Rightarrow 0.85I_a\angle(90° + \theta) = 0.7311\angle 111.44° \Rightarrow \tilde{I}_a = 0.86\angle 21.44°$ pu and pf = 0.931 leading (consumer notation).

$$\begin{aligned}S &= P + jQ = VI_a^* = 1(0.86\angle - 21.44°) = 0.8 - j0.314 \Rightarrow Q\\ &= -0.314\,\text{pu}.\end{aligned} \quad (10.250)$$

The phasor diagram is shown in Fig. 10.71.

(b)

$$\begin{cases} Q = \dfrac{VE_{a\_new}\cos\delta_{new} - V^2}{X_s} = 0 \\ \text{For constant power } E_{a\_new}\sin\delta_{new} = \text{constant} = E_a\sin\delta \end{cases} \quad (10.251)$$

$$\Rightarrow \begin{cases} V = E_{a\_new}\cos\delta_{new} \\ E_{a\_new}\sin\delta_{new} = (1)\sin 42.84° \end{cases} \Rightarrow \begin{cases} E_{a\_new}\cos\delta_{new} = 1 \\ E_{a\_new}\sin\delta_{new} = 0.68 \end{cases} \Rightarrow \tan\delta_{new}$$

$= 0.68$

$\Rightarrow \delta_{new} = 34.2°$ and $E_{a\_new} = 1/\cos\delta_{new} = 1.21$ pu $\Rightarrow \tilde{E}_{a\_new} = 1.21\,\text{pu}\angle 34.2°$.

Application Example 10.20: Steady-State Stability of Synchronous Generator Supplying Power to an Infinite Bus Based on Equal Area Approach

A 50 Hz synchronous generator is supplying power through two reactive parallel transmission lines with a line reactance of $X_{line} = 0.4$ pu each to an infinite bus, as

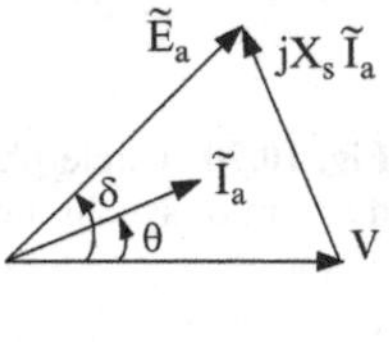

Fig. 10.71 Phasor diagram

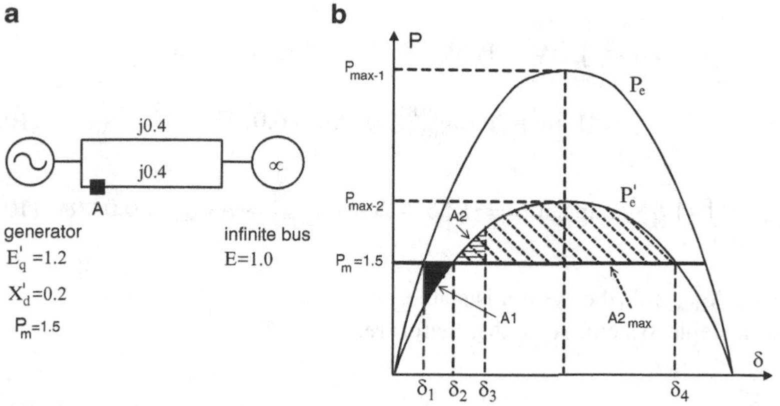

Fig. 10.72 Power system: (**a**) one-line diagram, (**b**) power-angle curves (all values are in per unit)

shown in Fig. 10.72. Before circuit breaker A opens at time t_1 the (power/torque) angle is δ_1. Determine:

(a) The value of P_{max-1} corresponding to the $P_e - \delta$ characteristics before the circuit breaker is opened and the rotor-torque angle δ_1.
(b) The value of P_{max-2} corresponding to the $P'_e - \delta$ characteristic after the circuit breaker is opened, and rotor-torque angles δ_2 and δ_4.
(c) Areas A1 and $A2_{max}$. Will the system be stable after the breaker action? Why?
(d) Assuming the system is stable, use area A2 to compute the angle δ_3.

Note: area A1 is from δ_1 to δ_2, area $A2_{max}$ is from δ_2 to δ_4, area A2 is from δ_2 to δ_3.

Solution

(a)

$$X_{eq} = \frac{0.4(0.4)}{0.4 + 0.4} = 0.2\,\text{pu}. \tag{10.252}$$

$$P_e = \frac{EE'_q}{X'_d + X_{eq}} \sin\delta = \frac{1.2(1)}{0.2 + 0.2} \sin\delta \Rightarrow P_e = 3\sin\delta \Rightarrow P_{max-1} = 3\,\text{pu}, \tag{10.253}$$

$$\Rightarrow 3\sin\delta_1 = 1.5 \Rightarrow \delta_1 = 30^\circ = 0.524\,\text{rad}.$$

(b)

$$P'_e = \frac{1.2(1)}{0.2 + 0.4} \sin\delta \Rightarrow P'_e = 2\sin\delta \Rightarrow P_{max-2} = 2\,\text{pu}, \tag{10.254}$$

$$\Rightarrow P'_e = P_m = 2\sin\delta_2 = 1.5 \Rightarrow \delta_2 = 48.6^\circ = 0.848\,\text{rad},$$

$$\Rightarrow \delta_4 = \pi - \delta_2 = 131.4^\circ = 2.293\,\text{rad}.$$

(c)

$$A_1 = \int_{\delta_1}^{\delta_2} (P_m - P_e)d\delta = \int_{0.524}^{0.848} (1.5 - 2\sin\delta)d\delta$$
$$= [1.5\delta + 2\cos\delta]|_{0.524}^{0.848} \Rightarrow A_1 = 0.0773, \quad (10.255)$$

$$A_{2max} = \int_{\delta_2}^{\delta_4} (1.5 - 2\sin\delta)d\,\delta = [1.5\delta + 2\cos\delta]|_{2.293}^{0.848} \Rightarrow A_{2max} = 0.0478. \quad (10.256)$$

Since $A_{2max} > A_1$ the system is stable.

(d) For a stable system, $A_1 = A_2$, therefore:

$$\int_{\delta_2}^{\delta_3} (P_m - P'_e)d\delta = 0.0773 \Rightarrow \int_{0.848}^{\delta_3} (1.5 - 2\sin\delta)d\delta = 0.0773, \quad (10.257)$$

$$\Rightarrow 1.5\delta_3 + 2\cos\delta_3 = 2.818 \Rightarrow \delta_3 = 1.218\,\text{rad} = 69.8°.$$

Application Example 10.21: Steady-State Parallel Operation of Two Synchronous Generators Supplying Power to a Passive Load

Two 3-phase, Y-connected synchronous generators have per phase generated voltages of $\tilde{E}_{a\_1} = 120\,V\angle 10°$ and $\tilde{E}_{a\_2} = 120\,V\angle 20°$ at no-load, and reactances of $X_{s\_1} = j5\ \Omega$/phase and $X_{s\_2} = j8\ \Omega$/phase, respectively. They are connected in parallel to a load impedance of $Z_L = (4 + j3)\ \Omega$/phase. Compute:

(a) Per phase terminal voltage $\tilde{V}_a$ of both generators operating in parallel.
(b) Armature currents for generator 1 ($\tilde{I}_{a\_1}$) and generator 2 ($\tilde{I}_{a\_2}$).
(c) Power supplied by each generator P_1 and P_2.
(d) The total output power of both generators P_{out}.

Solution

(a) Using the circuit diagrams of Fig. 10.73, find the Norton equivalent circuit ($I_{Norton} = I_N$ and $Z_{Thevenin} = Z_{TH}$) with respect to terminals ab.

$$\tilde{I}_N = \tilde{I}_{a\_1} + \tilde{I}_{a\_2} = \frac{120\angle 10°}{j5} + \frac{120\angle 20°}{j8} = 24\angle 80° + 15\angle 70°$$
$$= 38.83\,A\angle 76.14°. \quad (10.258)$$

$$\frac{1}{Z_{TH}} = \frac{1}{j5} + \frac{1}{j8} + \frac{1}{j(4 + j3)} \Rightarrow Z_{TH} = 2.11\ \Omega\angle 70.2°. \quad (10.259)$$

$$\tilde{V}_a = \tilde{V}_{TH} = \tilde{I}_N Z_{TH} = 38.83\angle 76.14°(2.11\,\angle 70.2°) = 82\,V\angle 5.94°. \quad (10.260)$$

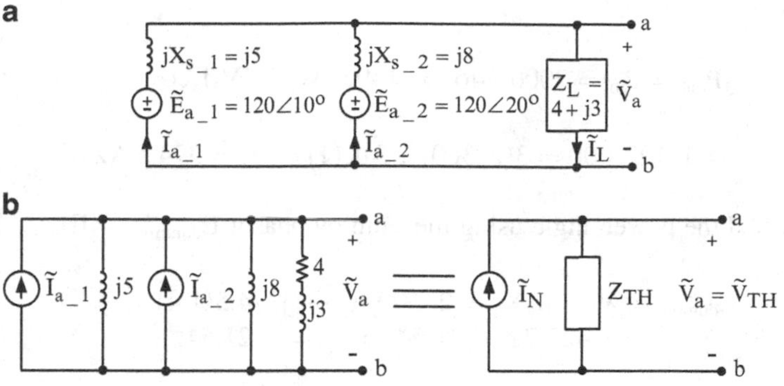

Fig. 10.73 (**a**) Circuit diagram and (**b**) the simplified Norton equivalent circuit

(b)

$$\begin{cases} \tilde{V}_a = \tilde{E}_{a\_1} + j\tilde{I}_{a\_1}X_1 \Rightarrow 82\angle 5.94^\circ = 120\angle 10^\circ + j5\tilde{I}_{a\_1} \\ \tilde{V}_a = \tilde{E}_{a\_2} + j\tilde{I}_{a\_2}X_2 \Rightarrow 82\angle 5.94^\circ = 120\angle 20^\circ + j8\tilde{I}_{a\_2} \end{cases} \tag{10.261}$$

$$\Rightarrow \begin{cases} \tilde{I}_{a\_1} = 9.36\,\mathrm{A}\angle -51.17^\circ. \\ \tilde{I}_{a\_2} = 7.31\,A\angle -32.06^\circ. \end{cases}$$

(c)

$$\begin{cases} P_1 = 3V_aI_{a\_1}\cos\theta_1 = 3(82)(9.36)\cos(51.7 - 5.93) = 1{,}621\,\mathrm{W}. \\ P_2 = 3V_aI_{a\_2}\cos\theta_2 = 3(82)(97.31)\cos(32.06 - 5.93) = 1{,}614.5\,\mathrm{W}. \end{cases} \tag{10.262}$$

(d)

$$P_{out} = P_1 + P_2 = 3{,}236\,\mathrm{W}. \tag{10.263}$$

Application Example 10.22: Steady-State Stability Limit of a Synchronous Motor Connected to Infinite Bus

A 2,000 hp, 2,300 V, unity power factor, 3-phase, Y-connected, 60 Hz salient pole synchronous motor has the direct-axis synchronous reactance of $X_d = 1.95\ \Omega$/phase and the quadrature-synchronous reactance of $X_q = 1.4\ \Omega$/phase. The motor is supplied from an infinite bus at rated voltage and frequency (1 hp = 746 W). Neglect all losses and compute:

(a) Armature current $\tilde{I}_a$.
(b) Power/torque angle $\delta$and excitation or induced voltage $\tilde{E}_a$.
(c) Maximum three-phase power P_{max}that the motor can deliver.
(d) Use the equal-area approach to find the maximum load angle δ_{max} at which motor can operate before instability occurs.

Solution

(a)

$$P_{out} = P_{in} = 2000(746) = 1492\,\text{kW} = 3V_aI_a\cos\theta, \tag{10.264}$$

$$\Rightarrow 1,492,000 = 3(2,300/\sqrt{3})I_a(1) \Rightarrow I_a = 374.5\,\text{A}\angle 0°.$$

(b) First find the power angle using the dummy phasor $\tilde{E}_{dummy} = |E_{dummy}|\angle\,\delta$:

$$\begin{aligned}\tilde{E}_{dummy} &= V_t - jI_aX_q = 2,300/\sqrt{3} - j374.5(1.4)\\ &= 1,427.7\angle -21.54° \Rightarrow \delta = -21.54°.\end{aligned} \tag{10.265}$$

Then use the computed power angle to compute $\tilde{E}_a$:

$$|\tilde{E}_a| = V_t\cos\theta + X_dI_d = V_t\cos\theta + X_d[I_a\sin(\theta - \delta)], \tag{10.266}$$

$$\Rightarrow |\tilde{E}_a| = 2,300/\sqrt{3}\cos(-21.54) + 1.95(374.5)\sin(21.54) = 1,503.3\,\text{V} \Rightarrow \tilde{E}_a = 1,503.3\,\text{V}\angle -21.54°.$$

Note that $\tilde{E}_{dummy}$ and $\tilde{E}_a$ have the same phase angles but different magnitudes.

(c)

$$P_{phase} = \frac{VE_a}{X_d}\sin\delta + \frac{V^2(X_d - X_q)}{2X_dX_q}\sin 2\delta. \tag{10.267}$$

$$\Rightarrow P_{phase} = \frac{2,300(1,503.3)}{\sqrt{3}(1.95)}\sin\delta + \frac{2,300^2(0.55)}{2(1.95)(1.4)}\sin 2\delta$$

$$= 1,024\sin\delta + 178\sin 2\delta \text{ kW/phase.}$$

$$\frac{dP_{phase}}{dt} = 1,024\cos\delta_{max} + 356\cos 2\delta_{max} = 0 \Rightarrow \delta_{max}^{steady-state} = 73.2°. \tag{10.268}$$

$$P_{max-phase} = 1,024\sin 73.2 + 178\sin(2\times 73.2) = 1,078.8\,\text{kW/phase}$$
$$\Rightarrow P_{max-3phase} = 3,236\,\text{kW}.$$

(d) Use the Equal-Area concept (Fig. 10.74) with $P_{electrical} = P_{3phase} = 3,072\sin\delta + 534\sin 2\delta$, $P_m = 1,492\,\text{W}$, and $\delta = 21.54°$ to find δ_{max}:

$$A_1 = A_2$$

$$\Rightarrow 1,492(21.54)\frac{\pi}{180}$$

$$-\int_0^{21.54} P_{3\,phase}(\delta)d\delta = \int_{21.54}^{\delta_{max}} P_{3phase}(\delta)d\delta - 1,492(\delta_{max} - 21.54)\frac{\pi}{180}.$$

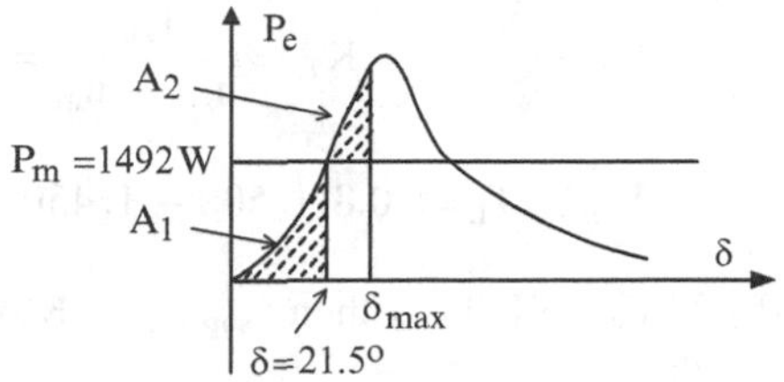

Fig. 10.74 Power-angle characteristic

$$\Rightarrow \delta_{max} = \delta_{max}^{transient} = 45.2°. \qquad (10.269)$$

Note that the maximum value of δ for transient operation ($\delta_{max}^{transient} = 45.2°$) is smaller that its corresponding value under steady-state condition ($\delta_{max}^{steady-state} = 73.2°$).

Application Example 10.23: Speed Control of Induction Motor with Constant Flux

A 3-phase, 50 Hz, 4-pole, 440 V (line-to-line) induction motor drive has a rated speed and a rated torque of 1,450 rpm and 40 Nm, respectively. The maximum air-gap flux density is maintained constant at its rated value. The motor is supplying a load with the linear torque-speed characteristic $T_L = (K_L n_m + 12)$Nm, such that it equals the rated motor torque at the rated speed. The motor draws 8 A (per phase) at a power factor of 0.8 lagging (consumer notation) at rated voltage and load conditions and its stator resistance is $R_s = 1.3\ \Omega$.

(a) Compute the value of K_L.
(b) To maintain flux density at its rated value, the applied voltage magnitude $\tilde{V}_a$ depends linearly on the frequency f of the applied voltages, except for the offset due to the resistance R_S; that is.

$$V_a = \frac{V_{a,rated} - R_S I_{a,rated}}{f_{rated}}(f) + R_S I_a.$$

The drive is started with a magnitude voltage of 50 V (line-to-line), such that the starting torque is $T_{st} = K_{st} T_{rated}$. Compute K_{st} and frequency of the starting voltage f_{start}.
(c) Calculate the required frequency f_{new} for a speed of $n_{m\_new} = 1{,}000$ rpm.

Solution

(a)

$$n_{syn} = \frac{120f}{P} = \frac{120(50)}{4} = 1,500 \text{ rpm}, \qquad (10.270)$$

$$\begin{cases} T_L = K_L n_m + 12, \\ T_{em} = K_T(n_{syn} - n_m), \end{cases} \qquad (10.271)$$

$$\Rightarrow K_T = \frac{T_{em}}{n_{syn} - n_m} = \frac{40}{1,500 - 1,450} = 0.8,$$

$$T_{em} = T_L \Rightarrow 0.8(1,500 - 1,450) = K_L(1,400) + 12 \Rightarrow K_L = 0.02.$$

(b) If $T_{st} = K_{st}T_{rated}$ then $n_{slip,start} = K_{st}n_{slip,rated} = K_{st}\,(1,500 - 14,450) = 50\,K_{st}$ then

$$\begin{cases} n_{syn} = \dfrac{120f}{P} \Rightarrow f = \dfrac{P(n_{syn})}{120}, \\ V_a = \dfrac{V_{a,rated} - R_S I_{a,rated}}{f_{rated}}(f) + R_S I_a. \end{cases} \quad (10.272)$$

Substituting the starting conditions, we have:

$$\begin{cases} f_{st} = \dfrac{P\,(n_{slip,start})}{120} = \dfrac{4(50\,K_{st})}{120} = 1.67\,K_{st}. \\ V_{a,st} = \dfrac{V_{a,rated} - R_S I_{a,rated}}{f_{rated}}(f_{st}) + R_S I_{a,st}. \end{cases}$$

$$\begin{cases} f_{st} = 1.67\,K_{st}. \\ V_{a,st} = \dfrac{V_{a,rated} - R_S I_{a,rated}}{f_{rated}}(f_{st}) + R_S(K_{st}I_a). \end{cases}$$

$$\Rightarrow \frac{50\sqrt{2}}{\sqrt{3}} = \frac{(440\sqrt{2}/\sqrt{3}) - 1.3(8)(\sqrt{2})(0.8)}{50}(1.67\,K_{st}) + 1.3(8)(\sqrt{2})(0.8)K_{st}$$

$$\Rightarrow K_{st} = 1.75,$$

and $f_{st} = 1.67K_{st} = 2.9\,\text{Hz}$.

(c) If $n_m = 1,000$ rpm then

$$0.8(n_{syn} - n_m) = 0.02n_m + 12 \Rightarrow n_{syn} = \frac{0.82n_m + 12}{0.8} = 1,040\,\text{rpm},$$

$$\text{and } n_{syn} = \frac{120f}{P} \Rightarrow f = \frac{P\,n_{syn}}{120} = 34.67\,\text{Hz}.$$

Application Example 10.24: Starting Torque Control of Induction Motor with Constant Flux

A 3-phase, 50 Hz, 4-pole, 440 V (line-to-line) induction motor drive has a rated speed and a rated torque of 1,440 rpm and 45 Nm, respectively. The maximum air-gap flux density is maintained constant at its rated value. The motor is supplying the load with the linear torque-speed characteristics $T_L = (A\,n_m + B)$ Nm, such that it equals the rated torque at rated speed.

(a) Plot the linear portion of the motor torque-speed characteristics for f = 50 Hz, indicating the slope, starting and rated operating points.
(b) Compute A and B if the motor speed drops to 855.6 rpm at f = 30 Hz.
(c) The drive is started with a voltage magnitude of 44 V (line-to-line). What is the starting torque T_{start} in terms of the rated torque T_{rated}? What is the frequency of the starting voltage f_{start}?

Solution

(a)

$$n_{syn} = \frac{120f}{P} = \frac{120(50)}{4} = 1,500\,\text{rpm} \Rightarrow n_{slip} = 1,500 - 1,440 = 60\,\text{rpm},$$

$$T_{em} = K_T(n_{syn} - n_m) \Rightarrow K_T = 45/60 = 0.75,$$

$$T_{em} = 0.75(1,500 - n_m). \tag{10.273}$$

The linear portion of the motor torque-speed characteristics is plotted in Fig. 10.75.

(b)

$$T_{em} = T_L \Rightarrow 0.75(n_{syn} - n_m) = An_m + B, \tag{10.274}$$

$$\begin{cases} f = 50\,\text{Hz},\ n_{syn} = 1,500\,\text{rpm},\ n_m = 1,440\,\text{rpm} \\ f = 30\,\text{Hz},\ n_{syn} = 900\,\text{rpm},\ n_m = 855.6\,\text{rpm} \end{cases}$$

$$\Rightarrow \begin{cases} 0.75(1,500 - 1,440) = 1,440A + B \\ 0.75(900 - 855.6) = 855.6A + B \end{cases} \Rightarrow \begin{cases} A = 0.02, \\ B = 16.2. \end{cases}$$

(c) Assume $T_{st} = A\, T_{rated}$ then $n_{slip,st} = A\, n_{slip,rated} = (1,500 - 1,440)A = (60 \cdot A)$ rpm. Substituting the starting conditions in (10.273), we get:

$$\begin{cases} f_{st} = \dfrac{n_{slip,st}}{60} \times \dfrac{P}{2} = \dfrac{60A}{60} \times \dfrac{4}{2} \\ V_a = \dfrac{V_{a,rated} - R_s I_{a,rtaed}}{f_{rated}}(f) + R_s I_{a,st} \end{cases} \Rightarrow \begin{cases} f_{st} = 2A \\ V_a = \dfrac{V_{a,rated} - R_s I_{a,rtaed}}{50}(f) + R_s(AI_{a,rated}) \end{cases}$$

$$\Rightarrow \begin{cases} f_{st} = 2A \\ 44(\sqrt{2}/\sqrt{3}) = \dfrac{440(\sqrt{2}/\sqrt{3}) - 1.3(8)(\sqrt{2})(0.8)}{50}(2A) + 1.3(8)(\sqrt{2})(0.8) \end{cases}$$

$$\Rightarrow A = 1.4,\ r \Rightarrow \begin{cases} f_{st} = 2A \\ T_{st} = AT_{rated} \end{cases} \Rightarrow \begin{cases} f_{st} = 2.8\ \text{Hz}, \\ T_{st} = 1.44T_{rated}. \end{cases}$$

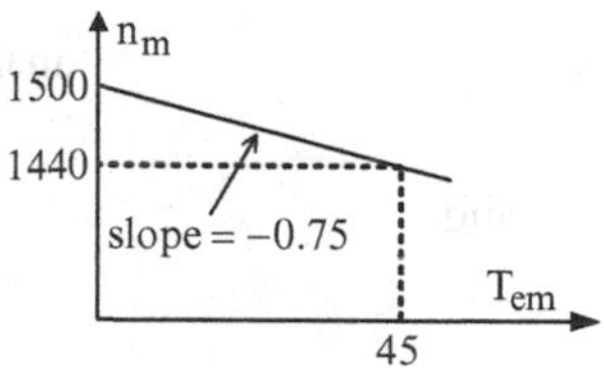

Fig. 10.75 Torque-speed characteristic

Application Example 10.25: Transient Step-Torque Control of Induction Motor with Constant Flux and dq Variables [60]

A 2-pole, 3-phase induction machine requires the magnetizing current $I_m = 4$ A to establish the rated air-gap flux density.

(a) Compute the required 3-phase currents at $t = 0^-$ to build up the rated flux.
(b) The desired step torque at $t = 0^+$ requires a step change in $i_{sq} = 10$ A. Calculate the 3-phase currents at $t = 0^+$ which results in the step torque while maintaining the rated flux density in the air gap.
(c) If the three-phase currents are set to $i_a(0^+) = 4$ A, $i_b(0^+) = 5$ A and $i_c(0^+) = -9$ A, calculate the new values of $i_{sd}(0^+)$ and $i_{sq}(0^+)$.

Solution

(a)

$$\begin{cases} i_a(0^-) = I_m = 4\,\text{A}, \\ i_b(0^-) = -I_m/2 = -2\,\text{A}. \\ i_c(0^-) = -I_m/2 = -2\,\text{A}. \end{cases} \quad (10.275)$$

(b)

$$I_{m,s,rated} = (3/2)I_{m,rated} = (3/2)(4) = 6\,\text{A}. \quad (10.276)$$

$$i_{sd}(0^-) = i_{sd}(0^+) = \sqrt{(2/3)}\,I_{m,s,rated} = \sqrt{(2/3)}(6) = 4.9\,\text{A}. \quad (10.277)$$

$$\Rightarrow \begin{cases} i_{sd}(0^+) = 4.9\,\text{A}. \\ i_{sq}(0^+) = 10\,\text{A} \end{cases} \Rightarrow i_s(0^+) = 4.9 + j10 = 11.13\,\text{A}\angle 63.9°.$$

Therefore,

$$\begin{cases} i_a(0^+) = (2/3)(11.13)\cos 63.9° = 3.26\,\text{A}. \\ i_b(0^+) = (2/3)(11.13)\cos(63.9° - 120°) = 4.138\,\text{A}. \\ i_c(0^+) = (2/3)(11.13)\cos(63.9° - 240°) = -7.40\,\text{A}. \end{cases} \quad (10.278)$$

(c)

$$\begin{cases} i_a(0^+) = (2/3)X\cos\Phi = 4\,\text{A} \\ i_b(0^+) = (2/3)X\cos(\Phi - 120°) = 5\,\text{A} \\ i_c(0^+) = (2/3)X\cos(\Phi - 240°) = -9\,\text{A} \end{cases} \Rightarrow \frac{\cos\Phi}{\cos(\Phi - 120°)} = \frac{4}{5}$$

$$\Rightarrow \frac{\cos\Phi}{\cos\Phi\cos 120° + \sin\Phi\sin 120°} = \frac{4}{5}$$

$$\Rightarrow \tan\Phi = \frac{5/4 - \cos 120°}{\sin 120°} \Rightarrow \Phi = 63.6°$$

and

$$X = \frac{4}{(2/3)\cos\Phi} = 13.5.$$

Therefore,

$$i_s(0^+) = i_{sd}(0^+) + j\, i_{sq}(0^+) = (6 + j\,12.1)\,\text{A} \Rightarrow \begin{cases} i_{sd} = 6\,\text{A}. \\ i_{sq} = 12.1\,\text{A}. \end{cases}$$

Single-phase induction machines: According to the IEEE Standard Dictionary of Electrical and Electronic Terms [61] a "single-phase" motor is a rotating machine that converts single-phase alternating current electric power into mechanical power, or that provides mechanical force or torque. The term "single-phase" therefore does not refer to the number of winding phases (main phase and auxiliary phase) the machine may have. This term is therefore a generic expression encompassing one-phase (main winding only) as well as nonsymmetrical two-phase (main-phase and auxiliary-phase windings) machines as well. There are two approaches for the design of such machines based on:

- Rotating-field theory
- Cross-field theory

Both of these methods are used in practice in industrial design offices, and each of them has advantages and disadvantages. There are four commonly used types of one- and two-phase motors:

- *Split-Phase Motor* has no external impedance in either main-or auxiliary-phase winding. Its auxiliary-phase winding is energized only during starting and its winding resistance is usually high.
- *Capacitor-Start Motor* has a capacitor connected in series with an auxiliary-phase winding which is energized only during the starting period.
- *Permanent-Split Capacitor Motor* is a capacitor motor using the same value of effective capacitance for both starting and running operations. The capacitor is connected in series with the auxiliary-phase winding.
- *Capacitor-Start, Capacitor-Run Motor* is a capacitor motor using two capacitors. One capacitor (with lower capacitance) is for running operation and both are connected in parallel for starting. The starting capacitor (with the higher capacitance and usually of the electrolytic type) is disconnected after starting. During the starting period both capacitors are paralleled and connected in series with the auxiliary-phase winding.

Efficiency optimization of single-phase induction machines: The optimal design of the motor dimensions, the capacitance of the run capacitor, the winding distribution, and the choice of the electrical steel are the most important sources for the improvement of the efficiency of modern single-phase induction motors for given performance and material cost constraints [62–66]. Figure 10.76 illustrates the efficiency η (full lines), the material cost C (dashed lines) in US$ (including capacitors) and all constraints in the (X_1, X_2) plane where X_1 is the stator-bore diameter and X_2 is the stator-core length.

Power quality of electric machines: With the deployment of variable-speed drives and the use of voltage-source and current-source inverters as well as rectifiers

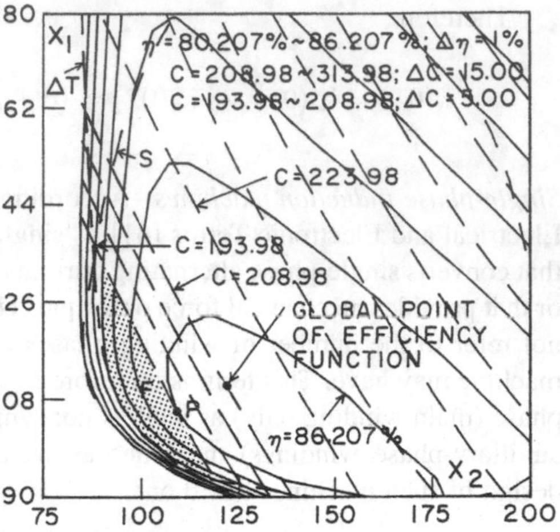

Fig. 10.76 Efficiency η in %, material cost C in US$ (including capacitors) and all constraints in the (X_1, X_2) plane (From Fuchs, E. F.; Huang, H.; Vandenput, A. J.; Höll, J.; Zak, Z.; Appelbaum, J.; Erlicki, M.: *Optimization of Induction Motor Efficiency, Volume 2: Single-Phase Induction Motors*, Publication of the Electric Power Research Institute, Palo Alto, California, EPRI EL-4152-CCM, May 1987, 450 pages, [66])

the power quality of electric machines gains importance. Power quality problems of machines are related to harmonic voltages and currents, harmonic losses, harmonic torques, generation of electromagnetic fields, non-periodic interference phenomena, asymmetric operation of three-phase machines, operation with DC currents due to asymmetric gating of semiconductor converters, and over-temperatures reducing their lifetime [23] and influence the reliability of power system protection [67].

10.6 Summary

The analysis of rotating and linear motion electric machines is based on Faraday's, Ampere's, Ohm's, and Kirchhoff's laws as well as the work of Fourier and the conservation of energy leading to magnetic force expressions. The application of these laws to electromechanical energy converters leads to nonlinear (differential) equations due to the product of independent variables and the nonlinear B-H characteristics of iron materials employed to increase and direct the magnetic flux within a magnetic device. Numerical methods based on the approximation of partial differential equations assist in the nonlinear magnetic field solutions. At the beginning of the chapter various flux patterns as they occur within the different electric machines are presented to give the reader a visualization of the various occurring magnetic fields. The elementary machine is used to explain winding factors and the generation of standing, alternating and rotating waves. The torque production in single- and three-phase machines is discussed next which leads to the conclusion that poly-phase machines lead to constant torques only, as invented by Tesla. Thereafter, the theory is applied to three-phase synchronous and induction machines as well as permanent-magnet machines including brushless DC machines.

The construction elements of the three-phase induction machine are highlighted through drawings. Doubly fed induction machines, in particular doubly fed induction generators (DFIG) as employed in variable-speed wind- and hydro-power plants, and single-phase induction machines conclude the discussion in this chapter. Efficiency optimization for given performance constraints and costs, and power quality issues of electric machines are addressed. Application examples highlight the concepts presented.

10.7 Problems

Problem 10.1: *Rotating fields in AC (induction, synchronous) machines*

A three-phase, two-pole induction motor is excited by balanced three-phase $f = 60$ Hz currents as described by $i_a = I_m \cos\omega t$, $i_b = I_m \cos(\omega t\text{-}120°)$, $i_c = I_m \cos(\omega t\text{-}240°)$. The mmfs of phases a, b, c can be written as $F_a = A_1 \cos\theta i_a$, $F_b = A_1 \cos(\theta\text{-}120°) i_b$, $F_c = A_1 \cos(\theta\text{-}240°) i_c$.

(a) Calculate the angular velocity $d\theta/dt = \dot{\theta}$ and the rotational direction of the total mmf $F_{total} = F_a + F_b + F_c$.

(b) What is the effect on the rotating (total) mmf wave, F_{total}, if two of the phase connections are interchanged? That is, $F_a = A_1 \cos\theta i_a$, $F_b = A_1 \cos(\theta\text{-}120°) i_c$, $F_c = A_1 \cos(\theta\text{-}240°) i_b$. Perform similar calculations as in (a).

Problem 10.2: *Construction of magnetomotive forces (mmfs)*

(a) Draw the magnetomotive force (mmf) of the stator winding of Fig. 10.77, where **x** (into sheet) represents an mmf of –100 At and • (out of sheet) represents +100 At.

(b) Draw the mmf of the rotor winding of Fig. 10.78 provided $N_1 = N_{10} = N_6 = N_5 =$ 4 turns, $N_2 = N_9 = N_7 = N_4 = 6$ turns, $N_8 = N_3 = 9$ turns, and $I_f = 10$ A.

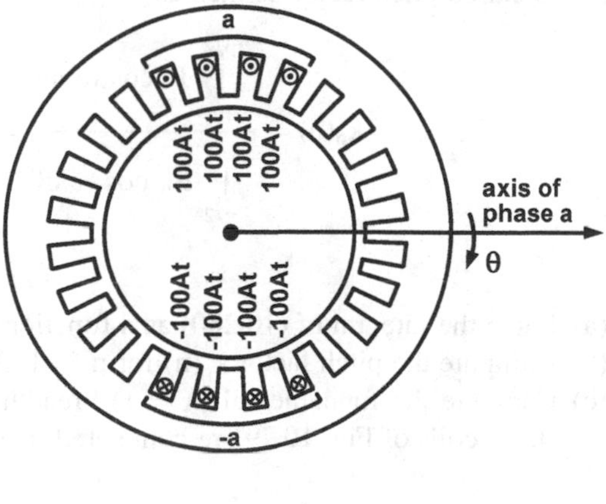

Fig. 10.77 MMF of one phase of a distributed two-pole, three-phase stator winding with full-pitch coils

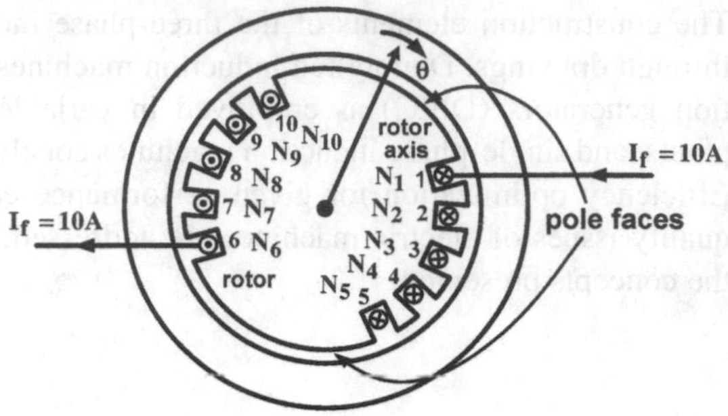

Fig. 10.78 MMF of a distributed two-pole, rotor winding

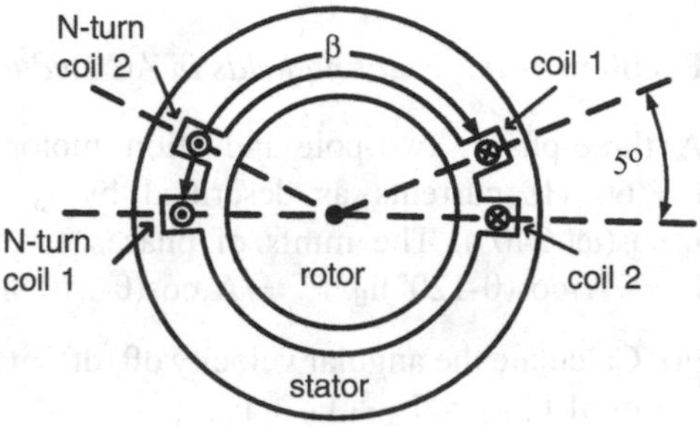

Fig. 10.79 Two short-pitched (fractional-pitch) stator coils connected in series

Problem 10.3: *Computation of winding factor $k_w = k_d \cdot k_p$*

Figure 10.79 shows two short-pitched stator coils of a two-pole machine. The coil sides are $\beta = 160°$ apart. For an air gap radial flux distribution of the form

$$B_{radial} = \sum_{n=odd} B_n \cos n\theta, \tag{10.279}$$

where n = 1 corresponds to the fundamental space harmonic, n = 3 to the third spatial harmonic, etc., the flux linkage of each coil is the integral of B_{radial}, over the surface spanned by that coil. Thus for the nth space harmonic, the ratio of the flux linkage of the fractional-pitch coil to that of the full-pitch coil is given by the so-called pitch factor defined as

$$k_p(n) = \frac{\int_{-\beta/2}^{\beta/2} B_n \cos n\theta d\theta}{\int_{-\pi/2}^{\pi/2} B_n \cos n\theta d\theta} = \frac{\int_{-\beta/2}^{\beta/2} \cos n\theta d\theta}{\int_{-\pi/2}^{\pi/2} \cos n\theta d\theta}. \tag{10.280}$$

(a) Solve the integral of (10.280) as a function of n and β.
(b) Compute the pitch factor $k_p(n)$ for n = 1, 3, 5, 7....
(c) Compute the fundamental (n = 1) breadth (or distribution) factor, if the two stator coils of Fig. 10.79 are connected in series.

Problem 10.4: *Construction of magnetomotive forces (mmfs), and calculation of winding factors*

(a) Draw the magnetomotive force (mmf) of the four coils of the stator winding of Fig. 10.80, where x represents an mmf of −100 At and • represents +100 At.
(b) Figure 10.80 shows four short-pitched stator coils of a two-pole machine. The coil sides are $\beta = 150°$ apart. For an air-gap radial flux distribution of the form $B_{radial} = \sum_{n=odd} B_n \cos n\theta$, where n = 1 corresponds to the fundamental space harmonic, n = 3 to the third spatial harmonic, etc., the flux linkage of each coil is the integral of B_{radial}, over the surface spanned by that coil. Thus for the *n*th space harmonic, the ratio of the flux linkage of the fractional-pitch coil to that of the full-pitch coil is given by the so-called pitch factor defined in (10.280) of Problem 10.3. Solve this integral as a function of n and β. Compute the pitch factor $k_p(n)$ for n = 1, 3, 5, 7, ... compute the fundamental (n = 1) breadth (or distribution) factor if the four stator coils of Fig. 10.80 are connected in series, and calculate the winding factor $k_{w\_stator} = k_b \cdot k_p$ of this stator winding for n = 1.
(c) Draw the mmf of the rotor winding of Fig. 10.81 provided $N_1 = N_6 = N_7 = N_{12} = 6$ turns, $N_2 = N_5 = N_8 = N_{11} = 7$ turns, $N_3 = N_4 = N_9 = N_{10} = 8$ turns, and $I_f = 10$ A

Problem 10.5: *Torque and power of AC (synchronous) machines*

A 4-pole 60 Hz three-phase, synchronous generator has a rotor length of $\ell = 3.9$ m, a rotor diameter $D^r_{out} = 1.12$ m and an air-gap (one-sided) length of g = 6.2 cm. The rotor winding consists of $N_{pole} = 55$ turns per pole with a winding factor of

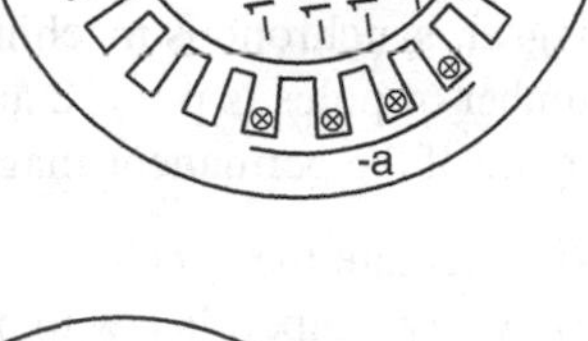

Fig. 10.80 MMF of one phase of a distributed two-pole, three-phase stator winding with four short-pitch coils

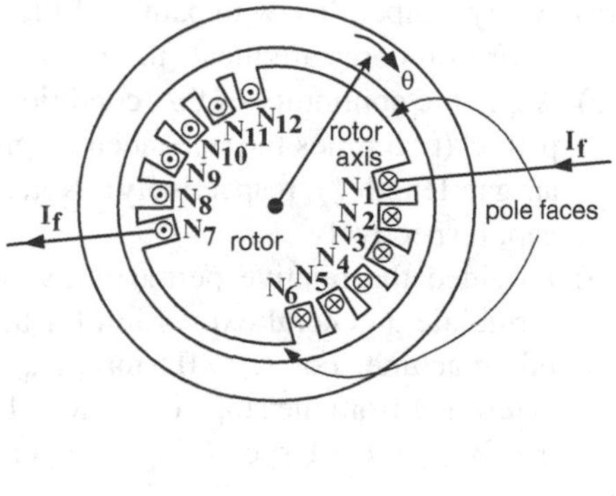

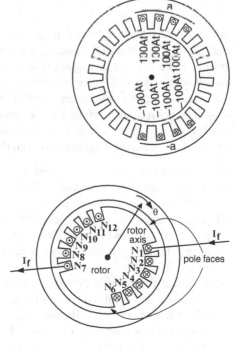

Fig. 10.81 MMF of a distributed two-pole, rotor winding

$k_{w\_rotor} = 0.89$. The maximum fundamental air-gap flux density is 1.0 T and the rotor winding (field) current is $I_f = 2{,}900$ A. Calculate:

(a) The maximum torque in newton-meters.
(b) The power output in megawatts which can be supplied by the machine under these conditions.

Problem 10.6: *Synchronous speed, mmfs, torque and power of synchronous machine*

A 4-pole, f = 60 Hz, $|\tilde{V}_{L\text{-}L\,rat}| = 12470$ V, three-phase, synchronous generator has the rotor length of $\ell = 4$ m, the rotor diameter $D^r_{out} = 1$ m, and an air-gap (one-sided) length of g = 6 cm. The rotor winding consists of $N_{pole} = 60$ turns per pole with a winding factor of $k_{w\_rotor} = 0.90$. The winding factor of the stator winding is $k_{w\_stator} = 1.0$. For generator operation we assume $E = 1.05\frac{V_{L-L\_rat}}{\sqrt{3}}$. The maximum fundamental air-gap flux density is 0.8 T and the rotor winding (field) current is $I_f = 3{,}000$ A.

(a) Calculate the synchronous speed $n_s^{(rpm)}$.
(b) What is the number of turns per phase N_{ph}?
(c) Calculate the mmf of the rotor F_r.
(d) Calculate the maximum torque in newton-meters.
(e) What is the angular velocity (frequency) ω_m of the rotor shaft?
(f) Determine the power output in megawatts which can be supplied by the machine under these conditions.

Problem 10.7: *Design of a permanent-magnet machine (motor or generator) [18–20]*

In Chap. 5 (Example 5.14 and homework problem 5.10) we performed a PSpice analysis of a brushless DC motor drive. Such a drive consists of a semiconductor inverter and a permanent-magnet motor. While in Chap. 5 we concentrated on the overall system we will focus now on the design/analysis of the permanent-magnet machine. See Application Example 10.11. A P = 50 kW, $V_{L\text{-}L} = 240$ V permanent magnet, synchronous machine has the cross-section of Figs. 10.27 and 10.28. The number of poles is p = 12 and the rated frequency is f = 60 Hz. The B-H characteristic of the permanent-magnet material (NdFeB) is shown in Fig. 6.37 of Chap. 6.

(a) Calculate the synchronous speed $n_s^{(rpm)}$ of the motor.
(b) Apply Ampere's law to path C of Figs. 10.27 and 10.28 assuming the stator and rotor iron cores are ideal ($\mu_r \to \infty$).
(c) Apply the continuity of flux condition for the areas A_m and A_g perpendicular to path C (two times the one-sided magnet length ℓ_m and two times the one-sided air gap length ℓ_g, respectively). Note, subscripts m and g stand for magnet and gap, respectively.
(d) Provided the relative permeability of the stator and rotor cores is $\mu_r \to \infty$, formulate a general expression for the load line which can be associated with this machine. For $\ell_g = 0.5$ mm, $A_m = 30{,}000$ mm^2, and $A_g = 35{,}000$ mm^2, determine from the slope of the load line the (one-sided) magnet length ℓ_m such that $B_{mo} = 1.0$ T specifying the operating point on the magnet characteristic.

(e) Plot the load line in a plot similar to that of Fig. 6.38b. Is this permanent-magnet material operated at the point of the maximum energy product?
(f) What is the recoil permeability μ_R of the NdFeB material?
(g) Compute the fluxes ϕ_{max}, $\phi_{totalmax} = 2\phi_{max}$, and the number of turns N per phase provided the induced stator phase voltage is $E_{ph} \approx \frac{V_{L-L}}{\sqrt{3}} = \frac{240\,V}{\sqrt{3}} = 138.57\,V$, the winding factor is $k_{w\_stator} = 0.8$, and the iron-core stacking factor is $k_s = 0.94$.

Problem 10.8: *Design of a permanent-magnet, three-phase synchronous machine (motor or generator)*

A $P = 50$ kW, $V_{L\text{-}L} = 240$ V permanent magnet, three-phase synchronous machine has the cross-section of Figs. 10.27 and 10.28. The number of poles is $p = 12$ and the rated frequency is $f = 240$ Hz. The B-H characteristic of the permanent-magnet material samarium-cobalt (SaCo) is shown in Fig. 6.37 of Chap. 6.

(a) Calculate the synchronous speed $n_s^{(rpm)}$ of the motor.
(b) Apply Ampere's law to path C of Figs. 10.27 and 10.28 assuming the stator and rotor iron cores are ideal ($\mu_r \to \infty$).
(c) Apply the continuity of flux condition for the areas A_m and A_g perpendicular to path C (two times the one-sided magnet length ℓ_m and two times the one-sided air gap length ℓ_g, respectively). Note, subscripts m and g stand for magnet and gap, respectively.
(d) Provided the relative permeability of the stator and rotor cores is $\mu_r \to \infty$, formulate a general expression for the load line which can be associated with this machine. For $\ell_g = 1$ mm, $A_m = 40{,}000$ mm^2, and $A_g = 45{,}000$ mm^2 determine from the slope of the load line the (one-sided) magnet length ℓ_m such that the samarium-cobalt (SaCo) magnet material (Fig. 6.37) is minimized.
(e) Plot the load line in a plot similar to that of Fig. 6.38b.
(f) What is the recoil permeability μ_R of the SaCo material?
(g) Compute the fluxes ϕ_{max}, $\phi_{totalmax} = 2\phi_{max}$, and the number of turns N_{ph} per phase provided the induced stator phase voltage is $E_{ph} \approx \frac{V_{L-L}}{\sqrt{3}} = \frac{240\,V}{\sqrt{3}} = 138.57\,V$, the winding factor is $k_{w\_stator} = 0.9$, and the iron-core stacking factor is $k_s = 0.94$.

Problem 10.9: *Permanent-magnet machine for a hybrid electric car*

A $P_{out} = 30$ hp, $V_{L\text{-}Lrms} = 120$ V permanent-magnet, synchronous machine (either motor or generator) has the cross-section of Figs. 10.27 and 10.28. The number of poles is $p = 30$ and the rated frequency is $f = 60$ Hz. The B-H characteristic of the permanent-magnet material (NdFeB) is shown in Fig. 6.37 of Chap. 6.

(a) Calculate the synchronous speed $n_s^{(rpm)}$ of the machine.
(b) Provided the relative permeability of the stator and rotor cores is $\mu_r \to \infty$ find an expression for the load line for which $\ell_g = 1$ mm, $A_g = 12{,}000$ mm^2, $\ell_m = 3$ mm and $A_m = 10{,}000$ mm^2.
(c) Draw the load line and the NdFeB characteristic and find the operating point (B_{mo}/H_{mo}) of this machine. Note: Approximate values for B_{mo} and H_{mo} are sufficient.
(d) Is this permanent-magnet material operated at the point of maximum energy product?

(e) Compute the recoil permeability μ_R of the NdFeB material.
(f) Compute the fluxes ϕ_{max}, $\phi_{totalmax}$ and the number of turns N per phase provided $E_{phrms} \approx V_{L-Lrms}/\sqrt{3}$, the winding factor is $k_{w\_stator} = 0.85$ and the iron-core stacking factor $k_s = 0.96$.

Problem 10.10: *Design of a 150 kVA, 800 Hz generator for an aircraft power supply*

A six-pole (p = 6), $S_{rated} = 150$ kVA, $V_{L\text{-}Lrated} = 480$ V, f = 800 Hz, three-phase synchronous generator has a rotor axial length of $\ell = 0.5$ m and a rotor diameter $D_{rotor\_out} = 0.3$ m (see, for example, the cross-section of Fig. 10.3d). The rotor winding consists of N_{pole}=20 turns per pole with a winding factor of k_{w_rotor}= 0.8. The peak (maximum) value of the fundamental air-gap flux density is $B_{max} = 0.8$ T and the rotor winding (field) current is $I_f = 10$ A.

(a) Calculate the peak mmf F_{rotor_peak} measured in [At].
(b) Calculate the maximum torque T_{fld_max} in [newton-meters].
(c) Determine the mechanical angular velocity ω_m in [rad/s] and the associated rotational speed n_m in [rpm].
(d) Compute the maximum output power P_{max} in [kW] the alternator can supply.

Problem 10.11: *Instantaneous and average torques*

An electromagnetic system has two windings and a uniform air gap. The self- and mutual inductances are given by (θ is the angle between the two axes of windings): $L_{11} = 6 + 1.5\cos(2\theta)$ H, $L_{22} = 4 + \cos(2\theta)$ H, and $L_{12} = L_{21} = M = 5\cos(\theta)$ H. The windings are connected to DC voltage sources. The first winding absorbs the current $i_1 = 15$ A and the second winding draws the current $i_2 = 4$ A. The instantaneous developed torque and the stored energy (as a function of θ) may be written in the following forms:

$$T_d = A_1 + A_2\sin(\theta) + A_3\sin(2\theta)\ [\text{Nm}]$$
$$W_f = B_1 + B_2\sin(\theta) + B_3\cos(\theta) + B_4\sin(2\theta) + B_5\cos(2\theta)\ [\text{J}].$$

a) Determine the values of A_1, A_2, and A_3.
b) Determine the values of B_1, B_2, B_3, B_4, and B_5.
c) For $\theta = 30^o$ find the magnitude and direction of torque applied to the rotor.

Problem 10.12: *Steady-state torque developed by a 3-phase induction motor*

A 415 V, 60 Hz, 8-pole, 855 rpm, Y-connected, 3-phase induction motor has stator resistance of 0.4 Ω, rotor copper loss of 96 W, and rotational (friction, windage) loss of 24 W. Neglecting core impedance (R_c, X_m) and assume the power factor is 0.6 lagging (consumer notation). Determine:

(a) The slip.
(b) The stator current.
(c) The output power and the efficiency.
(d) The load torque.

Problem 10.13: *Starting current of a 3-phase induction motor*

A 3-phase, Δ-connected, 4-pole, 60 Hz induction motor is rated at 25 hp and 460 V (line-to-line). Thevenin's equivalent circuit parameters are $V_{TH} = 440$ V (line-to-line) or $V_{TH\_L\text{-}N} = 254.04$ V (phase-to-neutral) and phase impedance $Z_{TH} = 1.2\Omega\ \angle 60°$. The rotor phase parameters (referred to the stator) are $R'_2 = 0.30\ \Omega$ and $X'_2 = 0.50\ \Omega$. Compute:

(a) Starting torque T_{start} and the corresponding motor speed n_{start} (in rpm).
(b) The starting current $|\tilde{I}_{start}|$ (assume $|\tilde{I}_{rotor}| = 0.7\ |\tilde{I}_{stator}|$).
(c) To limit the starting current, motor windings are connected in Y for starting. What are the new parameter values of V_{TH_new}, Z_{TH_new}, R'_{2_new}, and X'_{2_new}?
(d) For the conditions of part (c), compute the new value of the starting current $|\tilde{I}_{start\_new}|$

Problem 10.14: *Variation of starting torque of a 3-phase induction motor with terminal voltage and rotor resistance*

A 230 V, 3-phase, Δ-connected, four-pole, 60 Hz induction motor operates at a full-load speed of 1,700 rpm. The developed (steady-state) torque is T = 8 Nm. Starting torque, maximum torque and the speed corresponding to maximum torque are $T_{start} = 10$ Nm, $T_{max} = 15$ Nm, and $n_{max} = 1{,}400$ rpm, respectively. Assume $X_m \rightarrow \infty$

(a) Determine the new value of developed torque T_{new} if the supply voltage is decreased by 15%.
(b) Determine the new value of starting torque $T_{start,new}$ if the supply voltage is increased by 10%.
(c) Determine the new value of maximum torque $T_{max,new}$ if rotor resistance is doubled.
(d) Determine the new motor speed corresponding to maximum torque $n_{max,new}$ if rotor resistance is doubled.

Problem 10.15: *Synchronous generator connected to infinite bus*

A 3-phase synchronous generator is connected to an infinite bus. The infinite bus voltage and the generated voltage are $\tilde{V}_a = 1.0\,\text{pu}\,\angle 0°$ and $\tilde{E}_a = 1.0\,\text{pu}\,\angle 42.84°$, respectively. The synchronous reactance is $X_s = 0.85$ pu and armature resistance R_a is neglected.

(a) Determine (power/torque) angle δ, compute armature current $\tilde{I}_a$, power factor pf (generator notation), real power P, and reactive power Q. Draw the phasor diagram.
(b) If the prime mover torque is maintained constant at a value corresponding to P = 0.8 pu, compute the required generated voltage $\tilde{E}_{a\_new}$ at unity power factor condition. What is the new value of power/torque angle δ_{new}?

Problem 10.16: *Stability of a synchronous motor connected to infinite bus*

A synchronous motor is connected to an infinite bus via a short feeder with purely reactive impedance. The power-angle curve for transient condition is $1.3 \sin \delta'$ pu. With the motor operating initially unloaded, a shaft load of 1.0 pu is suddenly applied.

(a) Draw the transient $P' - \delta'$ characteristic showing area A1 and area A2 which can be used to determine the marginal stability limit.
(b) Compute the torque angle δ'.
(c) Compute area A1.
(d) Compute area A2.
(e) Does the motor remain in synchronism? Why or why not?

Problem 10.17: *Stability of a synchronous motor connected to infinite bus*

A synchronous motor whose input at rated operating conditions is 10 MVA is connected to an infinite bus over a short feeder whose impedance is purely reactive. The motor is rated at 60 Hz and 600 rpm. The power-angle curve for transient condition is $1.56 \sin \delta'$, where the amplitude is in pu on a 10 MVA base. With the motor operating initially unloaded a 10 MW shaft load is suddenly applied.

(a) Draw the transient $P' - \delta'$ characteristic showing area A1 and area A2 (used for margin of stability), and compute the torque angle δ_{ss}.
(b) Compute areas A1 and A2.
(c) Does the motor remain in synchronism with the infinite bus?
(d) Find the largest shaft load P_{m_max} that can be suddenly applied without loss of synchronism. What is the torque angle δ_{max} for this condition?

References

1. Fuchs, E. F.; McNaughton, G. A.: "Comparison of first-order finite difference and finite element algorithms for the analysis of magnetic fields, Part I: Theoretical analysis," *IEEE Transactions on Power Apparatus and Systems*, May 1982, Vol. PAS-101, No. 5, pp. 1170–1180.
2. McNaughton, G. A.; Fuchs, E. F.: "Comparison of first-order finite difference and finite element algorithms for the analysis of magnetic fields, Part II: Numerical results," *IEEE Transactions on Power Apparatus and Systems*, May 1982, Vol. PAS-101, No. 5, pp. 1181–1201.
3. Fuchs, E. F.; McNaughton, G. A.: "Properties of orthogonal and triangular grids for the analysis of magnetic fields based on the finite difference and finite element methods, Part I: Theoretical analysis," *Acta Technica*, March/April 1982, No. 2, pp. 168–205.
4. McNaughton, G. A.; Fuchs, E. F.; Siegl, M.: "Properties of orthogonal and triangular grids for the analysis of magnetic fields based on the finite difference and finite element methods, Part II: Numerical results based on Gaussian elimination," *Acta Technica,* March/April 1982, No. 2, pp. 206–238.
5. Fuchs, E. F.; Siegl, M.: "Properties of orthogonal and triangular grids for the analysis of magnetic fields based on the finite difference and finite element methods, Part III: Comparison of iterative solutions," *Acta Technica*, May/June 1982, No. 3, pp. 261–290.
6. Fuchs, E. F.; Chang, L. H.; Appelbaum, J.; Moghadamnia, S.: "Sensitivity of transformer and induction motor operation to power system's harmonics," *Topical Report*, Prepared for the US Department of Energy DOE-RA-50150-18, April 1983.
7. Say, M. G.: *Alternating Current Machines*, John Wiley and Sons, New York, 1983.
8. Weh, H.; Jiang, J.: "Berechnungsgrundlagen für Transversalflußmaschinen," *Archiv für Elektrotechnik*, Vol. 71, pp.187–198, Springer Verlag, 1988.

9. Teltsch, M.: *Adjustable-Speed Drive for an Electric Car with Large Rated Torque at Low Rated Speed*, M.S. Thesis, University of Colorado at Boulder, Boulder, CO, 1997.
10. Fuchs, E. F.; Pohl, G.: "Computer generated polycentric grid design and novel dynamic acceleration of convergence for the iterative solution of magnetic fields based on the finite difference method," *IEEE Transactions on Power Apparatus and Systems*, August 1981, Vol. PAS-100, No. 8, pp. 3911–3920.
11. Younglove, B. L.; Johnson, S. D.: *User's Manual for Program Fields*, Vol. 1, Department of Electrical and Computer Engineering, University of Colorado, Boulder, CO 80309, January 6, 1983.
12. Williams, R. W.: *Numerical Determination of the Flux Distribution in Salient-Pole Aircraft Alternators*, M.S. Thesis, University of Colorado, Boulder, CO 80309, 1974.
13. Fuchs, E. F.; Senske, K.: "Comparison of iterative solutions of the finite difference method with measurements as applied to Poisson's and the diffusion equations," *IEEE Transactions on Power Apparatus and Systems*, August 1981, Vol. PAS-100, No. 8, pp. 3983–3992.
14. Fuchs, E. F.: *Numerical Determination of Synchronous, Transient, and Subtransient Reactances of a Synchronous Machine*, Ph.D. Thesis, University of Colorado, Boulder, Colorado, 1970.
15. Fuchs, E. F.; Erdelyi, E. A.: "Nonlinear theory of turbogenerators, Part I: Magnetic fields at no-load and balanced loads," *IEEE Transactions on Power Apparatus and Systems*, March/April 1973, Vol. PAS-92, No. 2, pp. 583–591.
16. Fuchs, E. F.; Erdelyi, E. A.: "Nonlinear theory of turbogenerators, Part II: Load-dependent synchronous reactances," *IEEE Transactions on Power Apparatus and Systems*, March/April 1973, Vol. PAS-92, No. 2, pp. 592–599.
17. Fuchs, E. F.; Rosenberg, L. T.: "Analysis of an alternator with two displaced stator windings," *IEEE Transactions on Power Apparatus and Systems*, November/December 1974, Vol. PAS-93, No. 6, pp. 1776–1786.
18. Fardoun, A. A.; Fuchs, E. F.; Huang, H.: "Modeling and simulation of an electronically commutated permanent-magnet machine drive system using PSpice," *IEEE Transactions on Industry Applications*, July/August 1994, Vol. 30, No. 4, pp. 927–937.
19. Yildirim, D.: *Commissioning of 30 kVA Variable-Speed, Direct-Drive Wind Power Pl*ant, Ph. D. Thesis, University of Colorado at Boulder, May 1999.
20. Yildirim, D.; Fuchs, E. F.; Batan, T.: "Test results of a 20 kW direct-drive, variable-speed wind power plant," *Proceedings of the International Conference on Electrical Machines,* Istanbul, Turkey, Sept. 2–4, 1998.
21. Erdelyi, E. A.; Fuchs, E. F.; Binkley, D. H.: "Nonlinear magnetic field analysis of DC machines, Parts I, II and III," *IEEE Transactions on Power Apparatus and Systems*, Sept./Oct. 1970, Vol. PAS–89, No. 7 pp. 1546–1554.
22. Batan, T.: *Real-Time Monitoring and Calculation of the Derating of Single-Phase Transformers under (Non)Sinusoidal Operations*, Ph. D. Thesis, University of Colorado at Boulder, May 1999.
23. Fuchs, E. F., Masoum, M. A. S.: *Power Quality in Power Systems and Electrical Machines.* Elsevier, Academic Press, New York, 2008, 638 p. ISBN: 978-0-12-369536-9
24. Fuchs, E. F.; Schraud, J.; Fuchs, F. S.: "Analysis of critical-speed increase of induction machines via winding reconfiguration with solid-state switches," *IEEE Trans. on Energy Conversion,* Vol. 23, No. 3, Sept. 2008, pp. 774–780
25. Schraud, J.; Fuchs, E. F.; Fuchs, H. A.: "Experimental verification of critical-speed increase of induction machines via winding reconfiguration with solid-state switches," *IEEE Trans. on Energy Conversion*, Vol. 23, No. 2, June 2008, pp. 460–465
26. Fitzgerald, A. E.; Kingsley, Ch. Jr.; Umans, S. D.: *Electric Machinery*, 5th Edition, McGraw-Hill Publishing Company, New York, N.Y., 1990
27. Tesla, N.: "A new system of alternate current motors and transformers," *American Institute of Electrical Engineers*, May 1888.

28. Neidhöfer, G.: "Early three-phase power," *IEEE Power and Energy Magazine*, Sept./Oct. 2007, pp. 88–100.
29. Fuchs, E. F.; Masoum, M. A. S.: "Torques in induction machines due to low-frequency voltage/current harmonics," *International Journal of Power and Energy Systems,* Vol. 28, Issue 2, 2008, pp. 212–221.
30. Fuchs, E. F.: "Alternating current machine with increased torque above and below rated speed for hybrid/electric propulsion systems," *U.S. Patent Number* 8, 183, 814, May 22, 2012.
31. Fuchs, E.F.; Myat, M H.: "Speed and torque range increases of electric drives through compensation of flux weakening," *Proceedings of the 20th International Symposium on Power Electronics, Electrical Drives, Automation, and Motion,* Pisa, Italy, June 14–16, 2010.
32. White, D. C.; Woodson, H. H.: *Electromechanical Energy Conversion*, John Wiley & Sons, Inc., New York, N.Y., 1959.
33. Concordia, C. D.: *Synchronous Machines, Theory and Performance*, Schenectady, N.Y., General Electric Company, 1951.
34. Frank, U.V.: *The Feasibility of Ultra-High Speed Motors at 40,000* min^{-1} *and 75 kW with Significantly Reduced Rotor Losses*, M.S. Thesis, University of Colorado at Boulder, 1981.
35. Fuchs, E. F.; Frank, U. V.: "High-speed motors with reduced windage and eddy current losses, Part I: Mechanical design," *etzArchiv* 5 (1983) No. 1, pp. 17–23, and "Part II: Magnetic design," *etzArchiv* 5 (1983) No. 2, pp. 55–62.
36. Dubey, G. K.: (1989) *Power Semiconductor Controlled Drives*, Prentice Hall, Englewood Cliffs, N.J. 07632, 1989.
37. Fuchs, E. F.; Hanna, W. J.: "Measured efficiency improvements of induction motors with thyristor/triac controllers", *IEEE Transactions on Energy Conversion*, Vol. 17, No. 4, Dec. 2002, pp. 437–444; Vol. 19, No. 3, Sept. 2004, pp. 647–648.
38. Appelbaum, J.; Fuchs, E. F.; White, J. C.: "Optimization of three-phase induction motor design, Part I: Formulation of the optimization technique," *IEEE Transaction on Energy Conversion*, Sept. 1987, Vol. EC- 2, No. 3, pp. 407–414.
39. Appelbaum, J.; Khan, I. A.; Fuchs, E. F.; White, J. C.: "Optimization of three-phase induction motor design, Part II: The efficiency and cost of an optimal design," *IEEE Transaction on Energy Conversion*, Sept. 1987, Vol. EC- 2, No. 3, pp. 415–422.
40. Fei, R.; Fuchs, E. F.; Huang, H.: "Comparison of two optimization techniques as applied to three-phase induction motor design," *IEEE Transaction on Energy Conversion*, Dec. 1989, Vol. EC- 4, pp. 651–660.
41. Fuchs, E. F.; Chang, L. H.; Appelbaum, J.: "Magnetizing current, iron losses and forces of three-phase induction machines at sinusoidal and nonsinusoidal terminal voltages, Part I: Analysis," *IEEE Transaction on Power Apparatus and Systems*, Nov. 1984, Vol. PAS-103, No. 11, pp. 3303–3312.
42. Fuchs, E. F.; Roesler, D. J.; Chang, L. H.: "Magnetizing current, iron losses and forces of three-phase induction machines at sinusoidal and nonsinusoidal terminal voltages, Part II: Results," *IEEE Transaction on Power Apparatus and Systems*, Nov. 1984, Vol. PAS-103, No. 11, pp. 3313–3326.
43. Fuchs, E. F.; Appelbaum, J.; Khan, I. A.; Höll, J.; Frank, U. V.: "Optimization of induction motor efficiency, Volume 1: Three-phase induction motors," *Publication of the Electric Power Research Institute*, Palo Alto, California, EPRI EL-4152-CCM, July 1985, 336 pages.
44. Fuchs, E. F.; Poloujadoff, M.; Neal, G. W.: "Starting performance of saturable three-phase induction motors," *IEEE Transactions on Energy Conversion*, September 1988, Vol. EC-3, No. 3, pp. 624–635.
45. Courtesy of Siemens Corporation.
46. Li, S.; Haskew, T. A.; Challoo, R.: "Steady-state characteristic study for integration of DFIG wind turbines into transmission grid," *International Journal of Emerging Electric Power Systems*, Vol. 10, issue 1, 2009 Article 7, pp. 1–29.
47. Kling, W. L.; Slootweg, J. G.: "Wind turbines as Power Plants," *Proceedings of the IEEE/ Cigré Workshop on Wind Power and the Impacts on Power Systems*, 17–18 June 2002, Oslo, Norway.

48. Muller, S.; Deicke, M.; De Doncker, R. W.: "Doubly Fed Induction Generator Systems for Wind Turbines," *IEEE Industry Applications Magazine*, Vol. 8, No. 3, 26–33, May/June 2002.
49. Slootweg, J. G.; de Haan, S. W. H.; Polinder H.; Kling, W. L.: "General model for representing variable speed wind turbines in power system dynamics simulations," *IEEE Transactions on Power Systems*, Vol. 18, No. 1, February 2003.
50. Freris, L. L.: *Wind Energy Conversion System*, Upper Saddle River, NJ: Prentice Hall, 1990.
51. Miller, N. W.; Price W. W.; Sanchez-Gasca, J. J.: "Dynamic modeling of GE 1.5 and 3.6 wind turbine-generators," *GE Power Systems*, October 27, 2003.
52. Pena, R.; Clare, J. C.; Asher, G. M.: "Doubly fed induction generator using back-to-back PWM converters and its application to variable speed wind energy generation," *IEE Proc.-Electr. Power Appl.*, Vol. 143, No 3, May 1996.
53. Leonhard, W.: *Control of Electrical Drives*, Berlin, Germany: Springer-Verlag, 1996.
54. Mohan, N.: *Advanced Electric Drives – Analysis, Modeling and Control Using Simulink*, MN: Minnesota Power Electronics Research & Education, ISBN 0-9715292-0-5, 2001.
55. Erlich, I.; Brakelmann, H.: "Integration of wind power into the German high voltage transmission grid," *Proceedings of 2007 IEEE PES General Meeting*, 24–28 June 2007, Tampa, FL, USA.
56. Youssef, R. D.: "Integration of offshore wind farms into the local distribution network," available from http://www.berr.gov.uk/publications/index.html.
57. Li, S.; Sinha, S.: "A simulation analysis of double-fed induction generator for wind energy conversion using PSpice," *Proceedings of 2006 IEEE PES General Meeting, 18–22 June 2006, Montréal, Québec Canada.*
58. Simões, M. G.; Farret, F. A.: *Alternative Energy Systems: Design and Analysis with Induction Generators*, 2nd Edition, CRC, Boca Raton, 2008
59. http://www.vennemann-online.de/papers/Vennemann2010.pdf
60. Mohan, N.: *Electrical Drives, An Integrative Approach*, Monpere Publication, 2003.
61. *IEEE Standard Dictionary of Electrical and Electronics Terms, IEEE Std 100–1977.*
62. Huang, H.; Fuchs, E. F.; Zak, Z.: "Optimization of single-phase induction motor design, Part I: Formulation of the optimization technique," *IEEE Transaction on Energy Conversion*, June 1988, Vol. EC-3, No. 2, pp. 349–356.
63. Huang, H.; Fuchs, E. F.; White, J. C.: "Optimization of single-phase induction motor design, Part II: The maximum efficiency and minimum cost of an optimal design," *IEEE Transaction on Energy Conversion*, June 1988, Vol. EC-3, No. 2, pp. 357–366.
64. Huang, H.; Fuchs, E. F.; White, J. C.: "Optimal placement of the run capacitor in single-phase induction motor designs," *IEEE Transaction on Energy Conversion*, Sept. 1988, Vol. EC-3, No. 3, pp. 647–652.
65. Fuchs, E. F.; Vandenput, A. J.; Höll, J.; White, J. C.: "Design analysis of capacitor-start, capacitor-run single-phase induction motors," *IEEE Transaction on Energy Conversion*, June 1990, Vol. EC- 6, pp. 327–336.
66. Fuchs, E. F.; Huang, H.; Vandenput, A. J.; Höll, J.; Zak, Z.; Appelbaum, J.; Erlicki, M.: "Optimization of induction motor efficiency, Volume 2: Single-phase induction motors," *Publication of the Electric Power Research Institute*, Palo Alto, California, EPRI EL-4152-CCM, May 1987, 450 p.
67. Fuller, J. F.; Fuchs, E. F.; Roesler, D. J.: "Influence of harmonics on power system distribution protection," *IEEE Transaction on Power Delivery*, April 1988, Vol. 3, No. 2, pp. 546–554.

Chapter 11
Mechanical Loads

All components of the generic structure of the electromechanical drive system of Fig. 2.1 have been discussed, except the mechanical gear and the mechanical load. In this chapter the electrical torque T produced by the motor will be related to the mechanical load (L) torque T_L and to the angular velocity ω_m as transmitted to the mechanical load via the gear box. It is important to realize that the electrical torque characteristic $(T - \omega_m)$ of a drive motor must be compatible with the mechanical torque characteristic $(T_L - \omega_m)$ of the load. The compatibility of these two characteristics will be discussed in the first section, where the necessary and sufficient conditions for stable steady state operating points will be derived.

11.1 Steady-State Stability Criterion

According to the fundamental torque equation (see Fig. 2.2),

$$T = T_L + J\frac{d\omega_m}{dt}, \qquad (11.1)$$

where T is the developed torque of the motor measured in Nm, T_L is the load (resistive) torque (referred to the motor shaft) measured in Nm, J is polar moment of inertia of the motor–load system referred to the motor shaft measured in kgm^2, and ω_m is instantaneous angular velocity of the motor shaft measured in rad/s.

The equilibrium speed of a motor–load system is obtained when the motor torque T equals the load torque T_L. This is the speed at which the drive will normally operate in steady-state provided it is a speed of stable equilibrium. Equilibrium speed will be viewed as the stable speed provided that the operation will be restored to this speed after any small departure from it due to any disturbance in the motor–load system. The stability of an equilibrium point [1–3] can be readily investigated by using the concept of steady-state stability. In this concept, the stability of an equilibrium point is evaluated from the steady-state speed–torque curves of the motor and the load. It is assumed that any departure from the equilibrium point, due to any disturbance, will be along these curves. This in effect means that the motor is assumed to be in

E.F. Fuchs and M.A.S. Masoum, *Power Conversion of Renewable Energy Systems*,
DOI 10.1007/978-1-4419-7979-7_11, © Springer Science+Business Media, LLC 2011

electrical equilibrium for all stable operating points. The basis of this assumption is that the electrical time constant τ_a of the motor is usually negligible as compared to its mechanical time constant τ_m.

At a given equilibrium operating point, let the motor torque, load torque, and drive angular velocity be denoted by T_e, T_{Le}, and ω_{me}, respectively, where the subscript "e" stands for equilibrium. Then

$$T_e = T_{Le} \quad \text{and} \quad \frac{d\omega_{me}}{dt} = 0. \tag{11.2}$$

A disturbance in the supply, load, or any part of the drive will cause perturbations in the motor torque, load torque, and drive speed. At a time t, measured from the instant the disturbance is caused, let these perturbations be denoted by ΔT, ΔT_L, and $\Delta\omega_m$. Hence, at time t, motor torque, load torque, and motor angular velocity will be $(T_e + \Delta T)$, $(T_{Le} + \Delta T_L)$, and $(\omega_{me} + \Delta\omega_m)$, respectively. Now from (11.1)

$$J\frac{d(\omega_{me} + \Delta\omega_m)}{dt} + (T_{Le} + \Delta T_L) - (T_e + \Delta T) = 0. \tag{11.3}$$

Substitution from (11.2) gives

$$J\frac{d(\Delta\omega_m)}{dt} + \Delta T_L - \Delta T = 0. \tag{11.4}$$

This differential equation provides the relation between small perturbations around an equilibrium point. For small perturbations, the speed–torque curves of the motor and load can be assumed to be straight lines. Thus

$$\Delta T = \left(\frac{dT}{d\omega_m}\right)\Delta\omega_m \quad \text{and} \quad \Delta T_L = \left(\frac{dT_L}{d\omega_m}\right)\Delta\omega_m. \tag{11.5}$$

where $(dT/d\omega_m)$ and $(dT_L/d\omega_m)$ are the slopes of the steady-state speed–torque curves of the motor and the load at the operating point under consideration, respectively. Substituting (11.5) into (11.4) and rearranging the terms gives

$$J\frac{d(\Delta\omega_m)}{dt} + \left[\frac{dT_L}{d\omega_m} - \frac{dT}{d\omega_m}\right]\Delta\omega_m = 0. \tag{11.6}$$

This is a first-order linear differential equation. If the initial deviation in angular velocity at $t = 0$ is $(\Delta\omega_m)_0$, then the solution of (11.6) will be

$$\Delta\omega_m = (\Delta\omega_m)_0 e^{-\frac{1}{J}\left(\frac{dT_L}{d\omega_m} - \frac{dT}{d\omega_m}\right)t}. \tag{11.7}$$

An equilibrium point will be stable when $\Delta\omega_m$ approaches 0 as t approaches infinity ($t \to \infty$). For this to happen, the exponent in (11.7) must be negative. This gives

$$\left(\frac{dT_L}{d\omega_m} - \frac{dT}{d\omega_m}\right) > 0 \quad \text{or} \quad \left(\frac{dT_L}{d\omega_m} > \frac{dT}{d\omega_m}\right). \tag{11.8}$$

Equation (11.8) suggests that for an increase in speed, the load torque must exceed the motor torque so that deceleration takes place and the operation returns to the equilibrium speed. Similarly, for a decrease in speed, the motor torque must exceed the load torque so that acceleration occurs and the operation returns to the equilibrium speed.

Let us examine the equilibrium points A and B, which are obtained when an induction motor drives the load with the torque T_{L1}, as shown in Fig. 11.1. Let us first examine point A for steady-state stability. A small increase in angular velocity makes the load torque greater than the motor torque. Deceleration occurs and the operation is restored to point A. Similarly, a small decrease in angular velocity causes the motor torque to exceed the load torque. Acceleration occurs and the operation is restored to point A. Thus, A is a stable equilibrium point. Let us examine the stability of the equilibrium point B. A small increase in angular velocity causes the motor torque to exceed the load torque. Acceleration takes place and the operating point moves away from B. Similarly, a small decrease in angular velocity makes the load torque greater than the motor torque, causing deceleration and the operating point to drift away from B. Thus B is an unstable equilibrium point.

Let us now consider the equilibrium point C which is obtained when the motor drives the load torque T_{L2} which has the characteristic similar to the fan load (Fig. 11.7a). Examination of point C shows that it is a stable equilibrium point. Note that points B and C lie on the same part of the motor speed–torque curve. However, C provides the stable operation but not B. This shows that the stability of an equilibrium point depends not on either the motor characteristic alone or the load characteristic alone but on the relative nature of the two and their slopes at the point of intersection.

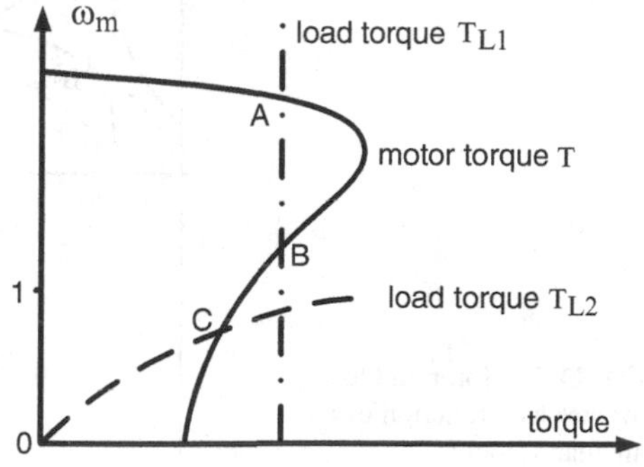

Fig. 11.1 Steady-state stability of equilibrium points

Notice that the part of the motor characteristic on which points B and C are situated has a positive slope. Such a motor characteristic gives an unstable operation with most loads (e.g., load torque T_{L1}), and therefore, it is sometimes called a statically unstable characteristic.

Application Example 11.1: Steady-State Stability of Induction Machines with Harmonic Torque [4–8]

Determine the stability of the points A to L as shown in Fig. 11.2. It will be sufficient to identify each operating point either with (S) for stable or with (U) for unstable.

Solution

A (S), B(S), C(U), D(U), E(S), F(U), G(U), H(S), I(U), J(S), K(S), L(U).

Application Example 11.2: Steady-State Stability of Electric Machines

In Fig. 11.3 the machine torque is given by the full (____) line, the load torque T_{LA} is given by the dashed (- - - - - -) line, the load torque T_{LB} is given by the

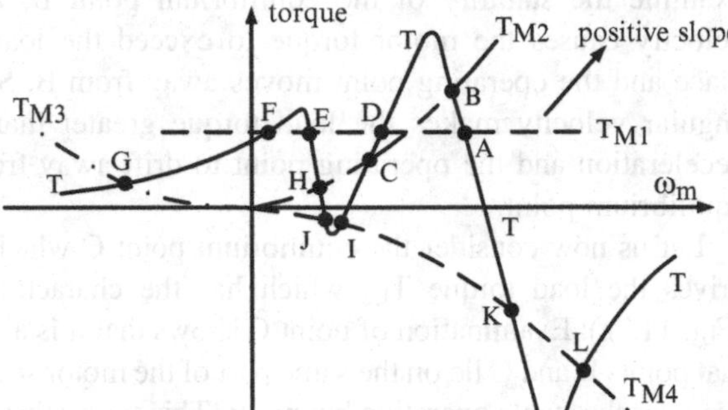

Fig. 11.2 Induction machine torque T and load torques T_{M1}, T_{M2}, T_{M3}, T_{M4} as a function of the angular velocity

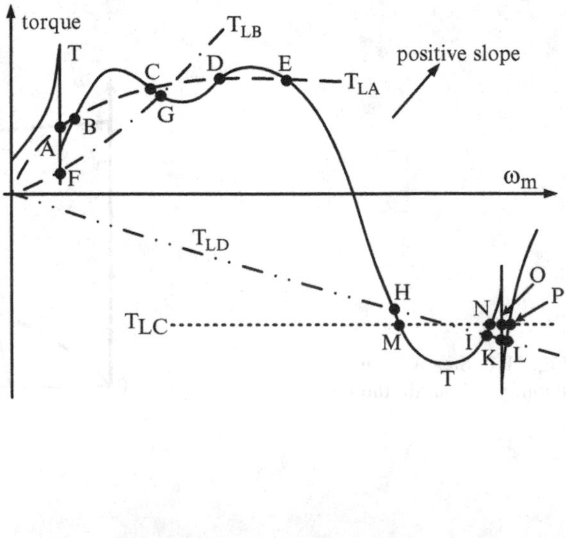

Fig. 11.3 Motor and load torques as a function of the angular velocity

dashed-dotted (· - · - · - · -) line, the load torque T_{LC} is given by the dotted (· · · · ·) line, and the load torque T_{LD} is given by the doubly dotted–dashed (·· - ·· -) line.
Find the stable operating points for load torques T_{LA} T_{LB}, T_{LC}, and T_{LD}.

Solution

Stable operating points for T_{LA}: A, C, E; stable operating points for T_{LB}: F, G; stable operating points for T_{LC}: M, O; stable operating points for T_{LD}: H, K.

Application Example 11.3: Steady-State Stability of a Pump Drive

A centrifugal pump has the torque characteristic $T_L = (\omega_m)^2$, and is driven by a motor (either of the DC or induction type) with the torque characteristic

$$T = \frac{(1 - \omega_m)}{[1 + (1 - \omega_m)^2]}.$$

(a) Plot T_L and T as a function of ω_m. Note ω_m, T, and T_L are expressed in per unit (pu).
(b) Find (e.g., graphically or iteratively) the steady-state angular velocity $\omega_m = \omega_{m\ \text{steady-state}} < \omega_{m0}$, where $\omega_{m0} = 1$ pu is the angular velocity of the motor at no-load (torque T = 0).
(c) Show that for $0 < \omega_{m\ \text{steady-state}} < \omega_{m0}$ the operating point is inherently stable from a steady-state point of view.

Solution

(a) Figure 11.4 shows the plots of T and T_L as a function of ω_m.
(b) The steady-state operating point is obtained from $T = T_L$ or

$$\omega_m^2 = \frac{(1 - \omega_m)}{[1 + (1 - \omega_m)^2]}$$

resulting in $\omega_{m\ \text{steady-state}} = 0.5916$ pu.
(c) The steady-state stability criterion (11.8) is

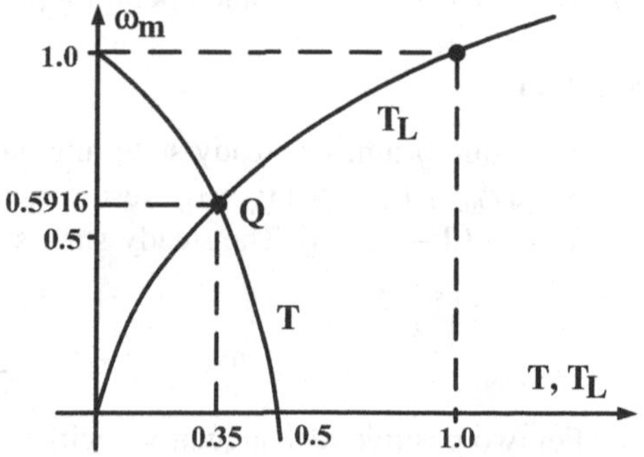

Fig. 11.4 Coordinate system for plotting ω_m as a function of T and T_L in per unit (pu)

$$\left(\frac{dT_L}{d\omega_m} > \frac{dT}{d\omega_m}\right).$$

From $T_L = (\omega_m)^2$ and

$$T = \frac{(1-\omega_m)}{[1+(1-\ \omega_m)^2]}$$

one can write the differentials

$$\frac{\partial T_L}{\partial \omega_m} = 2\omega_m,$$

$$\frac{\partial T}{\partial \omega_m} = \frac{-[1+(1-\omega_m)^2]+(1-\omega_m)\cdot 2\cdot(1-\omega_m)}{[1+(1-\ \omega_m)^2]^2}.$$

For $\omega_{m\ \text{steady-state}} = 0.5916$ pu one obtains the relation

$$\left.\frac{\partial T_L}{\partial \omega_m}\right|_{\omega m=0.5916} = 2\cdot 0.5916 = 1.1832 > \frac{-[1+0.16679]+2\cdot 0.16679}{[1+0.16679]^2}$$
$$= -1.10208 = \left.\frac{\partial T}{\partial \omega_m}\right|_{\omega m=0.5916}.$$

From this follows that introducing $\omega_m = \omega_{m\ \text{steady-state}} = 0.5916$ pu into (11.8) the inequality is satisfied. That is, the operating point Q of Fig. 11.4 is stable.

Application Example 11.4: Steady-State Stability of a Motor Drive

A motor drive develops the torque $T = a\omega_m + b$ and is loaded by the torque $T_L = c\,\omega_m^2 + d$, where a, b, c, and d are positive real constants.

(a) Find the steady-state angular velocities as a function of the above constants. What relation must exist between these constants for the drive to have two positive real angular velocities?
(b) Will the drive have stable operating points as obtained in part (a)?

Solution

(a) The equilibrium or steady-state angular velocities are obtained from $T = T_L$ or $(a\,\omega_m + b) = (c\,\omega_m^2 + d)$ resulting in the second-order equation $c\,\omega_m^2 - a\,\omega_m + (d - b) = 0$. The steady-state solutions are

$$\omega_m = \frac{a \pm \sqrt{a^2 - 4c(d-b)}}{2c}.$$

For two positive real angular velocities the relation $a^2 \rangle 4c(d-b)$ must be valid.

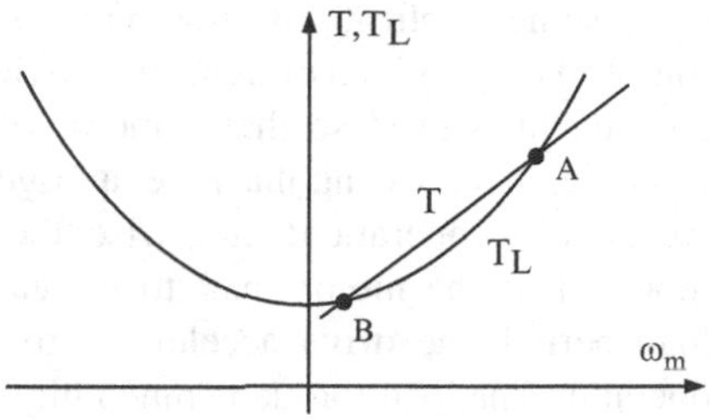

Fig. 11.5 Graphical representation of the two operating points A and B

(b) The partial derivatives are $\partial T_L/\partial\,\omega_m = 2c\omega_m$, $\partial T/\partial\,\omega_m = a$ resulting in the two differentials as a function of the above constants $\partial T_{LA}/\partial\,\omega_m = a + \sqrt{a^2 - 4c(d-b)}$, and $\partial T_{LB}/\partial\,\omega_m = a - \sqrt{a^2 - 4c(d-b)}$. Point A is a stable operating point and point B is unstable. The motor and load torques and the two operating points are depicted in Fig. 11.5.

11.2 Mechanical Load Torques

Equation (11.1) shows that the torque developed by the motor T is counterbalanced by a load torque T_L and the dynamic torque $J(d\omega_m/dt)$. The torque component $J(d\omega_m/dt)$ is called the inertia torque because it is present only during transient operations.

The drive accelerates or decelerates depending on whether T is greater or less than T_L. During the acceleration period, the motor should supply not only the load torque but also an additional component $J(d\omega_m/dt)$ to overcome the inertia of the drive. In applications having a load with large inertia, such as trains, the motor torque must exceed the load torque by a large amount to get adequate acceleration. Similarly, in applications requiring fast response, the motor torque should be maintained at the highest value and the motor–load system should be designed to have the lowest inertia. When the speed increases, the kinetic energy of the drive given by $E_{kin} = (1/2)J\,\omega_m^2$ also increases, and, therefore, in addition to the energy supplied to the load, the motor should also supply the kinetic energy. During the deceleration period, the dynamic torque $J(d\omega_m/dt)$ changes sign, and thus assists the motor torque T in maintaining the motion of the drive by extracting energy from the stored kinetic energy. When the load has a high inertia, the motor must produce a large braking torque (negative T) to get adequate deceleration. When fast response is required, the braking torque must be maintained at the highest value and the motor–load system must be designed with the lowest possible inertia.

When, for a short time, the load torque T_L exceeds the maximum torque capability of the motor running at a given speed, deceleration occurs and the dynamic torque assists the motor torque in maintaining the motion.

In some applications, involving a large torque of relatively short duration followed by a no-load or light-load period of sufficient duration, the dynamic torque component is used so that a motor of smaller rating can be employed [9]. For example, in a pressing machine, a large torque of short duration is required during the pressing operation; otherwise the torque is nearly zero. A flywheel [9] is mounted on the motor shaft to increase the equivalent inertia J. During the no-load period, the drive accelerates to store kinetic energy. During the pressing operation, the load torque is much higher compared to the motor torque. Deceleration occurs, producing an inertia torque, which together with the motor torque are able to produce the torque required for the pressing operation. In the absence of the flywheel, the motor will need to supply the entire torque required for the pressing operation, and therefore the motor rating has to be much higher.

11.2.1 Components of the Load Torque T_L

The load torque T_L can be further divided into the following components:

1. *Frictional torque T_F.* Friction will be present at the motor shaft and also in various parts of the load. The frictional torque T_F is the equivalent value of the various (dry or Coulomb friction, viscous friction, and standstill friction) frictional torques referred to the motor shaft.
2. *Windage torque T_W.* When a motor runs, the air in the air gap generates a torque opposing the motion.
3. *Torque required to do the useful mechanical work, T_M.* The nature of this torque depends on the type of load. It may be constant and independent of speed, it may be some function of speed, it may be time-invariant or time-variant, and its nature may also vary with the change in the load's mode of operation [9].

The variation of the frictional torque with angular velocity is shown in Fig. 11.6a. Its value at standstill is higher than its value at slightly above zero angular velocity. Friction at zero speed is called standstill friction. For the drive to start, the motor torque should at least exceed the standstill frictional torque. The frictional torque can be resolved into three components as shown in Fig. 11.6b: standstill (T_S), dry or Coulomb (T_C), and viscous (T_V) frictional torques.

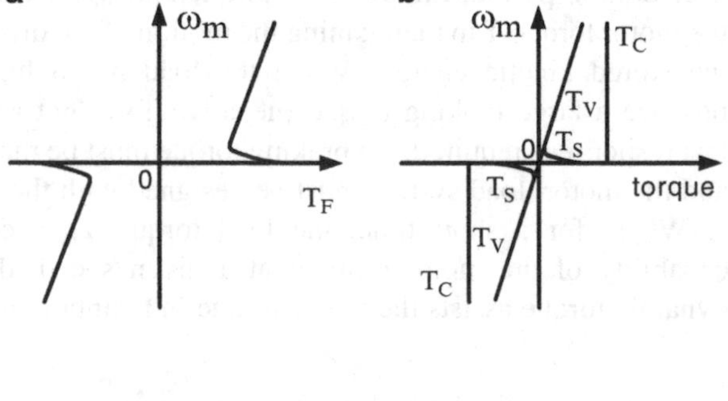

Fig. 11.6 Frictional torque and its components T_S, T_C, and T_V

The component T_V which varies linearly with speed is called viscous frictional torque and is given by the following equation

$$T_V = B\omega_m, \tag{11.9}$$

where B is the viscous or fluid friction coefficient.

The component T_C which is independent of speed is known as Coulomb friction or dry friction. A third component T_S accounts for the additional torque present at standstill. Since T_S is present only at standstill, it is not taken into account in a dynamic analysis.

The windage torque T_W which is (mostly) proportional to speed squared, is given by the following equation

$$T_W = C_0(\omega_m)^2. \tag{11.10}$$

where C_0 is a constant.

From the preceding discussion, for finite angular velocities one obtains the load torque

$$T_L = T_M + B\omega_m + T_C + C_0(\omega_m)^2. \tag{11.11}$$

In many applications $(T_C + C_0(\omega_m)^2)$ is very small compared to $B\omega_m$ and negligible compared to T_M. To simplify the analysis, the term $(T_C + C_0(\omega_m)^2)$ is approximately accounted for by updating the operating point-dependent value of the viscous friction coefficient, B. With this approximation, from (11.1)

$$T = J\frac{d\omega_m}{dt} + T_M + B\omega_m. \tag{11.12}$$

When T_M and T are either constant or proportional to angular velocity, (11.12) will be a first-order linear differential equation which can be solved either analytically or by Mathematica.[10]/Matlab [11].

11.2.2 Some Common Load Torques

It is interesting to learn about the speed–torque requirements of specific applications. Figure 11.7 shows speed–torque plots for some applications. In the case of centrifugal pumps, blowers, fans, and other loads involving the turbulent flow of fluid, the load–torque relationship varies as the square of the angular velocity (or any other exponential function) as shown in Fig. 11.7a. This represents the windage torque T_W given by (11.10). The windage force/torque is also an important component at high speeds in traction applications (e.g., trains, cars).

The variation of the traction load torque with angular velocity, excluding the torque due to gravity, is shown in Fig. 11.7b. It is applicable to trains and road

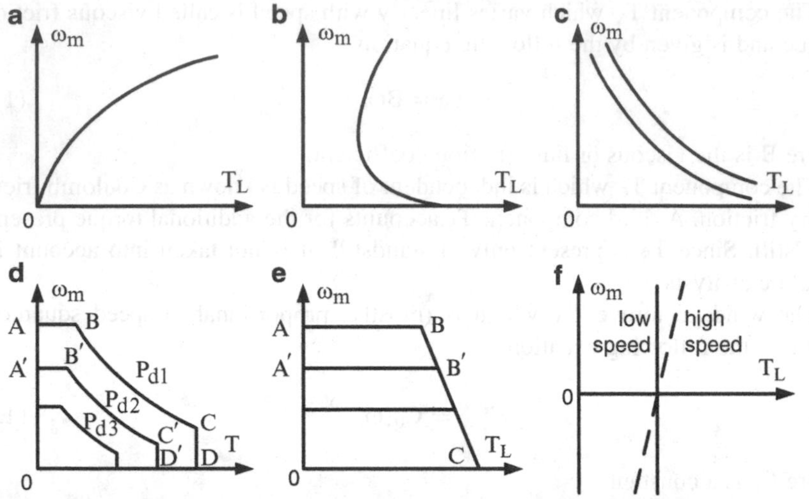

Fig. 11.7 Some examples of load torques as a function of the angular velocity ω_m: (**a**) fan and centrifugal pumps, (**b**) traction excluding gravity, (**c**) coiler drives, (**d**) diesel-electric locomotive, (**e**) excavators, (**f**) hoist

vehicles. It is comprised of the windage (T_W), viscous friction (T_V), Coulomb friction (T_C), and standstill (T_S) torques.

For example Fig. 11.7c illustrates the angular velocity–torque curves for applications where the driving motor is required to operate at constant power. One such example is a coiler drive which is used in steel strips/sheets, paper, and plastic mills. Figure 11.7d shows the required motor angular velocity–torque characteristics of a (diesel) electric locomotive drive where the maximum torque and angular velocity are limited. The motor angular velocity–torque requirements for excavators are depicted in Fig. 11.7e. The purpose of the excavator is to dig earth. While digging, it may come across a rock. The motor will then simply stop. The crane–hoist characteristics are shown in Fig. 11.7f. In a low-speed hoist, the torque is mainly due to gravity, which is constant and independent of angular velocity. In high-speed hoists, the viscous friction and windage torques also form an appreciable portion of the load torque.

11.2.3 Classification of Load Torques

Various load torques can be classified into two broad categories:

1. Active load torques.
2. Passive load torques.

Load torques which have the potential to drive the motor under equilibrium conditions are called active load torques. Such load torques usually retain their sign when the direction of the drive rotation is changed. Torque due to the force of gravity and torques due to tension, compression, and torsion undergone by an elastic body come under this category.

Load torques which always oppose the motion and change their sign on the reversal of motion are called passive load torques. Torques due to friction, cutting, etc. are in this category.

Application Example 11.5: Starting of a DC Motor

For the starting of (any type of) motors with a rating of larger than about $P_{out} = 1$ hp, a starting circuit must be used so that the starting current can be limited to an acceptable value.

(a) Sketch for a DC motor an appropriate starting circuit, where for the limitation of the motor armature current $I_{a\_starting}$ during start-up a two (or three or more)-step procedure is used.

(b) The starting of a DC motor with a two-step procedure is to be analyzed based on the solution of a first-order differential equation as follows:
Hint: For parts (b1) and (b2) you may want to introduce two different coordinate systems for the time axis t and t_{new}, as illustrated in Fig. 11.8.

(b1) At time $t = 0$ the motor torque $T = 100$ Nm, the useful mechanical load torque $T_M = 40$ Nm, and the viscous frictional torque $T_v = B\omega_m$ where $B = 2$ Nm/(rad/s) are applied at the shaft. Find the solution $\omega_m(t)$ of the differential equation $T = T_M + T_V + J(d\ \omega_m/dt)$, where $J = 1$ Nm/(rad/s^2) and calculate the time t_1 when 95% of the change in angular velocity $\omega_m(t_1)$ has been reached (see Fig. 11.8). Calculate the angular velocity $\omega_m(t_1)$ at time t_1, that is, $0.95 \cdot \omega_{m\_ss}$.

(b2) At time $t = t_1$ the motor torque $T = 160$ Nm, the useful mechanical load torque $T_M = 40$ Nm, and the viscous frictional torque $T_V = B\omega$ where $B = 2$ Nm/(rad/s) are applied at the shaft. Find the solution $\omega_m(t)$ of the differential equation $T = T_M + T_V + J(d\ \omega_m/dt)$, where $J = 1$ Nm/(rad/s^2) and calculate the time t_2 when 95% of the change in angular velocity ω_{m_tnew}

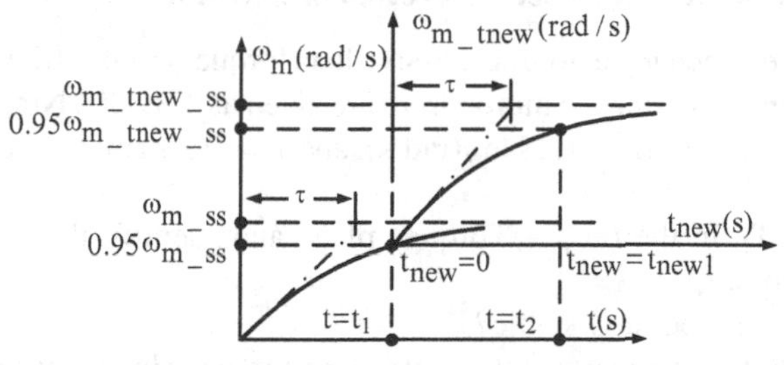

Fig. 11.8 Qualitative sketch of the angular velocity versus time diagram of a DC motor during start-up based on the two-step procedure. Note that ω_{m_ss} is the angular velocity which will be reached at $t \to \infty$ after the first load step has occurred, and $\omega_{m\_tnew\_ss}$ is the angular velocity which will be reached at $t \to \infty$ after the second load step has occurred

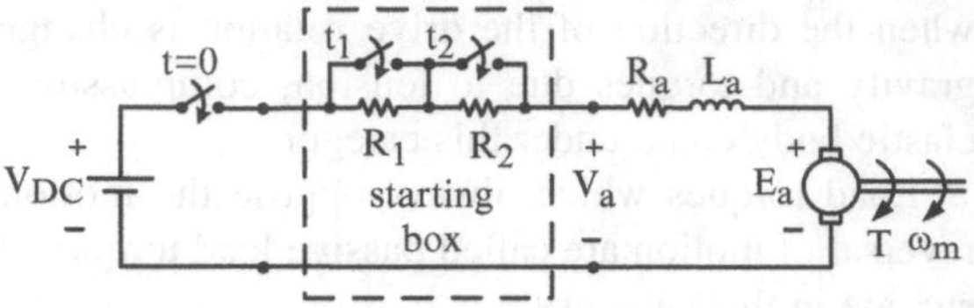

Fig. 11.9 Starting circuit for DC motor

has been reached (see Fig. 11.8), that is, $0.95{\cdot}\omega_{m\_tnew\_ss}$. Calculate the final steady-state (ss) angular velocity $\omega_m(t_2)$ at time t_2, that is, $\omega_{m\_final\_ss} = \omega_m(t_2) = 0.95\omega_{m\_ss} + 0.95\omega_{m\_tnew\_ss}$.

Solution

(a) A starting circuit for a DC motor is shown in Fig. 11.9 where a two-step procedure is used.

(b) (b1) The differential equation to be solved is $T = T_M + T_V + J(d\,\omega_m/dt)$ or $100 = 40 + 2\omega_m + d\omega_m/dt$ which results in $0.5(d\omega_m/dt) + \omega_m = 30$ with the initial condition $\omega_m(t = 0) = 0$ and the time constant $\tau = J/B = 0.5$ s. The solution of this differential equation is for the initial condition $\omega_m(t = 0) = 0$ $\omega_m(t) = 30(1 - e^{-t/\tau})$ resulting with $0.95\ \omega_{m\_ss} = 0.95{\cdot}30$ rad/s in $0.95 \cdot 30 = 30(1 - e^{-t_1/\tau})$ or $t_1 = \tau\{-\ln(0.05)\} = 1.4979$ s. The angular velocity at time t_1 is $\omega_m(t_1) = 0.95{\cdot}\omega_{m\_ss} = 28.5$ rad/s.

(b2) Introducing a new coordinate system $(\omega_{m\_tnew} - t_{new})$ one obtains the differential equation $160 = 40 + 2\omega_{m\_tnew} + d\omega_{m\_tnew}/dt_{new}$ or $60 = \omega_{m\_tnew} + \tau{\cdot}d\omega_{m\_tnew}/dt_{new}$ with the initial condition $\omega_{m\_tnew}(t_{new} = 0) = 28.5$ rad/s in the time t_{new} coordinate system. The solution of the above differential equation is $\omega_{m\_tnew}(t_{new}) = 60(1 - e^{-t_{new}/\tau}) + 28.5e^{-t_{new}/\tau}$ resulting in $0.95 \cdot 60 = 60(1 - e^{-t_{new1}/\tau}) + 28.5e^{-t_{new1}/\tau}$ or $t_{new1} = 1.1757$s and $t_2 = t_1 + t_{new1} = 1.4979$ s + 1.1757 s = 2.6736 s. The final steady-state (ss) angular velocity is $\omega_{m\_final\_ss} = 0.95\omega_{m\_ss} + 0.95\omega_{m\_tnew\_ss} = 28.5$ rad/s + 57 rad/s = 85.5 rad/s.

Application Example 11.6: Speed Reversal of a Motor

The motor torque, mechanical torque, passive load torque, viscous friction coefficient, and polar moment of inertia of a motor drive are given as $T = 500$ Nm, $T_M = 50$ Nm, $T_L = (T_M + 5\omega_m)$ Nm, $B = 1.2$ Nm/(rad/s), and $J = 0.3$ kgm^2, respectively.

(a) Plot T and T_L in the (ω_m – Torque) plane and identify the steady-state (ss) operating point Q.

(b) Is this operating point Q stable?

Hint: For part c you may want to introduce the coordinate system $(\omega_{m\_tnew} - t_{new})$ as illustrated in Fig. 11.10.

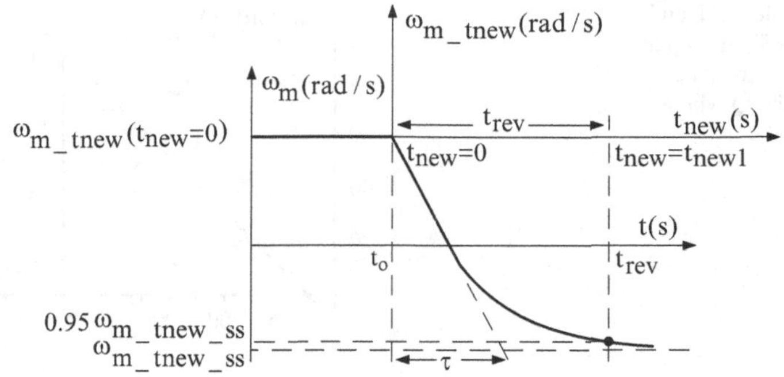

Fig. 11.10 Definition of speed reversal time

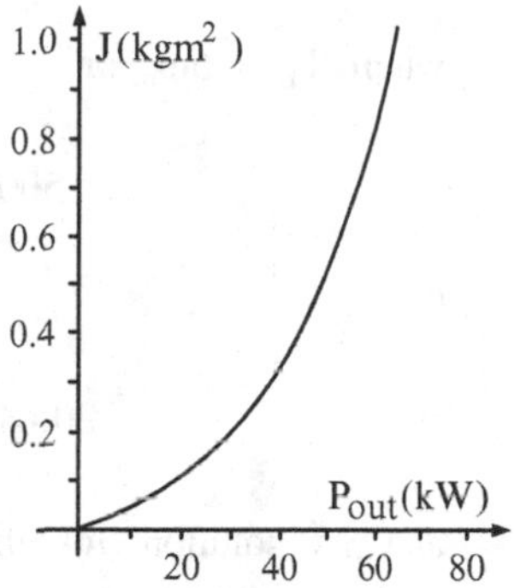

Fig. 11.11 Inertia-output power relationship for standard DC motors

(c) Calculate the time it takes for speed reversal, t_{rev}, from positive angular velocity $\omega_{m\_tnew}(t_{new} = 0)$ to negative angular velocity $0.95\omega_{m\_tnew\_ss}$ as illustrated in Fig. 11.10 provided the negative motor torque is $T = -500$ Nm, the useful mechanical torque is $T_M = 0$, and $T_L = 5\omega_{m\_tnew}$, that is, there is only viscous friction present.
(d) What is the rated power of the motor for the condition of (a)?
(e) Note the inertia-power relationship (Fig. 11.11) for commonly available motors. How could you reduce the inertia sufficiently in order to achieve a speed reversal time of $t_{rev\_new} = 0.1$ s?

Solution

(a) Figure 11.12 shows the plot of T and T_L in the (ω_m − Torque) plane where ω_{m_rat} at operating point Q is $\omega_{m\_rat} = 90$ rad/s.
(b) The steady-state (ss) operating point Q is stable due to $(\partial T_L/\partial \omega_m)\rangle(\partial T/\partial \omega_m)$, where $(\partial T_L/\partial \omega_m)\rangle 0$ and $\partial T/\partial \omega_m = 0$.
(c) The equation of motion is

Fig. 11.12 Plot of T and T_L in the (ω_m – Torque) plane with stable steady-state operating point Q where ω_{m_rat}=90 rad/s

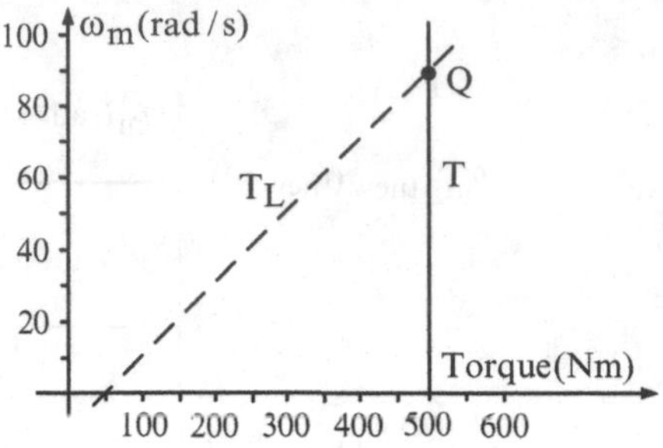

$$T = J\frac{d\omega_{m\_tnew}}{dt_{new}} + T_L + B\,\omega_{m\_tnew},$$

where $T_L = 5\omega_m$ or

$$-500 = (0.3\frac{d\omega_{m\_tnew}}{dt_{new}} + 6.2\,\omega_{m\_tnew})$$

or

$$-80.645 = (0.0484\frac{d\omega_{m\_tnew}}{dt_{new}} + \omega_{m\_tnew})$$

and its solution for the initial condition $\omega_{m\_tnew}(t_{new} = 0) = 90$ rad/s is $\omega_m(t) = -80.645(1 - e^{-t_{new}/\tau}) + 90e^{-t_{new}/\tau}$ where the time constant is $\tau = J/B = 0.3/6.2$ s $= 0.0484$ s.

The change in the angular velocity is $0.95 \cdot \omega_{m\_tnew\_ss} = 0.95 \cdot (-80.645$ rad/s). Introducing this value into the above solution one gets $0.95 \cdot (-80.645) = \{-80.645(1 - e^{-t_{new1}/\tau}) + 90e^{-t_{new1}/\tau}\}$ or $-76.61 = (-80.645 + 170.645e^{-t_{new1}/\tau})$ resulting in $t_{new1} = t_{rev} = 0.18$ s.

(d) The rated power is with T = 500 Nm and $\omega_{m\_rat} = 90$ rad/s $P = T\omega_m = 500 \cdot 90 = 45$ kW which has an inertia (approximately according to Fig. 11.11) of $J = 0.3$ kgm^2.

(e) Reduce the inertia constant J by a factor of 2, that is, $J_{new} = 0.15$ kgm^2 resulting in a reversal time of $t_{rev\_new} = 0.09$ s. Redesign the motor so that it has a smaller diameter and a longer axial length.

Application Example 11.7: Starting Time t_{start} and Reversing Time $t_{reverse}$ of a DC Motor Drive with a Current Source

The variable-speed drive of Fig. 11.13a consists of DC current source I_a, separately excited DC motor and load. The DC current source permits that the torque T of the motor to be adjusted to any desired value.

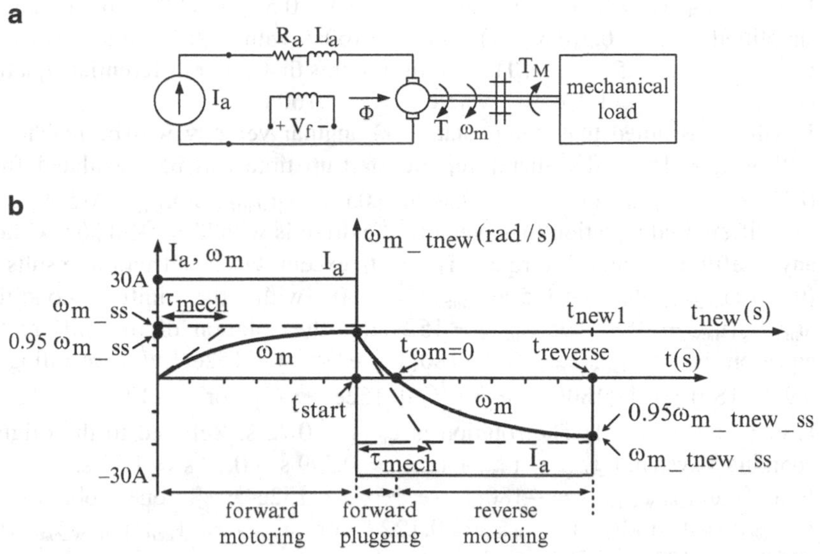

Fig. 11.13 (**a**) Variable-speed drive. (**b**) Qualitative sketch current of I_a and angular velocity ω_m as a function of time defining t_{start}, $t_{reverse}$ and $t_{\omega m=0}$

(a) Calculate the motor torque $T = (k_e\Phi)\ I_a$ (in Nm) for $(k_e\Phi) = 3$ Vs/rad and $I_a = 30$ A (see Fig. 11.13a). Derive an expression for the angular velocity $\omega_m(t)$ during starting of the drive, provided the load torque is $T_L=\{T_M+B\omega_m(t)\}$ expressed in Nm, where $T_M = 10$ Nm, the axial moment of inertia for the entire drive (motor and load) referred to the motor shaft is $J = 0.1\ \text{Nm/rad/s}^2$, the viscous damping coefficient is $B = 0.5$ Nm/rad/s, and the mechanical time constant is $\tau_{mech} = J/B$.

(b) Calculate the time t_{start} taken for the angular velocity to change from 0 to 95% of the steady-state (ss) angular velocity ω_{m_ss} for forward motoring (see Fig. 11.13b).

(c) At time t_{start} the drive is being reversed through $I_a = -30$ A (see Fig. 11.13b). Compute the angular velocity $\omega_m(t)$ during speed reversal provided $T_L = B\omega_m(t)$ expressed in Nm, that is, speed reversal occurs at no-load, where the useful mechanical torque is $T_M = 0$, and only viscous friction is present. Calculate the time $t_{reverse}$ taken for the angular velocity to change from $0.95\omega_{m\_ss}$ to $0.95\omega_{m\_tnew\_ss}$, where $\omega_{m\_tnew\_ss}$ is the steady-state angular velocity for reverse motoring.

(d) Determine the time $t_{\omega m=0}$ when the motor drive reaches zero angular velocity $\omega_m(t_{\omega m=0}) = 0$ after speed reversal has been initiated and forward plugging occurs.

Solution

(a) The motor torque is $T = 3{\cdot}30$ Nm $= 90$ Nm. The expression for the angular velocity $\omega_m(t)$ during starting is given by the differential equation

$J(d\omega_m/dt) + T_L = T$ or $(0.1(d\omega_m/dt) + 10 + 0.5\omega_m) = 90$, which can be simplified to $(0.2(d\omega_m/dt) + \omega_m) = 160$ with the time constant $\tau = J/B = 0.1/0.5 = 0.2$ s. The solution of this first-order differential equation is $\omega_m(t) = \omega_{m\_ss}(1 - e^{-t/\tau})$, where $\omega_{m\_ss} = 160$ rad/s.

(b) It will be assumed that steady-state (ss) angular velocity will be reached at $0.95\omega_{m\_ss} = 152$ rad/s, therefore, the start-up time can be calculated from $0.95\ \omega_{m\_ss} = \omega_{m\_ss}(1 - e^{-t_{start}/\tau})$ as $\ln(0.05) = -t_{start}/\tau$ or $t_{start} = 0.599$ s.

(c) The differential equation for reversing the drive is with $T = -90$ Nm without any useful mechanical torque ($T_M = 0$) except viscous friction results in $(0.1(d\ \omega_{m\_tnew}/dt_{new}) + 0.5\ \omega_{m\_tnew}) = -90$ with the initial condition $\omega_{m\_tnew}(t_{new} = 0) = 0.95\ \omega_{m\_ss} = 152$ rad/s the solution of this differential equation is $\omega_{m\_tnew}(t_{new}) = \{-180(1 - e^{-t_{new}/\tau}) + 152e^{-t_{new}/\tau}\}$ resulting in $0.95(-180) = \{-180(1 - e^{-t_{new1}/\tau}) + 152e^{-t_{new1}/\tau}\}$ or $-171 = \{-180 + (180 + 152)e^{-t_{new1}/\tau}\}$. The solution is $t_{new1} = 0.72$ s. Referred to the original coordinate system $t_{reverse} = t_{start} + t_{new1} = 0.599\ s + 0.72\ s = 1.32$ s.

(d) From $\omega_{m\_tnew}(t_{new}) = -180(1 - e^{-t_{new}/\tau}) + 152e^{-t_{new}/\tau}$ one obtains at $\omega_{m\_tnew} = 0$ rad/s $t_{new\_\omega m=0} = 0.122$ s or $t_{\omega m=0} = (t_{start} + t_{new\_\omega m=0}) = (0.599 + 0.122) = 0.721$ s.

Application Example 11.8: Starting and Reversing of an Electric Machine

A motor driving a load must be started. To limit the motor current during start-up a two-step procedure is used.

(a) At time $t = 0$ the motor torque $T = 100$ Nm and load torque $T_L = (T_M + B\omega_m)$ Nm, where $T_M = 40$ Nm, and $B = 1$ Nm/(rad/s) are applied at the shaft. Find the solution $\omega_m(t)$ of the differential equation $T = T_L + J(d\ \omega_m/dt)$, where $J = 1$ Nm/(rad/s^2) and calculate the time t_1 when 95% of the change in angular velocity has been reached (see Fig. 11.14). Calculate the angular velocity $\omega_m(t_1) = 0.95\ \omega_{m\_ss}$ at time t_1.

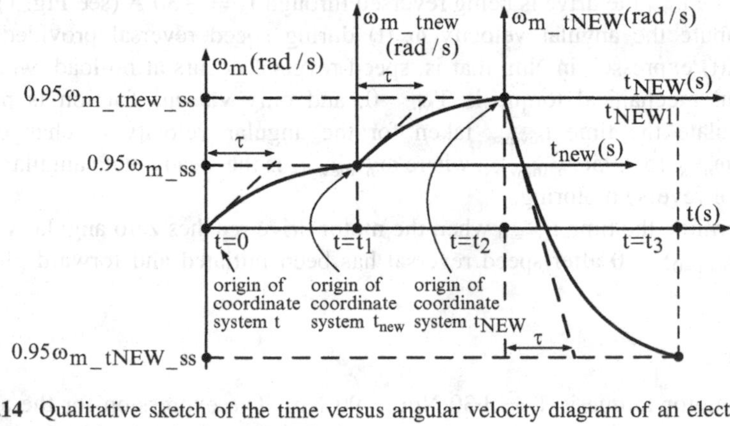

Fig. 11.14 Qualitative sketch of the time versus angular velocity diagram of an electric motor during start-up and reversal of angular velocity

Hint: for parts (b) and (c) you may want to introduce new coordinate systems $(\omega_{m\_tnew} - t_{new})$ and $(\omega_{m\_tNEW} - t_{NEW})$, respectively.

(b) At time $t = t_1$ the motor torque $T = 160$ Nm, the mechanical torque $T_M = 40$ Nm, and the load torque $T_L = (T_M + B \cdot \omega_{m\_tnew})$ Nm, where $B = 1$ Nm/(rad/s), are applied at the shaft. Find the solution $\omega_{m\_tnew}(t)$ of the differential equation $T = T_L + J(d\,\omega_{m\_tnew}/dt_{new})$, where $J = 1$ Nm/(rad/s^2) and calculate the time t_2 when 95% of the change in angular velocity has been reached (see Fig. 11.14). Calculate the angular velocity $\omega_m(t_2)$ at time t_2.

(c) The reversal of the motor occurs with the load torque $T_L = B\omega_{m\_tNEW}$ Nm where $B = 1$ Nm/(rad/s), the mechanical torque $T_M = 0$, and the motor torque is $T = -80$ Nm. Find the solution $\omega_{m\_tNEW}(t)$ of the differential equation $T = T_L + J(d\,\omega_{m\_tNEW}/dt_{NEW})$, where $J = 1$ Nm/(rad/s^2). Calculate the reversal time t_{NEW1} when 95% of the change in angular velocity has been reached (see Fig. 11.14). Calculate the angular velocity $\omega_m(t_3)$ at time t_3.

Solution

(a) The solution of the differential equation $T = T_L + J(d\omega_m/dt)$ or $60 = \omega_m + (d\omega_m/dt)$ for the initial condition $\omega_m(t = 0) = 0$ is with the time constant $\tau = J/B = 1$ s and $\omega_m(t) = 60(1 - e^{-t/\tau}) + \omega_m(t = 0)e^{-t/\tau} = 60(1 - e^{-t/\tau})$. The startup time t_1 is obtained from $0.95 \cdot 60 = 60(1 - e^{-t_1/\tau})$ or $t_1 = 2.9957$ s. The angular velocity at time t_1 is $\omega_m(t_1) = 0.95\omega_{m\_ss} = 57$ rad/s.

(b) Introducing the new coordinate system t_{new} one can write the differential equation $120 = \omega_{m\_new} + (d\,\omega_{m\_new}/dt_{new})$ with the solution, taking into account the initial condition $\omega_{m\_new}(t_{new=0}) = \omega_m(t_1) = 57$ rad/s, $\omega_{m\_tnew}(t_{new}) = \{120(1 - e^{-t_{new}/\tau}) + 57e^{-t_{new}/\tau}\}$ resulting in $0.95 \cdot 120 = 120(1 - e^{-t_{new1}/\tau}) + 57e^{-t_{new1}/\tau}$ or $t_{new1} = 2.351$ s and $t_2 = t_1 + t_{new1} = 2.996$ s $+ 2.351$ s $= 5.347$ s. The angular velocity at t_2 is $0.95\omega_{m\_ss} + 0.95\omega_{m\_tnew\_ss} = 57 + 114 = 171$ rad/s.

(c) The introduction of the additional new coordinate system t_{NEW} leads with the differential equation $(-80) = \omega_{m\_tNEW} + (d\omega_{m\_tNEW}/dt_{NEW})$ and the initial condition $\omega_{m\_tNEW}(t_{NEW=0}) = 0.95\omega_{m\_tnew\_ss} = 114$ rad/s to the solution $\omega_{m\_tNEW}(t_{NEW}) = \{-80(1 - e^{-t_{NEW}/\tau}) + 114e^{-t_{NEW}/\tau}\}$. The reversal time t_{NEW1} can be calculated from $0.95(-80) = \{-80(1 - e^{-t_{NEW1}/\tau}) + 114e^{-t_{NEW1}/\tau}\}$ leading to $t_{NEW1} = 3.88156$ s or $t_3 = 5.347$ s $+ 3.88156$ s $= 9.229$ s. The angular velocity at t_3 is $0.95\omega_{m\_tNEW}(t_{NEW1}) = 0.95\omega_{m\_tNEW\_ss} = -76$ rad/s.

11.3 Mechanical Gears

Until now in this text a direct drive between motor coupling and mechanical load has been assumed. However, in many applications some kind of gear, possibly involving a transformation of rotation to translation or linear motion, as in a rotating motor driving a vehicle, will be introduced.

Inertia and friction may be referred through ideal gears, just as inductance and resistances are referred to one side or the other of an ideal transformer. Thus, if, in Fig. 11.15, N_1 and N_2 are the number of teeth on the gear wheels, then

$$\omega = \frac{N_2}{N_1}\omega_2 \quad \text{and} \quad T_L = \frac{N_1}{N_2}T_2. \tag{11.13}$$

but

$$T_2 = J_2\frac{d\omega_2}{dt} + B_2\omega_2 = J_2\frac{N_1}{N_2}\frac{d\omega}{dt} + B_2\frac{N_1}{N_2}\omega = \frac{N_2}{N_1}T_L \tag{11.14}$$

and the above relation solved for T_L

$$T_L = J_2\left(\frac{N_1}{N_2}\right)^2\frac{d\omega}{dt} + B_2\left(\frac{N_1}{N_2}\right)^2\omega = J\frac{d\omega}{dt} + B\omega, \tag{11.15}$$

where $J = J_2(N_1/N_2)^2$ is the equivalent of the inertia constant J_2 referred to the drive (motor) shaft, and $B = B_2(N_1/N_2)^2$ is the equivalent of viscous damping coefficient B_2 referred to the drive (motor) shaft.

An alternative to gearing is the belt drive illustrated in Fig. 11.16, in which D_1 is the diameter of the driving pulley and D_2 is that of the driven pulley. If slippage is ignored or a toothed belt is used, (11.13)–(11.15) apply to the pulley drive if D_1 and D_2 are substituted for N_1 and N_2, respectively. Among the disadvantages of belt, as opposed to gear, drives are (a) slippage and (b) the possibility of oscillations due to the elasticity of the belt. In many applications [12–15] such as electric road vehicles

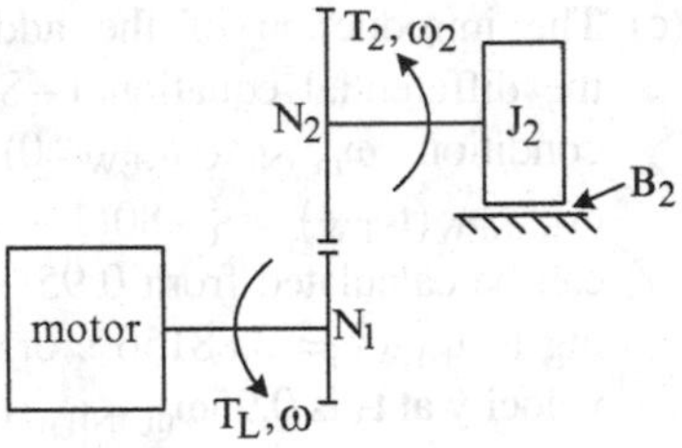

Fig. 11.15 Gear drive

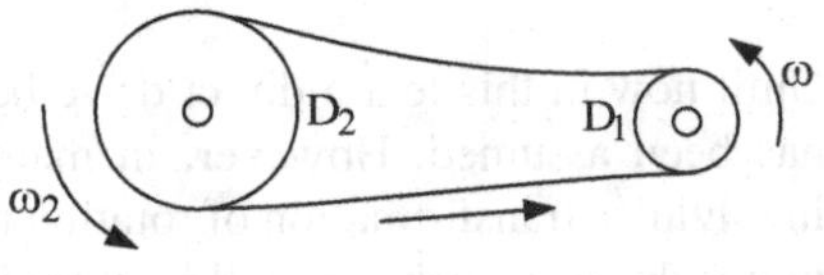

Fig. 11.16 Belt drive

no mechanical gear is desired in order to reduce the weight of the vehicle and thus increase its mileage. Similar considerations are valid for wind-power plants [16, 17].

11.4 Summary

The steady-state stability criterion of a motor drive linking motor torque T and load torque T_L as a function of the angular velocity ω_m is introduced. The various frictional and inertia torques are discussed, and the Newton torque law is applied to simple drive configurations such as starting and reversing a motor drive. The discussion of the effect of mechanical gears on angular velocity and torques concludes the discussion of this chapter.

11.5 Problems

Problem 11.1: *Approximation of nonlinear motor (torque–angular velocity) characteristic*

A drive load has the normalized torque–angular velocity characteristic $T_{load} = 0.25$ pu, that is, the torque is independent of the angular velocity ω_m. This load has a polar moment of inertia of $J_{load} = 1\ kgm^2$. The load is driven by a motor with the per-unit torque–angular velocity characteristic of

$$T_{motor} = (1 - \omega_m)/\{1 + (1 - \omega_m)^2\}\ pu, \tag{11.16}$$

where ω_m is also measured in pu. The polar moment of inertia of the motor is $J_{motor} = 1\ kgm^2$.

(a) Sketch T_{load} and T_{motor} in the (ω_m – T) plane.
(b) Find the steady-state angular velocity $\omega_{m\_steady\text{-}state}$; does steady-state stability exist at $\omega_{m\ steady\text{-}state}$? Note ω_m varies between $0 \le \omega_m \le 1.0$ pu.
(c) The nonlinear motor torque–angular velocity characteristic of (see Application Example 11.3) can be approximated by the linear function

$$T_{motorlinear} = 0.5(1 - \omega_m), \tag{11.17}$$

Sketch this approximation in the above-mentioned (ω_m – T) plane.
(d) Calculate and plot $\omega_m(t)$ for the linearized torque $T_{motor\ linear}$ within the time $0 \le t \le 100$ s; you may assume that the motor drive is at rest at t = 0 and has reached $\omega_{m+} = +\omega_{m0} = 0.5$ pu after 100 s.
(e) After the drive has reached about steady-state speed at t = 100 s calculate the reversal time t_{rev}^{new} if the drive needs to reach 95% of $\omega_{m-} = -\omega_{m0} = -0.5$ pu.

Problem 11.2: *Steady-state operation of a pump drive*

A pump has the torque characteristic $T_{pump} = 10 + 2\omega_m$, a polar moment of inertia of $J_{pump} = 1\ kgm^2$ and is driven by a motor with the torque characteristic $T = \omega_m + 100$ with a polar moment of inertia of $J_{motor} = 1\ kgm^2$.

(a) Find the steady-state angular velocity ω_{m0}; does steady-state stability exist at ω_{m0}?
(b) Calculate the time for the angular velocity to change from 50% of equilibrium angular velocity to 95% of equilibrium angular velocity.

Problem 11.3: *Redesign of a motor drive*

The rated speed and rated torque of a motor (either AC or DC) is $n_{rated} = 780$ rpm and $T_{rat} = 500$ Nm, respectively. In general, the motor torque–angular velocity characteristic can be approximated by $T = (2{,}500 - 25\omega_m)$ Nm, while the load torque–angular velocity characteristic is $T_L = (50 + 5\omega_m)$ Nm.

(a) What is the rated output power P_{out_rat}?
(b) Identify the steady-state operating point P_1 in Fig. 11.17; calculate ω_{mP1} and $T_{P1}=T_{LP1}$.
(c) Is this operating point P_1 stable?
(d) The speed is increased from 390 rpm at time $t = 0$ to 780 rpm at time $t \to \infty$. Provided the viscous friction coefficient and the polar moment of inertia are given as $B = 1$ (Nm/rad/s) and $J = 0.6\ kgm^2$, respectively, and the motor torque T is limited (independent of speed) by the solid-state converter to $T = 540$ Nm, calculate the time Δt it takes to reach the speed of $n_{m95\%} = 390$ rpm + 0.95 (780 − 390) rpm.
(e) How could you redesign this drive if $\Delta t = 0.5$ s is required? Support your reasoning through calculations.

Problem 11.4: *Operation in quadrants 1 and 3*

A drive load has the torque–speed characteristic $T_{load} = 1.11\omega_m$ Nm, which is valid in the first and third quadrants of the $(\omega_m - T)$ plane. That is, the torque is

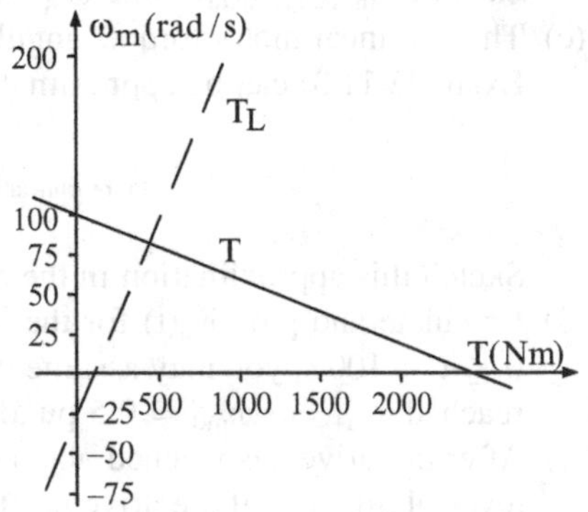

Fig. 11.17 $(\omega_m - T)$ plane

proportional to the angular velocity ω_m. This load has a polar moment of inertia of $J_{load} = 1\ kgm^2$. The load is driven by a motor with the torque–angular velocity characteristic of $T = (1{,}000 - 10\,\omega_m)$ Nm in quadrant 1, and $T = (-1{,}000 - 10\omega_m)$ Nm in quadrant 3 of the $(\omega_m - T)$ plane, where ω_m is measured in rad/s. The polar moment of inertia of the motor is $J_{motor} = 1\ kgm^2$.

(a) Sketch T_{load} and T in the first and third quadrants of the $(\omega_m - T)$ plane.
(b) Find the steady-state angular velocities ω_{mss+} (1st quadrant), ω_{mss-} (3rd quadrant) and the corresponding steady-state torques T_{ss+}, T_{ss-} at rated operation where $T_{load} = T$.
(c) Calculate and plot $\omega_m(t)$ within the time $0 \le t \le 1$ s; you may assume that the motor drive is at rest at $t = 0$ and has reached $\omega_m = \omega_{mss+}$ after $t_1 = 1$ s.
(d) After the drive has reached about steady-state angular velocity ω_{mss+} at $t_1 = 1$ s (or $t_2 = 0$ s, note there are two coordinate systems) calculate the reversal time t_{2rev}, provided the drive must reach an angular velocity ω_m of 95% of ω_{mss-} calculated in part (b).

Problem 11.5: *Drive coupled to overhauling load*

A 240 V, 1,000 rpm, 180 A DC separately excited motor has an armature resistance of 0.05 Ω, and iron-core including rotational losses of 1,200 W. It is coupled to an overhauling load with a torque of $T_L = -(200 + 2\omega_m)$.

(a) Compute the rated torque of this machine T_{rated}.
(b) Plot T_L and the rated motor torque T_{rated} in the $(\omega_m - T)$ plane.
(c) What is the rated angular velocity $\omega_{m\ rated}$?
(d) For the above-mentioned regeneration load torque would you expect $\omega_{m\ regen}$ to be larger or smaller than $\omega_{m\ rated}$, provided the maximum motor torque is $T_{max} \le -410$ Nm?
(e) Calculate angular velocity $\omega_{m\ regen}$ at which the motor can hold the load T_L by regenerative braking.
(f) Find $I_{a\ regen}$ and T_L for $\omega_{m\ regen}$ as obtained in (e).

Problem 11.6: *DC machine drive operating in motoring and regeneration regions*

A separately excited $P_{outrat} = 40$ kW DC machine (motor) has a rated terminal voltage of $V_t = 230$ V, a rated speed of 780 rpm, and an armature resistance of $R_a = 0.05\ \Omega$. The (machine torque–angular velocity) characteristic is $T = (2{,}500 - 25\omega_m)$ Nm and the (load torque–angular velocity) characteristic is $T_L = (-1{,}500 + 5\,\omega_m)$

(a) What is the input power P_{in}, the copper loss $P_{loss} = I_a^2 R_a$, the armature current I_a, and the efficiency η of this motor, provided iron-core losses, friction, and windage losses can be neglected?
(b) Calculate $(k_e\Phi)$ of this machine.
(c) Plot T and T_L in the $(\omega_m - T)$ plane of Fig. 11.18 and identify the steady-state operating point P_2; calculate ω_{mP2} and $T_{P2} = T_{LP2}$.

Fig. 11.18 ω_m – T plane

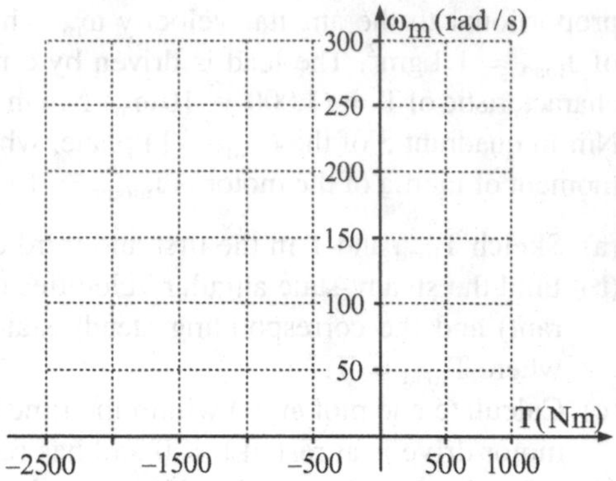

(d) Is this operating point P_2 stable?
(e) What is the output power of the drive at P_2?
(f) Calculate I_a, and E_a for this operating point.
(g) Make an assessment whether this machine is overloaded; you may neglect all losses but the copper losses.
(h) Find an operating point P_3 within the regeneration region (by adjusting the motor torque characteristic according to some control algorithm) so that the copper losses correspond to the rated copper losses.

Problem 11.7: *Start-up time t_{start} and stopping time t_{stop} of a drive*

(a) Derive an expression for the angular velocity $\omega_m(t)$ during starting of a motor–load set provided the motor torque is T = 100 Nm, the load torque is $T_L = T_M = 0.5 \cdot \omega_m(t)$, and the polar moment of inertia of the entire set – referred to the motor shaft – is J = 0.1 Nm/(rad/s^2). Calculate the time t_{start} taken for the angular velocity to change from 0 to 95% of the steady-state angular velocity $\omega_{m\ steady\ state}$. Sketch $\omega_m(t)$ for the start-up interval indicating the mechanical time constant τ_m.
(b) After t_{start} has elapsed the motor is removed from the terminal voltage (T = 0). The load torque remains $T_L = T_M = 0.5 \cdot \omega_m(t)$. Calculate the time t_{stop} taken for the angular velocity to change from $0.95 \cdot \omega_{m\ steady\ state}$ to $0.01 \cdot \omega_{m\ steady\ state}$. Sketch$\cdot\omega_m(t)$ for the stopping interval indicating the mechanical time constant τ_m.

References

1. Dubey, G. P.: *Power Semiconductor Controlled Drives*, Prentice Hall, Englewood Cliffs, NJ, 07632, 1989.
2. Leonhard, W.: *Control of Electrical Drives*, Springer, Berlin, 1985
3. Dewan, S. B.; Slemon, G. R.; Straughen, A.: *Power Semiconductor Drives*, Wiley, New York, 1984.

4. Heller, B.; Hamata, V.: *Harmonic Field Effects in Induction Machines*, Academia, Publishing House of Czechoslovak Academy of Sciences, Prague, 1977.
5. Oberretl, K.: "The field-harmonic theory of the squirrel cage motor taking multiple armature reaction and parallel winding branches into account," *Archiv für Elektrotechnik*, 49, pp. 343–364, 1965.
6. Oberretl, K.: "Field-harmonic theory of slip-ring motor taking multiple armature reactions into account," *Proceedings of the IEE*, 117, pp. 1667–1674, 1970.
7. Oberretl, K.: "Influence of parallel winding branches, delta connection, coil pitching, slot opening, and slot skew on the starting torque of squirrel-cage motors," *Elektrotech. Z.*, 86A, pp. 619–627, 1965.
8. Oberretl, K.: "General field-harmonic theory for three-phase, single-phase and linear motors with squirrel-cage rotor, taking multiple armature reaction and slot openings into account," *Archiv für Elektrotechnik,* 76 (1993), Part I: Theory and method of calculation, pp. 111–120, and Part II: Results and comparison with measurements, pp. 201–212.
9. Fuchs, E. F.; Masoum, M. A. S.: *Power Quality in Power Systems and Electrical Machines*, Elsevier/Academic Press, 638 pages. ISBN: 978-0-12-369536-9, February 2008.
10. http://www.wolfram.com/
11. http://www.mathworks.com/
12. Fuchs, E. F.; Schraud, J.; Fuchs, F. S.: "Analysis of critical-speed increase of induction machines via winding reconfiguration with solid-state switches," *IEEE Transactions on Energy Conversion*, 23(3), pp. 774–780, 2008.
13. Schraud, J.; Fuchs, E. F.; Fuchs, H. A.: "Experimental verification of critical-speed increase of induction machines via winding reconfiguration with solid-state switches," *IEEE Transactions on Energy Conversion*, 23(2), pp. 460–465, 2008.
14. Fuchs, E. F.; Myat, M. H.: "Speed and torque range increases of electric drives through compensation of flux weakening," *Proceedings of the 20th International Symposium on Power Electronics, Electrical Drives, Automation, and Motion*, Pisa, Italy, June 14–16, 2010.
15. Fuchs, E. F.: "Alternating current machine with increased torque above and below rated speed for hybrid/electric propulsion systems", *U.S. Patent Number 8*, 183, 814, May 22, 2012.
16. Yildirim, D.; Fuchs, E. F.; Batan, T.: "Test results of a 20 kW, direct drive, variable speed wind power plant," *Proceedings of the International Conference on Electrical Machines, Istanbul*, Turkey, pp. 2039–2044, September 2–4, 1998.
17. Yildirim, D.: *Commissioning of 30 kVA Variable-Speed, Direct-Drive Wind Power Plant*, PhD Dissertation, University of Colorado at Boulder, February 14, 1999.

Chapter 12
Analyses and Designs Related to Renewable Energy Systems

12.1 Introduction

Chapters 1–11 including the examples are appropriate for an introductory undergraduate course on energy conversion and renewable energy. The following examples in Chap. 12 encompass entire energy systems which include transformer, electric machine systems/drives, centralized (conventional) power system, distribution system analysis with/without power factor compensation, design of power plants and renewable energy storage facilities, photovoltaic, wind, and fuel-cell systems as well as examples associated with the smart grid. This very broad menu of examples can be the topics of a graduate course where the above-mentioned undergraduate course is a prerequisite.

As has been pointed out in Chap. 1 the rationale for this book is best characterized by a recent publication of the New IEEE-USA National Energy Policy Recommendation [1] stressing that the electric power system is an integral tool for the remaking of America. The clean and efficient production, transmission and use of electricity can play a key role in addressing the challenges of economic revitalization, climate change and breaking our addiction to foreign oil. Both established and new technologies must be applied at unprecedented scale and on an accelerated schedule. Bold actions and substantial investments will be required, in particular

1. Increasing energy efficiency
2. Breaking our addiction to oil by transforming transportation
3. Encouraging pollution-free and green electric power supply and
4. Building a stronger and smarter electrical energy supply infrastructure.

On a regional and local basis steps have been taken to achieve the above-mentioned goals with the Midwestern utility Xcel Energy's SmartGridCity Project [2]. Smart Grid is a $100 million investment by Xcel Energy and its partners to improve the City of Boulder's electric distribution system to combine traditional and new technology that provides real-time information about consumption and new ways to control the flow of electric power and natural gas supplies. It will give consumers the ability to monitor their power use and hopefully induce power saving measures.

E.F. Fuchs and M.A.S. Masoum, *Power Conversion of Renewable Energy Systems*,
DOI 10.1007/978-1-4419-7979-7_12,

This may result in the reduction of power consumption during peak-power times and the increase of power consumption during low-power demand periods. About 52,000 residences within Boulder will be equipped with so-called smart meters where the power use can be monitored via the Internet. In addition several substations will be equipped with power monitoring devices.

While this monitoring approach may lead to energy conservation, it may not be sufficient to reduce CO_2 consumption in the (coal-fired) power plants supplying within Colorado about 80% of the electrical energy. What is needed is the installation of renewable energy sources and their integration within the existing distribution and transmission system: this calls for short-term and long-term energy storage systems in the hundreds of MWh range. Short–term storage systems must be deployable or accessible within a few 60 Hz cycles (e.g., battery, super-capacitor, fuel cell storage plants, and variable-speed hydro-power plants) while long-term storage systems (e.g., constant-speed hydro, and compressed air) require about 6 min for deployment/access. Figures 12.1–12.4 depict the CO_2 emissions of countries, the CO_2 emissions per capita of countries, the CO_2 emissions of tons-force per thousands of dollars of gross national product of countries, and CO_2 emissions concern of different countries, respectively.

To address some of the issues involved, this chapter provides application examples that will highlight conventional and renewable energy problems and their solutions which may contribute to the remaking of power systems:

- Efficiency of a gas turbine of a cogeneration power plant
- Frequency control of an isolated power plant (islanding operation)
- Frequency control of an interconnected power system broken into two areas each having one generator

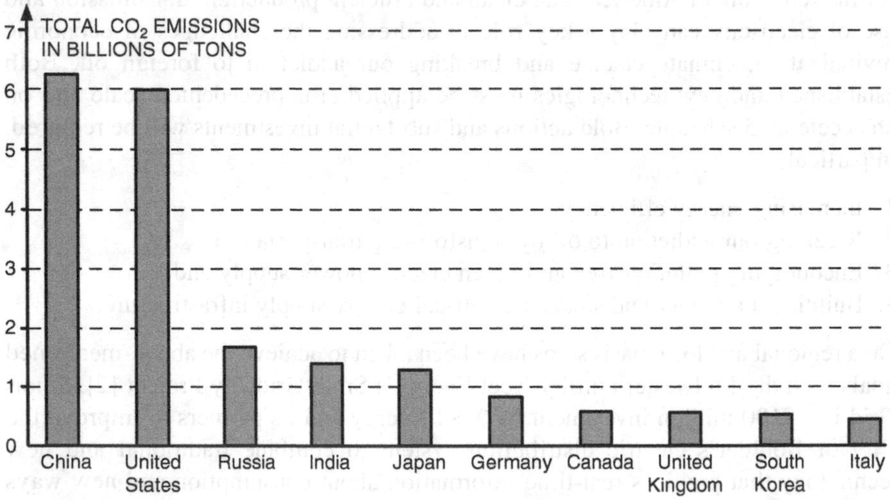

Fig. 12.1 CO_2 emissions of countries. Courtesy of U.S. Energy Information Administration [3], published in http://money.cnn.com/galleries/2009/news/0912/gallery.global_warming/index.html

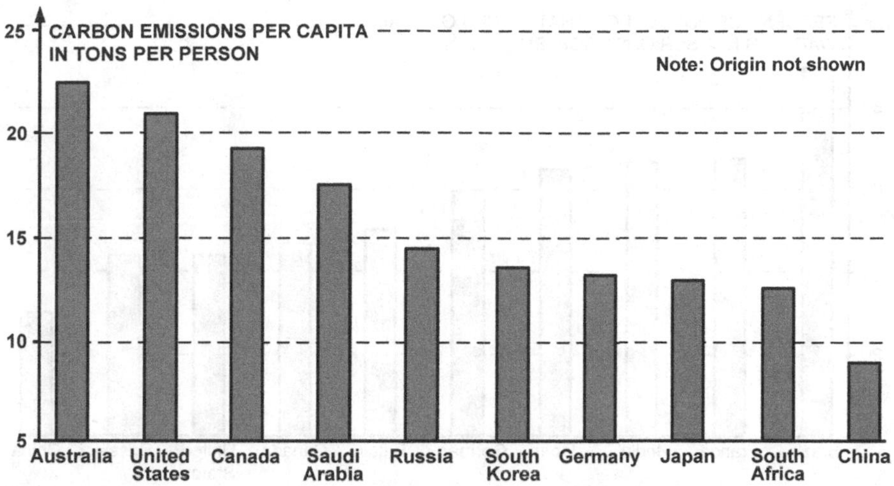

Fig. 12.2 CO_2 emissions per capita of countries. Courtesy of U.S. Energy Information Administration [3], published in http://money.cnn.com/galleries/2009/news/0912/gallery.global_warming/index.html

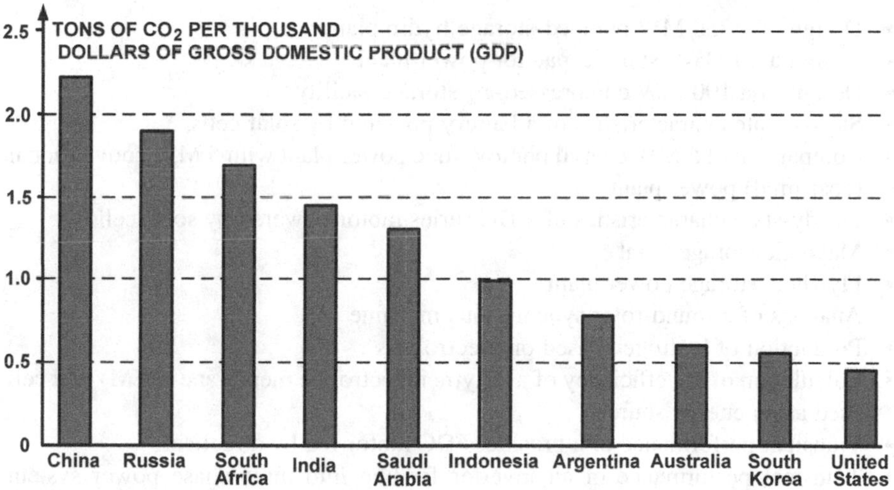

Fig. 12.3 CO_2 emissions of tons-force per thousands of dollars of gross national product of countries. Courtesy of U.S. Energy Information Administration [3], published in http://money.cnn.com/galleries/2009/news/0912/gallery.global_warming/index.html

- Frequency variation within an interconnected power system as a result of two load changes and structure of a smart/micro grid
- Fundamental power flow
- Fundamental power flow and reactive power control

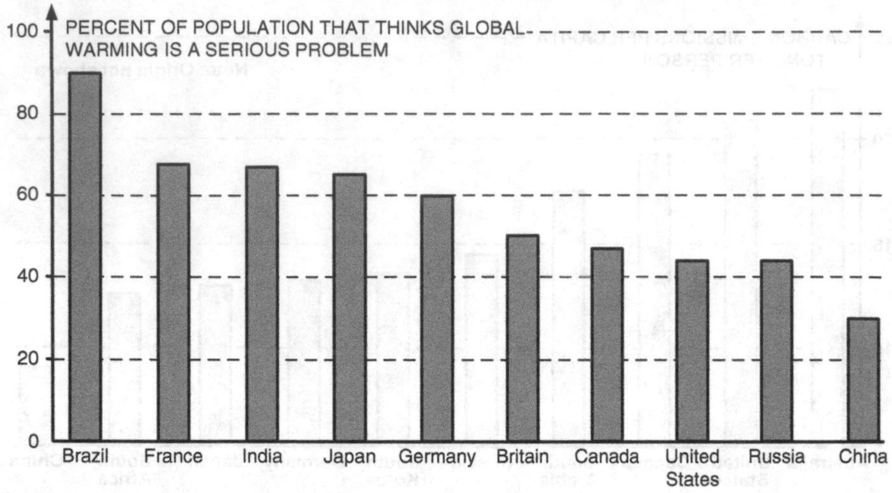

Fig. 12.4 CO_2 emissions concern of different countries. Courtesy of U.S. Energy Information Administration [3], published in http://money.cnn.com/galleries/2009/news/0912/gallery.global_-warming/index.html

- Design of a 250 MW pumped-storage hydro plant
- Design a 10 MWh super-capacitor power plant
- Design of a 100 MW compressed-air storage facility
- Steady-state characteristics of a battery powered by solar cells
- Comparison of 5 MW central photovoltaic power plant with 5 MW conventional (coal-fired) power plant
- Steady-state characteristics of a DC series motor powered by solar cells
- Magnetic storage plant
- Flywheel storage power plant
- Analysis of a round-rotor synchronous machine
- Production of hydrogen based on electrolysis
- Calculation of the efficiency of a polymer electrolyte membrane (PEM) fuel cell used as an energy source
- Transient performance of a brushless DC motor fed by a fuel cell
- Transient performance of an inverter feeding into three-phase power system when supplied by a fuel cell
- Calculation of load current $\tilde{I}_{AB\_\Delta}$ in phase of Δ-load and $\tilde{I}_{AB_Y}$in Y-load
- AC power flow and complex power calculation through single-phase feeder with load power-factor correction
- Stator and rotor frequencies in a 3-phase induction machine
- Efficiency of a 3-phase induction motor
- Design of a 5 MW variable-speed wind power plant operating at an altitude of 1,600 m
- Design of a 10 kW wind power plant at sea level and at an altitude of 1,500 m

- Design of a stand-alone wind-power plant with a compressed-air storage facility for a farm
- CO_2 generation of a coal-fired power plant
- Determination of the sequence-component equivalent circuits and matrices $Z_{bus}^{(1)}$, $Z_{bus}^{(2)}$, and $Z_{bus}^{(0)}$ of a two-generator system, including their application to short-circuit calculations
- Load management and load shedding in an emergency or operation with renewable sources

12.2 Electric Machine Systems/Drives

The design of electric drives has gained renewed interest with the emerging needs of electric/hybrid automobiles where high starting torque, speed control above and below rated speed and regeneration ability are required. Although the DC machine is not anymore used for electric car propulsion it serves as an excellent role model and can be emulated by AC drives such as brushless DC machine and induction machine drives. This section presents a review of DC, synchronous and induction machine analyses.

Application Example 12.1: Design of a 50 kW Drive for an Electric Automobile

An electric car with a total weight of 1.2 tons-force (including load) is supposed to travel with the energy content E = 25 kWh of one battery charge 120 miles. This results in a figure of merit (FM) of FM = 5.76. The breakdown of weight components is as follows: the weight of the car – without electric drive components and load – is 500 kg-force, the four persons and stowage represent the load and amount to 400 kg-force, the 25 kWh battery weighs 125 kg-force for 100% depth of discharge (DoD) or 250 kg-force for a 50 kWh battery where the depth of discharge is DoD = 50%. The electric drive including the inverter has a specific power of 2 kW/kg-force resulting for a rated output power of P_{out} = 50 kW of the electric drive in 25 kg-force. In addition, the mechanical gear has a weight of 25 kg-force. The electric DC motor of Fig. 12.5 has to provide during uphill travel the rated output torque of T_{rated} = 500 Nm at an angular velocity of ω_{m_rated} = 100 rad/s. The armature resistance is R = 0.05 Ω and the armature inductance L is large enough so that the current ripple due to pulse-width modulation (PWM) can be neglected.

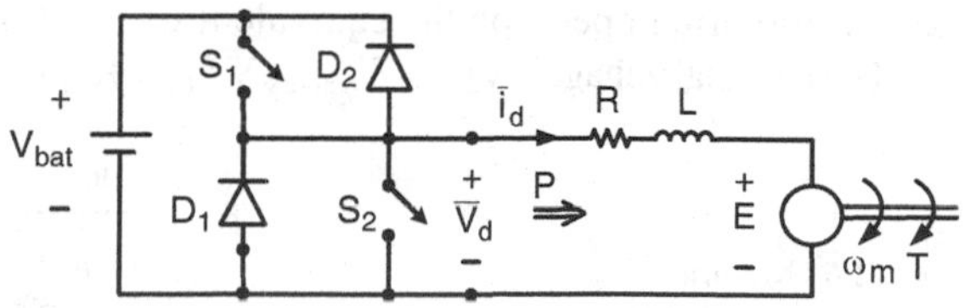

Fig. 12.5 Two-quadrant DC motor drive. For motoring $P=P_{motoring}$ is positive, and for regeneration (braking) $P=P_{regeneration}$ is negative

(a) Verify the figure of merit (FM).
(b) If the torque-current relation is $T = \text{const} \cdot \bar{i}_d$, where const = 5.0 Nm/A compute for the rated torque the rated current i_{drated}.
(c) If the induced voltage-angular velocity relation is $E = \text{const} \cdot \omega_m$, where const = 5.0 V/(rad/s), compute the rated induced voltage E_{rated} for the rated angular velocity.
(d) If the battery voltage is $V_{bat} = 550$ V compute t_{on}^{VGS1} of the step-down converter required for rated motor operation provided $f_{switching} = 10$ kHz.
(e) What is the efficiency of the DC drive $\eta_{motoring} = \frac{P_{out}}{P_{in}} = \frac{P_{out}}{P_{out} + P_{loss}}$ for rated operation? Assume that the iron-core and windage/friction losses of the motor are 20% of the rated ohmic losses of the motor, and the losses of the step-down DC-to-DC converter and the battery are negligible.
(f) When traveling downhill the electric motor operates as a brake or as a generator and the regenerated power of the DC machine is $P_{regeneration} = -20$ kW at the induced voltage of the DC machine $E_{regeneration} = 200$ V and the battery voltage of 550 V. Compute the on time t_{on}^{VGS2} of the step-up converter required for regeneration.
(g) Compute the current $\bar{i}_{dregeneration}$ for the condition of (e). Does this current exceed the rated current value determined in (b)?
(h) What is the efficiency $\eta_{regeneration} = \frac{(-P_{regeneration})}{(-P_{regeneration}) + P_{loss}}$ of the DC machine drive for the operating condition of part (f)? Assume that the iron-core and friction/windage losses of the motor are 15% of the ohmic losses of the machine acting as a generator and the losses of the step-up DC-to-DC converter and the battery are negligible. Would a DC machine be used for an actual electric automobile drive?

Solution

(a) The figure of merit for an electric automobile is defined as

$$FM = \frac{(\text{weight in tons} - \text{force}) \cdot (\text{distance in miles})}{(\text{energy used at wheel from battery})} = \frac{(1.2) \cdot (120)}{25\,\text{kWh}} = 5.76.$$

Note that the value of FM depends upon the driving style (e.g., high speed vs. low speed), the geographic properties of path driven (e.g., flat vs. hilly), the location of the path driven (e.g., in town with frequent stops vs. cross country with few stops), and the type of use of the battery (discharging only vs. discharging with regeneration due to braking).
(b) The rated current is $i_{d\_rated} = \frac{T_{rated}}{5} = 100$ A.
(c) The rated induced voltage is $E_{rated} = 5 \cdot 100 = 500$ V.
(d) For motoring operation the equivalent circuit Fig. 12.6 is valid.
The terminal voltage is $v_d = \delta_{motoring} V_{bat} = Ri_d + E = 0.05 \cdot 100 + 500 = 505$ V

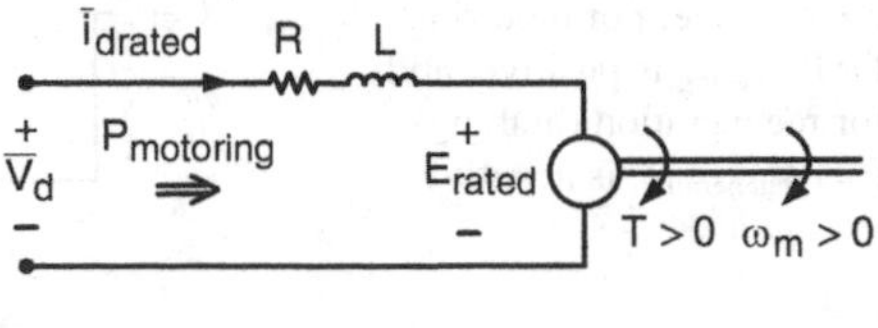

Fig. 12.6 Motoring operation, note that $P_{motoring}$ is positive

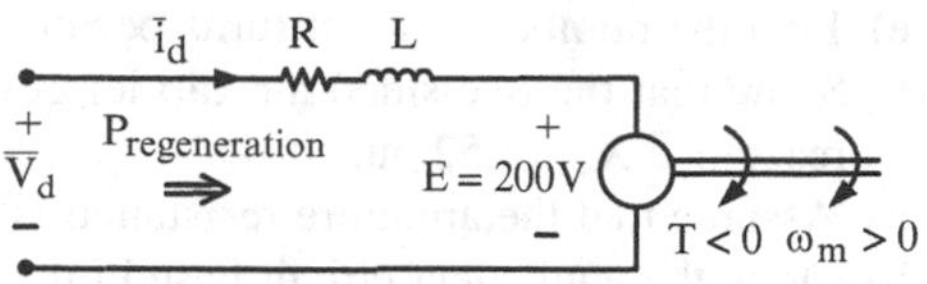

Fig. 12.7 Regeneration/braking operation, note that $P_{regeneration}$ is negative

The on-time of the DC–to–DC step-down converter is

$$t_{on}^{VGS1} = T_{switching}\delta_{motoring} = \frac{\delta_{motoring}}{f_{switching}} = \frac{\bar{v}_d}{V_{bat}f_{switching}} = \frac{505}{550 \cdot 10k} = 0.092ms.$$

(e) The efficiency at rated motor operation is
$\eta_{motoring} = \frac{P_{out}}{P_{in}} = \frac{P_{out}}{P_{out} + P_{loss}}$, where the loss is $P_{loss} = (\bar{i}_d)^2R + 0.2(\bar{i}_d)^2R =$ $600W$ resulting in the efficiency of $\eta_{motoring}$=0.988.

(f) For regeneration the equivalent circuit Fig. 12.7 is valid.
The terminal current is $\bar{i}_d = \frac{-20,000}{\bar{v}_d} = \frac{\bar{v}_d - E}{R}$ from which one obtains the second-order equation $(\bar{v}_d)^2 - 200(\bar{v}_d) + 1,000 = 0$ resulting in the two solutions $\bar{v}_{d1} = 194.87\,V$ and $\bar{v}_{d2}$= 5.13 V. The second solution is not feasible because it results in a very large armature current and in very large losses.
The duty cycle for motoring is $\delta_{motoring} = \frac{\bar{v}_d}{V_{bat}} = 0.3543$ and for regeneration operation $\delta_{regeneration}$= 1 $\delta_{motoring}$ = 0.6457.
The on-time of the DC–to–DC step-up converter is
t_{on}^{VGS2}=$T_{switching}\ \delta_{regeneration}$=$\frac{\delta_{regeneration}}{f_{switching}}$ = 0.06457 ms.

(g) The current for regeneration is $\bar{i}_{d1} = \frac{-20,000}{194.87} = -102.63$ A. This current is about the same as the rated current and can be tolerated.

(h) The efficiency is $\eta_{regeneration} = \frac{(-P_{regeneration})}{(-P_{regeneration}) + P_{loss}}$, where the losses are $P_{loss} = (\bar{i}_d)^2R + 0.15(\bar{i}_d)^2R = 605.64\,W$ resulting in $\eta_{regeneration}$ =0.971.
The DC machine would be replaced by either an induction, synchronous or brushless DC machine. A most recent publication [4] presents important data for electric cars which are/will be available in the near future.

Application Example 12.2: Analysis of a Round-Rotor (Non-Salient-Pole) Synchronous Machine Using Generator Notation for Power-Factor Definition

A three-phase synchronous machine operated as a generator has the electrical data $S = 100$ MVA, $V_{L-L} = 10{,}000$ V, p = 2 poles, $B_{sr} = B_{max} = 0.7$ T, $I_f = 5{,}000$ A, f = 60 Hz, and rotor diameter D = 1 m.

(a) Determine the rated speed $n_{s\ rated}$, the rated angular velocity $\omega_{s\ rated}$, and the rated stator current $I_{ph\ rated}$.
(b) Calculate the axial ideal and actual core lengths ℓ_i and ℓ_{actual}, respectively. These lengths can be determined based on either the utilization factor (Esson's number [5, 6]) C = 7.0 kWmin/m^3 at cosΦ=1.0 or the model laws [7]. Note that these two concepts can be applied to any type of electric machine.
(c) Calculate the rated torque T_{rated}.
(d) Determine the number of rotor turns per pole (N_r/p).

(e) Find the number of stator turns per phase $N_{s\ phase}$.
(f) Show that the one-sided air gap length $g = 0.08$ m results in the synchronous reactance $X_s = 1.52$ pu.
(g) Assume that the armature resistance is $R_a = 0.05 \cdot X_s$.
(h) Draw the equivalent circuit based on generator notation.
(i) Plot the phasor diagram for a $\cos\Phi = 0.8$ lagging power factor (generator notation) and compute E_a and δ.

Solution

(a) $n_{s\ rated} = (120 \cdot f)/p = (120 \cdot 60)/2 = 3{,}600$ rpm, $\omega_{s\ rated} = (2\pi \cdot ns)/60 = 376.98$ rad/s, $V_{ph\ rated} = V_{L-L\ rated}/\sqrt{3} = 10{,}000/\sqrt{3} = 5.77$ kV, $I_{ph\ rated} = (S_{rated}/3)/V_{ph\ rated} = 100/3)/5.77 = 5.77$ kA.

(b) The ideal core length is

$$\ell_i = \frac{P_{out}}{C \cdot D^2 \cdot n_s} = \frac{100 \cdot 10^3\,\text{kW}}{7\frac{\text{kW min}}{\text{m}^3} \cdot (1\,\text{m}^2) \cdot 3{,}600\,\frac{1}{\text{min}}} = 3.97\,\text{m},$$

and the actual core length is with the iron-core stacking factor $k_{fe} = 0.95$

$$\ell = \frac{\ell_i}{k_{fe}} = \frac{3.97\,\text{m}}{0.95} = 4.18\,\text{m}.$$

(c) The rated torque is

$$T_{fld\,rated} = \frac{P_{out\,rated}}{\omega_{s\,rated}} = \frac{100 \cdot 10^6}{376.98}\,\text{Nm} = 265.3\,\text{kNm}.$$

(d) For an assumed torque angle of $\delta_{r\ rated} = 30°$ one obtains the rotor magnetomotive force

$$F_r = \frac{T_{fld\,rated}}{(-)\frac{p}{2}\frac{\pi \cdot D \cdot \ell}{2} B_{sr} \sin\ \delta_{r\,rated}} = \frac{265.3\,\text{kNm}}{1 \cdot \frac{\pi \cdot 1 \cdot 4.18}{2} 0.7 \cdot 0.5} = 115.45\,\text{kAt}.$$

With the rotor winding factor $k_{wr} = 0.9$ and the field current $I_f = I_r$ one obtains the number of rotor turns per pole

$$N_{pole} = \left(\frac{N_r}{p}\right) = \frac{F_r}{\frac{4}{\pi} k_{wr} I_f} = \frac{115.45\,\text{kAt}}{\frac{4}{\pi} 0.9 \cdot 5000} = 20.15\ \text{turns}.$$

The total number of turns of the field (rotor) winding is $N_r = 2N_{pole} \approx 40$ turns.

(e) The number of turns of the stator winding per phase is

$N_{s\,phase} = \frac{E}{4.44 \cdot f \cdot k_{ws} \phi_{max}}$, where $E = 1.05 \cdot V_{phase\ rated} = 6.06$ kV, $k_{ws} = 0.9$, $\Phi_{max} = B_{max}$ (area$_p$), with area$_p = \pi \cdot r \cdot \ell_i = 6.24\,\text{m}^2$ follows $\Phi_{max} = 4.37$ Wb/pole or $N_{s\,phase} = \frac{6.06\,k}{4.44 \cdot 60 \cdot 0.9 \cdot 4.37} = 5.78$ turns or $N_{s\ phase} \approx 6$ turns.

(f) The stator magnetomotive force is $F_s = \frac{4}{\pi} k_{ws} \frac{N_{s\,phase}}{p} i_s$, with $H_s = F_s/g$, $B_s = \mu_o H_s = \mu_o (F_s/g)$, $i_s(t) = I_{smax}\cos(\omega t + \Phi)$ follows for the flux $\phi_{max} = (\mu_o \frac{4}{\pi} k_{ws} \frac{N_{s\,phase} I_{smax}}{g \cdot p} area_p)$ or the self-inductance of phase "a" $L_{aa} = \mu_o \frac{4}{\pi} k_{ws} \frac{(N_{s\,phase})^2}{g \cdot p} area_p$. With g=0.08 m one obtains
$L_{aa} = \mu_o \frac{4}{\pi} 0.9 \frac{(6)^2}{0.08 \cdot 2} 6.24 = 20.22 \cdot 10^{-4} H = 2.022 mH$. With Concordia [8] follows $X_s = X_d = X_{aa} + X_{ab} \approx 2X_{aa}$ or $X_s = 2L_{aa}\ \omega = 2 \cdot 2.202$ mH 377 rad/s $= 1.52\ \Omega$. The base current and base impedance are $I_{phase\,base} = \frac{S_{phase\,base}}{V_{phase\,base}} = \frac{100\,MVA/3}{10{,}000/\sqrt{3}} = 5.77\,kA$, and $Z_{base} = \frac{V_{phase\,base}}{I_{phase\,base}} = \frac{5.77kV}{5.77kA} = 1\ \Omega$, respectively, resulting in the per-unit synchronous reactance $X_s^{pu} = \frac{X_s}{Z_{base}} = 1.52$ pu.

(g) The armature resistance is $R_a = 0.05 \cdot X_s = 0.076$ pu.

(h) The equivalent circuit is shown in Fig. 12.8 based on generator notation.

(i) The phasor diagram for a $\cos\Phi = 0.8$ lagging power factor (generator notation) is depicted in Fig. 12.9 and E_a and δ are computed as follows:

$$|\tilde{E}_a| = \sqrt{(|\tilde{V}_a| + |\tilde{V}_{Xs}| \sin\phi + |\tilde{V}_{Ra}| \cos\phi)^2 + (|\tilde{V}_{Xs}| \cos\phi - |\tilde{V}_{Ra}| \sin\phi)^2},$$

$$\text{and} \quad \delta = \tan^{-1}\left\{\frac{(|\tilde{V}_{Xs}| \cos\phi - |\tilde{V}_{Ra}| \sin\phi)}{(|\tilde{V}_a| + |\tilde{V}_{Xs}| \sin\phi + |\tilde{V}_{Ra}| \cos\phi)}\right\}.$$

With $|\tilde{V}_a| = 5770$ V or 1 pu, $|\tilde{I}_a| = 5770$ A or 1 pu, $\phi = -36.87°$, $|\tilde{V}_{Xs}| = X_s|\tilde{I}_a| = 1.52$ pu, and $|\tilde{V}_{Ra}| = R_s|\tilde{I}_a| = 0.05$ pu follows $|\tilde{E}_a| = 2.28$ pu and $\delta = 31.18°$.

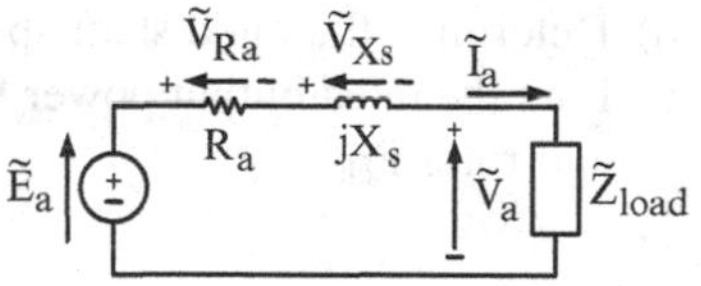

Fig. 12.8 Equivalent circuit of a non-salient-pole synchronous generator based on generator notation for power-factor definition

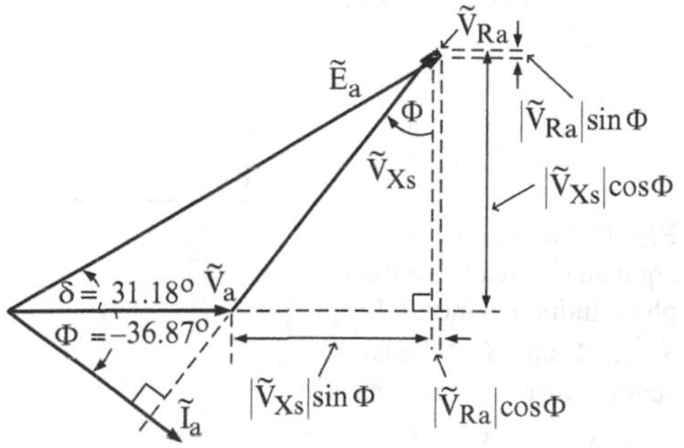

Fig. 12.9 Phasor diagram for $\cos\Phi = 0.8$ lagging power factor

Application Example 12.3: Stator (f_s) and Rotor (f_r) Frequencies in a 3-Phase Squirrel-Cage Induction Machine

A 3-phase induction machine operates as a motor and has the output power of $P_{out} = 20$ hp, the terminal voltage of $|\tilde{V}_{L-L}| = 240$ V, the frequency of the stator voltages/currents of f_s=60 Hz, and the rated (nominal) speed $n^{rpm}_{m\_rated} = 873$ rpm.

(a) Find the number of poles p of this machine for the stator frequency $f_s = 60$ Hz, and determine the slip s_{rated} for rated (nominal) operation.
(b) What is the rated frequency f_r of the rotor currents and rotor voltages?
(c) What is the output power P_{out} expressed in kW?

Solution

(a) The synchronous speed of the stator field is $n_s^{rpm} = (120 \cdot f_s)/p$ and one obtains for various pole numbers and $f_s = 60$ Hz the values of Table 12.1.
The corresponding slip values are given by $s = \frac{n_s^{rpm} - n^{rpm}_{m\_rated}}{n_s^{rpm}}$ and listed in Table 12.1. The slip of an induction motor must be for stable steady-state operation in the range $0 \leq s \leq 0.10$ and thus the rated slip is $s = s_{rated} = 0.03 = 3\%$, and the pole number is $p = 8$.
(b) The rated rotor frequency is $f_r = s{\cdot}f_s$ or $f_{r\_rated} = s_{rated}{\cdot}f_s = 0.03{\cdot}60$ Hz $= 1.8$ Hz.
(c) P_{out} =20 hp·0.746 kW/hp = 14.92 kW.

Application Example 12.4: Efficiency of a 3-Phase Squirrel-Cage Induction Motor

A 3-phase, $V_{L-L} = 480$ V, $f_s = 60$ Hz, $p = 6$ pole induction motor (Fig. 12.10) has the following parameters: $R_s = 0.2\ \Omega$, $X_s = 0.4\ \Omega$, $R_r' = 0.1\ \Omega$, $X_r' = 0.3\ \Omega$, $X_m \rightarrow \infty$, and the rated slip $s = s_{rated} = 3\%$.

(a) Determine the rated synchronous speed $n^{rpm}_{s\_rated}$.
(b) Determine the rated shaft speed $n^{rpm}_{m\_rated}$.
(c) Find the rated output power P_{out_rated}, and the copper (cu) losses in stator P_{cu_s}, and rotor P_{cu_r} .

Table 12.1 Synchronous speed n_s^{rpm} and slip s as a function of pole numbers p

p [–]	$n_s^{rpm} = 120 \cdot f_s/p$ [rpm]	s[–]
2	3,600	0.76
4	1,800	0.52
6	1,200	0.27
8	900	0.03
10	720	–0.22

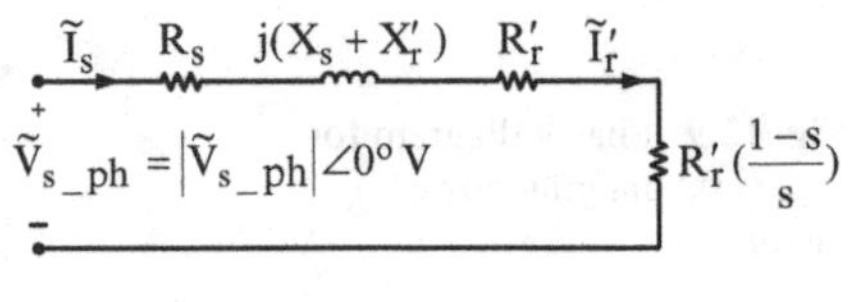

Fig. 12.10 Per-phase equivalent circuit of a three-phase induction motor. In $\tilde{V}_{s\_ph}$, R_s and X_s "s" means stator and "ph" means phase in $\tilde{V}_{s\_ph}$, otherwise "s" is the slip

(d) Compute the rated input power P_{in_rated} neglecting the fixed (core, friction and windage) losses, and compute the rated (power) efficiency η_{rated}.

Solution

(a) For p = 6 poles the synchronous speed is (see Table 12.1) $n_s^{rpm} = 1,200$ rpm.

(b) The rated shaft speed is $n_{m\_rated}^{rpm} = n_s^{rpm}(1-s) = 1,164$ rpm.

(c) The output power is $P_{out} = 3|\tilde{I}_r'|^2 R_r' \frac{(1-s)}{s}$ where the current is defined as $\tilde{I}_r' = \frac{\tilde{V}_{s\_ph}}{[R_s + \frac{R_r'}{s} + j(X_s + X_r')]}$. With $\tilde{V}_{s\_ph} = \frac{480}{\sqrt{3}}\angle 0°V = 277.14\angle 0°V$ one obtains $\tilde{I}_r' = 76.95\angle -11.21°A$ and the rated output power $P_{out\_rated} = 57.44$ kW. The copper losses in stator and rotor are $P_{loss} = P_{cu\_s} + P_{cu\_r} = 3(I_r')^2 (R_s + R_r') = 5.33$ kW.

(d) The rated input power is now $P_{in\_rated} = P_{out\_rated} + P_{loss} = 62.77$ kW resulting in the rated efficiency of $\eta_{rated} = P_{out\_rated}/P_{in\_rated} = 91.51\%$.

Application Example 12.5: Non-salient Pole (Round-Rotor) Synchronous Motor Using Consumer Notation for Power-Factor Definition

A three-phase (m = 3), four-pole (p = 4) non-salient pole (round-rotor) synchronous machine (either motor or generator) has the parameters $X_s = 2$ pu and $R_a = 0.05$ pu. It is operated as a motor with I_a=1.0 pu and (the phase-voltage) $V_a = 1$ pu at an overexcited (leading current based on the consumer notation, where the current flows into the machine) power factor of $\cos\Phi = 0.8$ leading.

The so called per unit system (pu) is frequently used to avoid large numbers in your calculations. One defines base (rated phase) values $V_a = V_{base} = 24$ kV, $I_a = I_{base} = 1.4$ kA, and the base impedance $Z_{base} = V_a/I_a = 17.14\ \Omega$. This motor cannot develop any starting torque at f = 60 Hz. For this reason this machine is used in variable–speed drives, where an inverter can output currents and voltages at f ≈0 Hz, for example the TGV (train grand vitesse) in France. Another application is that of a generator in power plants – indeed 95% of the total installed power in the US (e.g., 1000 GW) is generated with such machines.

(a) Draw the per-phase equivalent circuit of this machine.
(b) What is the total rated apparent power S measured in MVA?
(c) What is the total rated real power P measured in MW and what is the total rated reactive power Q measured in MVAr?
(d) Draw a phasor diagram with the voltage scale of 1.0 pu ≡ 1.5 in. (or about 4 cm), and the current scale of 1.0 pu ≡ 1.25 in. (or about 3 cm), or any other convenient scale so that the phasor diagram fits on one page.
(e) From this phasor diagram determine the per-phase induced voltage E_a and the torque angle δ.
(f) Calculate the rated speed (in rpm) of this machine at f = 60 Hz.
(g) Calculate the rated angular velocity $\omega_s = \omega_m$ (in rad/s) of this machine at f = 60 Hz.
(h) Derive and calculate – based on the phasor diagram neglecting $R_a \ll X_s$ – the total output power $P_{output} \approx 3(E_a V_a \sin\delta)/X_s$.

(i) Find the output torque T_{out} in Nm.
(j) Determine the total input electrical power $P_{in} = 3V_a I_a \cos\Phi$.
(k) Calculate the total machine loss $P_{loss} = 3R_a(I_a)^2$.
(l) Repeat the above analysis for an underexcited (lagging current based on consumer notation, where the current flows into the motor) power factor of $\cos\Phi = 0.8$ lagging, and $I_a = 0.5$ pu.

Solution

(a) The per-phase equivalent circuit is depicted in Fig. 12.11.

The per-phase base values are $V_{base} = 24$ kV, $I_{base} = 1.4$ kA resulting in $Z_{base} = V_{base}/I_{base}) = 24$ kV/1.4 kA $= 17.143\ \Omega$ the synchronous reactance $X_s = 34.28\ \Omega$ and the armature resistance $R_a = 0.857\ \Omega$.

(b) The rated apparent power is $S_{rat} = 3|\tilde{V}_{a\_rat}||\tilde{I}_{a\_rat}| = 100.8$ MVA.
(c) $P_{rat} = S_{rat}\cos\Phi = 80.64$ MW, and $Q_{rat} = S_{rat}\sin\Phi = 60.48$ MVAr.
(d) $\Phi = \cos^{-1}(0.8) = 0.6435°$ leading, with $|\tilde{V}_{a\_rat}| = 1.5$ in., $|\tilde{I}_{a\_rat}| = 1.25$ in., $|\tilde{V}_R| = |\tilde{I}_{a\_rat}| \cdot R_a = 1.4\,\text{kA} \cdot 0.05 \cdot 17.143 = 1,199.9\,\text{V}$, corresponding to 0.075 in.

$|\tilde{V}_X| = |\tilde{I}_{a\_rat}| \cdot X_s = 1.4\,\text{kA} \cdot 2 \cdot 17.143 = 48,000\,\text{V}$, corresponding to 3 in.

Figure 12.12 shows the phasor diagram for a leading power factor of 0.8 (consumer notation).

(e) The magnitude of the induced voltage is

$$|\tilde{E}_a| = \sqrt{(|\tilde{V}_a| - |\tilde{V}_R|\cos\phi + |\tilde{V}_X|\sin\phi)^2 + (|\tilde{V}_X|\cos\phi + |\tilde{V}_R|\sin\phi)^2} = 2.71\,\text{pu},$$

and the (torque) angle of the induced voltage is $\cos\delta = \{|\tilde{V}_a| - |\tilde{V}_R|\cos\phi + |\tilde{V}_X|\sin\phi\}/E_a = 0.7982$, that is, $\delta = 0.6465\,\text{rad}$ or $37.04°$.

(f) The synchronous speed is $n_s = (120\ f)/p = 1,800$ rpm.

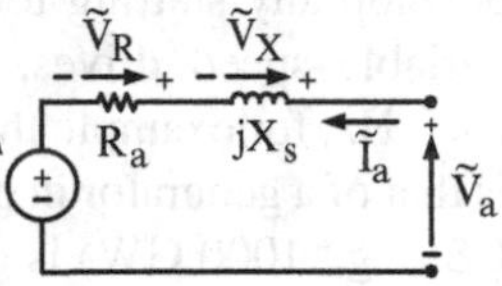

Fig. 12.11 Per-phase equivalent circuit of non-salient pole synchronous motor based on consumer notation for power-factor definition

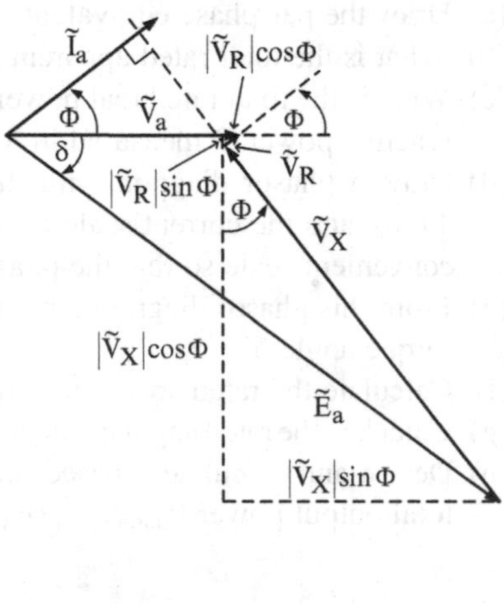

Fig. 12.12 Phasor diagram for a leading power factor of 0.8 (consumer notation)

(g) The angular velocity/frequency is $\omega_s = (2\pi n_s)/60 = 188.7$ rad/s.

(h) The complex power is defined as $\tilde{S} = P + jQ = 3\tilde{V}_a\tilde{I}_a^*$, with $\tilde{I}_a = \frac{\tilde{V}_a - \tilde{E}_a}{jX_s} = \frac{|\tilde{V}_a|\angle 0° - |\tilde{E}_a|\angle\delta}{jX_s}$ follows $\tilde{S} = \frac{3|\tilde{V}_a|\angle 0°}{X_s}(|\tilde{V}_a|\angle 90° - |\tilde{E}_a|\angle(90° - \delta))$

$$\tilde{S} = \frac{3|\tilde{V}_a|}{X_s}\{|\tilde{V}_a|[\cos(90°) + j\sin(90°)] - |\tilde{E}_a|[\cos(90° - \delta) + j\sin(90° - \delta)]\}$$

$$\tilde{S} = \frac{3|\tilde{V}_a|}{X_s}\{j|\tilde{V}_a| + |\tilde{E}_a|\sin\delta - j|\tilde{E}_a|\cos\delta\}$$

$$= j\frac{3|\tilde{V}_a|^2}{X_s} + \frac{3|\tilde{V}_a||\tilde{E}_a|}{X_s}\sin\delta - j\frac{3|\tilde{V}_a||\tilde{E}_a|}{X_s}\cos\delta$$

or the real input (=output) power – because the losses are neglected – is $P = \frac{3|\tilde{V}_a||\tilde{E}_a|}{X_s}\sin\delta$ and the corresponding reactive power $Q = \frac{3|\tilde{V}_a|^2}{X_s} - \frac{3|\tilde{V}_a||\tilde{E}_a|}{X_s}\cos\delta$.

$P = \frac{3|\tilde{V}_a||\tilde{E}_a|}{X_s}\sin\delta = (32.71 \cdot 1 \cdot 0.602)/2 = 2.444\,\text{pu}$, or

$P = 2.444 \cdot V_{base} \cdot I_{base} = 82.1$ MW, and $Q = 58.62$ kVAr.

(i) The torque developed is $T = P/\omega_s = 435.11$ kNm.

(j) $P_{in} = 80.64$ MW.

(k) $P_{loss} = 3 \cdot R_a|\tilde{I}_{a\_rat}|^2 = 3 \cdot 0.05 \cdot 17.143 \cdot (1.4)^2 = 5.04\,\text{MW}$.

(l) The results for a lagging (consumer notation) power factor of $\cos\Phi = 0.8$ and $I_f = 0.5$ pu are listed below:

$$S = 3|\tilde{V}_{a\_rat}|\left|\frac{I_{a\_rat}}{2}\right| = 3 \cdot 24\,\text{kV} \cdot 0.7\,\text{kA} = 50.4\,\text{MVA},$$

$P = S{\cdot}\cos\Phi = 50.4$ MW$\cdot$0.8 = 40.32 MW, and $Q = -30.24$ kVAr, $\Phi = -36.87°$ lagging,

$$|\tilde{V}_R| = \left|\frac{\tilde{I}_{a\_rat}}{2}\right| \cdot R_a = 0.7\,\text{kA} \cdot 0.05 \cdot 17.143 = 600\,\text{V}$$

corresponding to 0.0375 in.

$$|\tilde{V}_X| = \left|\frac{\tilde{I}_{a\_rat}}{2}\right| \cdot X_s = 0.7\,\text{kA} \cdot 2 \cdot 17.143 = 24{,}000\,\text{V}$$

corresponding to 1.5 in. Figure 12.13 shows the phasor diagram for a lagging power factor of 0.8 (consumer notation).

From the phasor diagram one obtains $|\tilde{E}_a| = 1.25$ in. corresponding to a voltage of 20,000 V, $n_s = 1{,}800$ rpm, $\delta = 65°$, $\omega_s = 188.7$ rad/s, $P = 38.05$ MW, $Q = -32.65$ kVAr, $T = 201.65$ kNm, and $P_{loss} = 1.26$ MW.

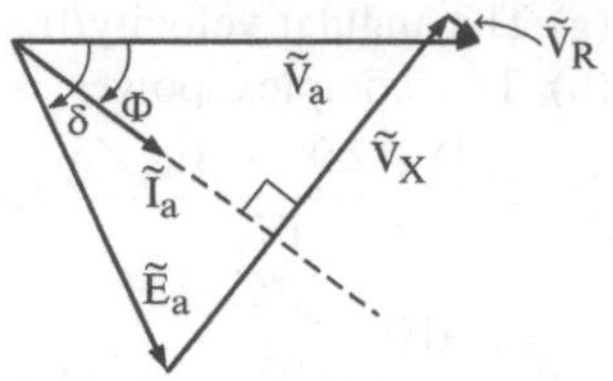

Fig. 12.13 Phasor diagram for a lagging power factor of 0.8 (consumer notation)

12.3 Centralized (Conventional) Power System

Within conventional power systems operation, cogeneration plants play an important role for the increase of efficiency. Related to the joint operation of central power stations in parallel with cogeneration/renewable plants is the frequency control either for islanding or interconnected operation. Load changes induce frequency variations which must be minimized by connecting either spinning reserve or additional storage plants. A prerequisite for the frequency control at steady-state is the Newton–Raphson power flow algorithm on which most commercially available load-flow software programs are based. The voltage control is achieved by compensation of the reactive power, load management, and load shedding.

Application Example 12.6: Carnot Cycle, Adiabatic Process, and Carnot Efficiency of a Gas Turbine of a Cogeneration Power Plant

The University of Colorado at Boulder operates two gas turbine-driven synchronous generators combined with a cogeneration plant providing heating and cooling to the campus facilities as shown in Fig. 12.14. Gas turbines operate based on the Brayton cycle which is related to the Carnot cycle. Note the temperatures in this figure are given in degrees Fahrenheit. The gas turbines as employed in Fig. 12.14 are depicted in Fig. 12.15.

The following data were provided by Mr. J.E. Hanlon of the University of Colorado at Boulder: gas costs = \$7–15/mmbtu; coal costs = \$1.50/mmbtu; #2 Diesel oil costs = \$18/mmbtu; electricity costs = \$0.065/kWh; operational break-even point = \$6/mmbtu (\$5/mmbtu if maintenance costs are included); Xcel costs of production = \$0.038/kWh; wind generation cost of production = \$0.05/kWh; photovoltaic cost of production = \$0.20/kWh; typical house uses 20 mmbtu/yr of natural gas; plant uses 20 mmbtu/10min; CTG@15MW or boiler at 100% = 150mmbtu/hr; typical house uses 10,000 kWh/yr; plant uses 6,500,000 kWh/yr; plant is capable of generating 270,000,000 kWh/yr; Campus electrical use = 21 MW; Brayton-cycle efficiency = 38%; cogeneration efficiency = 55%; boiler flame temperature = 3,500°F.

(a) Determine the Carnot efficiency $\eta_{carnot} = (T_1 - T_2)/T_1$, of one gas turbine assuming an adiabatic process [9], that is, there are no heat losses within the turbine.

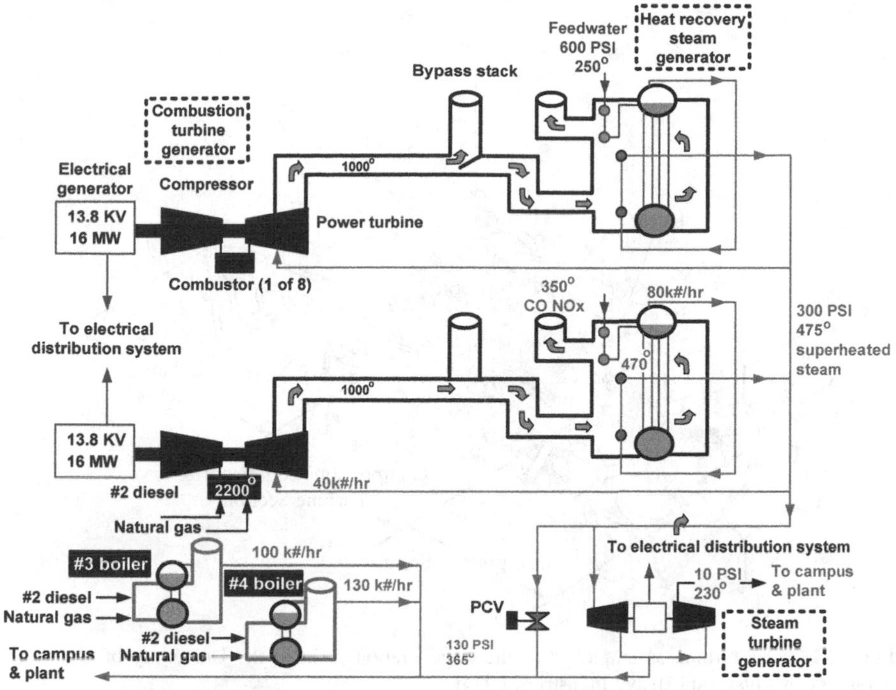

Fig. 12.14 Cogeneration plant at the University of Colorado (Courtesy of University of Colorado at Boulder, prepared by J.E Hanlon)

(b) In practice the turbine cannot be completely insulated, and there is a loss of heat of about 10%. Therefore, the overall efficiency of the gas turbine is $\eta_{\text{gas turbine}} = 0.9 \cdot \eta_{\text{carnot}}$.

(c) If the synchronous generator has an efficiency of $\eta_{\text{generator}} = 0.9$, what is the overall efficiency of the generating set (turbine+ generator) $\eta_{\text{overall}} = \eta_{\text{generator}} \cdot \eta_{\text{gas turbine}}$?

(d) If the power-plant transformer and the associated transmission and distribution network has an efficiency of $\eta_{\text{trans.+distribution}} = 0.85$, what is the overall efficiency of the power plant including transmission and distribution $\eta_{\text{overall power plant}}$?

(e) Compute the work $W = \dfrac{K(V_f^{1-\gamma} - V_i^{1-\gamma})}{1-\gamma}$ [joules] done by the natural gas provided $V_i = 16{,}000\ \text{m}^3$ $V_f = 103{,}500\ \text{m}^3$, $\gamma = 1.32$, and the constant $K = 10.20 \cdot 10^{11}$ [J/m^3].

(f) The energy content of natural gas is $E_{\text{natural gas}} = 1{,}021$ BTU/cubic foot. Compute the required volume of natural gas. Provided one cubic foot of natural gas costs either \$0.005 (US price for utilities) or \$0.01 (Europe) how much is the fuel cost for 1 kWh generated by the University power plant for both cases?

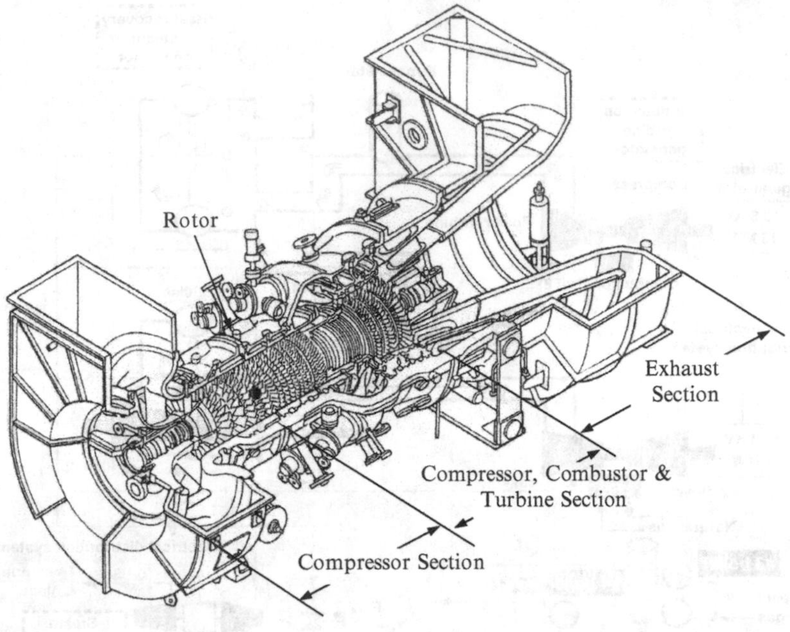

Fig. 12.15 Gas turbine as employed in the cogeneration plant at the University of Colorado (Courtesy of Mitsubishi Heavy Industries, LTD)

(g) The wage costs for power plant personnel are estimated to be 30% of the fuel costs. What is the generation cost for one kWh of electricity of the University power plant?
Hint: For the calculation of the Carnot efficiency of the gas turbine you may use the software available on the Internet address: http://hyperphysics.phy-astr.gsu.edu/hbase/thermo/adiab.html#c3

(h) How will the electricity cost per kWh be reduced due to the fact that it is a cogeneration plant providing heating and cooling power to University campus buildings?

Solution

(a) Carnot efficiency is obtained based on the degrees Fahrenheit of Fig. 12.14 and converted to Kelvin (K)

$T_{1\_Fahrenheit} = 2{,}200°F$ or $T_1 = (5/9)2{,}200 + 255.38 = 1{,}477.60$ K, and $T_{2\_Fahrenheit} = 1{,}000°$ F or $T_1 = (5/9)1{,}000 + 255.38 = 810.94$ K resulting in the Carnot efficiency of

$$\eta_{Carnot} = \frac{T_1 - T_2}{T_1} = \frac{14,77.60 - 810.94}{1,477.60} = 0.451.$$

(b) The gas turbine efficiency is $\eta_{gas\ turbine} = 0.9 \cdot \eta_{carnot} = 0.90 \cdot 0.451 = 0.406.$

(c) The overall efficiency of the turbine generator set is $\eta_{overall} = \eta_{generator} \cdot \eta_{gas\text{-}turbine} = 0.9 \cdot 0.406 = 0.3654$.

(d) The efficiency of the power plant including transmission and distribution $\eta_{overall\ power\ plant} = \eta_{trans.+distribution} \cdot \eta_{overall} = 0.85 \cdot 0.3654 = 0.311$.

(e) The work done by the natural gas is

$$W = \frac{K(V_f^{1-\gamma} - V_i^{1-\gamma})}{1-\gamma} = \frac{10.20 \cdot 10^{11}(103,500^{-0.32} - 16,000^{-0.32})}{-0.32}$$
$$= 0.647 \cdot 10^{11}\,\mathrm{Ws}, \text{ or } W = 6.47 \cdot 10^4/3,600\,\mathrm{MWh} = 17.9\,\mathrm{MWh}.$$

Taking the generator efficiency of 0.90 into account the power delivered to the power system is $P_{out} = 17.9 \cdot 0.9 = 16.1$ MW.

(f) The energy content of natural gas is $E_{natural\ gas} = 1{,}021$ BTU/(ft^3). Note 1 Wh = 3.413 BTU or 1 MWh = 3.413 10^6 BTU, which results in $W = 17.9 \cdot 3.413 \cdot 10^6$ BTU = $61.09 \cdot 10^6$ BTU. 1 cubic foot of natural gas contains 1,021 BTU, therefore, the required gas volume is

$$Vol_{cubic\_foot} = \frac{61.09 \cdot 10^6\,\mathrm{BTU}}{1021\,\mathrm{BTU/cubic\ foot}} = 59.84\,\mathrm{k\ cubic\ foot}.$$

The overall plant efficiency is 0.311, therefore, the required gas volume is

$$Vol_{required} = \frac{59.84\,\mathrm{k}}{0.311} = 192.40\,\mathrm{k\ cubic\ foot}.$$

1 cubic foot of natural gas costs \$0.005 thus the total cost in the US is $cost_{US} = 192.40\,\mathrm{k} \cdot 0.005 = \$962$ and 1 kWh generated costs

$$Cost_{US}^{1\,kWh} = \frac{\$962}{17.9\,\mathrm{MWh}} = \$0.054.$$

The cost in Europe is $Cost_{Europe}^{1\,kWh} = \$\,0.11$.

(g) $Cost_{US+labor}^{1\,kWh} = 1.3 \cdot \$0.054 = \$0.07$, without any tax and interest payments.

(h) The cost per kWh will be reduced if the heating and cooling power of the University campus is taken into account (beyond the scope of this example).

Application Example 12.7: Frequency Control of an Isolated Power Plant (Islanding Operation)

Figure 12.16 illustrates the block diagram of governor, prime mover (steam turbine) and rotating mass & load of a turbo generator set [10, 11].

For the frequency change $\Delta\omega$ per change in generator output power ΔP, that is, $R = \frac{\Delta\omega}{\Delta P}\,\mathrm{pu} = 0.01\,\mathrm{pu}$, the frequency-dependent load change $\Delta P_{L1}|_{frequ}$ per frequency change $D = \frac{\Delta P_{L1}|_{frequ}}{\Delta\omega} = 0.8\,\mathrm{pu}$, step-load change $\Delta P_L(s) = \frac{\Delta P_L}{s}\,\mathrm{pu} = \frac{0.2}{s}$,

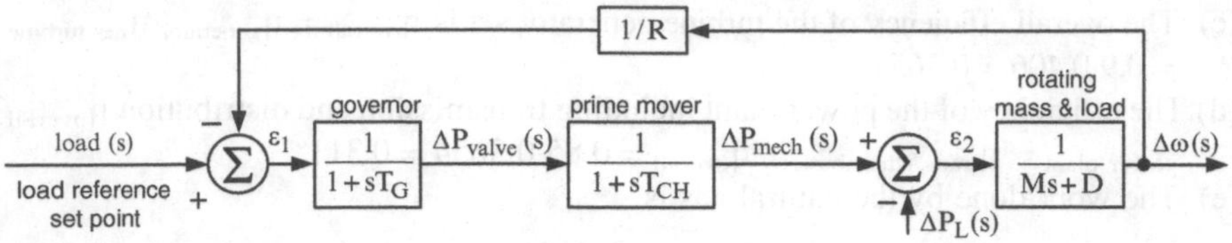

Fig. 12.16 Block diagram of governor, prime mover and rotating mass & load, where D corresponds to the frequency dependent load and $\Delta P_L(s)$ is the frequency independent load

angular momentum of steam turbine and generator set M = 4.5, base apparent power S_{base} = 500 MVA, governor time constant T_G = 0.01 s, valve changing/charging time constant T_{CH} = 1.0 s, load reference set point load(s) = 1.0:

(a) Derive for Fig. 12.16 $\Delta\omega_{steady\ state}$ by applying the final value theorem. You may assume load reference set point load(s) = 0.8 pu, and $\Delta P_L(s) = \frac{\Delta P_L}{s}$ pu $= \frac{0.2}{s}$ pu. For the nominal frequency f* = 60 Hz calculate the frequency f_{new} after the load change has taken place.
(b) List the ordinary differential equations and the algebraic equations of the block diagram of Fig. 12.16.
(c) Use either Mathematica or Matlab to establish steady-state conditions by imposing a step function for load reference set point load(s) $= \frac{1}{s}$ pu and run the program with a zero-step load change $\Delta P_L = 0$ (for 5 s) in order to establish the equilibrium condition without load step. After 5 s impose a step-load change of $\Delta P_L(s) = \frac{\Delta P_L}{s}$ pu $= \frac{0.2}{s}$ pu to find the transient response of Δω(t) for a total of 25 s.
(d) Use Mathematica or Matlab to establish steady-state conditions by imposing a step function for load reference set point load(s) $= \frac{1}{s}$ pu and run the program with a zero-step load change $\Delta P_L = 0$ (for 5 s) as in part c). After 5 s impose a step-load change of $\Delta P_L(s) = \frac{\Delta P_L}{s}$ pu $= \frac{-0.2}{s}$ pu to find the transient response of Δω(t) for a total of 25 s.

Solution

(a) Define the transfer functions $F_1 = \frac{1}{1+s\tau_G}$, $F_2 = \frac{1}{1+S\tau_{CH}}$, $F_3 = \frac{1}{Ms + D}$, and $F_4 = \frac{1}{R}$. From the block diagram one obtains $\Delta\omega(s) = F_3 \cdot \Delta P_{mech}(s) - F_3 \cdot \Delta P_L(s)$, where $\Delta P_{mech}(s) = F_1F_2\{\text{load}(s) - F_4\Delta\omega(s)\}$ resulting in $\Delta\omega(s) = \dfrac{F_1F_2F_3\text{load}(s) - F_3\Delta P_L(s)}{1 + F_1F_2F_3F_4}$ or

$$\Delta\omega(s) = \frac{\dfrac{\text{load}(s)}{(1 + s\tau_G)(1 + s\tau_{CH})(Ms + D)} - \dfrac{\Delta P_L(s)}{(Ms + D)}}{\left\{1 + \dfrac{1}{(1 + s\tau_G)(1 + s\tau_{CH})(Ms + D)R}\right\}}.$$

At steady state load(s) = 0.8 pu and $\Delta P_L(s) = \frac{\Delta P_L}{s}$ pu $= \frac{0.2}{s}$ pu one obtains with the finite-value theorem the droop characteristic in the frequency-load coordinate system $\Delta\omega|_{\text{steady state}} = \frac{-\Delta P_L}{(D+\frac{1}{R})}$, or for the given parameters $\Delta\omega|_{steady\,state} = \frac{-0.2}{(0.8+\frac{1}{0.01})} = -1.98\cdot10^{-3}\,pu = -0.198\%$ with $f_{\text{rated}} = f* = 60\,\text{Hz}$

one obtains the new frequency $f_{\text{new}} = f^* - 0.00198\cdot 60$ Hz = 59.88 Hz.

(b) The algebraic and differential equations of the block diagram are:

$$\varepsilon_1 = \left\{\text{load} - \frac{\Delta\omega}{R}\right\},\quad \Delta P_{\text{valve}} + \tau_G\frac{d(\Delta P_{\text{valve}})}{dt} = \varepsilon_1,\quad \Delta P_{\text{mech}} + \tau_{CH}\frac{d(\Delta P_{\text{mech}})}{dt} = \Delta P_{\text{valve}},$$

$$\varepsilon_2 = \{\Delta P_{\text{mech}} - \Delta P_L\},\quad \text{and}\quad \Delta\omega D + M\frac{d(\Delta\omega)}{dt} = \varepsilon_2.$$

(c) The results based on Mathematica are shown in Fig. 12.17. The Mathematica input program is listed below.

```
R=0.01;
d=0.8;
M=4.5;
Tg=0.01;
Tch=1;
Lr=1;
DPL[t_]:=If[t<5, 0, 0.2];
ic1=Dw[0]==0;
ic2=DPmech[0]==0;
ic3=DPvalve[0]==0;
E1[t_]:=Lr-Dw[t]/R;
E2[t_]:=DPmech[t]-DPL[t];
eqn1=Dw'[t]==(1/M)*(E2[t]-d*Dw[t]);
eqn2=DPmech'[t]==(1/Tch)*(DPvalve[t]-DPmech[t]);
eqn3=DPvalve'[t]= = (1/Tg)*(E1[t]-DPvalve[t]);
```

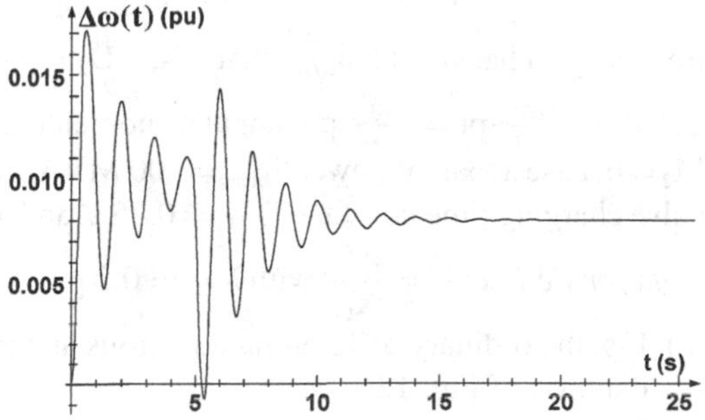

Fig. 12.17 Angular frequency change Δω(t) for a positive load-step function

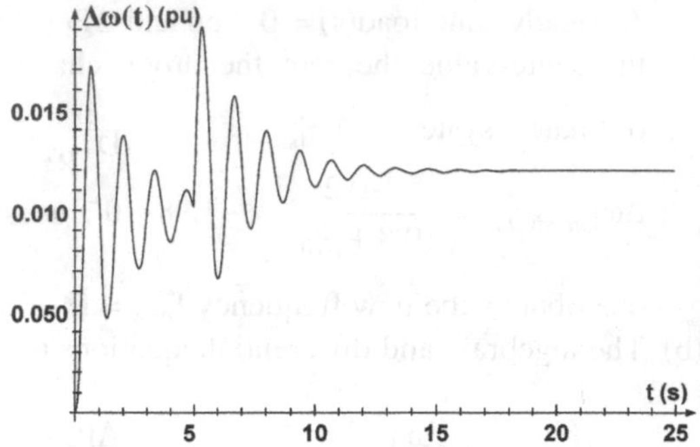

Fig. 12.18 Angular frequency change Δω(t) for a negative load-step function

```
sol=NDSolve[{eqn1, eqn2, eqn3, ic1, ic2, ic3}, {Dw[t], DPmech[t], DPvalve
   [t]},{t, 0, 25},
MaxSteps -> 100000];
Plot[Dw[t]/.sol,{t, 0, 25}, PlotRange -> All, AxesLabel ->{"t(s)", "Dw[t](pu)"}]
```

(d) The results based on Mathematica are shown in Fig. 12.18. Replace in above input program DPL[t_]:=If[t<5, 0, 0.2]; by DPL[t_]:=If[t<5, 0, –0.2];

Application Example 12.8: Frequency Control of an Interconnected Power System Broken into Two Areas Each Having One Generator

Figure 12.19 shows the block diagram of two generators interconnected by a transmission tie line [10, 11].

Data for generation set (steam turbine and generator) #1: For the frequency change ($\Delta\omega_1$) per change in generator output power (ΔP_1), that is, $R_1 = \frac{\Delta\omega_1}{\Delta P_1}$ pu $= 0.01$ pu (e.g., coal-fired plant), the frequency-dependent load change ($\Delta P_{L1}|_{frequ}$) per frequency change ($\Delta\omega_1$), that is, $D_1 = \frac{\Delta P_{L1}|_{frequ}}{\Delta\omega_1} = 0.8$ pu, step-load change $\Delta P_{L1}(s) = \frac{\Delta P_{L1}}{s}$ pu $= \frac{0.2}{s}$ pu, angular momentum of steam turbine and generator set $M_1 = 4.5$, base apparent power $S_{base} = 500$ MVA, governor time constant $T_{G1} = 0.01$ s, valve charging time constant $T_{CH1} = 0.5$ s, and load $ref_1(s) = 0.8$ pu.

Data for generation set (steam turbine and generator) #2: For the frequency change ($\Delta\omega_2$) per change in generator output power (ΔP_2), that is, $R_2 = \left(\frac{\Delta\omega_2}{\Delta P_2}\right)$ pu $= 0.02$ pu (e.g., coal-fired plant), the frequency-dependent load change $\left(\Delta P_{L2}|_{frequ}\right)$ per frequency change ($\Delta\omega_2$), that is, $D_2 = \frac{\Delta P_{L2}|_{frequ}}{\Delta\omega_2} = 1.0$ pu, step-load change $\Delta P_{L2}(s) = \frac{\Delta P_{L2}}{s}$ pu $= \frac{-0.2}{s}$ pu, angular momentum of steam turbine and generator set $M_2 = 6$, base apparent power $S_{base} = 500$ MVA, governor time constant $T_{G2} = 0.02$ s, valve charging time constant $T_{CH2} = 0.75$ s, and load $ref_2(s) = 0.8$ pu.

Data for tie line: $T = \dfrac{377}{X_{tie}}$ with $X_{tie} = 0.2$ pu

(a) List the ordinary differential equations and the algebraic equations of the block diagram of Fig. 12.19.

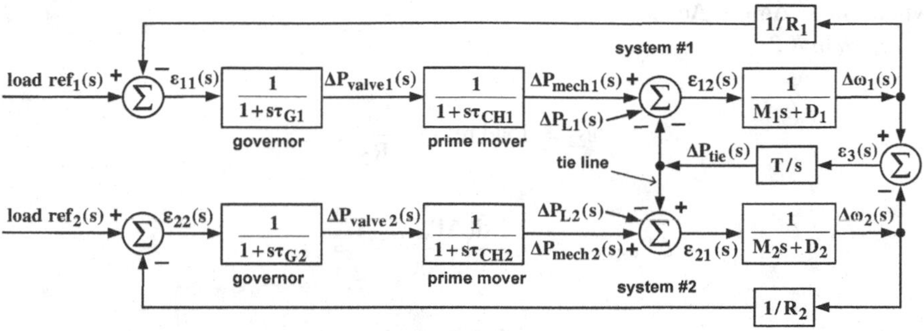

Fig. 12.19 Block diagram of two interconnected generators through a tie (transmission) line, where D_1 and D_2 correspond to the frequency dependent loads and $\Delta P_{L1}(s)$ and $\Delta P_{L2}(s)$ are the frequency independent loads

(b) Use either Mathematica or Matlab to establish steady-state conditions by imposing a step function for load $\text{ref}_1(s) = \frac{0.8}{s}$ pu, load $\text{ref}_2(s) = \frac{0.8}{s}$ pu and run the program with a zero step-load changes $\Delta P_{L1} = 0$, $\Delta P_{L2} = 0$ (for 5 s) in order to establish the equilibrium condition. After 5 sec impose step-load changes $\Delta P_{L1}(s) = \frac{\Delta P_{L1}}{s}\text{pu} = \frac{0.2}{s}$ pu, and after 7 s impose $\Delta P_{L2}(s) = \frac{\Delta P_{L2}}{s}\text{pu} = \frac{-0.2}{s}$ pu to find the transient responses $\Delta\omega_1(t)$ and $\Delta\omega_2(t)$ for a total of 100 s for $R_1 = 0.01$ pu and $R_2 = 0.02$ pu.

(c) Repeat part b) for $R_1 = 0.5$ pu, (e.g., wind-power plant), and $R_2 = 0.01$ pu (e.g., coal-fired plant) for

$$\Delta P_{L1}(s) = \frac{\Delta P_{L1}}{s}\text{pu} = \frac{0.1}{s}\text{pu} \quad \text{and } \Delta P_{L2}(s) = \frac{\Delta P_{L2}}{s}\text{pu} = \frac{-0.1}{s}\text{pu}$$

Solution

(a) Differential and algebraic equations

System # 1:

$$\varepsilon_{11} = \text{load ref}_1 - \frac{\Delta\omega_1}{R_1},$$

$$\Delta P_{\text{valve\_1}} + \tau_{G1}\frac{d(\Delta P_{\text{valve\_1}})}{dt} = \varepsilon_{11},$$

$$\Delta P_{\text{mech\_1}} + \tau_{CH1}\frac{d(\Delta P_{\text{mech\_1}})}{dt} = \Delta P_{\text{valve\_1}},$$

$$\varepsilon_{12} = \Delta P_{\text{mech\_1}} - \Delta P_{L1} - \Delta P_{\text{tie}},$$

$$\Delta\omega_1 D_1 + M_1\frac{d(\Delta\omega_1)}{dt} = \varepsilon_{12}.$$

Coupling (tie, transmission) network:

$$\frac{1}{T}\frac{d(\Delta P_{\text{tie}})}{dt} = \varepsilon_3,$$

where $\varepsilon_3 = \Delta\omega_1 - \Delta\omega_2$.

System # 2:

$$\varepsilon_{22} = \text{load ref}_2 - \frac{\Delta\omega_2}{R_2},$$

$$\Delta P_{valve\_2} + \tau_{G2}\frac{d(\Delta P_{valve\_2})}{dt} = \varepsilon_{22},$$

$$\Delta P_{mech\_2} + \tau_{CH2}\frac{d(\Delta P_{mech\_2})}{dt} = \Delta P_{valve\_2},$$

$$\varepsilon_{21} = \Delta P_{mech\_2} - \Delta P_{L2} + \Delta P_{tie},$$

$$\Delta\omega_2 D_2 + M_2\frac{d(\Delta\omega_2)}{dt} = \varepsilon_{21}.$$

(b) The results based on Mathematica are shown in Figs. 12.20a, b. The Mathematica input program is listed below.

```
R1=0.01;
d1=0.8;
M1=4.5;
Tg1=0.01;
Tch1=0.5;
Lr1=0.8;
DPL1[t_]: = If [t<5, 0, 0.2];
R2=0.02;
d2=1.0;
M2=6;
Tg2=0.02;
Tch2=0.75;
Lr2=0.8;
```

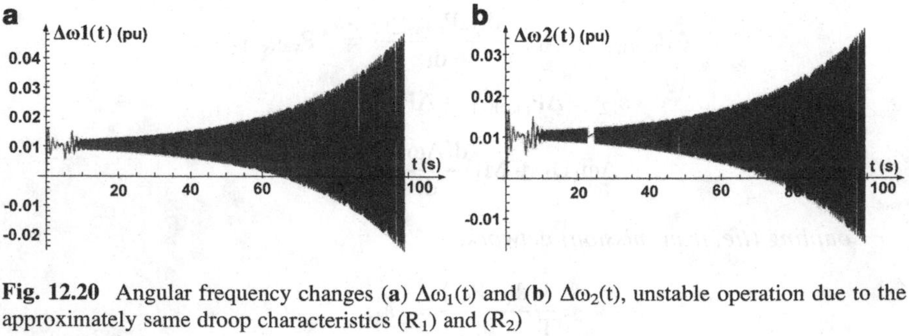

Fig. 12.20 Angular frequency changes (**a**) $\Delta\omega_1(t)$ and (**b**) $\Delta\omega_2(t)$, unstable operation due to the approximately same droop characteristics (R_1) and (R_2)

```
DPL2[t_]: = If [t<7, 0, −0.2];
Xtie=0.2;
Tie=377/Xtie;
ic1=Dw1[0]==0;
ic2= DPmech1[0]==0;
ic3=DPvalve1[0]==0;
ic4=DPtie[0]==0;
ic5=Dw2[0]==0;
ic6= DPmech2[0]==0;
ic7=DPvalve2[0]==0;
E11[t_]:=Lr1-Dw1[t]/R1;
E12[t_]:=DPmech1[t]-DPL1[t]-DPtie[t];
E3[t_]: = Dw1[t]-Dw2[t];
E22[t_]: = Lr2-Dw2[t]/R2;
E21[t_]: = DPmech2[t]-DPL2[t]+DPtie[t];
eqn1=Dw1'[t] = = (1/M1)*(E12[t]-d1*Dw1[t]);
eqn2=DPmech1'[t] = = (1/Tch1)*(DPvalve1[t]-DPmech1[t]);
eqn3=DPvalve1'[t] = = (1/Tg1)*(E11[t]-DPvalve1[t]);
eqn4=DPtie'[t] = = Tie*E3[t];
eqn5=Dw2'[t] = = (1/M2)*(E21[t]-d2*Dw2[t]);
eqn6=DPmech2'[t] = = (1/Tch2)*(DPvalve2[t]-DPmech2[t]);
eqn7=DPvalve2'[t] = = (1/Tg2)*(E22[t]-DPvalve2[t]);
sol=NDSolve[{eqn1, eqn2, eqn3, eqn4, eqn5, eqn6, eqn7, ic1, ic2, ic3, ic4, ic5,
   ic6, ic7},{Dw1[t], DPmech1[t], DPvalve1[t], DPtie[t], Dw2[t], DPmech2[t],
   DPvalve2[t] },{t, 0, 100}, MaxSteps -> 100000];
Plot[Dw1[t]/.sol,{t, 0, 100}, PlotRange -> All, AxesLabel ->{"t(s)", "Dw1[t]
   (pu)"}]
Plot[Dw2[t]/.sol,{t, 0, 100}, PlotRange -> All, AxesLabel ->{"t(s)", "Dw2[t]
   (pu)"}]
```

(c) Replace in above input program R1=0.01; by R1=0.5; replace DPL1[t_]: = If [t<5, 0, 0.2]; by DPL1[t_]: = 0.1; replace R2=0.02; by R2=0.01; and replace DPL2[t_]: = If [t<7, 0, −0.2]; by DPL2[t_]: = −0.1. The results based on Mathematica are shown in Figs. 12.21a, b.

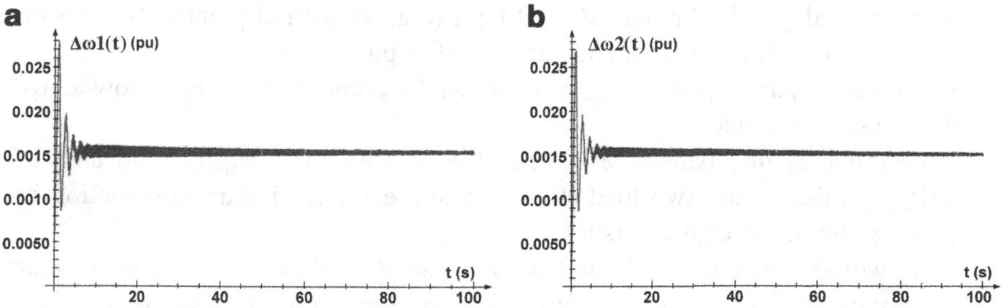

Fig. 12.21 Angular frequency change $\Delta\omega_1(t)$ and $\Delta\omega_2(t)$, stable operation due to the different droop characteristics (R_1) and (R_2)

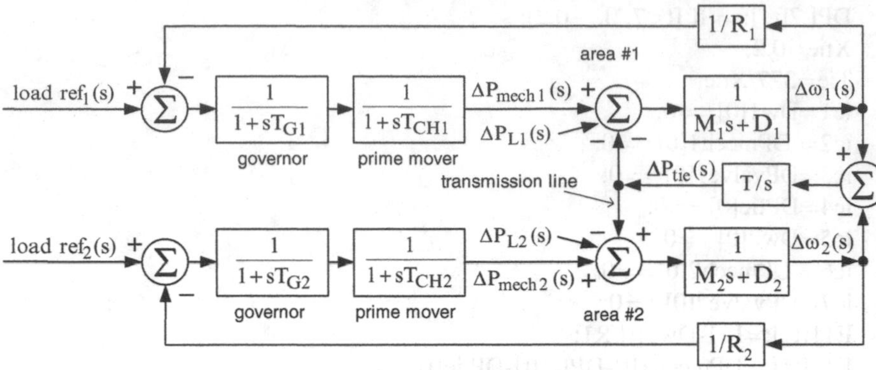

Fig. 12.22 Interconnected power system consisting of area #1 and area #2, where D_1 and D_2 correspond to the frequency dependent loads and $\Delta P_{L1}(s)$ and $\Delta P_{L2}(s)$ are the frequency independent loads

Application Example 12.9: Frequency Variation Within an Interconnected Power System as a Result of Two Load Changes

A block diagram [10, 11] of two interconnected areas of a power system (e.g., area #1 and area #2) is shown in Fig. 12.22. The two areas are connected by a single transmission line. The power flow over the transmission line will appear as a positive load to one area and an equal but negative load to the other, or *vice versa*, depending upon the direction of power flow.

(a) For steady-state operation show that with $\Delta\omega_1 = \Delta\omega_2 = \Delta\omega$ the change in the angular velocity (which is proportional to the frequency f) is

$$\Delta\omega = \text{function of } (\Delta P_{L1}, \Delta P_{L2}, D_1, D_2, R_1, R_2), \text{and} \tag{12.1}$$

$$\Delta P_{tie} = \text{function of the same parameters as in (12.1).} \tag{12.2}$$

(b) Determine values for $\Delta\omega$ (12.1), ΔP_{tie} (12.2) and the new frequency f_{new}, where the nominal (rated) frequency is $f_{rated} = 60$ Hz, for the parameters: load $ref_1 = 0.2$ pu = constant1, load $ref_2 = 0.8$ pu = constant2, $R_1 = 0.05$ pu (e.g., natural-gas fired plant), $R_2 = 0.1$ pu (e.g., coal-fired plant), $D_1 = 0.8$ pu, $D_2 = 1.0$ pu, $\Delta P_{L1} = 0.2$ pu, and $\Delta P_{L2} = -0.3$ pu.

(c) For a base apparent power $S_{base} = 1{,}000$ MVA compute the power flow across the transmission line.

(d) How much is the load increase/decrease in area #1 (ΔP_{mech1}) and area #2 (ΔP_{mech2}) due to the two load steps? Illustrate the load sharing/reduction by plotting the droop characteristics.

(e) How would you change R_1 and R_2 in case R_2 belongs to a wind or solar (photovoltaic) power plant operating at its maximum power point (and cannot accept any significant load increase due to the two load steps) and R_1 belongs to a natural-gas fired plant which can accept additional load? Illustrate the load sharing by plotting the corresponding droop characteristics.

(f) Provided the natural-gas fired plant is replaced by a short–term and a long-term storage plant illustrate the load sharing by plotting the corresponding droop characteristics of the (intermittently operating but always operating at peak power) photovoltaic plant ($R_1 = 10$), the short-term storage plant ($R_2 = 0.01$, can be put on-line within a 60 Hz cycle) and long-term storage plant ($R_3 = 10$, can be put on line within 6–10 min).

Solution

(a) $\Delta P_{mech1} - \Delta P_{L1} - \Delta P_{tie} = \Delta\omega D_1$, $\Delta P_{mech2} - \Delta P_{L2} + \Delta P_{tie} = \Delta\omega D_2$, $\Delta P_{mech1} = -\frac{\Delta\omega}{R_1}$, $\Delta P_{mech2} = -\frac{\Delta\omega}{R_2}$, from these equations follows for the tie line power change

$\Delta P_{tie} = -\Delta P_{L1} - \Delta\omega(D_1 + \frac{1}{R_1})$, and $\Delta P_{tie} = \Delta P_{L2} + \Delta\omega(D_2 + \frac{1}{R_2})$. With these relations one obtains for the change in angular frequency $\Delta\omega = \frac{(-\Delta P_{L1} - \Delta P_{L2})}{(D_1 + D_2 + \frac{1}{R_1} + \frac{1}{R_2})}$, and the change in the tie-line power

$$\Delta P_{tie} = \Delta P_{L2} + (D_2 + \frac{1}{R_2})\frac{(-\Delta P_{L1} - \Delta P_{L2})}{(D_1 + D_2 + \frac{1}{R_1} + \frac{1}{R_2})}.$$

(b) The change in the angular frequency is $\Delta\omega = \frac{(-0.2+0.3)}{(0.8+1.0+20+10)} = 0.003145\,\text{pu} = \Delta f\,[\text{pu}]$, resulting in the new frequency $f_{new} = f_{rated}\,(1+\Delta f) = 60 \cdot 1.003145\,\text{Hz} = 60.1887\,\text{Hz}$.

(c) The change in the tie line power is $\Delta P_{tie} = -0.3 + (1 + 10)\frac{(-0.2+0.3)}{(0.8+1.0+20+10)} = -0.2654\text{pu}$ or with the base apparent power of 1,000 MVA the real power change is $\Delta P_{tie} = -265.4\,\text{MW}$.

(d) The changes in the mechanical powers are $\Delta P_{mech1} = \frac{-0.003145}{0.05} = -0.0629\,\text{pu}$ or $\Delta P_{mech1} = -62.9\,\text{MW}$, $\Delta P_{mech2} = \frac{-0.003145}{0.1} = -0.03145\,\text{pu}$ or $\Delta P_{mech1} = -31.45\,\text{MW}$.

Figure 12.23 illustrates the sharing (reduction) of the additional power among a natural-gas fired (e.g., $R_1 = 5$) plant and a coal-fired (e.g., $R_2 = 10$) plant causing a frequency increase.

(e) Figure 12.24 illustrates the sharing (increase) of the additional power among a natural-gas fired plant (e.g., $R_1 = 0.01$) and a wind power plant (e.g., $R_2 = 50$) causing a frequency decrease.

(f) Figure 12.25 illustrates the sharing (increase) of the additional power among a short-term storage plant (e.g., $R_2 = 0.01$), a long-term-storage plant (e.g., $R_3 = 10$), and a photovoltaic-power plant (e.g., $R_1 = 20$) causing a frequency decrease. One obtains a stable frequency control where the short-term storage plant compensates the intermittent power output of the PV plant and the plant with spinning reserve (natural-gas fired plant) is replaced by a long-term

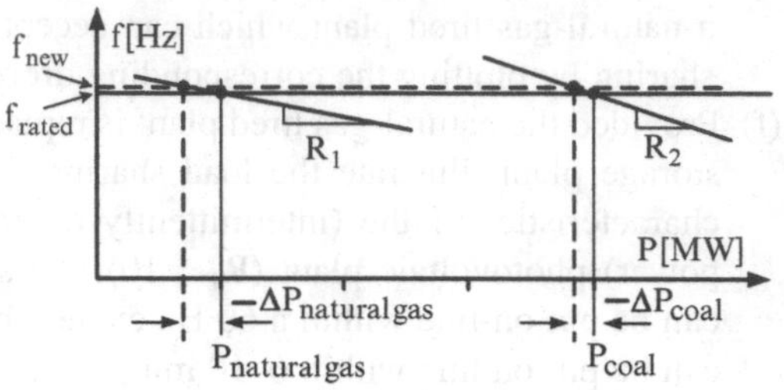

Fig. 12.23 Drooping characteristics of coal-fired (e.g., R_2=10) plant and natural-gas fired (e.g., R_1=5) plant, accommodating reduced demand of power

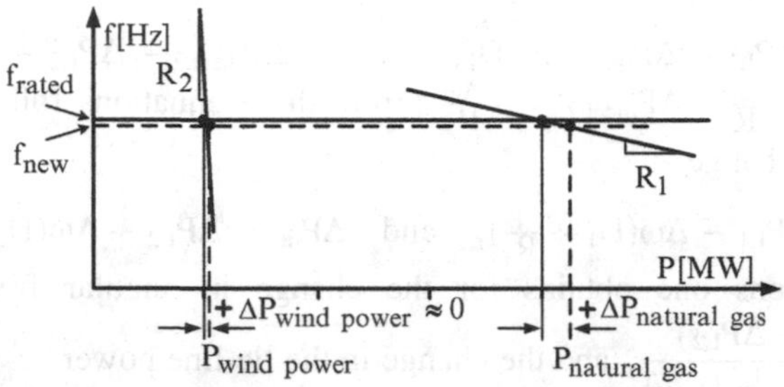

Fig. 12.24 Drooping characteristics of natural-gas fired (e.g., R_1=0.01) plant and wind power (e.g., R_2=50) plant, accommodating additional demand/increase of power, where the wind-power plant is operated at peak power and cannot deliver additional power

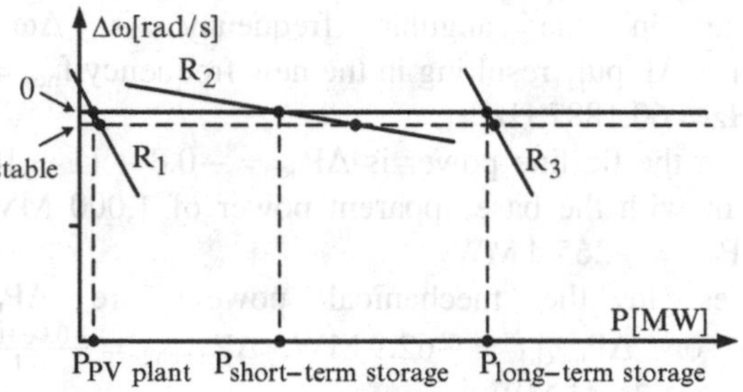

Fig. 12.25 Drooping characteristics of the short-term storage plant (e.g., R_2=0.01), the long-term storage plant (e.g., R_3=10) plant, and intermittently operating photovoltaic-power (e.g., R_1=20) plant, accommodating additional demand/increase of power, where the photovoltaic-power plant is operated at peak power and cannot deliver additional power

storage plant. The long-term storage plant is connected all the time to the power system and serves therefore as frequency leader. The PV plant and the short-term storage plant may operate intermittently only.

References [12, 55] describe the operation of a smart/micro grid consisting of natural-gas fired plant – serving as frequency leader – a long-term storage plant, a wind-power plant with its dedicated short-term storage plant, and a photovoltaic-power plant with its associated short-term storage plant. Figure 12.26 illustrates the block

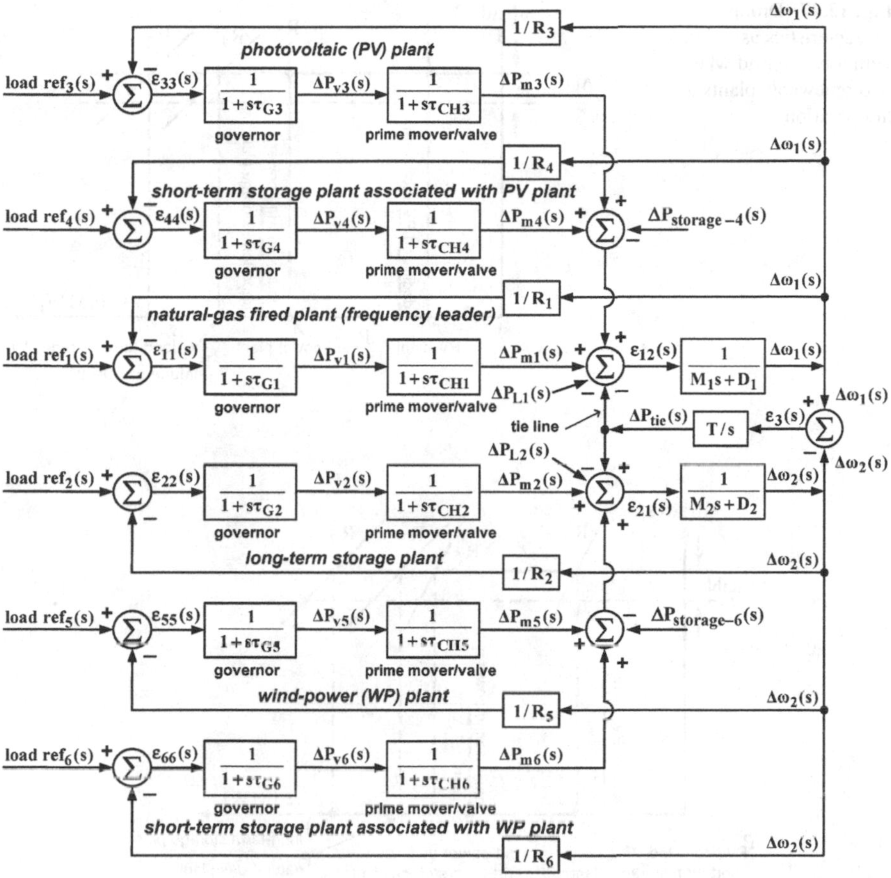

Fig. 12.26 Structure of a smart/micro grid

diagram in the frequency domain. Figure 12.27 shows the corresponding droop characteristics for frequency/load control when natural-gas plant, long-term storage plant and the two renewable plants are connected to the system, while Fig. 12.28 illustrates the operation when the two renewable plants are disconnected and instead the two short-term storage plants have replaced the two renewable plants.

Application Example 12.10: Fundamental Power Flow

To the two-bus power system of Fig. 12.29, apply the Newton–Raphson load-flow analysis technique [11].

(a) Find the fundamental bus admittance matrix in polar form $\bar{Y}_{bus}$.
(b) Assume that bus #1 is the swing or slack bus, and bus #2 is a PQ bus. Make an initial guess for the bus voltage vector $\bar{X}^{(o)}$ and evaluate the initial mismatch power $\bar{W}^{(o)}$.

Fig. 12.27 Droop characteristics associated with a micro grid when two renewable plants are in operation

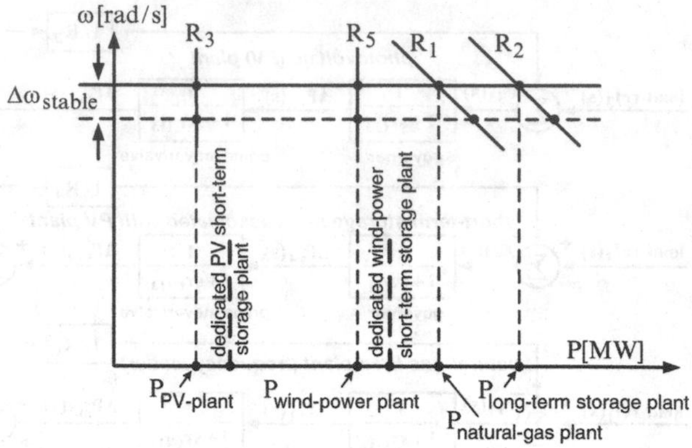

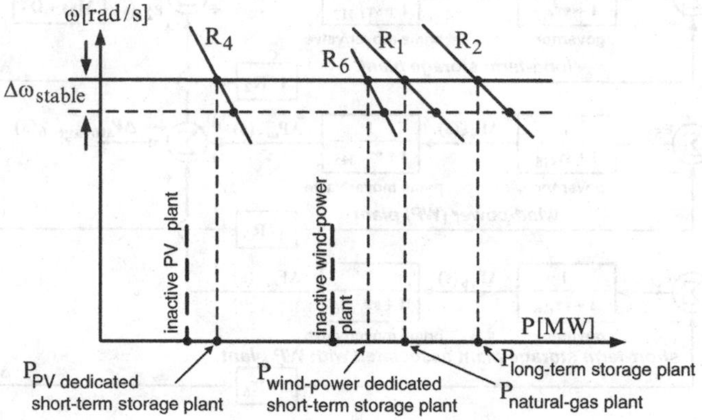

Fig. 12.28 Droop characteristics associated with a smart/micro grid when two short-term storage plants are in operation

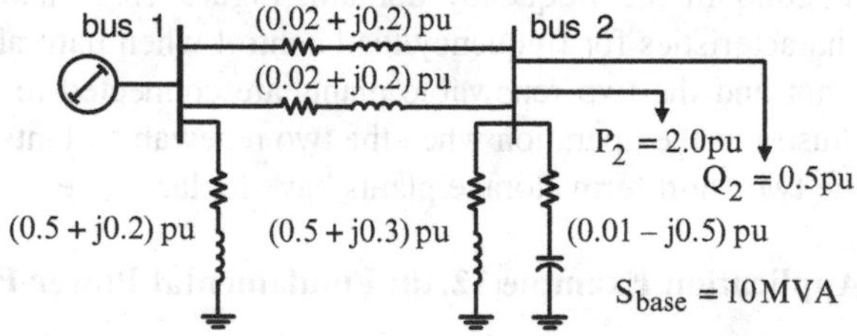

Fig. 12.29 Two-bus system

(c) Find the Jacobian $\bar{J}^{(o)}$ for this power system configuration, and compute the bus voltage correction vector $\Delta\bar{X}^{(o)}$.

(d) Update the bus voltage vector $\bar{X}^{(1)} = \bar{X}^{(o)} - \Delta\bar{X}^{(o)}$, and recompute the updated mismatch power vector $\overline{W}^{(1)}$.

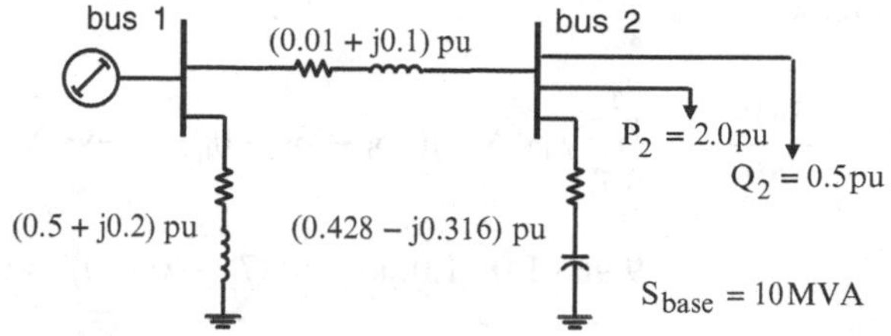

Fig. 12.30 Simplified two-bus system

Solution

The network of Fig. 12.29 is simplified as shown in Fig. 12.30.

(a) The $\bar{Y}_{bus}$ matrix is in polar form: $\bar{Y}_{bus} = \begin{vmatrix} 10.93\angle-75.63^\circ & 9.95\angle 95.7^\circ \\ 9.95\angle 95.7^\circ & 9.13\angle-74.1^\circ \end{vmatrix}$.

(b) Initial guess for the bus voltage vector $\bar{X}^{(o)}$ and evaluation of the initial mismatch power $\overline{W}^{(o)}$:

The bus voltages are defined as $\tilde{V}_1 = |\tilde{V}_1|e^{j\delta 1} = |\tilde{V}_1|\angle\,\delta_1 = 1\angle 0^o$ which is the swing or slack bus, and $\tilde{V}_2 = |\tilde{V}_2|e^{j\delta 2} = |\tilde{V}_2|\angle\,\delta_2$. In all subsequent calculations the per unit (pu) magnitudes are represented by $V_i = |\tilde{V}_i|$.
$\bar{X}^{(o)} = (\delta_2,\ V_2)^t = (0,\ \ 1)^t$, the mismatch power vector is defined as $\Delta\overline{W}^{(o)} = (\Delta P_2,\ \Delta Q_2)^t$,
where $\Delta P_i = P_i + \sum_{j=1}^{2} y_{ij} V_j V_i \cos(-\theta_{ij} - \delta_j + \delta_i)$,

$$\Delta P_2 = P_2 + y_{21} V_1 V_2 \cos(-\theta_{21} - \delta_1 + \delta_2) + y_{22} V_2 V_2 \cos(-\theta_{22} - \delta_2 + \delta_2)$$

$$\Delta P_2 = 2 + 9.95 \cdot 1.0 \cdot 1.0 \cos(-95.7^\circ - 0 + 0) + 9.13 \cdot 1.0 \cdot 1.0 \cos(74.1 - 0 + 0) \\ = 3.512\,\text{pu},$$

$$\Delta Q_i = Q_i + \sum_{j=1}^{2} y_{ij} V_j V_i \sin(-\theta_{ij} - \delta_j + \delta_i),$$

$$\Delta Q_2 = Q_2 + y_{21} V_1 V_2 \sin(-\theta_{21} - \delta_1 + \delta_2) + y_{22} V_2 V_2 \sin(-\theta_{22} - \delta_2 + \delta_2)$$

$$\Delta Q_2 = 0.5 + 9.95 \cdot 1.0 \cdot 1.0 \sin(-95.7^\circ - 0 + 0) + 9.13 \cdot 1.0 \cdot 1.0 \sin(74.1 - 0 \\ + 0) \\ = -0.616\,\text{pu},$$

therefore

$$\Delta\bar{W}^{(o)} = (3.512 - 0.616)^t.$$

(c) Jacobian $\bar{J}^{(o)}$ and bus voltage correction vector $\Delta\bar{X}^{(o)}$ are

$$\bar{J}^{(o)} = \begin{vmatrix} \dfrac{\partial \Delta P_2}{\partial \delta_2} & \dfrac{\partial \Delta P_2}{\partial V_2} \\ \dfrac{\partial \Delta Q_2}{\partial \delta_2} & \dfrac{\partial \Delta Q_2}{\partial V_2} \end{vmatrix}$$

where

$$\frac{\partial \Delta P_2}{\partial \delta_2} = -\sum_{\substack{j=1\\ \neq 2}}^{2} y_{ij} V_i V_j \sin(\delta_i - \delta_j - \theta_{ij}) = -y_{21} V_2 V_1 \sin(-\theta_{21} - \delta_1 + \delta_2)$$

$$= -9.95 \cdot 1.0 \cdot 1.0 \sin(-95.7^\circ - 0 + 0) = 9.90,$$

$$\frac{\partial \Delta P_2}{\partial V_2} = -\sum_{\substack{j=1\\ \neq 2}}^{2} y_{ij} V_i V_j \cos(\delta_i - \delta_j - \theta_{ij}) + 2V_i y_{ii} \cos(-\theta_{ii})$$

$$= y_{21} V_1 \cos(-\theta_{21} - \delta_1 + \delta_2) + 2V_2 y_{22} \cos(-\theta_{22})$$
$$= 9.95 \cdot 1.0 \cos(-95.7^\circ - 0 + 0) + 2 \cdot 1.0 \cdot 9.13 \cos(74.1^\circ) = 4.01,$$

$$\frac{\partial \Delta Q_2}{\partial \delta_2} = \sum_{\substack{j=1\\ \neq 2}}^{2} y_{ij} V_i V_j \cos(\delta_i - \delta_j - \theta_{ij}) = y_{21} V_2 V_1 \cos(-\theta_{21} - \delta_1 + \delta_2)$$

$$= 9.95 \cdot 1.0 \cdot 1.0 \cos(-95.7^\circ - 0 + 0) = -0.98823,$$

$$\frac{\partial \Delta Q_2}{\partial V_2} = \sum_{\substack{j=1\\ \neq 2}}^{2} y_{ij} V_i V_j \sin(\delta_i - \delta_j - \theta_{ij}) + 2V_i y_{ii} \sin(-\theta_{ii})$$

$$= y_{21} V_1 \sin(-\theta_{21} - \delta_1 + \delta_2) + 2V_2 y_{22} \sin(-\theta_{22})$$
$$= 9.95 \cdot 1.0 \sin(-95.7^\circ - 0 + 0) + 2 \cdot 1.0 \cdot 9.13 \sin(74.1^\circ) = 7.661,$$

$$\bar{J}^{(o)} = \begin{vmatrix} 9.90 & 4.01 \\ -0.988 & 7.661 \end{vmatrix}, \quad \text{and} \quad [\bar{J}^{(o)}]^{-1} = \begin{vmatrix} 9.90 & 4.01 \\ -0.988 & 7.661 \end{vmatrix},$$

$$\Delta\bar{X}^{(o)} = [\bar{J}^{(o)}]^{-1} \Delta\bar{W}^{(o)} = \begin{vmatrix} 9.90 & 4.01 \\ -0.988 & 7.661 \end{vmatrix} \begin{vmatrix} 3.512 \\ -0.616 \end{vmatrix} = \begin{vmatrix} 0.368 \text{ rad} \\ -0.033 \text{ pu} \end{vmatrix},$$

or its transpose $\Delta\bar{X}^{(o)} = (21.1^\circ, -0.033\,\text{pu})^t$.

(d) Updated bus voltage vector $\bar{X}^{(1)} = \bar{X}^{(o)} - \Delta\bar{X}^{(o)}$ and recomputed (updated) mismatch power vector $\overline{W}^{(1)}$ are

$$\bar{X}^{(1)} = \bar{X}^{(o)} - \Delta\bar{X}^{(o)} = \begin{vmatrix} 0 \\ 1 \end{vmatrix} - \begin{vmatrix} 21.1 \\ -.033 \end{vmatrix} = \begin{vmatrix} -21.1^\circ \\ 1.033\text{pu} \end{vmatrix},$$

or its transpose $[\Delta\bar{X}^{(1)}] = (-21.1^\circ,\ 1.033\,\text{pu})^t$.

$$\Delta P_i = P_i + \sum_{j=1}^{2} y_{ij} V_j V_i \cos(-\theta_{ij} - \delta_j + \delta_i),$$

$$\Delta P_2 = P_2 + y_{21} V_1 V_2 \cos(-\theta_{21} - \delta_1 + \delta_2) + y_{22} V_2 V_2 \cos(-\theta_{22} - \delta_2 + \delta_2)$$

$$\Delta P_2 = 2 + 9.95 \cdot 1.0 \cdot 1.033 \cos(-95.7^\circ - 0 - 21.1) + 9.13 \cdot (1.033)^2 \cos(74.1 - 21.1 + 21.1)$$
$$= 0.0348 \text{ pu},$$

$$\Delta Q_i = Q_i + \sum_{j=1}^{2} y_{ij} V_j V_i \sin(-\theta_{ij} - \delta_j + \delta_i),$$

$$\Delta Q_2 = Q_2 + y_{21} V_1 V_2 \sin(-\theta_{21} - \delta_1 + \delta_2) + y_{22} V_2 V_2 \sin(-\theta_{22} - \delta_2 + \delta_2)$$

$$\Delta Q_2 = 0.5 + 9.95 \cdot 1.0 \cdot 1.033 \sin(-95.7^\circ - 0 - 21.1) + 9.13 \cdot (1.033)^2 \sin(74.1 - 21.1 + 21.1)$$
$$= 0.695 \text{ pu},$$

therefore $\Delta \overline{W}^{(1)} = (0.0348, \quad 0.695)^t$.

Application Example 12.11: Fundamental Power Flow and Reactive Power Control

The objective of this example is to get acquainted with fundamental power flow concepts, such as admittance matrix, initial guesses for voltages, evaluation of mismatch powers, establishment of the Jacobian matrix, and the effect of inductive and capacitive loads on the voltage control of buses.

Fundamental load flow with inductive load: To the three-bus power system of Fig. 12.31, apply the Newton–Raphson load flow analysis technique [11] provided switch S_1 is closed and switch S_2 is open.

(a) Find the fundamental bus admittance matrix Y_{bus} in polar form.
(b) Assume that bus #1 is the swing or slack bus, and buses #2 and 3 are PQ buses. Make an initial guess for the bus voltage vector $\overline{X}^{(0)}$ and evaluate the initial mismatch power $\Delta \overline{W}^{(0)}$.
(c) Find the Jacobian $\bar{J}^{(o)}$ for this power system configuration and compute the bus voltage correction vector $\Delta \bar{X}^{(0)}$.
(d) Update the bus voltage vector $\bar{X}^{(1)} = \bar{X}^{(0)} - \Delta \bar{X}^{(0)}$ and re-compute the updated mismatch power vector $\Delta \overline{W}^{(1)}$.

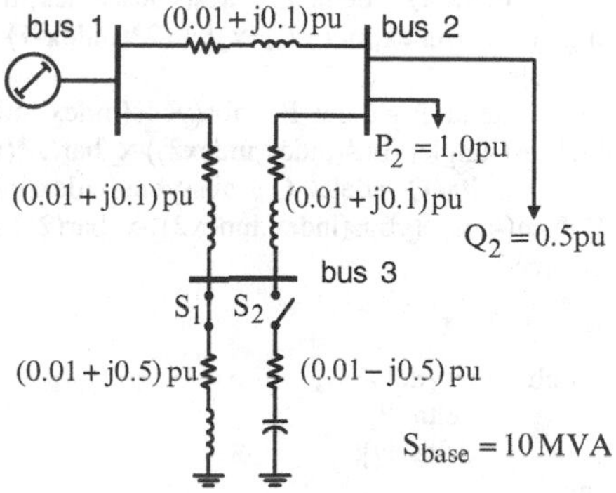

Fig. 12.31 Three-bus system where the real and reactive demand powers at bus 3 are P_3=0 and Q_3=0, respectively

(e) *Fundamental load flow with capacitive load*: Repeat all parts (a, b, c, and d) for the condition when switch S_1 is open and switch S_2 is closed.

For the solution of this problem you may use the following Matlab program:

Matlab program for fundamental power flow with 3 buses

```
clear all;
clc;

%define ybus and P&Q vectors... to save entry work only upper triangular needs to be entered
ybus = [19.9*exp(-1.47*i) -9.95*exp(-1.47*i) -9.95*exp(-1.47*i);
        0 19.9*exp(-1.47*i) -9.95*exp(-1.47*i);
        0 0 21.89*exp(-1.48*i)];

P_vec = [1 0];  %P[bus2 bus3 bus4...]
Q_vec = [.5 0];  %Q[bus2 bus3 bus4...]

%constructs the full ybus matrix
ybus = triu(ybus,1) + transpose(triu(ybus,1)) + ybus.*eye(size(ybus,1))

%construcs the initial guess vector and delta w vector
x_bar = [];
for index = 1:size(ybus,1)-1
  x_bar = [x_bar;
        0;
        1];
end
%prints the initial guess vector
x_bar

for iteration_num = 1:2
%constructs the delta-w column vector
delta_w = [];
for index = 2:size(ybus,1)
  delta_P = P_vec(index-1);
  delta_Q = Q_vec(index-1);
  for index2 = 1:size(ybus,1)
    if index2 == 1
        delta_P = delta_P + abs(ybus(index,index2))*x_bar(2*(index-1))*cos(-
angle(ybus(index,index2))+x_bar(2*(index-1)-1));
        delta_Q = delta_Q + abs(ybus(index,index2))*x_bar(2*(index-1))*sin(-
angle(ybus(index,index2))+x_bar(2*(index-1)-1));
    else
        delta_P = delta_P + abs(ybus(index,index2))*x_bar(2*(index-1))*x_bar(2*(index2-
1))*cos(-angle(ybus(index,index2))-x_bar(2*(index2-1)-1)+x_bar(2*(index-1)-1));
        delta_Q = delta_Q + abs(ybus(index,index2))*x_bar(2*(index-1))*x_bar(2*(index2-
1))*sin(-angle(ybus(index,index2))-x_bar(2*(index2-1)-1)+x_bar(2*(index-1)-1));
    end
  end

  delta_w = [delta_w;
          delta_P;
          delta_Q];
end
```

```
    J_mat = [];
    for index1 = 2:size(ybus,1)
        P_mattemp = [];
        Q_mattemp = [];
        for index2 = 2:size(ybus,1)
            Pdtemp = 0;
            Pvtemp = 0;
            Qdtemp = 0;
            Qvtemp = 0;
            if index1 == index2  %if diagonal of matrix
               for index3 = 1:size(ybus,1) %summation for diagonal case
                  if index3 ~= index1  %skips i=j
                     if index3 == 1  %handles v1=1 and d1=0
                        Pdtemp = Pdtemp - (abs(ybus(index1,index3))*x_bar(2*(index1-1))*sin(-
    angle(ybus(index1,index3))+x_bar(2*(index1-1)-1)));
                        Pvtemp = Pvtemp + (abs(ybus(index1,index3))*cos(-
    angle(ybus(index1,index3))+x_bar(2*(index1-1)-1)));
                        Qdtemp = Qdtemp + (abs(ybus(index1,index3))*x_bar(2*(index1-1))*cos(-
    angle(ybus(index1,index3))+x_bar(2*(index1-1)-1)));
                        Qvtemp = Qvtemp + (abs(ybus(index1,index3))*sin(-
    angle(ybus(index1,index3))+x_bar(2*(index1-1)-1)));
                     else
                  Pdtemp = Pdtemp - (abs(ybus(index1,index3))*x_bar(2*(index1-
1))*x_bar(2*(index3-1))*sin(-angle(ybus(index1,index3))-x_bar(2*(index3-1)-1)+x_bar(2*(index1-
1)-1)));
                  Pvtemp = Pvtemp + (abs(ybus(index1,index3))*x_bar(2*(index3-1))*cos(-
angle(ybus(index1,index3))-x_bar(2*(index3-1)-1)+x_bar(2*(index1-1)-1)));
                  Qdtemp = Qdtemp + (abs(ybus(index1,index3))*x_bar(2*(index1-
1))*x_bar(2*(index3-1))*cos(-angle(ybus(index1,index3))-x_bar(2*(index3-1)-
1)+x_bar(2*(index1-1)-1)));
                  Qvtemp = Qvtemp + (abs(ybus(index1,index3))*x_bar(2*(index3-1))*sin(-
angle(ybus(index1,index3))-x_bar(2*(index3-1)-1)+x_bar(2*(index1-1)-1)));
                end
            end

        end

        Pvtemp = Pvtemp + 2*x_bar(2*(index1-1))*abs(ybus(index1,index1))*cos(-
angle(ybus(index1,index1)));
        Qvtemp = Qvtemp + 2*x_bar(2*(index1-1))*abs(ybus(index1,index1))*sin(-
angle(ybus(index1,index1)));

      else  %k will never = 1 so no need to handle v1 and d1
         Pdtemp = abs(ybus(index1,index2))*x_bar(2*(index1-1))*x_bar(2*(index2-1))*sin(-
angle(ybus(index1,index2))-x_bar(2*(index2-1)-1)+x_bar(2*(index1-1)-1));
         Pvtemp = abs(ybus(index1,index2))*x_bar(2*(index1-1))*cos(-angle(ybus(index1,index2))-
x_bar(2*(index2-1)-1)+x_bar(2*(index1-1)-1));
         Qdtemp = -abs(ybus(index1,index2))*x_bar(2*(index1-1))*x_bar(2*(index2-1))*cos(-
angle(ybus(index1,index2))-x_bar(2*(index2-1)-1)+x_bar(2*(index1-1)-1));
         Qvtemp = abs(ybus(index1,index2))*x_bar(2*(index1-1))*sin(-angle(ybus(index1,index2))-
x_bar(2*(index2-1)-1)+x_bar(2*(index1-1)-1));
      end
      P_mattemp = [P_mattemp Pdtemp Pvtemp];
      Q_mattemp = [Q_mattemp Qdtemp Qvtemp];
   end
```

```
J_mat = [J_mat;
      P_mattemp;
      Q_mattemp];
end

%Prints the important values
delta_w
J_mat
delta_x_bar = J_mat^-1*delta_w
%solves for the new x_bar
x_bar = x_bar - delta_x_bar
end
```

Solution

(a) Find the fundamental bus admittance matrix $\bar{Y}_{bus}$ in polar form, S_1 closed and S_2 open:

$$y_{11} = y_{22} = \frac{1}{0.01 + j0.1} + \frac{1}{0.01 + j0.1} = 19.9007\angle -1.471 \text{ rad},$$

$$y_{33} = \frac{1}{0.01 + j0.1} + \frac{1}{0.01 + j0.1} + \frac{1}{0.01 + j0.5} = 21.8953\angle -1.4784 \text{ rad},$$

$$y_{12} = y_{21} = y_{13} = y_{31} = y_{23} = y_{32} = -\frac{1}{0.01 + j0.1} = 9.9503\angle 1.6705 \text{ rad},$$

$$\bar{Y}_{bus} = \begin{vmatrix} 19.9007\angle -1.471 & 9.9503\angle 1.6705 & 9.9503\angle 1.6705 \\ 9.9503\angle 1.6705 & 19.9007\angle -1.471 & 9.9503\angle 1.6705 \\ 9.9503\angle 1.6705 & 9.9503\angle 1.6705 & 21.8953\angle -1.4784 \end{vmatrix}.$$

(b) Initial guess (without swing-bus voltage) for the bus voltage vector $\bar{X}^{(0)}$ and evaluation of the initial mismatch power $\Delta\overline{W}^{(0)}$:

The bus voltages are defined as $\tilde{V}_1 = |\tilde{V}_1| e^{j\delta 1} = |\tilde{V}_1|\angle\delta_1 = 1\angle 0°$ which is the swing or slack bus, and $\tilde{V}_2 = |\tilde{V}_2| e^{j\delta 2} = |\tilde{V}_2|\angle\delta_2$ etc. In all subsequent calculations the per-unit (pu) magnitudes are represented by $V_i = |\tilde{V}_i|$.

$\bar{X}^{(o)} = (\delta_2, V_2, \delta_3, V_3)^t = (0, 1, 0, 1)^t$

$\Delta\overline{W}^{(o)} = (P_2 + F_{r,2},\ Q_2 + F_{i,2},\ P_3 + F_{r,3},\ Q_3 + F_{i,3})^t$, where

$$\Delta P_2 = P_2 + F_{r,2} = P_2 + \sum_{j=1}^{3} y_{2j} V_j V_2 \cos(-\theta_{2j} - \delta_j + \delta_2)$$

$$\begin{aligned} \Delta P_2 &= 1 + 9.9503 \cdot 1.0 \cdot 1.0 \cos(-1.6705 - 0 + 0) + 19.9007 \cdot 1.0 \\ &\quad \cdot 1.0 \cos(1.471 - 0 + 0) + 9.9503 \cdot 1.0 \cdot 1.0 \cos(-1.6705 - 0 + 0) \\ &= 1.0019, \end{aligned}$$

$$\Delta P_3 = P_3 + F_{r,3} = P_3 + \sum_{j=1}^{3} y_{3j} V_j V_3 \cos(-\theta_{3j} - \delta_j + \delta_3)$$

$$\Delta P_3 = 0 + 9.9503 \cdot 1.0 \cdot 1.0\cos(-1.6705 - 0 + 0) + 9.9503 \cdot 1.0 \cdot 1.0 \times \cos(-1.6705 - 0 + 0) + 21.8953 \cdot 1.0 \cdot 1.0\cos(1.4784 - 0 + 0) = 0.03929,$$

$$\Delta Q_2 = Q_2 + F_{i,2} = Q_2 + \sum_{j=1}^{3} y_{2j} V_j V_2 \sin(-\theta_{2j} - \delta_j + \delta_2)$$

$$\Delta Q_2 = 0.5 + 9.9503 \cdot 1.0 \cdot 1.0\sin(-1.6705 - 0 + 0) + 19.9007 \cdot 1.0 \cdot 1.0 \times \sin(1.471 - 0 + 0) + 9.9503 \cdot 1.0 \cdot 1.0\sin(-1.6705 - 0 + 0) = 0.5,$$

$$\Delta Q_3 = Q_3 + F_{i,3} = Q_3 + \sum_{j=1}^{3} y_{3j} V_j V_3 \sin(-\theta_{3j} - \delta_j + \delta_3)$$

$$\Delta Q_3 = 0 + 9.9503 \cdot 1.0 \cdot 1.0\sin(-1.6705 - 0 + 0) + 9.9503 \cdot 1.0 \cdot 1.0 \times \sin(-1.6705 - 0 + 0) + 21.8953 \cdot 1.0 \cdot 1.0\sin(1.4784 - 0 + 0) = 2.0.$$

$$\Delta \overline{W}^{(0)} = (1.0019\text{pu},\ 0.5\text{pu},\ 0.03929\text{pu},\ 2.0\text{pu})^t.$$

(c) Jacobian $\bar{J}^{(o)}$ for this power system configuration and computation of the bus voltage correction vector $\Delta\bar{X}^{(0)}$ are

$$\bar{J}^{(o)} = \begin{vmatrix} \frac{\partial \Delta P_2}{\partial \delta_2} & \frac{\partial \Delta P_2}{\partial V_2} & \frac{\partial \Delta P_2}{\partial \delta_3} & \frac{\partial \Delta P_2}{\partial V_3} \\ \frac{\partial \Delta Q_2}{\partial \delta_2} & \frac{\partial \Delta Q_2}{\partial V_2} & \frac{\partial \Delta Q_2}{\partial \delta_3} & \frac{\partial \Delta Q_2}{\partial V_3} \\ \frac{\partial \Delta P_3}{\partial \delta_2} & \frac{\partial \Delta P_3}{\partial V_2} & \frac{\partial \Delta P_3}{\partial \delta_3} & \frac{\partial \Delta P_3}{\partial V_3} \\ \frac{\partial \Delta Q_3}{\partial \delta_2} & \frac{\partial \Delta Q_3}{\partial V_2} & \frac{\partial \Delta Q_3}{\partial \delta_3} & \frac{\partial \Delta Q_3}{\partial V_3} \end{vmatrix}$$

The entries of the Jacobian are:

$J_{11} = \frac{\partial \Delta P_2}{\partial \delta_2}$ for $i=2$, $j=1$ and 3, $j \neq i$

$$J_{11} = -y_{21} V_2 V_1 \sin(\delta_2 - \delta_1 - \theta_{21}) - y_{23} V_2 V_3 \sin(\delta_2 - \delta_3 - \theta_{23})$$

$$J_{11} = -9.9503 \cdot 1.0 \cdot 1.0\sin(0 - 0 - 1.6705) - 9.9503 \cdot 1.0 \cdot 1.0\sin(0 - 0 - 1.6705) = 19.80177,$$

$J_{12} = \frac{\partial \Delta P_2}{\partial V_2}$ for $i=2$, $j=1$ and 3, $j=i$

$$J_{12} = y_{21}V_1\cos(\delta_2 - \delta_1 - \theta_{21}) + y_{23}V_3\cos(\delta_2 - \delta_3 - \theta_{23}) + 2V_2y_{22}\cos(-\theta_{22})$$

$$J_{12} = 9.9503 \cdot 1.0\cos(0 - 0 - 1.6705) + 9.9503 \cdot 1.0\cos(0 - 0 - 1.6705). + 2 \cdot 1.0 \cdot 19.9007\cos(1.471) = 1.9845,$$

$$J_{13} = \frac{\partial \Delta P_2}{\partial \delta_3} \text{ for } i = 2 \text{ and } k = 3$$

$$J_{13} = y_{23}V_2V_3\sin(\delta_2 - \delta_3 - \theta_{23}) = 9.9503 \cdot 1.0 \cdot 1.0\sin(0 - 0 - 1.6705) = -9.90088,$$

$$J_{14} = \frac{\partial \Delta P_2}{\partial V_3} \text{ for } i = 2 \text{ and } k = 3$$

$$J_{14} = y_{23}V_2\cos(\delta_2 - \delta_3 - \theta_{23}) = 9.9503 \cdot 1.0\cos(0 - 0 - 1.6705) = -0.99044,$$

$$J_{21} = \frac{\partial \Delta Q_2}{\partial \delta_2} \text{ for } i = 2,\ j = 1 \text{ and } 3,\ j \neq i$$

$$J_{21} = y_{21}V_2V_1\cos(\delta_2 - \delta_1 - \theta_{21}) + y_{23}V_2V_3\cos(\delta_2 - \delta_3 - \theta_{23})$$

$$J_{21} = 9.9503 \cdot 1.0 \cdot 1.0\cos(0 - 0 - 1.6705) + 9.9503 \cdot 1.0 \cdot 1.0\cos(0 - 0 - 1.6705) = -1.98088,$$

$$J_{22} = \frac{\partial \Delta Q_2}{\partial V_2} \text{ for } i = 2,\ j = 1 \text{ and } 3,\ j = i$$

$$J_{22} = y_{21}V_1\sin(\delta_2 - \delta_1 - \theta_{21}) + y_{23}V_3\sin(\delta_2 - \delta_3 - \theta_{23}) + 2V_2y_{22}\sin(-\theta_{22})$$

$$J_{22} = 9.9503 \cdot 1.0\sin(0 - 0 - 1.6705) + 9.9503 \cdot 1.0\sin(0 - 0 - 1.6705). + 2 \cdot 1.0 \cdot 19.9007\sin(1.471) = 19.7844,$$

$$J_{23} = \frac{\partial \Delta Q_2}{\partial \delta_3} \text{ for } i{=}2 \text{ and } k{=}3$$

$$J_{23} = -y_{23}V_2V_3\cos(\delta_2 - \delta_3 - \theta_{23}) = -9.9503 \cdot 1.0 \cdot 1.0\cos(0 - 0 - 1.6705) = 0.99044,$$

$$J_{24} = \frac{\partial \Delta Q_2}{\partial V_3} \text{ for } i = 2 \text{ and } k = 3$$

$$J_{24} = y_{23}V_2\sin(\delta_2 - \delta_3 - \theta_{23}) = 9.9503 \cdot 1.0\sin(0 - 0 - 1.6705) = -9.90088,$$

$$J_{31} = \frac{\partial \Delta P_3}{\partial \delta_2} \text{ for i} = 3 \text{ and k} = 2$$

$$J_{31} = y_{32}V_3V_2\sin(\delta_3 - \delta_2 - \theta_{32}) = 9.9503 \cdot 1.0 \cdot 1.0\sin(0 - 0 - 1.6705) \\ = -9.90088,$$

$$J_{32} = \frac{\partial \Delta P_3}{\partial V_2} \text{ for i} = 3 \text{ and k} = 2$$

$$J_{32} = y_{32}V_3\cos(\delta_3 - \delta_2 - \theta_{32}) = 9.9503 \cdot 1.0\cos(0 - 0 - 1.6705) = -0.99044,$$

$$J_{33} = \frac{\partial \Delta P_3}{\partial\, \delta_2} \text{ for i} = 3,\ \text{j} = 1 \text{ and } 2$$

$$J_{33} = -y_{31}V_3V_1\sin(\delta_3 - \delta_1 - \theta_{31}) - y_{23}V_3V_2\sin(\delta_3 - \delta_2 - \theta_{32})$$

$$J_{33} = -9.9503 \cdot 1.0 \cdot 1.0\sin(0 - 0 - 1.6705) - 9.9503 \cdot 1.0 \cdot 1.0\sin(0 - 0 \\ - 1.6705) \\ = 19.80177,$$

$$J_{34} = \frac{\partial \Delta P_3}{\partial V_3} \text{ for i} = 3,\ \text{j} = 1 \text{ and } 2,\ \text{j} = \text{i}$$

$$J_{34} = y_{31}V_1\cos(\delta_3 - \delta_1 - \theta_{31}) + y_{32}V_2\cos(\delta_3 - \delta_2 - \theta_{32}) + 2V_3y_{33}\cos(-\theta_{33})$$

$$J_{34} = 9.9503 \cdot 1.0\cos(0 - 0 - 1.6705) + 9.9503 \cdot 1.0\cos(0 - 0 - 1.6705) + 2 \\ \cdot 1.0 \cdot 21.8953\cos(1.471) \\ = 2.05946,$$

$$J_{41} = \frac{\partial \Delta Q_3}{\partial \delta_2} \text{ for i} = 3 \text{ and k} = 2$$

$$J_{41} = -y_{32}V_3V_2\cos(\delta_3 - \delta_2 - \theta_{32}) = -9.9503 \cdot 1.0 \cdot 1.0\cos(0 - 0 - 1.6705) \\ = 0.99044,$$

$$J_{42} = \frac{\partial \Delta Q_3}{\partial V_2} \text{ for i} = 3 \text{ and k} = 2$$

$$J_{42} = y_{32}V_3\sin(\delta_3 - \delta_2 - \theta_{32}) = 9.9503 \cdot 1.0\sin(0 - 0 - 1.6705) = -9.90088,$$

$$J_{43} = \frac{\partial \Delta Q_3}{\partial\, \delta_3} \text{ for i} = 3,\ \text{j} = 1 \text{ and } 2$$

$$J_{43} = y_{31}V_3V_1\cos(\delta_3 - \delta_1 - \theta_{31}) + y_{32}V_3V_2\cos(\delta_3 - \delta_2 - \theta_{32})$$

$$\begin{aligned}J_{43} &= 9.9503 \cdot 1.0 \cdot 1.0 \cos(0-0-1.6705) + 9.9503 \cdot 1.0 \cdot 1.0 \cos(0-0\\ &\quad -1.6705)\\ &= -1.98088,\end{aligned}$$

$$J_{44} = \frac{\partial \Delta Q_3}{\partial V_3} \text{ for i = 3, j = 1 and 2, j = i}$$

$$\begin{aligned}J_{44} &= y_{31} V_1 \sin(\delta_3 - \delta_1 - \theta_{31}) + y_{23} V_2 \sin(\delta_3 - \delta_2 - \theta_{32})\\ &\quad + 2V_3 y_{33} \sin(-\theta_{33})\end{aligned}$$

$$\begin{aligned}J_{44} &= 9.9503 \cdot 1.0 \sin(0-0-1.6705) + 9.9503 \cdot 1.0 \sin(0-0-1.6705)\\ &\quad + 2 \cdot 1.0 \cdot 21.8953 \sin(1.4784)\\ &= 23.80205.\end{aligned}$$

Therefore, the Jacobian and its inverse are:

$$\bar{J}^{(o)} = \begin{vmatrix} 19.80177 & 1.9845 & -9.90088 & -0.99044\\ -1.98088 & 19.7844 & 0.99044 & -9.90088\\ -9.90088 & -0.99044 & 19.80177 & 2.05946\\ 0.99044 & -9.90088 & -1.98088 & 23.80205 \end{vmatrix}$$

$$[\bar{J}^{(o)}]^{-1} = \begin{vmatrix} 0.0667 & -0.0064 & 0.0334 & -0.0028\\ 0.0063 & 0.0632 & 0.0026 & 0.0263\\ 0.0334 & -0.0028 & 0.0668 & -0.0055\\ 0.0026 & 0.0263 & 0.0053 & 0.0526 \end{vmatrix}.$$

$$\Delta\bar{X}^{(0)} = [\bar{J}^{(o)}]^{-1} \cdot \Delta\bar{W}^{(o)},$$

or its transpose $\Delta\bar{X}^{(0)} = (0.059,\ 0.0906,\ 0.0237,\ 0.121)^t$.

(d) Updating the bus voltage vector $\bar{X}^{(1)} = \bar{X}^{(0)} - \Delta\bar{X}^{(0)}$ and re-computing the updated mismatch power vector $\Delta\overline{W}^{(1)}$.
$\overline{X}^{(1)} = \overline{X}^{(0)} - \Delta\overline{X}^{(0)} = (-0.059 \text{ rad},\ 0.9094 \text{ pu}, -0.0237 \text{ rad},\ 0.879 \text{ pu})^t$.
Note this bus-voltage vector represents the voltage magnitudes and phase angles

$$V_1 = |\tilde{V}_1| = 1.0 \text{ pu},\ \delta_1 = 0 \text{ rad},\quad V_2 = |\tilde{V}_2| = 0.9094 \text{ pu},\ \delta_2 = -0059 \text{ rad},$$

$$V_3 = |\tilde{V}_3| = 0.879 \text{ pu},\ \delta_3 = -0.0237 \text{ rad}.$$

$$\Delta\overline{W}^{(1)} = (P_2 + F_{r,2},\ Q_2 + F_{i,2},\ P_3 + F_{r,3},\ Q_3 + F_{i,3})^t,$$

$$\Delta P_2 = P_2 + F_{r,2} = P_2 + \sum_{j=1}^{3} y_{2j} V_j V_2 \cos(-\theta_{2j} - \delta_j + \delta_2)$$

$$\begin{aligned}\Delta P_2 = {} & 1 + 9.9503 \cdot 1.0 \cdot 0.9094 \cos(-1.6705 - 0 - 0.059) + 19.9007 \\ & \cdot (0.9094)^2 \cos(1.471 + 0.059 - 0.059) + 9.9503 \cdot 0.879 \\ & \cdot 0.9094 \cos(-1.6705 + 0.0237 - 0.059) \\ = {} & 0.1391,\end{aligned}$$

$$\Delta P_3 = P_3 + F_{r,3} = P_3 + \sum_{j=1}^{3} y_{3j} V_j V_3 \cos(-\theta_{3j} - \delta_j + \delta_3)$$

$$\begin{aligned}\Delta P_3 = {} & 0 + 9.9503 \cdot 1.0 \cdot 0.879 \cos(-1.6705 - 0 - 0.0237) + 9.9503 \cdot 0.9094 \\ & \cdot 0.879 \cos(-1.6705 + 0.059 - 0.0237) + 21.8953 \cdot 0.879 \\ & \cdot 0.879 \cos(1.4784 + 0.0237 - 0.0237) \\ = {} & -0.0982,\end{aligned}$$

$$\Delta Q_2 = Q_2 + F_{i,2} = Q_2 + \sum_{j=1}^{3} y_{2j} V_j V_2 \sin(-\theta_{2j} - \delta_j + \delta_2)$$

$$\begin{aligned}\Delta Q_2 = {} & 0.5 + 9.9503 \cdot 0.9094 \cdot 1.0 \sin(-1.6705 - 0 - 0.059) + 19.9007 \\ & \cdot (0.9094)^2 \sin(1.471 + 0.059 - 0.059) + 9.9503 \cdot 0.879 \\ & \cdot 0.9094 \sin(-1.6705 + 0.0237 - 0.059) \\ = {} & 0.05952,\end{aligned}$$

$$\Delta Q_3 = Q_3 + F_{i,3} = Q_3 + \sum_{j=1}^{3} y_{3j} V_j V_3 \sin(-\theta_{3j} - \delta_j + \delta_3)$$

$$\begin{aligned}\Delta Q_3 = {} & 0 + 9.9503 \cdot 1.0 \cdot 0.879 \sin(-1.6705 - 0 - 0.0237) + 9.9503 \cdot 0.9094 \\ & \cdot 0.879 \sin(-1.6705 + 0.059 - 0.0237) + 21.8953 \cdot 0.879 \\ & \cdot 0.879 \sin(1.4784 + 0.0237 - 0.0237) \\ = {} & 0.2277.\end{aligned}$$

$$\Delta\overline{W}^{(1)} = (0.1391\ \text{pu},\ 0.05952\ \text{pu}, -0.0982\ \text{pu},\ 0.2277\ \text{pu})^t.$$

(e) *Fundamental load flow with capacitive load*: Repeat all parts (a, b, c, and d) for the condition when switch S_1 is open and switch S_2 is closed. This part is to be completed by the reader. The results are as follows:

$$\overline{X}^{(1)} = \overline{X}^{(0)} - \Delta\overline{X}^{(0)} = (-0.07371,\ 1.04699, -0.05251,\ 1.15398)^t,$$

$$\Delta\overline{W}^{(1)} = (-0.07694\ \text{pu},\ 0.01057\ \text{pu},\ 0.00644\ \text{pu},\ 0.27511\ \text{pu})^t.$$

Conclusion: By connecting either inductive or capacitive loads to the individual buses the voltage levels can be controlled.

Application Example 12.12: CO_2 Generation of a Coal-Fired Power Plant

A coal-fired power plant has an efficiency of η_{plant}=30%. It is known that 1 (kg-force) = 1 kp = 1 kg m/s^2 coal contains 8.2 kWh of energy.

(a) Calculate the amount (weight) of CO_2 (measured in kg-force) released for each 1 kWh energy generated by the coal-fired plant.
(b) On the average one person uses 300 kWh per month (for lighting, cooking, heating and air conditioning). How many tons-force (= 1,000 kg-force) CO_2 are produced by 300 million persons per year?

Solution

(a) 1 kWh of electricity requires the weight of $(1/8.2)$ kg-force = 0.122 kg-force of coal for 100% efficiency of the power plant. For a 30% efficiency the required coal per 1 kWh is $(0.122/0.30)$ kg-force = 0.4065 kg-force of coal. Coal or carbon (C) has an atomic weight of 12 and oxygen (O) has the atomic weight of 16. From this follows that for the oxidation/burning of coal, that is, C+2O→ CO_2 12+2·16 = 44 units in terms of atomic weights of CO_2 is produced. If the weight of 0.407 kg-force of coal is burned then a weight of $0.407(44/12)$ = 1.49 kg-force of CO_2 is generated. In summary, the generation of 1 kWh of electricity produces the weight of 1.49 kg-force of CO_2.
(b) The total energy consumed by 300 million persons is $E_{total} = 300$ kWh·12·300·10^6 = 1.08·10^{12} kWh per year, or $CO_{2\ total}$ = 1.49 (kg-force)/kWh·1,080·10^9 kWh = 1.61·10^{12} kg-force = 1.61·10^9 tons-force per year.

Application Example 12.13: Calculation of Current in Phase of Y-Load and in Phase of Δ-Load

Find the load currents $\tilde{I}_{aA}$, $\tilde{I}_{bB}$, $\tilde{I}_{cC}$, $\tilde{I}_{A\_Y}$, and $\tilde{I}_{AB\_\Delta}$ in the balanced three-phase circuit of Fig. 12.32.

Solution

Define the phase voltages as $\tilde{V}_{an} = 208\angle 0°\text{V}$, $\tilde{V}_{bn} = 208\angle -120°\text{V}$, $\tilde{V}_{cn} = 208\angle -240°\text{V}$ this results in the line-to-line voltage $\tilde{V}_{ab} = 208 \cdot \sqrt{3}\angle 30°\text{V} = 360\angle 30°\text{V}$.

Replacing the Δ-load by an equivalent Y-load yields -based on the Δ-Y transformation- the load impedance

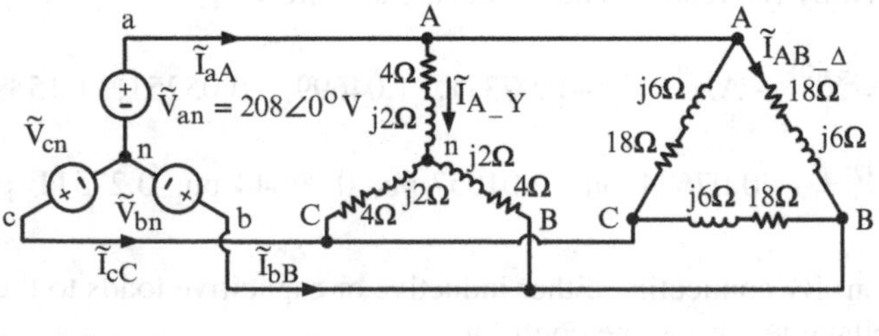

Fig. 12.32 Y- and Δ loads

$$Z_{eq\_Y} = \frac{1}{3} Z_\Delta = \frac{(18 + j6)}{3} \Omega = (6 + j2)\Omega.$$

With $Z_Y = (4 + j2)\Omega$ the parallel impedance of Z_{eq_Y} and Z_Y is

$$Z_p = \frac{(4 + j2)(6 + j2)}{(4 + j2 + 6 + j2)} = 2.63\angle 23.2°\Omega.$$

The phase current $\tilde{I}_{aA}$ supplied by $\tilde{V}_{an} = 208\angle 0°V$ is

$$\tilde{I}_{aA} = \frac{208\angle 0°V}{2.63\angle 23.2°\Omega} = 79.09\angle - 23.2°A.$$

Correspondingly, $\tilde{I}_{bB} = 79.09\angle - 143.2°A$, and $\tilde{I}_{cC} = 79.09\angle - 263.2°A$.
The phase current in the Y-load is

$$\tilde{I}_{A\_Y} = \frac{\tilde{V}_{an}}{Z_Y} = \frac{208\angle 0°V}{(4 + j2)\Omega} = 46.51\angle - 26.57°A$$

and the phase current in Δ-load is

$$\tilde{I}_{AB\_\Delta} = \frac{\tilde{V}_{ab}}{Z_\Delta} = \frac{208 \cdot \sqrt{3}\angle 30°V}{(18 + j6)\Omega} = 18.99\angle 11.57°A.$$

Application Example 12.14: AC Power Flow and Complex Power Calculation Through Single-Phase Feeder with Load

Figure 12.33 illustrates a source voltage $\tilde{V}_{source}$, a transmission line with the impedance $\tilde{Z}_{line} = (0.1 + j0.2)\Omega$, a load with the real power $P_{load} = 100$ kW, and a load voltage $\tilde{V}_{load} = 277\angle 0°V$ with a load power factor of $PF_{load} = \cos\theta_{load} = 0.8$ lagging (consumer notation).

(a) Determine $\tilde{I}_{load}$.
(b) Determine the complex power $\tilde{S}_{load}$ at the load.
(c) Determine the transmission-line losses P_{line_loss}.
(d) Find the source voltage $\tilde{V}_{source}$ that the utility must supply.
(e) Calculate the power factor at the source $\cos\theta_{source}$, and determine the complex power at the source.

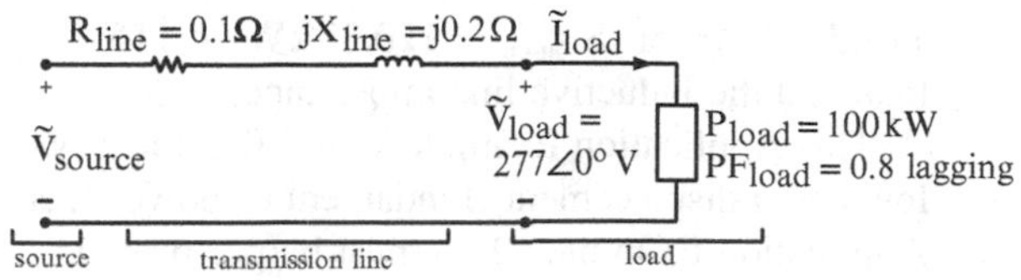

Fig. 12.33 AC power flow through single-phase feeder

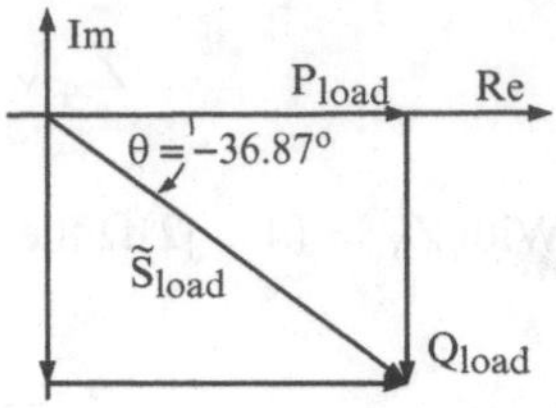

Fig. 12.34 Real power P_{load}, reactive power Q_{load}, and complex power $\tilde{S}_{load}$

Solution

(a) The load current at a lagging power factor of 0.8 is

$$\tilde{I}_{load} = \frac{P_{load}}{\tilde{V}_{load}\cos\theta_{load}} \angle -36.87° = 451.26\angle -36.87° \text{A}.$$

(b) The complex power at the load is $\tilde{S}_{load} = P_{load} + jQ_{load}$, where $Q_{load} = P_{load} \cdot \tan(-36.87°) = -74.99$ kVAr or $\tilde{S}_{load} = 100\text{kW} - j74.99\text{kVAr} = 124.998\angle -36.87°\text{kVA}$. These powers are illustrated in Fig. 12.34.

(c) Transmission line loss is $P_{line\_loss} = |\tilde{I}_{load}|^2 R_{line} = 20.36\,\text{kW}$

(d) The utility must supply the voltage $\tilde{V}_{source} = \tilde{I}_{load}(R_{line} + jX_{line}) + \tilde{V}_{load} = (367.24 + j45.13)\text{V} = 370\angle 7.0°\text{V}$

(e) The total supplied real power by the utility is $P_{source} = P_{line\_loss} + P_{load} = 20.36$ kW + 100 kW = 120.36 kW, and the total supplied reactive power is $Q_{source} = Q_{line\_loss} + Q_{load}$, where the reactive line loss is $Q_{line\_loss} = |\tilde{I}_{load}|^2 X_{line} = -40.73$ kVAr. Note this reactive loss is negative due to the (line) inductor consuming reactive power. Thus $Q_{total} = -40.73$ kVAr − 74.99 kVAr = −115.72 kVAr and the complex power at the source is $\tilde{S}_{source} = P_{source} + jQ_{source} = 120.36\text{kW} - j115.72\text{kVAr} = 166.97\text{kVA}\angle -43.88°$.

The power factor at the source is now

$$\cos\theta_{source} = \frac{P_{source}}{S_{source}},$$

where S_{source} is the apparent power or the magnitude of the complex power. Therefore, the source power factor is $\cos\theta_{source} = 0.72$ lagging.

An alternative formulation of the real and reactive powers at the source is given by

$P_{source} = \text{Re}\{\tilde{V}_{source}\tilde{I}^*_{load}) = 120.36\,\text{kW}$, and $Q_{source} = \text{Im}\{\tilde{V}_{source}\tilde{I}^*_{load}) = 115.72\,\text{kVAr}$ or $\tilde{S}_{source} = 120.37\,\text{kW} - j115.71\,\text{kVAr}$ due to the inductive load and the inductive line impedance.

This Application Example shows that the power factor at the source is very low and a displacement (fundamental) power factor correction as discussed in Application Example 12.16 must be performed.

Application Example 12.15: Load Management and Load Shedding

A block diagram of two interconnected areas of a power system (e.g., area #1 is fed by a natural gas-fired plant and area #2 is supplied by a wind-power plant) is shown in Fig. 12.35.

Part a: For steady-state operation show that with $\Delta\omega_1 = \Delta\omega_2 = \Delta\omega$ the (change in)

(a1) ΔP_{mech1} is a function of (Δload ref$_1$), $\Delta\omega$, and R_1,
(a2) ΔP_{mech2} is a function of (Δload ref$_2$), $\Delta\omega$, and R_2,
(a3) ΔP_{tie} is a function of (Δload ref$_2$), $\Delta\omega$, and R_2, D_2, and ΔP_{L2},
(a4) the angular velocity $\Delta\omega$ is a function of (Δload ref$_1$), (Δload ref$_2$), ΔP_{L1}, ΔP_{L2}, D1, D_2, R_1, R_2.

Part b: For the load commands or reference values (Δload ref$_1$) = 0.2 pu, (Δload ref$_2$) = 0.8 pu, (given by the control and dispatch center), the load demands are $\Delta P_{L1} = 0.5$ pu, $\Delta P_{L2} = 0.5$ pu, and the fixed parameters are $R_1 = 1$ pu (e.g., natural gas-fired plant), $R_2 = 5$ pu (e.g., wind-power plant), $D_1 = 0.5$ pu, $D_2 = 0.7$ pu,

(b1) Calculate the steady-state angular velocity $\Delta\omega_{ss\_1}$. Note, $\Delta\omega_{ss\_1}$ is defined in Fig. 12.36,
(b2) Calculate Δf_{ss-1} and f_{ss-1}, where the nominal (rated) frequency is $f_{rated} = 60$ Hz,
(b3) Calculate ΔP_{mech2} and ΔP_{tie} provided the base apparent power is S_{base}=10 MVA.

Part c: For the load reference values (Δload ref$_1$) = 0.2 pu, (Δload ref$_2$) = 0.0 pu, (given by the control and dispatch center when there is less wind, the wind-power plant has no output power), the load demands are $\Delta P_{L1} = 0.5$ pu, $\Delta P_{L2} = 0.5$ pu, and the fixed parameters $R_1 = 1$ pu (e.g., natural gas-fired plant), $R_2 = 5$ pu (e.g., wind-power plant), $D_1 = 0.5$ pu, $D_2 = 0.7$ pu,

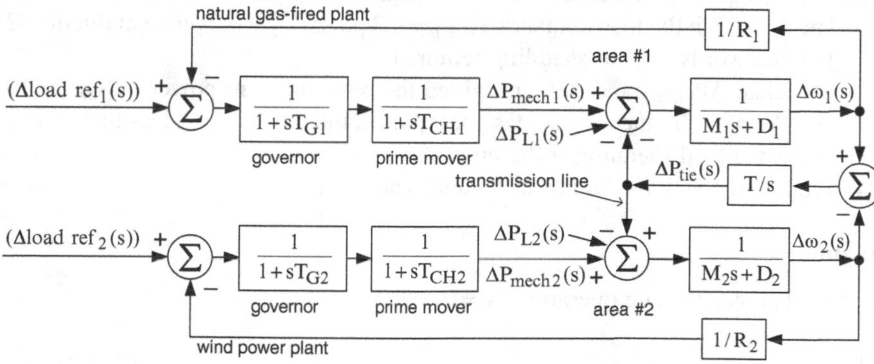

Fig. 12.35 Interconnected micro-power system (S_{base}=10 MVA) consisting of a natural gas-fired plant (area #1) and a wind power plant (area #2), where D_1 and D_2 correspond to the frequency dependent loads and $\Delta P_{L1}(s)$ and $\Delta P_{L2}(s)$ are the frequency independent loads

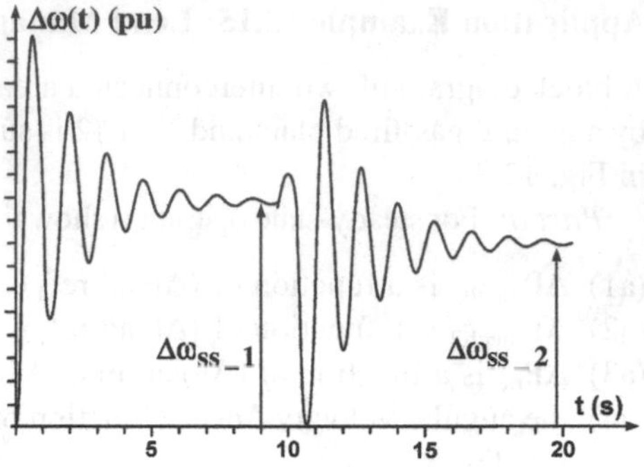

Fig. 12.36 Qualitative change of the steady-state angular frequency from $\Delta\omega_{ss\_1}$ to $\Delta\omega_{ss\_2}$ due to a change in the load reference 2, (Δload ref$_2$) at about t=10 s from 0.8 pu to 0.0 pu

(c1) Calculate the steady-state angular velocity $\Delta\omega_{ss\_2}$. Note, $\Delta\omega_{ss\_2}$ is defined in Fig. 12.36
(c2) Calculate Δf_{ss-2} and f_{ss-2}, where the nominal (rated) frequency is $f_{rated} = 60$ Hz
(c3) Calculate ΔP_{mech2} and ΔP_{tie} provided the base apparent power is $S_{base} = 10$ MVA
(c4) Is the frequency decrease acceptable, (e.g., is Δf_{ss-2} less than $\pm 1\%$), and can the load demands $\Delta P_{L1} = 0.5$ pu, $\Delta P_{L2} = 0.5$ pu be maintained?

Part d: For the load reference values (Δload ref$_1$) = 0.2 pu, (Δload ref$_2$) = 0.0 pu, (given by the control and dispatch center when there is less wind, the wind-power plant has no output power), the load demands are $\Delta P_{L1} = 0.2$ pu, $\Delta P_{L2} = 0.2$ pu, and the fixed parameters $R_1 = 1$ pu (e.g., natural gas-fired plant), $R_2 = 5$ pu (e.g., wind-power plant), $D_1 = 0.5$ pu, $D_2 = 0.7$pu,

(d1) Calculate the steady-state angular velocity $\Delta\omega_{ss\_3}$.
(d2) Is the frequency decrease acceptable, (e.g., is Δf_{ss-3} less than $\pm 1\%$)? Calculate f_{ss-3}. Can the load demands $\Delta P_{L1} = 0.2$ pu, $\Delta P_{L2} = 0.2$ pu be maintained? In other words, is load shedding required?
(d3) Calculate ΔP_{mech2} and ΔP_{tie} provided the base apparent power is $S_{base} = 10$ MVA. Is the frequency decrease acceptable, that is, is Δf_{ss-3} less than $\pm 1\%$, and is the load shedding sufficient?
(d4) Propose a way how the power balance can be achieved.

Solution

Part a): For steady-state operation (s = 0) one obtains

(a1)

$$\Delta P_{mech1} = \Delta loadref_1 - (\Delta\omega / R_1), \tag{12.3}$$

(a2)

$$\Delta P_{mech2} = \Delta loadref_2 - (\Delta\omega/R_2), \quad (12.4)$$

(a3) $\Delta\omega D_2 = \Delta P_{mech2} - \Delta P_{L2} + \Delta P_{tie}$,

$$\Delta P_{tie} = \Delta\omega D_2 - \Delta P_{mech2} + \Delta P_{L2}. \quad (12.5)$$

Equation (12.4) into (12.5) yields

$$\Delta P_{tie} = -\Delta loadref_2 + \Delta P_{L2} + \Delta\omega(1/R_2 + D_2). \quad (12.6)$$

(a4)

$$\Delta\omega D_1 = \Delta P_{mech1} - \Delta P_{L1} - \Delta P_{tie}. \quad (12.7)$$

Equation (12.3) into (12.7) results in

$$\Delta P_{tie} = \Delta loadref_1 - \Delta P_{L1} - \Delta\omega(1/R_1 + D_1). \quad (12.8)$$

Equation (12.6) equals (12.8) gives

$$\Delta\omega = \frac{(\Delta loadref_1 + \Delta loadref_2 - \Delta P_{L1} - \Delta P_{L2})}{(D_1 + D_2 + 1/R_1 + 1/R_2)}. \quad (12.9)$$

Part b): For steady-state operation and the given dispatch center load commands one obtains

(b1) $\Delta\omega_{ss-1} = \frac{(0.2+0.8-0.5-0.5)}{(0.5+0.7+1+0.2)} = 0\,\text{pu}$,
(b2) $\Delta f_{ss-1}[\text{pu}] = \Delta\omega_{ss-1}[\text{pu}]$ or $f_{ss-1} = f_{rated}(1 + \Delta f_{ss-1}) = 60\,\text{Hz}$.
(b3) $\Delta P_{mech2} = 0.8 - 0 = 0.8\,\text{pu}$ or $\Delta P_{mech2} = 8\,\text{MW}$, $\Delta P_{tie} = -0.8 + 0.5 + 0 = -0.3\,\text{pu}$ or $\Delta P_{tie} = -3\,\text{MW}$. The power balance for area #2 is illustrated in Fig. 12.37.

Part c): For steady-state operation and the given dispatch center load commands one obtains

(c1) $\Delta\omega_{ss-2} = \frac{(0.2+0-0.5-0.5)}{(0.5+0.7+1+0.2)} = -0.333\,\text{pu}$,
(c2) $\Delta f_{ss-2}[\text{pu}] = \Delta\omega_{ss-2}[\text{pu}]$ or $f_{ss-2} = f_{rated}(1 - 0.333) = 40\,\text{Hz}$. This low frequency is not acceptable.
(c3) $\Delta P_{mech2} = 0 + 0.333/5 = 0.0666\,\text{pu}$ or $\Delta P_{mech2} = 0.666\,\text{MW}$, $\Delta P_{tie} = 0 + 0.5 - 0.333\,(1/5 + 0.7)\,0 = 0.2\,\text{pu}$ or $\Delta P_{tie} = 2\,\text{MW}$. The power

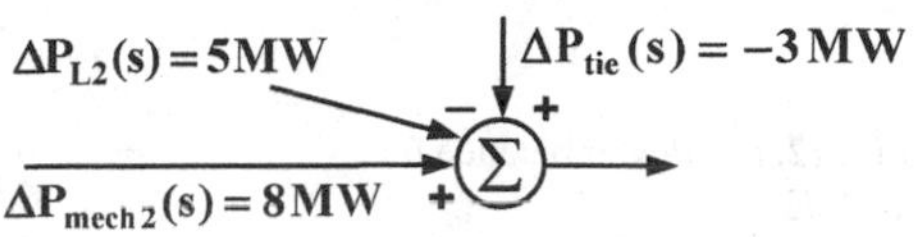

Fig. 12.37 Power balance for area #2

distribution for area #2 is illustrated in Fig. 12.38 and it is not acceptable, there is no power balance.

Part d): For steady-state operation and the given dispatch center load commands one obtains

(d1) $\Delta\omega_{ss-3} = \frac{(0.2+0-0.2-0.2)}{(0.5+0.7+1+0.2)} = -0.08333\,\text{pu}$,

(d2) $\Delta f_{ss-3}[\text{pu}] = \Delta\omega_{ss-3}[\text{pu}]$ or $f_{ss-3} = f_{rated}(1 - 0.08333) = 55$ Hz. This low frequency is not acceptable.

(d3) $\Delta P_{mech2} = 0 + 0.08333/5 = 0.01666\,\text{pu}$ or $\Delta P_{mech2} = 0.1666\,\text{MW}$, $\Delta P_{tie} = 0 + 0.2 - 0.08333(1/5 + 0.7) = 0.125\,\text{pu}$ or $\Delta P_{tie} = 1.25\,\text{MW}$. The power distribution for area #2 is illustrated in Fig. 12.39 and it is not acceptable, there is no power balance.

(d4) Power balance and rated frequency can be achieved by setting $D_2 = 0$, and $\Delta P_{L1} = \Delta P_{L2} = 0.1\,\text{pu}$. Based on these constraints one obtains $\Delta\omega_{ss-4} = \frac{(0.2+0-0.1-0.1)}{(0.5+0+1+0.2)} = 0.0\,\text{pu}$ or $\Delta f_{ss-4}[\text{pu}] = \Delta\omega_{ss-4}[\text{pu}]$ or $f_{ss-4} = f_{rated}(1 - 00) = 60$ Hz. The rated frequency is restored. $\Delta P_{mech2} = 0$, $\Delta P_{tie} = 0 + 0.1 + 0 = 0.1\,\text{pu}$, or $\Delta P_{tie} = 1\,\text{MW}$. The power balance – illustrated in Fig. 12.40 – is restored.

Conclusions: There is a need for additional load shedding ($D_2 = 0$, and $\Delta P_{L2} = \Delta P_{L1} = 0.1\,\text{pu}$), where D_2 corresponds to a frequency dependent load and ΔP_{L1} and ΔP_{L2} are frequency independent loads. Both adjustments can be done by the Control and Dispatch Center.

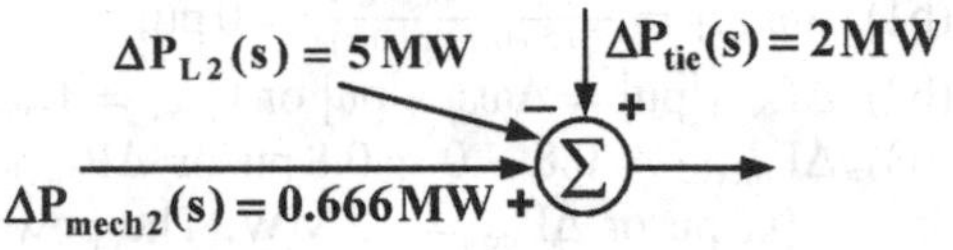

Fig. 12.38 Power imbalance for area #2

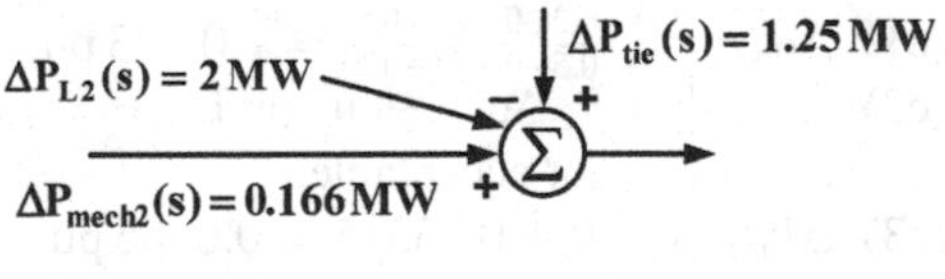

Fig. 12.39 Power imbalance for area #2

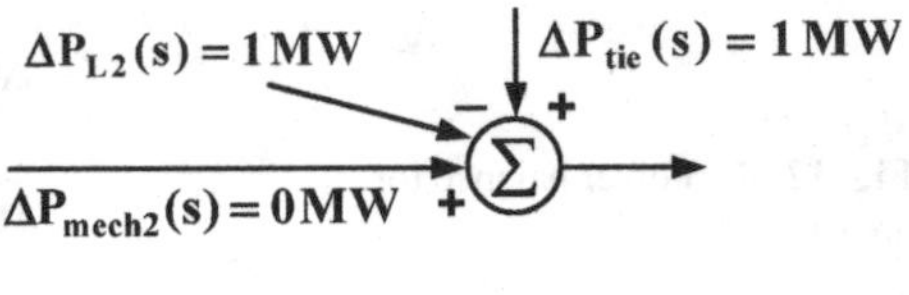

Fig. 12.40 Power balance for area #2

12.4 Distribution System Analysis with/without (Fundamental or Displacement) Power-Factor Compensation

At the distribution system level the power factor correction is very important for voltage control and the increase of the efficiency of the delivery of power. The minimization of the effects of faults based on short-circuit calculations are a prerequisite for the reliability of a power system. Therefore, this subsection covers the short-circuit calculations based on the symmetrical components of voltages and currents.

Application Example 12.16: Displacement (Fundamental or Displacement) Power-Factor Correction

A single-phase load (Fig. 12.41) is connected to a f = 60 Hz voltage $\tilde{V}_{load} = 240\,V\angle 0°$ V. The load consumes a (real) power of $P_{load} = 22$ kW at a power factor of $PF_{load} = \cos\theta_{load} = 0.6$ lagging (consumer notation). Determine the capacitor value C needed to raise the load power factor to $PF_{load\_corrected} = \cos\theta_{load\_corrected} = 0.9$ lagging (consumer notation).

Solution

The load current at a lagging power factor of 0.6 is $\tilde{I}_{load} = \frac{P_{load}}{\tilde{V}_{load}\cos\theta_{load}} \angle -53.13° = 152.78\angle -53.13°$A. This current is depicted in Fig. 12.42 in the complex plane.

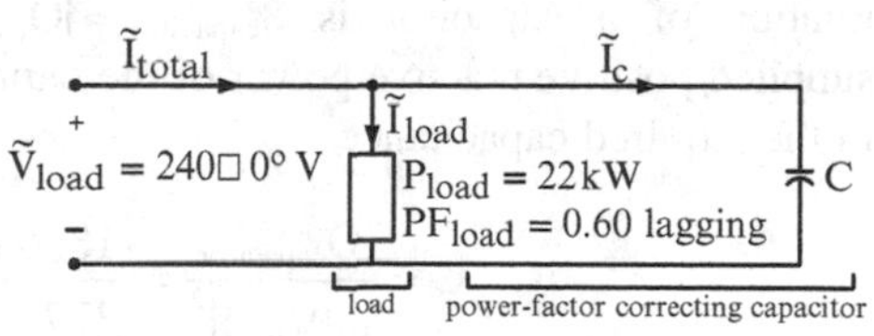

Fig. 12.41 Displacement (fundamental) power-factor correction (consumer notation)

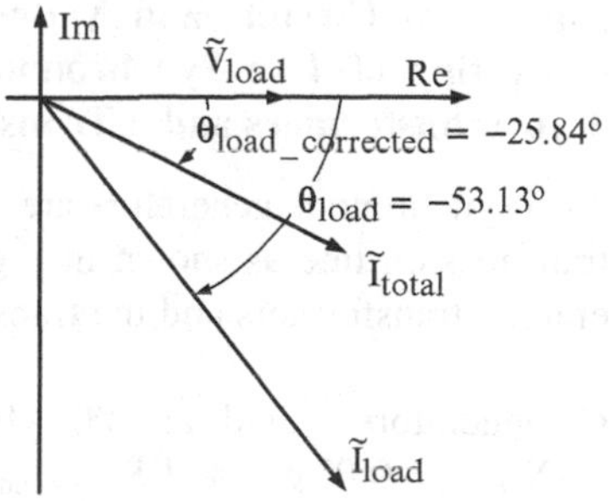

Fig. 12.42 Voltage $\tilde{V}_{load}$ and current $\tilde{I}_{load}$ in complex plane

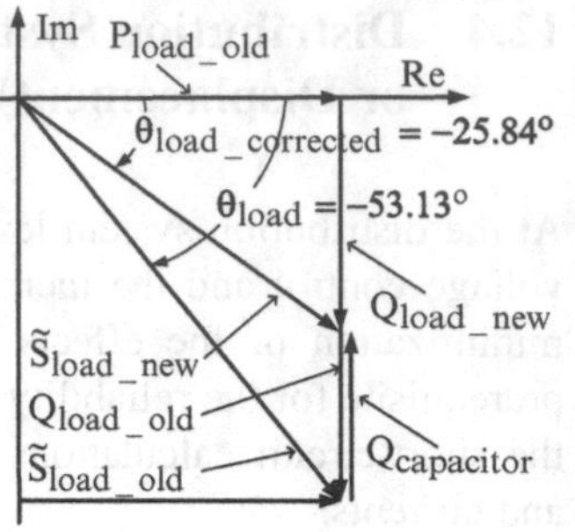

Fig. 12.43 Real power P_{load}, reactive power Q_{load}, and complex power $\tilde{S}_{load}$ in complex plane

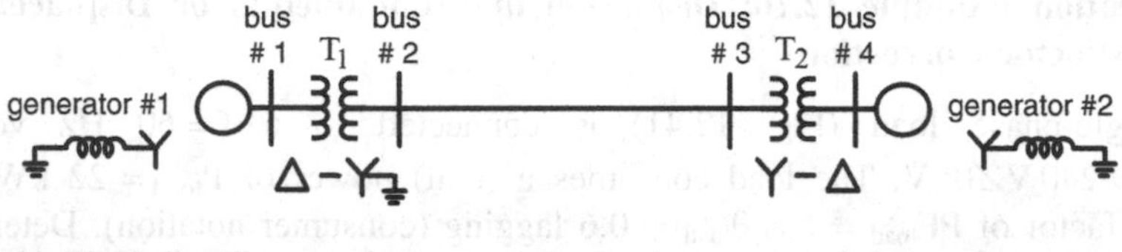

Fig. 12.44 Single-line diagram of a two-generator system connected via transformers and transmission line

The reactive power at a power factor of 0.6 lagging (called old) is $Q_{load\_old} = P_{load\_old} \cdot \tan(-53.13°) = -29.33$ kVAr. Thus the old complex power is $\tilde{S}_{old} = P_{load\_old} + Q_{load\_old} = 22\,\text{kW} - 29.33\,\text{kVAr}$.

The corresponding reactive power is shown in the complex plane in Fig. 12.43.

At the increased power factor of 0.9 lagging (called new) the reactive power is $Q_{load\_new} = P_{load\_old} \cdot \tan(-25.84°) = -10.65$ kVAr. This results in the complex power $\tilde{S}_{capacitor} = j(Q_{load\_new} - Q_{load\_old}) = j(-10.65\,\text{kVAr} + 29.33\,\text{kVAr}) = j\,18.68\,\text{kVAr}$ which must be compensated by the capacitor. The reactive or complex power relation of a capacitor is $\tilde{S}_{capacitor} = jQ_{capacitor} = j\,\omega C\left|\tilde{V}_{load}\right|^2$ or the delivered/supplied/positive reactive power of the capacitor is $Q_{capacitor} = \omega C\left|\tilde{V}_{load}\right|^2$ resulting in the required capacitance

$$C = \frac{Q_{capacitor}}{\omega\left|\tilde{V}_{load}\right|^2} = \frac{18,680\,\text{VAr}}{377 \cdot (240)^2} = 0.86\,\text{mF}.$$

Application Example 12.17: Determination of the Sequence-Component Equivalent Circuits and Matrices $Z_{bus}^{(1)}$, $Z_{bus}^{(2)}$, and $Z_{bus}^{(0)}$ of a Power System Consisting of Two Synchronous Generators Which Are Interconnected by Two Transformers and a Transmission Line

Two synchronous generators are connected by three-phase transformers through a transmission line as shown in Fig. 12.44. The ratings and reactances of the generators, transformers and the transmission line are:

- Generators 1 and 2: 600 MVA, 34.5 kV, $X_d'' = X_{1\_gen} = X_{2\_gen} = 0.16$ pu, $X_{0gen} = 0.03$ pu, and $X_{n\_grounding\_coil} = 0.05$ pu

- Transformer T_1 600 MVA, (34.5 kV connected in Δ)/(345 kV connected in Y), and $X_{transformer\ leakage} = X_{T1} = 0.04$ pu. Note that the Y neutral of transformer T_1 is grounded
- Transformer T_2 600 MVA, (34.5 kV connected in Δ)/(345 kV connected in Y), and $X_{transformer\ leakage} = X_{T2} = 0.04$ pu. Note that the Y neutral of transformer T_2 is not grounded
- On a chosen base of $S_{base} = 600$ MVA, $V_{L-L\ base} = 345$ kV the transmission line (L) reactances are $X_{1L} = X_{2L} = 0.20$ pu and $X_{0L} = 0.50$ pu
- Draw the zero-, positive-, and negative-sequence networks of Fig. 12.44.
- Determine the zero, positive, and negative-sequence bus impedance (reactance) matrices by means of the Z_{bus} building algorithm [13] and Kron's matrix reduction technique [13].

Solution

(a) Figures 12.45, and 12.46 show the zero-sequence, negative- and positive-sequence networks, respectively, based on the procedure of [13]. Note buses 5 and 6 are auxiliary buses or fictitious nodes.

(b) Zero-sequence impedance (reactance) network is given in Table 12.2, and the negative and positive-sequence impedance (reactance) network is given in Table 12.3.

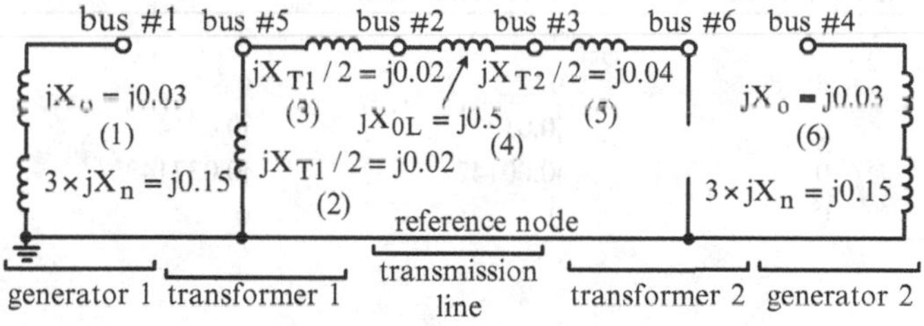

Fig. 12.45 Zero-sequence network

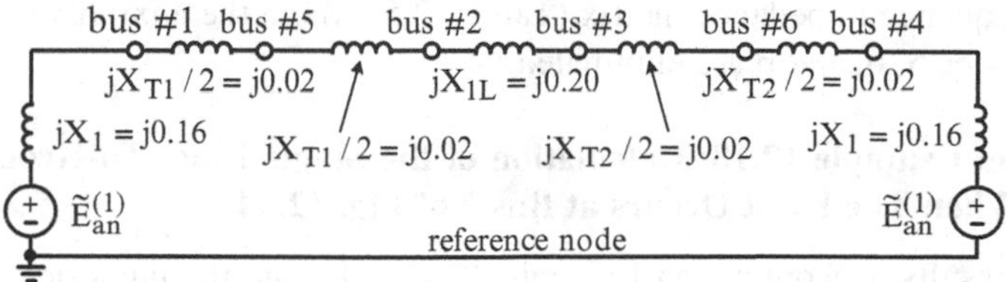

Fig. 12.46 Negative- and positive-sequence network. Note, for negative-sequence network the voltage $\tilde{E}_{an}^{(1)} = 0$

Table 12.2 Zero-sequence reactance matrix. Buses 5 and 6 are auxiliary buses/fictitious nodes

Bus	1	5	2	3	6	4
1	j0.18	0	0	0	0	0
5	0	j0.02	j0.02	j0.02	j0.02	0
2	0	j0.02	j0.04	j0.04	j0.04	0
3	0	j0.02	j0.04	j0.54	j0.54	0
6	0	j0.02	j0.04	j0.54	j0.58	0
4	0	0	0	0	0	j0.18

Table 12.3 Positive- and negative-sequence reactance matrices. Buses 5, 6 and p are auxiliary buses/fictitious nodes

Bus	1	5	2	3	6	4	p
1	j0.16	j0.16	j0.16	j0.16	j0.16	j0.16	j0.16
5	j0.16	j0.18	j0.18	j0.18	j0.18	j0.18	j0.18
2	j0.16	j0.18	j0.20	j0.20	j0.20	j0.20	j0.20
3	j0.16	j0.18	j0.20	j0.40	j0.40	j0.40	j0.40
6	j0.16	j0.18	j0.20	j0.40	j0.42	j0.42	j0.42
4	j0.16	j0.18	j0.20	j0.40	j0.42	j0.44	j0.44
p	j0.16	j0.18	j0.20	j0.40	j0.42	j0.44	j0.60

Table 12.4 The reduced zero-sequence impedance matrix $Z_{bus}^{(0)}$

Bus	1	2	3	4
1	j0.18	0	0	0
2	0	j0.0193	j0.00142	0
3	0	j0.00142	j0.037143	0
4	0	0	0	j0.18

Application of Kron's matrix reduction technique results in the zero-sequence impedance matrix (Table 12.4) where the auxiliary or fictitious buses or nodes 5 and 6 are eliminated.

The use of Kron's matrix reduction technique results in the positive- and negative-sequence impedance matrix (Table 12.5) where the auxiliary or fictitious buses or nodes 5, 6, and p are eliminated.

Application Example 12.18: Calculation of the Single Line- To-Ground Fault Current When The Fault Occurs at Bus 2 of Fig. 12.44

Using the results of Application Example 12.17 calculate the numerical values of the (subtransient) sequence-fault currents $\tilde{I}_{fa}^{(1)}$, $\tilde{I}_{fa}^{(2)}$, and, $\tilde{I}_{fa}^{(0)}$ [13], where "f" stands for fault and the fault impedance is $Z_f = j0.05$ pu.

Table 12.5 The reduced positive- and negative-sequence impedance matrix $Z_{bus}^{(1)} = Z_{bus}^{(2)}$

Bus	1	2	3	4
1	j0.01774	0	0	0
2	0	j0.01825	j0.0017	0
3	0	j0.0017	j0.01825	0
4	0	0	0	j0.01774

Solution

$$\tilde{I}_{fa}^{(1)} = \tilde{I}_{fa}^{(2)} = \tilde{I}_{fa}^{(0)} = \frac{\tilde{E}_{an}^{(1)}}{Z_{22}^{(1)} + Z_{22}^{(2)} + Z_{22}^{(0)} + 3Z_f}$$

$$= \frac{1\angle 0^\circ}{j0.01825 + j0.01825 + j0.0193 + j0.15} = 4.86\text{pu}\angle -90^\circ.$$

Application Example 12.19: Calculation of the Line-To-Line Fault Current When the Fault Occurs at Bus 2 of Fig. 12.44

Using the results of Application Example 12.17 calculate the numerical values of the (subtransient) sequence-fault currents $\tilde{I}_{fa}^{(1)}$, $\tilde{I}_{fa}^{(2)}$, and $\tilde{I}_{fa}^{(0)}$ [13], where "f" stands for fault and the fault impedance is $Z_f = j0.05$ pu.

Solution

$$\tilde{I}_{fa}^{(1)} = -\tilde{I}_{fa}^{(2)} = \frac{\tilde{E}_{an}^{(1)}}{Z_{22}^{(1)} + Z_{22}^{(2)} + Z_f} = \frac{1\angle 0^\circ}{j0.01825 + j0.01825 + j0.05}$$

$$= 11.56\,\text{pu}\angle -90^\circ, \quad \text{and} \quad \tilde{I}_{fa}^{(0)} = 0.$$

Application Example 12.20: Operation of Custom Power Device with Unbalanced and Distorted System Voltage and Load Current

Flexible AC Transmission System (FACTS) controllers are static or power electronic-based devices incorporated in power systems to enhance controllability (e.g., voltage, reactive power) and increase real-power transfer capacity. FACTS devices are capable of instantaneous individual or simultaneous control of V, P and Q. Custom Power devices are FACTS controllers that can also improve the quality of the electric power (e.g., compensate harmonics, imbalance, sags, swells, flickers). Figure 12.47 shows an ideal Custom Power device, located at the point of common coupling (PCC), represented by the combination of a voltage-source converter (injecting series voltage v_c) and a current-source converter (injecting shunt current i_c). Shunt compensator (i_c) forces the system current i_s to become balanced (harmonic-free) and sinusoidal, while series compensator (v_c) ensures balanced, sinusoidal and regulated load voltage.

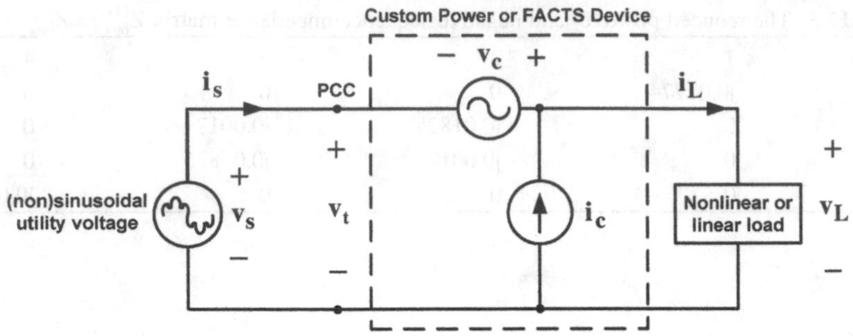

Fig. 12.47 Single-phase representation of ideal Custom Power or FACTS device in distorted three-phase power system

The three-phase system voltages and load currents are unbalanced and contain harmonics. The imbalance or distortion of the three-phase system (Fig. 12.47) consists of positive-, negative-, and zero-sequence fundamental and harmonic components

$$v_s(t) = v_{s+}(t) + v_{s-}(t) + v_{s0}(t) + \sum v_{sh}(t), \tag{12.10}$$

where subscripts "+", "−" and "0" refer to the positive-, negative-, and zero-sequence fundamental components, respectively. $v_{s+}(t) = V_{s+}\sin(\omega t + \phi_+)$, $v_{s-}(t) = V_{s-}\sin(\omega t + \phi_-)$ and $v_{s0}(t) = V_{so}\sin(\omega t + \phi_0)$ are positive-, negative-, and zero-sequence fundamental frequency components of the system voltage, respectively. $\sum v_{sh}(t) = \sum V_{sh}\sin(h\omega t + \phi_h)$ represents the harmonics of the voltage, h is harmonic order, ϕ_+, ϕ_-, ϕ_0, and ϕ_h are the corresponding voltage phase angles.

The distorted nonlinear load current (Fig. 12.47) is expressed as

$$i_L(t) = i_{L+}(t) + i_{L-}(t) + i_{L0}(t) + \sum i_{Lh}(t), \tag{12.11}$$

where $i_{L+}(t) = I_{L+}\sin(\omega t + \delta_+)$, $i_{L-}(t) = I_{L-}\sin(\omega t + \delta_-)$, and $i_{L0}(t) = I_{L0}\sin(\omega t + \delta_o)$ are the fundamental frequency positive-, negative-, and zero-sequence components of the load current, respectively, and $\sum i_{Lh}(t) = \sum I_{Lh}\sin(h\omega t + \delta_h)$ represents harmonics of the load current. δ_+, δ_-, δ_0, and δ_h are the corresponding current phase angles.

(a) Derive the equation of the series-compensation voltage $v_c(t)$ for a sinusoidal, balanced and harmonic-free load voltage $v_L(t) = V_L\sin(\omega t + \phi_+)$.
(b) Write the equation of the shunt-compensation current $i_c(t)$ for a sinusoidal, balanced and harmonic-free system current $i_s(t)$ and a displacement (fundamental) power factor angle of θ_L (between the positive-sequences of voltage and current at the load terminals, $\theta_L = \delta_+ - \phi_+$).
(c) For the conditions of part b, find the equation of the sinusoidal system current.

Solution

(a) The series converter will need to compensate the following components of voltage:

$$v_c(t) = v_L(t) - v_s(t)$$
$$= (V_L - V_{s+})\sin(\omega t + \phi_+) - v_{s-}(t) - v_{s0}(t) - \sum v_{sh}(t). \qquad (12.12)$$

The control system should automatically control the series converter so that its generated voltage at its output terminals is $v_c(t)$ and matched with the above equation.

(b) It is usually desired to have a certain phase angle (displacement power factor angle), θ_L, between the positive-sequences of voltage and current at the load terminals

$$\theta_L = (\delta_+ - \phi_+) \Rightarrow \delta_+ = (\phi_+ + \theta_L). \qquad (12.13)$$

Substituting the above (12.13) into (12.12) and simplifying yields

$$i_L(t) = I_{L+}\sin(\omega t + \phi_+)\cos\theta_L + [I_{L+}\cos(\omega t + \phi_+)\sin\theta_L + I_{L-}(t)$$
$$+ I_{L0}(t) + \sum i_{Lh}(t)]. \qquad (12.14)$$

It is clear that the output current of the shunt converter should be controlled and must assume the waveshape as specified by the 2nd, 3rd, 4th, and 5th terms of the above equation. That is,

$$i_c(t) = I_{L+}\cos(\omega t + \phi_+)\sin\theta_L + i_{L-}(t) + i_{L0}(t) + \sum i_{Lh}(t). \qquad (12.15)$$

(c) The system (feeder) current has the sinusoidal waveform

$$i_s(t) = i_L(t) - i_c(t) = (I_{L+}\cos\theta_L)\sin(\omega t + \phi_+). \qquad (12.16)$$

Equations (12.12) and (12.15) demonstrate the principles of an ideal Custom Power device for systems with voltage and current harmonics. If these equations are dynamically implemented by the device controller, terminal load voltage and system current will be sinusoidal. In addition, system voltage and current will be in phase if θ_L=0 (e.g., unity displacement power factor), and neither positive nor negative reactive power will be supplied by the source.

Application Example 12.21: Operation of FACTS Controller with Unbalanced Sinusoidal System Voltages and Load Currents

In Application Example 12.20, the Custom Power device (Fig. 12.47) is replaced with a FACTS controller and we assumed in Application Example 12.20 that system voltages and load currents are unbalanced but not distorted (e.g., no harmonics).

(a) Show that for a balanced load voltage (with positive-sequence magnitude of $V_{L+} = V_L\angle 0°$) and the condition of no positive-sequence (real) power absorption of the series compensator the following relations hold

$$V_{c0} = -V_{s0},\ V_{c-} = -V_{s-},\ |V_{c+}|^2 - 2a|V_L|\,|V_{c+}| + |V_L|^2 - |V_{S+}|^2 = 0, \quad (12.17)$$

where (a+jb) is the unit vector that is perpendicular to I_{s+}.

(b) Show that for balanced system currents and the condition of no positive-sequence (real) power absorption of the shunt compensator the following conditions hold

$$I_{c0} = I_{L0}\,,\quad I_{c+} = 0,\quad I_{c-} = I_{L-}. \quad (12.18)$$

(c) In addition to the conditions of parts a) and b), the shunt compensator is to supply β= 20% of the average load reactive power ($Q_{L\text{-avg}}$). Show that the new conditions for the shunt compensator are

$$I_{c0} = I_{L0}\,,\ |I_{c+}| = \beta\frac{|Q_{L-avg}|}{|V_L|},\quad I_{c-} = I_{L-}, \quad (12.19)$$

where $\beta = 0.2$ (=20%) is the percentage of load reactive power provided by the FACTS controller.

Solution

(a) From the structure of the ideal FACTS (Fig. 12.47) device one obtains

$$v_L(t) = v_s(t) + v_c(t). \quad (12.20)$$

To obtain balanced voltages at the load terminals, $v_c(t)$ must cancel the imbalance of the system voltages, therefore

$$V_{c0} = -V_{s0},\quad V_{c-} = -V_{s-}, \quad (12.21)$$

which implies that V_L is of positive sequence.

The positive-sequence component of the load voltage (V_L) should be set to the desired regulated voltage while its phase angle depends upon the load displacement (fundamental) power factor.

To define an injected positive-sequence voltage ($|V_{c+}|$) such that the series compensator does not require any positive-sequence (real) power (since i_s flows through the series compensator), v_{c+} must have a phase difference of 90° with i_{s+}. This implies

$$V_{L+} = V_{s+} + V_{c+}(a + jb), \qquad (12.22)$$

where (a+jb) is the unit vector that is perpendicular to I_{s+}. Assuming $V_{L+} = V_L\angle 0°$, (12.22) results in the following second-order equation

$$|V_{c+}|^2 - 2a\,|V_L|\;|V_{c+}| + |V_L|^2 - |V_{S+}|^2 = 0. \qquad (12.23)$$

If the desired regulated voltage level V_L is achievable, this quadratic equation will have two real solutions for $|V_{c+}|$. The minimum solution should be selected because it will result in less cost (e.g., smaller rating) of the FACTS device

(b) From the structure of the ideal FACTS (Fig. 12.47) device one obtains

$$i_L(t) = i_s(t) + i_c(t). \qquad (12.24)$$

To obtain balanced system currents, the injected current $i_c(t)$ must become zero and one gets for the sequence components of the load current

$$I_{c0} = I_{L0}, \quad I_{c+} = 0, \quad I_{c-} = I_{L-}. \qquad (12.25)$$

This will ensure that a positive-sequence current (equal to the positive-sequence load current) is only passing through the system.

(c) The FACTS device can also supply part of the load reactive power through its shunt compensator (converter). To do this (12.25) must be modified while keeping in mind that the zero- and negative-sequence components of the load current must also be canceled by the shunt converter. The new conditions for the injected shunt currents are now

$$I_{c0} = I_{L0}, \quad I_{c+} \neq 0, \quad I_{c-} = I_{L-}. \qquad (12.26)$$

We can divide the average component of the load reactive power between the shunt converter, the series converter and the system. Due to the fact that the average values of the active and the reactive powers of the shunt converter depend only on the positive-sequence current and its real power consumption (or absorption) must be zero, one gets

$$\begin{cases} |V_{L+}|\,|I_{c+}| \cos\,\theta = 0 \;\;\Rightarrow\;\; \theta = \pm 90^o \;\; \text{(because } |V_{L+}| \neq 0 \text{ and } |I_{c+}| \neq 0\text{).} \\ |V_{L+}|\;|I_{c+}|\sin\,\theta = \beta\,(Q_{L-avg}), \end{cases} \qquad (12.27)$$

where θ is the angle between V_{L+} and I_{c+}, scalar β indicates the percentage of reactive power supplied by the shunt converter, and $Q_{L\text{-}avg}$ is the average of the load reactive power.

For inductive loads (e.g., negative $Q_{L\text{-}avg}$), compensator (converter) current I_{c+} must lag V_L by 90°; to satisfy (12.27) and one can write

$$|I_{c+}| = \beta \frac{|Q_{L-avg}|}{|V_L|}. \tag{12.28}$$

12.5 Design of Renewable Storage Power Plants

In the past pumped-hydro plants and compressed-air facilities were used for energy storage. These plants need a lead time of about 6–10 min before they can be put on-line. It appears that these storage plants must be augmented by supercapacitor, battery, magnetic and flywheel storage plants, which can be brought on-line within a few 60 Hz cycles. Variable-speed pump hydroplants can increase its output within a 60 Hz cycle. An attempt is made to analyze some of these different storage plants with their advantages and disadvantages including their payback periods.

Application Example 12.22: Design of a 250 MW Pumped-Storage Hydro-Power Plant

A pumped-storage hydro-power plant (Fig. 12.48) is to be designed [14–18] for a rated power $P_{rated} = 250$ MW and a rated energy capacity 1,500 MWh per day. It consists of an upper and a lower reservoir with a water capacity C_{H2O} each. In addition there must be an emergency reserve for 625 MWh. Water evaporation must be taken into account and is 10% per year for each reservoir, and the precipitation per year is 20 in. The maximum and minimum elevations of the upper reservoir are 1,050 ft and 1,000 ft, respectively. The maximum and minimum elevations for the lower reservoir are 200 ft and 150 ft, respectively. The water turbine is of the Francis type and it is coupled with a salient-pole synchronous machine with p = 24 poles, which can be used as a generator for generating electricity by releasing the water from the upper reservoir to the lower one, and as a motor for pumping the water from the lower reservoir to the upper one. A capacity factor of 100% can be assumed: in a real application the capacity factor may vary between 70 and 90%.

(a) If the power efficiencies of the water turbine, the synchronous generator, and the Δ-Y transformer are $\eta_{turbine} = 0.8$, $\eta_{synchronous\ machine} = \eta_{Y-\Delta\ transformer} = 0.95$, respectively, compute the required turbine input power $P_{turbine}^{required}$.

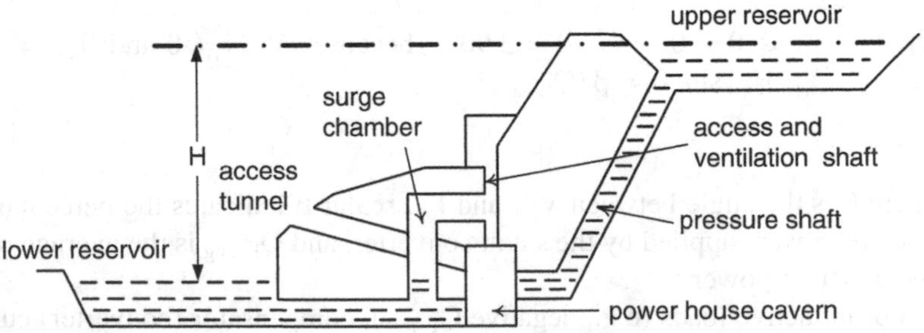

Fig. 12.48 Pumped-storage hydro-power plant for peak-power generation (requires start-up time of about 6 min)

(b) Provided the head of the water is $H = 850$ ft, the frictional losses between water and pipe amount to 15%, and the water flow measured in cubic feet per second is $Q = 6{,}000$ cfs, compute the mechanical power available at the turbine input P_{kW} [16–18].
(c) How does $P^{required}_{turbine}$ compare with P_{kW}?
(d) Compute the specific speed N_q [16–18]. Is the selection of the Francis turbine justified?
(e) What other types of water turbines exist?
(f) What is the amount of water the upper or lower reservoirs must hold to generate $E = (1{,}500\ \text{MWh} + 25\ \text{MWh}) = 2{,}125$ MWh per day during an 11.3 h period?
(g) Is the given precipitation per year sufficient to replace the evaporated water? If not, what is the required "rain-catch" area to replace the yearly water loss through evaporation?
(h) The pumped-storage plant delivers the energy $E = (1{,}500\ \text{MWh} + 625\ \text{MWh}) = 2{,}125$ MWh per day for which customers pay \$0.20/kWh due to peak-power generation. What is the payback period of this pumped storage plant if the construction price is \$3,000 per installed power capacity of 1 kW, the cost for pumping is \$0.03/kWh, and the interest rate is 3%?

Solution

(a) Computation of the turbine input power $P^{required}_{turbine}$:

$$P^{required}_{turbine\_input} = \frac{P_{rated}}{\eta_{turbine} \cdot \eta_{synchronous\_machine} \cdot \eta_{\Delta-Y\_transformer}} = 346.25\ \text{MW}$$

(b) Computation of the mechanical power available at the turbine input P_{kW}:

$$P_{kW} = \frac{H \cdot Q \cdot W \cdot 0.746}{550} = \frac{722.5 \cdot 6000 \cdot 62.4 \cdot 0.746}{550} = 366.9\ \text{kW},$$

where H is measured in ft, Q is measured in cubic feet per second (cfs), W is the weight of 1 cubic foot of water of 62.4 lb-ft/(ft)^3, 0.746 kW = 1 hp, and 550 is a constant converting ft·(lb-force)/s to horse power (hp).

(c) How does $P^{required}_{turbine}$ compare with P_{kW}?

$$P^{required}_{turbine\_input} \leqslant P_{kW}.$$

(d) Computation the specific speed. Is the selection of the Francis turbine justified?
Synchronous speed of generator $n_s = \frac{120 \cdot f}{p} = 300$ rpm, $Q = 6000 \cdot 0.0283 = 169.8\ \text{m}^3/\text{s}$, $H_{eff} = 722.5 \cdot 0.3048 = 220.22$ m. Specific speed of a water turbine is now $N_q = n_s \frac{Q^{0.5}}{H_{eff}^{0.75}} = 68.38$, which means a Francis turbine is satisfactory.

(e) What other types of water turbines exist?
Pelton wheel, Kaplan turbine, bulb-type turbine, mixed-flow turbine, propeller turbine, Gorlov turbine.

(f) The amount of water the upper or lower reservoirs must hold: $Q = 6{,}000$ cfs, therefore, the water required during one day is $Capacity_{H2O} = 6{,}000 \cdot 0.0283 \cdot 3600 \cdot 11.3 = 6.908\ 10^6\ m^3$, this requires two (an upper and lower) reservoirs of 20 m depth, 588 m long and 588 m wide, each. As is well known the water inlets at the upper and lower reservoirs will be in the middle of the reservoir depths in order to prevent that sand and rocks enter the turbine and the pump. This is to say the 20 m depths are effective depths and the actual depths might be more than twice as large, say 50 m.

(g) Required "rain-catch" area: $Capacity_{H2O\ evaporation} = 6.908 \cdot 10^5\ m^3$, $Capacity_{H2O} + Capacity_{H2O\ evaporation} = 6.908 \cdot 10^6 + 6.908 \cdot 10^5 = 7.599 \cdot 10^6\ m^3$. For 20 in. of precipitation one gets per-year $Capacity_{H2O\ precipitation} = 20 \cdot 0.0254\ (588)^2 = 1.756 \cdot 10^5\ m^3$. Note, $Capacity_{H2O\ evaporation}$ exceeds $Capacity_{H2O\ precipitation}$ from this follows that a rain-catch area is required. At the time of commissioning the plant the additional rain-catch area is required to fill the upper reservoir; some run-off will be required after the commissioning of the plant. The additional water to be supplied initially by the rain-catch area is $Capacity_{H2O\ additional} = Capacity_{H2O} + Capacity_{H2O\ evaporation} - Capacity_{H2O\ precipitation} = 7.599 \cdot 10^6\ m^3 - 1.756 \cdot 10^5\ m^3 = 7.42 \cdot 10^6\ m^3$ requiring a rain-catch area of $(area)_{rain\text{-}catch} = 7.42 \cdot 10^6 / (20 \cdot 0.0254) = 14.61 \cdot 10^6\ m^2 = 14.61\ km^2$.

(h) Payback period of pumped storage plant
Earnings $= \$2125 \cdot 10^3 \cdot 0.20 \cdot 365y = 155.125 \cdot 10^{6y}$,
expenses $= \$750 \cdot 10^6 + (2125 \cdot 10^3 \cdot 0.03 \cdot 365y)/(0.95 \cdot 0.95 \cdot 0.80 \cdot 0.85) = \$750 \cdot 10^6 + \$37.92y \cdot 10^6$.
Payback period in y years without interest payments: $y = 750 \cdot 10^6/(155.125 \cdot 10^6 - 37.92 \cdot 10^6) = 6.4$ years
Payback period with interest payments: $155.125y = 750(1.03)^y + 37.92y$ resulting in about $y = 8$ years.

Application Example 12.23: Design a 10 MWh Supercapacitor Power Plant

One of the major problems in using wind and solar energy is energy storage. Wind/solar energy may not be available when it is needed, for example, a wind farm could lose as much as 60 MW within a minute. For example, there are several scenarios of how the power change of 60 MW per minute can be mitigated through complementary, albeit more expensive power sources: one of them is the combination of a compressed-air power plant with a supercapacitor plant for bridging the time from when the wind power plant output decreases (60 MW per minute) to when the compressed-air storage (CAES) plant [19, 20] can take over. A CAES plant requires a start-up time of about 6 min. To bridge this 6 min start-up time required for a 100 MW compressed-air power plant, supercapacitors or ultracapacitors are proposed to provide up to 100 MW during a 6 min interval amounting to a required energy storage of 10 MWh. Inverters fed from supercapacitors can deliver power within milliseconds to the power system replacing the lost power of 60 MW per minute almost instantaneously. This combination of CAES plant and supercapacitors as bridging energy sources can be employed for peak-power operation as well as for improving power quality by preventing short brown/blackouts. Figure 12.49 depicts the block diagram of such a supercapacitor plant consisting of wind turbine, mechanical gear, synchronous generator, 3-phase transformer, 3-phase rectifier, supercapacitor bank, three-phase inverter, 3-phase transformer, and power system.

Recommended voltages are: output line-to-line voltage of generator $V^g_{L-L} = 677$ V, input DC voltage of inverter $V_{DC} = 600$ V, output line-to-line voltage of inverter $V^i_{L-L} = 240$ V, power system voltage $V^s_{L-L} = 480$ V, the two transformers are of the $\Delta - Y$ type connected to the generator and of the Y-Δ type connected to the power system.

(a) Draw a detailed circuit diagram depicting transformers, rectifier, capacitor, and inverter in more detail as compared to Fig. 12.49 and specify the powers, voltages and currents provided 10 rectifier/capacitor/inverter combinations are connected in parallel: this will improve the overall efficiency of the plant, because some of the components can be disconnected, if wind power generation is low. The efficiency for each and every component is about $\eta = 0.96$.

(b) Determine the specifications of the Y-Δ three-phase transformer between inverter and power system.

Design of PWM (pulse-width-modulated) three-phase current controlled voltage–source inverter feeding power into the utility system via Y– Δ transformer

The inverter circuit of Fig. 12.50 is to be analyzed with PSpice, where the DC voltage is $V_{DC} = 600$ V, output voltage of inverter $V^i_{L-L} = 240$ V, power

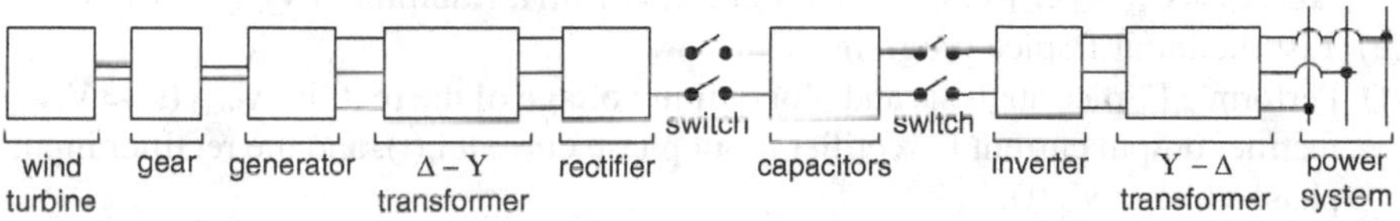

Fig. 12.49 Block diagram for charging and discharging supercapacitor bank (requires start-up time of a few 60 Hz cycles)

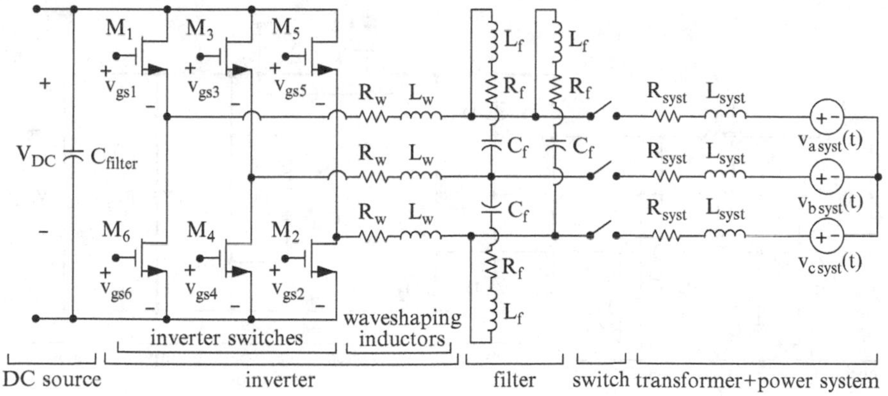

Fig. 12.50 PWM (pulse-width-modulated) three-phase current controlled voltage–source inverter

system voltage $V^{s}_{L-L} = 480$ V, and the switching frequency of the IGBTs is $f_{switch} = 7.2$ kHz. The DC input capacitance is $C_{filter} = 1{,}000$ μF, the components connected at the output of the inverter bridge are $L_w = 1$ mH, $R_w = 10$ mΩ, $L_f = 45$ μH, $C_f = 10.3$ μF, $R_f = 10$ mΩ; transformer and power system inductance and resistance per phase are $L_{syst} = 265$ μH, and $R_{syst} = 50$ mΩ, respectively, the power system (phase) voltages referred to the secondary (Y) side of the Y-Δ transformer are $v_{asyst}(t) = 196\mathrm{V}\sin\omega t$, $v_{bsyst}(t) = 196\mathrm{V}\sin(\omega t - 120°)$, and $v_{csyst}(t) = 196\mathrm{V}\sin(\omega t - 240°)$.

(c) Calculate with PSpice (you may ignore the output filter because of the 64-node limit of the PSpice software program) and plot output phase voltage and phase current of the PWM inverter provided the output voltages are sinusoidal/co-sinusoidal as given above by the power system voltages. List the PSpice input program.

(d) Subject the output current of the inverter to a Fourier analysis.

Design of a controlled three-phase rectifier

Figure 12.51 shows the three-phase rectifier of Fig. 12.49 with six diodes and one self-commutated switch, an insulated-gate bipolar transistor (IGBT). The load resistance is $R_{load} = 3.19$ Ω. The nominal input phase voltages of the rectifier can be assumed to be $v_{an} = 738\mathrm{V}\sin\omega t$, $v_{bn} = 738\mathrm{V}\sin(\omega t - 120°)$, and $v_{cn} = 738\mathrm{V}\sin(\omega t - 240°)$ resulting at a duty cycle of 50% in a DC output voltage of 600 V. The IGBT is gated with a switching frequency of 3 kHz and you may assume a duty cycle of $\delta = 50\%$. $C_{f1} = 200$ μF, $L_{f1} = 90$ μH, $C_{f2} = 50$ μF, ideal diodes $D_1 - D_6$ and D_f, $v_{gs} = 100$ V magnitude, $R_{sd} = R_{ss} = 10$ Ω, $C_{sd} = C_{ss} = 0.1$ μF, $C_f = 1{,}000$ μF, $L_f = 1$ mH, resulting in $V_{load} = 600$ V.

(e) List the input PSpice program

(f) Perform a PSpice analysis and plot output voltage of the rectifier $v_{load}(t) \approx V_{DC}$, rectifier output current I_r, rectifier input phase current $i_r(t)$, and the rectifier input phase voltage $v^{r}_{ph}(t)$.

(g) Subject the rectifier input current $i_r(t)$ to a Fourier analysis.

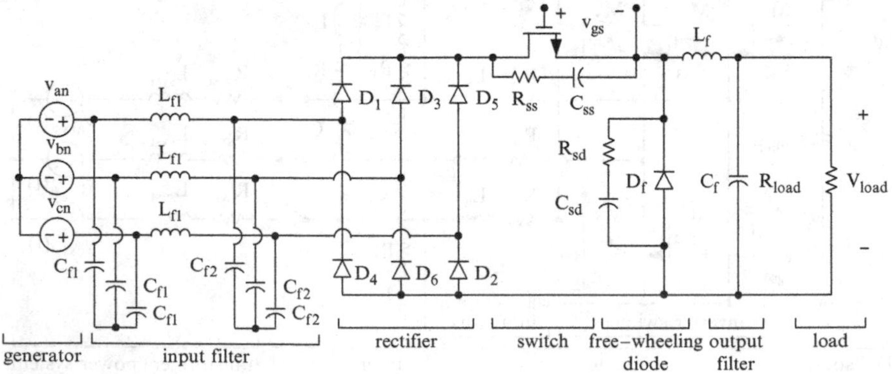

Fig. 12.51 Controlled three-phase rectifier

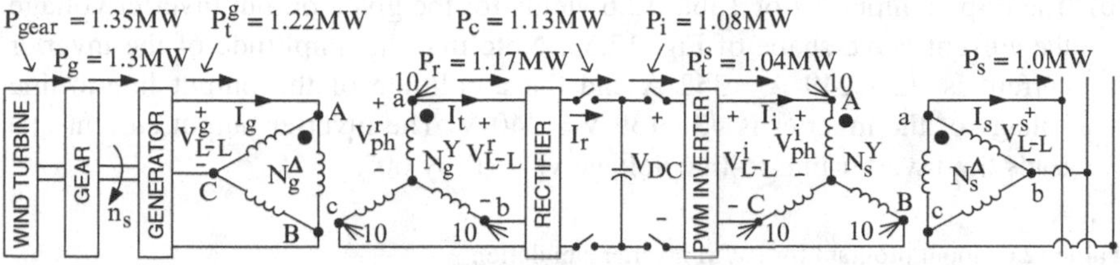

Fig. 12.52 Circuit diagram depicting transformers, rectifier, capacitor, and inverter in more detail as compared to Figure 12.49 supplying P_s=1.0 MW to the power system, that is, 100 such plants must be connected in parallel to supply 100 MW

(h) Determine the specifications of the three-phase transformer between (wind) generator and rectifier.
(i) Determine the overall costs of this power plant if a specific cost of \$4000/kW installed capacity is assumed. Note that a coal-fired plant has a specific cost of \$2,000/kW installed capacity.

Solution

(a) The detailed circuit diagram depicting transformers, rectifier, capacitor, and inverter in more detail as compared to Fig. 12.49 is shown in Fig. 12.52 with a rated output power of $P_s = 1.0$ MW. For the generation of 100 MW 100 such plants must be connected in parallel. Assuming a unity power factor the output powers of the various components are:

Inverter-side transformer output power: $P_s = 1.0$ MW, PWM inverter output power: $P_t^s = 1.04$ MW, supercapacitor output power: $P_i = 1.08$ MW, rectifier P_c output power $= 1.13$ MW, generator-side transformer output power: $P_r = 1.17$ MW, generator output power: $P_t^g = 1.22$ MW, gear output power: $P_g = 1.3$ MW, wind turbine output power: $P_{gear} = 1.35$ MW as listed in Fig. 12.52.

(b) The input voltage of inverter of $V_{DC} = 600$ V results with a modulation index of $m = 0.655$ in the output phase voltage of the inverter

$$V_{ph}^i = \frac{m \cdot V_{DC}}{2\sqrt{2}} = \frac{0.655 \cdot 600}{2\sqrt{2}} = 139\,\text{V},$$

or $V_{L-L}^i = \sqrt{3} V_{ph}^i = 240$V. The transformation ratio of the inverter-side Y-Δ transformer is $\frac{N_s^\Delta}{N_s^Y} = \frac{V_{L-L}^s}{V_{ph}^i} = \frac{480}{139} = 3.45$ resulting in a current fed into the utility system of $I_s = \frac{P_s}{\sqrt{3} V_{L-L}^s} = \frac{1\text{MW}}{\sqrt{3} \cdot 480} = 1,202.8$ A. The output current of one inverter is $I_i = \frac{P_t^s}{3 \cdot V_{ph}^i \cdot 10} = \frac{1.04\text{MW}}{3 \cdot 139 \cdot 10} = 249$A.

(c) The PSpice input list of Table 12.6 yields for the given output inverter voltage the current wave shape of Fig. 12.53. Note that the amplitude of the inverter current is $\sqrt{2} \cdot 249\ \text{A} = 352\ \text{A}$ and the amplitude of the output line-to-line voltage of the inverter is $\sqrt{3} \cdot 139\ \text{V} = 240\ \text{V}$. The inverter output current $i_i(t)$ leads the inverter line-to-line voltage $v^s_{L-L}(t)$ by 60°.

Table 12.6 Input program for PWM inverter simulation

*Current-controlled PWM voltage-source	vtrial 5 0 pulse(−10 10 0 69u 69u
* inverter with P-controller	+ 1u 139u)
*DC voltage supply	* Gating signals for upper MOSFETs
Vsuppl 2 0 600	* as result of comparison between
* IGBT switches	* triangular waveform and error signal
msw1 2 11 10 10 qfet	xgs1 14 5 11 10 comp
dsw1 10 2 diode	xgs2 24 5 21 20 comp
msw2 2 21 20 20 qfet	xgs3 34 5 31 30 comp
dsw2 20 2 diode	* Gating for lower MOSFETs
msw3 2 31 30 30 qfet	egs4 41 0 poly(1) (11,10) 50 −1
dsw3 30 2 diode	egs5 51 0 poly(1) (21,20) 50 −1
msw4 10 41 0 0 qfet	egs6 61 0 poly(1) (31,30) 50 −1
dsw4 0 10 diode	* Filter removed, as node limit is 64
msw5 20 51 0 0 qfet	* lfi1 16 15b 45u
dsw5 0 20 diode	* lfi2 26 25b 45u
msw6 30 61 0 0 qfet	* lfi3 36 35b 45u
dsw6 0 30 diode	* rfi1 15b 15c 0.1
L_W1 10 15 1m	* rfi2 25b 25c 0.1
L_W2 20 25 1m	* rfi3 35b 35c 0.1
L_W3 30 35 1m	* cfi1 15c 26 10.3u
* Resistors or voltage sources required	* cfi2 25c 36 10.3u
*to measure current resistors used as shunt	* cfi3 35c 16 10.3u
R_W1 15 16 10m	* Power system parameters
R_W2 25 26 10m	RM1 16 18 50m
R_W3 35 36 10m	LM1 18 19 265u
* Voltages represent reference currents	Vout1 19 123 sin(0 196 60 0 0 −30)
vref1 12 0 sin(0 352 60 0 0 0)	RM2 26 28 50m
vref2 22 0 sin(0 352 60 0 0 −120)	LM2 28 29 265u
vref3 32 0 sin(0 352 60 0 0 −240)	Vout2 29 123 sin(0 196 60 0 0 −150)
*Voltages represent load currents, measured	RM3 36 38 50m
* with shunts	LM3 38 39 265u
eout1 13 0 15 16 100	Vout3 39 123 sin(0 196 60 0 0 −270)
eout2 23 0 25 26 100	* Model of comparator: v1−v2, vgs
eout3 33 0 35 36 100	.subckt comp 1 2 9 10
*Error signals are (vref – eout)	rin 1 3 2.8k
rdiff1 12 13a 1k	r1 3 2 20meg
rdiff2 22 23a 1k	e2 4 2 3 2 50
rdiff3 32 33a 1k	r2 4 5 1k
cdiff1 12 13a 1u	d1 5 6 zenerdiode1
cdiff2 22 23a 1u	d2 2 6 zenerdiode2

(continued)

Table 12.6 (continued)

cdiff3 32 33a 1u	e3 7 2 5 2 1
rdiff4 13a 13 1k	r3 7 8 10
rdiff5 23a 23 1k	c3 8 2 10n
rdiff6 33a 33 1k	r4 3 8 100k
ecin1 14 0 12 13a 2	e4 9 10 8 2 1
ecin2 24 0 22 23a 2	.model zenerdiode1 D (Is=1p BV=0.1)
ecin3 34 0 32 33a 2	.model zenerdiode2 D (Is=1p BV=50)
* Program list is continued in column 2	.ends comp
	* Program list is continued below table

```
*Models
.model qfet nmos(level=3 gamma=0 kappa=0 tox=100n rs=42.69m kp=20.87u l=2u w=2.9
+ delta=0 eta=0 theta=0 vmax=0 xj=0 uo=600 phi=0.6 vto=3.487 rd=0.19 cbd=200n pb=0.8
+ mj=0.5 cgso=3.5n cgdo=100p rg=1.2 is=10f)
.model diode d(is=1p)
* Options
.options abstol=0.01m chgtol=0.01m reltol=50m vntol=1m itl5=0 itl4=200
.tran 5u 350m 320m 5u uic
.probe
.four 60 99 I(LM1)
.end
```

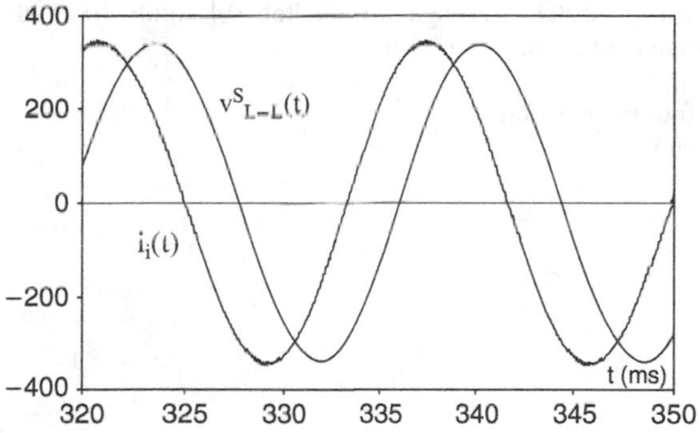

Fig. 12.53 Inverter output current for the given (by power system) inverter output line-to line voltage: inverter output current $i_I(t)$ leads the inverter line-to-line voltage $v^s_{L-L}(t)$ by 60°

(d) Fourier analysis of inverter output current: all current harmonics are negligibly small.

(e) List of the input PSpice program for the rectifier simulation is given in Table 12.7.

(f) Perform a PSpice analysis and plot output voltage of the rectifier $v_{load}(t) \approx V_{DC}$, rectifier output current I_r, rectifier input phase current $i_r(t)$, and the rectifier input phase voltage $v^r_{ph}(t)$ Figure 12.54 depicts the input and output voltages and currents of the rectifier.

(g) Subject the rectifier input current $i_r(t)$ to a Fourier analysis.

Table 12.7 Input program for rectifier simulation

*Controlled three-phase rectifier	D4 10 5 ideal	Rf2 8 10 10meg
*with self-commutated switch, duty	D5 7 9 ideal	*MOSFET
cycle = *50%, rectifier phase rms voltage	D6 10 6 ideal	MOS 9 15 12 12
*V_{rect_ph} = 522V, frequency of generator	Df 10 12 ideal	SMM
voltage *f_{gen} = 60 Hz	*input filter	*resistor and
VAN 1 0 sin(0 738 60 0 0 0)	Lfa 1 5 90u	capacitor
VBN 2 0 sin(0 738 60 0 0 −120)	Lfb 2 6 90u	Rss 9 11 10
VCN 3 0 sin(0 738 60 0 0 −240)	Lfc 3 7 90u	Css 11 12 0.1u
* Gating signal 3 kHz	Cfa 1 4 200u	Rsd 12 13 10
vg 15 12 pulse(0 30 10u 0n 0n 166.6u 333u)	Cfb 2 4 200u	Csd 13 10 0.1u
*diodes	Cfc 3 4 200u	*output
D1 5 9 ideal	Rf1 4 10 10meg	Lf 12 14 0.001
D2 10 7 ideal	Cf2a 5 8 50u	Cf 14 10 1000u
D3 6 9 ideal	Cf2b 6 8 50u	Rload 14 10 3.19
* continue program list	Cf2c 7 8 50u	*continue program
	*continue program list	list *below table
*in column #2	* in column #3	

```
*Model for MOSFET
.model SMM NMOS(level=3 gamma=0 kappa=0 tox=100n rs=0 kp=20.87u l=2u w=2.9
+ delta=0 eta=0 theta=0 vmax=0 xj=0 uo=600 phi=0.6 vto=0 rd=0 cbd=200n pb=0.8
+ mj=0.5 cgso=3.5n cgdo=100p rg=0 is=10f)
*diode model
.model ideal d(is=1p)
.options abstol=10u chgtol=10p reltol=0.1 vntol=100m itl4=200 itl5=0
.tran 0.5u 350m 300m 0.5m
.probe
.four 60 60 I(Lfa)
.end
```

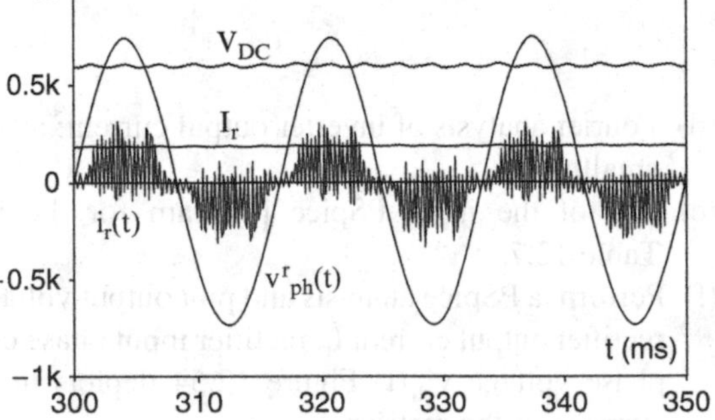

Fig. 12.54 Rectifier input and output voltages and currents

Dominant harmonic amplitudes including the fundamental are:
1st: 105.4 A; 5th: 17.7 A; 7th: 11.2 A; 36th: 12.8 A; 37th: 23.0 A; 43rd: 17.3 A; 45th: 18.2 A; 49th: 84.2 A; 51st: 75.7 A; 55th: 10 A. Note, to get rms values these values must be divided by $\sqrt{2}$. That is, the fundamental rms value is 74.54 A.

(h) Determine the specifications of the three-phase transformer between (wind) generator and rectifier.
The rectifier output current is

$$I_r = \frac{P_c}{V_{DC} \cdot 10} = \frac{1.13\,\text{MW}}{600 \cdot 10} = 188\,\text{A}$$

and results with $V_{DC} = 600$ V in a load resistor of $R_{load} = 3.19\ \Omega$. The PSpice simulation requires for a duty cycle of 50% with a load resistor $R_{load} = 3.19\ \Omega$ an input AC voltage of $V_{ph}^{r} = 522$ V or an amplitude of 738V. With the given generator voltage of $V_{L-L}^{g} = 677$ V one obtains the generator-side transformer ratio of $\frac{N_g^{\Delta}}{N_g^{Y}} = \frac{677}{522} = 1.3$. If the rectifier and the generator are matched with respect to their voltages one does not need a generator-side transformer. The input current to the rectifier is approximately [note there are harmonics as indicated in part g)]

$$I_t = \frac{P_r}{3 \cdot V_{ph}^{r} \cdot 10} = \frac{1.17\text{MW}}{3 \cdot 522 \cdot 10} = 74.7\text{A},$$

this compares well with the rms value of the fundamental of part g), that is, 105.4/1.414 = 74.54 A. The input current to the generator-side transformer at unity power factor is

$$I_g = \frac{P_t^{g}}{\sqrt{3}V_{L-L}^{g}} = \frac{1.22\,\text{MW}}{\sqrt{3}\cdot 677} = 1{,}040.5\,\text{A}.$$

(i) Determination of the overall costs of this power plant if a specific cost of \$4,000/kW installed power capacity is assumed. Note that a coal-fired plant has a specific cost of \$2,000/kW installed capacity.
The cost of the 100 MW super-capacitor plant is $\$400\cdot 10^{6}$.

Application Example 12.24: Design of a 100 MW Compressed-Air Storage Facility

Design a compressed-air energy storage (CAES) plant for $P_{out\_generator} = P_{rated} = 100$ MW, $V_{L-L} = 13{,}800$ V, and $\cos\Phi = 1$ which can deliver rated power for 2 h. The overall block diagram of such a plant [19, 20] is shown in Fig. 12.55.

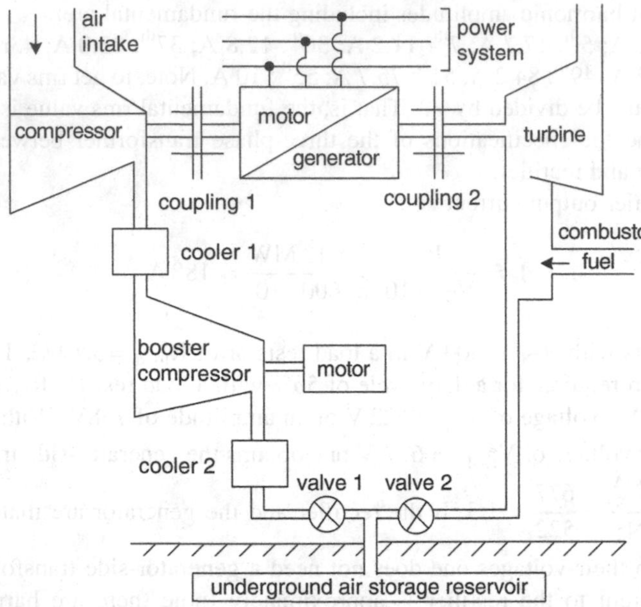

Fig. 12.55 Compressed-air storage (CAES) power plant requiring a start-up time of a few (e.g., 6) minutes

It consists of a compressor, cooler, booster compressor, booster cooler, booster-compressor three-phase, 2-pole induction motor with a 1:2 mechanical gear to increase the compressor speed, underground air-storage reservoir, combustor fired either by natural gas or oil or coal, a modified (without compressor) gas turbine, and a three-phase, 2-pole synchronous generator/motor.

Design data:

Air inlet temperature of compressor: 50°F ≡ 283.16 K at ambient pressure 1 atm ≡ 14.696 psi ≡ 101.325 kPa(scal).
Output pressure of compressor: 11 atm ≡ 161 psi ≡ 1,114.5 kPa.
Output temperature of booster cooler: 120°F ≡ 322.05 K at 1,000 psi ≡ 6,895 kPa.
Output temperature of combustor: 1,500°F ≡ 1,089 K at 650 psi ≡ 4,482 kPa.
Output temperature of gas turbine: 700°F ≡ 644 K at ambient pressure 1 atm ≡ 14.696 psi ≡ 101.325 kPa.
Generation operation: 2 h at 100 MW.
Re-charging (loading) operation of underground air storage reservoir: 8 h at 20 MW.
Capacity of air storage reservoir: $3.5 \cdot 10^6$ ft^3 ≡ $97.6 \cdot 10^3$ m^3 ≡ (46 m × 46 m × 46 m).
Start-up time: 6 min.
Turbine and motor/generator speed: $n_{s\ mot/gen} = 3{,}600$ rpm
Compressor speed: $n_{s\ comp} = 7{,}000$ rpm.
Cost per 1 kW installed power capacity: $2,000.

(a) Calculate the Carnot efficiency of the gas turbine

$$\eta_{\text{carnot gas turbine}} = \eta_{\text{compressor}} \cdot \eta_{\text{turbine}} = \frac{T_1 - T_2}{T_1}.$$

(b) Note that the compressor of a gas turbine has an efficiency of $\eta_{\text{compressor}} = 0.50$. In this case the compressor is not needed because pressurized air is available from the underground reservoir. The absence of a compressor increases the overall efficiency of the CAES, if "free" wind power is used for charging the reservoir. Provided the synchronous machine (motor/generator set) has an efficiency of $\eta_{\text{generator}} = 0.9$ calculate

b1) The efficiency of the turbine without the compressor $\eta_{\text{compressor}}$,

b2) $P_{\text{out turbine}}$,

b3) $P_{\text{out combustor}}$,

b4) For a heat rate of the combustor of 5,500 BTU/kWh, determine the input power of the combustor required (during 1 h) $P_{\text{in combustor}}$.

b5) The two air compressors (charging the reservoir during 8 h) including the two coolers require an input power of 20 MW (whereby wind energy is not free) calculate the overall efficiency of the CAES (during 1 h of operation)

$$\eta_{\text{overall}}^{\text{wind energy is not free}} = \frac{P_{\text{out\_generator}}}{P_{\text{incombustor}} + P_{\text{2\_compressors+2\_coolers}}}$$

b6) Compute the overall efficiency if wind energy is free, that is,

$$\eta_{\text{overall}}^{\text{wind energy is free}} = \frac{P_{\text{out\_generator}}}{P_{\text{in combustor}}}$$

(c) What is the overall construction cost of this CAES plant, if the construction price is $2,000 per installed power capacity of 1 kW?

(d) The CAES plant delivers the energy E = 200 MWh per day (during 2 h) for which customers pay $0.25/kWh due to peak-power generation. What is the payback period of this CAES plant, if the cost for pumping (loading, recharging) is neglected (free wind power), the cost of 1 $(\text{ft})^3$ of natural gas is $0.005, and the interest rate is 3%?

(e) Repeat the analysis for a heat rate of 4,000 BTU/kWh at a turbine efficiency of 0.9.

Hint: For the calculation of the Carnot efficiency of the gas turbine and of the compressors you may use the software available on the Internet address:
http://hyperphysics.phy-astr.gsu.edu/hbase/thermo/adiab.html#c3

Solution

(a) The Carnot efficiency of the gas turbine is

$$\eta_{\text{carnot gas turbine}} = \eta_{\text{compressor}} \cdot \eta_{\text{turbine}} = \frac{T_1 - T_2}{T_1} = \frac{1{,}089\,\text{K} - 644\,\text{K}}{1{,}089\,\text{K}} = 0.41.$$

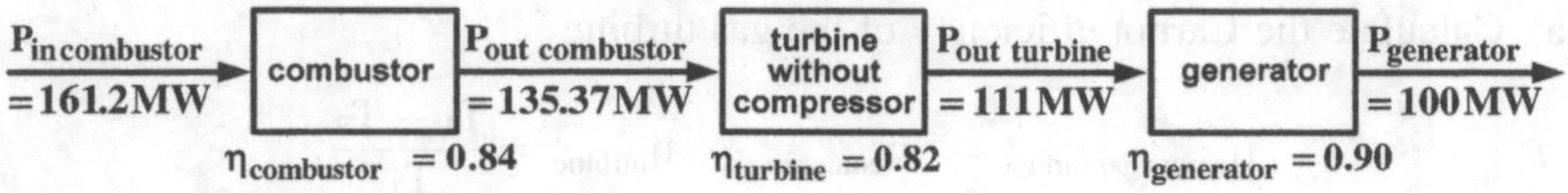

Fig. 12.56 Output powers and the associated efficiencies of generator, gas turbine without compressor, and combustor

b1) The efficiency of the gas turbine without the compressor $\eta_{\text{compressor}}$ is

$$\eta_{\text{turbine}} = \frac{\eta_{\text{carnot gas turbine}}}{\eta_{\text{compressor}}} = \frac{0.41}{0.5} = 0.82.$$

b2)
$$P_{\text{outturbine}} = \frac{P_{\text{generator}}}{\eta_{\text{generator}}} = \frac{100\,\text{MW}}{0.9} = 111\,\text{MW}$$

b3)
$$P_{\text{out combustor}} = \frac{P_{\text{out turbine}}}{\eta_{\text{turbine}}} = \frac{111\,\text{MW}}{0.82} = 135.37\,\text{MW}$$

b4) For a heat rate of the combustor of 5,500 BTU/kWh, determine the input power of the combustor required (during 1 h) $P_{\text{in combustor}}$. Figure 12.56 shows the output powers and the associated efficiencies of generator, gas turbine without compressor, and combustor.

The heat rate = 5,500 BTU/kWh indicates the thermal input energy to the combustor required to generate 1 kWh of electrical energy, that is, 5,500 BTU = 1.612 kWh of thermal energy is required to generate 1 kWh of electrical output energy.

The input energy delivered within 1 h to the combustor is identical to the input power of the combustor

$$P_{\text{in combustor}} = P_{\text{out\_generator}} \left(\frac{1.612\,\text{kWh}}{1\,\text{kWh}} \right) = 100\,\text{MW} \frac{1.612\,\text{kWh}}{1\,\text{kWh}} = 161.2\,\text{MW}.$$

The efficiency of the combustor is.

$$\eta_{\text{combustor}} = \frac{135.37}{161.2.} = 0.84.$$

b5) Two air compressors (charging the reservoir during 8 h) including the two coolers require an input power of 20 MW (whereby wind energy is not free) calculate the overall efficiency of the CAES (during 1 h of operation).

The energy required during an 8 h period for the two compressors and two coolers is $E_{2\text{ compressors}+2\text{ coolers}} = 20\,\text{MW} \cdot 8\,\text{h} = 160\,\text{MWh}$. If we want to deliver 100 MW during 1 h then the power required is $P_{2\ \text{compressors+}}$ $_{2\ \text{coolers}} = 80$ MW. The overall efficiency is

$$\eta_{\text{overall}}^{\text{wind energy is not free}} = \frac{P_{\text{out\_generator}}}{P_{\text{in combustor}} + P_{\text{2\_compressors+2\_coolers}}} = \frac{100\,\text{MW}}{(161.2 + 80)\,\text{MW}}$$
$$= 0.415.$$

b6) If wind energy is free, then

$$\eta_{\text{overall}}^{\text{wind energy is free}} = \frac{P_{\text{out\_generator}}}{P_{\text{in combustor}}} = \frac{100\,\text{MW}}{161.2\,\text{MW}} = 0.62.$$

(c) The construction cost of this CAES plant is \$200 M.
(d) Earnings per day by discharging the CAES
Earning = 200,000 kWh·\$0.25 = \$50k/day.
Natural gas cost: 5,500 BTU = $1.6117 \cdot 10^3$ Wh or 1 Wh = 3.413 BTU resulting in 1 MWh = $3.413 \cdot 10^6$ BTU.
The fuel energy delivered to the combustor is $E_{\text{natural gas}} = 161.2$ MWh·2·$3.413 \cdot 10^6$ BTU/MWh = $1{,}100 \cdot 10^6$ BTU. $1(\text{ft})^3$ of natural gas contains 1,021 BTU, and the volume of natural gas required is $\text{Vol}_{\text{natural gas}} = \frac{1,100 \cdot 10^6}{1,021} (\text{ft})^3 = 1.08 \cdot 10^6\ (\text{ft})^3$ within a 2 h period or per day. The fuel costs are then $\text{Cost}_{\text{fuel}} = 1.08 \cdot 10^6 \cdot \$0.005 = \$5400$ for 2 h or per day.
Payback period:
$\$(1.03)^y 1{,}00{,}000$ kW·\$2,000/kW+\$5,400·365·y = \$50,000·365·y or $(1.03)^y$ 200 = 16.28·y resulting in approximately y = 30 years.
(e) This part is to be completed by the reader.

Application Example 12.25: Magnetic Storage Plant

The fact that wind power plants can change their power output relatively quickly (e.g., 60 MW per minute) and compressed-air power plants have a start-up time of 6 min calls for bridging power sources such as a superconducting magnetic storage system or flywheel storage which can take over energy production within a few 60 Hz cycles.

Design a magnetic storage system which can provide for about 6 min power of 100 MW, that is, energy of 10 MWh. A coil of radius R = 10 m, height of h = 6 m, and N = 3,000 turns is shown in Fig. 12.57. The magnetic field strength H inside such a coil is axially directed, essentially uniform [21], and equal to H=Ni/h. The coil is enclosed by a superconducting box of two times its volume. The magnetic field outside the superconducting box can be assumed to be negligible and you may assume a permeability of free space for the calculation of the stored energy. Calculate the radial pressure in newtons per square meter acting on the sides of the coil for constant current I_0=8,000 A.

Fig. 12.57 Superconducting coil (start-up time of a few 60 Hz cycles)

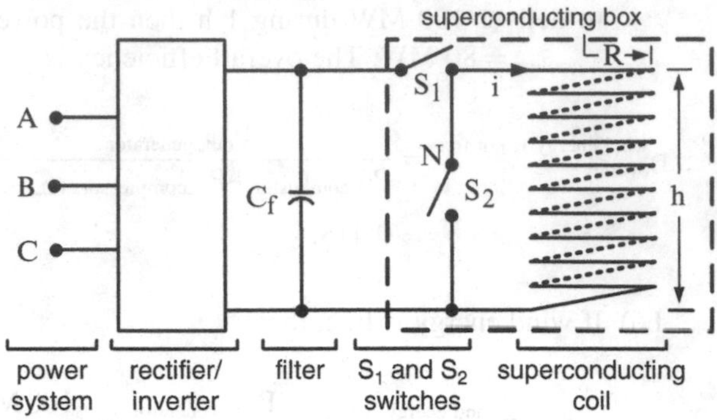

(a) Calculate the energy E_{stored} stored in the magnetic field (either H or B).
(b) Determine the radial pressure p_{radial} acting on the coil.
(c) Devise an electronic circuit which enables the controlled extraction of the energy from this storage device: S_1 is ON and S_2 is OFF for charging the superconducting coil, S_1 is OFF and S_2 is ON for the storage time period, and S_1 is ON and S_2 is OFF for discharging the superconducting coil.

Solution

(a) Energy stored in magnetic field

The magnetic field intensity is $H = \left(\frac{N \cdot I_o}{h}\right) = \left(\frac{3{,}000 \cdot 8{,}000}{6}\right)$ At/m $= 4 \cdot 10^6$ At/m, and the magnetic flux density becomes $B = \mu_o H = 4\pi \cdot 10^{-7} \cdot 4 \cdot 10^6 T = 5.026 T$. The volume of the coil including the superconducting box is $Vol = 2 \cdot (\pi \cdot r^2 \cdot h) = 2(\pi(10)^2 \cdot 6) = 3.77 \cdot 10^3 m^3$.

The energy stored in the coil is then $E_{stored} = \frac{1}{2} \int_{volume\ V} H \cdot B dV = \frac{1}{2} 4 \cdot 10^6 \cdot 5.026 \cdot 3.77 \cdot 10^3 = 37.89 \cdot 10^9\ Ws = 10.53\ MWh$.

(b) Radial pressure acting on coil

The stored (co)magnetic energy as a function of i and r for this linear problem is

$$E_{stored} = W'_{fld}(i, r) = \frac{1}{2} \int_{volume\ V} H \cdot B dV = \frac{1}{2} \int_{volume\ V} \mu_o H^2 dV = \frac{1}{2} \mu_o \left(\frac{N \cdot i}{h}\right)^2 \pi \cdot r^2 \cdot h$$
$$= \frac{1}{2} \mu_o \frac{N^2 i^2}{h} \pi \cdot r^2.$$

The radial force due to the magnetic field is

$$f_{fld}(i, r_o) = \left.\frac{\partial W'_{fld}(i, r)}{\partial r}\right|_{r=r_o} = \frac{\mu_o \pi \cdot r_o N^2 i^2}{h}.$$

The radial pressure is $p = \frac{f_{fld}(i, r_o)}{(\text{surface area of coil})} = \frac{f_{fld}(i, .r_o)}{2\pi r_o h} = \frac{\mu_o N^2 i^2}{2 \cdot h^2} = \frac{4\pi \cdot 10^{-7}(3,000)^2(8,000)^2}{2 \cdot 36} = 1.0053 \cdot 10^7 \text{N/m}^2$ or $p = 1457.5$ psi.

(c) To be completed by reader.

Application Example 12.26: Flywheel Storage Power Plant

Design a flywheel storage system which can provide a power of 100 MW for about 6 min, that is, energy of 10 MWh. The flywheel power plant consists (see Fig. 12.58) of a flywheel, mechanical gear, synchronous machine, inverter-rectifier set and a step-up transformer. The individual components of this plant must be designed as follows:

(a) Possible flywheel configurations

For the flywheels (made from steel) based on a spherical ($R_{1o} = 1.5$ m), cylindrical ($R_{1o} = 1.5$ m, $h = 0.9$ m), and wheel-type (with four spokes) configurations as shown in Fig. 12.59 ($h = 0.9$ m, $R_{1o} = 1.5$ m, $R_{1i} = 1.3$ m, $R_{2o} = 0.50$ m, $R_{2i} = 0.10$ m, $b = 0.2$ m), compute the ratio of inertia (J) to the weight (W) of the flywheel, that is (J/W). You may assume that the flywheel has magnetic bearings; see internet address: http://www.skf.com/portal/skf_rev/home

The axial moment of inertia of a rotating sphere is:

$$J_{sphere} = \frac{8}{15} \cdot \pi \cdot \gamma \cdot R_{1o}^5 \left[\text{kgm}^2\right] \tag{12.29}$$

where the mass density of iron/steel is

$$\gamma_{steel} = 7.8 \left[\text{kg/dm}^3\right] \tag{12.30}$$

The axial moment of inertia of a rotating cylinder is:

$$J_{cylinder} = \frac{1}{2} \cdot \pi \cdot \gamma \cdot h \cdot R_{1o}^4 \left[\text{kgm}^2\right]. \tag{12.31}$$

The axial moment of inertia of a rotating wheel-type configuration with 4 spokes is

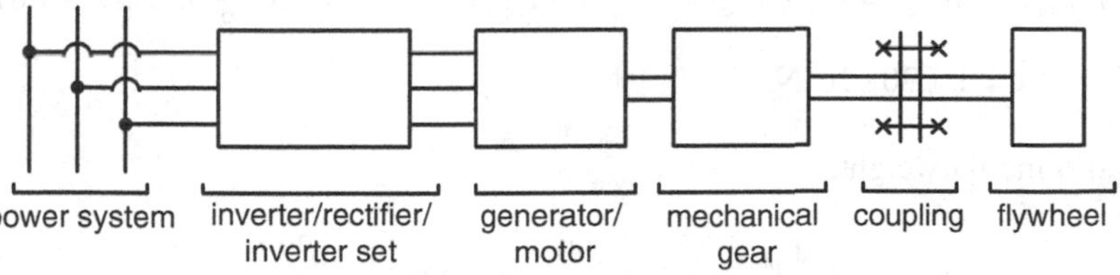

Fig. 12.58 Flywheel power plant (start-up time of a few 60 Hz cycles)

Fig. 12.59 Three flywheel configurations

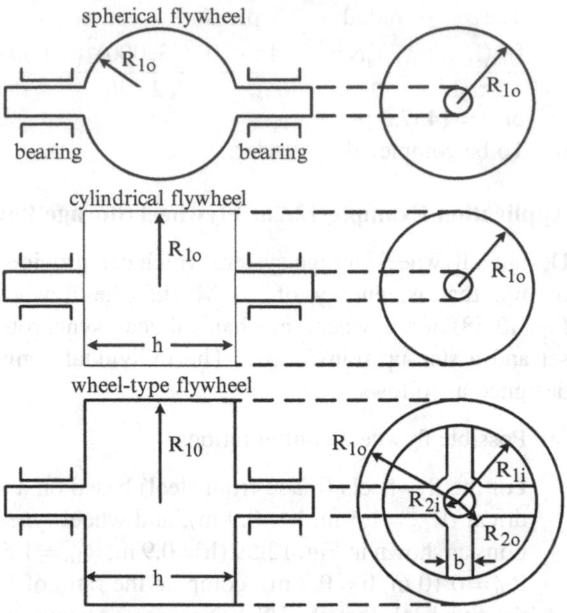

$$J_{wheel} = \frac{1}{2}\cdot \pi \cdot \gamma \cdot h \cdot (R_{1o}^4 - R_{1i}^4) + \frac{4}{3}\cdot h \cdot b \cdot \gamma \cdot (R_{1i}^3 - R_{2o}^3) + \frac{1}{2}\cdot \pi \cdot \gamma \cdot h \cdot (R_{2o}^4 - R_{2i}^4)\ [kgm^2]. \quad (12.32)$$

(b) For the given values $h = 0.9$ m, $R_{1o} = 1.5$ m, $R_{1i} = 1.3$ m, $R_{2o} = 0.50$ m, $R_{2i} = 0.10$ m, and $b = 0.2$ m calculate for the wheel-type configuration the stored energy $E_{stored\ rated}$ provided the flywheel rotates at $n_{flywheel\ rated} = 21{,}210$ rpm.

Solution

(a) Axial moment of inertia of rotating sphere:

$$J_{sphere} = \frac{8}{15}\cdot \pi \cdot \gamma \cdot R_{1o}^5 = \frac{8}{15}\pi \cdot 7.86 \cdot 10^3 (1.5)^5 = 100 \cdot 10^3 kgm^2,$$

weight of sphere:

$$W_{sphere} = m \cdot g = (\gamma \cdot V) g = \gamma \frac{4}{3}\pi \cdot R_{1o}^3 \cdot g = 7.86 \cdot 10^3 \frac{4}{3}\pi (1.5)^3 \cdot 9.81 = 1{,}090 \cdot 10^3 N,$$

ratio inertia/weight:

$$\left(\frac{J_{sphere}}{W_{sphere}}\right) = \frac{100}{1{,}090} = 0.0917 \frac{kgm^2}{N}.$$

Axial moment of inertia of rotating cylinder:

$$J_{cylinder} = \frac{1}{2} \cdot \pi \cdot \gamma \cdot h \cdot R_{1o}^4 = \frac{1}{2} \pi \cdot 7.86 \cdot 10^3 \cdot 0.9 \cdot (1.5)^4 = 56.25 \cdot 10^3 kgm^2,$$

weight of cylinder:

$$W_{cylinder} = m \cdot g = (\gamma \cdot V)g = \gamma \cdot \pi \cdot R_{1o}^2 \cdot h \cdot g = 7.86 \cdot 10^3 \cdot 0.9 \cdot \pi(1.5)^2 \cdot 9.81$$
$$= 490.51 \cdot 10^3 N,$$

ratio inertia/weight:

$$\left(\frac{J_{cylinder}}{W_{cylinder}}\right) = \frac{56.25}{490.51} = 0.1147 \frac{kgm^2}{N}.$$

Axial moment of inertia of wheel-type configuration: $J_{wheel} = \frac{1}{2} \cdot \pi \cdot \gamma \cdot h \cdot (R_{1o}^4 - R_{1i}^4) + \frac{4}{3} \cdot h \cdot b \cdot \gamma \cdot (R_{1i}^3 - R_{2o}^3) + \frac{1}{2} \cdot \pi \cdot \gamma \cdot h \cdot (R_{2o}^4 - R_{2i}^4)$,

$$J_{wheel} = \frac{1}{2} \pi \cdot 7.86 \cdot 10^3 \cdot 0.9[(1.5)^4 - (1.3)^4] + \frac{4}{3} 0.9 \cdot 0.2 \cdot 7.86 \cdot 10^3 [(1.3)^3$$
$$- (0.5)^3] + \frac{1}{2} \pi \cdot 7.86 \cdot 10^3 \cdot 0.9[(0.5)^4 - (0.1)^4]$$
$$= 29.117 \cdot 10^3 \, kgm^2,$$

weight of wheel type configuration:

$$W_{wheel} = (\gamma \cdot V)g$$
$$= \gamma \cdot [\pi \cdot h(R_{1o}^2 - R_{1i}^2) + 4h \cdot b(R_{1i} - R_{2o}) + \pi h(R_{2o}^2 - R_{2i}^2)] \cdot g,$$

$$W_{wheel} = 7.86 \cdot 10^3 \cdot [\pi \cdot 0.9(1.5^2 - 1.3^2) + 4 \cdot 0.9 \cdot 0.2(1.3 - 0.5) + \pi 0.9$$
$$(0.5^2 - 0.1^2)] \cdot 9.81 = 218.753 \cdot 10^3 N,$$

ratio inertia/weight:

$$\left(\frac{J_{wheel}}{W_{wheel}}\right) = \frac{29.117}{218.75} = 0.133 \frac{kgm^2}{N}.$$

Clearly the wheel-type flywheel is the best solution due to the greatest axial moment to weight ratio.

(b) Stored energy at rated speed:
The rated angular velocity is

$$\omega_{flywheel\_rated} = 2\pi\left(\frac{n_{flywheel\_rated}}{60}\right) = 2,221.04 \, rad/s,$$

$$E_{stored\_rated} = \frac{1}{2} J_{wheel}(\omega_{flywheel})^2 = \frac{1}{2} 29.117 \cdot 10^3 (2,221.04)^2 = 71.82 \cdot 10^9 \text{ Ws}$$
$$= 19.95 \text{ MWh}$$

Note the flywheel must not be completely discharged. The angular velocity at 9.95 MWh is

$$9.95 \text{ MWh} = \frac{1}{2} 29.117 \cdot 10^3 (\omega_{flywheel\_minimum})^2 \quad \text{or}$$

$$\omega_{flywheel\_minimum} = \sqrt{\frac{2 \cdot 9.95 \text{ MWh}}{29.117 \cdot 10^3}} = 1,568.57 \text{ rad/s}.$$

Application Example 12.27: Analysis of a Pumped-Storage Hydro-Power Plant

Partial design data of a pumped-storage hydro-power plant (see Fig. 12.48) are available. Complete/calculate the missing information. Note that for generation a Francis turbine [16–18], and for pumping a separate (dedicated) pump is used.

Partial available rated design data of this hydro-power plant are as follows:
Synchronous generator

Rated synchronous generator output power: $P_{generator}$=
Power factor of synchronous generator: cosΦ=0.9 lagging (overexcited based on generator notation)
Apparent output power of synchronous generator: $S_{generator}$=
Efficiency of synchronous generator: $\eta_{generator}$= 0.982
Synchronous speed: n_s=600 rpm
Number of poles: p=
Rated frequency: f=60 Hz
Terminal voltage: $V_{L-L} = 12.4$ kV

Y/Δ Three-phase transformer

Apparent output power of transformer: $S_{transformer}$=
Efficiency of transformer: $\eta_{transformer} = 0.995$
Primary voltage (machine side): $V_{L-Lprimary} = 12.4$ kV
Secondary voltage (system side): $V_{L-Lsecondary} = 115$ kV

Pipeline

Head: H = 292 m
Efficiency of pipeline due to frictional losses $\eta_{pipeline} = 0.969$
Effective head: H_{eff} =

Operational times

Generation time at rated generator power $P_{generator}$: $t_{generation}^{discharge} = 6.25$ h/day
Pumping time at rated pump power P_{pump} : $= t_{pumping}^{charge} = 11.25$ h/day

Francis turbine

Water flow: $Q_{turbine} = 36\ m^3/s$
Rated output power of turbine power: $P_{turbine\_out} =$
Efficiency of turbine: $\eta_{turbine} = 0.89$
Speed of turbine: $n_s = 600$ rpm

Pump

Water flow: $Q_{pump} = 20\ m^3/s$
Rated pump output power = 67,024 hp = 50 MW.
Efficiency of pump: $\eta_{pump} = 0.89$
Speed of pump: $n_s = 600$ rpm

(a) How many poles p has the synchronous generator?
(b) Determine the effective head of the pipeline H_{eff}.
(c) Calculate the input power P_{kW} [16–18] of the water turbine, and the rated output power $P_{turbine\_out}$ of the water turbine.
(d) Calculate the rated real output power $P_{generator\text{-}rated}$ of the synchronous generator.
(e) Determine the rated apparent output power $S_{generator}$ of the synchronous generator.
(f) Find the apparent output power of transformer $S_{transformer}$.
(g) Calculate the minimum required water content of the upper reservoir $C_{H_2O}^{discharge}$ and that of the lower reservoir $C_{H_2O}^{charge}$.
(h) Calculate the specific speed N_q [16–18]. Is the use of a Francis turbine justified?
(i) Determine the efficiency of this pumped-storage plant for discharging $\eta_{discharge}$ and for charging η_{charge}, and then compute the overall efficiency $\eta_{overal} = \eta_{discharge} \cdot \eta_{charge}$.
(j) Calculate the phase current of the generator $I_{phase\_generator}$.
(k) Determine the base impedance Z_{base} of the generator.
(l) Determine the turns ratio of the Y/Δ transformer (N_p/N_s), where the Y-side has N_p turns and the Δ-side has N_s turns.
(m) Determine the payback period if the construction cost is $4,000/ kW installed power capacity and the utility can sell 1 kWh during the peak-period times for $0.20 and buy 1 kWh during the off-period times for $0.06. You may neglect interest payments.
Hint: 1 ft = 0.3048 m.

Solution

(a) Synchronous speed of generator is $n_s = \frac{120 \cdot f}{p}$ or the pole number is $p = \frac{120 \cdot f}{n_s} = 12$ poles.
(b) $H_{eff} = \eta_{pieline} \cdot H = 282.95$ m.
(c) The input power of water turbine [16–18] with $Q_{turbine} = 36\ m^3/s = 1{,}271.19\ ft^3/s$ and $H_{eff} = 282.96\ m = 928.31$ ft is

$$P_{kW} = \frac{H_{eff} \cdot W \cdot 0.746 \cdot Q_{turbine}}{550} = \frac{928.31 \cdot 62.4 \cdot 0.746 \cdot 1271.19}{550} = 99,880\,\text{kW}.$$

Output power of water turbine $P_{turbine\_out} = P_{kW} \cdot \eta_{turbine} = 99{,}880 \cdot 0.89 = 88.80$ MW.

(d) $P_{generator\text{-}rated} = P_{turbine\_out} \cdot \eta_{generator} = 88.89\ \text{MW} \cdot 0.982 = 87.23\ \text{MW}$.

(e) $S_{generator} = P_{generator\text{-}rated} / \cos\Phi = 87.23\text{MW}/0.9 = 96.92\ \text{MVA}$.

(f) $S_{transformer} = S_{generator} \cdot \eta_{transformer} = 96.92\ \text{MVA} \cdot 0.995 = 96.44\text{MVA}$.

(g) $C_{H_2O}^{discharge} = t_{generation}^{discharge}\ Q_{turbine} \cdot 3{,}600 = 6.25 \cdot 36 \cdot 3{,}600 = 810 \cdot 10^3\ \text{m}^3$, and $C_{H_2O}^{charge} = t_{pumping}^{charge}\ Q_{pump} \cdot 3{,}600 = 11.25 \cdot 20 \cdot 3{,}600 = 810 \cdot 10^3\ \text{m}^3$.

(h) $$N_q = N \frac{Q_{turbine}^{0.5}}{H_{eff}^{0.75}} = \frac{600\sqrt{36}}{282.95^{0.75}} = 52.18.$$

From this follows that a Francis turbine is acceptable [16–18].

(i) $\eta_{discharge} = \eta_{pipeline} \cdot \eta_{turbine} \cdot \eta_{generator} \cdot \eta_{transformer} = 0.843$, $\eta_{charge} = \eta_{transformer} \cdot \eta_{motor} \cdot \eta_{pump} \cdot \eta_{pipeline} = 0.843$, and $\eta_{overall} = \eta_{discharge} \cdot \eta_{charge} = 0.843 \cdot 0.843 = 0.71$.

(j) $I_{phase\_generator} = 3V_{ph} \cdot I_{ph}$, where

$$I_{ph\_generator} = \frac{S_{generator}}{3V_{ph\_generator}} = \frac{96.92\ \text{MVA}}{3 \cdot 12.4\,\text{kV}/\sqrt{3}} = 4512.5\ \text{A}.$$

(k)

$$Z_{base} = \frac{V_{ph}}{I_{ph}} = \frac{12,400/\sqrt{3}}{4,512.5} = 1.587\ \Omega.$$

(l) From Fig. 12.60 follows the turns ratio

$$\frac{N_p}{N_s} = \frac{12,400/\sqrt{3}}{115,000} = 0.0623$$

or

$$\frac{N_s}{N_p} = 16.06 \approx 16\ \text{turns}.$$

(m) The construction cost is $87.23 \cdot 10^3$ kW·\$4000/kW = $\$348.92 \cdot 10^6$, the earnings per day are $87.23 \cdot 10^3$ kW·6.25h·\$0.20/kWh=\$109.04k/day, the expenses for

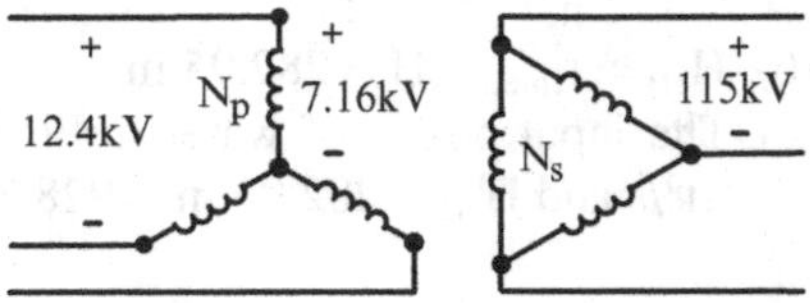

Fig. 12.60 Y-Δ transformer

pumping per day are 56,180 kW·11.25h·$0.06/kWh=$37.92k/day, and the payback period is then obtained from the equation $348.92·10^6+$37.92·10^3·365·y = 109.04·10^3·365·y or y = 13.5 years.

12.6 Photovoltaic Power Plants

Various concepts for photovoltaic power plants are presented: the central power-station type, the roof-top type (see Problem 3.4) and the configuration where the storage device (e.g., battery) is matched with the characteristic of the solar plant without taking recourse to a maximum-power tracker. Solar arrays can be also matched with electric motors driving water pumps. This is a very robust configuration where the electronic components are minimized.

Application Example 12.28: Steady-State Characteristics of a Battery Powered by Solar Cells

Figure 12.61 shows the circuit diagram of a battery directly connected to a solar generator (array) and Fig. 12.62 depicts the 12 V lead-acid battery and solar-power plant characteristics. The operating points (intersections of battery and solar panel

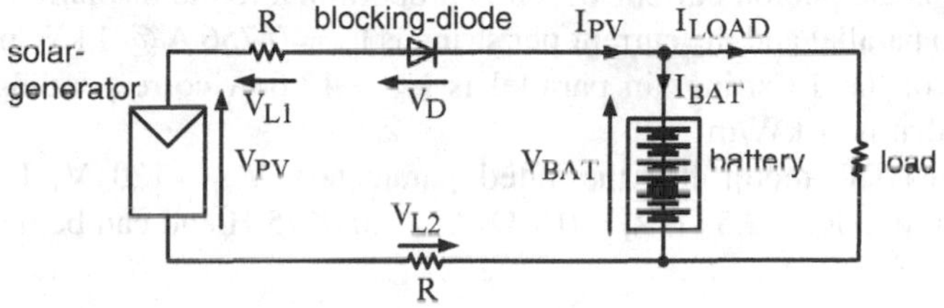

Fig. 12.61 Circuit diagram of lead-acid battery directly connected to solar power plant . V_{PV} is the panel voltage

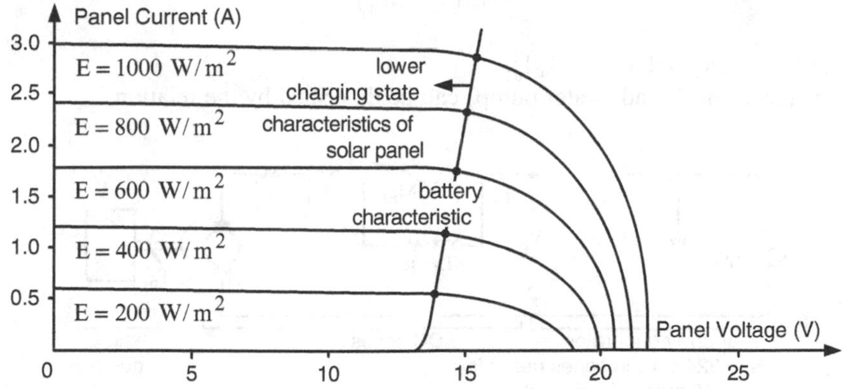

Fig. 12.62 Operating points of a lead-acid battery directly connected to a solar plant (requires start-up time of a few 60 Hz cycles). E is the insolation

characteristics) can be made rather close to the maximum power points of the solar power plant [22]. Note that a 12 V lead-acid battery has a rated voltage of $V_{BAT} = E_0 = 12.6$ V, if fully charged. A voltage of $V_{BAT} = E_0 = 11.7$ V if almost empty, and a voltage of $V_{BAT} = V_{BAToc} + V_{BATover} = 14$ V if fully charged, that is, the voltage drops across the electrochemical capacitance C_B and the nonlinear capacitance C_b ($V_{BATover}$) of battery cells are included (see Chap. 3).

Application Example 12.29: Steady-State Characteristics of a DC series Motor Powered By Solar Cells

The performance of a DC series motor supplied by a solar cell array (Fig. 12.63) is to be analyzed [23], where load matching (solar array, DC series motor and mechanical load) plays an important role. In this case there is no need for a peak-power tracker because the load matching provides inherent peak-power tracking for steady-state operation.

The solar array is governed by the relation

$$V_g = -0.9 I_g + \frac{1}{0.0422} \ln\left(1 + \frac{I_{phg} - I_g}{0.0081}\right), \tag{12.33}$$

where V_g is the output voltage, and I_g is the output current of the solar array. The current I_{phg} is the photon current which is proportional to the insolation. There are 18 strings in parallel and the current per string is $I_{ph} = 0.756$ A @ 1 kW/m^2. That is, the current of the 18 strings in parallel is $I_{phg} = 13.6$ A corresponding to 100% insolation, that is, 1 kW/m^2.

The series DC motor has the rated parameters $V_{rat} = 120$ V, $I_{rat} = 9.2$ A, $n_{rat} = 1{,}500$ rpm, $R_a = 1.5\ \Omega$, $R_f = 0.7\ \Omega$, $M_{af} = 0.0675$ H and can be described by the relation

$$\omega = \frac{V_g - (T_m/M_{af})^{1/2} R}{M_{af}(T_m/M_{af})^{1/2}}, \tag{12.34}$$

where $R = R_a + R_f$ and $T_m = M_{af} I_a^2$.

The mechanical load (water pump) can be described by the relation

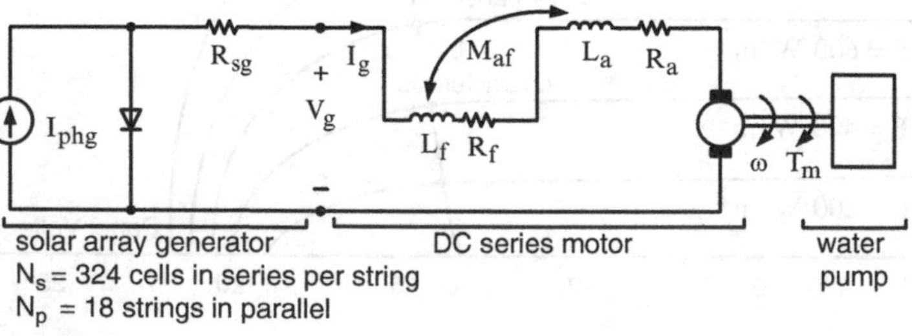

Fig. 12.63 Solar array powering a DC series motor which drives an (irrigation) water pump

$$T_m = T_{m1} = A + B\,\omega + T_{L1} = 4.2 + 0.002387\ \omega, \qquad (12.35)$$

where $A = 0.2$ Nm, $B = 0.0002387$ Nm/(rad/s), and $T_{L1} = 4.0$ Nm.

(a) Plot the solar array characteristic I_g vs. V_g for three insolation levels of 100%, 80%, and 60% using (12.33) with I_{phg} as parameter.
(b) Plot the DC series motor torque relation (12.34) and the water pump torque relation (12.35) in the angular velocity (ω) versus torque (T_m) plane
(c) Using the relations $V_m = V_g = M_{af} I_a \omega + R I_a$, and $T_m = M_{af} I_a^2$, where $I_a = I_g$ plot the load line T_{m1} in the $I_g - V_g$ plane. Is there a good load matching?
(d) Calculate the input power of the series DC motor P_{in} from V_{rat} and I_{rat}.
(e) Calculate the power delivered to the pump by the solar array P_{solar} at 100%, 80%, and 60% insolation.
(f) How does P_{solar} compare with the input power of the series DC motor P_{in} for the three different insolation levels?

Solution

(a) The total solar array current generated (g) is at 100% insolation $I_{phg} = 0.756$ A·18=13.6A, at 80% insolation $I_{phg} = 0.8 \cdot 0.757 \cdot 18 = 10.9$A, and at 60% insolation $I_{phg} = 0.6 \cdot 0.756 \cdot 18 = 8.16$A.

Introducing $I_{phg} = 13.6$ A into (12.33) yields Table 12.8. Correspondingly one gets for $I_{phg} = 10.9$ A and $I_{phg} = 8.16$ A Tables 12.9 and 12.10, respectively.

The $I_g - V_g$ characteristics of Tables 12.8, 12.9, and 12.10 are graphically displayed in Fig. 12.64

(b) With (12.34) and (12.35) one obtains the relations

$$I_g = I_a = \sqrt{\frac{T_m}{M_{af}}}, \qquad (12.36)$$

Table 12.8 $I_g - V_g$ characteristic for 100% insolation

I_g[A]	0	2	4	6	8	10	12	13	13.6
V_g[V]	175.98	170.42	164.14	156.81	147.78	135.53	114.58	90.63	≈0

Table 12.9 $I_g - V_g$ characteristic for 80% insolation

I_g[A]	0	2	4	6	8	10	10.5	10.9
V_g[V]	170.75	164.14	156.32	146.42	132.22	102.84	83.43	≈0

Table 12.10 $I_g - V_g$ characteristic for 60% insolation

I_g[A]	0	2	4	6	8	8.17
V_g[V]	163.89	155.43	144.35	127.06	64.67	≈0

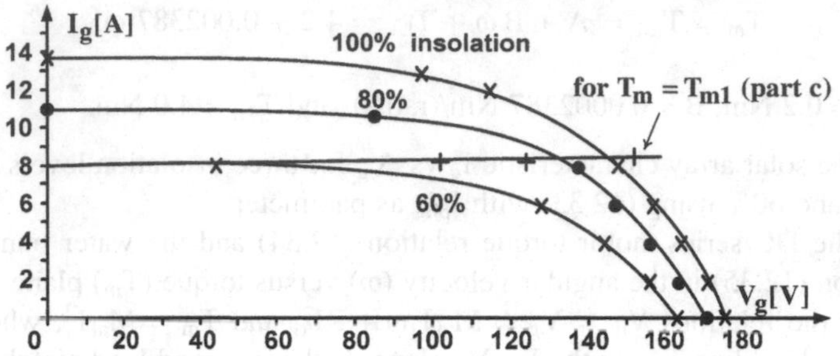

Fig. 12.64 I_g-V_g characteristics for insolation levels of 100%, 80% and 60%, and the mechanical pump torque T_{m1} as requested in part c)

Table 12.11 I_g-V_g, and $T_m-\omega$ characteristics for 100% insolation

I_g[A]	0	2	4	6	8	10	12	13	13.6
V_g[V]	175.98	170.42	164.14	156.81	147.78	135.53	114.58	90.63	≈0
T_m[Nm]	0	0.27	1.08	2.43	4.34	6.75	9.72	11.41	–
ω[rad/s]	∞	1229.8	575.3	354.6	240.7	168.2	108.9	70.69	–

Table 12.12 I_g-V_g, and $T_m-\omega$ characteristics for 80% insolation

I_g[A]	0	2	4	6	8	10	10.5	10.9
V_g[V]	170.75	164.14	156.32	146.42	132.22	102.84	83.43	≈0
T_m[Nm]	0	0.27	1.08	2.43	4.34	6.75	7.44	–
ω[rad/s]	∞	1199.6	546.3	328.9	212.3	119.8	85.12	–

Table 12.13 I_g-V_g, and $T_m-\omega$ characteristics for 60% insolation

I_g[A]	0	2	4	6	8	8.17
V_g[V]	163.89	155.43	144.35	127.06	64.67	≈0
T_m[Nm]	0	0.27	1.08	2.43	4.34	–
ω[rad/s]	∞	1118.74	502.0	281.1	87.2	–

$$T_m = M_{af} I_g^2, \tag{12.37}$$

$$\omega = \frac{V_g - I_g R}{M_{af} I_g}. \tag{12.38}$$

Introducing the values of Tables 12.8, 12.9, and 12.10 into (12.36), (12.37) and (12.38) one obtains the values or Tables 12.11, 12.12, and 12.13 for insolation values of 100%, 80% and 60%, respectively.

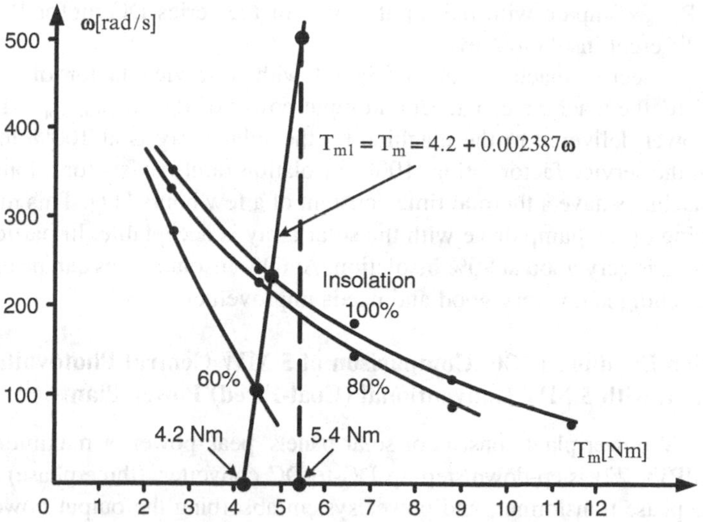

Fig. 12.65 The DC series motor and the water pump torque ω -T_m relation

Table 12.14 Operation of series DC motor with T_m=T_{m1}

I_g [A]	6	8	8.2	8.3	8.4	8.5
V_g [V]	–	44.34	96.13	123.47	151.8	181.2

The DC series motor torque relation as specified by Tables 12.11, 12.12, and 12.13, and the water pump torque relation as given by (12.35) are plotted in Fig. 12.65 in the angular velocity (ω) versus torque (T_m) plane.

(c) Using the relations $V_m = V_g = M_{af}I_a\omega + RI_a$, and $T_m = M_{af}I_a^2$, where I_a=I_g, the load line T_{m1} is plotted in the I_g–V_g plane of Fig. 12.64.
At steady-state operation one obtains for T_m=T_{m1}

$$V_g = \frac{(0.0675)^2 I_g^3 - 0.0675 \cdot 4.2 I_g}{0.002387} + 2.2 I_g = 1.908(I_g)^3 - 116.57 I_g.$$

Table 12.14 represents some of the functional values which are plotted in Fig. 12.64.

The load matching as seen from Fig. 12.64 is very good for 80% insolation only.

(d) Calculation of the input power of the series DC motor P_{in} from V_{rat} and I_{rat}: The rated input power is $P_{in\ rat} = V_{rat} I_{\ rat} = 120$ V·9.2 A = 1.104 kW.

(e) Calculation of the power delivered to the pump by the solar array P_{solar} at 100%, 80%, and 60% insolation: The input power of the DC motor is for 100, 80, and 60% insolation $P_{in\ 100\%} = 145$ V·6 A = 1.25 kW, $P_{in\ 80\%} = 132.$V · 8.5 A = 1.12 kW, and $P_{in\ 60\%}$ =44.34 V·8 A = 0.355 kW, respectively.

(f) Does P_{solar} compare with the input power of the series DC motor P_{in} for the three different insolation levels?

Most electric machines are designed with a service factor of 1.15 and, therefore, the machine can accept an input power of $P_{in\ service\ factor} = 1.27$ kW. The power delivered to the machine by the solar array is at 100% insolation within the service factor rating. 100% insolation rarely exists for a longer time and machines have a thermal time constant of a few hours [11]. This means the matching of the pump drive with the solar array is acceptable. In particular the matching is very good at 80% insolation. At 60% insolation, as can be expected, the matching is not very good and needs improvement.

Application Example 12.30: Comparison of 5 MW Central Photovoltaic (PV) Power Plant with 5 MW Conventional (Coal-Fired) Power Plant

A central PV power plant consists of solar panels, peak-power or maximum power tracker (MPT) [22], step-down/step-up DC-to-DC converter, (three-phase) inverter, Y-Δ three-phase transformer, and power system absorbing the output power of the transformer $P_{out\ transformer} = 5$ MW. Note: 5 MW can provide power to 1,800 residences.

Performance of 5 MW PV plant:

(a) Draw a block diagram of the above-mentioned components.
(b) If the power efficiencies of the maximum power tracker, step-down/step-up DC-to -DC converter, inverter, and transformer are $\eta = 90\%$ each, what is the required maximum power output of the solar array ($P_{solar\ array\ max}$) for a transformer output power of $P_{out\ transformer} = 5$ MW?
(c) Provided the efficiency of the solar cells is $\eta_c = 15\%$, what is the insolation power ($P_{Qs\ max}$) required?
(d) The solar array consists of 43,000 solar panels each having an area of $(0.8 \times 1.60)\ m^2$. What is the total area of all solar panels ($area_{total\ array}$), and what is the maximum insolation required (Q_s measured in kW/m^2) at the location of this photovoltaic plant?
(e) What is the payback period (in years) if 1 kW installed output power capacity of the 5 MW PV plant costs $4,000 provided the average price of 1 kWh during the future 15 years is $0.20? Note: The fuel costs are zero and the operational costs are negligible; you may assume 6 h operation of the plant per day at 80% of power capacity (0.8×5 MW $= 4$ MW).

Some useful data pertaining to a 5 MW PV plant:
Construction timeframe: 1 year
Payback period: 10–20 years, depending upon rebates
Warranty: 25 years limited warranty
Exposure to hail, snow and wind loading: no damage whatsoever
efficiency (photovoltaic): 10–15% reduction of efficiency through aging: 0.5–0.75% per year
power factor: unity-power factor, there is no power-factor control
Number of AC/DC inverters required: 10–20

Peak-power tracking: on-line search for maximum power
Bypass diodes: 18 solar cells are bypassed by one diode, that is, for 72 solar cells in series four bypass diodes are used
Performance of 5 MW conventional (coal-fired) plant:

(f) For a 5 MW conventional (coal-fired) power plant the cost for 1 kW installed power capacity is $2,000. What is the payback period if the utility has to pay for 1 kWh fuel and operational costs of $0.075, and the utility charges its customers $0.20/kWh? You may assume 24 h operation of the plant per day at 80% of power capacity (0.8×5 MW = 4 MW).

Solution

(a) Block diagram of PV plant is shown in Fig. 12.66.
(b) Maximum output power of solar array

$$P_{\text{solar array max}} = \frac{P_{\text{out transformer}}}{\eta_t \cdot \eta_i \cdot \eta_{con} \cdot \eta_{MPT}} = \frac{5\text{ MW}}{(0.9)^4} = 7.62\,\text{MW}.$$

(c) The required insolation power at a solar cell efficiency of $\eta_{cell} = 0.15$: $P_{Qsmax} = \frac{P_{\text{solar array max}}}{\eta_{cell}} = \frac{7.62\text{ MW}}{0.15} = 50.81$ MW.
(d) Required solar array area

$$\text{area}_{\text{total array}} = 43{,}000 \cdot 0.8 \cdot 1.6 = 55.04 \cdot 10^3\,\text{m}^2$$

resulting in the maximum insolation level

$$Q_{smax} = \frac{P_{Qsmax}}{\text{area}_{\text{total array}}} = \frac{50.81 \cdot 10^3\,\text{kW}}{55.04 \cdot 10^3\,\text{m}^2} = 0.92\,\text{kW/m}^2.$$

(e) Payback period of the PV plant in years
Construction cost: $\text{cost}_{1kW} = \$4$k per 1 kW installed power capacity leads to $\text{cost}_{5MW} = \$20$ M.

The earnings are based on $\text{price}_{1kWh} = \$0.20$. At 80% power capacity = 4 MW the earnings per y years are earning = 4k·0.20·365·6·y = \$1.752M·y or the payback period is y = 20/1.752 = 11.42 years neglecting any tax, rebates or interest payments.
(f) Payback period of the coal-fired plant in years

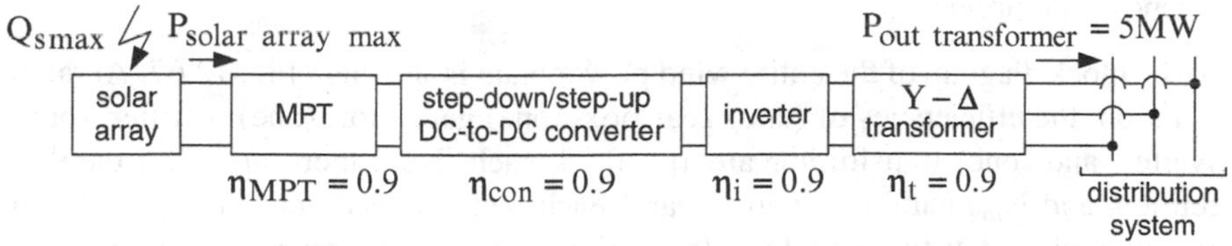

Fig. 12.66 Block diagram of 5 MW PV plant

Construction cost $cost_{1kW} = \$2k$ per 1 kW installed power capacity leads to $cost_{5MW} = \$10M$.

There are fuel and operational costs for y years: cost $_{fuel+operation} = \$0.075 \cdot 4k \cdot 365 \cdot 24 \cdot y = \$2.628M \cdot y$.

The earnings are based on $price_{1\ kWh} = \$0.20$. At 80% power capacity = 4 MW the earnings per y years are earning = $4k \cdot 0.20 \cdot 365 \cdot 24 \cdot y = \$7.01M \cdot y$ or the payback period is $y = 10/4.382 = 2.28$ years neglecting any (carbon) tax or interest payments. Note this payback period would increase if there were a carbon tax. On the other hand, the solar plant does not generate during a 6 h period 4 MW because of varying insolation.

12.7 Wind Power Plants

The application of wind power generation is demonstrated by designing a large plant for central power station applications and small plants for stand-alone micro-grids. In most applications the variable-speed concept is utilized requiring inverters/rectifiers for the generation of the appropriate output voltages.

Application Example 12.31: Design of a 5 MW Variable-Speed Wind-Power Plant to Operate at an Altitude of 1,600 m

Design a wind-power plant feeding the rated power $P_{out} = 5$ MW at unity power factor into the three-phase distribution system at a line-to-line voltage of $V_{L-L\ system} = 12.47$ kV. The wind-power plant consists of

- One (Y-grounded/Δ) three-phase, step-up transformer, N_{inv} parallel-connected three-phase PWM inverters with an input voltage of $V_{DC} = 600\ V + 600\ V = 1{,}200$ V, where each inverter delivers an output AC current $I_{phase\_inverter}$ at unity power factor $\cos\Phi = 1$ to the low-voltage winding of the transformer (e.g., the angle Φ between the line-to-neutral voltage of the inverter $V_{L-n\_inverter}$ and the phase current of the inverter $I_{phase\_inverter}$ is zero). Note that the midpoint of the DC inverter voltage is grounded or represents a virtual ground
- N_{rect} (=N_{inv}) three-phase rectifiers, each one equipped with one self-commutated switch and six diodes operating at a duty cycle of $\delta = 50\%$. Note that each rectifier feeds one inverter
- One synchronous generator
- One mechanical gear and
- One wind turbine

The block diagram of the entire wind power plant is shown in Fig. 12.67. At rated operation, the efficiencies of (one) gear box, (one) generator, (one) rectifier, (one) inverter, and (one) transformer are $\eta = 0.95$, each. Note there are N_{rect} parallel rectifiers and N_{inv} parallel inverters, and each rectifier feeds one inverter. At an operation of less than rated load the efficiencies will be smaller. The parallel

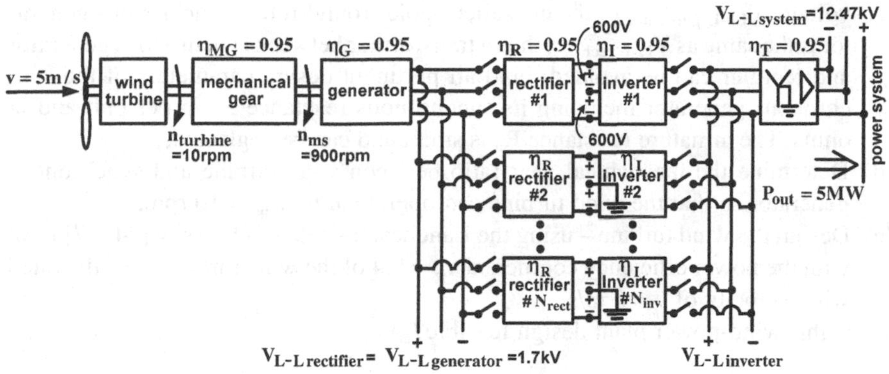

Fig. 12.67 Block diagram of a 5 MW wind-power plant feeding power into the utility distribution system

configuration of inverters and rectifiers permits an efficiency increase, because at light and medium loads only a few inverters and rectifiers must be operated and some can be disconnected. Rated operation is when all rectifiers and inverters are connected.

(a) For rated operation determine the output power of each component, e.g., transformer, inverters, rectifiers, generator, gear box, and wind turbine.
(b) Determine for a modulation index of $m = 0.8$ the inverter output line-to-neutral voltage $V_{L-n\_ inverter}$ such that the inverter can deliver at unity power factor an approximately sinusoidal current to the transformer. N_{inv} commercially available three-phase PWM inverters are connected in parallel and one inverter has the output power rating $P_{inverter} = 500$ kW. What is the number of inverters required? Determine $I_{phase\_inverter}$ of one inverter, the resulting output current of all N_{inv} inverters, that is, $\Sigma\ I_{phase\_inverter}$, and the transformer ratio $a = (N_s/N_p)$ of the (Y-grounded/Δ) step-up transformer, where N_p is the # of turns of the grounded Y per phase and N_s is the # of turns of the Δ per phase. For your calculations you may assume (one) ideal transformer and an ideal power system, where all resistances and leakage inductances are neglected. However, the resistances of the transformer are taken into account in the efficiency calculation of the transformer. Why do we use a Y-grounded/ Δ transformer configuration?
(c) Each of the N_{inv} inverters is fed by one three-phase rectifier with one self-commuted PWM-operated switch (IGBT) and 6 diodes, and all N_{rect} three-phase rectifiers are fed by (one) three-phase synchronous generator with $p = 8$ poles at $n_s = n_{ms} = 900$ rpm, $C = 3.5$ kWmin/m^3 [5, 6], $D_{rotor} = 1.0$ m, $f = 60$ Hz, $B_{sr} = B_{max} = 0.7$ T, $k_{ws} = k_{wr} = 0.8$, $\delta_r = 30°$, $N_f = 800$ or $(N_f/p) = 100$, $k_{fe} = 0.95$, and $g = 0.01$ m The input line-to-line voltage $V_{L-L\ rectifier} = 1.7$ kV of one three-phase rectifier results at a duty cycle of $0 < \delta < 1.0$ in the DC output voltage (of one rectifier, including filter) $V_{DC} = 1{,}200$ V. The line-to-line output

voltage $V_{L-L\ generator}$ of the nonsalient-pole (round-rotor) synchronous generator is the same as $V_{L-L\ rectifier}$; thus a transformer between synchronous generator and rectifier can be avoided. Find all pertinent design parameters of the synchronous generator including its synchronous reactance X_s in per unit and in ohms. The armature resistance R_a is small and can be neglected,

(d) Determine the mechanical gear ratio between wind turbine and synchronous generator so that the wind turbine can operate at $n_{turbine} = 10$ rpm.

(e) Design the wind turbine – using the Lanchester–Betz–Joukowsky [24–27] limit with the power efficiency coefficient $c_p = 0.4$ of the wind turbine – for the rated wind velocity of $v = 5$ m/s.

(f) Is this wind-power plant design feasible?

Solution

(a) The output powers of the power plant components are:

$$P_{out}^{transformer} = 5\,\text{MW}, \qquad P_{out}^{N\_inverters} = \frac{P_{out}^{transformer}}{0.95} = 5.26\,\text{MW},$$

$$P_{out}^{N\_rectifiers} = \frac{P_{out}^{N\_inverters}}{0.95} = 5.54\,\text{MW}, \qquad P_{out}^{generator} = \frac{P_{out}^{N\_rectifiers}}{0.95} = 5.83\,\text{MW},$$

$$P_{out}^{gear} = \frac{P_{out}^{generator}}{0.95} = 6.14\,\text{MW}, \qquad P_{out}^{wind\_turbine} = \frac{P_{out}^{gear}}{0.95} = 6.46\,\text{MW}.$$

(b) The output phase (line-to-neutral) voltage of the inverter is as a function of the input DC voltage

$$V_{L-N\_inverter} = m\frac{\frac{V_{DC}}{2}}{\sqrt{2}} = 0.8\frac{1200/2}{\sqrt{2}} = 340\,\text{V}$$

The number of inverters required is

$$N_{inverter} = \frac{P_{out}^{N\_inverters}}{P_{inverter}} = \frac{5,260\,\text{kW}}{500\,\text{kW}} \approx 11 \text{ inverters.}$$

The phase current of one inverter is

$$I_{phase\_inverter} = \frac{P_{inverter}/3}{V_{L-N\_inverter}} = \frac{500\,\text{kW}/3}{340\,\text{V}} = 490\,\text{A},$$

and the resulting output current of the 11 inverters is $\Sigma I_{phase\_inverter} = 11 \cdot I_{phase\_inverter} = 5.392$ kA or the total inverter phase current can be obtained from

$$\frac{P_{out}^{N\_inverters}/3}{V_{L-N\_inverter}} = \frac{5,260\,\text{kW}/3}{340\,\text{V}} = 5.16\,\text{kA}.$$

The transformer ratio is

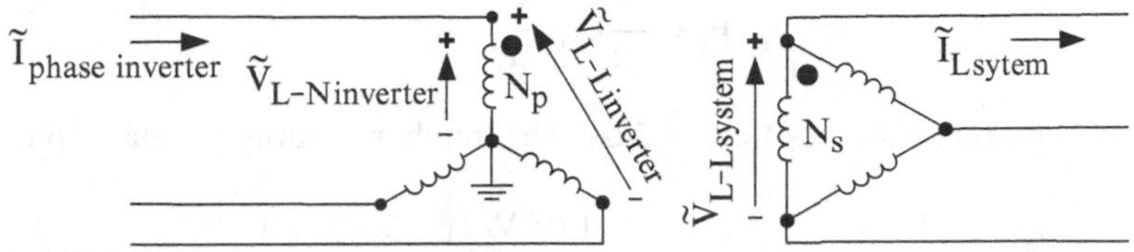

Fig. 12.68 Y-grounded/Δ transformer

$$a = \frac{N_s}{N_p} = \frac{V_{L-L\,system}}{V_{L-N\_inverter}} = \frac{12.47\,k}{340} = 37.$$

The transformer connection is illustrated in Fig. 12.68.
The Y-grounded /Δ transformer is used to avoid zero-sequence components entering the power system.

(c) The line-to-line voltage of the rectifier is $V_{L-L\,rectifier} = 1.7\,kV = V_{L-L\,generator}$. The phase voltage of the generator connected in Y is $V_{phase\,generator} = \frac{V_{L-L\,generator}}{\sqrt{3}} = \frac{1.7\,kV}{\sqrt{3}} = 982\,V$, and the (synchronous) angular velocity of the generator is $\omega_s = \omega_{ms} = \frac{2\pi \cdot 900}{60} = 94.245\,rad/s$. The generator output power is $P_{out}^{generator} = 5.83\,MW = C \cdot D^2 \cdot L_i \cdot n_{ms}$ or the ideal length of the generator stator core is

$$L_i = \frac{P_{out}^{generator}}{C \cdot D^2 \cdot n_{ms}} = \frac{5,830\,kW}{3.5 \cdot 1 \cdot 900} = 1.85\,m$$

and the actual length of the core is then

$$L_{actual} = \frac{L_i}{k_{fe}} = 1.95\,m$$

with the generator torque

$$T_{rated} = \frac{P_{out}^{generator}}{\omega_{ms}} = \frac{5,830\ kW}{94.245\,rad/s} = 61.86\,kNm.$$

With the area per pole

$$area_p = \frac{2\pi \cdot D_{rotor} \cdot L_i}{2 \cdot p} = \frac{2\pi \cdot 1 \cdot 1.85}{2 \cdot 8} = 0.727\,m^2$$

one obtains for the rotor magnetomotive force (mmf)

$$F_r = \frac{(-)T_{rated}}{(\frac{p}{2})^2 \cdot area_p \cdot B_{sr} \cdot \sin\ \delta_r} = \frac{61.86\,k}{(4)^2 \cdot 0.727 \cdot 0.7 \cdot 0.5} = 15.2\,kAt.$$

Or with

$$F_r = \frac{4}{\pi} k_{wr} (\frac{N_f}{p}) \cdot I_f$$

the field current becomes I_f=149.22A. The number of turns per stator phase is

$$N_{s\_phase} = \frac{E_{s\_phase}}{4.44 \cdot f \cdot k_{ws} \cdot B_{sr} \cdot area_p} = \frac{1.05\, V_{L-Lgenerator}/\sqrt{3}}{4.44 \cdot 60 \cdot 0.8 \cdot 0.7 \cdot 0.727} \approx 10 \text{ turns/phase}$$

The self-inductance of phase "a" is

$$L_{aa} = \frac{\mu_o \frac{4}{\pi} k_{ws} (N_{s\_phase})^2 area_p}{p \cdot g} = \frac{4\pi \cdot 10^{-7} \frac{4}{\pi} \cdot 0.8 \cdot 100 \cdot 0.727}{8 \cdot 0.01} = 1.16\,\text{mH}$$

or the self-reactance X_{aa}=2πf·L_{aa}=0.437 Ω. The synchronous reactance is approximately according to Concordia [8]

$X_s \approx 2X_{aa} = 0.875\,\Omega$. With the base values V_{base_phase} = 982 V, P_{base_phase} = S_{base_phase} = 1.94 MW, I_{base_phase} = 1.979 kA Z_{base_phase} = 0.4962 Ω one obtains the per unit synchronous reactance of $X_s^{pu} = 1.76$ pu. This value is acceptable from a transient stability point of view.

(d) The gear ratio is ratio = 900/10 = 90.

(e) The maximum efficiency of a wind turbine is [24–27] $c_p = \eta_{max} = 16/27 = 0.59259$, and the maximum output energy/work of the wind turbine is therefore [24–27] $W_{max}|_{cp=0.59259} = \frac{8}{27} \cdot \rho \cdot A \cdot v^3$.

The air density at 1,600 m altitude is ρ=1.04852(kg-force)/m^3 resulting in $W_{max}|_{cp=0.59259} = \frac{8}{27} \cdot 1.04825 \cdot A \cdot (5)^3 = 6.46 \cdot 10^3$ kW or in the area A swept by the wind turbine rotor blades

$$A = \frac{6.46 \cdot 10^6}{\frac{8}{27}(1.0485) \cdot (5)^3} = 166,352\,\text{m}^2$$

or a turbine blade radius (length) of

$$R = \sqrt{\frac{A}{\pi}} = \sqrt{\frac{166,352}{\pi}} = 230\,\text{m}$$

For an efficiency of c_p = 0.4 one obtains the required power relation

$$W_{max}|_{cp=0.4} = W_{max}|_{cp=0.59259} \frac{0.59259}{0.4} = 9.57\,\text{MW}$$

or the required area swept by the rotor blades

$$A_{required} = \frac{9.57\,\text{MW}}{\frac{8}{27}(1.0485) \cdot (5)^3} = 246,446\,\text{m}^2$$

or a rotor blade length of

$$R_{required} = \sqrt{\frac{246,446}{3.1414}} = 280\,\text{m}$$

(f) This long length of a rotor blade ($R_{required}$) is not feasible. The rotor-blade length can be reduced for the given output power of 5 MW by eliminating the gear -as has been demonstrated by the Enercon Co and the Siemens Co of Germany based on permanent-magnet synchronous machine designs- and permitting a larger rated wind speed (e.g., v = 8 m/s).

Application Example 12.32: Design of a 10 kW Wind-Power Plant at Sea Level and at an Altitude of 1,500 m

Design a wind-power plant feeding $P_{out}^{transformer} = 10\,\text{kW}$ into the distribution system at a line-to-line voltage of $V_{L-L}^{system} = 12.47$ kV where some of the components (e.g., generator, inverter, transformer) are designed for unity power factor cosΦ = 1. The wind-power plant consists of wind turbine, gear, generator, rectifier and a three-phase Y-Δ transformer between three-phase PWM inverter and distribution system, where the inverter has the DC input voltage $V_{DC} = 600$ V and delivers rated output AC current $I_{ph}^{inverter}$ to the Y-side of the transformer. All components have an efficiency of 95%.

(a) Determine the output power of all components of the wind-power plant. Choose the inverter output line-to-line voltage $V_{L-L}^{inverter}$ such that the inverter can deliver an approximately sinusoidal rated current to the transformer. You may assume a modulation index of m = 0.65.
(b) Determine the design data of the (Y-grounded/Δ) inverter-power system transformer, whereby the Δ-side of the transformer is connected to the power system.
(c) The three-phase rectifier with one self-commuted PWM-operated switch (IGBT) is fed by a three-phase synchronous generator. Choose the input line-to-line voltages $V_{L-L}^{rectifier}$ of the three-phase rectifier such that the DC output voltage of the rectifier (including filter) is $V_{DC} = 600$ V assuming a duty ratio of $\delta = 50\%$.
(d) Choose the generator line-to-line output voltages $V_{L-L}^{generator}$ such that they match with the required input line-to-line voltage of the rectifier$V_{L-L}^{rectifier}$; thus a transformer between generator and rectifier can be avoided. Design the synchronous generator for a per-unit synchronous reactance of $X_s^{pu} = 1.5$pu.
(e) Design the mechanical gear between wind turbine and synchronous generator (with p = 2 poles at n_s=3,600 rpm, C = 1.3 kW min/m^3 [5, 6], $D_{rotor} = 0.2$ m so that a wind turbine can operate at half a revolution per second ($n_{turbine} = 30$ rpm).
(f) The wind turbine – using the Lanchester–Betz–Joukowsky Limit [24–27] for the maximum energy efficiency of a wind turbine as a guideline- - is to be

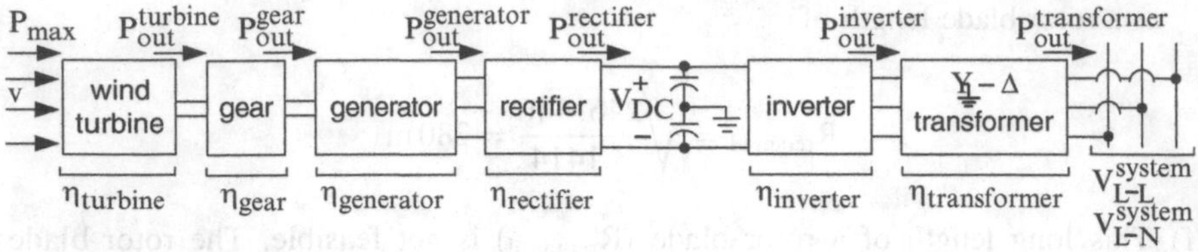

Fig. 12.69 Block diagram of wind-power plant

designed for the rated wind velocity of $v = 10$ m/s. Thereafter, the wind turbine design should be adjusted for a wind turbine efficiency coefficient of $c_p = 0.30$.

Solution

(a) Figure 12.69 illustrates the block diagram of the wind power plant with the follwing outputs powers: $P_{out}^{transformer} = 10\,\text{kW}$,

$$P_{out}^{inverter} = \frac{P_{out}^{transformer}}{0.95} = 10.53\,\text{kW},$$

$$P_{out}^{rectifier} = \frac{P_{out}^{inverter}}{0.95} = 11.1\,\text{kW},$$

$$P_{out}^{generator} = \frac{P_{out}^{rectifier}}{0.95} = 11.7\,\text{kW},$$

$$P_{out}^{gear} = \frac{P_{out}^{generator}}{0.95} = 12.3\,\text{kW},$$

$$P_{out}^{windturbine} = \frac{P_{out}^{gear}}{0.95} = 12.9\,\text{kW}.$$

The line-to-neutral output voltage of the inverter for a modulation index of $m = 0.65$ is

$$V_{L-n}^{inverter} = m\frac{V_{DC}}{2\sqrt{2}} = 0.65\frac{600}{2\sqrt{2}} = 138\,\text{V}.$$

The connection diagram of the inverter, DC source and power system with the associated grounding is depicted in Fig. 12.70.

(b) The Y-grounded/Δ inverter-power system transformer winding diagram is shown in Fig. 12.71, and the associated phasor diagrams are depicted in Fig. 12.72.

If $\tilde{V}_{a-n}^{inverter} = \left|\tilde{V}_{a-n}^{inverter}\right|\angle 0°$ then $\tilde{V}_{a-b}^{inverter} = \left|\tilde{V}_{a-b}^{inverter}\right|\angle 30° = \sqrt{3}\left|\tilde{V}_{a-n}^{inverter}\right|\angle 30°$, and $\tilde{V}_{A-C}^{system} = \left|\tilde{V}_{A-C}^{system}\right|\angle 0°$. The line-to-neutral voltage of the power system is $\tilde{V}_{A-N}^{system} = \left|\tilde{V}_{A-N}^{system}\right|\angle 0° = \frac{\tilde{V}_{A-C}^{system}}{\sqrt{3}}\angle 30°$. The turns ratio of the transformer is

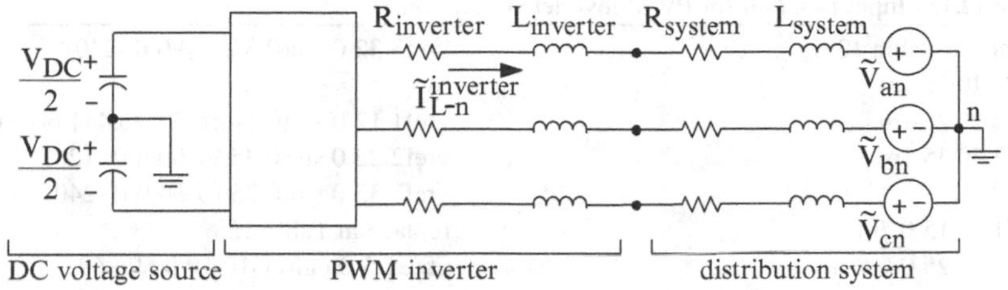

Fig. 12.70 DC voltage source, inverter and power system (PSpice circuit) where the ground of the inverter input and the power system neutral n might be connected to different ground

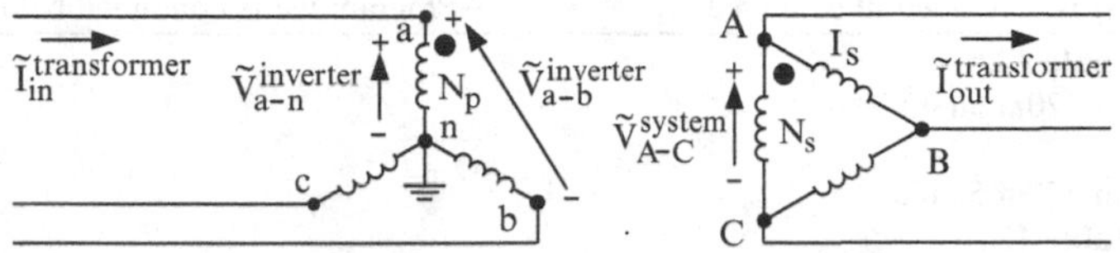

Fig. 12.71 Y-grounded/Δ transformer between inverter and power system

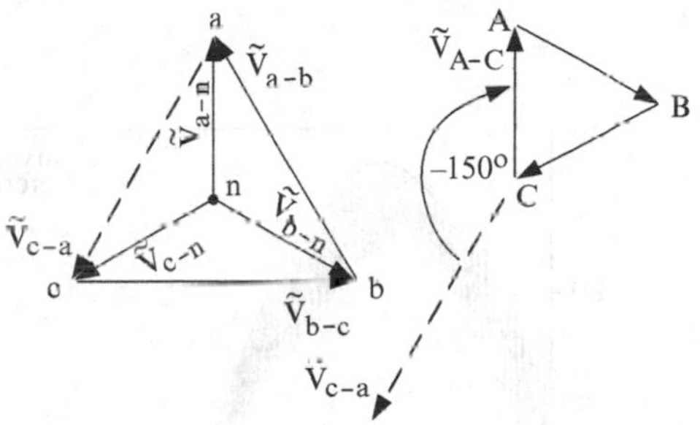

Fig. 12.72 Phasor diagram of voltages for Y-grounded/Δ transformer between inverter and power system

$\frac{N_s}{N_p} = \frac{\tilde{V}_{A-C}^{system}}{\tilde{V}_{a-n}^{inverter}} = \frac{12.47\,k}{138} = 90.36 \approx 90$. The output current of the transformer (see Fig. 12.71) is $\left|\tilde{I}_{out}^{transformer}\right| = \frac{P_{out}^{transformer}}{\sqrt{3}\left|\tilde{V}_{L-L}^{system}\right|} = \frac{10{,}000}{\sqrt{3}\cdot 12470} = 0.463\,A$, and the input current of the transformer is $\left|\tilde{I}_{in}^{transformer}\right| = \frac{P_{out}^{inverter}}{\sqrt{3}\left|\tilde{V}_{L-L}^{inverter}\right|} = \frac{10{,}530}{\sqrt{3}\cdot 239} = 25.44\,A$.

The Pspice program for the inverter is given in Table 12.15.

The calculated inverter current and its reference current are shown in Fig. 12.73.

(c) The block diagram for the rectifier is depicted in Fig. 12.74, where the output current is $I_{out}^{rectifier} = \frac{P_{out}^{rectifier}}{V_{DC}} = \frac{11{,}100}{600} = 18.5\,A$, the corresponding input current is obtained from $P_{out}^{generator} = 11.7\,kW = \sqrt{3}\left|\tilde{V}_{L-L}^{rectifier}\right|\left|\tilde{I}_{in}^{rectifier}\right|$.

The PSpice program for the rectifier listed below yields for the input voltage of the rectifier $\left|\tilde{V}_{L-L}^{rectifier}\right| = 623.52\,V$ an input rectifier current of

Table 12.15 Input program for PWM inverter simulation

```
Replace in Table 12.6
L_W1 10 15 1m
L_W2 20 25 1m
L_W3 30 35 1m
by
L_W1 10 15 0.5m
L_W2 20 25 0.5m
L_W3 30 35 0.5m
replace in Table 12.6
vref1 12 0 sin(0 352 60 0 0 0)
vref2 22 0 sin(0 352 60 0 0 -120)
* Program list is continued in column 2
vref3 32 0 sin(0 352 60 0 0 -240)
by
vref1 12 0 sin(0 {sqrt (2)*25.44} 60 0 0 0)
vref2 22 0 sin(0 35.97 60 0 0 -120)
vref3 32 0 sin(0 35.97 60 0 0 -240)
replace in Table 12.6
vtrial 5 0 pulse(-10 10 0 69u 69u
+ 1u 139u)
by
vtrial 5 0 pulse(-10 10 0 86.5u 86.5u
+ 0.6u 173.6u)
* Program list is continued below table
```

```
replace in Table 12.6
.tran 5u 350m 320m 5u uic
by
.tran 5u 275m 250m 5u uic
replace in Table 12.6
.four 60 99 I(LM1)
by
.four 60 I(LM1)
```

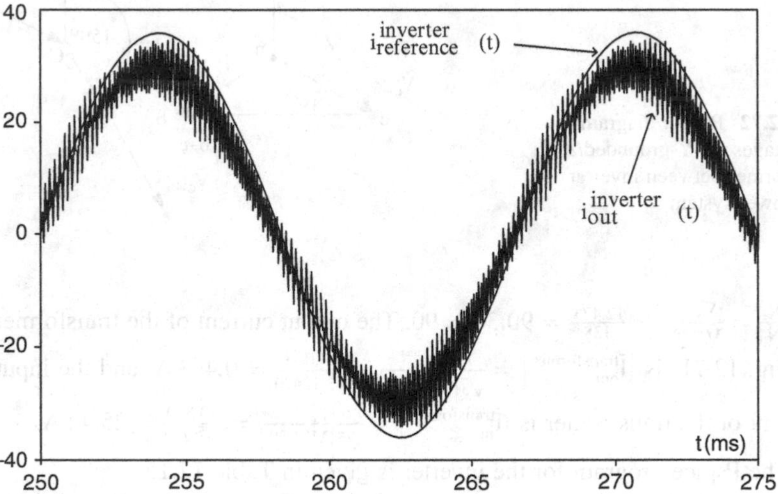

Fig. 12.73 Reference and actual PWM inverter output currents

$\left|\tilde{I}_{in}^{rectifier}\right| = \frac{11,700}{\sqrt{3}\cdot 623.52} = 10.83\,\text{A}$. The PSpice program for rectifier is given in Table 12.16.

The calculated rectifier output voltage and current as well as its input voltage are shown in Fig. 12.75.

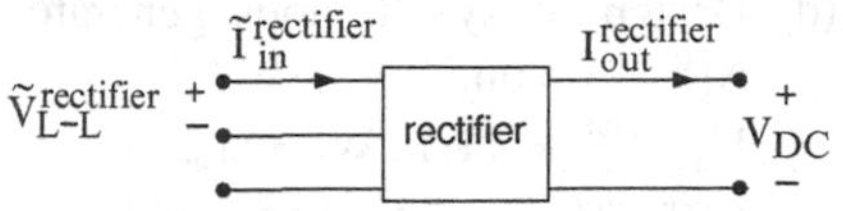

Fig. 12.74 Block diagram of rectifier

Table 12.16 Input program for rectifier simulation

Replace in Table 12.7	by	replace in Table 12.7
VAN 1 0 sin(0 738 60 0 0 0)	VAN 1 0 sin(0 510 60 0 0 0)	Rload 14 10 3.19
VBN 2 0 sin(0 738 60 0 0 −120)	VBN 2 0 sin(0 510 60 0 0 −120)	by
VCN 3 0 sin(0 738 60 0 0 −240)	VCN 3 0 sin(0 510 60 0 0 −240)	Rload 14 10 32.5
* continue program list	*continue program list	*continue program list *below table
*in column #2	* in column #3	

replace in Table 12.7
.options abstol=10u chgtol=10p reltol=0.1 vntol=100m itl4=200 itl5=0
by
.options abstol=10m chgtol=10m reltol=0.1 vntol=100m itl4=200 itl5=0
replace in Table 12.7
.tran 0.5u 350m 300m 0.5m
by
.tran 5u 90m 0m uic
replace in Table 12.7
.four 60 60 I(Lfa)
by
.four 60 I(Lfa)

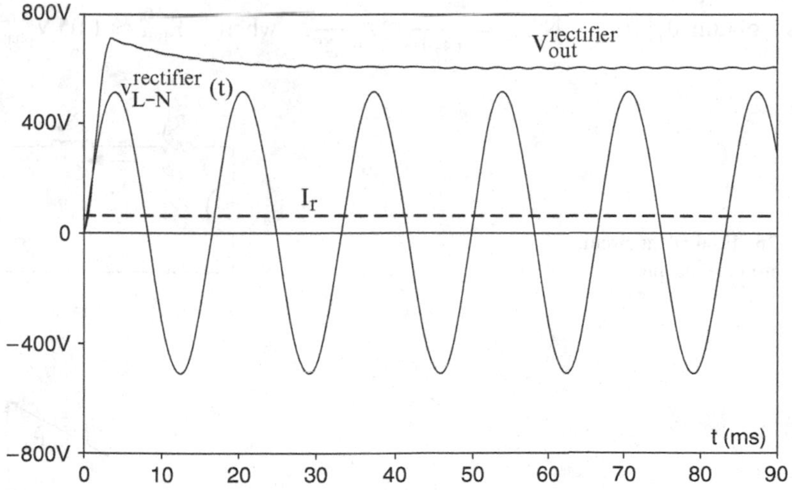

Fig. 12.75 Calculated output voltage $V_{out}^{rectifier}$ and current $I_r = I_{out}^{rectifier}$ of rectifier, and line-to-neutral input voltage of rectifier $v_{L-N}^{rectifier}(t)$

(d) Design of synchronous generator for a per-unit synchronous reactance of $X_s^{pu} = 1.5\,\text{pu}$.

$P_{out}^{generator} = 11.7\,\text{kW} = S_{base}$, $I_{phase}^{generator} = \left|\tilde{I}_{in}^{rectifier}\right| = 10.83\,\text{A} = I_{base}$,

$V_{phase}^{generator} = 360\,\text{V} = V_{base}$,

$$Z_{base} = \frac{V_{base}}{I_{base}} = \frac{360\,\text{V}}{10.83\,\text{A}} = 33.24\,\Omega,$$

$$X_s^{\Omega} = X_s^{pu} \cdot Z_{base} = 49.86\,\Omega$$

$$P_{loss}^{generator} = P_{out}^{gear} - P_{out}^{generator} = (12.3 - 11.7)\,\text{kW} = 600\,\text{W} = 3(I_{phase}^{generator})^2 R_a$$

$$R_a^{\Omega} = \frac{600\,\text{W}}{3(10.83)^2} = 1.70\,\Omega,$$

$$R_a^{pu} = \frac{R_a^{\Omega}}{Z_{base}} = 0.051\,\text{pu}$$

The equivalent circuit of the synchronous machine is illustrated in Fig. 12.76. The phasor diagram for unity power factor is shown in Fig. 12.77. The ideal length of the generator is

$$L_i = \frac{P_{out}^{generator}}{(D_{rotor})^2 \cdot C \cdot n_s} = \frac{11.7\,\text{kW}}{(0.2)^2 \cdot 1.3 \cdot 3,600} = 0.0625\,\text{m}.$$

The actual length including the insulation of the core laminations is $L_{actual} = \frac{L_i}{K_{fe}} = 0.066\,\text{m}$. The number of turns per phase of the stator winding is obtained from $N_{phase}^{stator} = \frac{E_{rms}}{4.44 \cdot f \cdot K_w^{stator} \cdot \Phi_{max}^{stator}}$, where $E_{rms} \approx 1.05\, V_{phase}^{generator}$

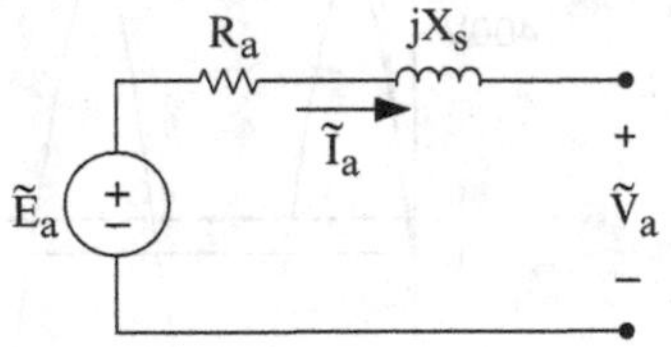

Fig. 12.76 Equivalent circuit of synchronous machine

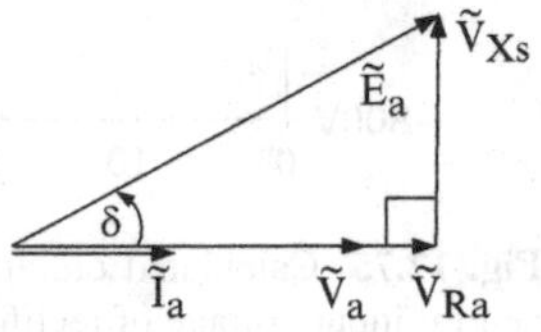

Fig. 12.77 Phasor diagram for unity power factor

$= 1.05 \cdot 360\,V = 378\,V$, the stator winding factor is estimated to be $k_w^{stator} = 0.8$, and $B_{max} = 0.7\,T$ resulting in the stator flux of

$$\Phi_{max}^{stator} = \frac{B_{max} \cdot area_p}{p} = \frac{0.7 \cdot 2\pi \cdot (\frac{D_{rotor}}{2})L_i}{2} = 0.01374\,Wb/pole.$$

The area per pole is $area_p = \frac{\Phi_{max}^{stator}}{B_{max}} = 0.01963\,m^2$ and $N_{phase}^{stator} = 129$ turns/phase. The synchronous reactance is now $X_s = X_{ab} + X_{aa} \approx 2X_{aa}$ yielding the self-inductance $L_{aa} = \frac{X_{aa}}{2\,\pi \cdot f} = 0.06613\,H = 66.13\,mH$.
Determination of the one-sided air gap: The self-inductance is

$$L_{aa} = \frac{\mu_o \frac{4}{\pi} k_w^{stator} (N_{phase}^{stator})^2 \cdot area_p}{p \cdot g}$$

or the air gap length

$$g = \frac{\mu_o \frac{4}{\pi} k_w^{stator} (N_{phase}^{stator})^2 \cdot area_p}{p \cdot L_{aa}} = \frac{4\pi \cdot 10^{-7} \frac{4}{\pi} 0.8 \cdot (129)^2 \cdot 0.01963}{2 \cdot 0.06613} = 0.00315\,m$$
$$= 3.15\,mm$$

Determination of rotor (field) excitation: The electric torque is

$$T_{fld} = -\frac{P_{out}^{generator}}{\omega_s} = -\frac{11.7\,kW}{3,600 \frac{2\pi}{60}} = -31.04\,Nm = -\frac{p}{2} area_p\,B_{max}\,F_r \cdot \sin\,\delta_r$$

with $B_{max} = 0.7$ T and $\delta_r = 30°$ one obtains the rotor mmf

$$F_r = -\frac{T_{fld}}{(p/2) \cdot area_p \cdot B_{max} \cdot \sin\,\delta_r} = \frac{31.04}{1 \cdot 0.01963 \cdot 0.7 \cdot \sin(30°)} = 4.516\,kAt$$

The rotor mmf as a function of the total rotor turns N_f is $F_r = \frac{4}{\pi} k_w^{rotor} (\frac{N_f}{p}) I_f$. With the estimated rotor winding factor $k_w^{rotor} = 0.8$ the product of number of rotor turns times the field current becomes $(N_f I_f) = F_r \cdot p \cdot \frac{\pi}{4} \frac{1}{k_w^{rotor}} = 4.516k \cdot 2 \cdot \frac{\pi}{4} \cdot \frac{1}{0.8} = 8,866\,At$. Selecting $N_f = 100$ turns one obtains a field excitation current of $I_f = 88.66$ A.

(e) Design of mechanical gear is based on the gear ratio $= (n_s/n_{turbine}) = 3,600/30 = 120$.

(f) The air density is different at sea level and at an altitude of 1,500 m. Table 12.17 illustrates the air density ρ as a function of the altitude.

Table 12.17 Air density ρ as a function of the altitude

Altitude [m]	−500	0 (sea level)	500	1,000	1,500	2,000	2,500	3,000
ρ[kg/m^3]	1.2854	1.2255	1.1677	1.112	1.0583	1.0067	0.957	0.9092

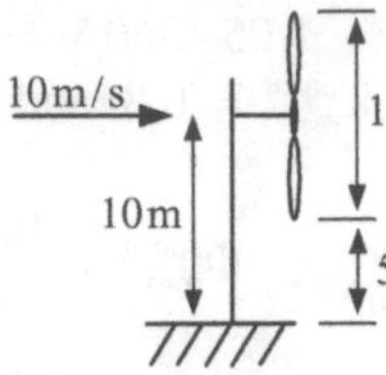

Fig. 12.78 Design dimensions of the wind-power plant

The maximum efficiency of a wind turbine is $c_p = \eta_{WT} = 16/27 = 0.59259$ and the maximum energy/work developed by the wind turbine rotor is $W_{max} = \frac{8 \cdot A \cdot v^3 \cdot \rho}{27}$ where A is the rotor sweep area, v the free stream wind velocity, and ρ the air density [24–27]. From this follows for the sweep area at sea level $A_{sea\ level} = \frac{W_{max} \cdot 27}{8 \cdot v^3 \cdot \rho} = \frac{12{,}900 \cdot 27}{8 \cdot 1{,}000 \cdot 1.2255} = 35.53\,m^2 = \pi \cdot R^2_{sea\ level}$ or $R_{sea\ level} = 3.36\,m$. For a c_p=0.3 one obtains the maximum power of the wind turbine $P_{max} = W_{max}(0.59259/0.30) = 25.48$ kW or a wind sweep area of $A^{required}_{sea\ level} = \frac{P_{max} \cdot 27}{8 \cdot V^3 \cdot \rho} = \frac{25{,}480 \cdot 27}{8 \cdot 1{,}000 \cdot 1.2255} = 70.18\,m^2 = \pi \cdot R^{2\,required}_{sea\,level}$ or $R^{required}_{sea\,level} = 4.73\,m$.

For an altitude of 1,500 m the corresponding values are

$$A_{1{,}500m} = \frac{W_{max} \cdot 27}{8 \cdot v^3 \cdot \rho} = \frac{12{,}900 \cdot 27}{8 \cdot 1{,}000 \cdot 1.0583} = 41.1\,m^2 = \pi \cdot R^2_{1{,}500\,m}$$

or $R_{1{,}500\,m} = 3.62\,m$. For a c_p=0.3 one obtains the maximum power of the wind turbine $P_{max} = W_{max}(0.59259/0.30) = 25.48$ kW or a wind sweep area of

$$A^{required}_{1{,}500\,m} = \frac{P_{max} \cdot 27}{8 \cdot v^3 \cdot \rho} = \frac{25{,}480 \cdot 27}{8 \cdot 1{,}000 \cdot 1.0583} = 81.26\,m^2 = \pi \cdot R^{2 required}_{1{,}500\,m}$$

or $R^{required}_{1{,}500\,m} = 5.09\,m$. The design of the wind tower is illustrated in Fig. 12.78.

Application Example 12.33: Design of a Stand-Alone Wind-Power Plant With a Compressed-Air Storage Facility for a Farm

The block diagram for a stand-alone power plant is shown in Fig. 12.79, consisting of a compressed-air storage facility (wind turbine, gear, compressor with cooler, gas turbine and synchronous generator feeding a rectifier and an inverter). The wind turbine is directly coupled with the synchronous generator, gas turbine, cooler, and compressor through a mechanical gear. During normal operation the wind turbine provides (per day) during an 8-h timeframe (on the average) 5 kW for the farm via synchronous generator, rectifier and inverter. In addition on the average 5 kW is delivered to charge the compressed-air storage tank up to 500 psi (pounds per square inch) at an ambient temperature of 70°C during an 8-h period per day. The cooler's and compressor's energy is supplied by the wind turbine through the synchronous generator via a common shaft. During no-wind conditions the energy is provided by the compressed-air storage tank, a pressure reduction valve from 500 psi to 250 psi, the gas turbine/combustor and the synchronous generator, rectifier and inverter.

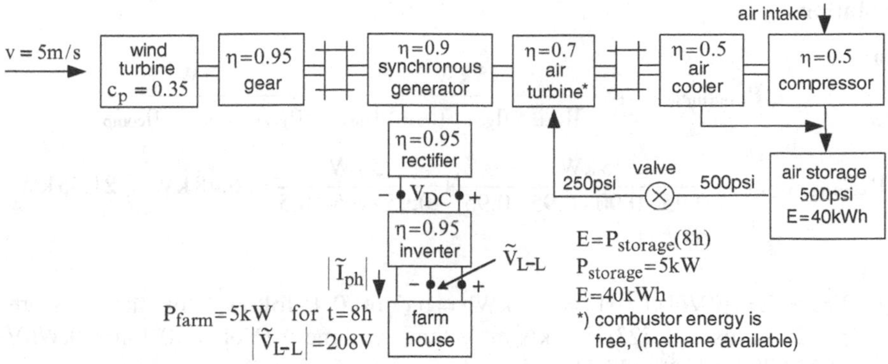

Fig. 12.79 Wind-power plant combined with compressed-air power plant for a farm delivering during wind conditions simultaneously P_{farm}=5 kW to the farm and $P_{storage}$= 5 kW to the compressed-air storage facility. During no-wind conditions the compressed-air facility delivers P_{farm}=5 kW to the farm

(a) For the given efficiencies compute the output power of the wind turbine $P_{wind\ turbine\_out}$ delivering $P_{farm} = 5$ kW to the farm and at the same time $P_{storage} = 5$ kW to the compressed-air storage facility. Note that the efficiency of the cooler and that of the compressor (see Fig. 12.79) takes into account the additional energy required to cool and to compress the air for the air storage.
(b) Compute the volume of the storage tank (Vol) if the stored energy in the compressed-air storage facility is $E = p(\mathrm{Vol})/(\gamma - 1)$, where for air $\gamma = C_p/C_v = 1.4$, (Vol) is the storage volume, $E = 5$ kW·8h = 40 kWh, and the pressure is p = 250 psi (250 psi is the worst case condition, while 500 psi is the best case condition).
(c) During periods when there is no wind, the compressed-air storage supplies to the farm $P_{farm} = 5$ kW via the gas turbine, synchronous generator, rectifier and inverter. How long can energy be provided by the compressed-air tank? Assume that the pressure reduction valve maintains a uniform pressure of 250 psi for the air entering the gas turbine, where the air is heated by combustor.
(d) For $c_p = 0.35$ determine the required radius R of the wind turbine blades at an altitude of 1,500 m.
(e) Calculate the payback period provided the avoided cost for 1 kWh is $0.20, and the construction cost is $8,000 per 1 kW installed power capacity ($P_{farm} + P_{storage} = P_{total} = 10$ kW). For heating up the air in the combustor methane is available on the farm at no cost. No transmission line to the isolated farm has to be built which represents an avoided construction cost of $10,000 per mile of transmission line; assume an avoided line length of 2 miles. Neglect any interest payments.

Hints: 1 N = 1 kg·m/s^2, 1 W = 1 kg·m^2/s^3, 1 h = 3,600 s, and 1 psi = 6.895 kPa = 6.895 kN/m^2 @ 70°C.

Solution

(a)

$$P_{\text{wind turbine\_out}} = \frac{5\,\text{kW}}{\eta_{gear}\cdot\eta_{gen}\cdot\eta_{rect}\cdot\eta_{inv}} + \frac{5\,\text{kW}}{\eta_{gear}\cdot\eta_{cool}\cdot\eta_{comp}}$$

$$P_{\text{wind turbine\_out}} = \frac{5\,\text{kW}}{0.95\cdot 0.90\cdot 0.95\cdot 0.95} + \frac{5\,\text{kW}}{0.95\cdot 0.5\cdot 0.5} = 6.48\,\text{kW} + 21.05\,\text{kW}$$
$$= 27.53\,\text{kW}.$$

(b) From $E = p(Vol)/(\gamma-1) = 40$ kWh $= p\cdot(Vol)/0.4$ follows for the pressure $p = 250$ psi $= 1{,}723.75$ kN/m^2 the volume $(Vol) = (0.4\cdot 40$ kWh)/ (1,723.75 kN/m^2) = 33.415 m^3.

(c) $E_{farm} = 40$ kWh·0.7·0.90·0.95·0.95 = 22.74 kWh resulting in a time during which 5 kW can be supplied $t = E_{farm}/P_{farm} = 4.55$ h.

(d) The wind turbine input power at an air density of $\rho = 1.0583$(kg-force)/m^3 corresponding to an altitude of 1,500 m is obtained [24–27] from $\pi R^2 = A = \frac{P_{\text{wind turbine\_out}}\cdot 27}{8V^3\rho}$, where R is the ideal wind rotor radius and v the wind velocity. From $\pi R^2 = \frac{27.53\,\text{k}\cdot 27}{8(5)^3 1.0583}$ follows $R_{ideal} = R = 14.95$ m or taking $c_p = 0.35$ into account yields [24–27] for $P_{required} = P_{\text{wind turbine \_out}}\cdot(0.593/0.35) = 46.64$ kW or $\pi R^2_{required} = \frac{41.69\,\text{k}\cdot 27}{8(5)^3 1.0583}$ follows $R_{required} = 19.46$ m.

(e) The construction cost are \$8k/kW·10 kW = \$80k.

The avoided cost is (5 kW·8 h+5 kW·4.55 h)·365·\$0.20 = \$4,581 per year, and the avoided investment cost (not to pay for a 2 mile line) is \$20 k.

One obtains from the relation (\$80k – \$20k) = \$4.581k·y the payback period y = 13.1 years.

12.8 Fuel Cell Systems

Fuel cells are a promising substitute for batteries and therefore the behavior of such energy storage devices are explored feeding brushless DC machines as used in electric automobiles, and supplying energy to inverters which are used for distributed generation (DG) supplying the necessary power to the grid to control the frequency.

Application Example 12.34: Production of Hydrogen Based on Electrolysis and its Application with Respect to Electric Cars with Fuel Cells

A well-known way of splitting water (H_2O) is by electrolysis. A container (trough) contains pure (distilled) water, and two electrodes, - - the anode (A) and cathode (C) consisting of either platinum or nickel – connected to a DC current source. Pure H_2O is a poor conductor; the addition of H_2SO_4 to the water makes the solution conducting. Although the solution of H_2O and H_2SO_4 heats up - due to the losses caused by the current flowing from anode to cathode- at the anode oxygen O_2 and at the cathode hydrogen H_2 is accumulated. The efficiency of splitting H_2O based on electrolysis is about $\eta_{electrolysis} = 80\%$, that is, 20 % of the energy is converted to heat within the electrolysis process. The energy density of hydrogen is about 2.28 times that of gasoline ($E_{gasoline} = 12.3$ kWh/kg-force), that is, $E_{H2} = 28$ kWh/kg-force. {Note 1 l (dm^3) of water has a mass m of 1 kg at 4°C, and exerts a (force or) weight of 1 kg-force (or 1 kp) on a scale}. The additional energy loss due to the production of distilled water can be deduced from the definition of 1 kcal. Definition of unit 1 kcal: The energy of 1 kcal is required to heat 1 kg-force of H_2O by 1°C (this is approximately true only because the energy required depends upon the temperature range. It is exactly true heating pure water from 14.5 to 15.5°C). Note that 1 kcal = 4.186 kWs.

(a) The energy of a $P_{out} = 5$ MW wind-power plant can be used to generate hydrogen during 8 h of operation per day during 365 days per year. The construction cost of the 5 MW plant, and the generation of pure water including the electrolysis equipment is $4,000/kW installed power capacity. How much hydrogen energy (expressed in MWh) can be obtained during one year taking into account the above-described losses?
(b) Calculate the payback period provided one kg-force of hydrogen (weight) can be sold for $4.00. For your payback period calculation you may neglect interest for borrowing the money for the construction cost.
(c) How many "equivalent-gasoline gallons" of hydrogen can be produced during one year?
(d) Provided a car owner (on average) travels 15,000 miles per year with an average "mileage" of 40 miles/(gallon "equivalent gasoline"), how many car owners can get the "equivalent gasoline gallons of hydrogen" from the 5 MW plant?

Solution

(a) The hydrogen that can be produced within one year is given by the total energy available from the wind turbine $E_{total} = (5$ MW)(8 h/day)(365 days/year) = 14,600 MWh. The efficiency of electrolysis reduces this energy to $E_{useful} = 0.8\ E_{total} = 11{,}680$ MWh. The energy content of 1 kg of hydrogen is $E_{H2} = 28$ kWh/(kg-force), thus the amount of hydrogen produced is $H_2 = E_{useful}/E_{H2} = 417{,}143$ (kg-force). The atomic weight of hydrogen (H_2) is 2 and that of water (H_2O) is 18, therefore the amount of H_2O required for the electrolysis is $H_2O = 417{,}143/(2/18) = 3.754 \cdot 10^6$ kg-force. The energy loss $E_{distillation}$ due to the production of distilled water is obtained from $E_{1kg-force\ H_2O} = (90°C)(4.186\,kWs/°C)(1\,h/3{,}600\,s) = 0.10465\,kWh$ as $E_{distillation} = (3.754 \cdot 10^6$ kg-force)(0.10465 kWh/kg-force) = 392.9 MWh. The energy available for the electrolysis is now $E_{H2} = E_{useful} - E_{distillation} = 11{,}680$ MWh − 392.9 MWh = 11,287 MWh.
(b) The payback period – neglecting interest payments – is based on the earnings per year and the construction cost. For $H_2 = E_{H2}/(E_{H2\ per\ kg\text{-}force}) = (11{,}287$

$\cdot 10_6$ Wh)/(28 kWh/kg-force) = 403,107 kg-force per year, the earnings per year are Earnings/year = (403,107 kg-force/year)($4.00/kg-force) = $1,612,428/year. The construction cost is $C_{construction}$ = ($4,000)(5,000kW) = $20,000,000 resulting in a payback period of y = ($20,000,000/$1,612,428) = 12.4 years.

(c) With $E_{H2} = 11.287 \cdot 10^6$ kWh, the specific weight of gasoline of $\gamma_{gasoline} = 0.72$ kg-force/l = 2.736 kg-force/gallon, and the energy content of gasoline $E_{gasoline}$ = 12.3 kWh/kg-force the weight of gasoline in kg-force is $Weight_{gasoline} = 11.287 \cdot 10^6$ kWh/(12.3 kWh/kg-force) = 917,642 kg-force or the equivalent gallons of gasoline produced per year are $Gallons_{gasoline}$ = 917,642 kg-force/(2.736kg-force/gallon) = 335,396 equivalent gallons of gasoline per year.

(d) Each car owner uses (15,000 miles/year)/(40 miles/gallon) = 375 gallons/year, that is, (335,396 gallons/year)/(375 gallons/year/owner) = 895 owners. In other words the 5 MW wind-power plant produces enough hydrogen to meet the needs of 895 car owners.

Application Example 12.35: Calculation of the Efficiency of a Polymer Electrolyte Membrane (PEM) Fuel Cell Used as an Energy Source

A PEM fuel cell has the parameters:

Performance: output power = P_{rat} = 2,400 W[1)], output current = I_{rat} = 92 A[1)], DC voltage range = V_{rat} = 25–50 V, operating lifetime: T_{life} = 1,500 h[2)].

Fuel: composition=C=99.99% dry gaseous hydrogen, supply pressure = p = 10–250 PSIG, consumption = V = 37 SLPM[3)].

Operating environment: ambient temperature = t_{amb} = 3°C to 30°C, relative humidity = HR = 0–95%, location = indoors and outdoors[4)].

Physical: length × width × height = 56 × 50 × 33 cm, weight = W = 26 kg-force.

Emissions: liquid water = H_2O = 1.74 l maximum per hour.

1. Beginning of life, sea level, rated temperature range
2. CO destroys the proton exchange membrane
3. At rated power output, SLPM ≡ standard liters per minute (standard flow)
4. Unit must be protected from inclement weather, sand and dust.
   (a) Calculate the power efficiency of a PEM fuel cell in two different ways.
   (b) Find the specific power density of this PEM fuel cell expressed in W/kg-force.
   (c) How does this specific power density compare with that of a lithium-ion battery?

   Hints:

1. The nominal energy density of hydrogen is 28 kWh/(kg-force), which is significantly larger than that of gasoline (12.3 kWh/kg-force). This makes hydrogen a desirable fuel for automobiles,
2. The (weight) density of hydrogen is $\gamma_{hydrogen}$ = 0.0899g-force/l
3. The oxygen atom has eight electrons, eight protons and eight neutrons

Solution

(a) *Atomic-weight method*: Atomic weights of hydrogen (H), oxygen (O), and water (H_2O) is 1, 16, and 18, respectively. If 1.74 l of water are emitted then the weight of hydrogen consumed is $\text{weight}_{H2} = (2/18)(1.74\text{kg-force/h}) = 0.19333$ kg-force/h and the corresponding hydrogen input energy provided is $E_{H2} = (0.19333$ kg-force/h)(28 kWh/kg-force) = 5.41 kWh/h which corresponds to an input power of $P_{in} = 5.4133$ kW. The output power is $P_{out} = 2.4$ kW resulting in a power efficiency of $\eta = P_{out}/P_{in} = 0.443$.
Hydrogen-density method: The input power of hydrogen is $P_{in} = (\gamma_{hydrogen})$(standard flow/h)($E_{H2}$/kg-force) = ($0.0899 \cdot 10^{-3}$ kg-force/l)(37·60 l/h)(28 kWh/kg-force) = 5.59 kWh/h, or $P_{in} = 5.59$ kW yielding a power efficiency of $\eta = P_{out}/P_{in} = 0.43$.

(b) The specific power density per unit of weight is power density = P_{out}/weight = 2,400W/(26kg-force) = 92.3 W/kg-force.

(c) The power density of a lithium-ion battery is 300 W/kg-force. This indicates that the PEM fuel cell is relatively heavy for a given output power.

Application Example 12.36: Transient Performance of a Brushless DC Motor Fed by a Fuel Cell

Replace the battery (with the voltage $V_{DC} = 300$ V) of Fig. 5.49 of Application Example 5.14 of Chap. 5 by the equivalent circuit of the fuel cell as described in Fig. 2 of the paper by Wang et al. [28]. You may assume that in Fig. 2 of the Wang et al. paper the voltage changes $E = 300\ V \pm 30$ V, where the superimposed rectangular voltage ± 30 V has a period of $T \pm 30V = 1$ s. This voltage variation of a fuel cell as a function of time reflects the nonuniform DC voltage generation of a fuel cell. The remaining parameters of the fuel cell equivalent circuit can be extrapolated from Table III of the paper by Wang et al. (e.g., $E = V_{DC} = 300$ V, $R_{conc} = 0.41\ \Omega$, $R_{act} = 6.0\ \Omega$, $R_{ohmic} = 1.42\ \Omega$, $C_{fc} = 0.02$ F).

(a) Perform a PSpice analysis for the phase angles $\theta = -60°, -30°, 0°, 30°, 60°$ and plot phase A machine current i_{MA} from 0.6 to 1.3 ms.

(b) Perform a PSpice analysis for the phase angles $\theta = 60°$ and plot phase A machine current i_{MA} from1 s to 2 s.

Solution

(a) A part of the PSpice input program with .cir extension is listed below and the simulation plots are shown in Figs. 12.80–12.84.

Replace the following five statements in the PSpice program list of Application Example 5.14.

```
*Brushless DC motor drive.
vplus 1 0 150
vminus 2 0 –150
```

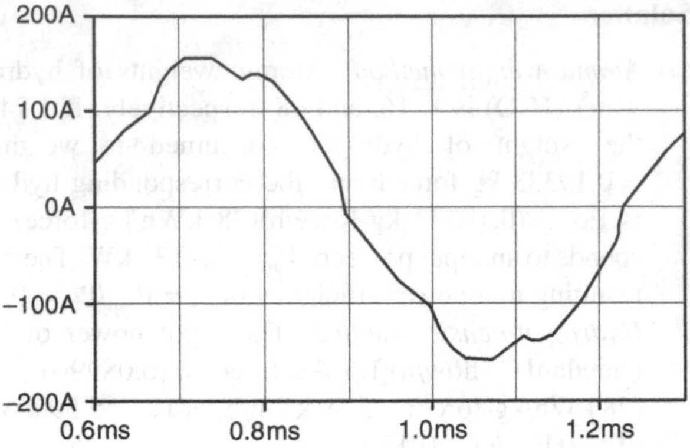

Fig. 12.80 PSpice simulation of the phase A machine current i_{MA} for the phase angle θ= −60° from 0.6 ms to 1.3 ms

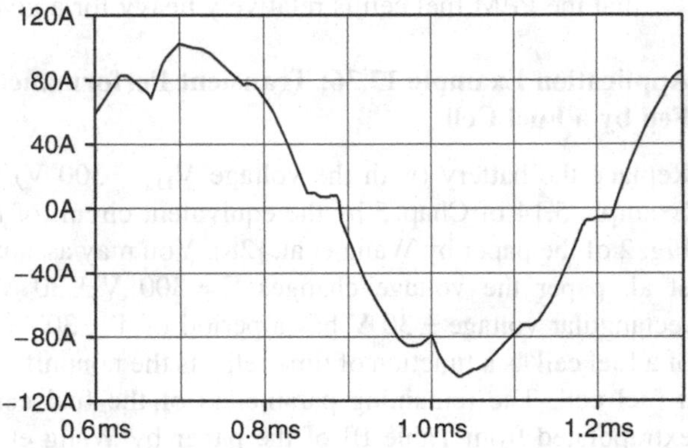

Fig. 12.81 PSpice simulation of the phase A machine current i_{MA} for the phase angle θ= −30° from 0.6 ms to 1.3 ms

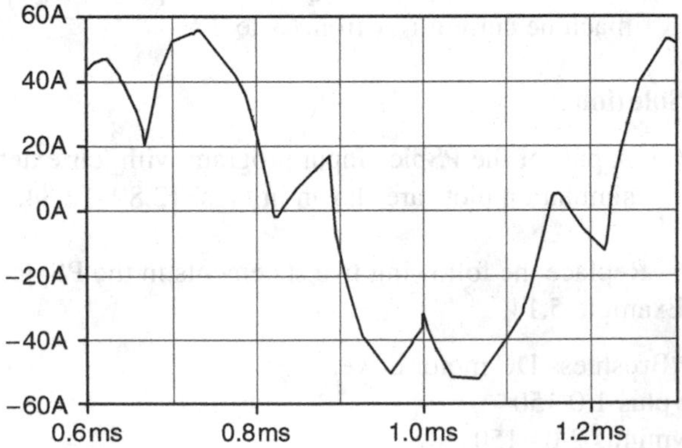

Fig. 12.82 PSpice simulation of the phase A machine current i_{MA} for the phase angle θ= 0° from 0.6 ms to 1.3 ms

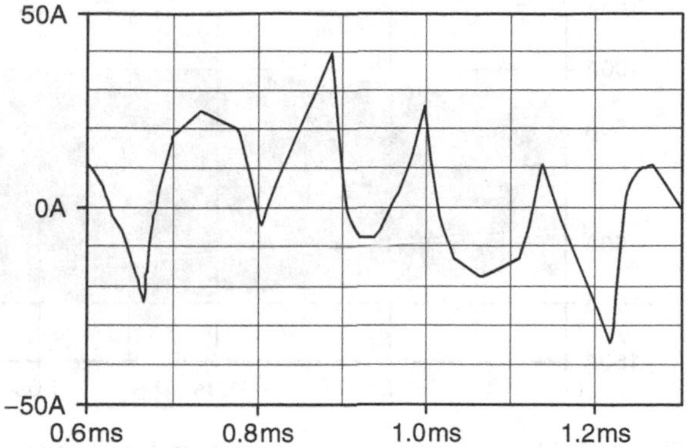

Fig. 12.83 PSpice simulation of the phase A machine current i_{MA} for the phase angle $\theta = 30°$ from 0.6 ms to 1.3 ms

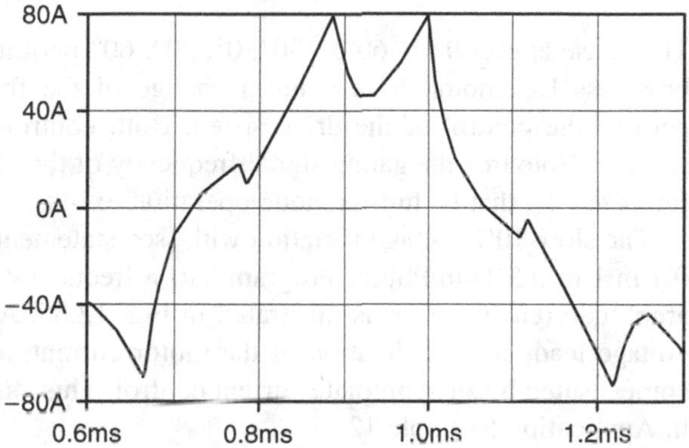

Fig. 12.84 PSpice simulation of the phase A machine current i_{MA} for the phase angle $\theta = 60°$ from 0.6 ms to 1.3 ms

```
c1 0 1 100u
c2 0 2 100u
```

by

```
* Brushless DC motor drive.
VDC 3 2 300
Vpm 3a 3 PULSE (30 -30 0 1m 1m 0.5 1)
Rconc 3a 4 0.41
Ract 4 5 6.0
Cfc 3a 5 0.02
Rohmic 5 1 1.42
Rlarge 2 0 1000Meg
```

The above plots show that the torque control of a brushless DC machine is performed by changing the phases of the stator currents which in turn changes the amplitude of the stator currents and the maximum flux density within the machine.

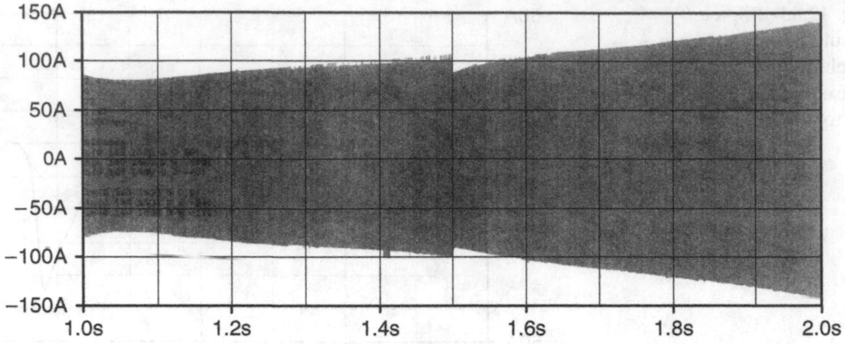

Fig. 12.85 PSpice simulation of the phase A machine current i_{MA} for the phase angle $\theta = 60°$ from 1 s to 2 s. This oscillogram illustrates the envelope of the machine current

The phase angles $\theta = -60°, -30°, 0°, 30°, 60°$ permit to control the torque of the brushless DC motor drive while a change of the frequency f_1 (e.g., 1,500 Hz) permits the control of the drive speed. Both controls are independent from one another. Note that the gating signal frequency of the MOSFETs corresponds to the frequency f_1, that is, full-on mode operation exists.

The slow 10% voltage variation with [see statement Vpm 3a 3 PULSE (30 −30 0 1 m 1 m 0.5 1) in PSpice program list] a frequency of 1 Hz changes the current amplitude (envelope) as is illustrated in Fig. 12.85. Any variation of the fuel-cell voltage leads to a modulation of the motor current, and this modulation must be compensated by an appropiate current control. This variation will be also discussed in Application Example 12.37.

Application Example 12.37: Transient Performance of an Inverter Feeding into Three-Phase Power System when Supplied by a Fuel Cell

Replace the DC source of Fig. 5.54 of Application Example 5.15 of Chap. 5 by the equivalent circuit of the fuel cell as described in Fig. 2 of the paper by Wang et al. [28].

(a) You may assume that in Fig. 2 of this paper the voltage changes $E = V_{DC} = 600$ V $\pm$ 30 V (that is, vfuel = 600 V, and vp = $\pm$ 30 V) where the superimposed rectangular voltage vp = $\pm$ 30 V has a period of T $\pm$ 30 V = 1 s. This voltage variation of a fuel cell as a function of time reflects the nonuniform DC voltage generation of a fuel cell. The remaining parameters of the fuel cell equivalent circuit can be extrapolated (e.g., $R_{act} = 6.81\ \Omega$, $R_{ohmic} = 1.42\ \Omega$, $C_{fuel} = 0.02$ F) from Table III of the paper by Wang et al. [28] for $E = V_{DC} = 600$ V. Plot the inverter reference current (56.6 A amplitude) and the inverter output current from 200 to 270 ms.
(b) Repeat part (a) with vfuel = 390 V.
(c) Repeat part (a) and plot the current from 1 to 2 s.

Solution

(a) A part of the PSpice input program with .cir extension is listed below and the simulation plots are shown in Figs. 12.86–12.88.

Replace the following two statements in the PSpice program list of Application Example E5.15 with

proportional (P) control.

*DC voltage supply

Vsuppl 2 0 380

by

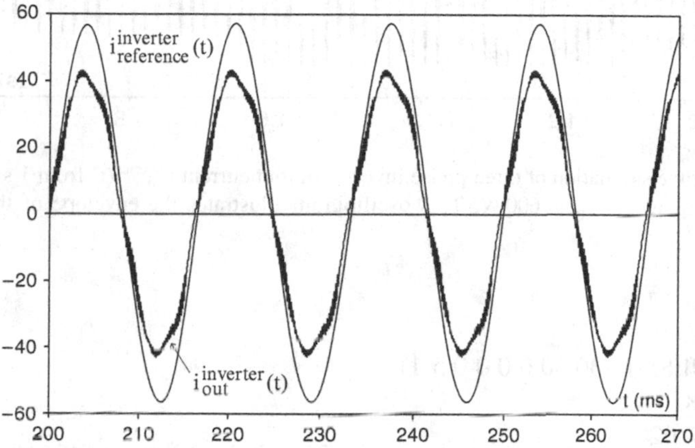

Fig. 12.86 PSpice simulation of three-phase inverter output current $i_{out}^{inverter}(t)$ and reference current $i_{reference}^{inverter}(t)$ with amplitude of 56.6 A from 200 ms to 270 ms at an input DC voltage of v_{fuel}=600 V

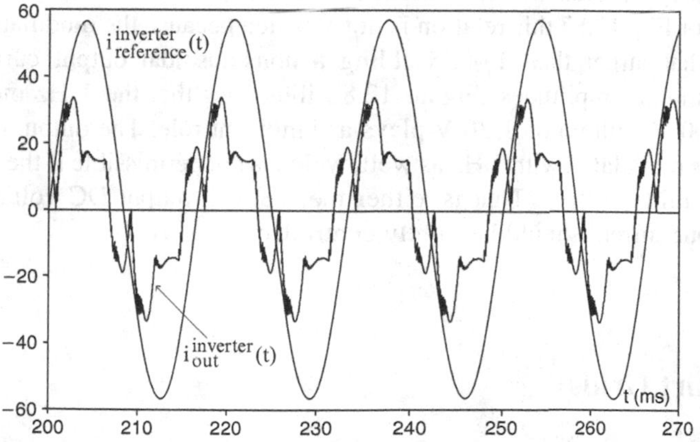

Fig. 12.87 PSpice simulation of three-phase inverter output current $i_{out}^{inverter}(t)$ and reference current $i_{reference}^{inverter}(t)$ with amplitude of 56.6 A from 200 ms to 270 ms at at an input DC voltage of v_{fuel}=390 V

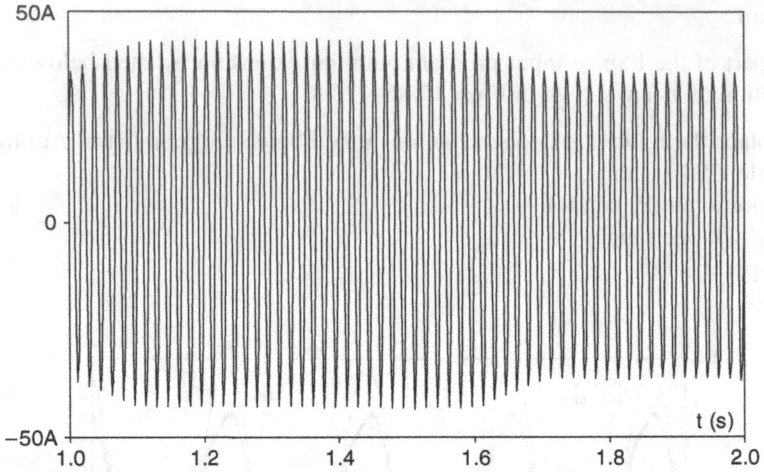

Fig. 12.88 PSpice simulation of three-phase inverter output current $i_{out}^{inverter}(t)$ from 1 s to 2 s at at an input DC voltage of v_{fuel}=600 V. This oscillogram illustrates the envelope of the inverter current

```
* Fuel cell.
vfuel b c 600
vp c 0 PULSE (−30 30 0 0 0 0.5 1)
rac a b 6.81
cfuel a b 0.02
rohmic 2 a 1.42
```

Figure 12.86 illustrates that the inverter relation $V_{rms\_out} = m\frac{V_{DC}}{2\sqrt{2}}$ for the inverter AC output voltage $V_{rms\_out} = 138.61V$ as a function of the inverter DC input voltage V_{DC}=600 V for a modulation index of m=0.653 is satisfied resulting in an approximate sinusoidal inverter output current with constant fundamental amplitude, while for Fig. 12.87 this relation is not satisfied because the modulation index m is somewhat larger than 1.00, yielding a nonsinusoidal output current with constant harmonic amplitudes. Figure 12.88 illustrates that the 1 Hz modulation of vfuel = 600 V with vp of ± 30 V plays a significant role. The output current of the inverter is modulated with 1 Hz as well, which is not permissible if the current is fed into the utility system. That is, either the fuel cell output DC voltage or the inverter output current should be tightly controlled.

12.9 Smart Grids

To date the concept of the smart grid is not well defined: different engineers and corporations have different ideas how a smart grid should look like. However, smart grid technologies are presently undergoing rapid development in an effort to modernize power grids to cope with the increasing energy demands of the future [29]. High speed bi-directional communications networks will provide the framework for real-time monitoring and control of transmission, distribution and end-user consumer assets for effective coordination and usage of available energy resources. Furthermore, integration of computer automation into all levels of power network operations, especially at the distribution and consumer level (e.g., smart meters), enables smart grids to rapidly self regulate and heal, improve system power quality, reliability and security, and more efficiently manage energy delivery and consumption.

One present-day important concept relates to the plug-in electric vehicle (PEV) [30] which is expected to become an integral part of smart grids in the near future. Furthermore, smart appliances for households such as air-conditioning, dishwashers, clothes dryers, washing machines, as well as PEVs, could "talk" to the grid and decide how best to operate and automatically schedule their activities at strategic times based on available generation. The idea is to use the hybrid/electric car as a load if the renewable energy is plentiful and to supply energy to the grid if there is a lack of renewable energy due to the intermittent nature of photovoltaic and wind-power sources. Important components of a smart grid are the renewable energy storage facilities discussed in Sects. 12.5–12.8.

With the rapid development of smart grid technology, it is anticipated that PEVs and smart appliances will have a large impact on the operation of low-voltage residential networks [29–34]. Electric utilities are becoming concerned about the potential burdening issues of local distribution circuits (e.g., transformers, and cables) due to additional stresses imposed by public PEV charging stations and multiple domestic PEV charging activity. Such operations may lead to overloading at peak residential hours, poor power quality, decreased reliability, outages and increased operational costs. Unlike the conventional small domestic loads with predictable daily load curves, the ratings of PEVs are relatively large, their daily load curves (e.g., charging times) are unpredictable and their penetration in some residential systems might rapidly increase within the next few years. Furthermore, it is reasonable to assume that most uncoordinated (random) PEV charging will take place in the evenings during the peak residential load hours.

Application Example 12.38: Impact of Uncoordinated Plug-In Electric Vehicle Charging on Residential and Distribution Systems

To illustrate the impact of Plug-in Electric Vehicle (PEV) on distribution and residential networks, consider the 145 bus system of Fig. 12.89a. It consists of six low voltage 19-bus 415 V residential networks connected to the 31-bus 22 kV distribution system. The typical residential daily load curve of Fig. 12.89b is assumed for the residential loads where the peak occurs approximately at 6 pm. System line and load parameters are listed in Tables 12.18–12.21 [31]. Each PEV is rated at 4 kW and the available battery energy is 14.4 kWh; the size of the battery depends upon the permissible depth of discharge (DoD). Assume three constant current charging rates corresponding to charging durations of 6 h

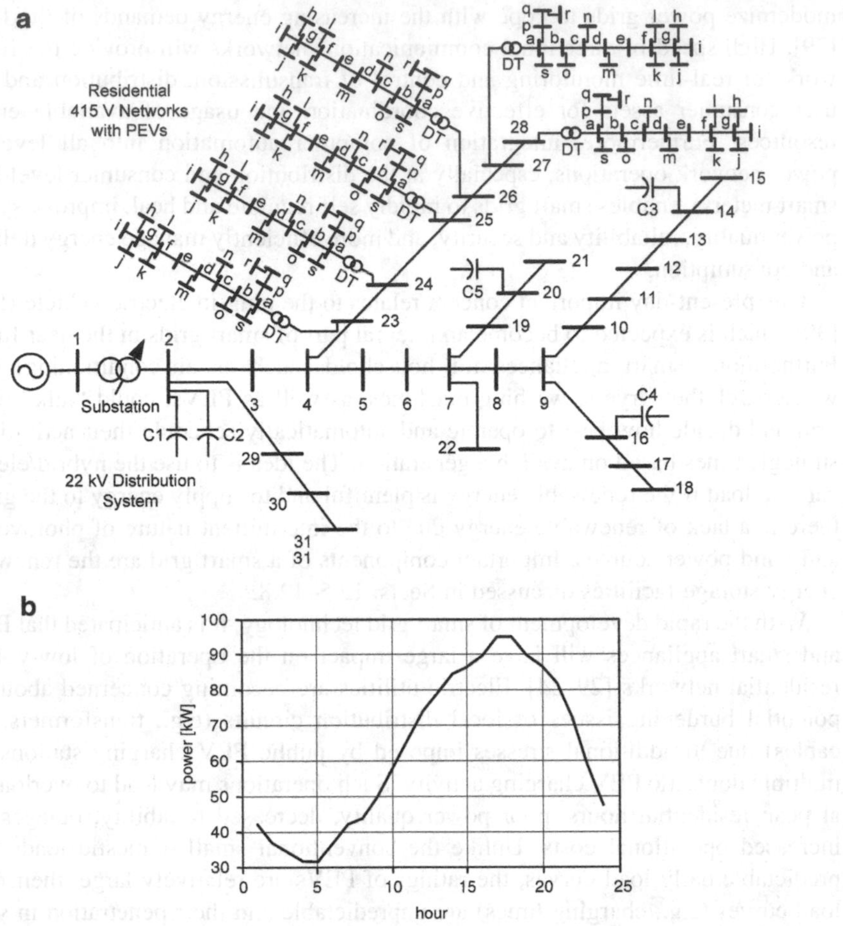

Fig. 12.89 The 145-bus system consisting of six low voltage and 19-bus 415 V residential networks connected to the 31-bus 22 kV distribution system; (**a**) system diagram, (**b**) residential daily load curve for all PQ loads, where P stands for real power and Q for reactive power [32]

($P_{charger}$ = 14.4 kWh/6h = 2.4 kW), 4 h ($P_{charger}$ = 14.4 kWh/4h = 3.6 kW) and 1 h ($P_{charger}$ = 14.4(0.8) = 11.5 kW, assuming 80% battery capacity during fast charging due power limitations [32]). Consider random locations of PEVs in the six residential networks with two penetration levels of 21 and 79% corresponding to 6(19)(0.21) = 24 and 6(19)(0.79) = 90 PEVs, respectively. Vehicles will be randomly plugged in during three charging zones; green or dotted line (5 pm to 8 am), blue or full line (5 pm to 2 am) and red or dashed line (during the peak load; 5 pm to 11 pm). The harmonic distortions of the PEV loads are not considered and only the fundamental frequency component is to be included.

Table 12.18 Line parameters of the 31-bus 22 kV distribution system (Fig. 12.89)

The 31-bus 22 kV distribution system [32]							
Line				Line			
From bus	To bus	Line resistance R [Ω]	Line reactance X [Ω]	From bus	To bus	Line resistance R [Ω]	Line reactance X [Ω]
1	2	0.090	1.553	16	17	1.374	0.774
2	3	0.279	0.015	17	18	1.374	0.774
3	4	0.444	0.439	7	19	0.864	0.751
4	5	0.864	0.751	19	20	0.864	0.751
5	6	0.864	0.751	20	21	1.374	0.774
6	7	1.374	0.774	7	22	0.864	0.751
7	8	1.374	0.774	4	23	0.444	0.439
8	9	1.374	0.774	23	24	0.444	0.439
9	10	1.374	0.774	24	25	0.864	0.751
10	11	1.374	0.774	25	26	0.864	0.751
11	12	1.374	0.774	26	27	0.864	0.751
12	13	1.374	0.774	27	28	1.374	0.774
13	14	1.374	0.774	2	29	0.279	0.015
14	15	1.374	0.774	29	30	0.279	0.015
9	16	0.864	0.751	30	31	1.374	0.774

Table 12.19 Line parameters of the 19 bus 415V residential networks (Fig. 12.89)

The 19 bus 415 V residential networks [32]							
Line				Line			
From bus	To bus	Line resistance R [Ω]	Line reactance X [Ω]	From bus	To bus	Line resistance R [Ω]	Line reactance X [Ω]
a	b	0.0415	0.0145	f	l	1.3605	0.1357
b	c	0.0424	0.0189	d	m	0.140	0.0140
c	d	0.0444	0.0198	c	n	0.7763	0.0774
d	e	0.0369	0.0165	b	o	0.5977	0.0596
e	f	0.0520	0.0232	a	p	0.1423	0.0496
f	g	0.0524	0.0234	p	q	0.0837	0.0292
g	h	0.0005	0.0002	q	r	0.3123	0.0311
g	i	0.2002	0.0199	a	s	0.0163	0.0062
g	j	1.7340	0.1729	Distribution transformer reactance			0.0654
f	k	0.2607	0.0260				

(a) Simulate the system using the Newton–Raphson load flow in Matlab environment. For the two PEV penetration levels (21 and 79%), the three charging durations (6, 4 and 1 h) and the three charging zones (green, blue and red), compute the ratio of peak system losses to peak-load demand

Table 12.20 Load parameters of the 31 bus 22 kV distribution system (Fig. 12.89)

The 31 bus 22 kV distribution system [32]							
Bus	Bus type	P [kW]	Q [kVAr]	Bus	Bus type	P [kW]	Q [kVAr]
2	PQ	0.1305	0.0430	18	PQ	0.1195	0.03925
5	PQ	0.2345	0.0770	19	PQ	0.1080	0.0355
10	PQ	0.0472	0.0155	20	PQ	0.16825	0.05525
12	PQ	0.0840	0.0275	21	PQ	0.1240	0.04075
13	PQ	0.1645	0.0540	22	PQ	0.05175	0.0170
14	PQ	0.1960	0.0645	29	PQ	0.22075	0.0725
15	PQ	0.1825	0.060	31	PQ	0.2210	0.0725
16	PQ	0.1195	0.03925				
17	PQ	0.1375	0.04525				

Table 12.21 Load parameters of the 19 bus 415V residential networks (Fig. 12.89)

The 19 bus 415 V residential networks [32]			
Bus	Bus type	P [kW]	Q [kVAr]
Bus "b" to "s"	PQ	3.5	2.42
Randomly selected buses based on PEV penetrations (21% or 79%)	PEV charger	3.6	0

($P_{loss}^{max}/P_{total}^{peak}$), maximum voltage deviation from the rated voltage (1 pu) at the worst bus (ΔV^{max}) and the maximum of all distribution transformer load currents.

(b) Plot the system losses and the maximum voltage deviation at the worst buses during the 24 h period for high penetration of PEVs within the three charging zones.

(c) Plot the system daily load curves without and with high penetration of PEVs. Plot and compare the load currents of all distribution transformers feeding the residential networks.

Solution

(a) Simulation results are compared in Table 12.22. The Matlab source codes for this Application Example are listed in [35].

(b) System losses and voltage profile at the worst buses over the 24 h period are plotted in Figs. 12.90a, b, respectively.

(c) The daily load curves and the distribution transformer currents are compared in Figs. 12.91a, b, respectively.

Table 12.22 Impact of PEV charge rates, charging zones and penetrations on the performance of the 145-bus system (Fig. 12.89) [32, 35]

		Low PEV penetration (21%)			High PEV penetration (79%)		
Charging rate	Charging zones	$\frac{P_{loss}^{max}}{P_{total}^{peak}}$* [%]	ΔV^{max}** [%]	I_{trans}^{max}*** [%]	$\frac{P_{loss}^{max}}{P_{total}^{peak}}$* [%]	ΔV^{max}** [%]	I_{trans}^{max}*** [%]
Normal Charge 6 h @ $P_{chg,max} = 2.4$ kW	5pm–8am	2.146	12.658	0.812	2.298	16.913	0.976
	5pm–2am	2.198	12.909	0.859	2.732	18.715	1.130
	5pm–11pm	2.273	13.007	0.884	3.212	19.747	1.181
Medium Charge 4 h @ $P_{chg,max} = 3.6$ kW	5pm–8am	2.224	13.019	0.862	2.530	20.624	1.039
	5pm–2am	2.224	13.036	0.896	2.726	20.981	1.150
	5pm–11pm	2.265	13.148	0.930	3.649	22.755	1.343
Quick Charge (at 80% Capacity) 1 h @ $P_{chg,max} = 11.5$ kW	5pm–8am	2.244	13.022	0.908	3.030	36.907	1.522
	5pm–2am	2.199	13.633	1.018	3.130	39.445	1.455
	5pm–11pm	2.293	13.561	1.027	3.863	43.301	1.658

*) Ratio of peak-system losses to peak-load demand
**) Voltage deviation from the rated voltage (1 pu) at the worst bus
***) Maximum of all distribution transformer load currents

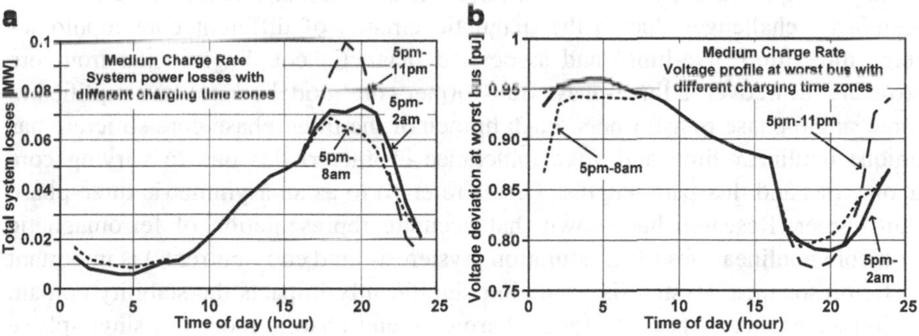

Fig. 12.90 Simulation results over the 24 h period for medium charge rate (4 h @ $P_{charger}$= 3.6kW) with randomly distributed charging of 90 PEVs (79% penetration) within the three time zones; (**a**) System losses, (**b**) Voltage profile at the worst buses [32]

12.10 Ferroresonance in Power Transformers

Ferroresonance is one of the longest recognized power quality disturbances in the history of AC power systems, spanning nearly a century of accumulated research [11, 36]. The symptoms of ferroresonance are regarded as most serious and

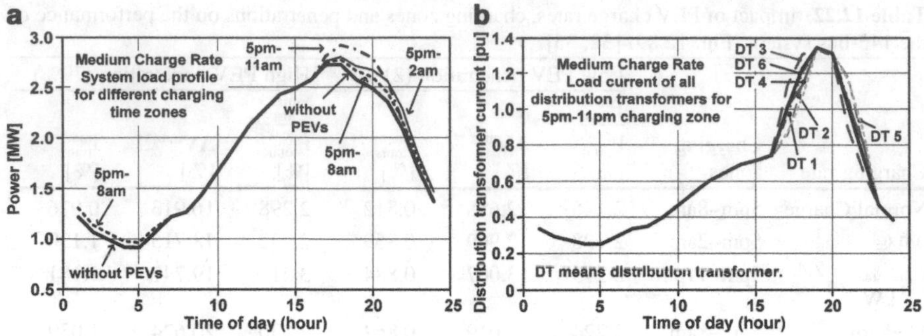

Fig. 12.91 Simulation results over the 24 h period for medium charge rate (4 h @ $P_{charger}$= 3.6kW) with randomly distributed charging of 90 PEVs (79% penetration) within the three time zones; (**a**) Daily load curve without and with PEVs, (**b**) Load current of the six distribution transformers serving residential networks [32]

typically result in large currents and over voltages that can reach in excess of 4 pu accompanied with severe waveform distortion.

Much progress has been made in the modeling and understanding of single-phase transformer ferroresonance. However, one of the weakest areas in ferroresonance research remains in the modeling of (a)symmetric three-phase transformers. Exact behavior of three-phase transformers under ferroresonance conditions is largely unexplored due to the lack of adequate core models. These models present significant challenges due to the magnetic circuits of different core topologies (e.g., three-limb, five-limb) and associated magnetic couplings in the iron-core structure. Moreover, a three-phase transformer core model cannot be simplified to three single-phase transformers. Each branch of the three-phase core structure has unique nonlinear limb and yoke reluctance characteristics due to varying core geometries and flux path lengths. This is referred to as an asymmetric three-phase transformer. Research has shown that accurate representation of ferromagnetic iron-core nonlinearities (e.g., saturation, hysteresis, and eddy-currents) is important in ferroresonance. Hysteresis formation significantly impacts the stability domain of ferroresonance, especially for subharmonic and chaotic modes in single-phase transformer core models [37, 38]. The common approach of approximating core nonlinearities with non-hysteretic single-value functions (e.g., piece-wise, exponential, and polynomials) has proven to be inadequate for ferroresonance studies. The impacts of dynamic hysteresis nonlinearities (e.g., major and minor loops) and magnetically coupled asymmetric legs on three-phase transformer ferroresonance are challenging issues that require more research.

There have been very few attempts at studying three-phase transformers under ferroresonance conditions. References [39–41] implement models considering core topology and employ single-value nonlinear functions for core effects. Analytical and numerical approaches are developed in [42, 43] to calculate ferroresonance modes. However, these approaches do not consider dynamic hysteresis effects with core asymmetry and associated magnetic leg couplings.

The occurrence of ferroresonance in single- and three-phase transformers are complicated issue that can be demonstrated through simulation studies using an accurate nonlinear model of power transformer taking into account the core non-linearities including saturation, magnetic coupling between phases, asymmetrical electromagnetic core structures and hysteresis loops. Next section introduces a relatively accurate and simple transformer model that includes saturation, electric and magnetic couplings with asymmetrical magnetic core structures.

12.10.1 The Nonlinear Electric and Magnetic Circuits of Power Transformer

An accurate and relatively simple nonlinear three-phase transformer model for (a) symmetrical, (un)balanced and (non)sinusoidal operation including harmonic distortions, DC biasing, sag/swell and ferroresonance is proposed and implemented in references [44–48]. It is based on the simultaneous solution of electric and magnetic equivalent circuits of three-phase, three-limb transformers. The model has also been tested for single-phase and three-phase, five-limb transformers. The nodal equations of the circuits are solved in time domain using iterative techniques such as Newton–Raphson. The electric circuit governs the electrical connections of the source, load and the transformer. For three-phase transformers, the magnetic circuit is necessary to represent the multiple flux paths, reluctances and magnetomotive forces within multi-legged iron core structures. It also incorporates the asymmetric magnetizing behavior of the core.

The electric circuit of Fig. 12.92a can be simulated by most software packages such as PSIM or PSPICE. Typically a nodal matrix of equations describing the circuit is formed and solved through iterative time-domain numerical techniques. The induced primary and secondary voltages are modeled as voltage sources controlled by the time derivative of the magnetic fluxes (Faraday's Law). This establishes a link between the magnetic and electric circuits.

The transformer model implements core-loss resistance R_{core} as a linear element in parallel with the induced primary voltage. Some models use nonlinear resistances to represent core losses because they are functions of harmonic voltage magnitudes and phase angles [49, 54]. However, according to [11] a linear R_{core} is a valid simplification with good agreement shown between model simulations and measured hysteresis loop areas. Likewise, winding resistances and leakage fluxes are assumed constant.

The electric circuit parameters, winding resistances, core loss resistances and leakage inductances can be estimated from three-phase, open-circuit and short-circuit tests.

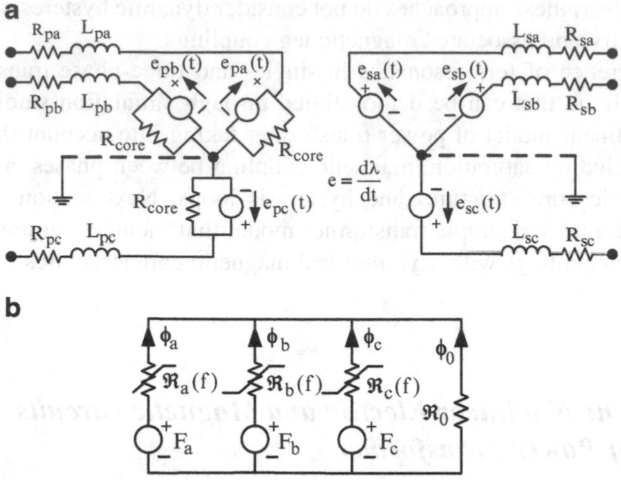

Fig. 12.92 Nonlinear model of three-phase three-leg transformer; (**a**) electric equivalent circuit, (**b**) magnetic equivalent circuit (From Masoum M.A.S.; Moses, P.S.; Masoum, A. S: Derating of asymmetric three-phase transformers serving unbalanced nonlinear loads, *IEEE Transactions on Power Delivery,* vol. 23, pp. 2033-2041, Oct. 2008, [45])

The circuit in Fig. 12.92b is an approximation of the equivalent circuit proposed by [50]. In this model, the seven reluctances of the core are reduced to three reluctance parameters that can be easily measured. This magnetic circuit can be programmed into PSPICE as an electrical circuit using the magnetic-electric duality principle.

The nonlinear reluctances are implemented in PSPICE as flux sources dependent on their own MMF drops

$$\Phi\{f_x\} = f_x\ \Re\{f_x\}^{-1}, \tag{12.39}$$

where f_x ($x = a$, b, and c) are mmfs developed in the limbs (of the phases "a", "b" and "c") by currents in the primary and secondary windings. This is another relationship linking the electric and magnetic circuits. Nonlinear reluctance functions can be fitted to each leg's magnetizing characteristic

$$\Phi(f_x) = \mathrm{sgn}(f_x)\cdot\ \alpha_x \log_{10}(\beta_x|f_x| + 1), \tag{12.40}$$

where the empirically determined parameters α_x and β_x are constants that shape the function to any measured saturation curve ($\Phi - f_x$). Note that α_x and β_x influence the vertical and horizontal scaling of $\Phi(f_x)$, respectively.

The measurement procedure to obtain the magnetic circuit parameters of an asymmetric three-limb transformer is described in [50]. The zero-sequence, open-circuit test determines the linear air path reluctance, $\Re_0$. The $\lambda - i$ hysteresis loops for each limb can be obtained by exciting a phase at a time and integrating the corresponding induced voltages.

12.10.2 Understanding Ferroresonance

Ferroresonance is known to be caused by the interaction of power system capacitances with nonlinear inductances associated with wound magnetic cores in power transformers and instrument voltage transformers. Many stable and unstable operating points (ferroresonance modes) are possible due to the constantly changing transformer inductances during its magnetizing and demagnetizing cycles. Sources of capacitances can include circuit breakers equipped with grading capacitors, shunt and series transmission line capacitances (overhead and underground) and stray capacitances in transformer windings, bushings, bus bars and feeders. Furthermore, initial conditions and system perturbations play an important role in ferroresonance behaviour whereby many operating modes for the same set of system parameters can be observed. Hence, different instances of AC voltage amplitude at breaker operation, load transients and residual core fluxes can initiate widely different ferroresonance behaviour.

To understand [11] the ferroresonance phenomena, consider a simple power circuit, consisting of a transformer connected to a power system as shown by Fig. 12.93a; its single-phase equivalent circuit is illustrated in Fig. 12.93b. The current is

$$|\tilde{I}| = \frac{|\tilde{V}_s|}{\sqrt{R^2 + (X_L - X_C)^2}}. \tag{12.41}$$

Even though switch S_2 is open, a current i(t) can still flow because the cable shunt capacitance C closes the circuit. The transformer is assumed to be unloaded but energized. The magnetizing current i(t) has to return to its source via the cable

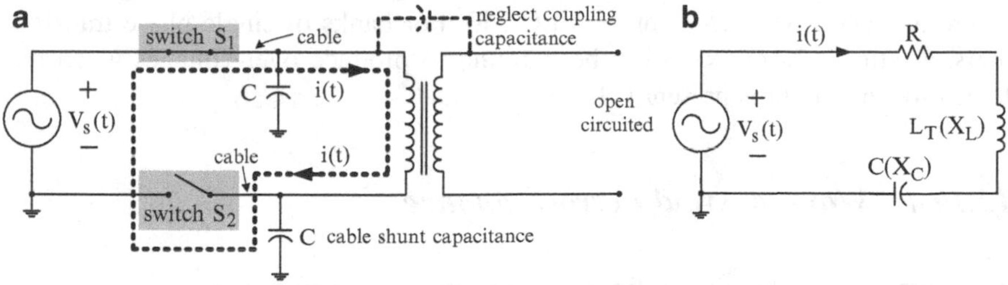

Fig. 12.93 Formation of ferroresonance in an unloaded single-phase transformer connected to power system via cable; (**a**) system diagram, (**b**) single-phase equivalent circuit, where C stands for cable and T for transformer

shunt capacitance C. Therefore, ferroresonance occurs whenever X_L and X_C are close to each other.

In power systems, the variations of X_L are mainly due to the nonlinear magnetizing inductance of transformers and most changes in X_C are caused by switching capacitor banks and cables. Figure 12.94 demonstrates [11] the formation of ferroresonance in power system due to unsynchronized or nonsimultaneous three-phase switching of three-phase no-load (delta-Y and Y-Y) transformers.

12.10.3 Power System Conditions Contributive to Ferroresonance

The important question is what conditions in an actual power system are contributive (conducive) to ferroresonance? Some conditions may not by themselves initiate the problem, but they are necessary so that when other conditions are present in a power system, ferroresonance can occur. There are four prerequisites for ferroresonance [11]

- The employment of cables with relatively large capacitances.
- The transformer connection.
- An unloaded transformer. Adding a load to the secondary side of the transformer will considerably reduce the possibility of ferroresonance due to the reflection of the load impedance Z_L to the primary side (by the square of the turns ratio) and limiting the circulating current.
- Having either one or two transformer phases connected. When the third phase is energized, the system will be balanced and the current will flow directly to and from the transformer, bypassing the capacitance to ground.

It should be noted that the mere presence of the ferroresonant circuit does not necessarily mean that the circuit involved will experience over-voltages due to this problem. The ratio of the capacitive reactance X_C to the transformer magnetizing reactance X_T of the ferroresonant circuit (X_C/X_T) will determine whether the circuit will experience over-voltages or not. Experience and simulations for a wide range of transformer sizes (15 kVA to 15 MVA) and line-to-ground (line-to-neutral) voltages ranging from 2.5 kV to 19.9 kV indicate that ferroresonant conditions with over-voltages above (1.1–1.25) pu will occur when the ratio of $X_C/X_T < 40$, where $X_T = \omega L$, $X_C = 1/\omega C$ or $1/\omega^2 LC < 40$. For banks of single-phase transformers, a ratio of $X_C/X_T < 25$ has been found to produce over-voltages when the ferroresonant circuit is present [11].

12.10.4 Ways to Avoid Ferroresonance

There are a few techniques to limit the probability of ferroresonance:

- The grounded-Y/grounded-Y transformer- when the neutral of the transformer is grounded, the capacitance to ground of the line (cable) is essentially bypassed,

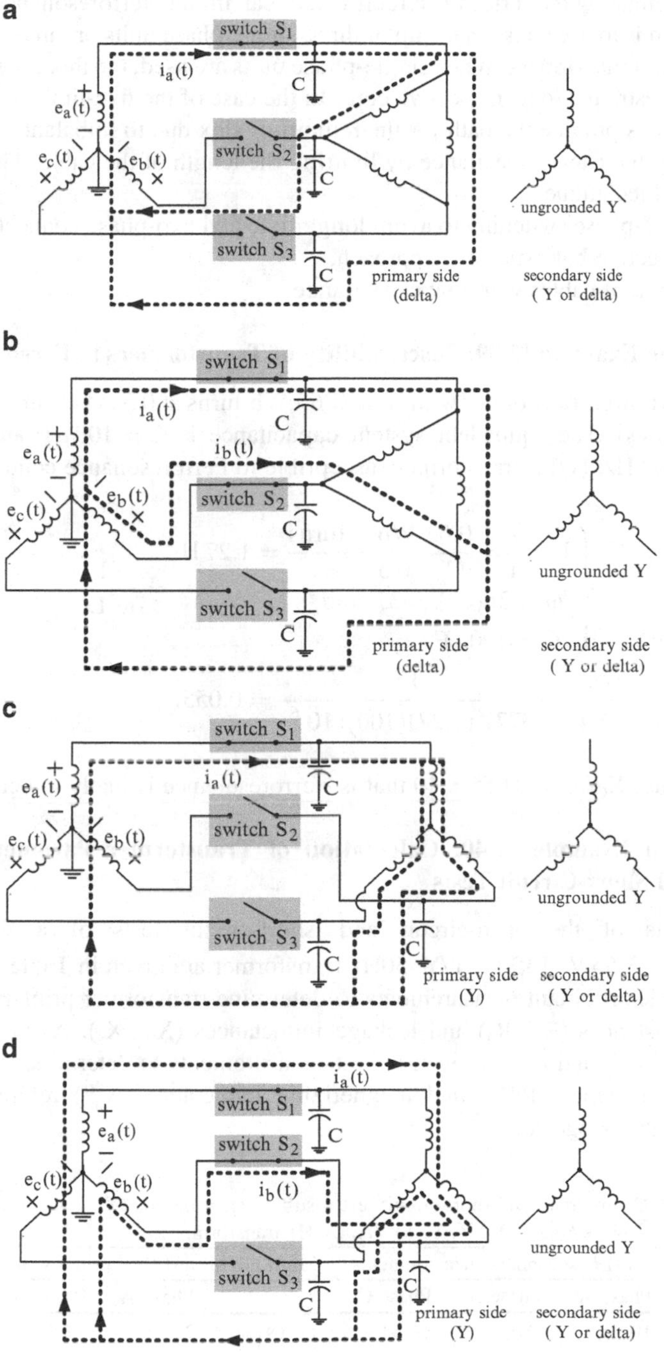

Fig. 12.94 Formation of resonance in three-phase power system; (**a**) delta-Y transformer and one phase energized, (**b**) delta-Y transformer and two phases energized, (**c**) Y-Y transformer and one phase energized, (**d**) Y-Y transformer and two phases energized

thus eliminating the series LC circuit which can initiate ferroresonance. Another approach is to use banks made up of three single-phase units or three-phase five-limb core transformers. When single-phase units are used, the flux for each phase will be restricted to its respective core. In the case of the five-limb core, the two extra limbs provide the path for the remaining flux due to imbalance.

- Limiting the cable capacitance by limiting the length of the cable. This is not a practical technique.
- Fast three-phase switching to avoid longer one and two-phase connections. This is an effective but expensive approach.
- Utilization of cables with low capacitance.

Application Example 12.39: Susceptibility of Transformers to Ferroresonance

Consider a power transformer with $\lambda = 0.63$ Wb-turns at 0.5 A (assume linear λ-i characteristics). The equivalent system capacitance is $C = 100\,\mu F$ and the frequency is 60 Hz. Is this transformer susceptible to Ferroresonance conditions?

Solution

$$\left\{\begin{array}{l} L = \frac{\lambda}{i} = \frac{0.63\ \text{Wb}-\text{turns}}{0.5\,\text{A}} = 1.27\,\text{H} \\ \omega = 2\pi(60) = 377\,\text{rad/s} \\ C = 100\mu F \end{array}\right. \Rightarrow \frac{1}{\omega^2 LC}$$

$$= \frac{1}{(377)^2(1.27)(100)(10^{6})} = 0.055. \tag{12.42}$$

Therefore, $X_C/X_T = 0.055 < 40$ that is, ferroresonance is likely to occur.

Application Example 12.40: Calculation of Transformer Parameters from Open- and Short-Circuit Tests

The results of the open-circuit and short-circuit tests of a three-phase $\sqrt{3}\,440\,\text{V}/\sqrt{3}\,55\,\text{V}$, 1 kVA, Y/Y, 50Hz transformer are given in Table 12.23.

Use the short-circuit measurements to determine transformer primary and secondary resistances (R_p, R_s) and leakage inductances (X_p, X_s). Assume a linear magnetic circuit and use the open-circuit measurements to determine transformer core loss resistance (R^{pri}_{core}) and magnetizing inductance (X^{pri}_m) refereed to the primary high voltage side.

Table 12.23 Open-circuit and short-circuit test results (performed in Y/Y configuration) for the three-phase $\sqrt{3}\,440\,\text{V}/\sqrt{3}\,55\,\text{V}$, Y/Y, 1 kVA, 50 Hz transformer

Open circuit test (LV secondary measurements)				Short circuit test (HV primary measurements)			
	Phase A	Phase B	Phase C		Phase A	Phase B	Phase C
P^{phase}_{oc} [W]	10.14	8.6	16.54	P^{phase}_{sc} [W]	27.21	27.54	27.91
I^{line}_{oc} [A]	0.359	0.247	0.382	I^{line}_{sc} [A]	1.255	1.274	1.286
V^{L-N}_{oc} [V]	55.01	55.00	54.99	V^{L-N}_{sc} [V]	22.31	22.30	22.30

Tables 12.24 Calculated parameters of the three-phase 440V/55V, 1 kVA, 50 Hz transformer based on open-circuit and short-circuit test results (Table 12.23)

From the short-circuit test				From the open-circuit test			
	Phase A	Phase B	Phase C		Phase A	Phase B	Phase C
θ_{sc} [°]	13.36	14.22	13.29	θ_{oc} [°]	59.11	50.72	38.06
R_p [Ω]	8.64	8.48	8.444	R^{pri}_{core} [kΩ]	19.10	22.51	11.70
mean	8.52 Ω/phase			mean	17.78 kΩ/phase		
R_s [Ω]	0.135	0.133	0.132	X^{pri}_{m} [kΩ]	11.43	18.41	14.95
mean	0.133 Ω/phase			mean	14.93 kΩ/phase		
X_p [Ω]	2.10	2.15	1.99				
mean	2.08 Ω/phase ⇒ $L_p = 6.62$ mH/phase						
X_s [Ω]	0.033	0.034	0.031				
mean	0.033 Ω/phase ⇒ $L_p = 0.103$ mH/phase						

Solution

From the open-circuit test:

$$\theta_{oc} = \cos^{-1}\left(\frac{P^{phase}_{oc}}{V^{L-N}_{oc} I^{line}_{oc}}\right) \Rightarrow \begin{cases} R^{pri}_{core} = \dfrac{V^{L-N}_{oc}}{I^{line}_{oc} \cos\theta_{oc}} \\ X^{pri}_{m} = \dfrac{V^{L-N}_{oc}}{I^{line}_{oc} \sin\theta_{oc}}. \end{cases}$$

$$\Rightarrow \begin{cases} R^{sec}_{core} = \left(\frac{440}{55}\right)^2 R^{pri}_{core}. \\ X^{sec}_{m} = \left(\frac{440}{55}\right)^2 X^{pri}_{m}. \end{cases} \tag{12.43}$$

From the short-circuit test:

$$\theta_{sc} = \cos^{-1}\left(\frac{P^{phase}_{sc}}{V^{L-N}_{sc} I^{line}_{sc}}\right) \Rightarrow \begin{cases} R_{eq} = R_p + R'_s = \dfrac{V^{L-N}_{sc}}{I^{line}_{sc}} \cos\theta_{sc}. \\ X_{eq} = X_p + X'_s = \dfrac{V^{L-N}_{sc}}{I^{line}_{sc}} \sin\theta_{sc}. \end{cases}$$

$$\Rightarrow \begin{cases} R_p = R_{eq}/2,\ R_s = (R_{eq}/2)(55/440)^2. \\ X_p = X_{eq}/2,\ X_s = (X_{eq}/2)(55/440)^2. \end{cases} \tag{12.44}$$

Using (12.43) and (12.44), transformer parameters are calculated and listed in Table 12.24.

Application Example 12.41: Ferroresonance in Three-Phase, Three-Limb Transformer

The ferroresonance study for this application example is performed for an unloaded Y-Y three-phase, three-limb transformer supplied by a three-phase source (Fig. 12.95). The transformer primary and secondary windings are rated at 440 V and 55 V (1 kVA), respectively, and operate at a nominal frequency of 50 Hz.

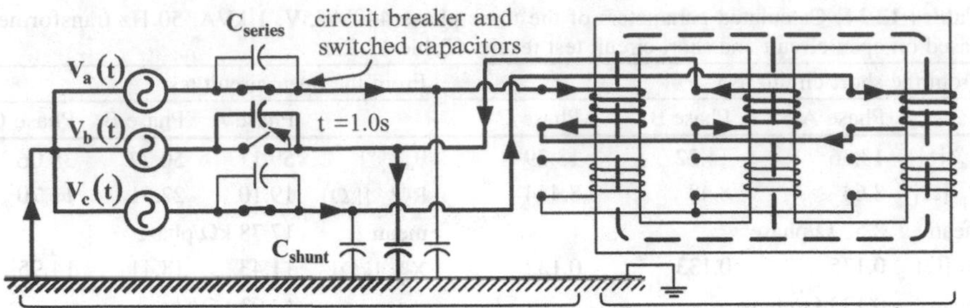

Fig. 12.95 Ferroresonance in an unloaded transformer fed through series and shunt capacitors (e.g., cable capacitance). The thicker lines indicate the ferroresonance path when circuit breaker in phase b is opened (From Moses, P.S.; Masoum, M.A.S.; Toliyat, H.A.: "Impacts of hysteresis and magnetic couplings on the stability domain of ferroresonance in asymmetric three-phase three-leg transformers," *IEEE Transactions on Energy Conversion*, [48])

Transformer primary and secondary parameters R_p, R_s, X_p, X_s and R_{core}^{pri} are calculated in Application Example 12.40. The constants α_x and β_x of the nonlinear reluctance, calculated from the measured ($\lambda - i$) hysteresis characteristic are $\alpha_a = 0.92$, $\alpha_b = 0.7$, $\alpha_c = 0.92$, $\beta_a = 100$, $\beta_b = 1{,}000$ and $\beta_c = 100$.

The ferroresonance circuit consists of series and shunt capacitances interacting with the magnetizing inductances of the wound transformer core limbs. C_{shunt} and C_{series} can originate from circuit breakers equipped with grading capacitors, shunt and series transmission line capacitances (overhead and underground cables), reactive power compensation capacitor banks, and lumped stray capacitances in transformer windings, bushings, bus bars and feeders.

In order for ferroresonance to occur, a system perturbation (e.g., switching transient) with initial conditions conducive to ferroresonance is usually necessary. For the system of Fig. 12.95, phase b circuit breaker is used to impose switching transients to initiate ferroresonance. There have been many such practical occurrences where unsynchronized/nonsimultaneous three-phase switching or single-phase circuit breaker operations resulted in one or two suddenly disconnected phases while the transformer is unloaded or lightly loaded, giving rise to ferroresonance [51]. For example, feeders employing single-phase circuit breakers or fuses can suddenly develop faults causing one of the phases to de-energize or to become disconnected. The switching transient and resulting unbalanced excitation of the transformer can lead to a ferroresonance path involving magnetizing inductances in series with capacitances. Furthermore, in unloaded or lightly loaded systems (e.g., rural distribution feeders), there may be insufficient damping for ferroresonance. The resulting large distorted currents and over-voltages can cause severe damage to networks.

(a) System simulation- using the nonlinear transformer model of Fig. 12.92 and the magnetizing functions of (12.39) and (12.40), simulate the system using PSpice. Show system diagrams in PSpice.
(b) Bifurcation Analysis- a useful and effective approach for identifying system parameters conducive to ferroresonance is through bifurcation analysis.

A system parameter (e.g., C_{shunt}) is selected as the bifurcation parameter and the primary phase voltages are studied. For each phase, the bifurcation diagram is constructed from each bifurcation parameter change by sampling the voltage waveform at the power frequency (50 Hz). For the fixed $C_{series} = 30$ pF, perform simulations for C_{shunt} ranging from 0.1 to 50 μF with a single-phase, open-phase condition occurring at t = 1 s in phase b. Plot the bifurcation diagram (e.g., sample points of the peak phase to neutral voltage of phase b (in pu) versus the values of C_{shunt}).

(c) Poincaré maps- the Poincaré maps are constructed in a similar way by sampling the phase-plane trajectory orbits. The resulting patterns can be interpreted for visual classification of ferroresonance modes. For the fundamental ferroresonance mode ($C_{series} = 30$ pF and $C_{shunt} = 0.6$ μF), plot the phase b time domain waveforms of transformer primary voltages, winding currents and core fluxes. Also plot the phase-plane trajectory and Poincaré maps.

(d) Repeat part (c) for the period-3 subharmonic ferroresonance ($C_{series} = 30$ pF, and $C_{shunt} = 10$ μF) and the chaotic ferroresonance ($C_{series} = 30$ pF, and $C_{shunt} = 38$ μF) modes.

Solution

(a) The PSpice circuit diagram is shown in Fig. 12.96 and source code is listed in Table 12.25

(b) The Bifurcation diagram of ferroresonance modes (phase b voltage) for fixed $C_{series} = 30$ pF and different C_{shunt} values is shown in Fig. 12.97. The transient period is ignored in the sampling process to only analyse the subsequent stable

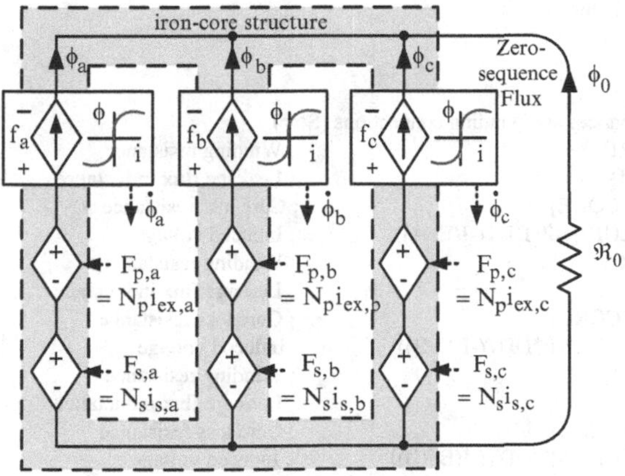

Fig. 12.96 Three-phase three-limb transformer magnetic equivalent circuit (From Moses, P.S.; Masoum, M.A.S.; Toliyat, H.A.: "Impacts of hysteresis and magnetic couplings on the stability domain of ferroresonance in asymmetric three-phase three-leg transformers," *IEEE Transactions on Energy Conversion*, [48, 52])

Table 12.25 PSpice source code for application example 12.41 [53]

```
*********** Transient analysis settings *********
.TRAN 0.2ms 5s 0.0s 0.2ms SKIPBP
**** Constants
.PARAM PI=3.14159265 FSEQ={2*PI/3} FREQ = 50 W0={2 * PI * FREQ} + PHASEDEG = {0}
+PHASE = {PHASEDEG * PI/180}
**** Electrical open and short circuit test parameters
.PARAM NP = 1 NS = {1/(440/55)}+ RPRI = 9.4229 RSEC = 0.1472 LPRI = 6.3411m LSEC = 0.0991m
+ RCORE = 14308  RZERO = 150
***** Define nonlinear magnetizing characteristic function
.FUNC H(X,K1,K2) {SGN(X)*(K1*LOG10(K2*ABS(X)+1))}
**** Primary phase voltage values (RMS and peak)
.PARAM URMS = {440} UMAX = {URMS*SQRT(2/3)}
**** Magnetizing reluctance characteristic function
+ K1A = 0.92 K1B = 0.7  K1C = 0.92 + K2A = 100  K2B = 1000  K2C =  100
ERAMP 1 0 VALUE {TABLE(TIME, 1m, 0, 0.5, 1)}        ;Ramping initial voltage
**** Single-phase switching transients
.MODEL Smod VSWITCH(Ron=1m Roff=1MEG Von=1V Voff=0V)
Sw1 810 10 501 0 Smod
Sw2 820 20 502 0 Smod
Sw3 830 30 503 0 Smod
ESW1   501 0 VALUE {1}
ESW2   502 0 VALUE  {TABLE(TIME,  0, 1,   1.0000,1,   1.00000001,0 )}
ESW3   503 0 VALUE  {1}
***************************** Electric equivalent circuit ************************
***** Three-phase voltage source functions connected to transformer primary
EGA 810 0 VALUE {UMAX*V(1)*COS(W0*TIME  +PHASE)}
EGB 820 0 VALUE {UMAX*V(1)*COS(W0*TIME-FSEQ+PHASE)}
EGC 830 0 VALUE {UMAX*V(1)*COS(W0*TIME+FSEQ+PHASE)}
**** CB grading capacitor
.PARAM Cseries = 0.03n
CGA 810 10 {Cseries}
CGB 820 20 {Cseries}
CGC 830 30 {Cseries}
**** Shunt capacitance
.PARAM Cshunt = 30.6u
CPA 10 0 {Cshunt}
CPB 20 0 {Cshunt}
CPC 30 0 {Cshunt}
**** Primary impedances and winding connections (Star)
RPRIA  10 12 {RPRI}                              ; Winding resistance
LPRIA 12 14 {LPRI}                               ; Leakage flux inductance
RCOREA 14 NP {RCORE}                             ; Core loss resistance
EPRIA  14 NP VALUE {NP*DDT(-I(EFA))}             ; Induced voltage
RPRIB  20 22 {RPRI}                              ; Winding resistance
LPRIB 22 24 {LPRI}                               ; Leakage flux inductance
RCOREB 24 NP {RCORE}                             ; Core loss resistance
EPRIB  24 NP VALUE {NP*DDT(-I(EFB))}             ; Induced voltage
RPRIC  30 32 {RPRI}                              ; Winding resistance
LPRIC 32 34 {LPRI}                               ; Leakage flux inductance
RCOREC 34 NP {RCORE}                             ; Core loss resistance
EPRIC  34 NP VALUE {NP*DDT(-I(EFC))}             ; Induced voltage
**** Primary neutral connection
* VNP NP 0 0      ;Solidly grounded (Enable this line for grounding the neutral)
RNP NP 0 {1/0}    ;Non-grounded
**** Secondary impedances and winding connections (Star)
```

(continued)

Table 12.25 (continued)

```
RSECA 40 42 {RSEC}
LSECA 42 44 {LSEC}
ESECA 44 NS VALUE {NS*DDT(-I(EFA))}; esa=Ns_dphi_a/dt
RSECB 50 52 {RSEC}
LSECB 52 54 {LSEC}
ESECB 54 NS VALUE {NS*DDT(-I(EFB))}; esb=Ns_dphi_b/dt
RSECC 60 62 {RSEC}
LSECC 62 64 {LSEC}
ESECC 64 NS VALUE {NS*DDT(-I(EFC))}; esc=Ns_dphi_c/dt
**** Secondary neutral connection
* VNS NS 0 0        ;Solidly grounded (Enable this line for grounding the neutral)
RNS NS 0 {1M}    ;Non-grounded
**** Secondary load (open-circuit)
.PARAM RLOAD = {1/0}
RLA 40 NS {RLOAD}
RLB 50 NS {RLOAD}
RLC 60 NS {RLOAD}
****************** Magnetic equivalent circuit for three-leg topology *******************
EFA 102   0 VALUE {NP*I(EPRIA)+NS*I(ESECA)}; Fa = Np_Ipea+Ns_Isa
GFA 102 115 VALUE {H(V(102, 115), K1A, K2A)}; phi_a
EFB 202   0 VALUE {NP*I(EPRIB)+NS*I(ESECB)}; Fb = Np_Ipeb+Ns_Isb
GFB 202 115 VALUE {H(V(202, 115), K1B, K2B)}; phi_b
EFC 302   0 VALUE {NP*I(EPRIC)+NS*I(ESECC)}; Fc = Np_Ipec+Ns_Isc
GFC 302 115 VALUE {H(V(302, 115), K1C, K2C)}; phi_c
RD 115 0 {RZERO}; Zero-sequence reluctance
**** Flux measurements
EFLUXA FLUXA 0 VALUE {-I(EFA)}
EFLUXB FLUXB 0 VALUE {-I(EFB)}
EFLUXC FLUXC 0 VALUE {-I(EFC)}
********************************** Output plotting *******************************
.PROBE V(10,NP) V(20,NP) V(30,NP) V(40,NS) v(50,NS) v(60,NS) ; primary/secondary phase voltages
+ I(LPRIA) I(LPRIB) I(LPRIC)                ; primary winding currents
+ I(LSECA) I(LSECB) I(LSECC)                ; secondary winding currents
+ I(RCOREA) I(RCOREB) I(RCOREC)             ; core loss current
+ I(EPRIA)  I(EPRIB)  I(EPRIC)              ; magnetizing current
+ V(EPRIA)  V(EPRIB)  V(EPRIC)              ; primary induced voltage
+ V([FLUXA]) V([FLUXB]) V([FLUXC])          ; flux linkage waveforms
.END
```

ferroresonance oscillations. This figure illustrates the existence of multiple ferroresonance modes in the bifurcation diagram.

(c) Figure 12.98 shows the time domain plots and the phase-plane trajectory for phase b for the fundamental ferroresonance mode. The circuit breaker transient perturbs the system oscillations which then settle to a stable attracting limit cycle.

(d) Figures 12.99 and 12.100 show the time domain plots, the phase-plane trajectory for period-3 subharmonic ferroresonance mode and chaotic ferroresonance mode, respectively.

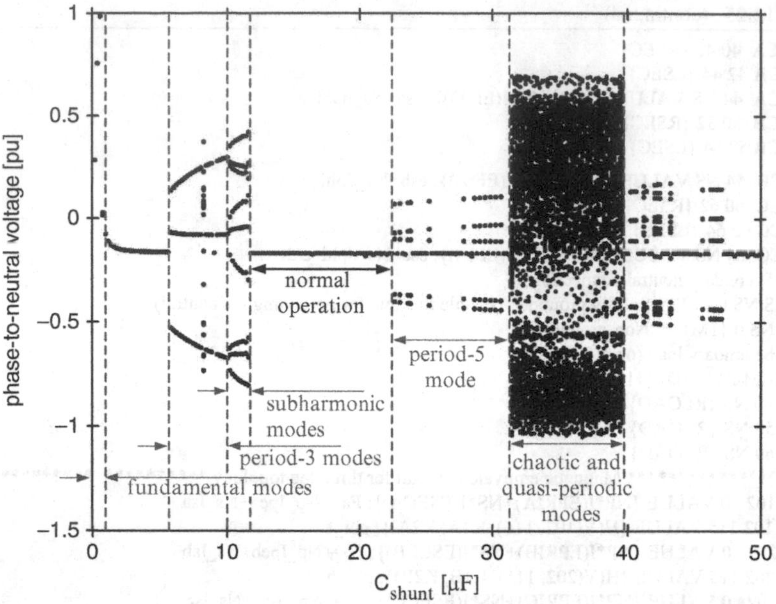

Fig. 12.97 Bifurcation diagram of ferroresonance modes (phase b voltage) for fixed C_{series}=30 pF and different C_{shunt} values (From Moses, P.S.; Masoum, M.A.S.: "Modeling ferroresonance in asymmetric three-phase power transformers," Australasian Universities Power Engineering Conference, AUPEC 2009, pp. 1-6, 2009 [52])

12.11 Summary

This chapter presents designs of electric automobiles with either battery or fuel cells as energy sources, cogeneration plants including the analysis of gas turbines, pumped-storage hydro plants, supercapacitor power plants, compressed-air storage facilities, magnetic storage plants, and flywheel storage power plants.

The transient frequency control of an isolated power plant (islanding operation), the frequency variation of an interconnected power system broken into two areas each having one generator, and the frequency change within an interconnected power system as a result of two load changes are analyzed. The utilization of intermittently operating renewable sources (e.g., PV, and wind) is possible if short-term and long-term energy storage facilities are employed as well. Short-term storage plants can supply additional power via appropriate droop characteristics within a few 60 Hz cycles, while long-term storage facilities can be put on line within 6–10 min and supply additional power with appropriate droop characteristics within minutes. Steady-state fundamental power flow and reactive power control augment the transient frequency/load analysis.

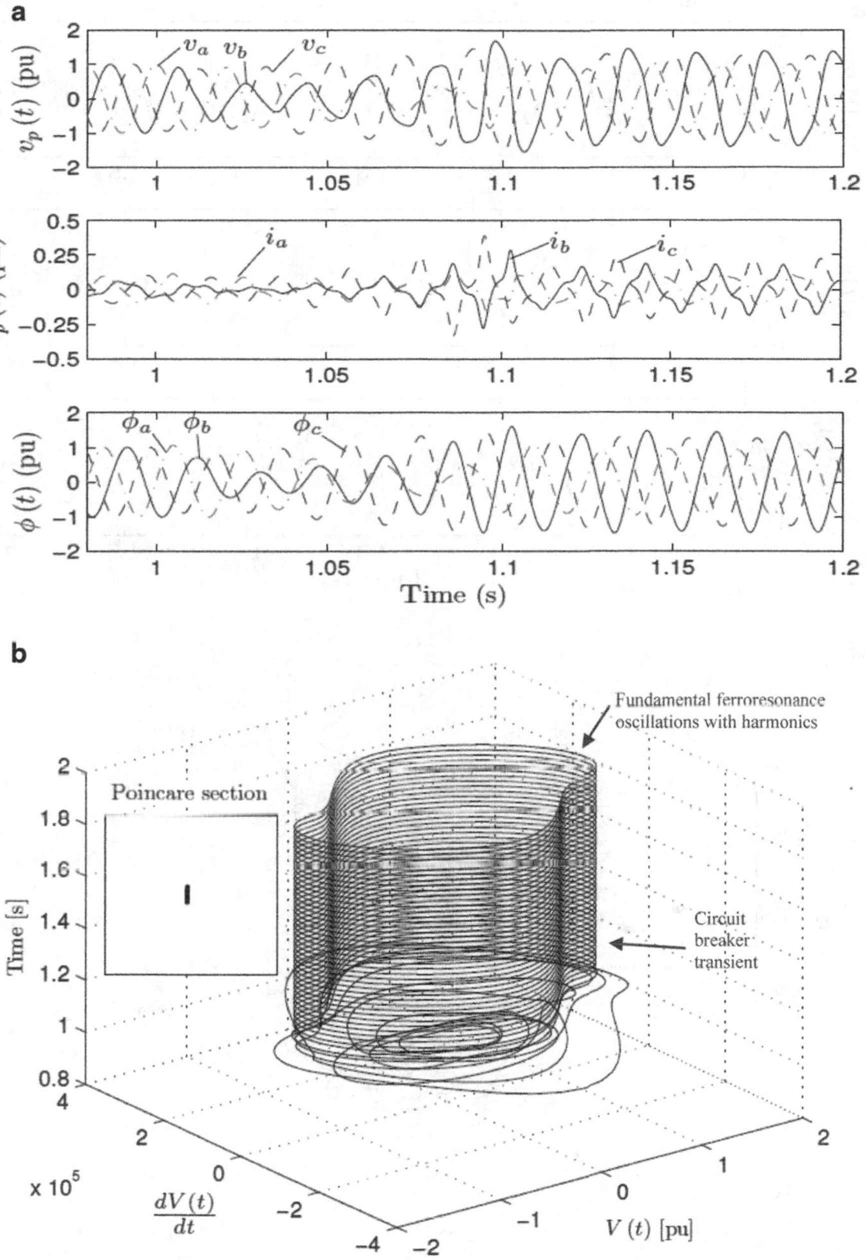

Fig. 12.98 (**a**) Time domain waveforms of transformer primary voltages, winding currents and core fluxes for fundamental ferroresonance mode (C_{series} =30 pF, C_{shunt}=0.6 μF), (**b**) Phase-plane trajectory (From Moses, P.S.; Masoum, M.A.S.: "Modeling ferroresonance in asymmetric three-phase power transformers," Australasian Universities Power Engineering Conference, AUPEC 2009, pp. 1-6, 2009 [52])

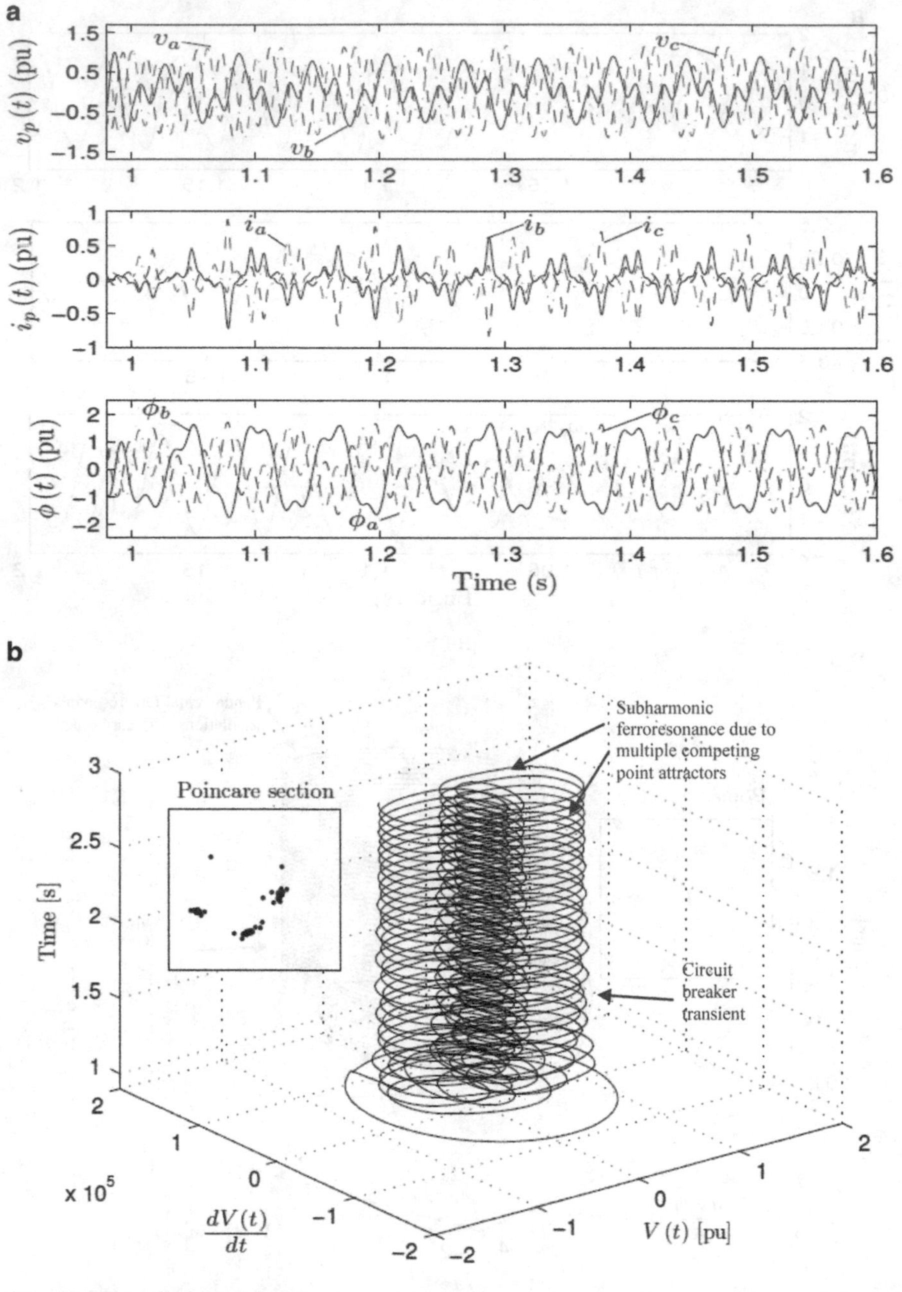

Fig. 12.99 (**a**) Time domain waveforms of transformer primary voltages, winding currents and core fluxes for period-3 subharmonic ferroresonance mode (C_{series} =30 pF, C_{shunt}=10 μF), (**b**) Phase-plane trajectory (From Moses, P.S.; Masoum, M.A.S.: "Modeling ferroresonance in asymmetric three-phase power transformers," Australasian Universities Power Engineering Conference, AUPEC 2009, pp. 1-6, 2009 [52])

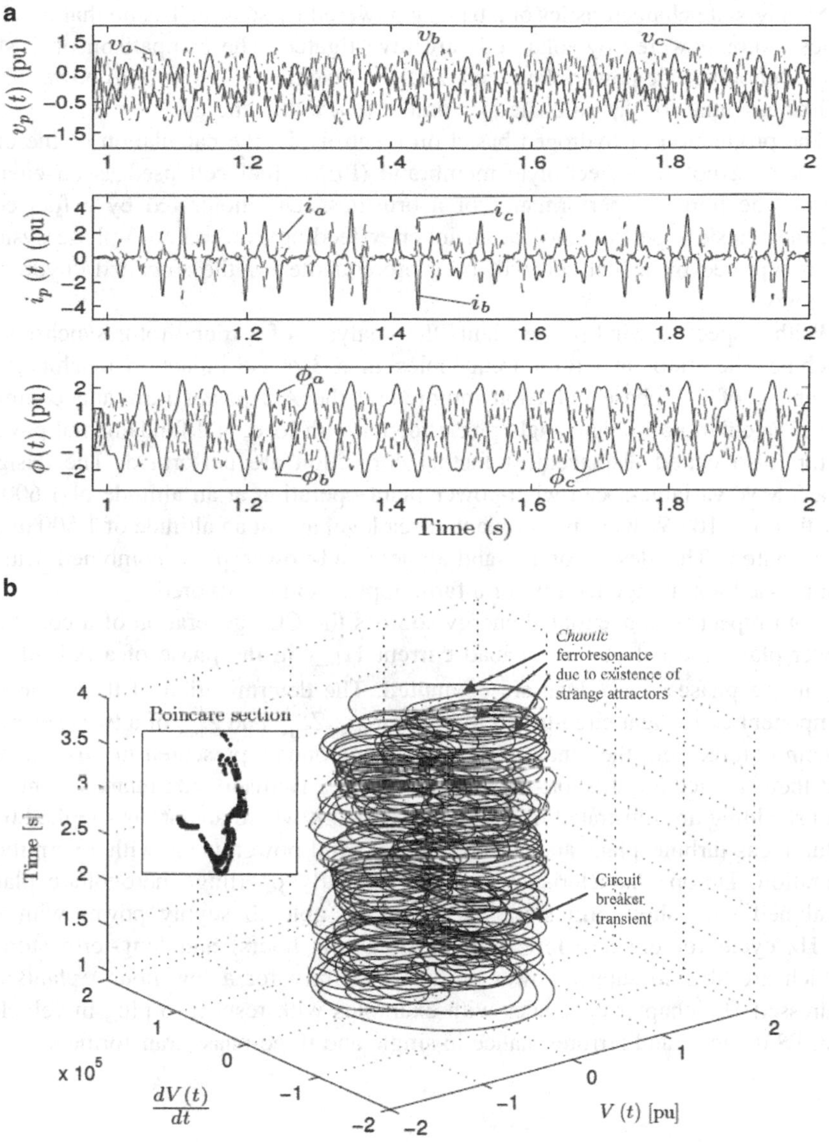

Fig. 12.100 (**a**) Time domain waveforms of transformer primary voltages, winding currents and core fluxes for chaotic ferroresonance mode (C_{series} =30 pF, C_{shunt}=38 μF), (**b**) Phase-plane trajectory (From Moses, P.S.; Masoum, M.A.S.: "Modeling ferroresonance in asymmetric three-phase power transformers," Australasian Universities Power Engineering Conference, AUPEC 2009, pp. 1-6, 2009 [52])

Steady-state characteristics of a battery powered by solar cells, and that of a DC series motor powered by solar cells are investigated. The comparison of 5 MW central photovoltaic power plant with 5 MW conventional (coal-fired) power plant will be essential for the rationale to promote renewable energy.

The production of hydrogen based on electrolysis, the calculation of the efficiency of a polymer electrolyte membrane (PEM) fuel cell used as an energy source, the transient performance of a brushless DC motor fed by a fuel cell, and the transient performance of an inverter feeding power into 3-phase system when supplied by a fuel cell will be a prerequisite for the so-called "hydrogen economy".

With respect to wind-power plants the analyses of a round-rotor synchronous machine, the stator and rotor frequencies in a 3-phase induction machine, the efficiency of a 3-phase induction motor, and the AC power flow and complex power calculation through single-phase feeder with load, and fundamental power-factor (also called displacement factor) correction are performed. The designs of a 5 MW variable-speed wind-power plant operating at an altitude of 1,600 m and that of a 10 kW wind-power plant at sea level and at an altitude of 1,500 m are investigated. The design of a stand-alone wind-power plant combined with a compressed-air storage facility for a farm application is explored.

With repect to conventional energy sources the CO_2 generation of a coal-fired power plant, the calculation of load current $\tilde{I}_{AB\_\Delta}$ in the phase of a Δ-load and $\tilde{I}_{A_Y}$in the phase of a Y-load are computed. The determination of the sequence-component equivalent circuits and matrices $Z_{bus}^{(1)}$, $Z_{bus}^{(2)}$, and $Z_{bus}^{(0)}$ of a two-generator system complement the renewable energy components presented in this chapter, and their application to short-circuit calculations. Demand-side management and load shedding are illustrated in a two-area network where area #1 is supplied by a natural gas-turbine plant and area #2 by a wind-power plant with intermittent operation. Droop characteristics for intermittently operting photovoltaic plants combined with short-term storage (which are able to supply power within a 60 Hz cycle for either a few minutes or a few hours) and long-term storage (which are able to supply power within 6–10 min for a few hours) plants are addressed.The chapter concludes with examples with respect to plug-in vehicles, FACTS devices, and ferroresonance in single and three-phase transformers

References

1. http://www.ieeeusa.org/policy/positions/energypolicy.pdf
2. http://birdcam.xcelenergy.com/sgc/media/pdf/smartgridcityhypothesisWhitePaper_July2008.pdf
3. http://money.cnn.com/galleries/2009/news/0912/gallery.global_warming/index.html
4. Dickerman, L.; Harrison, J.; "A new car, a new grid," *IEEE Power and Energy*, Vol. 8, No. 2, March/April 2010, pp. 55–61.
5. Bödefeld, Th.; Sequenz H.: *Elektrische Maschinen*, 6th Edition, Springer, Wien 1962.
6. Schuisky. W.: *Berechnung elektrischer Maschinen*, Springer, Wien, 1960.
7. Jordan H.; Klima, V.; Kovacs, K.P.: *Asynchronmaschinen*, Akademiai Kiado, Budapest 1975.

8. Concordia, C.: *Synchronous Machines*, John Wiley & Sons, New York, 1951.
9. Gerthsen, C.H.; Kneser, H.O.: *Physik*, 6th Edition, Springer, Berlin, 1960
10. Wood, A.J.; Wollenberg, B.F.: *Power Generation Operation & Control*, John Wiley & Sons, New York, 1984.
11. Fuchs, E.F.; Masoum M.A.S.: *Power Quality in Power Systems and Electrical Machines*, Elsevier, Amsterdam, 2008, 638 pages. ISBN: 978-0-12-369536-9.
12. Fuchs, E.F.; Fuchs, W.L.: "Complementary Control of Intermittently Operating Renewable Sources with Short- and Long-Term Storage Plants," Chapter in *Energy Storage*, InTech Open Access Publisher, University Campus STeP Ri, Slavka Krautzeka 83/A, 51000 Rijeka, Croatia, ISBN: 978-953-307-269-2, 2011.
13. Grainger, J.J.; Stevenson, W.D., Jr.: *Power System Analysis*, McGraw-Hill, New York, 1994.
14. 90 MW Pumpspeicherwerk Glems, Technische Werke der Stadt Stuttgart A.G., 1964.
15. http://www.tva.gov/sites/raccoonmt.htm
16. Cowdrey, J.: Boulder's municipal hydroelectric system, July 7, 2004.
17. Zipparro, V.J.; Hasen, H.: *Davis' Handbook of Applied Hydraulics*, 4th Edition, McGraw-Hill, New York, 1993, ISBN 0-70-073002-4.
18. American Society of Mechanical Engineers, *The Guide to Hydropower Mechanical Design*, HIC Publications, 1996, ISBN 0-9651765-0-9.
19. Mattick, W.; Haddenhorst, H.G.; Weber, O.; Stys, Z.S.: " Huntorf- the world's first 290 MW gas turbine air storage peaking plant," *Proceedings of the American Power Conference*, Vol. 37, 1975, pp. 322–330.
20. Vosburgh, K.G.: "Compressed air energy storage," *Journal of Energy*, Vol. 2, No. 2, March-April 1978, pp. 106–112.
21. Fitzgerald, A.F.; Kingsley Jr., Ch.; Umans, S.D.: *Electric Machinery*, 5th Edition, McGraw-Hill Publishing Company, New York, 1990.
22. Masoum, M.A.S.; Dehbonei, H.; Fuchs, E.F.: "Theoretical and experimental analyses of photovoltaic systems with voltage- and current-based maximum power point tracking," *IEEE Transaction on Energy Conversion*, Vol. 17, No. 4, December 2002, pp. 514–522.
23. Appelbaum, J.: "Starting and steady-state characteristics of DC motors powered by solar cell generators, *IEEE Transaction on Energy Conversion*, Vol. EC-1, No. 1, March 1986.
24. Bergey, K.H.: "The Lanchester-Betz limit", *Journal of Energy*, Vol. 3, No. 6, November–December, 1979.
25. van Kuik, G.A.M.: " The Lanchester-Betz-Joukowsky limit,"*Wiley Interscience*, February 14, 2007, (www.interscience.wiley.com) DOI: 10.1002/we.218.
26. Joukowsky, N.E.: "Windmill of the NEJ type," *Transactions of the Central Institute for Aero-Hydrodynamics of Moscow 1920*. Also published in *Joukowsky N.E, Collected Papers Vol. VI. The Joukowsky Institute for Aero-Hydrodynamics, Moscow:* Vol VI, 405–409, 1937 (in Russian). Also published in Gauthier-Villars et Cie. (eds). *Theorie Tourbillionnaire de l'Helice Propulsive, Quatrieme Memoire*. 123–146: Paris, 1929 (in French).
27. Joukowsky, N.E.: Fourth paper published in the *Transactions of the Office for Aerodynamic Calculations and Essays of the Superior Technical School of Moscow 1918 (in Russian)*
28. Wang, C.; Nehrir, M.H.; Shaw S.R.: "Dynamic models and model validation for PEM fuel cells using electrical circuits," *IEEE Transaction on Energy Conversion*, Vol. 20, No. 2, June 2005, pp 442–451.
29. Ipakchi, A.; Albuyeh, F.: "Grid of the future," *IEEE Power and Energy Magazine*, Vol. 7, pp. 52–62, 2009.
30. Saber, A.Y.; Venayagamoorthy, G.K.: One million plug-in electric vehicles on the road by 2015, *12th International IEEE Conference on Intelligent Transportation Systems ITSC '09*, 2009, pp. 1–7.
31. Masoum, M.A.S.; Moses, P.S.; Deilami, S.: Load management in smart grids considering harmonic distortion and transformer derating, *International Conference on Innovative Smart Grid Technologies (ISGT)*, Jan. 19–21, Gaithersburg, MD, USA, 2010, pp. 1–7.

32. Masoum, A.S.; Deilami, S.; Moses, P.S.; Abu-Siada A.: Impacts of battery charging rates of plug-in electric vehicle on smart grid distribution systems, *International Conference on Innovative Smart Grid Technologies (ISGT)*, Oct 10–13, Gothenburg, Sweden, 2010, pp. 1-6.
33. Moses, P.S.; Deilami, S.: Masoum, A.S.; Masoum, M.A.S.; Power quality of smart grids with plug-in electric vehicles considering battery charging profile, *International Conference on Innovative Smart Grid Technologies (ISGT)*, Oct 10–13, Gothenburg, Sweden, 2010, pp. 1–7.
34. Deilami, S.; Masoum, A.S.; Moses, P.S.; Masoum, M.A.S.: Real-time coordination of plug-in electric vehicle charging in smart grids to minimize power losses and improve voltage profile, *IEEE Transactions on Smart Grid*, Vol. 2, pp. 456–467, 2011.
35. Masoum, A.S.: Impact of coordinated and uncoordinated charging of plug-in electrical vehicles on smart grid, MS Thesis, Curtin University, WA, Australia, Nov. 2010.
36. Iravani, M.R.; Chaudhary, A.K.S.; Giesbrecht, W.J.; Hassan, I.E.; Keri, A.J.F.; Lee, K.C.; Martinez, J.A.; Morched, A.S.; Mork, B.A.; Parniani, M.; Sharshar, A.; Shirmohammadi, D.; Walling, R.A.; Woodford, D.A.: Modeling and analysis guidelines for slow transients. III. The study of ferroresonance, *IEEE Transactions on Power Delivery,* Vol. 15, pp. 255–265, 2000.
37. Lamba, H.; Grinfeld, M.; McKee, S.; Simpson, R.: Subharmonic ferroresonance in an LCR circuit with hysteresis, *IEEE Transactions on Magnetics,* Vol. 33, pp. 2495–2500, 1997.
38. Rezaei-Zare, A.; Iravani, R.; Sanaye-Pasand, M.: Impacts of transformer core hysteresis formation on stability domain of ferroresonance modes, *IEEE Transactions on Power Delivery,* Vol. 24, pp. 177–186, 2009.
39. Khorasani, P.G.; Deihimi, A.: A new modeling of Matlab transformer for accurate simulation of ferroresonance, *International Conference on Power Engineering, Energy and Electrical Drives (POWERENG)*, 2009, pp. 529–534.
40. Moses, P.S.; Masoum, M.A.S.: Modeling ferroresonance in asymmetric three-phase power transformers, in *Australasian Universities Power Engineering Conference (AUPEC)*, 2009, pp. 1–6.
41. Makarov, A.V. ; Komin, V.G.: The research of ferroresonant phenomena in electric circuits under open-phase operating conditions, *IEEE Russia Power Tech*, 2005, pp. 1–7.
42. Lesieutre, B.C.; Mohamed, J.A.; Stankovic, A.M.: Analysis of ferroresonance in three-phase transformers, *International Conference on Power System Technology, 2000 (PowerCon)*, 2000, Vol. 2, pp 1013–1018.
43. Tokic, A.; Madzarevic, V.; Uglesic, I.: Numerical calculations of three-phase transformer transients, *IEEE Transactions on Power Delivery,* Vol. 20, pp. 2493–2500, 2005.
44. Pedra, J.; Sainz, L.; Corcoles, F.; Lopez, R.; Salichs, M.: PSPICE computer model of a nonlinear three-phase three-legged transformer, *IEEE Transactions on Power Delivery,* Vol. 19, pp. 200–207, 2004.
45. Masoum, M.A.S.; Moses, P.S.; Masoum, A.S.: Derating of asymmetric three-phase transformers serving unbalanced nonlinear loads, *IEEE Transactions on Power Delivery,* Vol. 23, pp. 2033–2041, 2008.
46. Masoum, M.A.S.; Moses, P.S.: Impact of balanced and unbalanced DC bias on harmonic distortion generated by asymmetric three-phase three-leg transformers, *IET Electric Power Applications*, Vol. 4, No. 7, pp. 507-515, 2010.
47. Moses, P.S.; Masoum, M.A.S.; Toliyat, H.A.: Dynamic modeling of three-phase asymmetric power transformers with magnetic hysteresis: no-load and inrush conditions, *IEEE Transactions on Energy Conversion*, Vol. 26, pp. 1040–1047, 2010.
48. Moses, P.S.; Masoum, M.A.S.; Toliyat, H.A.: Impacts of hysteresis and magnetic couplings on the stability domain of ferroresonance in asymmetric three-phase, three-leg transformers, *IEEE Transactions on Energy Conversion*, Vol. 26, pp. 581–592, 2011.
49. Masoum, M.A.S.; Fuchs, E.F.; Roesler, D.J.: Large signal nonlinear model of anisotropic transformers for nonsinusoidal operation, II. Magnetizing and core-loss currents, *IEEE Transactions on Power Delivery,* Vol. 6, pp. 1509–1516, 1991.

50. Fuchs, E.F.; You, Y.: Measurement of lambda-i characteristics of asymmetric three-phase transformers and their applications, *IEEE Transactions on Power Delivery,* Vol. 17, pp. 983–990, 2002.
51. Ferracci, P.: Ferroresonance, *Group Schneider: Cahier No. 190,* pp. 1–28, March 1998.
52. Moses, P.S.; Masoum, M.A.S.: Modeling Ferroresonance in Asymmetric Three-Phase Power Transformers, Australasian Universities Power Engineering Conference, AUPEC 2009, pp. 1–6, 2009.
53. Moses, P.S.: private communication
54. Fuchs, E.F.; Masoum, M.A.S.; Roesler, D.J.: Large signal nonlinear model of anisotropic transformers for nonsinusoidal operation, I. Lambda-i characteristics, *IEEE Transactions on Power Delivery,* Vol. 6, pp. 1874–1886, 1991.
55. Fuchs, W.L.; Fuchs, E.F.: Frequency variations of power system due to switching of renewable energy sources, *International Conference on Renewable Energies and Power Quality (ICREPQ' 12),* Santiago de Compostela (Spain), 28th to 30th March, 2012, 6 pages.

Index

E.F. Fuchs and M.A.S. Masoum, *Power Conversion of Renewable Energy Systems*, DOI 10.1007/978-1-4419-7979-7, © Springer Science+Business Media, LLC 2011

L

M

N

O

P

R